Student Study and Solutions Companion

Calculus
Single Variable

Brian E. Blank and Steven G. Krantz

Prepared by Salvatrice Farinella Keating

Key College Publishing
Innovators in Higher Education

www.keycollege.com

in cooperation with

Springer

Brian E. Blank and Steven G. Krantz

Department of Mathematics
Washington University in St. Louis
St. Louis, MO 63130

Salvatrice Farinella Keating

Department of Mathematics/Computer Science
Eastern Connecticut State University
Willimantic, CT 06226

Key College Publishing was founded in 1999 as a division of Key Curriculum Press® in cooperation with Springer New York, LLC. We publish innovative texts and courseware for the undergraduate curriculum in mathematics and statistics as well as mathematics and statistics education. For more information, visit us at www.keycollege.com.

Key College Publishing
1150 65th Street
Emeryville, CA 94608
(510) 595-7000
info@keycollege.com
www.keycollege.com

Consultant: Chi-Keung Cheung

Development Editor: Kristin Burke
Production Director: Casey FitzSimons
Production Coordinator: Mae Lum
Project Managers: Maura Brown and Lynn Lustberg, Interactive Composition Corporation
Copyeditor: Kathy Carlyle
Proofreader: Interactive Composition Corporation
Composition, Illustration: Interactive Composition Corporation
Text Designers: Suzanne Montazer and Interactive Composition Corporation
Cover Designer: Jensen Barnes
Cover Photo Credit: St. Louis Arch, Missouri, USA: Getty Images/Charles Thatcher
Printer: Von Hoffman Corporation

Editorial Director: Richard J. Bonacci
General Manager: Mike Simpson
Publisher: Steven Rasmussen

Printed in the United States of America
10 9 8 7 6 5 4 3 2 1   10 09 08 07 06

ISBN: 1-931914-71-0

Contents

Introduction

Calculus is one of the milestones of human thought. Every well-educated person should be acquainted with the basic ideas of the subject. In today's technological world, in which more and more ideas are being quantified, knowledge of calculus has become essential to a broader cross-section of the population.

How to Successfully Use the *Student Study and Solutions Companion*

The *Student Study and Solutions Companion* has been carefully designed and integrated with *Calculus: Single Variable* by Brian E. Blank and Steve G. Krantz to help you succeed in your course. Used in conjunction with your textbook, the tips and hints found in the *Student Study and Solutions Companion* will develop your critical thinking skills and help you master the essential ideas and concepts in each chapter. The *Student Study and Solutions Companion* is organized into two parts: a study guide section and a solutions section. In addition, this companion offers an appendix of solutions to the study guide self-test exercises and a reference guide of mathematical facts and common formulas.

Section I: Study Guide

Section I: Study Guide corresponds to the ten chapters in your *Calculus: Single Variable* textbook. In this section, you will find the following resources:

- **A list of important concepts at the beginning of each section**. This feature will help you immediately identify the important points in each section, quickly assess your understanding, and review for quizzes and exams.
- **Additional Worked Examples in each section**. Additional examples offer you extra practice with concepts and skills and provide warnings about common student errors to avoid. The solutions to the examples are given in detail and serve as a guide to solving problems found in the textbook.
- **Self-Test Exercises in each section**. These exercises have been carefully written to offer you additional opportunities to evaluate your understanding and mastery of the key concepts in each section.
- **Terms introduced in the chapter**. Terms that appear in the textbook are highlighted in italics and presented through words and symbols. This companion uses the same notation as the textbook to facilitate your learning.

Section II: Textbook Solutions and Answers

Section II contains complete solutions to all the odd-numbered exercises in the textbook. You will also find answers to the Quick Quiz questions that appear at the end of each section in the textbook.

Appendix: Solutions to Study Guide Self-Test Exercises

The appendix contains complete solutions to the Self-Test Exercises found in the Study Guide. These solutions are provided to enhance and refine your understanding of the material.

Quick Reference Guide

In the back of this companion is a Quick Reference Guide that offers supplementary mathematical facts, common formulas, and integral tables. The Quick Reference Guide can easily offer you a helping hand when you need to look up a mathematical formula.

Don't forget to make use of the extensive Web resources available for *Calculus: Single Variable*. If you purchased a new textbook, your access code can be found inside the back cover. If you purchased a used textbook, you will also need to purchase an Access Code Package to access the site. Visit www.keycollege.com/online now to start the registration process.

We hope this Student Study and Solutions Companion will assist you as you begin your study of calculus.

Section I   Study Guide

Basics

1.1 Number Systems

Important Concepts

- Real Numbers
- The Number Line and Intervals
- The Triangle Inequality
- Approximation
- Floating Point Representations

Real Numbers

A *set*, denoted by an uppercase letter or the symbol { }, is a collection of objects. Each object in a set is called an *element of the set*; the symbol $\in$ is used to indicate that an object is in the collection.

There are *finite sets* and *infinite sets*. For both types of sets, either a description indicating the properties of the set or a listing of elements is used. The format $\{x : P(x)\}$, where $P(x)$ represents the properties of the set, is used in the description method. If the listing method is chosen, an ellipsis consisting of three consecutive dots is often used. For example, in a finite set with many elements, three dots may appear between the next-to-last element and the last element; this is done in order to avoid recording every element in the set or to indicate a number pattern. However, if some elements are listed and three dots appear either before the first element and/or after the last element, this indicates that the set is infinite and that the number pattern continues. These conventions are illustrated in the following examples of sets. Elements in a set are listed only once. A set with no elements is called the *empty* or *null set* and is denoted by the symbol ϕ.

Example 1 Using the listing method, describe set A with elements a_1, a_2, and a_3.

Solution Using the listing method, set A is described by $A = \{a_1, a_2, a_3\}$. ∎

In the text, the authors discuss number systems and real numbers. It is helpful to see how the subsets of real numbers are related. In the following material, we present some different but equivalent descriptions. Figure 1 presents a visual image of how these subsets of $\mathbb{R}$ are related.

The *real number system*, $\mathbb{R}$, is an infinite set that is comprised of the following infinite sets of numbers called *subsets*:

- Natural Numbers, $\mathbb{N}$
- Integers, $\mathbb{Z}$

- Rational Numbers, $\mathbb{Q}$
- Irrational Numbers, $\mathbb{R} - \mathbb{Q}$
- Real Numbers, $\mathbb{R}$

These subsets are defined as listed here:

The set of *natural numbers*, $\mathbb{N}$, is defined by $\mathbb{N} = \{0, 1, 2, 3, \ldots\}$.
The set of *integers*, $\mathbb{Z}$, is defined by $\mathbb{Z} = \{\ldots -3, -2, -1, 0, 1, 2, 3, \ldots\}$.
The set of *rational numbers*, $\mathbb{Q}$, is defined in two equivalent ways:

- $\mathbb{Q} = \{x : x = \dfrac{a}{b}, a, b \in Z, b \neq 0\}$. That is to say, a rational number x is a number that can be expressed as a fraction with a numerator and denominator that are both integers, but the denominator cannot be equal to the integer zero.
- $\mathbb{Q} = \{x : x$ is a terminating decimal or x is a non-terminating, repeating decimal$\}$. That is to say, a rational number x is a number that can be expressed as a decimal with digits after the decimal point that are either finite in length or infinite in length but repeat as a group.

The numbers $\dfrac{2}{5}$, 0.25, and $.33\bar{3}$ are examples of rational numbers. The bar above the three in the number $.33\bar{3}$ indicates that the digit three keeps repeating indefinitely.

The set of *irrational numbers*, $\mathbb{R} - \mathbb{Q}$, can also be defined two equivalent ways:

- $\mathbb{R} - \mathbb{Q} = \left\{x : x \neq \dfrac{a}{b}, a, b \in Z, b \neq 0\right\}$. That is, an irrational number x is a number that *cannot* be expressed as a fraction.
- $\mathbb{R} - \mathbb{Q} = \{x : x$ is a nonterminating, nonrepeating decimal$\}$. That is, an irrational number x is a number that *cannot* be expressed as a decimal with digits after the decimal point that are finite in length or infinite in length but repeat as a group.

The numbers $\sqrt{2}$ and π are examples of irrational numbers.

The set of *real numbers*, $\mathbb{R}$, is described as $\mathbb{R} = \{x : x \in \mathbb{Q}$ or $x \in \mathbb{R} - \mathbb{Q}\}$. That is to say, a real number x is either a rational number or an irrational number. The relationship amongst the subsets of $\mathbb{R}$ is illustrated in Figure 1.

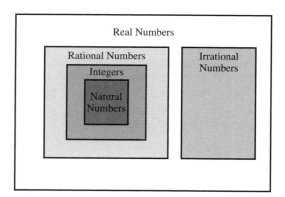

Figure 1

The figure illustrates that all natural numbers are integers, all integers are rational numbers, and all rational numbers are real numbers. Figure 1 also illustrates that all irrational numbers are real numbers and that irrational numbers are real numbers that are not rational numbers.

Example 2 Describe the set B of real numbers that are greater than or equal to 2.5 and that are less than or equal to 5.

Solution Set B is described by $B = \{x \in \mathbb{R} : 2.5 \le x \le 5\}$. ■

Example 3 Describe the set C of integers that are greater than or equal to 2.5 and that are less than 5.

Solution Set C is described by $C = \{3, 4\}$. ■

Absolute Value

It is very important that you understand the concept of absolute value. Absolute value notation often appears in computations. We emphasize the significance of absolute value in the material that follows.

The *absolute value of x*, denoted by $|x|$, is defined by $|x| = \begin{cases} x, & x \ge 0 \\ -x, & x < 0 \end{cases}$. When dealing with the absolute value of a real number x, only the number of units and not its sign is considered. Therefore, $|-3| = 3$, $|0| = 0$, and $\left|\dfrac{1}{2}\right| = \dfrac{1}{2}$. Absolute value plays a role when measuring distance; for example, $|u - v|$ represents the distance between u and v.

Absolute value also plays a role in dealing with inequalities: For real number N, $N > 0$,

- $|u| \le N$ implies $-N \le u \le N$ and, conversely, $-N \le u \le N$ implies $|u| \le N$.
- $|u| < N$ implies $-N < u < N$ and, conversely, $-N < u < N$ implies $|u| < N$.
- $|u| \ge N$ implies $u \le -N$ or $u \ge N$ and, conversely, $u \le -N$ or $u \ge N$ implies $|u| \ge N$.
- $|u| > N$ implies $u < -N$ or $u > N$ and, conversely, $u < -N$ or $u > N$ implies $|u| > N$.

The variable u in the previous expressions can either represent a single variable or a mathematical expression. Be careful when you evaluate equations. Some operations *reverse* the direction of the inequality symbol; multiplication or division by a negative value and taking reciprocals are some of the many operations that reverse the direction.

Self-Test Exercises

1. Describe the set D of real number defined by $D = \{x \in \mathbb{R} : |x| < 7\}$.
2. Describe the set E of rational numbers defined by $E = \{x \in \mathbb{Q} : |x| = 1\}$.

The Number Line and Intervals

The *number line* is a continuous and unending horizontal line where each "point" on the line is in one-to-one correspondence with the elements of $\mathbb{R}$; that is, each real number in $\mathbb{R}$ is uniquely represented by a single point on the number line (see Figure 2).

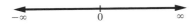

Figure 2

Intervals can be *bounded* or *unbounded*. The following tables summarize the types of intervals.

Type	Symbolic Representation	Definition	Description
bounded	(a, b)	$\{x : a < x < b\}$	open interval
bounded	$[a, b]$	$\{x : a \leq x \leq b\}$	closed interval
bounded	$[a, b)$	$\{x : a \leq x < b\}$	half open or half closed
bounded	$(a, b]$	$\{x : a < x \leq b\}$	half open or half closed

Type	Symbolic Representation	Definition
unbounded	(a, ∞)	$\{x : a < x\}$
unbounded	$[a, \infty)$	$\{x : a \leq x\}$
unbounded	$(-\infty, b)$	$\{x : x < b\}$
unbounded	$(-\infty, b]$	$\{x : x \leq b\}$
unbounded	$(-\infty, \infty)$	$\{x : -\infty < x < \infty\}$

You can find visual representations of these intervals in Figures 7 and 8 on pages 5 and 6 of the student text.

Example 4 Express the set $X = \{x : |2x + 4| < 3\}$ as an interval.

Solution

$$|2x + 4| < 3 \Rightarrow |2(x + 2)| < 3$$
$$\Rightarrow |2| \cdot |x + 2| < 3$$
$$\Rightarrow 2 \cdot |x + 2| < 3$$
$$\Rightarrow |x + 2| < \frac{3}{2}$$
$$\Rightarrow \frac{-3}{2} < (x + 2) < \frac{3}{2}$$
$$\Rightarrow \frac{-3}{2} - 2 < x < \frac{3}{2} - 2$$
$$\Rightarrow \frac{-7}{2} < x < \frac{-1}{2}$$

$\frac{-7}{2} < x < \frac{-1}{2}$ is the definition of all x in the open interval $\left(\frac{-7}{2}, \frac{-1}{2}\right)$. ∎

Example 5 Express the set $S = \{x : |2x + 3| \le 3\}$ as an interval.

Solution

$$|2x + 3| \le 3 \Rightarrow \left| 2\left(x + \frac{3}{2} \right) \right| \le 3$$

$$\Rightarrow |2| \cdot \left| x + \frac{3}{2} \right| \le 3$$

$$\Rightarrow 2 \cdot \left| x + \frac{3}{2} \right| \le 3$$

$$\Rightarrow \left| x + \frac{3}{2} \right| \le \frac{3}{2}$$

$$\Rightarrow \frac{-3}{2} \le \left(x + \frac{3}{2} \right) \le \frac{3}{2}$$

$$\Rightarrow \frac{-3}{2} - \frac{3}{2} \le x \le \frac{3}{2} - \frac{3}{2}$$

$$\Rightarrow \frac{-6}{2} \le x \le 0$$

$$\Rightarrow -3 \le x \le 0$$

$-3 \le x \le 0$ is the definition of all x in the closed interval $[-3, 0]$.   ∎

Self-Test Exercises

3. Solve for x: $|-5x - 25| \le 10$.
4. Does $|5x| < 2$ imply that $-0.4 < x < 0.4$? Substantiate your answer.

The Triangle Inequality

The *Triangle Inequality*, a useful estimation tool, states that for all real numbers a and b, $|a + b| \le |a| + |b|$.

Example 6 If the distance from points A to B is 5 and the distance from points B to C is 2, estimate the distance from points A to C.

Solution

$$|A - C| = |A - B + B - C|$$
$$= |(A - B) + (B + C)|$$

If you apply the Triangle Inequality with $a = (A - B)$ and $b = (B - C)$, then

$$|(A - B) + (B - C)| \le |A - B| + |B - C|$$
$$= 5 + 2$$
$$= 7$$

Therefore, the distance from points A to C, $|A - C|$, is less than or equal to 7.   ∎

Approximation

The difference between the exact value and the approximate value is called the *error*. If x represents the *exact value* and a represents the *approximate value*, then $|x - a|$ represents the *absolute error* and $\dfrac{|x - a|}{|x|}$ represents the *relative error*.

Rounding errors and truncation errors are common when making approximations. These errors often occur in calculations that are obtained by using a calculator. The statement "$|x - a| \leq 5 \times 10^{-(q+1)}$" indicates that *the approximation "a" agrees with exact value "x" to "q" decimal places*.

Floating Point Representations

In general, the *floating point decimal representation* of a nonzero real number x is $x = \pm 0.a_1 a_2 a_3 \ldots \times 10^p$ where $p \in \mathbb{N}$, $a_i \in \{0, 1, 2, 3, \ldots, 8, 9\}$, $i = 1, 2, 3$, and $a_1 \neq 0$. A number with a floating point decimal representation $x = (\pm 0.a_1 a_2 a_3 \ldots a_k)(10^p)$ is said to have k *significant digits*.

For example, in the statement $|x - a| \leq 5 \times 10^{-(q+1)}$, the expression $5 \times 10^{-(q+1)}$ can be rewritten as 0.5×10^{-q} in floating point representation.

Example 7   Identify the number of significant digits in the expression $7 \cdot 10^{-1}$.

Solution $7 \cdot 10^{-1} = 0.7 \times 10^0$; therefore, $a_k = 7$ with $k = 1$. This implies that there is one significant digit; it is the digit 7.   ■

Example 8   Express the number .0002156 in floating point decimal representation.

Solution   The floating-point decimal representation of .0002156 is 0.2156×10^{-3}.   ■

Self-Test Exercises

5.   Express the number 276.55589156 in floating point decimal representation.
6.   Identify the number of significant digits in the floating point decimal representation of 276.55589156.

1.2 Planar Coordinates and Graphing in the Plane

Important Concepts

- The Cartesian Plane $\mathbb{R} \times \mathbb{R}$
- Conics
- Regions in $\mathbb{R} \times \mathbb{R}$

The *Cartesian product of set X and Y*, denoted by $X \times Y$, is the set of all ordered pairs (x, y) where $x \in X$ and $y \in Y$. The order in which the elements appear is important. In the ordered pair, the first element, x, is an element in the first set listed in $X \times Y$ and the second element, y, is an element in the second set in $X \times Y$.

The *Cartesian plane* is defined as the Cartesian product of $\mathbb{R}$ with itself. Recall that $\mathbb{R}^2 = \mathbb{R} \times \mathbb{R} = \{(a, b) | a, b \in \mathbb{R}\}$. Earlier, we identified every unique point on the number line with a unique real number in a one-to-one fashion. Here, we identify every unique point in the Cartesian plane with a unique ordered pair (a, b) in a one-to-one fashion. Figure 1 illustrates the Cartesian plane.

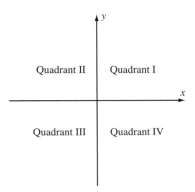

Figure 1

In the figure, the horizontal line that is referred to as x is called the x-axis. The x-axis is represented by the interval $(-\infty, \infty)$, which is a copy of the real number line. The vertical line, which is referred to as y, is called the y-axis. The y-axis is also represented by the interval $(-\infty, \infty)$, which is again a copy of the real number line. The intersection of the two axes is called the *origin*. The origin is represented by the ordered pair $(0, 0)$.

The two axes divide the *Cartesian* or *xy-plane* into four quadrants: Quadrant I, Quadrant II, Quadrant III, and Quadrant IV. *Quadrant I* is defined as the Cartesian product $X \times Y$ with x-values that are elements in the interval $(0, \infty)$ and with y-values that are elements in the interval $(0, \infty)$. *Quadrant II* is defined as the Cartesian product $X \times Y$ with x-values that are elements in the interval $(-\infty, 0)$ and with y-values that are elements in the interval $(0, \infty)$. *Quadrant III* is defined as the Cartesian product $X \times Y$ with x-values that are elements in the interval $(-\infty, 0)$ and with y-values that are elements in the interval $(-\infty, 0)$. *Quadrant IV* is defined as the Cartesian product $X \times Y$ with x-values that are elements in the interval $(0, \infty)$ and with y-values that are elements in the interval $(-\infty, 0)$.

The formulas for the midpoint of a line segment and distance are useful. If points P_1 and P_2 in the Cartesian plane are defined by $P_1 = (x_1, y_1)$, $P_2 = (x_2, y_2)$, then the formula for the *midpoint* $M_{\overline{P_1 P_2}}$ of line segment $\overline{P_1 P_2}$ is given by

$$M_{\overline{P_1 P_2}} = \left(\frac{x_1 + x_2}{2}, \frac{y_1 + y_2}{2} \right)$$

and the formula for the *distance between P_1 and P_2* (i.e., the length $|\overline{P_1 P_2}|$ of line segment $\overline{P_1 P_2}$) is given by $|\overline{P_1 P_2}| = \sqrt{(x_1 - x_2)^2 + (y_1 - y_2)^2}$. *The order of the coordinate values is consistent*; that is, the values of $P_1 = (x_1, y_1)$ each came first in the difference expressions. Note that

$$|\overline{P_1 P_2}| = \sqrt{(x_1 - x_2)^2 + (y_1 - y_2)^2} = \sqrt{(x_2 - x_1)^2 + (y_2 - y_1)^2}.$$

Example 1 Identify the quadrant in which point $P_1 = \left(\frac{-1}{10}, 1000 \right)$ is located.

Solution Quadrant II is defined as the Cartesian product $X \times Y$ with x-values that are elements in the interval $(-\infty, 0)$ and with y-values that are elements in the interval $(0, \infty)$. $\frac{-1}{10} \in (-\infty, 0)$ and $1000 \in (0, \infty)$. Therefore, point P_1 is located in Quadrant II. ∎

Example 2 Calculate the midpoint $M_{\overline{P_1 P_2}}$ of line segment $\overline{P_1 P_2}$ if $P_1 = (-1, 5)$ and $P_2 = \left(\frac{1}{2}, -3 \right)$.

Solution

$$M_{\overline{P_1 P_2}} = \left(\frac{x_1 + x_2}{2}, \frac{y_1 + y_2}{2} \right)$$
$$= \left(\frac{-1 + \frac{1}{2}}{2}, \frac{5 - 3}{2} \right)$$

$$\left(\frac{-1+\frac{1}{2}}{2}, \frac{5-3}{2}\right) = \left(\frac{\frac{-1}{2}}{4}, \frac{2}{2}\right)$$

$$\left(\frac{\frac{-1}{2}}{4}, \frac{2}{2}\right) = \left(\frac{-1}{4}, 1\right)$$

Therefore, the midpoint $M_{\overline{P_1 P_2}}$ of line segment $\overline{P_1 P_2}$ is $= \left(\frac{-1}{4}, 1\right)$.   ■

Self-Test Exercises

1. Identify points P_1 and P_2 on the Cartesian plane if $P_1 = (.1, .5)$ and $P_2 = (-1, 2)$.
2. Calculate $|\overline{P_1 P_2}|$, the length of the line segment $\overline{P_1 P_2}$, if $P_1 = (.1, .5)$ and $P_2 = (-1, 2)$.

Conics

Picture two empty ice-cream cones with the bottom points of each ice-cream cone touching. Think of a plane as a single sheet of paper. If the plane could slice through the cones at different angles without destroying them, the intersection of the plane and cone or cones would create different conics depending on the angle of the plane at impact. The geometric figures circle, ellipse, parabola, and hyperbola are known as *conics*.

If the plane meets a line through the center of the cone at a 90° angle, a *circle* is formed. If the plane intersects only one of the cones, either an *ellipse* or a *parabola* is formed. If the plane intersects both of the cones, a *hyperbola* is formed. These conic figures are illustrated in Figure 2.

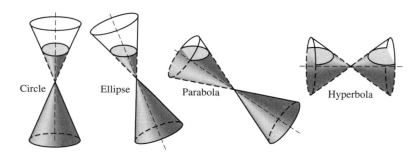

Circle   Ellipse   Parabola   Hyperbola

Figure 2

Figures 3, 4, and 5 offer more detailed views.

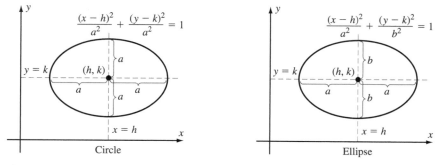

Figure 3

Figure 3 illustrates that the equation of the ellipse, $\dfrac{(x-h)^2}{a^2} + \dfrac{(y-k)^2}{b^2} = 1$, when $a = b$, results in the equation of the circle, $\dfrac{(x-h)^2}{a^2} + \dfrac{(y-k)^2}{a^2} = 1$. (The graph of the circle is *not* circular in appearance because the two axes are not in the same scale, but the equation is that of a circle.)

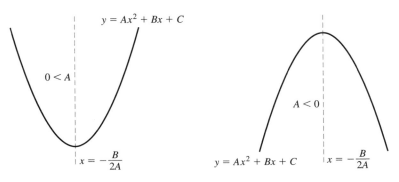

Figure 4

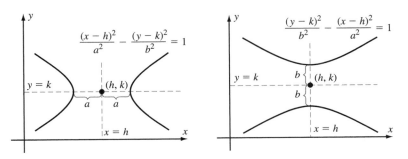

Figure 5

As indicated in Figures 3, 4, and 5, the equations of the conic figures in standard form are

- **Circle:** $\dfrac{(x-h)^2}{a^2} + \dfrac{(y-k)^2}{a^2} = 1$, which is equivalent to $(x-h)^2 + (y-k)^2 = a^2$.

- **Ellipse:** $\dfrac{(x-h)^2}{a^2} + \dfrac{(y-k)^2}{b^2} = 1$. The values a and b can be any real number.

- **Parabola:** $y = Ax^2 + Bx + C$, where $A \neq 0$. The vertex is the point $\left(\dfrac{-B}{2A}, f\left(\dfrac{-B}{2A}\right) \right)$, where

$$ f\left(\frac{-B}{2A}\right) = A\left(\frac{-B}{2A}\right)^2 + B\left(\frac{-B}{2A}\right) + C $$
$$ = \frac{-B^2 + 4AC}{4A} $$

and the equation of the axis of symmetry L is $x = \dfrac{-B}{2A}$.

- **Hyberbola:** $\dfrac{(x-h)^2}{a^2} - \dfrac{(y-k)^2}{b^2} = 1$ or $\dfrac{(y-k)^2}{b^2} - \dfrac{(x-h)^2}{a^2} = 1$.

Completing the Square is a useful algebraic technique used to express quadratic equations in standard form. This method is demonstrated in Example 3 in an equivalent but slightly different way than the method presented in the text.[1]

Example 3 Complete the square in x of $Ax^2 + Bx + C = 0$, $A \neq 0$, and solve for x.

Solution

$$\text{For } A \neq 0,\ Ax^2 + Bx + C = 0 \Rightarrow x^2 + \frac{B}{A}x + \frac{C}{A} = 0$$

$$\Rightarrow x^2 + \frac{B}{A}x = -\frac{C}{A}$$

$$\Rightarrow x^2 + \frac{B}{A}x + \left(\frac{1}{2}\cdot\frac{B}{A}\right)^2 = -\frac{C}{A} + \left(\frac{1}{2}\cdot\frac{B}{A}\right)^2$$

$$\Rightarrow \left(x + \frac{B}{2A}\right)^2 = \frac{B^2 - 4AC}{4A^2}$$

If we let $u = x + \dfrac{B}{2A}$ and take the square roots of $\left(x + \dfrac{B}{2A}\right)^2$ and $\dfrac{B^2 - 4AC}{4A^2}$, using the rule that $\sqrt{u^2} = |u|$, then $\left|x + \dfrac{B}{2A}\right| = \dfrac{\sqrt{B^2 - 4AC}}{2A} \Rightarrow x + \dfrac{B}{2A} = \dfrac{\pm\sqrt{B^2 - 4AC}}{2A} \Rightarrow x = \dfrac{-B \pm \sqrt{B^2 - 4AC}}{2A}$. ∎

For $A \neq 0$, $Ax^2 + Bx + C = 0 \Rightarrow x = \dfrac{-B \pm \sqrt{B^2 - 4AC}}{2A}$ is the statement of the *Quadratic Formula*. The solution of Example 3 is its derivation.

Example 4 Identify the conic with the equation given by $x^2 + 14x + 48 - y = 0$.

Solution

$$x^2 + 14x + 48 - y = 0 \Rightarrow x^2 + 14x + 48 = y$$

$$\Rightarrow x^2 + 14x + 49 - 1 = y$$

$$\Rightarrow x^2 + 14x + 49 = y + 1$$

$$\Rightarrow (x + 7)^2 = y + 1$$

$$\Rightarrow (x - (-7))^2 = (y - (-1))$$

$(x - (-7))^2 = (y - (-1))$ is the equation of a parabola that opens up, has the vertex point $(-7, -1)$, and has an axis of symmetry that satisfies the equation $x = -7$. ∎

---

[1] The method presented in the text is

$$Ax^2 + Bx + C = 0 \Rightarrow A\left(x^2 + \frac{B}{A}x\right) + C = 0 \Rightarrow A\left(x^2 + \frac{B}{A}x + \left(\frac{B}{2A}\right)^2\right) + C - A\left(\frac{B}{2A}\right)^2 = 0$$

$$\Rightarrow \left(x + \frac{B}{2A}\right)^2 + \frac{C}{A} - \left(\frac{B}{2A}\right)^2 = 0 \Rightarrow \left(x + \frac{B}{2A}\right)^2 = \left(\frac{B}{2A}\right)^2 - \frac{C}{A} \Rightarrow \left(x + \frac{B}{2A}\right)^2 = \frac{B^2 - 4AC}{4A^2}.$$

This is the same solution obtained in Example 3 above.

Self-Test Exercises

3. Identify the conic with the equation that is given by $x^2 - 2x - 14 + y^2 + 6y = 0$.
4. Identify the conic with the equation that is given by $x^2 - y^2 - \pi = 0$.

Regions in $\mathbb{R} \times \mathbb{R}$

Regions in $\mathbb{R} \times \mathbb{R}$ are usually described by inequalities. In the graph of a planar region, a *solid line* denotes that the points on the line are part of the solution (see Figure 14 on page 18 in the student text), and a *dotted line* indicates that the points on the line are not part of the solution (see Figure 13 on page 18 in the student text). When you are deciding which part of the region to shade, choose a test point within a region and substitute its x- and y-values into the inequality statement. If the result after substitution is true, shade the region that contains the selected test point. If the result after substitution is false, shade the region that does not contain the selected test point.

Example 5 Sketch $T = \{(x, y) : y \geq 2x\}$.

Solution The graph of the line $y = 2x$ will be a solid line because the property the ordered pairs (x, y) in the region must satisfy is $y \geq 2x$. A test point $(2, 10)$ gives a true result because $10 > (2)(2)$ is true. Therefore, the region to be shaded will be on the side that contains the test point $(2, 10)$ and is bounded by the graph of $y = 2x$. A sketch of the shaded regions appear in Figure 6.

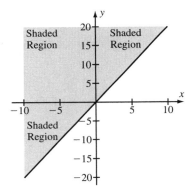

Figure 6 ■

Example 6 Sketch $U = \{(x, y) : y < 10\}$.

Solution The horizontal line $y = 10$ will be dotted because $y < 10$ is a strict inequality. There are no conditions on the variable x so its values can range from minus infinity to plus infinity. Choose $(0, 0)$ as a test point. The value $y = 0$ is certainly less than 10, so the region is bounded by the dotted line $y = 10$ and is located below this boundary. A sketch of the shaded regions appears in Figure 7.

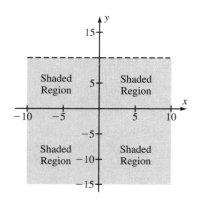

Figure 7

Self-Test Exercises

5. Sketch the region $F = \{(x, y) : y \leq x^2\}$.
6. Sketch the region $G = \{(x, y) : y \geq x^2\}$.

1.3 Lines and Their Slopes

Important Concepts

- Slopes
- Equations of Lines
- Least Squares Lines

Slopes

Suppose $A = (x_1, y_1)$ and $B = (x_2, y_2)$ in $\mathbb{R} \times \mathbb{R}$. Points A and B determine the straight line AB. Line AB has a *slope* $m_{\overline{AB}}$, which is defined in the following *equivalent* ways:

- $m_{\overline{AB}} = \dfrac{\Delta y}{\Delta x}$

- $= \dfrac{y_2 - y_1}{x_2 - x_1}$

- $= \dfrac{rise}{run}.$

If line AB is horizontal, then $m_{\overline{AB}} = 0$ because $\Delta y = 0$. If line AB is vertical, then $m_{\overline{AB}}$ is *undefined* because $\Delta x = 0$. If line AB is neither horizontal nor vertical, then its slope will either be positive (if the graph of line AB appears to go up from left to right) or its slope will be negative (if the graph of line AB appears to go down from left to right).

Example 1 If $A = (.5, 2)$ and $B = (3, .25)$, determine the slope, $m_{\overline{AB}}$, of line AB.

Solution
$$m_{\overline{AB}} = \frac{(.25 - 2)}{3 - .5}$$
$$= \frac{-1.75}{2.5}$$
$$= -.7$$

Therefore, the slope of line AB is $-.7$.

Example 2 If $A = (3, 4)$, $B = (x_b, y_b)$, and the slope of line AB is equal to 5, find the coordinate values of point B.

Solution

$$5 = m_{\overline{AB}} = \frac{(y_b - 4)}{(x_b - 3)} \Rightarrow 5(x_b - 3) = (y_b - 4)$$
$$\Rightarrow 5x_b - 15 = y_b - 4$$
$$\Rightarrow y_b = 5x_b - 11$$

Therefore, the coordinates of point B satisfy $B = (x_b, y_b) = (x_b, 5x_b - 11)$. The number of possibilities is infinite because x_b can have *any* real value. For example, if we choose $x_b = 1$, then $B = (1, 5 - 11) = (1, -6)$ and $m_{\overline{AB}} = 5$. Similarly, if we choose $x_b = 0$, then $B = (0, 0 - 11) = (0, -11)$ and $m_{\overline{AB}} = 5$. See if you can find any other values that work, or try to find *one* set $B = (x_b, y_b) = (x_b, 5x_b - 11)$ such that $m_{\overline{AB}} \neq 5$. ∎

Self-Test Exercises

1. Determine the slope of line AB if $A = (3, 4)$ and $B = (-2, 4)$.
2. Determine the slope of line AB if $A = (3, 4)$ and $B = (3, -4)$.

Equations of Lines

Slope $m_{\overline{AB}}$ appears in two formulas that correspond to the equation of line AB; they are the *Point-Slope Form* and the *Slope-Intercept Form*.

If $A = (x_a, y_a)$ is a point on line AB and $m_{\overline{AB}}$ is the slope of line AB, the Point-Slope Form of line AB is given by $y = m_{\overline{AB}}(x - x_a) + y_a$.

If $A = (x_a, y_a)$ is a point on line AB and $m_{\overline{AB}}$ is the slope of line AB, the Slope-Intercept Form of line AB is given by $y = m_{\overline{AB}} \cdot x + b$ where b is the y-intercept. The y-intercept is the point $(0, b)$.

The *Intercept Form* is a third formula for the equation of line AB. If a is the x-intercept, it is the point $(a, 0)$. If b is the y-intercept, it is the point $(0, b)$. The Intercept Form of line AB is given by $\frac{x}{a} + \frac{y}{b} = 1$, $a \neq 0$, $b \neq 0$.

Example 3 If $A = (1, -2)$ and $B = (-5, 2)$, express the equation of line AB using the Slope-Intercept Form.

Solution

$$m_{\overline{AB}} = \frac{(-2 - 2)}{(1 - (-5))} = \frac{-4}{6} = \frac{-2}{3}.$$

$A = (1, -2)$ is a point on line AB, so $x = 1$ and $y = -2$. Substitute the values of $m_{\overline{AB}}$, x, and y in the equation $y = m_{\overline{AB}} \cdot x + b$ to obtain $-2 = \frac{-2}{3} + b$. This implies that $b = \frac{-6 + 2}{3} = \frac{-4}{3}$.

Therefore, the equation of line AB using the Slope-Intercept Form is given by $y = \frac{-2}{3}x - \frac{4}{3}$. ∎

Lines AB and CD are *perpendicular* if their intersection forms $90°$ angles. Lines AB and CD are *parallel* if they never meet. Information about their slopes, $m_{\overline{AB}}$ and $m_{\overline{CD}}$, can identify whether lines AB and CD are either perpendicular or parallel.

- Lines AB and CD are *perpendicular* if and only if $m_{\overline{AB}} \cdot m_{\overline{CD}} = -1$.
- Lines AB and CD are *parallel* if and only if $m_{\overline{AB}} = m_{\overline{CD}}$.

Example 4 If the equation of line AB is $y + 2x - 7 = 0$ and the equation of line CD is $-2y + 2x - 8 = 0$, determine whether lines AB and CD are parallel, perpendicular, or neither.

Solution $y + 2x - 7 = 0$ implies that $y = -2x + 7$, so the slope of line AB is $m_{\overline{AB}} = -2$. $-2y + 2x - 8 = 0$ implies that $2y = 2x - 8$. This in turn implies that $y = x - 4$. Therefore, the slope of line CD is $m_{\overline{CD}} = 1$. Because the slopes of the two lines are not equal, the lines are not parallel. Because the product of their slopes is not equal to -1, the two lines are not perpendicular. Therefore, lines AB and CD are neither parallel nor perpendicular. ∎

Self-Test Exercises

3. Find an equation of line AB if line AB is perpendicular to line CD and passes through the point $(1, -5)$. The equation of line CD is given by $y = 3x + 10$.
4. Find the equation of line AB that passes through the points $(0, 5)$ and $(2, 1)$.

Least Squares Lines

Data gathered from real-world conditions may not lend itself to being expressed in a formula, so finding a *line of best fit* is a common problem. A method that is used quite often to find the line of best fit is the *Method of Least Squares*.

A line of best fit has slope. It has been established that given the $N + 1$ data points, $(x_0, y_0), (x_1, y_1), \ldots, (x_N, y_N)$, the *least squares line* through (x_0, y_0) is given by $y = m(x - x_0) + y_0$ where

$$m = \frac{(x_1 - x_0)(y_1 - y_0) + (x_2 - x_0)(y_2 - y_0) + \cdots + (x_N - x_0)(y_N - y_0)}{(x_1 - x_0)^2 + (x_2 - x_0)^2 + \cdots + (x_N - x_0)^2}.$$

Example 5 Calculate the least squares line through (x_0, y_0) if $(x_0, y_0) = (1, 2), (x_1, y_1) = (7, 5)$, and $(x_2, y_2) = (3, 3)$.

Solution

i	x_i	y_i	$x_i - x_0$	$y_i - y_0$	$(x_i - x_0)^2$	$(x_i - x_0)(y_i - y_0)$
1	7	5	6	3	36	18
2	3	3	2	1	4	2
Total					40	20

$m = \dfrac{20}{40} = \dfrac{1}{2}$, so the equation for the least squares line is $y = \dfrac{1}{2}(x - 1) + 2 = \dfrac{1}{2}x + \dfrac{3}{2}$.

Therefore, the equation of the least squares line through $(x_0, y_0) = (1, 2)$ is $y = \dfrac{1}{2}x + \dfrac{3}{2}$. ∎

Self-Test Exercise

5. Calculate the least squares line through (x_0, y_0) if $(x_0, y_0) = (2, -5)$, $(x_1, y_1) = (1, 15)$, and $(x_2, y_2) = (7, 1)$.

1.4 Functions and Their Graphs

Important Concepts

- Functions
- Graphs of Functions

Functions

A *function f* on a set S with values in T, $f: S \rightarrow T$, assigns a *unique* value $f(s) \in T$ to each $s \in S$. S is called the *domain of S*, and T is called the *range of f*. The *image of f* is the set $\{y = f(s) : s \in S\}$. This set is entirely contained in the set T; in other words, the image of f is a subset of T, the range of f. If the domain of S, the image of f and the range of f consist only of real numbers, then f is said to be *real-valued*.

Functions can be defined by formulas, data, or their graphs. Functions can act on a single variable or on several variables.

Example 1 f is defined by $\begin{array}{c} 2 \mapsto 4 \\ 4 \mapsto 8 \\ 6 \mapsto 12 \end{array}$. Is f a function? If so, state the domain and range of f. Substantiate your answer.

Solution Since our definitions of *domain* and *range* are defined for functions, we must first verify that f is a function. The rule f acts on the elements 2, 4, and 6 (the *input* elements). It assigns to each input element one unique element (an *output* element) in its image. Thus f is a function. The domain of f is the set $S = \{2, 4, 6\}$ and the range of f is the set $T = \{4, 8, 12\}$. We note that in this example the image of f is equal to the range of f. ■

A function f is a *sequence* if its domain is a finite or infinite set of consecutive integers. A sequence can be described by listing its values, $f_1, f_2, \ldots f_k, \ldots$, where $f_1 = f(1)$ is the first element of the list, $f_2 = f(2)$ is the second element of the list, and $f_k = f(k)$ is the kth element of the list for some k.

Example 2 Provide an example of a sequence.

Solution

$$1 \mapsto 4$$
$$2 \mapsto 8$$
$$3 \mapsto 12$$

is an example of a finite sequence with *domain* $S = \{1, 2, 3\}$, *range* $T = \{4, 8, 12\}$, which is described by the function $f(x) = 4x$, $x \in \{1, 2, 3\}$. Therefore, the elements of the sequence are $f_1 = 4$, $f_2 = 8$, and $f_3 = 12$. ■

Example 3 R is a rectangle whose length, ℓ, is equal to twice its width, w. Is the area A a function (of a single variable)? If so, identify the *domain* and *range* of A.

Solution Because we are dealing with a nondegenerate rectangle, its dimensions are positive real numbers. Area A is equal to the product of the length and width of rectangle R. But because $\ell = 2w$ we can think of A as either a rule where $A(\ell) = (\ell)\left(\dfrac{\ell}{2}\right) = \dfrac{\ell^2}{2}$ or a rule where $A(w) = (2w)(w) = 2w^2$. In either case, an input value is assigned to one output value, so A is a function of the variable w or the variable ℓ as was demonstrated. A finite product of positive real numbers is itself a positive real number. If $\mathbb{R}^+$ denotes the set of positive real numbers, then the domain of A is a subset of $\mathbb{R}^+$ and the range of A is also a subset of $\mathbb{R}^+$. ■

Self-Test Exercises

1. Is f defined by $f(x) = 3$ a function? Substantiate your answer.
2. Is f defined by $f(x) = \sqrt{1 - x^2}$, $x \in \mathbb{R}$ a function? Substantiate your answer.

Graphs of Functions

It is understood that it is best to use *rigor* to determine whether a rule f is a function or not; however, it is sometimes more expedient to visually determine whether a rule f is a function or not by inspecting its graph.

The graph of f is defined to be the set of all points (x, y) in the xy-plane for which x is in the *domain* of f and $y = f(x)$, i.e., *graph* of $f = \{(x, y) \mid x \in S, y = f(x)\}$ where S is the domain of f.

Based on the graph of f, the *Vertical Line Test* is very useful in determining whether or not f is a function. The statement of the Vertical Line Test is *if every vertical line drawn through a curve intersects that curve only once, then the curve is the graph of a function.*

Example 4 The *graph* of f, consisting of horizontal lines in Quadrants I and III, respectively, is shown in Figure 1. In Figure 1, it is understood that the x- and y-*axes* extend from minus infinity to plus infinity. The horizontal line $y = -1$ extends from minus infinity up to but not including zero, and the horizontal line $y = 1$ extends from zero to positive infinity.

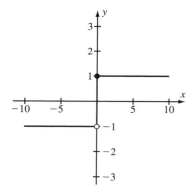

Figure 1

Is f a function? Substantiate your answer.

Solution We see that any vertical line intersects the graph of f only once; therefore, f is a function according to the Vertical Line Test. If more rigor is preferred, we remark that function f is defined by

$$f(x) = \begin{cases} 1, & x \geq 0 \\ -1, & x < 0 \end{cases}.$$

Any nonnegative x is uniquely assigned to 1, i.e., the point $(x, 1)$, and any negative x is uniquely assigned to -1, i.e., the point $(x, -1)$. So f is a function with domain $(-\infty, \infty)$ and range $\{-1, 1\}$. ■

Example 5 The graph of f consisting of a parabola that opens to the right is shown in Figure 2.

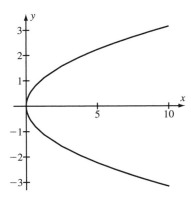

Figure 2

Is f a function? Substantiate your answer.

Solution The vertical line $x = 1$ intersects the given graph at two points: $(1, 1)$ and $(1, -1)$. Therefore, according to the Vertical Line Test, f is not a function. If we choose to be more rigorous, the graph in Figure 2 represents the rule $y^2 = x$. This implies that $y = \sqrt{x}$ or $y = -\sqrt{x}$. Since the *input* element x is assigned to both the positive square root of x and also the negative square root of x, *uniqueness* is not preserved, so f is not a function. ∎

Self-Test Exercises

3. The graph of f is shown in Figure 3.

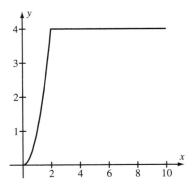

Figure 3

Is f a function? Substantiate your answer.

4.   The graph of f is given shown in Figure 4.

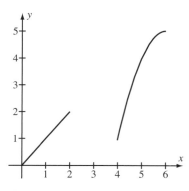

Figure 4

Is f a function? Substantiate your answer.

1.5 Combining Functions

Important Concepts

- Arithmetic Operations
- Functional Composition
- Translations

Arithmetic Operations

If c is a constant and functions f and g have the same domain X, then for $x \in X$ we create "new" functions by defining arithmetic operations on f and g in the following way.

$$(f + g)(x) = f(x) + g(x)$$
$$(f - g)(x) = f(x) - g(x)$$
$$(f \cdot g)(x) = f(x) \cdot g(x)$$
$$\left(\frac{f}{g}\right)(x) = \frac{f(x)}{g(x)}, g(x) \neq 0$$
$$(cf)(x) = c \cdot f(x)$$

A *polynomial function p* is defined by

$$p : x \mapsto p(x) = a_n x^n + a_{n-1} x^{n-1} + \cdots + a_1 x^1 + a_0, a_n \neq 0,$$

where n is a natural number. Polynomial functions are a very important class of function.

Example 1 If $f(x) = x^3$ and $g(x) = 2x$, define $(f + g)(x)$.

Solution Functions f and g have the same domain, which is the set of real numbers $\mathbb{R}$. Therefore, we can define $(f + g)(x)$ by $(f + g)(x) = f(x) + g(x) = x^3 + 2x$. The domain of $(f + g)(x)$ is $\mathbb{R}$. ■

Example 2 If $f(x) = x^3$ and $g(x) = 2x$, define $\left(\dfrac{f}{g}\right)(x)$.

Solution Functions f and g have the same domain, which is the set of real numbers $\mathbb{R}$. Therefore, we can define $\left(\dfrac{f}{g}\right)(x)$ by $\left(\dfrac{f}{g}\right)(x) = \dfrac{f(x)}{g(x)} = \dfrac{x^3}{2x} = \dfrac{x^2}{2}$, as long as $x \neq 0$. The domain of $\left(\dfrac{f}{g}\right)(x)$ is the set of all real numbers except 0. ∎

Self-Test Exercises

1. If $f(x) = 2x^4$, $g(x) = -3x^2$, and $c = -1$, define the function $(cf - g)(x)$.
2. If $f(x) = x$, $g(x) = x^5$, and $c = 10$, evaluate the function $(cf \cdot g)(5)$.

Functional Composition

If f and g are functions such that the range of f is entirely contained in the domain of g, then *g may be composed with f*. The statement *g may be composed with f* can also be stated as *the composition of g with f*. The composition of g with f is denoted by $g \circ f$, and it is defined by $(g \circ f)(x) = g(f(x))$. This explains why the range of f must be entirely contained in the domain of g.

The composition of functions is not commutative; that is, $g \circ f \neq f \circ g$ in general.

Example 3 If $f(x) = x^3$ and $g(x) = 2x$, define $(g \circ f)(x)$.

Solution $(g \circ f)(x) = g(f(x)) = g(x^3) = 2x^3$. Therefore, for given functions f and g, $(g \circ f)(x) = 2x^3$. ∎

Example 4 If $f(x) = x^3$ and $g(x) = 2x$, define $(f \circ g)(x)$.

Solution $(f \circ g)(x) = f(g(x)) = f(2x) = (2x)^3 = 8x^3$. Therefore, for given functions f and g, $(f \circ g)(x) = 8x^3$. ∎

Examples 3 and 4 illustrate that the composition of functions is *not* commutative.

Self-Test Exercises

3. If $f(x) = (x + 2)^2$, define $(f \circ f)(x)$.
4. If $f(x) = x - 3$ and $g(x) = 3 - x$, define $(f \circ g)(x)$.

Translations

A *translation of function f* is simply a movement of the graph of f to the right or left (i.e., a horizontal translation) or up or down (i.e., a vertical translation).

A *horizontal translation of f* is defined by $x \mapsto f(x + h)$:

- If $h > 0$, the horizontal shift is to the left, and the shift per point is $|h|$ units. (See Figure 3a on page 49 of the student text.)
- If $h < 0$, the horizontal shift is to the right, and the shift per point is $|h|$ units. (See Figure 3b on page 49 of the student text.)

A *vertical translation of f* is defined by $x \mapsto f(x) + k$:

- If $k > 0$, the vertical shift is up, and the shift per point is $|k|$ units. (See Figure 4a on page 49 of the text.)
- If $k < 0$, the vertical shift is down, and the shift per point is $|k|$ units. (See Figure 3b on page 49 of the text.)

Example 5 If $f(x) = x$, sketch the graphs of $f(x)$ and $f(x) + 2$ on the same set of axes.

Solution In Figure 1, the graph of $f(x) = x$ is identified as a solid line and the graph of $f(x) + 2$ is identified as a dotted line. A vertical translation or shift of two units per point is illustrated.

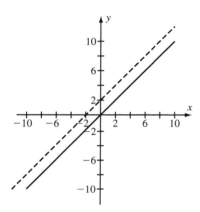

Figure 1   ■

Example 6 If $f(x) = -x^2$, sketch the graphs of $f(x)$ and $f(x - 10)$ on the same set of axes.

Solution In Figure 2, the graph of $f(x) = -x^2$ is identified as the solid curve and the graph of $f(x - 10)$ is identified as the dotted curve. A horizontal translation or shift to the right of ten units per point is illustrated.

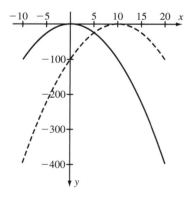

Figure 2   ■

Self-Test Exercises

5. If $f(x) = x^2 + 2x + 1$, describe the function $f(x + 5)$.
6. If $f(x) = x^2 + 2x + 1$, describe the function $f(x) + 5$.

1.6 Trigonometry

Important Concepts

- Types of Measure
- Trigonometric Functions

Types of Measure

The measurement of angles is based on rotation along the unit circle (center $= (0, 0)$, radius $= 1$) in the plane beginning with the positive x-axis. *Counterclockwise rotation* corresponds to positive angles, and *clockwise rotation* corresponds to negative angles.

Angles are measured either in *degrees* or in *radians*. One type of measurement can be converted to the other by using one of the following relationships:

$$\textbf{D:} \quad \text{angle in degrees} = \frac{180°}{\pi} \text{ times (angle in radians)}$$

$$\textbf{R:} \quad \text{angle in radians} = \frac{\pi}{180°} \text{ times (angle in degrees)}$$

If an angle is given in radians, substitute the value in equation **D** and solve for "angle in degrees." If the angle is given in degrees, substitute it in equation **R** and solve for "angle in radians."

Example 1 If the measurement of angle A in radians is $\frac{2\pi}{3}$, what is the measurement of angle A in degrees?

Solution Let x represent the degree measurement of angle A. Using equation **D**, angle in degrees $= \frac{180°}{\pi}$ times (angle in radians). Therefore,

$$x = \left(\frac{180°}{\pi}\right)\left(\frac{2\pi}{3}\right) = 120°. \qquad ■$$

Example 2 If the measurement of angle A in degree measure is $30°$, what is the measurement of angle A in radians?

Solution Let x represent the radian measurement of angle A. Then using equation **R**, angle in radians $= \frac{\pi}{180°}$ times (angle in degrees). Therefore,

$$x = \left(\frac{\pi}{180°}\right)(30°) = \frac{\pi}{6}. \qquad ■$$

Self-Test Exercises

1. Convert 57 degrees to radians.
2. Convert $\frac{\pi}{5}$ radians to degrees.

Trigonometric Functions

There are six *trigonometric functions*:

- $y = \sin(x)$: the *sine* function;
- $y = \cos(x)$: the *cosine* function;
- $y = \tan(x)$: the *tangent* function;
- $y = \csc(x)$: the *cosecant* function;
- $y = \sec(x)$: the *secant* function;
- $y = \cot(x)$: the *cotangent* function.

Consider the following triangle ABC in Figure 1. For the sake of simplicity, the triangle is placed in Quadrant I. The following explanation also applies to identical triangles that are placed in Quadrants II, III, and IV.

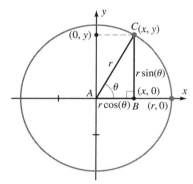

Figure 1

The six trigonometric functions will now be defined in relation to triangle ABC in Figure 1. $AB = x$ is the side *adjacent* to angle θ, $BC = y$ is the side *opposite* to angle θ, and $AC = r$ is the hypotenuse. We note that the circle in Figure 1 is of *arbitrary* radius r instead of the unit radius $r = 1$. This generalizes the formulae and Figure 7 on page 60 of the student text. There are six trigonometric functions:

- $y = \sin(\theta) = \dfrac{opposite}{hypotenuse} = \dfrac{y}{r}$; $\qquad y = \cos(\theta) = \dfrac{adjacent}{hypotenuse} = \dfrac{x}{r}$;

- $y = \csc(\theta) = \dfrac{hypotenuse}{opposite} = \dfrac{r}{y} = \dfrac{1}{\sin(\theta)}$; $\quad y = \sec(\theta) = \dfrac{hypotenuse}{adjacent} = \dfrac{r}{x} = \dfrac{1}{\cos(\theta)}$;

- $y = \tan(\theta) = \dfrac{opposite}{adjacent} = \dfrac{y}{x} = \dfrac{\sin(\theta)}{\cos(\theta)}$; $\quad y = \cot(\theta) = \dfrac{adjacent}{opposite} = \dfrac{x}{y} = \dfrac{1}{\tan(\theta)}$.

If we restrict ourselves to the unit circle where the radius $r = 1$, the trigonometric values for the angles can be found in the first column of the following table.

Angle	Sine	Cosine	Tangent	Cotangent	Secant	Cosecant
0	0	1	0	undef	1	undef
$\pi/6$	$1/2$	$\sqrt{3}/2$	$1/\sqrt{3}$	$\sqrt{3}$	$2/\sqrt{3}$	2
$\pi/4$	$\sqrt{2}/2$	$\sqrt{2}/2$	1	1	$\sqrt{2}$	$\sqrt{2}$
$\pi/3$	$\sqrt{3}/2$	$1/2$	$\sqrt{3}$	$1/\sqrt{3}$	2	$2/\sqrt{3}$
$\pi/2$	1	0	undef	0	undef	1
$2\pi/3$	$\sqrt{3}/2$	$-1/2$	$-\sqrt{3}$	$-1/\sqrt{3}$	-2	$2/\sqrt{3}$
$3\pi/4$	$\sqrt{2}/2$	$-\sqrt{2}/2$	-1	-1	$-\sqrt{2}$	$\sqrt{2}$
$5\pi/6$	$1/2$	$-\sqrt{3}/2$	$-1/\sqrt{3}$	$-\sqrt{3}$	$-2/\sqrt{3}$	2
π	0	-1	0	undef	-1	undef
$7\pi/6$	$-1/2$	$-\sqrt{3}/2$	$1/\sqrt{3}$	$\sqrt{3}$	$-2/\sqrt{3}$	-2
$5\pi/4$	$-\sqrt{2}/2$	$-\sqrt{2}/2$	1	1	$-\sqrt{2}$	$-\sqrt{2}$
$4\pi/3$	$-\sqrt{3}/2$	$-1/2$	$\sqrt{3}$	$1/\sqrt{3}$	-2	$-2/\sqrt{3}$
$3\pi/2$	-1	0	undef	0	undef	-1
$5\pi/3$	$-\sqrt{3}/2$	$1/2$	$-\sqrt{3}$	$-1/\sqrt{3}$	2	$-2/\sqrt{3}$
$7\pi/4$	$-\sqrt{2}/2$	$\sqrt{2}/2$	-1	-1	$\sqrt{2}$	$-\sqrt{2}$
$11\pi/6$	$-1/2$	$\sqrt{3}/2$	$-1/\sqrt{3}$	$-\sqrt{3}$	$2/\sqrt{3}$	-2

For all angles θ, $|\sin(\theta)| \le 1$ and $|\cos(\theta)| \le 1$.

$\sin^2(\theta) + \cos^2(\theta) = 1$ is a very important trigonometric identity. Other trigonometric identities are derived from it. The following table contains some useful trigonometric identities and formulas.

$$1 + \tan^2(\theta) = \sec^2(\theta)$$

$$1 + \cot^2(\theta) = \csc^2(\theta)$$

$$\sin(\theta + \phi) = \sin(\theta)\cos(\phi) + \cos(\theta)\sin(\phi) \qquad \text{Addition Formula}$$

$$\cos(\theta + \phi) = \cos(\theta)\cos(\phi) - \sin(\theta)\sin(\phi) \qquad \text{Addition Formula}$$

$$\sin(2\theta) = 2\sin(\theta)\cos(\theta) \qquad \text{Double Angle Formula}$$

$$\cos(2\theta) = \cos^2(\theta) - \sin^2(\theta) \qquad \text{Double Angle Formula}$$

$$\sin(-\theta) = -\sin(\theta) \qquad \text{Sine is an odd function.}$$

$$\cos(-\theta) = \cos(\theta) \qquad \text{Cosine is an even function.}$$

$$\sin^2\left(\frac{\theta}{2}\right) = \frac{1 - \cos(\theta)}{2} \qquad \text{Half-Angle Formula}$$

$$\cos^2\left(\frac{\theta}{2}\right) = \frac{1 + \cos(\theta)}{2} \qquad \text{Half-Angle Formula}$$

Example 3 Use the *Half-Angle Formula* to calculate the value of $\cos^2(45°)$.

Solution The Half-Angle Formula is given by

$$\cos^2\left(\frac{\theta}{2}\right) = \frac{1 + \cos(\theta)}{2}.$$

In this problem, $\frac{\theta}{2} = 45°$; therefore, $\theta = 90°$ and

$$\cos^2(45°) = \frac{1 + \cos(90°)}{2} = \frac{1 + 0}{2} = \frac{1}{2}.$$

Therefore, the value of $\cos^2(45°)$ is $\frac{1}{2}$. ■

Example 4 Use the trigonometric identity $\sin^2(\theta) + \cos^2(\theta) = 1$ to derive the trigonometric identity $1 + \cot^2(\theta) = \csc^2(\theta)$.

Solution $\sin^2(\theta) + \cos^2(\theta) = 1$ implies that

$$\frac{\sin^2(\theta)}{\sin^2(\theta)} + \frac{\cos^2(\theta)}{\sin^2(\theta)} = \frac{1}{\sin^2(\theta)}, \quad \sin^2(\theta) \neq 0,$$

which is equal to $1 + \cot^2(\theta) = \csc^2(\theta)$. We remark that if $\sin(\theta) = 0$, then $\sin^2(\theta) = 0$ and the expression "$\frac{1}{\sin^2(\theta)} = \frac{1}{0}$" is undefined. This remark substantiates the inclusion of $\sin^2(\theta) \neq 0$ in the statement above. ■

Self-Test Exercises

3. Use the trigonometric identity $\sin^2(\theta) + \cos^2(\theta) = 1$ to derive the trigonometric identity $1 + \tan^2(\theta) = \sec^2(\theta)$.
4. Derive the formula for $\sin(2A)$.

Limits

The Concept of Limit

Important Concepts

- Informal Definition of Limit
- One-Sided Limits
- Methods

Informal Definition of Limit

In this section, we will informally discuss the concepts of limit, obtaining limits from one side, and obtaining a limit using algebraic and graphical methods.

Our discussion is in the realm of real numbers; that is to say, a, b, and c are real numbers. $(a, c) \cup (c, b)$ is an open interval on which real-valued function f is defined and f is not necessarily defined at c. The mathematical statement $\lim_{x \to c} f(x) = \ell$ is read "$f(x)$ has a limit ℓ as x tends to c." Its equivalent mathematical statement, $f(x) \to \ell$ as $x \to c$, is read "$f(x)$ tends to ℓ as x tends to c." This simply means that if x gets close to the value c, but is not necessarily equal to c, then $f(x)$ gets close to the value ℓ. This concept is represented in Figure 1.

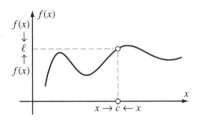

$f(x)$ has the limit ℓ as x tends to c.

Figure 1

If the limit ℓ exists, then it is unique and it is a finite real number. This means that there is only one limiting value and its value is a finite real number.

Example 1 Express the statement "$f(x) = \sin(x)$ tends to 1 as x tends to $\dfrac{\pi}{2}$" as a mathematical statement regarding limits.

Solution In the given statement, the value 1 represents the limiting value ℓ and the value $\dfrac{\pi}{2}$ represents the value c. Therefore, the statement $\lim_{x \to c} f(x) = \ell$ for the given statement becomes $\lim_{x \to \frac{\pi}{2}} (\sin(x)) = 1$. An equivalent statement is $\sin(x) \to 1$ as $x \to \dfrac{\pi}{2}$. ■

Self-Test Exercise

1. By examining its graph, determine whether $\lim\limits_{x \to 2} \dfrac{1}{x}$ exists.

One-Sided Limits

The statement that "a limit is unique" implies that the finite limiting value from the right is equal to the finite limiting value from the left. It is common in many "real-world applications" that a limit may not exist but a one-sided limit does exist. Therefore, we also discuss the concept of one-sided limit and related terminology.

The mathematical expression $\lim\limits_{x \to c^+} f(x) = \ell$ means that as x tends to c *through values larger than c*, then $f(x)$ gets close to ℓ. $\lim\limits_{x \to c^+} f(x) = \ell$ is often described as x approaching c from the right. Similarly, the mathematical expression $\lim\limits_{x \to c^-} f(x) = \ell$ means that as x tends to c *through values smaller than c*, then $f(x)$ gets close to ℓ. $\lim\limits_{x \to c^-} f(x) = \ell$ is often described as x approaching c from the left.

Example 2 If $f(x) = \begin{cases} 1 & x \geq 0 \\ -1 & x < 0 \end{cases}$, evaluate $\lim\limits_{x \to 0} f(x)$.

Solution Consider the graph of $f(x) = \begin{cases} 1 & x \geq 0 \\ -1 & x < 0 \end{cases}$ in the viewing window $[-3, 3] \times [-3, 3]$, as shown in Figure 2.

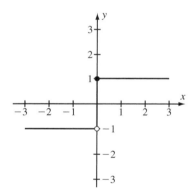

Figure 2

In Figure 2, observe that as zero on the x-axis is approached through values larger than zero, the associated y-values are all one. Similarly, observe that as zero on the x-axis is approached through values smaller than zero, the associated y-values are all negative one. These observations, in mathematical terms, are expressed as $\lim\limits_{x \to 0^+} f(x) = 1$ and $\lim\limits_{x \to 0^-} f(x) = -1$, respectively. Because the limit from the right is not equal to the limit from the left, we conclude that $\lim\limits_{x \to 0} f(x)$ does not exist. A different way to state this is "The limit obtained is not unique and, therefore, does not exist." ■

Example 3 If $f(x) = \dfrac{1}{x}$, evaluate $\lim\limits_{x \to 0^+} f(x)$.

Solution Observe the following values that mimic the method of approaching zero from the right.

Approach from the Right of Zero

$\frac{1}{100000000}$	$\frac{1}{10000000}$	$\frac{1}{1000000}$	$\frac{1}{100000}$	$\frac{1}{10000}$	$\frac{1}{1000}$	$\frac{1}{100}$	$\frac{1}{10}$	$\frac{1}{1}$	$0^+ \leftarrow x$
100000000	10000000	1000000	100000	10000	1000	100	10	1	$f(x)$

As we progress from right to left, the values of x become smaller. In fact, they are approaching zero but are not equal to zero. However, as we progress from right to left, the values of $f(x) = \dfrac{1}{x}$ become larger. As x gets closer to zero from the right, $f(x)$ increases without bound. Based on these observations, we conclude that $\lim\limits_{x \to 0^+} f(x) = \infty$; that is, the limit does not exist because the limiting value is not a finite real number. Graphical analysis also supports this conclusion. The graph of $f(x) = \dfrac{1}{x}$ in the viewing window $[0, 1] \times [0, 100]$ appears in Figure 3.

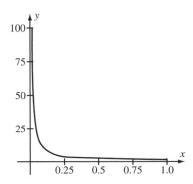

Figure 3

Self-Test Exercise

2. Evaluate $\lim\limits_{x \to 4} \sqrt{x}$, if it exists.

Algebraic Method

The Algebraic Method utilizes algebraic manipulation. If $f(x)$ is a rational function that is not defined at one or more expressions that appear in the denominator, when utilizing the Algebraic Method, one factors the numerator and denominator of rational function $f(x)$ and cancels all factors that appear in both the numerator and denominator. Once all of the common factors have been cancelled, the limit is calculated.

Example 4 If $f(x) = \dfrac{x^2 - 7x + 12}{x - 4}$, evaluate $\lim\limits_{x \to 4} f(x)$, if it exists.

Solution We observe that $f(x)$ is not defined at $x = 4$. Therefore, the Algebraic Method is applied.

$$f(x) = \frac{x^2 - 7x + 12}{x - 4} = \frac{(x - 3)(x - 4)}{(x - 4)} = x - 3, \text{ as long as } x \neq 4$$

We observe that $\lim\limits_{x \to 4} f(x) = \lim\limits_{x \to 4^+} f(x) = \lim\limits_{x \to 4^-} f(x) = 1$.

Graphical Method

If the graph of function $f(x)$ passes through the top or bottom of the boundaries of the viewing window, $[c - \delta, c + \delta] \times [\ell - \epsilon, \ell + \epsilon]$ where $c - \delta \leq x \leq c + \delta, \ell - \epsilon \leq y \leq \ell + \epsilon, \delta > 0$ and $\epsilon > 0$, then the value of δ is decreased until the graph of function $f(x)$ appears as a diagonal in this "new" viewing window. In other words, the graph of function $f(x)$ starts at a top (or bottom) corner of the viewing window and passes through the opposite on-the-diagonal bottom (or top) corner of the viewing window. (See Figures 4 and 5.) Then a vertical line through $x = c$ is drawn. The distance from the bottom of the viewing window to the point of intersection of the vertical line $x = c$ and the graph of function $f(x)$ is the value ℓ.

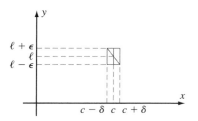

Figure 4   **Figure 5**

The Graphical Method is useful when a specific degree of accuracy is needed; in other words, this method is used to answer the question "How close to the value c does the value x have to be in order to have $f(x)$ be within a certain value, say .01, of ℓ?".

Example 5 For $f(x) = \dfrac{1}{x}$, $\lim\limits_{x \to 2} f(x) = \dfrac{1}{2}$. Find a positive number δ such that $f(x)$ stays within 0.01 of $\frac{1}{2}$ when x stays within δ of 2.

Solution In this example, we set $c = 2, \ell = \dfrac{1}{2}$, and $\epsilon = 0.01$. We experimented with different viewing windows until we obtained the graph in Figure 6 in the viewing window

$$[1.96, 2.04] \times [0.49, 0.51] = [2 - .04, 2 + .04] \times \left[\frac{1}{2} - .01, \frac{1}{2} + .01 \right].$$

Therefore, the value $\delta = .04$ suffices.

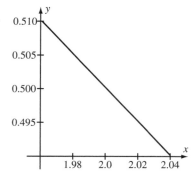

Figure 6

Self-Test Exercise

3. If $\lim\limits_{x \to 3}(x^2) = 9$, find a positive number δ such that $f(x) = x^2$ stays within 0.01 of 9 when x stays within δ of 3.

2.2 Limit Theorems

Important Concepts

- Formal Definition of Limit
- Formal Definition of One-Sided Limit
- Theorems

Formal Definition of Limit

An informal definition of *limit* was given in Section 2.1. In this section, we'll use a formal definition. This definition is often referred to as the ϵ-δ *definition of limit*.

Let f be a real-valued function that is defined in an open interval to the left of the real number c and in an open interval to the right of the real number c. The limit of $f(x)$ is equal to ℓ as x approaches c if for any $\epsilon > 0$ there exists a $\delta > 0$ such that $|f(x) - \ell| < \epsilon$ for all values of x such that $0 < |x - c| < \delta$. Figure 1 graphically illustrates the ϵ-δ *definition of limit*.

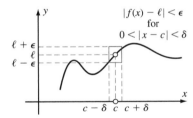

$f(x)$ has the limit ℓ as x tends to c.

Figure 1

Example 1 Demonstrate that the statement $\lim\limits_{x \to 1} x^2 - 5x + 6 = 2$ satisfies the ϵ-δ definition of limit.

Solution To demonstrate that the statement $\lim\limits_{x \to 1} x^2 - 5x + 6 = 2$ satisfies the ϵ-δ definition of limit, you need to produce a value for $\delta > 0$ that will satisfy, for $\epsilon > 0$, $|(x^2 - 5x + 6) - 2| < \epsilon$ whenever $0 < |x - 1| < \delta$. The following is a standard method to produce a value for δ that satisfies the ϵ-δ definition of limit.

Assume $\epsilon > 0$.

$$
\begin{aligned}
|(x^2 - 5x + 6) - 2| < \epsilon &\qquad \text{whenever } 0 < |x - 1| < \delta &\Leftrightarrow \\
|(x^2 - 5x + 4)| < \epsilon &\qquad \text{whenever } 0 < |x - 1| < \delta &\Leftrightarrow \\
|(x - 4)(x - 1)| < \epsilon &\qquad \text{whenever } 0 < |x - 1| < \delta &\Leftrightarrow \\
|(x - 4)||(x - 1)| < \epsilon &\qquad \text{whenever } 0 < |x - 1| < \delta.
\end{aligned}
$$

If $|x - 1| < 1$, then

$$-1 < x - 1 < 1 \Leftrightarrow 0 < x < 2 \Leftrightarrow -4 < x - 4 < -2.$$

This implies that $4 > -(x-4) > 2$ so $-(x-4) < 4$. Since

$$|-(x-4)| = |-1 \cdot (x-4)| = |-1| \cdot |x-4| = |x-4|,$$

we conclude that $|-(x-4)| < 4$, which is the same as $|x-4| < 4$. However, $|x-4| < 4$ and $|(x-4)||(x-1)| < \epsilon$ whenever $0 < |x-1| < \delta$ implies that $|(x-1)| < \dfrac{\epsilon}{4}$ whenever $0 < |x-1| < \delta$; therefore, δ is the smaller of the two values 1 and $\dfrac{\epsilon}{4}$.

This is expressed as $\delta = \min\left\{1, \dfrac{\epsilon}{4}\right\}$.   ■

Self-Test Exercises

1. Show that $\lim\limits_{x \to 2}(x^2 - 9) = -5$ satisfies the ϵ-δ definition of limit.
2. Show that $\lim\limits_{x \to 5}(2x + 6) = 16$ satisfies the ϵ-δ definition of limit.

Formal Definition of One-Sided Limit

We can also formalize the concept of a one-sided limit in the same manner as the ϵ-δ definition of limit.

Suppose f is a function defined in an interval just to the right of c. If for every $\epsilon > 0$ there is a $\delta > 0$ such that $|f(x) - \ell| < \epsilon$ for each x that satisfies $c < x < c + \delta$, then we say that f has a limit ℓ as a *limit from the right*, or f has *right limit* ℓ at c. Mathematically, this is expressed as $\lim\limits_{x \to c^+} f(x) = \ell$. Figure 2 illustrates the concept "f has *right limit* ℓ at c."

Suppose f is a function defined in an interval just to the left of c. If for every $\epsilon > 0$ there is a $\delta > 0$ such that $|f(x) - \ell| < \epsilon$ for each x that satisfies $c - \delta < x < c$, then we say that f has a limit ℓ as a *limit from the left*, or f has *left limit* ℓ at c. Mathematically, this is expressed as $\lim\limits_{x \to c^-} f(x) = \ell$. Figure 3 illustrates the concept of left limit.

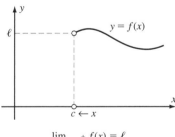

$\lim_{x \to c^+} f(x) = \ell$ $\lim_{x \to c^-} f(x) = \ell$

Figure 2   **Figure 3**

Example 2 If it exists, evaluate $\lim\limits_{x \to 2^+} f(x)$ if $f(x) = \begin{cases} |x-2| & x < -1 \\ 1 & -1 \le x < 2. \\ x-1 & 2 \le x \end{cases}$

Solution In the viewing window $[-5, 5] \times [0, 7]$, the graph of function $f(x)$ appears in Figure 4. Implementing the definition of $f(x) = x - 1$ for x greater than or equal to two, we see that $f(5) = 4$, $f(4) = 3$, $f(3) = 2$, $f(2.5) = 1.5$, $f(2.25) = 1.25$, $f(2.1) = 1.1$, $f(2.01) = 1.01$, $f(2.001) = 1.001$, $f(2.0001) = 1.0001$. These results illustrate that as $x \to 2^+$, $f(x) \to 1$ and is reflected in the graph of $f(x) = x$ (Figure 4) by the following observations. Each horizontal block has length .5 units and each vertical block is .2 units. Therefore, at $x = 5$, the vertical "y-distance" is 4; at $x = 4$, the vertical "y-distance" is 3; at $x = 3$, the vertical "y-distance" is 2; at $x = 2.5$, the vertical "y-distance" is 1.5; and at $x = 2$, the vertical "y-distance" is 1.

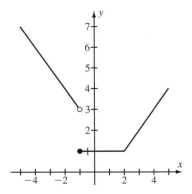

Figure 4

Using graphical analysis, we determine that $\lim_{x \to 2^+} f(x) = 1$. ◼

Theorems

For your convenience, we'll list several theorems that can be very useful when you are calculating limits. We'll also provide examples to illustrate applications of the theorems.

Theorem 1

Let $a < c < b$. A function f with a domain that contains a set of open intervals of the form $(a, c) \cup (c, b)$ cannot have two distinct limits at c.

Example 3 If $f(x) = \begin{cases} 1 & x \geq 0 \\ -1 & x < 0 \end{cases}$, evaluate $\lim_{x \to 0} f(x)$.

Solution The domain of $f(x)$ is $(-\infty, \infty)$ and it contains a set of open intervals of the form $(-3, 0) \cup (0, 3)$. Consider the graph of $f(x) = \begin{cases} 1 & x \geq 0 \\ -1 & x < 0 \end{cases}$ in the viewing window $[-3, 3] \times [-3, 3]$, which is shown in Figure 5.

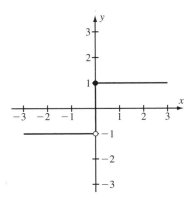

Figure 5

Graphical analysis shows that $\lim_{x \to 0^+} f(x) = 1$ but $\lim_{x \to 0^-} f(x) = -1$. Because the limit from the right is not equal to the limit from the left, we conclude that $\lim_{x \to 0} f(x)$ does not exist. A different way to state this is "The limit obtained is not unique and, therefore, does not exist." We remark that this solution illustrates an application of Theorem 1. As stated earlier, the domain of $f(x)$ contains a set of open intervals of the form $(-3, 0) \cup (0, 3)$. Theorem 1 states that f cannot have two distinct limits at $c = 0$, which agrees with our earlier answer. ∎

Theorem 2   If f and g are two functions, c is a real number, and $\lim_{x \to c} f(x)$ and $\lim_{x \to c} g(x)$ exist, then the following equations hold:

a. $\lim_{x \to c}(f + g)(x) = \lim_{x \to c}(f(x) + g(x)) = \lim_{x \to c} f(x) + \lim_{x \to c} g(x)$

   and $\lim_{x \to c}(f - g)(x) = \lim_{x \to c}(f(x) - g(x)) = \lim_{x \to c} f(x) - \lim_{x \to c} g(x)$

b. $\lim_{x \to c}(f \cdot g)(x) = \lim_{x \to c}(f(x)g(x)) = \left(\lim_{x \to c} f(x)\right)\left(\lim_{x \to c} g(x)\right)$

c. $\lim_{x \to c}\left(\dfrac{f}{g}\right)(x) = \lim_{x \to c}\left(\dfrac{f(x)}{g(x)}\right) = \dfrac{\lim_{x \to c} f(x)}{\lim_{x \to c} g(x)}$, provided $\lim_{x \to c} g(x) \neq 0$

d. $\lim_{x \to c}(\alpha \cdot f(x)) = \alpha \cdot \left(\lim_{x \to c} f(x)\right)$ for any constant α

Although Theorem 2 is defined using two functions (f and g), the results of parts (a) through (d) can be extended for a finite number of functions $f_1, f_2, \ldots, f_n$.

Theorem 3   If $p(x)$ is a polynomial, then $\lim_{x \to c} p(x) = p(c)$.

If $q(x)$ is also a polynomial and $q(c) \neq 0$, then $\lim_{x \to c}\left(\dfrac{p(x)}{q(x)}\right) = \dfrac{p(c)}{q(c)}$.

Theorem 4   Let n be a positive integer and c be a real number. Also assume that c is positive if n is even. Under these conditions, $\lim_{x \to c} \sqrt[n]{x} = \sqrt[n]{c}$.

Here are some examples to which these theorems apply.

Example 4 If $f(x) = x^2$ and $g(x) = -7x - 12$, verify that

$$\lim_{x \to 1}(f(x) + g(x)) = \lim_{x \to 1} f(x) + \lim_{x \to 1} g(x)$$

and

$$\lim_{x \to 2}(f(x) - g(x)) = \lim_{x \to 2} f(x) - \lim_{x \to 2} g(x).$$

Solution $(f(x) + g(x)) = (x^2) + (-7x - 12)$. Applying Theorem 3 to the polynomial $x^2 - 7x - 12$, we obtain

$$\lim_{x \to 1}(f(x) + g(x)) = 1^2 - (7)(1) - 12 = -18.$$

Applying Theorem 3 to the polynomial x^2 and once again to the polynomial $-7x - 12$, we obtain

$$\lim_{x \to 1} f(x) = \lim_{x \to 1}(x^2) = 1^2 = 1$$

and

$$\lim_{x \to 1} g(x) = \lim_{x \to 1}(-7x - 12) = (-7)(1) - 12 = -19.$$

Therefore,

$$\lim_{x \to 1} f(x) + \lim_{x \to 1} g(x) = 1 - 19 = -18 = \lim_{x \to 1}(f(x) + g(x)).$$

Similarly,

$$(f(x) - g(x)) = (x^2) - (-7x - 12) = x^2 + 7x + 12.$$

Applying Theorem 3 to the polynomial $x^2 + 7x + 12$, we obtain

$$\lim_{x \to 2}(f(x) - g(x)) = 2^2 + (7)(2) + 12 = 30.$$

Applying Theorem 3 to the polynomial x^2, and once again to the polynomial $7x + 12$, we obtain

$$\lim_{x \to 2} f(x) = \lim_{x \to 2}(x^2) = 2^2 = 4$$

and

$$\lim_{x \to 2} g(x) = \lim_{x \to 2}(-7x - 12) = (-7)(2) - 12 = -26.$$

Therefore,

$$\lim_{x \to 2} f(x) - \lim_{x \to 2} g(x) = 4 - (-26) = 30 = \lim_{x \to 2}(f(x) - g(x)). \qquad \blacksquare$$

Example 5 If $f(x) = \sqrt{x}$ and $g(x) = \dfrac{1}{x}$, verify that

$$\lim_{x \to 3}(f \cdot g)(x) = \left(\lim_{x \to 3} f(x)\right)\left(\lim_{x \to 3} g(x)\right).$$

Solution $(f \cdot g)(x) = (\sqrt{x})\left(\dfrac{1}{x}\right) = \dfrac{1}{\sqrt{x}}$. Applying Theorem 2c to $\dfrac{1}{\sqrt{x}}$, then Theorem 3 to the expression 1 and Theorem 4 to the expression $\sqrt{x}$, we obtain

$$\lim_{x \to 3}(f \cdot g)(x) = \lim_{x \to 3}\left(\frac{1}{\sqrt{x}}\right) = \frac{\lim_{x \to 3} 1}{\lim_{x \to 3}\sqrt{x}} = \frac{1}{\sqrt{3}}\lim_{x \to 3}(f \cdot g)(x) = \lim_{x \to 3}\left(\frac{1}{\sqrt{x}}\right) = \frac{\lim_{x \to 3} 1}{\lim_{x \to 3}\sqrt{x}} = \frac{1}{\sqrt{3}}.$$

Applying Theorem 4 to the expression $\sqrt{x}$ and applying Theorem 3 to the expression $\dfrac{1}{x}$, we obtain

$$\lim_{x \to 3} f(x) = \lim_{x \to 3}\sqrt{x} = \sqrt{3}$$

and

$$\lim_{x \to 3} g(x) = \lim_{x \to 3}\frac{1}{x} = \frac{1}{3}.$$

Therefore,

$$\left(\lim_{x \to 3} f(x)\right)\left(\lim_{x \to 3} g(x)\right) = (\sqrt{3})\left(\frac{1}{3}\right) = \frac{1}{\sqrt{3}} = \lim_{x \to 3}(f \cdot g)(x). \qquad \blacksquare$$

Example 6 If $f(x) = x^2 + 6x + 7$ and $g(x) = x - 7$, verify that $\displaystyle\lim_{x \to -2}\left(\frac{f}{g}\right)(x) = \frac{\lim_{x \to -2} f(x)}{\lim_{x \to -2} g(x)}$,

provided $\displaystyle\lim_{x \to -2} g(x) \neq 0$.

Solution $\dfrac{f(x)}{g(x)} = \dfrac{x^2 + 6x + 7}{x - 7}$. Applying Theorem 3 to $f(x)$ and to $g(x)$, we obtain

$$\lim_{x \to -2} f(x) = \lim_{x \to -2} (x^2 + 6x + 7) = (-2)^2 + 6(-2) + 7 = -1$$

and

$$\lim_{x \to -2} g(x) = \lim_{x \to -2} (x - 7) = -2 - 7 = -9.$$

Since $f(x)$ and $g(x)$ are the numerator and denominator, respectively, of the rational function $\dfrac{f(x)}{g(x)}$, we obtain from Theorem 3

$$\begin{aligned}
\lim_{x \to -2} \left(\frac{f(x)}{g(x)} \right) &= \lim_{x \to -2} \left(\frac{x^2 + 6x + 7}{x - 7} \right) \\
&= \frac{(-2)^2 + 6(-2) + 7}{-2 - 7} \\
&= \frac{-1}{-9} \\
&= \frac{1}{9}.
\end{aligned}$$

Therefore,

$$\frac{\lim\limits_{x \to -2} f(x)}{\lim\limits_{x \to -2} g(x)} = \frac{-1}{-9} = \frac{1}{9} = \lim_{x \to -2} \left(\frac{f(x)}{g(x)} \right). \qquad \blacksquare$$

Example 7 If $\alpha = 5$ and $f(x) = x$, verify that $\lim\limits_{x \to \frac{2}{3}} (\alpha f(x)) = \alpha \left(\lim\limits_{x \to \frac{2}{3}} f(x) \right)$.

Solution $\alpha f(x) = 5x$. Applying Theorem 3 to the polynomial function $5x$, we obtain

$$\lim_{x \to \frac{2}{3}} (\alpha f(x)) = \lim_{x \to \frac{2}{3}} 5x = (5) \left(\frac{2}{3} \right) = \frac{10}{3}.$$

Applying Theorem 3 to the polynomial function x, we obtain

$$\lim_{x \to \frac{2}{3}} (f(x)) = \lim_{x \to \frac{2}{3}} x = \frac{2}{3}.$$

Therefore,

$$\alpha \left(\lim_{x \to \frac{2}{3}} f(x) \right) = 5 \left(\frac{2}{3} \right) = \frac{10}{3} = \lim_{x \to \frac{2}{3}} (\alpha f(x)). \qquad \blacksquare$$

Example 8 If $p(x) = x^2 + 4x + 4$, $q(x) = x + 2$ and $c = 1$, verify that $\lim\limits_{x \to c} p(x) = p(c)$ and $\lim\limits_{x \to c} \left(\dfrac{p(x)}{q(x)} \right) = \dfrac{p(c)}{q(c)}$ graphically.

Solution Consider the graphs of $p(x)$, the horizontal line $y = 9$, and the vertical line $x = 1$, which appear in the window $[-2, 2] \times [0, 10]$ on the same set of axes, as shown in Figure 6.

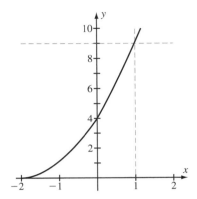

Figure 6

In the figure, we see that the graphs of $p(x)$, the horizontal line $y = 1$, and the vertical line $x = 1$ all intersect at the point $(1, 9)$. This implies that $p(1) = 9$. By examining the graph of $p(x)$, we can see that $\lim_{x \to 1} p(x) = \lim_{x \to 1}(x^2 + 4x + 4) = 9$. Therefore, $\lim_{x \to 1} p(x) = \lim_{x \to 1}(x^2 + 4x + 4) = 9 = p(1)$.

Now let's consider the graphs of $r(x) = \dfrac{p(x)}{q(x)}$, the horizontal line $y = 3$, and the vertical line $x = 1$. These graphs appear in the window $[-2, 2] \times [0, 4]$ on the same set of axes, which is shown in Figure 7.

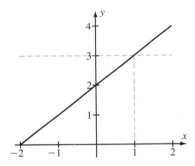

Figure 7

From the graph, we see that the graphs of the slanted line $r(x) = \dfrac{p(x)}{q(x)}$, the horizontal line $y = 3$, and the vertical line $x = 1$ all intersect at the point $(1, 3)$. This implies that $r(1) = 3$. By examining the graph of $r(x) = \dfrac{p(x)}{q(x)}$, we can see that $\lim_{x \to 1}\left(\dfrac{p(x)}{q(x)}\right) = 3$. Since $p(1) = 9$ and $q(1) = 1 + 2 = 3$, we have

$$\lim_{x \to 1}\left(\frac{p(x)}{q(x)}\right) = 3 = r(1) = \frac{p(1)}{q(1)}. \qquad \blacksquare$$

Example 9 If $c = 4$ and $n = 2$, verify that $\lim_{x \to c} \sqrt[n]{x} = \sqrt[n]{c}$ graphically.

Solution Recall that $\sqrt[2]{x} = \sqrt{x}$. Consider the graphs of $s(x) = \sqrt{x}$, the horizontal line $y = 2$, and the vertical line $x = 4$. These graphs appear in the window $[2, 6] \times [0, 4]$ on the same set of axes, as shown in Figure 8.

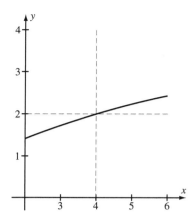

Figure 8

From the graph, we see that the graphs of $s(x) = \sqrt{x}$, the horizontal line $y = 2$, and the vertical line $x = 4$ all intersect at the point (4, 2). This implies that $s(4) = 2$. Note that $s(4) = \sqrt{4} = 2$. By examining the graph of $s(x)$, we can see that $\lim_{x \to 4} s(x) = 2$. Therefore,

$$\lim_{x \to 4} s(x) = \lim_{x \to 4} \sqrt{x} = \sqrt{4} = 2.$$ ∎

Theorem 5 If $\lim_{x \to c} g(x) = 0$ and if $\lim_{x \to c} f(x) \neq 0$, then $\lim_{x \to c} \left(\dfrac{f(x)}{g(x)} \right)$ does not exist.

We remark that if both $\lim_{x \to c} f(x) = 0$ and $\lim_{x \to c} g(x) = 0$, then $\lim_{x \to c} \left(\dfrac{f(x)}{g(x)} \right)$ may or may not exist.

Example 10 If $f(x) = x + 5$ and $g(x) = x^3$, verify graphically that $\lim_{x \to 0} \left(\dfrac{f(x)}{g(x)} \right)$ does not exist.

Solution $\lim_{x \to 0} g(x) = 0^3 = 0$. $\lim_{x \to 0} f(x) = 0 + 5 = 5 \neq 0$. Consider the graph of

$$r(x) = \frac{f(x)}{g(x)} = \frac{x + 5}{x^3}$$

in the window $[-5, 5] \times [-10{,}000, 10{,}000]$, as shown in Figure 9.

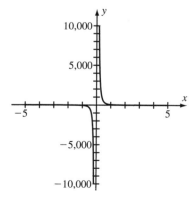

Figure 9

The graph implies that as $x \to 0^+$, the "y-distance" from the point on the x-axis to the graph increases vertically (in the positive direction) while as $x \to 0^-$, the "y-distance" from the point on the x-axis to the graph decreases vertically (in the negative direction). Therefore, $\lim\limits_{x \to 0} \left(\dfrac{f(x)}{g(x)} \right)$ does not exist because the right and left limits are not finite and are not equal. ■

Theorem 6 If f, g, and h are functions with domains that each contain $S = (a, c) \cup (c, b)$, $g(x) \le f(x) \le h(x)$ for all x in S, and $\lim\limits_{x \to c} g(x) = \lim\limits_{x \to c} h(x) = \ell$, then $\lim\limits_{x \to c} f(x) = \ell$. This theorem is referred to as the *Pinching Theorem*.

Example 11 If $f(x) = 2^x$, $g(x) = \dfrac{x}{2} + 1$ and $h(x) = 5x + 1$, calculate $\lim\limits_{x \to 0} 2^x$, both graphically and computationally.

Solution We shall first calculate this limit graphically. Functions $f(x) = 2^x$, $g(x) = \dfrac{x}{2} + 1$, and $h(x) = 5x + 1$ are given. By comparing their graphs in viewing window $[0, 1] \times [0, 6]$, as shown in Figure 10 (the graph of $h(x)$ is the dashed line, the graph of $g(x)$ is the dotted line, and the graph of $f(x)$ is the solid line), we see that for $0 \le x \le 1$, $g(x) \le f(x) \le h(x)$. Therefore,

$$\lim_{x \to 0^+} g(x) \le \lim_{x \to 0^+} f(x) \le \lim_{x \to 0^+} h(x),$$

which implies that $1 \le \lim\limits_{x \to 0^+} f(x) \le 1$. So $\lim\limits_{x \to 0^+} f(x) = 1$ by the Pinching Theorem. Now let's consider a second graph in the viewing window $[-1, 0] \times [-1, 2]$ (see Figure 11).

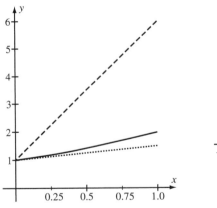

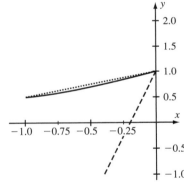

Figure 10   **Figure 11**

Comparing the graphs of the dashed line, the dotted line, and the solid line, we see that for $-1 \le x \le 0$, their respective graphs establish that $h(x) \le f(x) \le g(x)$. Therefore,

$$\lim_{x \to 0^-} h(x) \le \lim_{x \to 0^-} f(x) \le \lim_{x \to 0^-} g(x),$$

which implies that $1 \le \lim\limits_{x \to 0^-} f(x) \le 1$, so by the Pinching Theorem $\lim\limits_{x \to 0^-} f(x) = 1$. $\lim\limits_{x \to 0^-} f(x) = 1$ and $\lim\limits_{x \to 0^+} f(x) = 1$ establish that $\lim\limits_{x \to 0} f(x) = 1$.

We will now show that $\lim\limits_{x \to 0} f(x) = 1$ using non-graphical methods. Let's create a table of values that displays functional values for $f(x)$, $g(x)$, and $h(x)$ when $-1 \le x \le 1$.

Approach From the Right of Zero

.00001	.0001	.0010	.0100	.1000	1.0000	$0^+ \leftarrow x$
1.0001	1.0005	1.0005	1.0500	1.5000	6.0000	$h(x) = 5x + 1$
1.0000	1.0001	1.0007	1.0070	1.0718	2.0000	$f(x) = 2^x$
1.0000	1.0001	1.0005	1.0050	1.0500	1.5000	$g(x) = \frac{x}{2} + 1$

Approach From the Left of Zero

$x \to 0^-$	-1	$-.1$	$-.01$	$-.001$	$-.0001$	$-.00001$
$g(x) = \frac{x}{2} + 1$	.5000	.9500	.9950	.9995	1.0000	1.0000
$f(x) = 2^x$	.50000	.9930	.9931	.9993	.9999	1.0000
$h(x) = 5x + 1$	-4.0000	.5000	.9500	.9950	.9995	1.0000

Since $\lim_{x \to 0^-} f(x) = 1$ and $\lim_{x \to 0^+} f(x) = 1$, establish that $\lim_{x \to 0} f(x) = 1$. ∎

Theorem 7 For every real number c, $\lim_{t \to c} \sin(t) = \sin(c)$ and $\lim_{t \to c} \cos(t) = \cos(c)$.

Example 12 If $c = \dfrac{\pi}{4}$, show graphically that $\lim_{t \to c} \sin(t) = \sin(c)$ and $\lim_{t \to c} \cos(t) = \cos(c)$.

Solution Consider the graphs of the sine function, the cosine function, the vertical line $x = \dfrac{\pi}{4}$, and the horizontal line $y = \dfrac{\sqrt{2}}{2}$ in the viewing window $[0, 2] \times [-0.5, 1.5]$, as shown in Figure 12. (The gray curve is the graph of $y = \cos(x)$ and the black curve is the graph of $y = \sin(x)$ in Figure 12.)

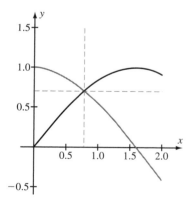

Figure 12

The graph establishes $\lim_{x \to \frac{\pi}{4}} \sin(x) = \dfrac{\sqrt{2}}{2} = \sin\left(\dfrac{\pi}{4}\right)$ and $\lim_{x \to \frac{\pi}{4}} \cos(x) = \dfrac{\sqrt{2}}{2} = \cos\left(\dfrac{\pi}{4}\right)$. ∎

Theorem 8 If t is measured in *radians*, then $\lim_{t \to 0} \left(\dfrac{\sin(t)}{t}\right) = 1$ and $\lim_{t \to 0} \left(\dfrac{1 - \cos(t)}{t^2}\right) = \dfrac{1}{2}$.

Self-Test Exercises

3. If $f(x) = 2x$ and $g(x) = x^3 + 5$, evaluate $\lim\limits_{x \to 3} \left(\dfrac{f}{g} \right)(x)$, if it exists.

4. Evaluate $\lim\limits_{x \to 1} \sqrt{4x}$, if it exists.

2.3 Continuity

Important Concepts

- Definitions of Continuity
- Theorems

Continuity at a Point

Suppose that a function f is defined on an open interval that contains the point c. We say that *f is continuous at c*, provided that $\lim\limits_{x \to c} f(x) = f(c)$.

The statement $\lim\limits_{x \to c} f(x) = f(c)$ implies that three conditions are satisfied:

1. f is defined at c, i.e., $f(c)$ exists.
2. $\lim\limits_{x \to c} f(x)$ exists.
3. $\lim\limits_{x \to c} f(x) = f(c)$.

Continuity of a function at a point is graphically illustrated in Figure 1.

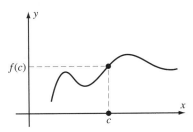

$f(x)$ has the limit $f(c)$ as x tends to c.

Figure 1

We only consider continuity of a function at points of its domain. If a point c is in the domain of f and f is not continuous at c, then we say that *f is discontinuous at c*. Point c is called a *point of discontinuity of f*.

Continuity on an Interval

When a function is continuous at every point in an interval, we say that the *function is continuous on the interval*.

Continuity on a Domain

When a function is continuous at every point in its domain, we say that the *function is a continuous function*.

An Equivalent Definition of Continuity

If $x = c + \Delta x$, then $x \to c$ is equivalent to $\Delta x \to 0$. With this notation, the definition of continuity at a point, $\lim_{x \to c} f(x) = f(c)$, is equivalent to the definition $\lim_{\Delta x \to 0} f(c + \Delta x) = f(c)$.

Continuous Extensions

Consider a function f that is continuous on $(a, c) \cup (c, b)$. If $\lim_{x \to c} f(x)$ exists, then we can extend the continuity of f to the entire interval (a, b) in the following way. Define

$$F(x) = \begin{cases} f(x) & \text{if} \quad x \neq c \\ \lim_{x \to c} f(x) & \text{if} \quad x = c \end{cases}.$$

The function $F(x)$ is called *a continuous extension of* f.

Functions $f(x)$ and $F(x)$ are not the same functions because their respective domains are not equal.

Example 1 Define a continuous extension of $f(x) = \dfrac{x^2 - 9}{x + 3}$.

Solution $\lim_{x \to -3} f(x) = \lim_{x \to -3} \dfrac{(x + 3)(x - 3)}{x + 3} = \lim_{x \to -3} (x - 3) = -6.$

Therefore, we define $F(x)$, a continuous extension of $f(x)$, by

$$F(x) = \begin{cases} f(x) & \text{if} \quad x \neq -3 \\ -6 & \text{if} \quad x = -3 \end{cases}.$$

One-Sided Continuity

Suppose f is a function with a domain that contains $[c, b)$. If $\lim_{x \to c^+} f(x) = f(c)$, then we say that *f is continuous from the right at c* or *f is right-continuous at c*. Figure 2 represents right-continuity at a point.

Suppose f is a function with a domain that contains $(a, c]$. If $\lim_{x \to c^-} f(x) = f(c)$, then we say that *f is continuous from the left at c* or *f is left-continuous at c*. Figure 3 represents left-continuity at a point.

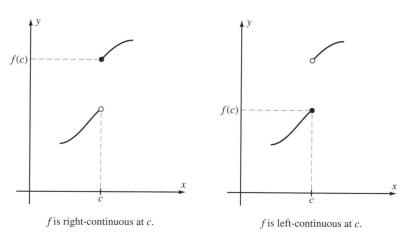

f is right-continuous at c. f is left-continuous at c.

Figure 2   **Figure 3**

If both $\lim_{x \to c^+} f(x)$ and $\lim_{x \to c^-} f(x)$ exist but are not equal, then f *is said to have a jump discontinuity at* c.

Example 2 If $f(x) = \begin{cases} 1 & if & x \geq 0 \\ -1 & if & x < 0 \end{cases}$, evaluate $\lim_{x \to 0} f(x)$.

Solution The domain of $f(x)$ contains a set of open intervals of the form $(-3, 0) \cup (0, 3)$.
Consider the graph of $f(x) = \begin{cases} 1 & if & x \geq 0 \\ -1 & if & x < 0 \end{cases}$ in the viewing window $[-3, 3] \times [-3, 3]$,
as shown in Figure 4.

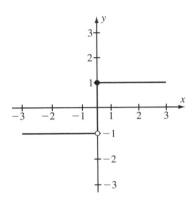

Figure 4

Both graphical and algebraic analysis show that $\lim_{x \to 0+} f(x) = 1$, but $\lim_{x \to 0-} f(x) = -1$.
Function f has a jump discontinuity at $x = 0$. ∎

Self-Test Exercises

1. If $f(x) = \begin{cases} 1 & if & x \geq 0 \\ -1 & if & x < 0 \end{cases}$, evaluate $\lim_{x \to 3} f(x)$.

2. If $f(x) = \begin{cases} 1 & if & x \geq 3 \\ x & if & x < 3 \end{cases}$, evaluate $\lim_{x \to 3+} f(x)$.

Theorems

Theorem 1

Let α be a constant.

a. If f and g are functions that are continuous at $x = c$, then so are functions $f + g$, $f - g$,
    $\alpha \cdot f$, and $f \cdot g$. If $g(c) \neq 0$, then function $\dfrac{f}{g}$ is also continuous at $x = c$.

b. If f and g are functions that are right-continuous at $x = c$, then so are functions $f + g$,
    $f - g, \alpha \cdot f$, and $f \cdot g$. If $g(c) \neq 0$, then function $\dfrac{f}{g}$ is also right-continuous at $x = c$.

c. If f and g are functions that are left-continuous at $x = c$, then so are functions $f + g$,
    $f - g, \alpha \cdot f$, and $f \cdot g$. If $g(c) \neq 0$, then function $\dfrac{f}{g}$ is also left-continuous at $x = c$.

Example 3 For which values of x is function $f \cdot g$ continuous if $f(x) = x^5$ and $g(x) = x - 3$?

Solution $(f \cdot g)(x) = (f(x))(g(x)) = (x^5)(x - 3) = x^6 - 3x^5$. Because $(f \cdot g)(x)$ is a
polynomial, we conclude that function $f \cdot g$ is continuous for all real x. This result substantiates
the statement of Theorem 1(a): $f(x) = x^5$ and $g(x) = x - 1$ are continuous functions for all
real x, therefore $(f \cdot g)(x) = x^6 - 3x^5$ is continuous for all real x. ∎

If the image of a function f is contained in the domain of a function g, then the composition of g with f is denoted as $(g \circ f)(x)$ and defined by $(g \circ f)(x) = g(f(x))$. The next theorem states that composition preserves continuity.

Theorem 2

Suppose that the image of a function f is contained in the domain of a function g. If f is continuous at c and g is continuous at $f(c)$, then the composed function $g \circ f$ is continuous at c.

Example 4 If $f(x) = 5x$ and $g(x) = \dfrac{1}{x}$, show that $(g \circ f)(x)$ is continuous at $x = 3$.

Solution $\lim\limits_{x \to 3}(f(x)) = 15 = f(3)$; therefore, $f(x)$ is continuous at $x = 3$. $\lim\limits_{x \to 15}(g(x)) = \dfrac{1}{15} = g(15)$; therefore, $g(x)$ is continuous at $x = f(3) = 15$.

$$(g \circ f)(x) = g(f(x)) = \frac{1}{f(x)} = \frac{1}{5x}. \lim_{x \to 3}(g(f(x))) = \frac{1}{15};$$ therefore, $(g \circ f)(x)$ is continuous at $x = 3$ in support of Theorem 2. ∎

Theorem 3

Suppose f is a continuous function with a domain that contains the closed interval $[a, b]$. Suppose further that $f(a) \neq f(b)$. For any number γ between $f(a)$ and $f(b)$, there is a number c between a and b such that $f(c) = \gamma$.

Theorem 3 is known as the *Intermediate Value Theorem*. This is an existence theorem.

Minimum and Maximum of a Function

Suppose f is a function with domain S. If there is a point α in S such that $f(\alpha) \leq f(x)$ for all x in S, then the point α is called *a minimum for the function f* and $m = f(\alpha)$ is called the *minimum value of f*.

If there is a point β in S such that $f(\beta) \geq f(x)$ for all x in S, then the point β is called *a maximum for the function f* and $M = f(\beta)$ is called the *maximum value of f*.

Theorem 4

If f is a continuous function that is defined on the closed interval $[a, b]$, then f has a minimum α in the interval $[a, b]$ and a maximum β in the interval $[a, b]$.

Theorem 4 is called the *Extreme Value Theorem*.

Example 5 Calculate the maximum and minimum values in $[2, 5]$ for function $f(x) = x$.

Solution The interval $[2, 5]$ is closed; therefore by the Extreme Value Theorem we know that a minimum value and a maximum value exist. The graph of $f(x) = x$ is a *straight* line passing through the origin in the first and third quadrants forming an angle of $45°$ with the positive x-axis. Therefore, for values in the closed interval $[2, 5]$, the endpoints of the line segment are $(2, 2)$ and $(5, 5)$. By evaluating the values of $f(x) = x$ at each endpoint, we obtain $f(2) = 2$ and $f(5) = 5$. For $2 < x < 5$, we have $2 < f(x) < 5$. Since the function is continuous with values on a closed and bounded interval, we conclude that the minimum value is 2 and the maximum value is 5 on the interval $[2, 5]$. ∎

Self-Test Exercises

3. With the help of a graphing calculator, calculate the maximum and minimum value in interval $[16, 25]$ for $f(x) = \sqrt{x}$.
4. If $f(x) = x^{1/3}$ and $\alpha = 8$, with the help of a graphing calculator, discuss the continuity of $(\alpha f)(x)$.

2.4 Infinite Limits and Asymptotes

Important Concepts

- Infinite-Valued Limits
- Vertical Asymptotes
- Limits at Infinity
- Horizontal Asymptotes

Infinite-Valued Limits

Let f be a function that is defined on an interval just to the left of c and also on an interval just to the right of c. Informally, we say if f becomes arbitrarily large and positive as x tends to c, then we write $\lim_{x \to c} f(x) = +\infty$. If f becomes arbitrarily large and negative without bound as x tends to c, then we write $\lim_{x \to c} f(x) = -\infty$.

However, a more rigorous definition of an infinite-value limit is given by the following. $\lim_{x \to c} f(x) = \infty$ if for any $N > 0$, there is a $\delta > 0$ such that $f(x) > N$ whenever $0 < |x - c| < \delta$. $\lim_{x \to c} f(x) = -\infty$ if for any $M > 0$, there is a $\delta > 0$ such that $f(x) < -M$ whenever $0 < |x - c| < \delta$. One-sided limits are defined similarly.

Example 1 Discuss $\lim_{x \to 2^+} \left(\dfrac{1}{(x - 2)} \right)$.

Solution As $x \to 2^+$, $(x - 2) \to 0^+$, so $\dfrac{1}{(x - 2)} \to +\infty$. ∎

Vertical Asymptotes

If a function f has a one-sided or two-sided infinite limit as x tends to c, the line $x = c$ is a *vertical asymptote of f*. A vertical asymptote appears in Figure 1.

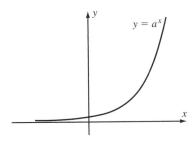

The graph of $y = a^x$ for $a > 1$

Figure 1

Example 2 For the function $f(x) = \dfrac{1}{(x - 2)}$, identify any vertical asymptotes if they exist.

Solution In Example 1, we saw that $\dfrac{1}{(x - 2)} \to +\infty$ as $x \to 2^+$. Based on the one-sided infinite limit, as $x \to 2^+$ we conclude that the line $x = 2$ is a vertical asymptote. ∎

Self-Test Exercise

1. Identify any vertical asymptotes for the function $f(x) = \dfrac{2}{x^2 - 5x + 6}$.

Limits at Infinity

Assume that the domain of f contains an interval of the form (a, ∞), and let α be a real number. If $f(x)$ approaches α when x becomes arbitrarily large and positive, we say that $\lim\limits_{x \to +\infty} f(x) = \alpha$.

Now assume that the domain of g contains an interval of the form $(-\infty, b)$, and let β be a real number. If $g(x)$ approaches β when x is negative and when x becomes arbitrarily large in absolute value, then we say that $\lim\limits_{x \to -\infty} g(x) = \beta$. Here is a more rigorous definition: $\lim\limits_{x \to +\infty} f(x) = \alpha$ if for any $\epsilon > 0$, there is an N such that $|f(x) - \alpha| < \epsilon$ whenever $N < x < \infty$. Also $\lim\limits_{x \to -\infty} g(x) = \beta$ if for any $\epsilon > 0$, there is an M such that $|g(x) - \beta| < \epsilon$ whenever $-\infty < x < M$.

Example 3 Calculate $\lim\limits_{x \to +\infty} \dfrac{1}{x}$.

Solution As $x \to +\infty$, $\dfrac{1}{x} \to 0$ because the numerator is fixed at the value 1 but the denominator increases without bound. Therefore, the value of the fraction gets smaller and smaller and the value tends to zero. ■

Example 4 Examine the limits at infinity for the function $f(x) = \dfrac{x^4 - 2x^3 + 7}{2x^4 + 9}$.

Solution

$$
\begin{aligned}
f(x) &= \frac{x^4 - 2x^3 + 7}{2x^4 + 9} \\
&= \frac{x^4 - 2x^3 + 7}{2x^4 + 9} \cdot \frac{\frac{1}{x^4}}{\frac{1}{x^4}} \\
&= \frac{1 - \frac{2}{x} + \frac{7}{x^4}}{2 + \frac{9}{x^4}} \to \frac{1}{2},
\end{aligned}
$$

as $x \to +\infty$ and also as $x \to -\infty$. The graph of $f(x)$ with full domain is shown in Figure 2 and the graph of $f(x)$ with restricted domain ($x = 100$ to $x = 2000000$) is shown in Figure 3. The limit "behavior" is visible in the graph of $f(x)$ with restricted domain.

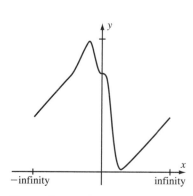

Figure 2

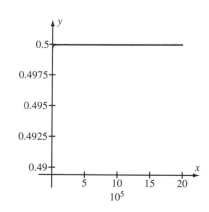

Figure 3   ■

Self-Test Exercise

2. If $f(x) = \dfrac{3}{x^2}$, evaluate $\lim\limits_{x \to +\infty} f(x)$ if it exists.

Horizontal Asymptotes

If either $\lim\limits_{x \to +\infty} f(x) = \alpha$ or $\lim\limits_{x \to -\infty} f(x) = \alpha$, then we say that the line $y = \alpha$ is a *horizontal asymptote of f.*

Example 5 Identify any horizontal asymptotes, if they exist, for $f(x) = x^{-1/2}$.

Solution $\lim\limits_{x \to +\infty} f(x) = \lim\limits_{x \to \infty} \dfrac{1}{\sqrt{x}} = 0$; therefore, a horizontal asymptote exists and it is the line $y = 0$. ∎

Self-Test Exercise

3. Identify any vertical and/or horizontal asymptotes, if they exist, for $f(x) = \dfrac{1}{2x}$.

2.5 Limits of Sequences

Important Concepts

- Sequences
- Theorems

Sequences

An *infinite sequence of real numbers* is a function f with a domain equal to the set of positive integers and a range equal to the set of real numbers. You can think of this function as an action on input, a positive integer, which produces an output, a real number, that is one of the terms of the sequence, as shown here:

$$Input \mapsto Function\ Action \mapsto Output$$
$$n \mapsto a \mapsto a(n) \equiv a_n$$

The graph of a sequence is discrete; that is, the graph of a sequence consists of points.

Convergence and Divergence

Suppose $\{a_n\}_{n=1}^{\infty}$ is a sequence and ℓ is a real number. We say the sequence has *limit ℓ* or *converges to ℓ* if for each $\epsilon > 0$ there is an integer N such that if $n \geq N$, then $|a_n - \ell| < \epsilon$.

The condition $|a_n - \ell| < \epsilon$ for $n \geq N$ is illustrated in Figure 1.

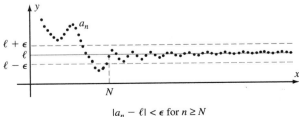

$|a_n - \ell| < \epsilon$ for $n \geq N$

Figure 1

When the sequence $\{a_n\}_{n=1}^{\infty}$ has limit ℓ, we write $\lim\limits_{n\to\infty} a_n = \ell$ or equivalently, $a_n \to \ell$ as $n \to \infty$.

A sequence that has a limit is *convergent*. If a sequence does not converge, then we say that it *diverges*, and we call it a *divergent sequence*.

Example 1 Discuss the convergence or divergence of the sequence $\{(-1)^n\}_{n=1}^{\infty}$.

Solution If n is an even positive integer, then $(-1)^n = 1$. However, if n is an odd positive integer, then $(-1)^n = -1$. Therefore, the values go back and forth from -1 to 1 to -1 and so forth. As a result, there is no unique limiting value and the sequence is divergent. ∎

The Tail of a Sequence

Informally, a key observation of the condition $|a_n - \ell| < \epsilon$ for $n \geq N$ is that *convergence depends only on what $\{a_n\}_{n=1}^{\infty}$ does when n is large*. If we fix any N, then the terms $a_{N+1}, a_{N+2}, a_{N+3}, \ldots$ are referred to as the *tail end of the sequence*. There are an infinite number of tail ends of a sequence because N is an arbitrary choice.

Some Special Sequences

If for any $M > 0$ there is an N such that $a_n > M$ for all indices $n \geq N$, then we write $\lim\limits_{n\to\infty} a_n = \infty$ and say that the sequence $\{a_n\}$ *tends to infinity*.

If for any $M < 0$ there is an N such that $a_n < M$ for all indices $n \geq N$, then we write $\lim\limits_{n\to\infty} a_n = -\infty$ and say that the sequence $\{a_n\}$ *tends to* $-\infty$.

Theorems

Theorem 1

Let r be any real number. Assume that $p > 0$.

    a. $\lim\limits_{n\to\infty} n^p = \infty$.

    b. $\lim\limits_{n\to\infty} \dfrac{1}{n^p} = 0$.

    c. If $|r| > 1$, then the sequence $\{r^n\}_{n=1}^{\infty}$ diverges and $\lim\limits_{n\to\infty} |r|^p = \infty$.

    d. If $|r| < 1$, then $\lim\limits_{n\to\infty} r^n = 0$.

    e. $\lim\limits_{n\to\infty} p^{1/n} = 1$.

Theorem 2

Suppose $\{a_n\}_{n=1}^{\infty}$ and $\{b_n\}_{n=1}^{\infty}$ are convergent sequences.

    a. $\lim\limits_{n\to\infty} (a_n \pm b_n) = \lim\limits_{n\to\infty} (a_n) \pm \lim\limits_{n\to\infty} (b_n)$

    b. $\lim\limits_{n\to\infty} (a_n \cdot b_n) = \left(\lim\limits_{n\to\infty} (a_n) \right) \left(\lim\limits_{n\to\infty} (b_n) \right)$

    c. $\lim\limits_{n\to\infty} \left(\dfrac{a_n}{b_n} \right) = \dfrac{\lim\limits_{n\to\infty} (a_n)}{\lim\limits_{n\to\infty} (b_n)}$, provided that $\lim\limits_{n\to\infty} (b_n) \neq 0$

    d. $\lim\limits_{n\to\infty} (\alpha \cdot a_n) = \alpha \cdot \left(\lim\limits_{n\to\infty} (a_n) \right)$ for any real number α

    e. $\lim\limits_{n\to\infty} (\alpha) = \alpha$ for any real number α

    f. $\lim\limits_{n\to\infty} (a_n)$ is unique.

    g. If $\lim\limits_{n\to\infty} a_n = \lim\limits_{n\to\infty} b_n$ and $a_n \leq c_n \leq b_n$ for all n, then $\{c_n\}$ converges to the same limit as $\{a_n\}_{n=1}^{\infty}$ and $\{b_n\}_{n=1}^{\infty}$.

Example 2 Calculate the limit of $a_j = \dfrac{2j^2 + 2}{j^3 + 2j^2 + 2}$ as $j \to \infty$.

Solution

$$a_j = \frac{2j^2 + 2}{j^3 + 2j^2 + 2}$$

$$= \frac{(2j^2 + 2)\,\frac{1}{j^3}}{(j^3 + 2j^2 + 2)\,\frac{1}{j^3}}$$

$$= \frac{\frac{2}{j} + \frac{2}{j^3}}{1 + \frac{2}{j} + \frac{2}{j^3}}$$

$$\lim_{j \to \infty} a_j = \frac{\displaystyle\lim_{j \to \infty}\left(\frac{2}{j} + \frac{2}{j^3}\right)}{\displaystyle\lim_{j \to \infty}\left(1 + \frac{2}{j} + \frac{2}{j^3}\right)} = \frac{0}{1} = 0 \qquad\blacksquare$$

Example 3 Calculate the limit of $a_j = \dfrac{10^j}{100^j}$ as $j \to \infty$.

Solution $a_j = \dfrac{10^j}{100^j} = \dfrac{10^j}{(10^2)^j} = \dfrac{10^j}{10^{2j}} = \dfrac{1}{10^{2j-j}} = \dfrac{1}{10^j}$. As $j \to \infty$, $10^j \to \infty$, so

$\dfrac{1}{10^j} \to 0$. Therefore, the limit exists and $\displaystyle\lim_{j \to \infty} a_j = 0$. $\qquad\blacksquare$

Geometric Series

Informally, you can think of any *series* as the sum of the terms of a sequence. A limit of the form $\displaystyle\lim_{N \to \infty}(1 + r + r^2 + \cdots + r^{N-1})$ is called a *geometric series*, which is frequently encountered in applications. The first term in the series is 1, the second term is equal to r times the first term, the third term is equal to r times the second term, and in general the nth term is equal to r times the $(n-1)$th term; r is called the *ratio* of the *geometric series*.

Theorem 3   Suppose that $r \neq 1$. For each positive integer N, $1 + r + r^2 + \cdots + r^{N-1} = \dfrac{r^N - 1}{r - 1}$. If $|r| < 1$,

then $\displaystyle\lim_{N \to \infty}(1 + r + r^2 + \cdots + r^{N-1}) = \dfrac{1}{r - 1}$.

Example 4 Calculate the sum of the series $1 + \dfrac{1}{4} + \dfrac{1}{16} + \dfrac{1}{64}$.

Solution The series $1 + \dfrac{1}{4} + \dfrac{1}{16} + \dfrac{1}{64}$ is a geometric series with the first term 1, ratio $\dfrac{1}{4}$, and $N - 1 = 3$. Therefore, its sum is equal to

$$\frac{r^N - 1}{r - 1} = \frac{\left(\frac{1}{4}\right)^4 - 1}{\left(\left(\frac{1}{4}\right) - 1\right)}$$

$$= \frac{\frac{1}{256} - \frac{256}{256}}{\frac{-3}{4}}$$

$$= \frac{85}{64} \approx 1.3281. \qquad\blacksquare$$

Composition with Continuous Functions

Theorem 4 Let $\{b_j\}_{j=1}^{\infty}$ be a sequence that converges to a limit ℓ. Suppose that $\{b_n\}$ is contained in the domain of a function f that is continuous at ℓ. Therefore, $\lim_{j\to\infty} f(b_j) = f\left(\lim_{j\to\infty} b_j\right) = f(\ell)$.

Example 5 Evaluate $\lim_{j\to\infty} \left(2 + \dfrac{5}{j^2}\right)^{10}$.

Solution Let $b_j = 2 + \dfrac{5}{j^2}$. $\lim_{j\to\infty}(b_j) = \lim_{j\to\infty}\left(2 + \dfrac{5}{j^2}\right) = 2 + 0 = 2 = \ell$, i.e., sequence $\{b_j\}_{j=1}^{\infty}$ converges to $\ell = 2$. Function $f(t) = t^{10}$ is continuous on the interval $(-\infty, \infty)$ and this interval contains $\ell = 2$. Therefore, we can apply Theorem 4 to obtain $\lim_{j\to\infty} f(b_j) = f\left(\lim_{j\to\infty} b_j\right) = f(2) = 2^{10} = 1024$. ∎

Self-Test Exercise

1. Identify the following series and calculate the requested value, if it exists.

$$\lim_{N\to\infty}\left(1 + \frac{1}{10} + \left(\frac{1}{10}\right)^2 + \cdots + \left(\frac{1}{10}\right)^{N-1}\right)$$

2.6 Exponential Functions

Important Concepts

- Precursors: Properties of Real Numbers
- Exponential Functions
- Applications

Monotonic Sequences

A sequence $\{a_n\}$ is said to be *increasing* if $a_n \leq a_{n+1}$ for every n. A sequence $\{b_n\}$ is said to be *decreasing* if $b_n \geq b_{n+1}$ for every n. Both types of sequences are said to be *monotone* or *monotonic*. Monotonic sequences are represented in Figure 1.

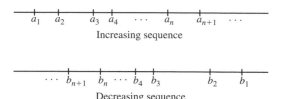

Figure 1

Example 1 Classify $\left\{\dfrac{1}{10^j}\right\}_{j=1}^{\infty}$.

Solution Let $b_j = \dfrac{1}{10^j}$. Since $10^{j+1} > 10^j$ for every $j \ge 1$, then $\dfrac{1}{10^j} > \dfrac{1}{10^{j+1}}$ for such j, i.e.,

we see that $b_j \ge b_{j+1}$. Therefore $\left\{\dfrac{1}{10^j}\right\}_{j=1}^{\infty}$ is a monotonic decreasing function.   ■

The Monotone Convergence Property of the Real Numbers

A sequence $\{a_n\}$ is *bounded above* if there is a real number U such that $a_n \le U$ for all n. The real number U is referred to as an *upper bound* for the sequence $\{a_n\}$, and we say that $\{a_n\}$ is *bounded above by U*.

A sequence $\{a_n\}$ is *bounded below* if there is a real number L such that $L \le a_n$ for all n. The real number L is referred to as a *lower bound* for the sequence $\{a_n\}$, and we say that $\{a_n\}$ is *bounded below by L*.

If $\{a_n\}$ is bounded both above and below, then we say that $\{a_n\}$ is *bounded*. If there is a number $M > 0$ such that $|a_n| \le M$ for all n, then we say that $\{a_n\}$ is *bounded* by M.

Theorem 1

If $\{a_n\}_{n=1}^{\infty}$ is a monotonic and bounded, then $\{a_n\}_{n=1}^{\infty}$ converges to some real number ℓ. If $\{a_n\}_{n=1}^{\infty}$ lies in a closed interval I, then ℓ belongs to I.

Theorem 1 is called the *Monotone Convergence Property*. It guarantees that a bounded increasing sequence actual converges.

Example 2 Investigate the convergence or divergence of $\left\{\dfrac{1}{10^j}\right\}_{j=1}^{\infty}$.

Solution Let $b_j = \dfrac{1}{10^j}$. In the previous example, Example 1, we established that $\left\{\dfrac{1}{10^j}\right\}_{j=1}^{\infty}$ is a monotonic (decreasing) function for every $j \ge 1$. Furthermore, there exists a real number $M = \dfrac{1}{10}$ such that $\left|\dfrac{1}{10^j}\right| \le \dfrac{1}{10} = M$ for all $j \ge 1$. Therefore, by the Monotone Convergence Theorem, $\left\{\dfrac{1}{10^j}\right\}_{j=1}^{\infty}$ converges to a limit.   ■

Limits of Recursive Sequences

If $b_n = f(b_{n-1})$ for some continuous function f and if $\{b_n\}$ converges to a point b that is in the domain of f, then the limit must be a root of the equation $f(x) = x$. A root of the equation $f(x) = x$ is called a *fixed point of f*.

Irrational Exponents

If a is a positive number and x is an irrational number, then we define a^x by $a^x = \lim\limits_{n\to\infty} a^{x_n}$ where $\{x_n\}$ is a sequence of rational number such that $x_n \to x$.

Exponential Functions

An *exponential function* is defined as a function of the form $y = a^x$ where either $a > 1$ or $0 < a < 1$ and x is a real number. The graph of $y = a^x$, $a > 1$, and x is a real number is shown in Figure 2.

Notice that in Figure 2, the graph of $y = a^x$, $a > 1$, has a y-intercept of one and as $x \to +\infty$, $y \to +\infty$; i.e., the function is increasing from left to right and has a y-intercept of 1.

The graph of $y = a^x$, $0 < a < 1$, and x is a real number is shown in Figure 3.

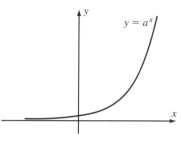

The graph of $y = a^x$ for $a > 1$

Figure 2

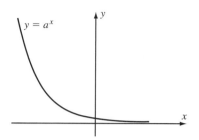

The graph of $y = a^x$ for $0 < a < 1$

Figure 3

Notice that in Figure 3, the graph of $y = a^x$, $0 < a < 1$, has a y-intercept of one and as $x \to +\infty$, $y \to 0$; i.e., the function is decreasing from left to right and has a y-intercept of 1.

Example 3 Describe the properties of the function $y = 2^x$.

Solution The function $y = 2^x$ is an exponential function with base $a = 2 > 1$. Therefore it is an increasing function with y-intercept equal to one. ∎

Self-Test Exercise

1. Describe the properties of the function $y = 2^{-x}$.

Applications

The following formulas are used to compute accumulated amounts when different types of interest are applied to funds initially deposited in an account.

If an amount of P dollars is put in the bank at r percent *simple interest per year*, then the amount in an account after *one year* is equal to $P + \dfrac{r}{100}P = P\left(1 + \dfrac{r}{100}\right)$ dollars.

However, if interest is *compounded n times per year*, then the accumulated amount after t years is given by the formula in Theorem 2.

Theorem 2 If the interest on principal P accrues at r percent *compounded n times per year*, then the value of the deposit *after t years* is $P\left(1 + \dfrac{r}{100n}\right)^{nt}$.

In addition to simple and/or compound interest calculations, there is a formula that calculates the accumulated amount when interest is continuously compounded. Continuous compounding is the limiting result of letting n tend to infinity. In this situation, the accumulated amount in an account is worth $\displaystyle\lim_{n \to \infty} P\left(1 + \dfrac{r}{100n}\right)^n$ *at the end of one year*.

Example 4 If $1000 is deposited in an account for 5 years and earns 5% interest compounded monthly, what is the accumulated amount?

Solution $A = P\left(1 + \dfrac{r}{100n}\right)^{nt} = 1000\left(1 + \dfrac{.05}{12}\right)^{(12)(5)} = \$1283.36$ ■

Theorem 3   There is a real number e such that $e = \lim\limits_{n \to \infty}\left(1 + \dfrac{1}{n}\right)^{n}$ and $e = \lim\limits_{n \to \infty}\left(1 - \dfrac{1}{n}\right)^{-n}$.

The function $y = e^x$ is an exponential function of the form $y = a^x$ where $a = e = 2.71828182845904\ldots$ is an *irrational number*.

Theorem 4   For every real number u, we have $e^u = \lim\limits_{x \to \infty}\left(1 + \dfrac{u}{x}\right)^{x}$.

Theorem 5   If the interest on principal P accrues at an annual rate of r percent, compounded continuously, then the money accumulated after t years is $Pe^{\frac{rt}{100}}$.

Exponential functions are used to model some types of growth and decay. The function $y = Ae^{kx}$ models *growth* if $k > 0$ and models *decay* if $k < 0$. In both instances, the term A in the function $y = Ae^{kx}$ is a constant. If $k > 0$, then k is called the *growth constant*. If $k < 0$, then k is called the *decay constant*.

Example 5 A culture of fruit flies doubled in 2 days. Assuming that the culture grows exponentially, how large will it be in 4 days?

Solution The applicable formula is $y = y(t) = Ae^{kt}$ with $k > 0$ because we are in a growth model. If A represents the original number of fruit flies and if in $t = 2$ days, the original amount doubles, then $y(t + 2) = 2y(t) = 2Ae^{kt}$. But $y(t + 2) = Ae^{k(t+2)} = Ae^{kt}e^{2k}$. Since both $2Ae^{kt}$ and $Ae^{kt}e^{2k}$ are equal to $y(t + 2)$, $2Ae^{kt} = Ae^{kt}e^{2k}$, which implies that $2 = e^{2k}$. Since $y(4) = Ae^{k4} = A(e^{2k})^2 = A(2)^2 = 4A$, we conclude that in 4 days, the culture will be four times its original size.   ■

Self-Test Exercises

2. If $1250 were deposited initially in an account that pays 2% interest compounded weekly, how much interest will be accumulated in 6 years?

3. Is the sequence $\left\{\dfrac{n}{n+1}\right\}_{n=1}^{\infty}$ convergent or divergent?

The Derivative

3.1 Rates of Change and Tangent Lines

Important Concepts

- Rates of Change
- Tangent and Normal Lines to a Curve at a Given Point

Average Rate of Change

Our knowledge of limits as presented in Chapter 2 will now be applied to rates of change.

The *average rate of change* of function $f(x)$ as x changes from c to $(c + \Delta x)$ is denoted as $\dfrac{\Delta f}{\Delta x}$, read "the change in f with respect to x," and is defined by $\dfrac{\Delta f}{\Delta x} = \dfrac{f(c + \Delta x) - f(c)}{\Delta x}$.

Example 1 Calculate the average rate of change from c to $(c + \Delta x)$ if $f(x) = x^2$.

Solution The average rate of change from c to $(c + \Delta x)$ is

$$\begin{aligned}
\frac{\Delta f}{\Delta x} &= \frac{f(c + \Delta x) - f(c)}{\Delta x} \\
&= \frac{(c + \Delta x)^2 - c^2}{\Delta x} \\
&= \frac{c^2 + 2c \cdot \Delta x + (\Delta x)^2 - c^2}{\Delta x} \\
&= \frac{2c \cdot \Delta x + (\Delta x)^2}{\Delta x} \\
&= \frac{\Delta x \cdot (2c + \Delta x)}{\Delta x} \\
&= 2c + \Delta x.
\end{aligned}$$

Instantaneous Rate of Change

The *instantaneous rate of change* of $f(x)$ at $x = c$ is defined by $\displaystyle\lim_{\Delta x \to 0} \left(\frac{f(c + \Delta x) - f(c)}{\Delta x} \right)$, provided this limit exists and is finite.

Example 2 Calculate the instantaneous rate of change at c if $f(x) = x^2$.

Solution In Example 1, we calculated the average rate of change from c to $(c+\Delta x)$ if $f(x) = x^2$ as $\dfrac{\Delta f}{\Delta x} = 2c + \Delta x$. The instantaneous rate of change of $f(x)$ at $x = c$ is

$$
\begin{aligned}
\lim_{\Delta x \to 0} \frac{\Delta f}{\Delta x} &= \lim_{\Delta x \to 0} (2c + \Delta x) \\
&= \lim_{\Delta x \to 0} 2c + \lim_{\Delta x \to 0} \Delta x \\
&= 2c + 0 \\
&= 2c.
\end{aligned}
$$

Instantaneous Velocity

Instantaneous velocity at $t = c$ is a special case of the *instantaneous rate of change* of $f(x)$ at $x = c$. When we are dealing with velocity, function f represents the *position of an object* and is denoted by the letter p rather than f, and the independent variable is t for time, rather than the independent variable x.

Therefore, if we represent the function by the letter p and the independent variable by t in the definition of instantaneous rate of change, we obtain a definition of instantaneous velocity.

Suppose the position of a body moving along a straight path is described by a function p of time t. The *instantaneous velocity* of the body at time c is given by $\lim\limits_{\Delta t \to 0} \left(\dfrac{p(c + \Delta t) - p(c)}{\Delta t} \right)$, provided that this limit exists and is finite.

Example 3 If the position of a body, p, at time t is given by $p(t) = t^2 + 4t + 4$ feet, where t is measured in seconds, calculate the instantaneous velocity at time 3 seconds.

Solution Let $v(t)$ denote instantaneous velocity (in ft/s) at time t.

$$
v(3) = \lim_{\Delta t \to 0} \left(\frac{p(3 + \Delta t) - p(3)}{\Delta t} \right).
$$

$$
\begin{aligned}
p(3 + \Delta t) &= (3 + \Delta t)^2 + 4(3 + \Delta t) + 4 \\
&= 9 + 6\Delta t + (\Delta t)^2 + 12 + 4\Delta t + 4 \\
&= 25 + 10\Delta t + (\Delta t)^2. \\
p(3) &= (3)^2 + 4(3) + 4 = 9 + 12 + 4 = 25.
\end{aligned}
$$

Therefore,

$$
\begin{aligned}
v(3) &= \lim_{\Delta t \to 0} \left(\frac{p(3 + \Delta t) - p(3)}{\Delta t} \right) \\
&= \lim_{\Delta t \to 0} \frac{(25 + 10\Delta t + (\Delta t)^2) - 25}{\Delta t} \\
&= \lim_{\Delta t \to 0} \frac{10\Delta t + (\Delta t)^2}{\Delta t} \\
&= \lim_{\Delta t \to 0} \frac{\Delta t(10 + \Delta t)}{\Delta t} \\
&= \lim_{\Delta t \to 0} (10 + \Delta t) \\
&= \lim_{\Delta t \to 0} (10) + \lim_{\Delta t \to 0} (\Delta t) \\
&= 10 + 0 \\
&= 10.
\end{aligned}
$$

Sums of Functions

In Section 2.2 of Chapter 2, we introduced some basic limit theorems such as "the limit of a finite sum of functions is equal to the finite sum of the limit of each function in the sum," and "the limit of a constant times a function is equal to the constant times the limit of the function." The following theorem is nothing more than an application of these limit theorems in our current framework of instantaneous rate of change.

Theorem 1

Suppose α and β are constants. If r and s are the instantaneous rates of change of $f(x)$ and $g(x)$ at a point $x = c$, then $\alpha \cdot r + \beta \cdot s$ is the instantaneous rate of change of $\alpha \cdot f(x) + \beta \cdot g(x)$ at $x = c$.

Theorem 1 is referred to as the *Sum Rule*.

Example 4 Calculate the instantaneous rate of change at $x = 2$ of $h(x) = 3f(x) + 2g(x)$ if $f(x) = 5x$ and $g(x) = x^2$.

Solution The instantaneous rate of change of $f(x)$ at $x = 2$ is

$$\lim_{\Delta x \to 0} \left(\frac{(f(2 + \Delta x)) - (f(2))}{\Delta x} \right) = \lim_{\Delta x \to 0} \left(\frac{(5(2 + \Delta x)) - (5(2))}{\Delta x} \right)$$

$$= \lim_{\Delta x \to 0} \left(\frac{(10 + 5\Delta x) - 10}{\Delta x} \right)$$

$$= \lim_{\Delta x \to 0} \left(\frac{5\Delta x}{\Delta x} \right)$$

$$= \lim_{\Delta x \to 0} (5)$$

$$= 5.$$

The instantaneous rate of change of $g(x)$ at $x = 2$ is

$$\lim_{\Delta x \to 0} \left(\frac{(g(2 + \Delta x)) - (g(2))}{\Delta x} \right) = \lim_{\Delta x \to 0} \left(\frac{(2 + \Delta x)^2 - (2^2)}{\Delta x} \right)$$

$$= \lim_{\Delta x \to 0} \left(\frac{(4 + 4\Delta x + (\Delta x)^2) - 4}{\Delta x} \right)$$

$$= \lim_{\Delta x \to 0} \left(\frac{4\Delta x + (\Delta x)^2}{\Delta x} \right)$$

$$= \lim_{\Delta x \to 0} \left(\frac{\Delta x (4 + \Delta x)}{\Delta x} \right)$$

$$= \lim_{\Delta x \to 0} (4 + \Delta x)$$

$$= \lim_{\Delta x \to 0} 4 + \lim_{\Delta x \to 0} \Delta x$$

$$= 4 + 0$$

$$= 4.$$

Therefore, according to the Sum Rule (Theorem 1), with $\alpha = 3$ and $\beta = 2$,

$$\lim_{\Delta x \to 0} \left(\frac{h(2 + \Delta x) - h(2)}{\Delta x} \right) = 3 \cdot (5) + 2 \cdot (4)$$

$$= 15 + 8$$

$$= 23.$$ ∎

Self-Test Exercises

1. If $f(x) = x^3$ and $g(x) = x$, calculate the average rate of change of $h(x) = f(x) - 2g(x)$ between $x = 1$ and $x = 2$.
2. If $f(x) = x^3$ and $g(x) = x$, calculate the instantaneous rate of change of $h(x) = f(x) - 2g(x)$ at $x = 1$.
3. If the position of an object at time t is given by $p(t) = t^2 + 3t + 1$, calculate the instantaneous velocity at time 2.

Tangent and Normal Lines to a Curve at a Given Point

The two pictures in Figure 1 represent various possibilities for a tangent line. The two pictures in Figure 2 illustrate that a tangent line can be visualized as a limiting function of a secant line as $\Delta x \to 0$.

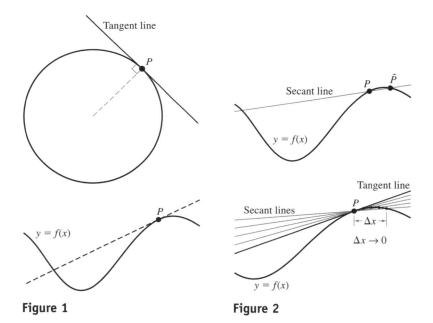

Figure 1 **Figure 2**

A *secant line* is a line that intersects a figure in two places. Informally, we think of the *slope of a secant line* at a point c as the change in y at c divided by the change in x at c. Formally, if the secant line passes through the points $P = (c, f(c))$ and $Q = (c + \Delta x, f(c + \Delta x))$, then its slope is calculated as

$$\frac{f(c + \Delta x) - f(c)}{(c + \Delta x) - c} = \frac{f(c + \Delta x) - f(c)}{\Delta x},$$

which has the same expression as that of the *average rate of change*.

Since a tangent line can be visualized as a limiting function of a secant line as $\Delta x \to 0$, we have the following result. Consider a function f defined on an open interval containing the point c. Suppose that

$$m = \lim_{\Delta x \to 0} \left(\frac{f(c + \Delta x) - f(c)}{\Delta x} \right)$$

exists and is finite. Then m is the *slope of the tangent line at* $P = (c, f(c))$. Furthermore, the equation of the tangent line to the graph of f at $P = (c, f(c))$ is given by $y = m \cdot (x - c) + f(c)$.

Example 5 Find the slope and equation of the line tangent to the graph of $f(x) = -x^2 + 6x$ at the point $(1, 5)$.

Solution The slope of the tangent line at the point $(1, 5)$ is given by

$$\begin{aligned}
m &= \lim_{\Delta x \to 0} \left(\frac{f(1 + \Delta x) - f(1)}{\Delta x} \right) \\
&= \lim_{\Delta x \to 0} \left(\frac{(-(1 + \Delta x)^2 + 6(1 + \Delta x)) - (-(1)^2 + 6(1))}{\Delta x} \right) \\
&= \lim_{\Delta x \to 0} \left(\frac{-1 - 2\Delta x - (\Delta x)^2 + 6 + 6\Delta x + 1 - 6}{\Delta x} \right) \\
&= \lim_{\Delta x \to 0} \left(\frac{4\Delta x - (\Delta x)^2}{\Delta x} \right) \\
&= \lim_{\Delta x \to 0} \left(\frac{\Delta x \, (4 - \Delta x)}{\Delta x} \right) \\
&= \lim_{\Delta x \to 0} (4 - \Delta x) \\
&= \lim_{\Delta x \to 0} 4 - \lim_{\Delta x \to 0} \Delta x \\
&= 4 - 0 \\
&= 4.
\end{aligned}$$

Therefore, the equation of the tangent line at the point $(1, 5)$ is

$$\begin{aligned}
y &= 4 \cdot (x - 1) + f(1) \\
&= 4x - 4 + 5 \\
&= 4x + 1.
\end{aligned}$$

You can easily check your answer graphically by drawing both the graph and the tangent line near the point $(1, 5)$ to see if they are tangential to each other (see Figure 3).

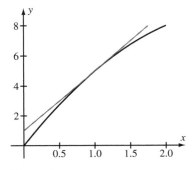

Figure 3

Normal Lines

A line L is said to be *perpendicular*, or *normal*, to a curve $y = f(x)$ at a point $P = (c, f(c))$ if L is perpendicular to the tangent line to the curve at the point $P = (c, f(c))$.

If the tangent line is the horizontal line $y = f(c)$, then the normal line is the vertical line $x = c$. For all other possibilities, the product of the slopes of the tangent line and the normal line at a point c is equal to negative one—i.e., $(m_{\text{tan}}) \cdot (m_{\text{normal}}) = -1$, or equivalently the slope of one of the two lines is equal to the negative reciprocal of the slope of the other line.

Example 6 If $y = 4x + 1$ is the equation of the tangent line to the curve $f(x) = -x^2 + 6x$ at the point $(1, 5)$, find the equation of the normal line to the curve at that point.

Solution The slope of the tangent line is 4. The slope of the normal line is the negative reciprocal of 4, i.e., $m_{normal} = \dfrac{-1}{4}$. Therefore, the equation of the normal line is

$$y = m \cdot (x - c) + f(c)$$
$$= \frac{-1}{4}(x - 1) + 5$$
$$= \frac{-1}{4}x + \frac{1}{4} + 5$$
$$= \frac{-1}{4}x + \frac{21}{4}.$$

Self-Test Exercises

4. Let $f(x) = -x^2 + 4x$.
    a. Find the slope and equation of the tangent line to the graph of f at the point $(2, 4)$.
    b. Find the equation of the normal line to the graph of f at the point $(2, 4)$.
5. If $f(x) = -4x + 9$, find the equation of the tangent line to the curve at the point $\left(-\dfrac{1}{4}, 10\right)$.

Corners, Cusps, and Vertical Tangent Lines

If $\displaystyle \lim_{\Delta x \to 0^+} \left(\frac{f(c + \Delta x) - f(c)}{\Delta x} \right) \neq \lim_{\Delta x \to 0^-} \left(\frac{f(c + \Delta x) - f(c)}{\Delta x} \right)$, but both the left-limit and right-limit are finite, then the function f has a *corner at* $P = (c, f(c))$ and does not have a tangent at P.

Example 7 Show that the function $f(x) = |x|$ has a corner at $(0, 0)$.

Solution

$$f(x) = |x| = \begin{cases} x, & \text{if } x \geq 0 \\ -x, & \text{if } x < 0 \end{cases}$$

$$\lim_{\Delta x \to 0^+} \left(\frac{f(0 + \Delta x) - f(0)}{\Delta x} \right) = \lim_{\Delta x \to 0^+} \frac{|\Delta x| - |0|}{\Delta x}$$
$$= \lim_{\Delta x \to 0^+} \frac{\Delta x - 0}{\Delta x}$$
$$= \lim_{\Delta x \to 0^+} 1$$
$$= 1$$

$$\lim_{\Delta x \to 0^-} \left(\frac{f(0 + \Delta x) - f(0)}{\Delta x} \right) = \lim_{\Delta x \to 0^-} \frac{|\Delta x| - |0|}{\Delta x}$$

$$= \lim_{\Delta x \to 0^-} \frac{-\Delta x - 0}{\Delta x}$$

$$= \lim_{\Delta x \to 0^-} -1$$

$$= -1$$

Therefore, the absolute value function has a corner at the point $(0, 0)$. This conclusion is substantiated by its graph in Figure 4.

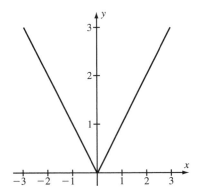

Figure 4 ∎

If $\lim_{\Delta x \to 0} \left(\frac{f(c + \Delta x) - f(c)}{\Delta x} \right) = +\infty$ or $-\infty$, then the definition of tangent line does not apply because the limit is not finite. In this case, a vertical line $x = c$ would intersect the graph of f at $P = (c, f(c))$.

Example 8 Show that $f(x) = x^{1/3}$ has a vertical "tangent" at the point $(0, 0)$.

Solution

$$\lim_{\Delta x \to 0^+} \left(\frac{f(0 + \Delta x) - f(0)}{\Delta x} \right) = \lim_{\Delta x \to 0^+} \left(\frac{(\Delta x)^{1/3} - 0}{\Delta x} \right)$$

$$= \lim_{\Delta x \to 0^+} \left(\frac{1}{(\Delta x)^{2/3}} \right)$$

$$= +\infty$$

$$\lim_{\Delta x \to 0^-} \left(\frac{f(0 + \Delta x) - f(0)}{\Delta x} \right) = \lim_{\Delta x \to 0^-} \left(\frac{(\Delta x)^{1/3} - 0}{\Delta x} \right)$$

$$= \lim_{\Delta x \to 0^-} \left(\frac{1}{(\Delta x)^{2/3}} \right)$$

$$= +\infty$$

Since $\lim_{\Delta x \to 0} \left(\frac{f(0 + \Delta x) - f(0)}{\Delta x} \right) = +\infty$, we conclude that $f(x) = x^{1/3}$ has a vertical tangent at $x = 0$. ∎

If $\lim_{\Delta x \to 0^+} \left(\frac{f(c + \Delta x) - f(c)}{\Delta x} \right) \neq \lim_{\Delta x \to 0^-} \left(\frac{f(c + \Delta x) - f(c)}{\Delta x} \right)$, but either the left-limit or right-limit is infinite, then we say that the point $P = (c, f(c))$ is a *cusp* of the graph of f.

Example 9 Show that $f(x) = (x^2)^{1/3}$ has a cusp at the point $(0, 0)$.

Solution Since $f(x) = (x^2)^{1/3} = x^{2/3}$, we have

$$\lim_{\Delta x \to 0^+} \left(\frac{f(0 + \Delta x) - f(0)}{\Delta x} \right) = \lim_{\Delta x \to 0^+} \left(\frac{(\Delta x)^{2/3} - 0}{\Delta x} \right)$$
$$= \lim_{\Delta x \to 0^+} \left(\frac{1}{(\Delta x)^{1/3}} \right)$$
$$= +\infty$$

and

$$\lim_{\Delta x \to 0^-} \left(\frac{f(0 + \Delta x) - f(0)}{\Delta x} \right) = \lim_{\Delta x \to 0^-} \left(\frac{(\Delta x)^{2/3} - 0}{\Delta x} \right)$$
$$= \lim_{\Delta x \to 0^-} \left(\frac{1}{(\Delta x)^{1/3}} \right)$$
$$= -\infty.$$

We conclude that $f(x)$ has a cusp at $x = 0$. This result is substantiated by the graph shown in Figure 5.

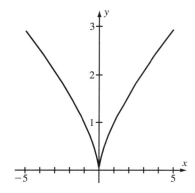

Figure 5

Self-Test Exercises

6. Show that the function $f(x) = |x - 1| = \begin{cases} x - 1 & \text{if } x \geq 1 \\ -x + 1 & \text{if } x < 1 \end{cases}$ has a corner at $(1, 0)$.
7. Show that $f(x) = x^{1/5}$ has a vertical "tangent" at the point $(0, 0)$.
8. Show that $f(x) = (x - 1)^{2/3}$ has a cusp at the point $(1, 0)$.

3.2 The Definition of the Derivative

Important Concepts

- Definition of the Derivative
- Continuity and Differentiability
- Visual Methods

Definition of the Derivative

The derivative of a function at a point in its domain is a limit value, provided the limit exists and is finite. We now define this concept more formally.

Let f be a function that is defined in an open interval that contains a point c. If $\lim_{\Delta x \to 0} \dfrac{f(c + \Delta x) - f(c)}{\Delta x}$ exists and is finite, then we say that f *is differentiable at c*. This limit is called the *derivative of the function at point c* and is denoted by $f'(c)$, or equivalently as $D(f)(c)$, $\dfrac{df}{dx}(c)$, $\dfrac{d}{dx} f(c)$, or $\dfrac{df}{dx}\bigg|_{x=c}$. The process of calculating $f'(c)$ is called *differentiation*. The expressions $\lim_{\Delta x \to 0} \dfrac{f(c + \Delta x) - f(c)}{\Delta x}$ and $\lim_{x \to c} \dfrac{f(x) - f(c)}{x - c}$ are equivalent definitions of $f'(c)$.

A *geometric interpretation of differentiation* is the function f is differentiable at c if and only if the graph of f has a tangent line at the point $(c, f(c))$. The geometric interpretation appears in Figure 1.

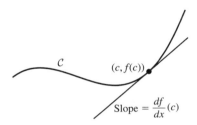

Figure 1

The Derived Function

If f has a derivative at every point of S, then we say that f *is differentiable on S* and we define the derivative $f'(x)$ as $f'(x) = \lim_{\Delta x \to 0} \dfrac{f(x + \Delta x) - f(x)}{\Delta x}$ for all x in S. The *derived function of f* or the *derivative of f* is f', equivalently denoted as $D(f)$, $\dfrac{df}{dx}$, and $\dfrac{d}{dx} f$. The domain of f' is a subset of the domain of f.

Example 1 Calculate $f'(1)$ if $f(x) = (x + 1)^3$.

Solution In this example, we will compute the derivative at $x = 1$ using the two equivalent definitions.

$$
\begin{aligned}
f'(1) &= \lim_{x \to 1} \frac{f(x) - f(1)}{x - 1} \\
&= \lim_{x \to 1} \frac{(x + 1)^3 - (1 + 1)^3}{x - 1} \\
&= \lim_{x \to 1} \frac{\left(x^3 + 3x^2 + 3x + 1\right) - (8)}{x - 1} \\
&= \lim_{x \to 1} \frac{x^3 + 3x^2 + 3x - 7}{x - 1} \\
&= \lim_{x \to 1} \frac{(x - 1)\left(x^2 + 4x + 7\right)}{x - 1} \\
&= \lim_{x \to 1} \left(x^2 + 4x + 7\right) \\
&= 12 + 0 + 0 \\
&= 12
\end{aligned}
$$

We can also compute the derivative using the other equivalent definition.

$$
\begin{aligned}
f'(1) &= \lim_{\Delta x \to 0} \frac{f(1 + \Delta x) - f(1)}{\Delta x} \\
&= \lim_{\Delta x \to 0} \frac{(1 + \Delta x + 1)^3 - 2^3}{\Delta x} \\
&= \lim_{\Delta x \to 0} \frac{(2 + \Delta x)^3 - 2^3}{\Delta x} \\
&= \lim_{\Delta x \to 0} \frac{\left(8 + 12(\Delta x) + 6(\Delta x)^2 + (\Delta x)^3\right) - 8}{\Delta x} \\
&= \lim_{\Delta x \to 0} \frac{12(\Delta x) + 6(\Delta x)^2 + (\Delta x)^3}{\Delta x} \\
&= \lim_{\Delta x \to 0} \frac{(\Delta x)\left(12 + 6(\Delta x) + (\Delta x)^2\right)}{\Delta x} \\
&= \lim_{\Delta x \to 0} \left(12 + 6(\Delta x) + (\Delta x)^2\right) \\
&= 12 + 0 + 0 \\
&= 12
\end{aligned}
$$

Both definitions of $f'(1)$ will give the same result. Of course, $f'(1)$ can also be written as $\dfrac{df}{dx}\Big|_{x=1}$, or $Df(1)$. ∎

Example 2 Calculate $f'(x)$ if $f(x) = x^3$.

Solution To compute $f'(x)$, we use the definition $f'(x) = \lim_{\Delta x \to 0} \dfrac{f(x + \Delta x) - f(x)}{\Delta x}$. Notice that $f(x + \Delta x) = (x + \Delta x)^3 = x^3 + 3x^2(\Delta x) + 3x(\Delta x)^2 + (\Delta x)^3$ and $f(x) = x^3$. We have

$$
\begin{aligned}
f'(x) &= \lim_{\Delta x \to 0} \frac{f(x + \Delta x) - f(x)}{\Delta x} \\
&= \lim_{\Delta x \to 0} \frac{(x^3 + 3x^2(\Delta x) + 3x(\Delta x)^2 + (\Delta x)^3) - (x^3)}{\Delta x} \\
&= \lim_{\Delta x \to 0} \frac{3x^2(\Delta x) + 3x(\Delta x)^2 + (\Delta x)^3}{\Delta x} \\
&= \lim_{\Delta x \to 0} \frac{(\Delta x)(3x^2 + 3x(\Delta x) + (\Delta x)^2)}{\Delta x} \\
&= \lim_{\Delta x \to 0} (3x^2 + 3x(\Delta x) + (\Delta x)^2) \\
&= 3x^2 + 0 + 0 \\
&= 3x^2.
\end{aligned}
$$
∎

Self-Test Exercises

1. Calculate $f'(2)$ if $f(x) = (x - 1)^2$.

2. Calculate $\dfrac{d}{dt}(t^{1/2})\Big|_{t=4}$.

3. Calculate $f'(x)$ if $f(x) = x^2 + 3x$.

Continuity and Differentiability

We shall now explain how the concepts of continuity and differentiability are related.

Recall the *geometric interpretation of differentiation*: the function f is differentiable at c if the graph of f has a well-defined tangent line at the point $(c, f(c))$. This happens only when the function f is continuous at the point $(c, f(c))$. If f is discontinuous at c, then there is no line that is tangent to the graph of f at the point $(c, f(c))$, as Figure 2 suggests.

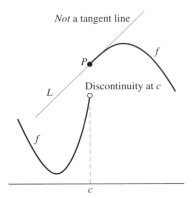

Figure 2

Theorem 1 If f is not continuous at a point c in its domain, then f is not differentiable at c. If f is differentiable at a point c in its domain, then f is continuous at c.

Theorem 1 formalizes the following diagram:

$$differentiability \implies continuity$$
[Differentiability implies continuity]

However,

$$continuity \not\Rightarrow differentiability$$
[Continuity does not imply differentiability]

Consider the graph of the function $y = |x|$, which appears in Figure 3.

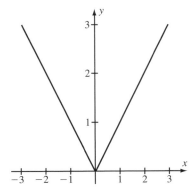

Figure 3

Although the viewing window is $[-3, 3] \times [0, 3]$, the function $f(x) = |x|$ is continuous for all real x. In particular, the function is continuous at $x = 0$ in its domain. However, $f(x) = |x|$ is *not* differentiable at $x = 0$ because, as the graph indicates, there is a sharp point at

$(0, 0)$. The derivative at $x = 0$ does not exist because $\lim\limits_{\Delta x \to 0^+} \dfrac{f(0 + \Delta x) - f(0)}{\Delta x} = 1$, but $\lim\limits_{\Delta x \to 0^-} \dfrac{f(x + \Delta x) - f(x)}{\Delta x} = -1$. This example illustrates that *continuity $\not\Rightarrow$ differentiability*.

Self-Test Exercise

4. Discuss the existence of $f'(2)$ for $f(x) = \dfrac{x^2 + 6}{x^2 + x - 6}$.

Visual Methods

Graphing utilities are powerful tools used to investigate differentiability. For example, Figure 3 clearly illustrates the concept of "*continuity does not imply differentiability.*"

In many simple cases, we can check the differentiability of a function f by inspecting its graph to see if tangent lines exist.

Example 3 Discuss the differentiability of the function $f(x) = \sin(x)$ using its graph.

Solution The graph of $f(x) = \sin(x)$ in the viewing window $[0, 2\pi] \times [-1, 1]$ appears in Figure 4.

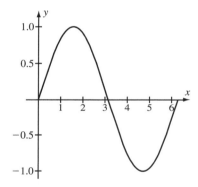

Figure 4

The function $f(x) = \sin(x)$ is continuous for all real x; $f(x) = \sin(x)$ is also a periodic function with period $p = 2\pi$. That means if you continue the viewing window either to the right of the point $(2\pi, 0)$ or to the left of the point $(0, 0)$ in "horizontal blocks" of length 2π, you will repeatedly get the same graph of $f(x) = \sin(x)$ that appears in Figure 4. We note that the graph of $f(x) = \sin(x)$ has no sharp corners, cusps, vertical tangents, or so forth. As a result, a tangent line exists at each and every point $(x, f(x))$ on the graph of the sine function. This implies that a derivative $f'(x)$ exists at each point x in the domain of $f(x) = \sin(x)$. ∎

Sometimes it is difficult to tell whether the tangent line exists from the graph of f. A more precise way to investigate graphically the differentiability of a function f at a point c is through the following procedures.

1. Define a new function $\phi(x) = \dfrac{f(x) - f(c)}{x - c}$ for $x \neq c$.
2. Graph this new function $\phi(x)$ near c.
3. Check if $\phi(x)$ can be extended to a continuous function at c. If it does, then f is differentiable at c. This is because $\lim\limits_{x \to c} \phi(x) = \lim\limits_{x \to c} \dfrac{f(x) - f(c)}{x - c} = f'(c)$.

Example 4 Discuss the differentiability of the function $f(x) = \begin{cases} \dfrac{\sin x}{x}, & \text{if } x \neq 0 \\ 1, & \text{if } x = 0 \end{cases}$ at $x = 0$ using an appropriate graph.

Solution To check if $f(x)$ is differentiable at $x = 0$, we consider the function

$$\phi(x) = \frac{f(x) - f(0)}{x - 0} = \frac{\left(\dfrac{\sin x}{x} - 1\right)}{x - 0} = \frac{\sin x - x}{x^2}$$

which is defined for $x \neq 0$. The graph of $\phi(x)$ appears in Figure 5. We see that $\phi(x)$ can be extended to a continuous function at $x = 0$. Therefore, we can believe that $f(x)$ is differentiable at $x = 0$.

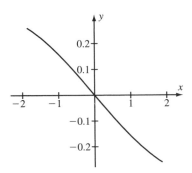

Figure 5

Example 3 establishes that the sine function is differentiable. The following theorem also establishes that if $f(x) = \sin(x)$ then $f'(x) = \cos(x)$.

Theorem 2 For each x, $\dfrac{d}{dx}\sin(x) = \cos(x)$ and $\dfrac{d}{dx}\cos(x) = -\sin(x)$.

Although not formally stated as a theorem, the following derivative formula, $\dfrac{d}{dx}x^n = n \cdot x^{n-1}$, for $n = -1, 0, 1, 2$ and 3, is a special case of the *Power Rule,* which will be discussed more fully in Section 3.4.

Example 5 Calculate the derivative of $f(x) = x^3$.

Solution Use the derivative rule $\dfrac{d}{dx}x^n = n \cdot x^{n-1}$ with $n = 3$ to obtain $f'(x) = \dfrac{d}{dx}x^3 = 3 \cdot x^{3-1} = 3x^2$, which is the same result we had in Example 2 earlier.

Self-Test Exercises

5. Discuss the differentiability of $f(x) = \cos(x)$, utilizing its graph.

6. Discuss the differentiability of $f(x) = \begin{cases} \sin(x), & \text{if } \quad x \geq \dfrac{\pi}{4} \\ \dfrac{2\sqrt{2}}{\pi}x, & \text{if } \quad x < \dfrac{\pi}{4} \end{cases}$ at $x = \dfrac{\pi}{4}$.

3.3 Rules for Differentiation

Important Concepts

- The Derivative of a Constant Function, Addition, Subtraction, and Scalar Multiplication of Functions
- The Derivative of Products and Quotients of Functions
- Numeric Differentiation

The Derivative of a Constant Function

A *constant function* f is defined by $f(x) = \alpha$, $\alpha \in \mathbb{R}$, for all x in the domain of f.

The derivative of a constant function is equal to zero. This means that if $f(x) = \alpha$, $\alpha \in \mathbb{R}$, for all x in the domain of f, then $f'(x) = 0$ for all x.

Example 1 Substantiate, using graphical analysis, that $f'(x) = 0$ if $f(x) = 3$.

Solution The graph of $f(x) = 3$ is a horizontal line through the point $(0, 3)$, which is parallel to the x-axis. (See Figure 1.) As the graph indicates, all tangent lines to points $(x, f(x))$ on the graph of $f(x) = 3$ coincide with the horizontal line. Horizontal tangents have slope of zero. The slope of a line tangent to the graph of a function at a point is equal to the derivative of the function at the point of tangency. Therefore, we conclude that $f'(x) = 0$ not only for the x-values in the graph in Figure 1, but for all x in the domain of function f because the graph remains a horizontal line through the point $(0, 3)$ for all real x.

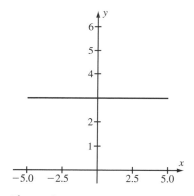

Figure 1

The Derivative of a Sum [or Difference] of Functions

The derivative of a sum [or difference] of functions is equal to the sum [or difference] of the individual derivatives. This means that *if $h(x) = (f \pm g)(x) = f(x) \pm g(x)$, then*

$$h'(x) = f'(x) \pm g'(x),$$

provided both $f'(x)$ and $g'(x)$ exist.

Example 2 Calculate the derivative of $h(x) = x^2 + 6$.

Solution Define two functions $f(x)$ and $g(x)$ by $f(x) = x^2$ and $g(x) = 6$. Apply the rule $\dfrac{d}{dx}x^n = n \cdot x^{n-1}$, for $n = 2$ to $f(x)$, and obtain $f'(x) = 2x$. Apply the rule "the derivative of a constant function is equal to zero" to $g(x)$ and obtain $g'(x) = 0$. Note that $h(x) = x^2 + 6 = f(x) + g(x)$; therefore, $h'(x) = f'(x) + g'(x) = 2x + 0 = 2x.$ ■

If $\alpha \in \mathbb{R}$ and $h(x) = (\alpha \cdot f)(x) = \alpha \cdot f(x)$, then $h(x)$ is a *scalar multiple of function $f(x)$*.

The Derivative of a Scalar Multiple of a Function

The derivative of a scalar multiple of a function f is equal to the scalar multiple of the derivative of the function f. This means that if $\alpha \in \mathbb{R}$ and $h(x) = (\alpha \cdot f)(x) = \alpha \cdot f(x)$, then

$$h'(x) = \alpha \cdot f'(x),$$

provided $f'(x)$ exists.

Example 3 Calculate the derivative of $h(x) = 3x$.

Solution Define function $f(x) = x$. Apply the rule $\dfrac{d}{dx}x^n = n \cdot x^{n-1}$, for $n = 1$ to $f(x)$, to obtain $f'(x) = 1$. Recall that $x^0 = 1$ for $x \neq 0$. Therefore, $h(x) = 3x = (3 \cdot f)(x)$, so $h'(x) = (3f)'(x) = 3 \cdot f'(x) = 3 \cdot (1) = 3.$ ■

The Derivative of a Linear Combination of Functions

If f and g are functions and if $\alpha \in \mathbb{R}$ and $\beta \in \mathbb{R}$, then *the linear combination of functions f and g* is defined by $\alpha \cdot f + \beta \cdot g$.

The derivative of a linear combination of functions is the linear combination of the derivatives of the function. This means that if $h(x) = (\alpha \cdot f + \beta \cdot g)(x) = \alpha \cdot f(x) + \beta \cdot g(x)$, then

$$h'(x) = \alpha \cdot f'(x) + \beta \cdot g'(x),$$

provided both $f'(x)$ and $g'(x)$ exist.

Self-Test Exercises

1. Calculate the derivative of $f(x) = 2x^3 + 5\sin x$.
2. Calculate $D(h)(-3)$ if $h(x) = 3x^3 + 5x^2$.

The Derivative of a Product of Functions

The derivative of a product of two functions is equal to the product of the second function and the derivative of the first function added to the product of the first function and the derivative of the second function. This means that if $h(x) = (f \cdot g)(x) = f(x) \cdot g(x)$, then

$$h'(x) = f'(x) \cdot g(x) + f(x) \cdot g'(x)$$
$$= \left(\frac{d}{dx}f(x)\right) \cdot g(x) + f(x) \cdot \left(\frac{d}{dx}g(x)\right)$$

provided both $f'(x)$ and $g'(x)$ exist.
    This rule is a theorem known as the *Product Rule*.

Example 4 Calculate the derivative of $h(x) = 3x$ using the Product Rule.

Solution Define the function $f(x) = 3$. Apply the rule that "the derivative of a constant function is zero" to $f(x)$ to obtain $f'(x) = 0$. Define function $g(x) = x$. Apply the rule $\frac{d}{dx}(x^n) = n \cdot (x^{n-1})$, for $n = 1$, to $g(x)$ to obtain $g'(x) = 1 \cdot x^0 = 1$. (Recall that $x^0 = 1$ for $x \neq 0$.) Because we think of $h(x) = 3x$ as the product of the two functions $f(x) = 3$ and $g(x) = x$, we can apply the Product Rule. Therefore,

$$h'(x) = f'(x) \cdot g(x) + f(x) \cdot g'(x)$$
$$= (0) \cdot (x) + (3) \cdot (1)$$
$$= 3.$$

∎

Example 5 Calculate the derivative of $h(x) = x^2 \sin x$.

Solution The function $h(x) = x^2 \sin x$ is the product of x^2 and $\sin x$. We can apply the product rule to obtain

$$h'(x) = \frac{d}{dx}(x^2 \cdot \sin x)$$
$$= \left(\frac{d}{dx}x^2\right) \cdot \sin x + x^2 \cdot \left(\frac{d}{dx}\sin x\right).$$

Recall that $\frac{d}{dx}x^2 = 2x^1 = 2x$ and $\frac{d}{dx}\sin x = \cos x$. We have

$$h'(x) = \left(\frac{d}{dx}x^2\right) \cdot \sin x + x^2 \cdot \left(\frac{d}{dx}\sin x\right)$$
$$= (2x) \cdot \sin x + x^2 \cdot (\cos x).$$

Please note that the most common mistake for beginners is to think that

$$\frac{d}{dx}(x^2 \cdot \sin x) = \left(\frac{d}{dx}x^2\right) \cdot \left(\frac{d}{dx}\sin x\right).$$

This is very wrong!

∎

The Derivative of a Reciprocal of a Function

The derivative of a reciprocal of a function is equal to minus the derivative of the function divided by the square of the function. This means that if $h(x) = \left(\frac{1}{f}\right)(x) = \frac{1}{f(x)}$, then

$$h'(x) = \frac{-f'(x)}{(f(x))^2},$$

provided $f'(x)$ exists and $f(x) \neq 0$.

Example 6 Calculate the derivative of $h(x) = \frac{1}{x}$ using the *Reciprocal Rule*.

Solution Define $f(x) = x$. Then we have $f'(x) = 1$. Since $h(x) = \frac{1}{x} = \frac{1}{f(x)}$, we have

$$h'(x) = \frac{-f'(x)}{(f(x))^2} = \frac{-1}{x^2}$$

using the Reciprocal Rule.

∎

The Derivative of a Quotient of Functions

The derivative of a quotient of two functions is equal to the product of the denominator and the derivative of the numerator minus the product of the numerator and the derivative of the denominator all over the square of the denominator. This means that if

$$h(x) = \left(\frac{f}{g}\right)(x) = \frac{f(x)}{g(x)},$$

then

$$h'(x) = \frac{g(x) \cdot f'(x) - f(x) \cdot g'(x)}{(g(x))^2},$$

provided both $f'(x)$ and $g'(x)$ exist and $g(x) \neq 0$. This is called the *Quotient Rule*.

Example 7 Calculate the derivative of $h(x) = \dfrac{\sin(x)}{\cos(x)}$ using the *Quotient Rule*.

Solution Theorem 2 in Section 3.2 states that if $f(x) = \sin(x)$, then $f'(x) = \cos(x)$, and if $g(x) = \cos(x)$, then $g'(x) = -\sin(x)$. Since

$$h(x) = \frac{\sin(x)}{\cos(x)} = \frac{f(x)}{g(x)},$$

according to the Quotient Rule, we have

$$h'(x) = \frac{g(x) \cdot f'(x) - f(x) \cdot g'(x)}{(g(x))^2}$$

$$= \frac{(\cos(x)) \cdot (\cos(x)) - (\sin(x)) \cdot (-\sin(x))}{(\cos(x))^2}$$

$$= \frac{(\cos(x))^2 + (\sin(x))^2}{(\cos(x))^2}$$

$$= \frac{1}{(\cos(x))^2}. \qquad\blacksquare$$

Because $\dfrac{\sin(x)}{\cos(x)} = \tan(x)$ and $\dfrac{1}{(\cos(x))^2} = \left(\dfrac{1}{\cos(x)}\right)^2 = (\sec(x))^2$, we have just demonstrated that *the derivative of the tangent function is equal to the square of the secant function.*

Self-Test Exercises

3. Calculate the derivative of $f(x) = x \sin(x)$.
4. Calculate the derivative of $f(x) = \dfrac{x}{\sin(x)}$.

Numeric Differentiation

The method for approximating the numerical value of $f'(c)$ is known as *numeric differentiation*. This method is convenient when dealing with functions for which calculation of the derivative is demanding and/or time-consuming. Numeric differentiation makes use of *difference quotients*.

Suppose that f is a function defined on (a, b) and differentiable for $c \in (a, b)$. When $h > 0$ and c is fixed, then the quotient

$$D_+ f(c, h) = \frac{f(c + h) - f(c)}{h}$$

is known as the *forward difference quotient*,

$$D_- f(c, h) = \frac{f(c) - f(c - h)}{h}$$

is known as the *negative difference quotient*, and

$$D_0 f(c, h) = \frac{f\left(c + \dfrac{h}{2}\right) - f\left(c - \dfrac{h}{2}\right)}{h}$$

is known as the *central difference quotient*.

Figure 2 visualizes these concepts.

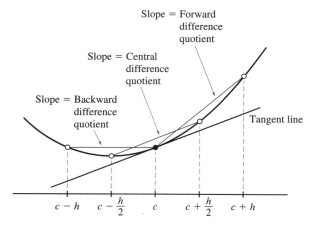

Figure 2

Although each of these difference quotients can be used to approximate $f'(c)$, the central difference quotient, $D_0 f(c, h)$, for a given value of h, gives the best approximation to $f'(c)$.

Choosing the best value of h with which to approximate $f'(c)$ is difficult. Two types of errors [*roundoff* $r(h)$ and *truncation* $t(h)$] can result when using $D_0 f(c, h)$ to approximate $f'(c)$.

Recall that *roundoff errors* occur when computation results in a loss of significant digits. *Truncation errors* result when the limit process is terminated at a particular value of h. The *total error* $e(h)$ is calculated as the sum of the roundoff error and the truncation error—i.e., $e(h) = r(h) + t(h)$. To ensure that both $r(h)$ and $t(h)$ are very small, let h decrease in value until either the desired accuracy is attained or the total error begins to increase. Figure 3 visualizes these errors.

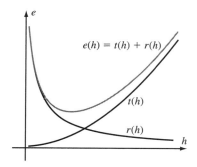

Figure 3

Example 8 Let $V(x) = \sqrt{2 + \cos(x)}$. Approximate $V'(0.4174)$ to four significant digits.

Solution Using $D_0V(0.4174, h) = \dfrac{V(0.4147 + h/2) - V(0.4147 - h/2)}{h}$, the following table presents the results of central difference quotient calculations.

h	0.1	0.01	0.001	0.0001	0.00001	0.000001
$D_0V(0.4174, h)$	−.11871043	−.1187356	−.118736	−.11874	−.11874	−.11874

Since the leftmost four digits of $D_0V(0.4174, h)$ agree when $h = 0.1, 0.01, 0.001, 0.0001$, 0.00001 and 0.000001 are equal to −0.1187, we can be reasonably certain that $V'(0.4174) = -0.1187$ to four significant digits. ∎

Self-Test Exercise

5. Let $V(x) = \sqrt{2 + x^2}$. Approximate $V'(0.4174)$ to three significant digits.

3.4 Differentiation of Some Basic Functions

Important Concepts

-   The Derivative of a Function Defined as a Power of the Variable x
-   Derivatives of Trigonometric Functions
-   The Derivative of the Natural Exponential Function

The Derivative of a Function Defined as a Power of the Variable x

As discussed in Section 3.2, the *Differentiation Rule* $\dfrac{d}{dx}x^n = n \cdot x^{n-1}$, for $n = -1, 0, 1, 2$, and 3, is a special case of the *Power Rule*, which is stated in the following theorem.

Theorem 1 If p is any real number, then $\dfrac{d}{dx}x^p = p \cdot x^{p-1}$.

Example 1 Calculate $\dfrac{d}{dx}(f(x))$ if $f(x) = \sqrt{x}$.

Solution $f(x) = \sqrt{x} = x^{1/2}$. Apply the Power Rule to calculate $\dfrac{d}{dx}(f(x))$.

$$\frac{d}{dx}(f(x)) = \left(\frac{1}{2}\right)x^{(1/2)-1}$$

$$= \left(\frac{1}{2}\right)x^{-1/2}$$

$$= \frac{1}{2\sqrt{x}}.$$
∎

Example 2 Calculate $\dfrac{d}{dx}\left(\dfrac{1}{x^3}\right)$.

Solution We can compute this derivative using the Quotient Rule, but the computation is easier if we use the Power Rule.

$$\begin{aligned}
\frac{d}{dx}\left(\frac{1}{x^3}\right) &= \frac{d}{dx}\left(x^{-3}\right)\\
&= (-3)\cdot\left(x^{-3-1}\right)\\
&= (-3)\cdot\left(x^{-4}\right)\\
&= -3x^{-4}\\
&= \frac{-3}{x^4}.
\end{aligned}$$

∎

The Power Rule is the basis for the following theorem. Recall that a *polynomial function* $p(x)$ *of degree n* is defined by

$$p(x) = a_n x^n + a_{n-1}x^{n-1} + \cdots + a_1 x^1 + a_0,$$

where $a_n \neq 0$ and n is a non-negative integer.

Theorem 2 If p is a polynomial function (of degree n), then its derivative p' is also a polynomial function with degree equal to $n-1$. If p is a constant function, then p' is identically equal to zero.

Example 3 Calculate $\dfrac{d}{dx}(f(x))$ if $f(x) = 2x^{10} - 5x^3 + 7x$.

Solution

$$\begin{aligned}
\frac{d}{dx}(2x^{10} - 5x^3 + 7x)) &= \frac{d}{dx}(2x^{10}) - \frac{d}{dx}(5x^3) + \frac{d}{dx}(7x)\\
&= 2\frac{d}{dx}(x^{10}) - 5\frac{d}{dx}(x^3) + 7\frac{d}{dx}(x).
\end{aligned}$$

According to the Power Rule, we have $\dfrac{d}{dx}(x^{10}) = 10x^9$, $\dfrac{d}{dx}(x^3) = 3x^2$, and $\dfrac{d}{dx}(x) = 1\cdot x^0 = 1$. Therefore,

$$\begin{aligned}
\frac{d}{dx}(2x^{10} - 5x^3 + 7x)) &= 2\frac{d}{dx}(x^{10}) - 5\frac{d}{dx}(x^3) + 7\frac{d}{dx}(x)\\
&= 2(10x^9) - 5(3x^2) + 7(1)\\
&= 20x^9 - 15x^2 + 7.
\end{aligned}$$

∎

Self-Test Exercises

1. Calculate $f'(x)$ if $f(x) = 25x^{10} - 3x + 2$.

2. Calculate $f'(x)$ if $f(x) = \dfrac{1}{\sqrt{x}}$.

Derivatives of Trigonometric Functions

The six trigonometric functions are the sine function, the cosine function, the tangent function, the cosecant function, the secant function, and the cotangent function. The derivatives of some of these functions were calculated earlier. For convenience, the derivatives of all six trigonometric functions are listed here.

$$\frac{d}{dx}(\sin(x)) = \cos(x) \qquad \frac{d}{dx}(\cos(x)) = -\sin(x) \qquad \frac{d}{dx}(\tan(x)) = (\sec(x))^2$$

$$\frac{d}{dx}(\csc(x)) = -\csc(x)\cot(x) \qquad \frac{d}{dx}(\sec(x)) = \sec(x)\tan(x) \qquad \frac{d}{dx}(\cot(x)) = -(\csc(x))^2$$

Example 4 Calculate $\dfrac{d}{dt}\left(\dfrac{1}{\cot(t)}\right)$.

Solution $\left(\dfrac{1}{\cot(t)}\right) = \tan(t)$, so $\dfrac{d}{dt}\left(\dfrac{1}{\cot(t)}\right) = \dfrac{d}{dt}(\tan(t)) = (\sec(t))^2$. ∎

Example 5 Calculate $\dfrac{d}{dx}(\tan(x) \cdot (\sqrt{x} - 3x))$.

Solution We will use the Product Rule here because we have to differentiate a product of two functions.

$$\frac{d}{dx}(\tan(x) \cdot (\sqrt{x} - 3)) = (\tan(x))' \cdot (\sqrt{x} - 3x) + (\tan(x)) \cdot (\sqrt{x} - 3x)'$$

$$= (\sec(x))^2 \cdot (\sqrt{x} - 3x) + (\tan(x)) \cdot (x^{1/2} - 3x)'$$

$$= (\sec(x))^2 \cdot (\sqrt{x} - 3x) + (\tan(x)) \cdot ((x^{1/2})' - (3x)')$$

$$= (\sec(x))^2 \cdot (\sqrt{x} - 3x) + (\tan(x)) \cdot \left(\frac{1}{2}x^{1/2-1} - 3\right)$$

$$= (\sec(x))^2 \cdot (\sqrt{x} - 3x) + (\tan(x)) \cdot \left(\frac{1}{2}x^{-1/2} - 3\right) \quad ∎$$

Self-Test Exercises

3. Calculate $\dfrac{d}{dt}\left(\dfrac{\cos(t)\tan(t)}{\sin(t)}\right)$.

4. Compute the derivative of $g(x) = \dfrac{\tan(x)}{\sqrt{x} - 3x}$.

The Derivative of the Natural Exponential Function

Recall that e^x is defined as $e^x = \lim\limits_{n\to\infty}\left(1 + \dfrac{x}{n}\right)^n$. The *natural exponential function* is defined as $y = e^x$. We note that the expressions e^x and $\exp(x)$ are equivalent.

Theorem 3   For every x, $\dfrac{d}{dx}(e^x) = e^x$.

Example 6 Calculate the derivative $(e^{2x})'$.

Solution $e^{2x} = e^{x+x} = (e^x) \cdot (e^x)$. Therefore, apply the Product Rule to the product $(e^x) \cdot (e^x)$ to calculate $(e^{2x})'$. We have

$$(e^{2x})' = (e^x)' \cdot (e^x) + (e^x) \cdot (e^x)'$$

$$= (e^x) \cdot (e^x) + (e^x) \cdot (e^x)$$

$$= 2(e^x) \cdot (e^x)$$

$$= 2e^{x+x}$$

$$= 2e^{2x}. \quad ∎$$

Self-Test Exercise

5. Calculate $f'(x)$ if $f(x) = \dfrac{e^x}{\sin(x)}$.

3.5 The Chain Rule

Important Concepts

*   The Chain Rule for Differentiating the Composition of Two Functions
*   The Chain Rule for Differentiating the Composition of Multiple Functions

The *Chain Rule* deals with differentiating functions that are defined as a composition of a finite number of other functions or itself.

The Chain Rule for Differentiating the Composition of Two Functions

Suppose that f and g are functions and that the domain of g contains the range of f. Then g *composed with* f, or *the composition of g with f*, is defined by $(g \circ f)(x) = g(f(x))$. In this definition, g is identified as the *outside function* and f is identified as the *inside function*. These identifications will be useful when the Chain Rule is defined. The composition of g with f is represented in Figure 1.

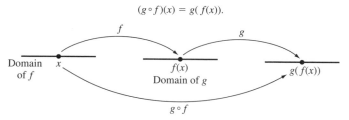

Figure 1

Theorem 1   Suppose the range of the function f is contained in the domain of the function g. If f is differentiable at c and if g is differentiable at $f(c)$, then $g \circ f$ is differentiable at c and is defined by

$$(g \circ f)'(c) = g'(f(c)) \cdot f'(c) = \left(\frac{dg}{du} \bigg|_{u=f(c)} \right) \cdot \left(\frac{df}{dx} \bigg|_{x=c} \right).$$

Theorem 1 is known as the *Chain Rule*. An informal definition of the Chain Rule is *the derivative of a function defined as the composition of two functions is equal to the derivative of the outside function evaluated at the inside function times the derivative of the inside function, provided these derivatives exist.*

Example 1 Use the Chain Rule to calculate $\dfrac{d}{dx}(\cos(5x^2))$.

Solution The function $\cos(5x^2)$ can be written as $\cos(5x^2) = (g \circ f)(x)$, where $g(u) = \cos(u)$ and $f(x) = 5x^2$. According to the Chain Rule,

$$\frac{d}{dx}(\cos(5x^2)) = \left(\frac{dg}{du}\bigg|_{u=f(x)} \right) \cdot \left(\frac{df}{dx} \right).$$

Since $\dfrac{dg}{du} = -\sin(u)$ and $\dfrac{df}{dx} = 10x$, we know that

$$\begin{aligned}
\frac{d}{dx}\cos(5x^2) &= \left(-\sin(u)|_{u=f(x)} \right) \cdot (10x) \\
&= -\sin(f(x)) \cdot 10x \\
&= -\sin(5x^2) \cdot 10x.
\end{aligned}$$

We can also use the informal definition of the Chain Rule mentioned above to calculate this derivative. In this example, the outside function is "cos" and the inside function is "$5x^2$." Now let us follow the statement of that informal definition:

"Derivative of the outside function" becomes "$-\sin$,"

"evaluated at the inside function" means "$-\sin(5x^2)$"

"times the derivative of the inside function" becomes "$-\sin(5x^2) \cdot (5x^2)' = -\sin(5x^2) \cdot (10x)$," which is the same answer as we obtained earlier. ■

The Chain Rule Applied to Powers

If the function g is defined by $g(u) = u^p$ and the function f is simply denoted as $f(x)$, then $(g \circ f)(x)$ is defined by $(g \circ f)(x) = g(f(x)) = (f(x))^p$. The outside function is g and the inside function is f. If the derivatives exist, $g'(u) = p \cdot u^{p-1}$; therefore, $g'(f(x)) = p \cdot (f(x))^{p-1}$ and the derivative of f is denoted as $f'(x)$. As such,

$$\begin{aligned}
\frac{d}{dx}(f(x))^p &= (g \circ f)'(x) \\
&= g'(f(x)) \cdot f'(x) \\
&= p \cdot (f(x))^{p-1} \cdot f'(x).
\end{aligned}$$

This statement is the Chain Rule applied to powers.

Example 2 Use the Chain Rule to calculate $(e^{2x})'$.

Solution Since $e^{2x} = (e^x)^2$, we can utilize the Chain Rule applied to powers, with $f(x) = e^x$ and $p = 2$. Therefore,

$$\begin{aligned}
\frac{d}{dx}(e^x)^2 &= (2(e^x)^{2-1}) \cdot (e^x)' \\
&= 2(e^x) \cdot (e^x)' \\
&= 2(e^x) \cdot (e^x) \\
&= 2e^{2x}.
\end{aligned}$$

■

The Chain Rule for Differentiating the Composition of Multiple Functions

If a function is defined as the composition of more than two functions, then calculating its derivative involves applying the Chain Rule several times, because the Chain Rule is defined for a function that is the composition of two functions.

Example 3 Use the Chain Rule to calculate $\dfrac{d}{dx} \tan(\cos(5x^2))$.

Solution The outside function is "$\tan(u)$," and the inside function is $\cos(5x^2)$. Since $\dfrac{d}{du} \tan u = \sec^2 u$, according to the Chain Rule, we have

$$\frac{d}{dx} \tan(\cos(5x^2)) = \left(\frac{d}{du} \tan u \bigg|_{u=\cos(5x^2)} \right) \cdot \left(\frac{d}{dx} \cos(5x^2) \right)$$

$$= \sec^2 u \big|_{u=\cos(5x^2)} \cdot \left(\frac{d}{dx} \cos(5x^2) \right)$$

$$= \sec^2(\cos(5x^2)) \cdot \left(\frac{d}{dx} \cos(5x^2) \right).$$

To finish the computation, we need to find $\dfrac{d}{dx} \cos(5x^2)$. Using the Chain Rule again, now the outside function is $\cos(u)$ and the inside function is $5x^2$. (See Example 1.) We obtain

$$\frac{d}{dx} \cos(5x^2) = (-\sin(u)\big|_{u=5x^2}) \cdot (10x)$$

$$= -\sin(5x^2) \cdot 10x.$$

Therefore,

$$\frac{d}{dx} \tan(\cos(5x^2)) = \sec^2(\cos(5x^2)) \cdot (-\sin(5x^2) \cdot 10x)$$

$$= -10x \cdot \sin(5x^2) \cdot \sec^2(\cos(5x^2)). \qquad ∎$$

Self-Test Exercises

1. Calculate the derivative of $f(x) = \sin(2x)$.
2. Calculate the derivative of $f(x) = (\sin(2x))^n$, $n = 0, 1, 2, \ldots$.
3. Calculate the derivative of $f(x) = \cos(\sin(2x))$.

3.6 Inverse Functions and the Natural Logarithm

Important Concepts

- Inverse Functions
- The Natural Logarithm

The Inverse of a Function

Let $f : S \rightarrow T$ be a function. We say that *f has an inverse (or f is invertible)* if there exists a function $g : T \rightarrow S$ such that $(f \circ g)(t) = t$ for all t in T and $(g \circ f)(s) = s$ for all s in S. If such a function g exists, then $g = f^{-1}$, the inverse of f.

Therefore, the domain of f is equal to the range of f^{-1}, and the range of f is equal to the domain of f^{-1}.

The Horizontal Line Test

If $f : S \rightarrow T$, then f is invertible if and only if for each t in T, the horizontal line $y = t$ intersects the graph of f exactly once.

Basic Rules for Finding Inverses

Suppose that function $f : S \rightarrow T$ and f is an invertible function with an explicit formula for $f(s) = t$ in T. To find f^{-1}, in the equation $f(s) = t$, solve this equation for s in terms of t. After the variable s is isolated on one side of the equation, the other side is the expression for $f^{-1}(t)$.

The expression $f^{-1}(t)$ is not the same as $(f(t))^{-1}$. The expression $(f(t))^{-1} = \dfrac{1}{f(t)}$, which is the reciprocal of $f(t)$, but $f^{-1}(t)$ is the expression for the inverse function of $f(t)$.

One-to-One Functions

If function $f : S \rightarrow T$ has the property that $f(s_1) \neq f(s_2)$ whenever $s_1 \neq s_2$, then we say that f *is one-to-one*, which is sometimes written as "f is 1-1." We note that f *is 1-1 if it takes different elements in its domain to different elements in its image.*

Onto Functions

If function $f : S \rightarrow T$ has the property that for every t in T there is at least one s in S for which $f(s) = t$, then we say that f *is onto.* We note that *function f is onto if its image is the same as its range.*

One-to-One and Onto Functions

A function $f : S \rightarrow T$ is both *one-to-one* and *onto* if for each t in T there exists exactly one s in S such that $f(s) = t$.

Theorem 1

A function f is invertible if and only if it is both one-to-one and onto.

Example 1 Find the inverse of function $f(x) = 3x + 5$.

Solution The graph of $f(x) = 3x + 5$ is a straight line with slope 3, x-intercept $\left(-\dfrac{5}{3}, 0\right)$, and y-intercept $(0, 5)$, as shown in Figure 1.

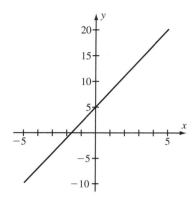

Figure 1

The *Horizontal Line Test* establishes that $f(x) = 3x + 5$ is *one-to-one*. The *Image of f* = $(-\infty, \infty)$ and the *Range of f* = $(-\infty, \infty)$, so function f is onto, therefore f is invertible. To calculate f^{-1}, we let $y = f(x)$ in the equation $f(x) = 3x + 5$, and solve for x. This will be the expression for $f^{-1}(y)$.

$$f(x) = 3x + 5$$
$$y = 3x + 5$$
$$x = \frac{y - 5}{3}$$
$$f^{-1}(y) = \frac{y - 5}{3}$$
$$= \frac{1}{3}y - \frac{5}{3}.$$

∎

We note that conventionally when graphing a function and its inverse on the same set of axes, the original function is expressed as a function of x, and its inverse is also expressed as a function of x. Thus, in the example $f(x) = 3x + 5$ and $g(x) = \frac{1}{3}x - \frac{5}{3}$ represent the original function $f(x)$ and its inverse $g(x)$.

The Graph of the Inverse Function

The graph of the function f^{-1} is the reflection in the line $y = x$ of the graph of f, if we draw both f and f^{-1} on the same graph with variable x.

Example 2 Show that the graph of f^{-1} is the reflection in the line $y = x$ of the graph of $f(x) = 3x + 5$.

Solution Recall that we found in Example 1 that $f^{-1}(y) = \frac{1}{3}y - \frac{5}{3}$. Now in order to draw both f and f^{-1} on the same graph as functions of the variable x, we need to rewrite $f^{-1}(x) = \frac{1}{3}x - \frac{5}{3}$. The graphs of $f^{-1}(x) = \frac{1}{3}x - \frac{5}{3}$, $f(x) = 3x + 5$, and $y = x$ appear in the graph in Figure 2 as the line with dashes, the line with small dots, and the solid line, respectively. The graph clearly shows that the graph of the inverse function is the reflection in the line $y = x$ of the function f.

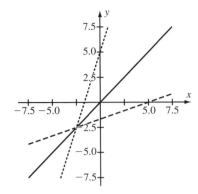

Figure 2 ■

Continuity and Differentiability of Inverse Functions

Theorem 3 Suppose S and T are open intervals in $\mathbb{R}$ and $f : S \to T$ is *one-to-one* and *onto* (and therefore, f^{-1} exists). If f is continuous on S, then f^{-1} is continuous on T. If f is differentiable on S and $f' \neq 0$ on S, then f^{-1} is differentiable on T.

Theorem 4 Let f be an invertible function defined on an open interval containing point s. Let $t = f(s)$. If f is differentiable at $s = f^{-1}(t)$ and if $\dfrac{df}{ds}$ is nonzero at this point, then

$$\frac{d}{dt}(f^{-1})\Big|_{t} = \frac{1}{\dfrac{df}{ds}\Big|_{s=f^{-1}(t)}}.$$

Theorem 4 is called the *Inverse Function Derivative Rule*.

Example 3 If $f(x) = 3x + 5$, calculate $\dfrac{d}{dy}(f^{-1}(y))$.

Solution $\dfrac{d}{dx}(f(x)) = 3$. Apply the Inverse Function Derivative Rule to obtain

$$\frac{d}{dy}(f^{-1}(y)) = \frac{1}{\dfrac{d}{dx}(f(x))} = \frac{1}{3}.$$

The inverse function was obtained in Example 1. Because $f^{-1}(y) = \dfrac{1}{3}y - \dfrac{5}{3}$ and $\dfrac{d}{dy}(f^{-1}(y))$ is equal to $\dfrac{1}{3}$, the results of the Inverse Function Derivative Rule are confirmed. ■

Example 4 If $f(s) = 2s^5 - 3$, use the Inverse Function Derivative Rule to calculate $\dfrac{d}{dt}(f^{-1}(t))$.

Solution We first note that $\dfrac{d}{ds}(f(s)) = \dfrac{d}{ds}(2s^5 - 3) = 10s^4$. The Chain Rule tells us that if $\dfrac{d}{ds}(f(s)) \neq 0$, then

$$\frac{d}{dt}(f^{-1}(t)) = \frac{1}{\dfrac{d}{ds}(f(s))}$$
$$= \frac{1}{10s^4}.$$

But since $t = f(s) = 2s^5 - 3$, we have $s = \left(\dfrac{t+3}{2}\right)^{1/5}$. Therefore,

$$\frac{d}{dt}(f^{-1}(t)) = \frac{1}{10s^4}$$
$$= \frac{1}{10\left(\dfrac{t+3}{2}\right)^{4/5}}$$
$$= \frac{2^{4/5}}{10(t+3)^{4/5}}. \qquad \blacksquare$$

Example 5 Assume that f is invertible and differentiable, with a differentiable inverse. Compute $(f^{-1})'(5)$, given that $f^{-1}(5) = 2$ and $f'(2) = 7$.

Solution The Inverse Function Derivative Rule tells us that

$$(f^{-1})'(5) = \frac{d}{dt}(f^{-1}(t))\Big|_{t=5}$$
$$= \frac{1}{\dfrac{d}{ds}(f(s))\Big|_{s=f^{-1}(5)}}$$
$$= \frac{1}{\dfrac{d}{ds}(f(s))\Big|_{s=2}}$$
$$= \frac{1}{f'(2)}$$
$$= \frac{1}{7}. \qquad \blacksquare$$

Self-Test Exercises

1. Calculate the inverse of $f(x) = 5 - \dfrac{x}{3}$.
2. Determine whether the function $f(x) = x^2$ is invertible.
3. Let $f(s) = 2 + s^3$. Use the Inverse Function Derivative Rule to find $\dfrac{d}{dt}(f^{-1}(t))$.

Increasing and Decreasing Functions

A function f is *increasing* if $f(s) < f(u)$ for $s < u$. *An increasing function is one-to-one.*

Example 6 Show that $f(x) = 3x + 5$ is an increasing function.

Solution Suppose x_1 and x_2 are elements in the domain of f and $x_1 < x_2$. Then $3x_1 < 3x_2$. This, in turn, implies that $3x_1 + 5 < 3x_2 + 5$, so $f(x_1) < f(x_2)$. Since the choices of x_1 and x_2 are arbitrary, this is true for all elements $x_1 < x_2$. ∎

A *decreasing function* is a function such that $f(s) > f(u)$ for $s < u$. A *decreasing function* is also one-to-one.

Example 7 Show that $f(x) = -3x + 5$ is a decreasing function.

Solution Suppose x_1 and x_2 are elements in the domain of f and $x_1 < x_2$. Then $-3x_1 > -3x_2$. This, in turn, implies that $-3x_1 + 5 > -3x_2 + 5$, so $f(x_1) > f(x_2)$. Since the choices of x_1 and x_2 are arbitrary, this is true for all elements $x_1 < x_2$. ∎

The Natural Logarithm Function

If $f(x) = e^x$, then $f^{-1}(x) = \ln(x)$. The function $y = \ln(x)$ is the *natural logarithm function*. Therefore, for every $s \in \mathbb{R}$ and $t > 0$, $\ln(e^s) = s$ and $e^{\ln(t)} = t$. In particular, if $s = 0$, then $\ln(1) = 0$; if $s = 1$, then $\ln(e) = 1$.

Basic Properties of the Natural Logarithm Function

The basic algebraic properties of the natural logarithm function are $\ln(xy) = \ln(x) + \ln(y)$, for $x, y > 0$ and $\ln(x^p) = p \cdot \ln(x)$.

Derivatives of Exponential Functions and the Natural Logarithm Function

In Section 3.4, the formula for the derivative of the function $y = e^x$ was given as $y' = e^x$. This gives rise to the formula for the derivative of the general exponential function $y = a^x$ for $a > 0$.

Theorem 2 If a is a positive constant, then $\dfrac{d}{dx}(a^x) = (a^x) \cdot (\ln(a))$.

Using the Inverse Function Derivative Rule, we have

Theorem 3 If $t > 0$, then $\dfrac{d}{dt}(\ln(t)) = \dfrac{1}{t}$.

Example 8 Calculate the derivative of $f(x) = 3^x \ln(x)$ for $x > 0$.

Solution Apply the Product Rule to $f(x) = 3^x \ln(x)$ and obtain

$$f'(x) = (3^x)' \cdot \ln(x) + 3^x \cdot (\ln(x))'$$
$$= (3^x \ln(3)) \cdot \ln x + 3^x \cdot \left(\frac{1}{x}\right).$$
∎

Self-Test Exercises

4. Calculate the derivative of $f(x) = (x^2 + 2)\ln(x)$ for $x > 0$.
5. Calculate the derivative of $f(x) = \cos(2^x)$.

3.7 Higher Derivatives

Important Concepts

- Higher Derivatives: Definition and Notation
- Velocity and Acceleration as Derivatives

Higher Derivatives: Definition and Notation

The derivative of the $(k-1)$th derivative of f is called the kth *derivative of f* and is *denoted by* $f^{(k)}$. The expressions $\dfrac{d^k f}{dx^k}$, $\dfrac{d^k}{dx^k}(f)$, and $D^k(f)$ are equivalent to the expression $f^{(k)}$.

Example 1 Calculate the second derivative of $f(x) = 5^x$.

Solution $f(x) = 5^x$ so $f'(x) = 5^x \cdot (\ln(5))$.

$$
\begin{aligned}
f''(x) &= \frac{d}{dx}(f'(x)) \\
&= \frac{d}{dx}((\ln(5)) \cdot 5^x) \\
&= (\ln(5)) \cdot \frac{d}{dx}(5^x) \\
&= (\ln(5)) \cdot ((\ln(5)) \cdot (5^x)) \\
&= 5^x (\ln(5))^2.
\end{aligned}
$$

■

Newton's Notation

If $p(t)$ is a function of the time variable t, then the first and second derivatives are represented by $\dot{p}(t)$ and $\ddot{p}(t)$. This notation is often used in physics and engineering.

Leibniz's Rule

The following theorem is known as *Leibniz's Rule for Second Derivatives*.

Theorem 1 If f and g are two times differentiable at c, then so is function $f \cdot g$ and $(f \cdot g)''(c) = f''(c) \cdot g(c) + 2f'(c) \cdot g'(c) + f(c) \cdot g''(c)$.

Example 2 Calculate the second derivative of $f(x) = 5^x \sin(x)$.

Solution According to the Leibniz's Rule, we have

$$(5^x \sin(x))'' = (5^x)'' \cdot (\sin(x)) + 2(5^x)' \cdot (\sin(x))' + (5^x) \cdot (\sin(x))''.$$

Since we have $(\sin(x))' = \cos(x)$, $(\sin(x))'' = -\sin(x)$, and from Example 1, $(5^x)' = 5^x \cdot (\ln(5))$, $(5^x)'' = 5^x \cdot (\ln(5))^2$, the second derivative becomes

$$
\begin{aligned}
(5^x \sin(x))'' &= 5^x \cdot (\ln(5))^2 \cdot (\sin(x)) + 2 \cdot 5^x \cdot (\ln(5)) \cdot (\cos(x)) + (5^x) \cdot (-\sin(x)) \\
&= 5^x \cdot \left(\sin(x) \cdot (\ln(5))^2 + 2 \cdot \ln(5) \cdot \cos(x) - \sin(x) \right).
\end{aligned}
$$

■

Self-Test Exercises

1. a. Calculate the second derivative of $f(x) = 3^x$.
    b. By considering 3^x as $(1) \cdot (3^x)$, use Leibniz's Rule for Second Derivatives to calculate the second derivative of $f(x) = 3^x$ again.
2. Calculate the third derivative of $f(x) = 3x^2 + 20x - 6$.

Velocity and Acceleration Expressed as Derivatives

If p is position and t is time, then *velocity*, $v(t) = p'(t)$, is the rate of change of position with respect to time. The rate of change of velocity with respect to time is *acceleration*, $a(t) = v'(t)$.

Example 3 If $p(t) = t^2 - 5t + 6$ represents position at time t, find the expression for acceleration.

Solution If $p(t) = t^2 - 5t + 6$, then $p'(t) = 2t - 5$. If $p'(t) = 2t - 5$, then $p''(t) = 2$. Since $a(t) = v'(t)$ and $v(t) = p'(t)$, we see that $a(t) = p''(t)$; therefore, the expression for $a(t)$ is $a(t) = 2$. ∎

Self-Test Exercises

3. The path of a projectile is given by the position function $p(t) = t^2 - 5t$. Find the expressions for velocity and acceleration.
4. If the velocity is given by $v(t) = 2t^2 - 5$, find the expression for the acceleration function. What is the acceleration at $t = 3$?

3.8 Implicit Differentiation and Related Rates

Important Concepts

- Implicit Definition of a Function and Implicit Differentiation
- Methods for Solving Related Rates Problems

Implicit Definitions of Function and Implicit Differentiation

If a function is defined by $y = f(x)$, then we say that y *is explicitly defined as a function of x*. However, if y is defined by an equation $F(x, y) = C$, then we say that y *is defined implicitly by the equation*. For example, by choosing $F(x, y) = y^{1/3} - xy^5$ and $C = 7$, we can say that y is defined implicitly by the equation $y^{1/3} - xy^5 = 7$.

Implicit Differentiation

If $F(x, y) = C$ is a given equation and if $P = (x_0, y_0)$ satisfies this equation, then we may find $\dfrac{dy}{dx}\bigg|_P$, if it exists, by differentiating the equation *without* first solving for y in terms of x.

When we do this, we treat y as though it were a differentiable function of x on an open interval centered at x_0. We say that we have *implicitly differentiated F* with respect to x.

Basic Steps for Differentiating Implicitly

1. If $F(x, y) = C$ is a given equation, then we proceed along the equation from left to right.
2. We implement the differentiation rules to each term on both sides of the equation. However, when differentiating the function of y, we calculate the derivative with respect to y first, then attach the expression $\dfrac{dy}{dx}$ to the result. The derivative of a function of x is calculated as usual.
3. Collect all the terms that contain the expression $\dfrac{dy}{dx}$ on one side of the equation, and collect all other terms on the other side of the equation.
4. Treat the expression $\dfrac{dy}{dx}$ as if it were a common factor.
5. Solve the equation for $\dfrac{dy}{dx}$. The result will be $\dfrac{dy}{dx}$ equals an expression that may contain both x and y.

Example 1 Differentiate the equation $y^{1/3} - xy^5 = 7$ implicitly.

Solution

Step 1: Differentiate the terms from left to right.

$$(y^{1/3})' - (xy^5)' = (7)'$$

Step 2: When differentiating $y^{1/3}$, we differentiate it first with respect to y, then attach to the result the expression $\dfrac{dy}{dx}$. Therefore, $(y^{1/3})'$ becomes $\left(\dfrac{d}{dy} y^{1/3}\right) \dfrac{dy}{dx} = \dfrac{1}{3} y^{-2/3} \dfrac{dy}{dx}$.

Treat the term xy^5 as the product of two functions, so the Product Rule will be applied. Therefore, $(xy^5)'$ becomes $x' \cdot y^5 + x \cdot (y^5)' = 1 \cdot y^5 + x \cdot 5y^4 \dfrac{dy}{dx}$.

On the right side of the equation, since 7 is a constant, we have $(7)' = 0$. Combining these results, the implicit differentiation of the given equation becomes

$$\frac{1}{3} y^{-2/3} \frac{dy}{dx} - \left(1 \cdot y^5 + x \cdot 5y^4 \frac{dy}{dx}\right) = 0.$$

Step 3: Collect all the terms that contain $\dfrac{dy}{dx}$ on one side of the equation:

$$\frac{1}{3} y^{-2/3} \frac{dy}{dx} - 1 \cdot y^5 - x \cdot 5y^4 \frac{dy}{dx} = 0$$

$$\frac{1}{3} y^{-2/3} \frac{dy}{dx} - x \cdot 5y^4 \frac{dy}{dx} = y^5$$

Step 4: Treat $\dfrac{dy}{dx}$ as if it were a common factor.

$$\left(\frac{1}{3} y^{-2/3} - x \cdot 5y^4\right) \frac{dy}{dx} = y^5$$

Step 5: Solve for $\dfrac{dy}{dx}$.

$$\frac{dy}{dx} = \frac{y^5}{\left(\frac{1}{3}y^{-2/3} - x \cdot 5y^4\right)}$$

$$= \frac{3y^5}{\left(y^{-2/3} - 15x \cdot y^4\right)}$$

$$= \frac{3y^{17/3}}{\left(1 - 15x \cdot y^{14/3}\right)}. \qquad \blacksquare$$

The Method of Implicit Differentiation can also be used to calculate higher derivatives.

Self-Test Exercise

1. Differentiate the expression $\dfrac{x^2}{y^5} - xy^{2/3} + y^6 = 10$ implicitly.

Methods for Solving Related Rates Problems

The Method of Implicit Differentiation is often used to solve related rates problems wherein two variables are related by means of an equation.

Basic Steps for Solving a Related Rates Problem

1. Identify the quantities (or functions) that vary, and identify the variable (which is often time) with respect to which the change in these quantities is taking place.
2. Establish a relationship (an equation) between the quantities isolated in step 1.
3. Differentiate the equation from step 2 with respect to the variable identified in step 1. Be sure to apply the Chain Rule carefully.
4. Substitute the numerical data, and solve for the unknown rate of change.

Example 2 A 5-foot ladder is leaning against a wall. If the bottom of the ladder slides away from the bottom of the wall at a rate of 3 ft/s, how fast is the top of the ladder sliding down the wall when the bottom of the ladder is 2 ft away from the wall?

Solution

Step 1: Let x be the distance from the bottom of the ladder to the bottom of the wall. Let y be the distance from the top of the ladder to the bottom of the wall. (See Figure 1.) $\dfrac{dx}{dt}$ is the rate at which the bottom of the ladder is sliding away from the wall; $\dfrac{dx}{dt} = 3$ ft/s. $\dfrac{dy}{dt}$ is the rate at which the top of the ladder is sliding down the wall; $\dfrac{dy}{dt}$ is unknown.

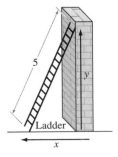

Figure 1

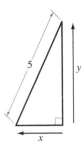

Figure 2

Step 2: A right triangle is formed by the ladder, which is the hypotenuse. The distance x from the bottom of the ladder to the bottom of the wall is the horizontal leg of the triangle, and the distance y from the top of the ladder to the bottom of the wall is the vertical leg of the triangle. (See Figure 2.) Because we are dealing with a right triangle, we can apply the Pythagorean Theorem to obtain $x^2 + y^2 = 5^2$.

Step 3: Differentiate $x^2 + y^2 = 5^2$ implicitly with respect to the variable t to obtain

$$2x\frac{dx}{dt} + 2y\frac{dy}{dt} = 0 \quad \text{so} \quad \frac{dy}{dt} = \frac{-x}{y}\frac{dx}{dt}.$$

Step 4: We know that $x = 2$ ft and $\dfrac{dx}{dt} = 3$ ft/s, so we just have to calculate the value of y when $x = 2$. We use the relationship $x^2 + y^2 = 5^2$ to obtain $2^2 + y^2 = 5^2$, which implies that $y^2 = 25 - 4 = 21$ so $y = \sqrt{21}$.

Therefore, when $x = 2$ and $y = \sqrt{21}$, we have

$$\frac{dy}{dt} = \frac{-x}{y}\frac{dx}{dt} = \left(\frac{-2}{\sqrt{21}}\right)(3) = \frac{-6}{\sqrt{21}} \text{ ft/s.} \qquad \blacksquare$$

Self-Test Exercise

2. A spherical balloon is losing air. Its diameter is decreasing at the rate of $\dfrac{1}{2}$ in/s. How fast is the volume changing when its radius is 5 in?

3.9 Differentials and Approximation of Functions

Important Concepts

- Approximation of Functions
- Applications

The Method of Increments

The definition $f'(c) = \lim\limits_{\Delta x \to 0} \dfrac{f(c + \Delta x) - f(c)}{\Delta x}$ implies that

$$f'(c) \approx \frac{f(c + \Delta x) - f(c)}{\Delta x}$$

when $\Delta x \neq 0$ and Δx is very small. Therefore, $f(c + \Delta x) \approx f(c) + f'(c) \cdot \Delta x$. If we define $\Delta f(c) = f(c + \Delta x) - f(c)$, then we approximate $\Delta f(c)$ by $\Delta f(c) \approx f'(c) \cdot \Delta x$. These steps describe the *Method of Increments*.

The approximation scheme known as the Method of Increments is visualized in Figure 1.

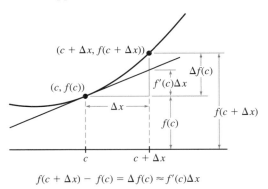

$$f(c + \Delta x) - f(c) = \Delta f(c) \approx f'(c)\Delta x$$

Figure 1

Example 1 Use the approximation $f(c + \Delta x) \approx f(c) + f'(c) \cdot \Delta x$ to estimate the value of $\sqrt{3.9}$.

Solution If $f(x) = x^{1/2}$, then $f'(x) = \dfrac{1}{2}x^{-1/2}$. Choose $c = 4$, so that $\Delta x = 3.9 - c = 3.9 - 4 = -0.1$. Using the suggested approximation $f(c + \Delta x) \approx f(c) + f'(c) \cdot \Delta x$, we obtain

$$
\begin{aligned}
\sqrt{3.9} = \sqrt{(4 - 0.1)} &= f(4 + (-0.1)) \\
&\approx f(4) + f'(4) \cdot (-0.1) \\
&= \sqrt{4} + \left(\frac{1}{2} \cdot 4^{-1/2} \right)(-0.1) \\
&= 2 + \left(\frac{1}{2} \right) \cdot \left(\frac{1}{4^{1/2}} \right)(-0.1) \\
&= 2 - .025 \\
&= 1.975
\end{aligned}
$$

Note that $\sqrt{3.9} = 1.9748417658$. ■

Applications

If we take $\Delta x = 1$ in the equation $f(c + \Delta x) \approx f(c) + f'(c) \cdot \Delta x$, we obtain $f(c + 1) \approx f(c) + f'(c)$. This approximation is often used in economics. For example, if $f(x)$ represents a quantity of goods produced as a function of the amount of labor x, then f is called the *product of labor*. The *marginal product of labor,* denoted as *MPL*, is the increment in production that is achieved by adding one or more units of labor.

Within this context, $f'(c) \approx f(c + 1) - f(c)$ and $MPL(c)$ plays the role of $f'(c)$. Other applications are *marginal cost* $C'(x)$, *marginal revenue* $R'(x)$, and *marginal profit* $P'(x)$.

The Tangent Line Approximation to *f* at *c*

In the approximation $f'(c) \approx \dfrac{f(c + \Delta x) - f(c)}{\Delta x}$ for $\Delta x \neq 0$ and Δx very small, if we define x as $x = (c + \Delta x)$, then we can approximate $f(x) \approx f(c) + f'(c) \cdot (x - c)$ for x close to c. The right side of this approximation is a linear function of x, which we denote by L. The graph of L is the tangent line to the graph of f at $(c, f(c))$.

The function $L(x) = f(c) + f'(c) \cdot (x - c)$ is called the *linearization of f at c*. The approximation $L(x) \approx f(x)$ for x close to c is called the *tangent line approximation to f at c* (or *best linear approximation*).

Important Linearizations

$(1 + u)^p \approx 1 + pu$ for values of u near 0 is a linearization that occurs often in calculus. Other important linearizations include $\sin(x) \approx x$, $\cos(x) \approx 1$, and $\tan(x) \approx x$.

Differentials

In the Method of Increments description, the approximation $\Delta f(c) \approx f'(c) \cdot \Delta x$ was derived for $\Delta f(c) = f(c + \Delta x) - f(c)$. When Δx becomes "infinitesimal," the approximation becomes equality. Denote the infinitesimal increments in x as dx. When this happens, we represent $\Delta f(c) = f(c + \Delta x) - f(c)$ as $df(x) = f'(x)dx$. This approximation is sometimes referred to as a *differential approximation*.

The rules of differentiation where c represents a constant expressed in differential form are listed in the following table.

$$
\begin{array}{ll}
d(c) = 0 & d(\sin(u)) = \cos(u)du \\
d(cu) = c\,du & d(\cos(u)) = -\sin(u)du \\
d(u + v) = du + dv & d(\tan(u)) = \sec^2(u)du \\
d(uv) = u\,dv = v\,du & d(\sec(u)) = \sec(u)\tan(u)du \\
d\left(\dfrac{u}{v}\right) = \dfrac{v\,du - u\,dv}{v^2} & d(\cot(u)) = -\csc^2(u)du \\
d\left(\dfrac{1}{v}\right) = \dfrac{-dv}{v^2} & d(\csc(u)) = -\csc(u)\cot(u)du \\
d(u^n) = nu^{n-1}du & d(\exp(u)) = \exp(u)du \\
d(\ln(u)) = \dfrac{du}{u} & d(a^u) = a^u \ln(a)du
\end{array}
$$

Self-Test Exercises

1. Approximate the value of $\sqrt{17}$.
2. Use the Method of Increments to estimate the value of $f(x) = x^{-1/3}$ at $x = 8.07$, using the known value at the initial point $c = 8$.

Applications of the Derivative

4.1 The Derivative and Graphing

Important Concepts

- Maxima and Minima
- Theorems

Local Maximum, Absolute Maximum, Local Minimum, and Absolute Minimum

An informal explanation of the difference between the adjectives "local" and "absolute" in the terms local maximum, absolute maximum, local minimum, and absolute minimum is whether the x-values related to $f(x)$ constitute *part of the domain* or constitute the *entire domain*.

A *local maximum* is the largest y-value identified from amongst *some* of the y-values, and an *absolute maximum* is the largest y-value identified from amongst *all* of the y-values. Similarly, a *local minimum* is the smallest y-value identified from amongst *some* of the y-values, and an *absolute minimum* is the smallest y-value identified from amongst *all* of the y-values.

The difference between "local maximum/minimum" and "absolute maximum/minimum" can also be defined formally. Let f be a function with domain S. Function f has a *local maximum* (or *relative maximum*) at the point c in S if there exists a $\delta > 0$ such that $f(x) \leq f(c)$ for all x in S such that $|x - c| < \delta$. Note that if the local maximum *occurs* at $x = c$ then the local maximum value *is* $f(c)$.

However, if $f(x) \leq f(c)$ for all x in S, then we say f has an *absolute maximum* (or *global maximum*) *at* c and $f(c)$ is the *absolute maximum value*.

Let f be a function with domain S. Function f has a *local minimum* (or *relative minimum*) at the point c in S if there exists a $\delta > 0$ such that $f(x) \geq f(c)$ for all x in S such that $|x - c| < \delta$. In this case, the local minimum *occurs* at $x = c$ and the local minimum value *is* $f(c)$.

However, if $f(x) \geq f(c)$ for all x in S, then we say that *f has an absolute minimum at* c and $f(c)$ *is the absolute minimum value for f*. These concepts are visualized in Figure 1.

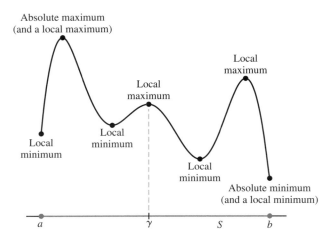

Figure 1

As Figure 1 shows, if f has a local maximum (minimum) at c, then f takes its greatest (smallest) value at c *only when compared with nearby points*. By contrast, an absolute maximum (minimum) at c is determined by comparing the value of f at all points of the function's domain. One can think of a local maximum point as a "regional high point," while the absolute maximum point is the "world high point."

Example 1 Locate all local maxima and minima, if they exist, for the function $f(x) = -(x-2)^2 + 10$ in the interval $(0, 3)$ by examining its graph.

Solution The graph of $f(x)$ in the interval $(0, 3)$ is shown in Figure 2.

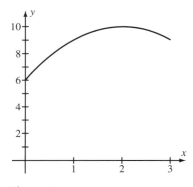

Figure 2

The local maximum value is 10, and it occurs at $x = 2$. There is no local minimum, as the next sentences explain. We do *not* state that the local minimum value is 6 because 6 is the y-value associated with $x = 0$ but $x = 0$ is *not* in the domain $(0, 3)$. Similarly we do not state the local minimum value is 9 because 9 is the y-value associated with $x = 3$, but $x = 3$ is not in the domain $(0, 3)$. ∎

Please note that if we change the interval in Example 1 from $(0, 3)$ to $[0, 3]$, then the answer will become:

"The local maximum value is 10, and it occurs at $x = 2$. A local minimum value is 6, and it occurs at $x = 0$. Another local minimum value is 9, and it occurs at $x = 3$."

Self-Test Exercises

1. Identify all local and absolute minima and maxima, if they exist, for $f(x) = 2x + 5$ in the interval $[-7, 10]$.

2. Identify all local and absolute minima and maxima, if they exist, for $f(x) = \sin(x)\cos(x)$ in the interval $\left(\dfrac{3\pi}{2}, 2\pi\right)$.

Theorems

The following theorems are important. You can use them to help locate maxima and minima, provided they exist.

Theorem 1 Let f be defined on an open interval that contains the point c. Suppose that f is differentiable at c. If f has a local extremum at c, then $f'(c) = 0$.

Theorem 1 is known as *Fermat's Theorem*.

Example 2 Use Fermat's Theorem to locate all local maxima and minima, if they exist, for the function $f(x) = -(x-2)^2 + 10$ in the interval $(0, 3)$.

Solution
$$
\begin{aligned}
f(x) &= -(x-2)^2 + 10 \\
&= -(x^2 - 4x + 4) + 10 \\
&= -x^2 + 4x - 4 + 10 \\
&= -x^2 + 4x + 6.
\end{aligned}
$$

$f'(x) = -2x + 4$ and $f'(x) = 0$ implies $-2x + 4 = 0$; therefore, $x = 2$. This means that $f'(2) = 0$. Although Fermat's Theorem does not guarantee the existence of local extrema at $x = 2$, it indicates that $x = 2$ is a candidate for local maxima and minima. We should examine values in the neighborhood of $x = 2$. We'll compare the values of f at $x = 2$, with its value at the nearby points $x = 2 - \delta$ and $x = 2 + \delta$, say, for $\delta = .01$. At $x = 2 - .01 = 1.99$, $f(1.99) = 9.9999$ and at $x = 2 + .01 = 2.01$, $f(2.01) = 9.9999$. Compare these results with $f(2) = 10$. We conclude that the local maximum value is 10, and it occurs at $x = 2$. ■

Theorem 2 Let f be a function that is continuous on $[a, b]$ and differentiable on (a, b). If $f(a) = f(b)$, then there exists a number c in (a, b) such that $f'(c) = 0$.

Theorem 2 is known as *Rolle's Theorem*.

Figure 3 visualizes Rolle's Theorem. Its geometric interpretation is "If a function f satisfies the hypotheses of Theorem 2, then there exists a number c in (a, b) at which the graph of f has a horizontal tangent line that is parallel to the line segment determined by the points $(a, f(a))$ and $(b, f(b))$."

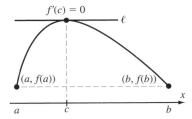

Figure 3

Example 3 Show that the hypotheses and conclusion of Rolle's Theorem are satisfied by function $f(x) = x^2 + 6$ on $[-2, 2]$.

Solution Function f is a polynomial, so it is continuous for all real x; in particular it is continuous for x in the interval $[-2, 2]$. Its derivative $f'(x) = 2x$ exists for all real x; in particular, it exists for x in the open interval $(-2, 2)$. The values $f(-2) = f(2) = 10$. Therefore, the hypotheses of Rolle's Theorem are satisfied. In Figure 4, the graph of $f(x)$ is an upright parabola and the graph of the constant function $y = 10$ is the dotted horizontal line. The graph of $y = 10$ is included to identify two specific points, $(-2, 10)$ and $(2, 10)$, on the parabola where $f(a) = f(b)$. A horizontal tangent to the parabola is possible only at the point $(0, 6)$. Therefore, there exists a value $c = 0$ in the open interval $(-2, 2)$ such that $f'(c) = 0$. The conclusion of Rolle's Theorem holds.

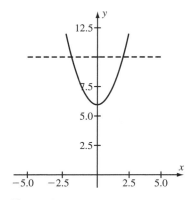

Figure 4

Theorem 3 If f is a function that is continuous on $[a, b]$ and differentiable on (a, b), then there exists a number c in (a, b) such that $f'(c) = \dfrac{f(b) - f(a)}{b - a}$.

Theorem 3 is known as the *Mean Value Theorem*, which is illustrated in Figure 5. When we compare Figure 3 and Figure 5, we see that Rolle's Theorem is a special case of the Mean Value Theorem.

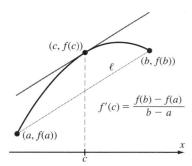

Figure 5

Example 4 Verify that the hypothesis of the Mean Value Theorem is satisfied for the function $f(x) = \sin(x)$ in the interval $\left[\dfrac{\pi}{4}, \dfrac{\pi}{2}\right]$. What can you conclude from the theorem in this case?

Solution $f(x) = \sin(x)$ is continuous in the interval $\left[\dfrac{\pi}{4}, \dfrac{\pi}{2}\right]$, and $f'(x) = \cos(x)$ exists in the open interval $\left(\dfrac{\pi}{4}, \dfrac{\pi}{2}\right)$. Since the hypothesis of the Mean Value Theorem is satisfied, the existence of a value c in $\left(\dfrac{\pi}{4}, \dfrac{\pi}{2}\right)$ such that $f'(c) = \dfrac{f\left(\dfrac{\pi}{2}\right) - f\left(\dfrac{\pi}{4}\right)}{\dfrac{\pi}{2} - \dfrac{\pi}{4}}$ is guaranteed. In this example, the conclusion of the Mean Value Theorem is "There exists a value c in $\left(\dfrac{\pi}{4}, \dfrac{\pi}{2}\right)$ such that $\cos(c) = \dfrac{1 - \dfrac{\sqrt{2}}{2}}{\dfrac{\pi}{4}} = \dfrac{4 - 2\sqrt{2}}{\pi}$." $\left(\text{Although the Mean Value Theorem does not tell us exactly what the value of } c \text{ is, we can find the value of } c \text{ using the inverse trigonometric function, namely, } c = \cos^{-1}\left(\dfrac{4 - 2\sqrt{2}}{\pi}\right) \approx 1.19, \text{ which is a value in the interval } \left(\dfrac{\pi}{4}, \dfrac{\pi}{2}\right).\right)$ ∎

Theorem 4 Let f be a differentiable function on an interval (α, β). If $f'(x) = 0$ for each x in (α, β), then f is a constant function.

Theorem 5 If F and G are differentiable functions such that $F'(x) = G'(x)$ for all x in (α, β), then there exists a constant C such that $F(x) = G(x) + C$ for all x in (α, β).

Self-Test Exercises

3. Let $f(x) = x^3 + 3x^2 + 3x + 5$. Use Fermat's Theorem (Theorem 1) to locate all possible candidates for local maxima and minima in the interval $(-20, 20)$. By examining the values of f in the neighborhood of each candidate point, determine if it is a local maximum, a local minimum, or neither.

4. Verify that the hypothesis of the Mean Value Theorem (Theorem 3) holds for $f(x) = 2x^4 - 32$ in $[-1, 3]$. State the conclusion of the Mean Value Theorem in this case. With the help of algebra, explicitly find all values of c that satisfy the conclusion of the Mean Value Theorem.

4.2 Maxima and Minima of Functions

Important Concepts

- Increasing and Decreasing Functions and the First Derivative
- The First Derivative Test for Extrema
- Critical Points

Increasing and Decreasing Functions

A function f is *increasing* on an interval I if $f(\alpha) < f(\beta)$ whenever α and β are points in I such that $\alpha < \beta$.

A function f is *decreasing* on an interval I if $f(\alpha) > f(\beta)$ whenever α and β are points in I such that $\alpha < \beta$.

Example 1 Is the function $f(x) = x^2$ increasing or decreasing on the interval $[3, 5]$?

Solution Consider the graph of $f(x) = x^2$ on the interval $[3, 5]$, as illustrated in Figure 1.

As we proceed from left to right, the numbers along the x-axis increase and the vertical distance from the x-axis to the graph of the function f increases. Since the y-values are associated with the vertical distance, we conclude that as $\alpha < \beta$ in the interval $[3, 5]$, then $f(\alpha) < f(\beta)$. Therefore, function f is increasing.

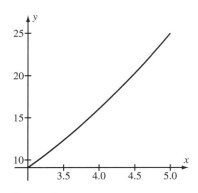

Figure 1

Theorem 1   If $f'(x) > 0$ for all x in interval I, then f is increasing on I. If $f'(x) < 0$ for all x in interval I, then f is decreasing on I.

Theorem 1 is useful; it gives a method to use the derivative to indicate whether function f is increasing or decreasing.

Example 2 Find all values x where the function $f(x) = x^3 - 6x^2 + 12x + 1$ is decreasing.

Solution $f(x) = x^3 - 6x^2 + 12x + 1$ implies that $f'(x) = 3x^2 - 12x + 12$. The function f decreases for all x where $f'(x) < 0$. The expression $3x^2 - 12x + 12$ factors into $3(x - 2)^2$. Since 3 is positive and the expression $(x - 2)^2 \geq 0$, we conclude that $f'(x) \geq 0$. Therefore, there are *no* values x where the function $f(x) = x^3 - 6x^2 + 12x + 1$ is decreasing.

Example 3 Determine the intervals on which the function $f(x) = x^3 - 7x^2$ is increasing or decreasing.

Solution $f(x) = x^3 - 7x^2$ implies that $f'(x) = 3x^2 - 14x$, which factors into $x(3x - 14)$. Therefore, $f'(x) = 0$ implies that $x = 0$ or $x = \dfrac{14}{3}$.

$f'(x) > 0$ implies that either $x > 0$ and $(3x - 14) > 0$ or $x < 0$ and $(3x - 14) < 0$. The first case, $x > 0$ and $(3x - 14) > 0$, implies that $x > \dfrac{14}{3}$. The second case, $x < 0$ and $(3x - 14) < 0$, implies that $x < 0$. We conclude that function f is *increasing* in the interval $(-\infty, 0)$ or the interval $\left(\dfrac{14}{3}, \infty\right)$.

$f'(x) < 0$ implies that either $x > 0$ and $(3x - 14) < 0$ or that $x < 0$ and $(3x - 14) > 0$. The first case, $x > 0$ and $(3x - 14) < 0$, implies that $0 < x < \dfrac{14}{3}$. The second case, $x < 0$ and $(3x - 14) > 0$, is impossible, so there is no solution. We conclude that function f is *decreasing* in the interval $\left(0, \dfrac{14}{3}\right)$. We can check these answers by drawing the graph of f on the interval $[-3, 8]$. See Figure 2.

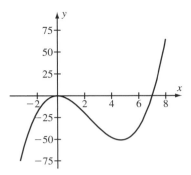

Figure 2

The First Derivative Test for Extrema

Theorem 2

Let f be a differentiable function on an open interval, and suppose that $f'(c) = 0$ at some point c inside the interval.

   a. If $f'(x) < 0$ for $x < c$ and $f'(x) > 0$ for $x > c$, then f has a local minimum at c.
   b. If $f'(x) > 0$ for $x < c$ and $f'(x) < 0$ for $x > c$, then f has a local maximum at c.
   c. If f' does not change sign at c even though $f'(c) = 0$, then f has neither a local minimum nor a local maximum at c.

Figure 3 illustrates the various possibilities described in Theorem 2.

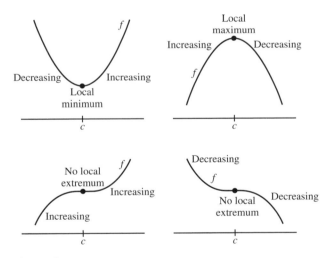

Figure 3

    Theorem 2 is known as the *First Derivative Test*; it is a powerful tool that helps determine if a differentiable function has a local minimum or maximum.

Example 4 Use the First Derivative Test to determine the values of any local extrema if $f(x) = x^3 - 7x^2$.

Solution Since f has domain the entire real line and f is differentiable at all x, the local extrema will be at points where $f' = 0$. Now, $f(x) = x^3 - 7x^2$ implies that $f'(x) = 3x^2 - 14x$, which factors into $x(3x - 14)$. Therefore, $f'(x) = 0$ implies that $x = 0$ or $x = \dfrac{14}{3}$. These points are the only candidates for local extrema.

Let us first examine the point $x = 0$. If x is just to the left of 0, for example, $x = -.0001$, then $x < 0$, and $(3x - 14) < 0$, so $f'(x) = x(3x - 14) > 0$ because it is a product of two negative numbers. If x is just to the right of 0, for example, $x = .0001$, then $x > 0$, and $(3x - 14) < 0$, so $f'(x) = x(3x - 14) < 0$ because it is a product of a positive and a negative number. The First Derivative Test tells us that a local maximum occurs at $x = 0$. The value of the local maximum is $f(0) = 0$.

Similarly, if x is just to the left of $\dfrac{14}{3}$, for example, $x = \dfrac{14}{3} - .0001$, then $x > 0$, and $(3x - 14) < 0$, so $f'(x) = x(3x - 14) < 0$. If x is just to the right of $\dfrac{14}{3}$, for example, $x = \dfrac{14}{3} + .0001$, then $x > 0$, and $(3x - 14) > 0$, so $f'(x) = x(3x - 14) > 0$. Therefore, by the First Derivative Test, a local minimum occurs at $x = \dfrac{14}{3}$. The value of the local minimum is $f\left(\dfrac{14}{3}\right) = \dfrac{-1372}{27} \approx -50.81$. ∎

Critical Points

Let c be a point in an open interval on which f is continuous. We call c a *critical point for f* if one of the following two conditions holds:

f is not differentiable at c.

or,

f is differentiable at c and $f'(c) = 0$.

As a consequence of the First Derivative Test, if the domain of f is an open interval, then the critical points are the only places where local extrema of f occur, provided they exist.

Example 5 Find all critical points of $f(x) = x^3 - 7x^2$ in the interval $(-10, 10)$.

Solution If we refer to the solution presented for Example 4, we see that $f(x)$ is a polynomial and, therefore, continuous in $(-10, 10)$ and that $f'(x)$ exists for all values of x. Therefore, the only critical points c are obtained from $f'(c) = 0$. These points are $c = 0$ and $c = \dfrac{14}{3}$. ∎

Example 6 Find and analyze the critical points of $f(x) = x^{1/3} + x^{4/3}$.

Solution If $x \neq 0$, then $f'(x) = \dfrac{1}{3}x^{-2/3} + \dfrac{4}{3}x^{1/3} = \dfrac{1}{3}x^{-2/3}(1 + 4x) = \dfrac{(1 + 4x)}{3x^{2/3}}$. f is not differentiable at $x = 0$ and $f' = 0$ at $x = \dfrac{-1}{4}$. Therefore, the critical points of f are $c = 0$ and $c = \dfrac{-1}{4}$.

At the critical point $c = 0$, if x is slightly left of 0, for example, $x = -.0001$, we have $f'(x) = \dfrac{\text{positive}}{\text{positive}} > 0$ and if x is slightly right of 0, for example, $x = .0001$, we have $f'(x) = \dfrac{\text{positive}}{\text{positive}} > 0$. By the First Derivative Test, a local extremum does not occur at $c = 0$.

At the critical point $c = \dfrac{-1}{4}$, if x is slightly left of $\dfrac{-1}{4}$, for example, $x = -\dfrac{1}{4} - .0001$, we have $f'(x) = \dfrac{\text{negative}}{\text{positive}} < 0$. If x is slightly right of $\dfrac{-1}{4}$, for example, $x = -\dfrac{1}{4} + .0001$, we have $f'(x) = \dfrac{\text{positive}}{\text{positive}} > 0$. By the First Derivative Test, there is a local minimum at $c = \dfrac{-1}{4}$. ∎

Self-Test Exercises

1. Find the intervals on which the function $f(x) = x^4 - 1$ is increasing and on which the function is decreasing.
2. Find all critical points of the function $f(x) = x^4 - 1$. Then use the First Derivative Test to determine whether each critical point is a local maximum, a local minimum, or neither.
3. Find all local extrema if $f(x) = x^3 - 12x^2$.
4. Find and analyze the critical points for the function $f(x) = x \cdot (x + 2)^{1/3}$.

4.3 Applied Maximum and Minimum Problems

Important Concepts

- Finding Extrema of a Continuous Function on a Closed Bounded Interval
- Maximum and Minimum Word Problems

Finding Extrema of a Continuous Function on a Closed Bounded Interval

The Extreme Value Theorem in Section 3 of Chapter 2 guarantees that f will have both an absolute minimum and an absolute maximum in $[a, b]$. The following strategy can be used to identify the absolute minimum value and the absolute maximum value of f in $[a, b]$.

To find the extrema of a continuous function f on a closed interval $[a, b]$, test the points in (a, b) where f is not differentiable, the points in (a, b) where f' exists and is equal to zero, and the endpoints a and b.

In other words, the points to be tested are the *critical points* and the *endpoints* of the interval.

Example 1 Determine the absolute maximum and absolute minimum of $f(x) = x^3 + 2x^2$ in the interval $[-5, 3]$.

Solution $f(x) = x^3 + 2x^2$ implies that $f'(x) = 3x^2 + 4x$. The derivative exists for all real numbers x, so the critical points c are obtained by setting the first derivative equal to zero. The expression $3x^2 + 4x$ factors into $x(3x + 4)$, so the critical points are 0 and $\dfrac{-4}{3}$. The values of $f(x)$ at the critical points and at the endpoints of the interval are $f(0) = 0$, $f\left(\dfrac{-4}{3}\right) \approx 1.19$, $f(-5) = -75$, and $f(3) = 45$. Therefore, the absolute minimum value is -75 and the absolute maximum value is 45. ∎

Maximum and Minimum Word Problems

The following six steps should be followed for every optimization word problem that involves a continuous function on a closed bounded interval.

1. Determine the function f to be maximized or minimized.
2. Identify the relevant variable or variables.
3. If f depends on more than one variable, find relationships amongst the variables that will allow substitution into the expression for f.

4. Determine the set of allowable values for the variable.
5. Find all places where the function is not differentiable, all places where the derivative is zero, and the endpoints. *These are the points to test.*
6. Use the First Derivative Test or, alternatively, simply substitute values to determine which of the points found in step 5 solves the maximum-minimum problem.

Example 1 A farmer has 100 feet of fencing with which to enclose a garden. If the garden is to be rectangular in shape, how large can the farmer make the garden and enclose it using the fencing he has?

Solution

1. The function to be maximized is the area A of the garden.
2. Let w represent the width in feet, and let ℓ represent the length in feet of the garden. (See Figure 1.) We have $A = \ell w$.

Figure 1

3. Since the total amount of fencing is 100, this implies that the perimeter of the garden is 100. We therefore have $100 = 2\ell + 2w$, or equivalently $100 - 2\ell = 2w$, so $50 - \ell = w$.

   Substituting this expression for w into the equation $A = \ell w$ produces the area function

$$A(\ell) = \ell(50 - \ell) = 50\ell - \ell^2.$$

4. The length of the garden cannot be a negative number; this means that $\ell \geq 0$. We cannot use more than 100 feet of fence for the perimeter; this means that $100 \geq 2\ell$, or equivalently, $\ell \leq 50$. Therefore, the range of values for ℓ is $0 \leq \ell \leq 50$.
5. The function $A(\ell) = 50\ell - \ell^2$ is differentiable everywhere and $A'(\ell) = 50 - 2\ell$. $A'(\ell) = 0$ implies that $\ell = 25$; it is the only critical point inside the interval $(0, 50)$. This critical point and the endpoints $\ell = 0$ and $\ell = 50$ are the candidates for the maximum we seek.
6. We calculate that $A(25) = 50(25) - (25)^2 = 625$, $A(0) = 50(0) - 0 = 0$ and $A(50) = 50(50) - 50^2 = 0$. Therefore, the maximum dimensions are $\ell = w = 25$ feet, and the garden's maximum area is 625 square feet. This conclusion makes sense. ■

Self-Test Exercises

1. If $f(x) = 2x + 5$, find all absolute minima and all absolute maxima for x in $[30, 35]$.
2. If $f(x) = \dfrac{1}{x}$, find all absolute minima and all absolute maxima for x in $[2, 5]$.
3. A farmer has 200 feet of fencing with which to enclose a garden. If the garden is rectangular in shape, how large a garden can the farmer enclose?
4. A farmer can spend $5000 to enclose a garden. The garden is rectangular in shape, with its perimeter bounded by two parallel stone walls and two parallel wooden fences. If it costs $5/ft to build a stone wall, and $3/ft to build a wooden fence, how large a garden can the farmer enclose?

4.4 Concavity

Important Concepts

- Curvature Test for Concavity
- Inflection Points and the Second Derivative Test

Let the domain of a differentiable function f contain an open interval I.

If f' (the slope of the tangent line to the graph of f) increases as x moves from left to right on I, then *the graph of f is concave up on I.* (See Figure 1.)

If f' (the slope of the tangent line to the graph of f) decreases as x moves from left to right on I, then *the graph of f is concave down on I.* (See Figure 2.)

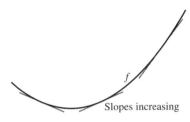

The graph of f is concave up.

Figure 1

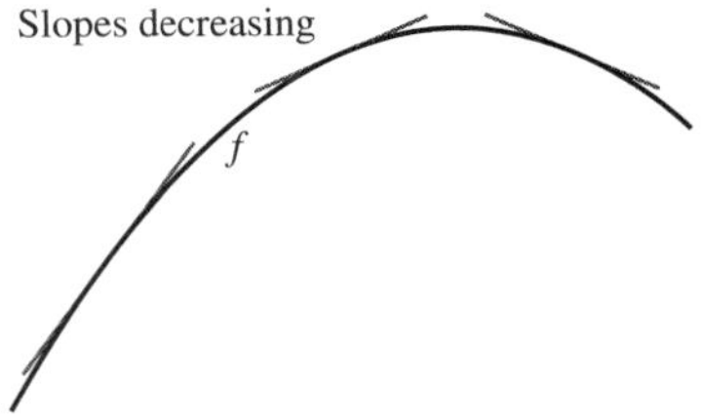

Slopes decreasing

The graph of f is concave down.

Figure 2

Example 1 Discuss the concavity of the function $f(x) = x^2 + 6$ on the interval $(-5, 5)$.

Solution We calculate that $f'(x) = 2x$. Function $f'(x) = 2x$ is an increasing function for all real x; in particular it is increasing on $(-5, 5)$. Therefore, the graph of f is concave up on $(-5, 5)$. The graph of $f(x) = x^2 + 6$ is illustrated in Figure 3.

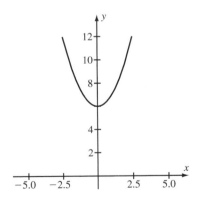

Figure 3

We can also use information about the second derivative to determine concavity.

Theorem 1   Suppose that the function f is twice differentiable on an open interval I.

a. If $f''(x) > 0$ for all x in I, then the graph of f is *concave up* on I.
b. If $f''(x) < 0$ for all x in I, then the graph of f is *concave down* on I.

Theorem 1 is called the *Second Derivative Test for Concavity.*

Example 2 Use the Second Derivative Test for Concavity to determine where $f(x) = -x^2 + 16$ is concave up or concave down in the interval $(-5, 5)$.

Solution $f(x) = -x^2 + 16$ implies that $f'(x) = -2x$ and $f''(x) = -2$. Therefore, $f''(x) < 0$ for all x, particularly for x in the interval $(-5, 5)$. By the Second Derivative Test for Concavity, function $f(x) = -x^2 + 16$ is concave down. ∎

Inflection Points and the Second Derivative Test

Let f be a continuous function on an open interval I. If the graph of f changes concavity (from positive to negative or from negative to positive) at a point c in I, then c is called a *point of inflection* or an *inflection point*.

As Figure 4 clearly shows, *inflection points are not the same as critical points*.

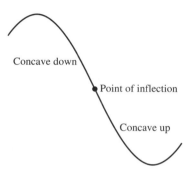

Concave down

Point of inflection

Concave up

Figure 4

Determining Points of Inflection

The following strategy can be used to identify inflection points:

1. Locate all points of I where $f''(x) = 0$ *or* where f'' is undefined.
2. At these points, check to see if f'' changes sign.

Example 3 Given that $f(x) = 3x^5 - 5x^4$, identify all the points of inflection of f.

Solution $f(x) = 3x^5 - 5x^4$ implies that $f'(x) = 15x^4 - 20x^3$ and $f''(x) = 60x^3 - 60x^2 = 60x^2(x - 1)$. We notice that f'' is defined everywhere and $f''(x) = 0$ at $x = 0$ and $x = 1$. Therefore, the points 0 and 1 are the only candidates for points of inflection. For those x that are just left of 0, for example, $x = -.001$, we have $x^2(x - 1) < 0$, and hence $f''(x) < 0$. For those x that are just right of 0, for example, $x = .001$, we have $x^2(x - 1) < 0$, and hence $f''(x) < 0$. These results indicate that f'' does not change sign at 0; therefore 0 is *not* a point of inflection. Following a similar procedure, we can check that f'' changes from negative to positive at $x = 1$. Therefore, 1 is the only point of inflection of f. ∎

The Second Derivative Test at a Critical Point

Theorem 2
Let f be twice differentiable (both f' and f'' exist) on open interval I containing a point c at which $f'(c) = 0$.

    a. If $f''(c) > 0$, then a local minimum occurs at $x = c$.
    b. If $f''(c) < 0$, then a local maximum occurs at $x = c$.
    c. If $f''(c) = 0$, no conclusion is possible *from this test*.

Theorem 2 is called the *Second Derivative Test for Extrema*.

Example 4 Use the Second Derivative Test for Extrema to examine the critical points of $f(x) = 3x^3 - 5x$.

Solution $f(x) = 3x^3 - 5x$ implies that $f'(x) = 9x^2 - 5$ and $f''(x) = 18x$. Both $f'(x)$ and $f''(x)$ exist for all real x. Therefore $f'(x) = 0$ implies that the critical points are $c = \sqrt{\dfrac{5}{9}} = \dfrac{\sqrt{5}}{3}$ and $c = -\sqrt{\dfrac{5}{9}} = -\dfrac{\sqrt{5}}{3}$. $f''\left(\dfrac{\sqrt{5}}{3}\right) = 18 \cdot \dfrac{\sqrt{5}}{3} > 0$ and $f''\left(-\dfrac{\sqrt{5}}{3}\right) = 18 \cdot \left(-\dfrac{\sqrt{5}}{3}\right) < 0$.

We conclude that by the Second Derivative Test for Extrema, a local minimum occurs at $c = \dfrac{\sqrt{5}}{3}$ and a local maximum occurs at $c = -\dfrac{\sqrt{5}}{3}$. ∎

Curvature

An informal explanation of the curvature of a function f is that curvature measures the amount of "bend" that the graph of f has at a given point.

A more formal explanation of curvature states:

Let C denote the graph of $y = f(x)$. For each x, let $\phi(x)$ be the angle that the tangent line to C at $(x, f(x))$ makes with the positive x-axis. The *curvature* $\kappa(x)$ of $y = f(x)$ at the point $(x, f(x))$ is equal to

$$\kappa(x) = \lim_{\Delta x \to 0} \frac{|\phi(x + \Delta x) - \phi(x)|}{\sqrt{(\Delta x)^2 + (\Delta y)^2}} = \lim_{\Delta x \to 0} \frac{|\phi(x + \Delta x) - \phi(x)|}{\sqrt{(\Delta x)^2 + (f(x + \Delta x) - f(x))^2}}.$$

Theorem 3
Suppose that $y = f(x)$ is twice differentiable. The curvature $\kappa(x)$ of the graph of f at the point $(x, f(x))$ is given by $\kappa(x) = \dfrac{|f''(x)|}{(1 + (f'(x))^2)^{3/2}}$.

Example 5 Discuss the curvature of $f(x) = x^5$ at the point $(2, 32)$.

Solution $f(x) = x^5$ implies that $f'(x) = 5x^4$ and $f''(x) = 20x^3$. The curvature of $f(x) = x^5$ at the point $(2, 32)$ is equal to $\kappa(2) = \dfrac{|f''(2)|}{(1 + (f'(2))^2)^{3/2}} = \dfrac{|160|}{(1 + (80)^2)^{3/2}} \approx .00031$. ∎

Self-Test Exercises

1. Determine on which intervals the function $f(x) = 5x^3 - 12x$ is concave up or concave down, find all critical points and points of inflection, and identify all local maxima and minima.
2. Given that the second derivative $f''(x) = (x - 1)^2 x$, find all points of inflection of f.
3. Calculate the curvature of $f(x) = x^2 - 6x$ at the point $(1, -5)$.

4.5 Graphing Functions

Important Concepts

- Curve Sketching
- Periodic Functions
- Skew Asymptotes
- Graphing Calculators and Software

Curve Sketching

Curve sketching is an important analytical tool. The following suggested checklist of steps will help you gather information so that you can sketch the graph of a function f.

Basic Steps in Sketching a Graph

1. Determine the domain and, if possible, the range of the function.
2. Find all horizontal and vertical asymptotes.
3. Calculate the first derivative, and find the critical points for the function.
4. Find the intervals on which the function is increasing or decreasing.
5. Calculate the second derivative, and find the intervals on which the function is concave up or concave down.
6. Identify all local maxima, local minima, and points of inflection.
7. Plot these points, as well as the y-intercept, if applicable. Plot any x-intercepts if they can be computed. Sketch the asymptotes.
8. Connect the points plotted in step 7, keeping in mind concavity, local extrema, and asymptotes.

Example 1 Identify all information needed to sketch the graph of $f(x) = \dfrac{1}{x^2}$.

Solution

Step 1: The domain of $f(x) = \dfrac{1}{x^2}$ is $(-\infty, 0) \cup (0, \infty)$. The range of $f(x) = \dfrac{1}{x^2}$ is $(0, \infty)$.

Step 2: The horizontal asymptote is $y = 0$. The vertical asymptote is $x = 0$.

Step 3: $f'(x) = \dfrac{-2}{x^3}$. f' is not defined at $x = 0$, but since 0 is not in the domain of f, there is *no* critical point for f.

Step 4: $f'(x) = \dfrac{-2}{x^3} > 0$, when x in $(-\infty, 0)$ and $f'(x) = \dfrac{-2}{x^3} < 0$, when x in $(0, \infty)$. Therefore, f is increasing on $(-\infty, 0)$ and decreasing on $(0, \infty)$.

Step 5: $f''(x) = \dfrac{6}{x^4}$ so $f''(0)$ is not defined, but 0 is not in the domain of f. Therefore, there isn't any point of inflection. Since $f''(x) = \dfrac{6}{x^4} > 0$ for all x in the domain, f is concave up there.

Step 6: There are no local extrema or points of inflection.

Step 7: There are no y-intercepts or x-intercepts. ∎

Periodic Functions

A function f is *periodic* if there is a positive number p such that $f(x + p) = f(x)$ for $x \in \mathbb{R}$.

The smallest positive number p for which $f(x + p) = f(x)$, $x \in \mathbb{R}$, is true is called the *period of* f.

Example 2 Determine the period of $f(x) = \sin(2x)$.

Solution The graph of $f(x) = \sin(2x)$ for $x \in [0, 2\pi]$ is shown in Figure 1.

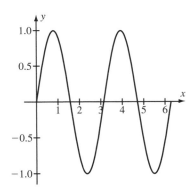

Figure 1

This suggests that the period p is π. Now, let us check this mathematically:

$$f(x + \pi) = \sin(2(x + \pi)) = \sin(2x + 2\pi) = \sin(2x) = f(x).$$

In general, to graph a periodic function f with a period p, we plot (over the interval $[0, p]$) and then translate this function. See Figure 2.

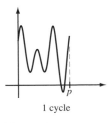

1 cycle   3 cycles

Figure 2

Skew Asymptotes

The line $y = mx + b$ is a *skew asymptote of f* if $\lim\limits_{x \to +\infty} ((mx + b) - f(x)) = 0$ or if $\lim\limits_{x \to -\infty} ((mx + b) - f(x)) = 0$.

The expressions *skew asymptote*, *slant asymptote*, and *oblique asymptote* are equivalent. Skew asymptotes often occur for rational functions where the degree of the numerator is one more than the degree of the denominator.

Example 3 Discuss the skew asymptote or asymptotes for $f(x) = \dfrac{x^2 + 3x + 5}{x}$.

Solution $f(x) = \dfrac{x^2 + 3x + 5}{x} = x + 3 + \dfrac{5}{x}$. The skew asymptote for $f(x) = \dfrac{x^2 + 3x + 5}{x}$ is $y = x + 3$. We can check this with:

$$\lim_{x \to \pm\infty} \left((x + 3) - \frac{x^2 + 3x + 5}{x} \right) = \lim_{x \to \pm\infty} \frac{-5}{x} = 0.$$

Self-Test Exercises

1. Identify all pertinent information necessary to sketch the graph of $f(x) = \dfrac{x}{x+5}$.
2. Determine the period of $f(x) = \cos(3x)$.
3. Discuss the skew asymptote of $f(x) = \dfrac{x^3 + 3x^2 + 5x}{x^2}$.

Graphing Calculators and Software

Although graphing calculators and plotting software are convenient tools, they have their limitations. They frequently distort graphs that are displayed on calculator or computer screens. By understanding curve sketching, you can compensate for those distortions and correct them.

4.6 L'Hôpital's Rule

Important Concept

- Indeterminate Forms and l'Hôpital's Rule

Indeterminate Forms and L'Hôpital's Rule

The symbols $\dfrac{0}{0}, \dfrac{\infty}{\infty}, (0)(\infty), \infty - \infty, 0^0, 1^\infty$, and ∞^0 are called *indeterminate forms*. l'Hôpital's Rule is a method for evaluating limits that result in an indeterminate form.

Theorem 1
Let $f(x)$ and $g(x)$ be differentiable functions on $(a, c) \cup (c, b)$. If $\lim\limits_{x \to c} f(x) = \lim\limits_{x \to c} g(x) = 0$, then $\lim\limits_{x \to c} \dfrac{f(x)}{g(x)} = \lim\limits_{x \to c} \dfrac{f'(x)}{g'(x)}$, provided $\lim\limits_{x \to c} \dfrac{f'(x)}{g'(x)}$ exists as a finite or infinite limit.

Theorem 1 is l'Hôpital's Rule for the indeterminate form $\dfrac{0}{0}$.

Example 1 Use l'Hôpital's Rule to evaluate $\lim\limits_{x \to 2} \dfrac{x^2 - 5x + 6}{x^2 - 4}$.

Solution $\lim\limits_{x \to 2} \dfrac{x^2 - 5x + 6}{x^2 - 4} = \dfrac{0}{0}$, so apply l'Hôpital's Rule.

$$\lim\limits_{x \to 2} \dfrac{x^2 - 5x + 6}{x^2 - 4} = \lim\limits_{x \to 2} \dfrac{2x - 5}{2x}$$
$$= \dfrac{-1}{4}$$

Theorem 2
Let $f(x)$ and $g(x)$ be differentiable functions on $(a, c) \cup (c, b)$. If $\lim\limits_{x \to c} f(x)$ and $\lim\limits_{x \to c} g(x)$ both exist and equal $+\infty$ or $-\infty$, then $\lim\limits_{x \to c} \dfrac{f(x)}{g(x)} = \lim\limits_{x \to c} \dfrac{f'(x)}{g'(x)}$, provided $\lim\limits_{x \to c} \dfrac{f'(x)}{g'(x)}$ exists as a finite or infinite limit.

Theorem 2 is l'Hôpital's Rule for the indeterminate form $\dfrac{\infty}{\infty}$.

Example 2 Use l'Hôpital's Rule to evaluate $\lim\limits_{x \to 0} \dfrac{\ln(|2x|)}{\ln(|3x|)}$.

Solution $\lim\limits_{x\to 0}\dfrac{\ln(|2x|)}{\ln(|3x|)} = \dfrac{-\infty}{-\infty}$. Therefore, we can employ l'Hôpital's Rule.

$$\lim\limits_{x\to 0}\frac{\ln(|2x|)}{\ln(|3x|)} = \lim\limits_{x\to 0}\frac{\dfrac{2}{2x}}{\dfrac{3}{3x}}$$
$$= \lim\limits_{x\to 0}\frac{x}{x}$$
$$= \lim\limits_{x\to 0}1$$
$$= 1.$$

Theorem 3 Let $f(x)$ and $g(x)$ be differentiable functions. If $\lim\limits_{x\to +\infty} f(x) = \lim\limits_{x\to +\infty} g(x) = 0$ or if $\lim\limits_{x\to +\infty} f(x) = \pm\infty$ and $\lim\limits_{x\to +\infty} g(x) = \pm\infty$, then $\lim\limits_{x\to +\infty}\dfrac{f(x)}{g(x)} = \lim\limits_{x\to +\infty}\dfrac{f'(x)}{g'(x)}$, provided $\lim\limits_{x\to +\infty}\dfrac{f'(x)}{g'(x)}$ exists as a finite or infinite limit. The same result holds for the limit as $x \to -\infty$.

Theorem 3 is l'Hôpital's Rule for limits at infinity.

Example 3 Use l'Hôpital's Rule to evaluate $\lim\limits_{x\to +\infty}\dfrac{\ln(x)}{2x}$.

Solution $\lim\limits_{x\to +\infty}\dfrac{\ln(x)}{2x} = \dfrac{+\infty}{+\infty}$. Therefore, we can employ l'Hôpital's Rule.

$$\lim\limits_{x\to +\infty}\frac{\ln(x)}{2x} = \lim\limits_{x\to +\infty}\frac{\dfrac{1}{x}}{2}$$
$$= \lim\limits_{x\to +\infty}\frac{1}{2x}$$
$$= 0$$

When the indeterminate forms 0^0, 1^∞, and ∞^0 result when calculating the limit of $\lim\limits_{x\to c}\dfrac{f(x)}{g(x)}$ (c can either be a finite real number or infinity), then the strategy to be used is to rewrite the functions $f(x)$ and $g(x)$ into equivalent expressions to which l'Hôpital's Rule can be applied. Some techniques utilize logarithms or expressing terms with a common denominator.

Example 4 Use l'Hôpital's Rule to evaluate $\lim\limits_{x\to 0} x^x$.

Solution Let $f(x) = x^x$. We have $\lim\limits_{x\to 0} f(x) = \lim\limits_{x\to 0} x^x = 0^0$. If we utilize logarithms, then $\ln(f(x)) = \ln(x^x) = x\ln(x)$, which gives us the intermediate form $(0)(\infty)$. But if we rewrite $x\ln(x)$ as $\dfrac{\ln(x)}{\dfrac{1}{x}}$, the intermediate form is $\dfrac{\infty}{\infty}$ and we can apply l'Hôpital's Rule.

$$\lim\limits_{x\to 0}(\ln(f(x))) = \lim\limits_{x\to 0}\frac{\ln(x)}{\dfrac{1}{x}}$$
$$= \lim\limits_{x\to 0}\frac{\dfrac{1}{x}}{-\dfrac{1}{x^2}}$$
$$= \lim\limits_{x\to 0}(-x)$$
$$= 0$$

This means that $\lim\limits_{x\to 0} f(x) = \lim\limits_{x\to 0} e^{\ln(f(x))} = e^{\left(\lim\limits_{x\to 0}\ln(f(x))\right)} = e^0 = 1.$

Self-Test Exercises

1. Evaluate $\lim\limits_{x \to \pi} \dfrac{\tan(x)}{(x - \pi)}$.
2. Evaluate $\lim\limits_{x \to 0} (2x)^{3x}$.

4.7 The Newton-Raphson Method

Important Concept

- The Newton-Raphson Method

The Newton-Raphson Method

The Newton-Raphson Method is used to find zeroes of equations for which no formulas exist. The fundamental idea behind the Newton-Raphson Method is the following. The graph of a differentiable function f is approximated near a point $(x_1, f(x_1))$ by the tangent line ℓ at the point $(x_1, f(x_1))$. If x_1 is near a point c where the graph of f crosses the x-axis, then the point x_2 at which the tangent line ℓ intersects the x-axis will be even closer. This is illustrated in Figure 1.

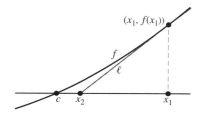

Figure 1

An algorithm needs to be developed to utilize the Newton-Raphson Method.

The Newton-Raphson Method (Algorithm)

If f is a differentiable function, then the $(j + 1)$th estimate x_{j+1} for a zero of f is obtained from the jth estimate x_j by the formula $x_{j+1} = x_j - \dfrac{f(x_j)}{f'(x_j)}$, provided that $f'(x_j) \neq 0$.

A loose rule of thumb is that in a successful application of the Newton-Raphson Method, the number of decimal places of accuracy doubles with each iteration. The technical description is that the Newton-Raphson Method is a *second-order process*.

The Newton-Raphson Method is often implemented using a computer algebra system.

Example 1 Use the Newton-Raphson Method to determine the value of $\sqrt{2}$ to within an accuracy of 10^{-7}.

Solution This problem is equivalent to finding the positive value where $f(x) = x^2 - 2$ vanishes. For a first estimate, we let $x_1 = 1.4$. Notice that $f'(x) = 2x$.

$$x_2 = x_1 - \frac{f(x_1)}{f'(x_1)}$$

$$= 1.4 - \frac{(1.4)^2 - 2}{2(1.4)}$$

$$= 1.414285714$$

$$x_3 = x_2 - \frac{f(x_2)}{f'(x_2)}$$

$$= 1.414285714 - \frac{(1.414285714)^2 - 2}{2(1.414285714)}$$

$$= 1.414213564$$

The value of $\sqrt{2} = 1.414213562$ was obtained using a calculator. Therefore, the value of x_3 agrees with $\sqrt{2}$ to seven places of accuracy. ∎

Self-Test Exercise

1. Use the Newton-Raphson Method to determine the value of $\sqrt{5}$ to within an accuracy of 10^{-7}.

4.8 Antidifferentiation and Applications

Important Concepts

- Antidifferentiation
- Applications

Antidifferentiation

Informally, the technique of antidifferentiation can be stated as follows. For a given function f, find another function F for which $F' = f$, if function F exists. In other words, antidifferentiation reverses the differentiation process.

More formally, we state:

Let f be a continuous function defined on an open interval I. If F is a function that is differentiable on I and satisfies $F' = f$ on I, then we call F an *antiderivative for f*.

The symbol for antidifferentiation is the integral sign, denoted by $\int$. The collection of all antiderivatives of f is denoted by $\int f(x)dx$. The expression $\int f(x)dx$ is called the *indefinite integral of f* and is equal to $\int f(x)dx = F(x) + C$, where C is called the *constant of integration*. The graph of $\int f(x)dx = F(x) + C$ is an infinite family of curves, one curve for each value of the real number C.

Indefinite Integrals of Powers of *x*

For any number $m \neq -1$, we have $\int x^m dx = \frac{1}{m+1} \cdot x^{m+1} + C$.

Example 1 Calculate the indefinite integral $\int x^{1/2}dx$.

Solution

$$\int x^{1/2}dx = \frac{x^{1/2+1}}{\frac{1}{2}+1} + C$$

$$= \frac{x^{3/2}}{\frac{3}{2}} + C$$

$$= \frac{2}{3}x^{3/2} + C.$$

Since $F' = f$, you can check your answer by differentiating

$$\frac{d}{dx}\left(\frac{2}{3}x^{3/2} + C\right) = \frac{2}{3}\cdot\frac{3}{2}x^{3/2-1} + 0 = x^{1/2}.$$ ■

Indefinite Integrals of Other Functions

Since $\dfrac{d}{dx}\sin(x) = \cos(x)$, $\dfrac{d}{dx}\cos(x) = -\sin(x)$, and $\dfrac{d}{dx}e^x = e^x$, we have

$$\int \cos(x)\,dx = \sin(x) + C,$$

$$\int \sin(x)\,dx = -\cos(x) + C,$$

$$\text{and } \int e^x dx = e^x + C.$$

Theorem 1 Let f and g be functions defined on an open interval I. Let F be an antiderivative for f, and let G be an antiderivative for g. Then

a. $\int(f(x) + g(x))dx = F(x) + G(x) + C$
b. $\int \alpha \cdot f(x)dx = \alpha \cdot F(x) + C$

Example 2 Calculate the indefinite integral $\int(3x + \sin(x))\,dx$.

Solution

$$\int(3x + \sin(x))dx = \int(3x)dx + \int(\sin(x))\,dx$$

$$= 3\int xdx + \int(\sin(x))\,dx$$

$$= \left(3\frac{x^2}{2} + C_1\right) + (-\cos(x) + C_2)$$

$$= \frac{3}{2}x^2 - \cos(x) + C, \text{ where } C = C_1 + C_2$$ ■

Velocity and Acceleration

Velocity is an antiderivative of acceleration. Position is an antiderivative of velocity.

Example 3 Find the expression for speed if a moving object accelerates from rest with the acceleration at time t given by $a(t) = \dfrac{5}{2}t^2$.

Solution Acceleration is equal to the first derivative of velocity. Speed is defined as the absolute value of velocity. Therefore, our strategy is to antidifferentiate $a(t) = \dfrac{5}{2}t^2$, evaluate the resulting constant of integration, and then express speed as the absolute value of the antiderivative of acceleration with its evaluated constant of integration.

$$
\begin{aligned}
v(t) &= \int v'(t)\, dt \\
&= \int a(t)\, dt \\
&= \int \frac{5}{2}t^2 \, dt \\
&= \frac{5}{2} \int t^2 \, dt \\
&= \left(\frac{5}{2}\right)\left(\frac{1}{3}\right) t^3 + C \\
&= \frac{5}{6}t^3 + C
\end{aligned}
$$

Since the object accelerates from rest, both velocity and acceleration are zero at time $t = 0$. We have $0 = v(0) = \dfrac{5}{6}(0)^3 + C = 0 + C$, which implies $C = 0$. Therefore, the expression for velocity is $v(t) = \dfrac{5}{6}t^3$, and the expression for speed is speed $= \left|\dfrac{5}{6}t^3\right| = \dfrac{5}{6}\left|t^3\right|$. ∎

Self-Test Exercises

1. Calculate the indefinite integral $\int \cos(x)\, dx$.
2. Calculate the indefinite integral $\int (5x^2 + \cos(x))\, dx$.
3. Find the expression for the position of a moving object that accelerates from rest at the origin with acceleration at time t given by $a(t) = \dfrac{5}{2}t^2$.

4.9 Applications to Economics

Important Concepts

- Profit Maximization
- Economic Lot Size and the Square Root Formula
- Average Cost
- Elasticity of Demand

In the realm of economics, applications often deal with the production of items, related production costs, revenue, profit, and so forth. Since we are dealing with functions that express these relationships, we can apply calculus to determine maxima and/or minima.

Profit Maximization

In business, a price p must be set in order to sell x items. If the price p is set too high, then fewer items will be sold. If the price p is set too low, then more items will be sold but perhaps at a loss. Therefore, a *demand function* $x = \mathcal{D}(p)$ determines the price p that a company must set in order to sell x items. Some authors also define $p = \mathcal{D}^{-1}(x)$ as a demand function. Equivalent representations of the demand equation appear in Figure 1.

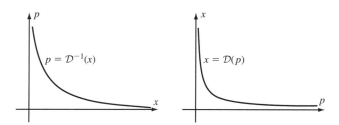

Figure 1

Cost is a very important factor in business. There are fixed costs and incremental costs. The *cost function* is a total of fixed cost plus incremental cost. Fixed costs play the role of a constant value in the cost equation, and the incremental costs are costs dependent on the number of items produced, so they play the role of a constant times the variable in the cost equation.

The *cost $C(x)$ of producing x units of a product* is given by $C(x) = C_0 + mx$, where C_0 represents *fixed costs that are independent of the number of units produced* and mx represents the marginal cost (m) of producing x units of the product. Therefore, the expression mx represents *incremental costs*.

The expression $C'(x)$ represents *marginal cost*. The *total revenue R* is the number of units sold times the price per unit (i.e., $R = xp$). The *profit P* is equal to revenue less cost ($P = R - C$). Therefore, we have $P = x\mathcal{D}^{-1}(x) - C(x)$ or $P = \mathcal{D}(p) \cdot p - C(\mathcal{D}(p))$. Typical cost, revenue, and profit plots are graphed as functions of p in Figure 2.

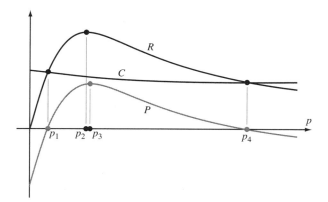

Figure 2

Please note that when maximizing the profit $P(x)$ or $P(p)$ (depends on which variable is used), we need to look for critical point $P'(x) = 0$ or $P'(p) = 0$, which means $R'(x) - C'(x) = 0$ or $R'(p) - C'(p) = 0$, equivalently $R'(x) = C'(x)$ or $R'(p) = C'(p)$.

Example 1 Suppose a restaurant attracts customers by setting a fixed cost to serve x customers for Sunday brunch at $C(x) = .04x^2 + x + 10$ dollars. The restaurant accountant has determined that to attract x customers to Sunday brunch the price should be $p(x) = 2.50 - .015x$ dollars. How many customers need to be served in order to maximize their profit?

Solution The demand function is $p(x) = 2.50 - .015x$, and the cost function is $C(x) = .04x^2 + x + 10$. Revenue $R(x) = x \cdot p(x) = x(2.50 - .015x) = 2.5x - .015x^2$. Therefore, the marginal revenue $R'(x)$ is

$$R'(x) = 2.5 - (.015)(2x)$$
$$= 2.5 - .03x.$$

The total cost is $C(x) = .04x^2 + x + 10$, so the marginal cost $C'(x)$ is

$$C'(x) = (.04)(2x) + 1$$
$$= .08x + 1.$$

To obtain the number, x, of customers needed to maximize the restaurant's profit, set marginal cost equal to marginal revenue and solve for x.

$C'(x) = R'(x)$ implies $.08x + 1 = 2.5 - .03x$. This implies $.08x + .03x = 2.5 - 1$, which implies $.11x = 1.5$. Therefore, $x = \dfrac{1.5}{.11} \approx 13.6$.

We conclude the restaurant needs to have 14 customers at Sunday brunch to maximize its profit. ∎

Example 1 illustrates the *Maximum Profit Principle*.

Maximum Profit Principle

Marginal revenue equals marginal cost at the production level that maximizes profit. Another classic economic problem is the problem of costs related to production and product storage. The economic lot size problem deals with the question, *What is the total number of production runs that minimize the total cost to the manufacturer?*

Economic Lot Size and the Square Root Formula

Let N denote the number of units produced each year in x production runs of equal size $\dfrac{N}{x}$. Let c be the start-up cost of each production run, and let μ be the marginal cost of each unit. The cost of each production run is $c + \mu\dfrac{N}{x}$, and the total production cost PC over the year is

$$PC(x) = x \cdot \left(c + \mu\frac{N}{x}\right) = cx + \mu N.$$ Figure 3 is the plot of the number of stored units as a function of time.

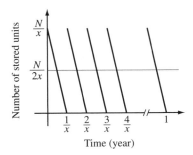

Figure 3

If σ is the annual cost of storing one unit, then the total storage cost SC is equal to the product of the annual cost of storing one unit and the average number of units in storage—i.e., $SC(x) = \sigma \dfrac{N}{2x}$.

The total cost TC of production and storage is $PC + SC$ or $TC(x) = cx + \mu N + \sigma \dfrac{N}{2x}$. If x_0 is the value of x that minimizes $TC(x)$, then the quantity $\dfrac{N}{x_0}$ is the *economic lot size*. Remember that $TC'(x) = c - \sigma \left(\dfrac{N}{2x^2}\right)$ and a critical point is located at $x_* = \sqrt{\dfrac{\sigma N}{2c}}$. Figure 4 shows a sketch of TC.

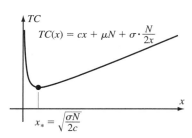

Figure 4

Example 2 A manufacturer expects to sell 25,000 copies of a computer game per year. The annual storage cost per package is $1.05 per year. The production start-up costs are $400, and the marginal cost is $2.00. How many production runs will minimize total costs?

Solution In this problem, $N = 25000$, $c = 400$, $\mu = 2$, and $\sigma = 1.05$. The total cost of production and storage is $TC(x) = cx + \mu N + \sigma \dfrac{N}{2x} = 400x + (2)(25000) + (1.05)\left(\dfrac{25000}{2x}\right)$ and the critical point is $x_* = \sqrt{\dfrac{\sigma N}{2c}} = \sqrt{\dfrac{(1.05)(25000)}{(2)(400)}} \approx 5.73$. $TC(5) = 54625$, $TC(6) = 54587.5$, and $TC(6) < TC(5)$; therefore, six production runs will minimize total costs. ∎

The following personal finance scenario is essentially the same as the economic lot size problem in economics. In this scenario, a retiree lives on the savings she has accumulated during her lifetime. The retiree earns interest on the money she has in a savings account, but the retiree must withdraw money from the account periodically to pay for expenses. A cost is associated with each withdrawal. *The goal of this personal finance problem is to determine the number of withdrawals that will minimize the associated costs.*

The Square Root Formula

We can make the following comparisons between the two problems mentioned:

Finance "Cash Holdings" ↔ Economics "Number of Stored Units"

Therefore, Figure 3 depicts the retiree's cash holdings as a function of time—assuming that the retiree lives on savings that earn $r\%$ annual interest, has N dollars in living expenses uniformly incurred during one year, and makes x number of equal withdrawals of $\dfrac{N}{x}$ dollars at equal intervals over the year. τ denotes the transaction cost of each withdrawal. The *average cash on hand* is $\dfrac{N}{2x}$, which results in an *annual loss* of $\left(\dfrac{r}{100}\right)\left(\dfrac{N}{2x}\right)$ in interest. This implies that the

$$transaction\ cost = x\tau + \frac{rN}{200x},$$ where τ denotes the transaction cost of each withdrawal.

The *square root formula* in economics, $x_* = \sqrt{\dfrac{rN}{200\tau}}$, is the value at which transaction total cost is minimized. Therefore, this problem in personal finance is essentially the same as the economic lot size problem. In simple terms, the problem to solve is, *How often should the retiree make withdrawals?*

Example 3 A retiree withdraws \$30,000 a year from her savings. If the transaction cost per withdrawal is \$5, and the retiree's savings earn 8% per year, then how many times per year should the retiree withdraw from her savings to cover expenses?

Solution In this problem, $N = 30000$, $\tau = 5$, and $r = .08$.

$$C = \text{Total transaction cost} = x\tau + \frac{rN}{200x} = 5x + \frac{.08(30000)}{200x} = 5x + \frac{12}{x}$$

$$x_* = \sqrt{\frac{rN}{200\tau}} = \sqrt{\frac{(.08)(30000)}{(200)(5)}} \approx 1.55$$

$C(1) = 17$ and $C(2) = 16$; therefore, the retiree should make two withdrawals per year. ■

The next application is one of minimizing average cost.

Average Cost

If the cost of producing x units of a commodity is $C(x)$, then the *average cost* $\overline{C}(x)$ of producing those units is the cost divided by the number of units or $\overline{C}(x) = \dfrac{C(x)}{x}$. If the fixed costs are c and the marginal cost is a constant μ, then the cost function $C(x)$ is given by

$$C(x) = \frac{(c + \mu x)}{x} = \mu + \frac{c}{x}$$

Average cost has a simple geometric interpretation that can be understood from Figure 5.

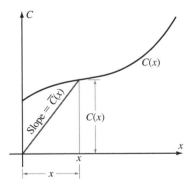

Figure 5

If a line is drawn from the origin to the point $(x, C(x))$, then the slope of this line segment is $\overline{C}(x) = \dfrac{C(x)}{x}$, which is the average cost. Moving from left to right along the curve $C(x)$, the slope $\overline{C}(x)$ initially decreases but eventually increases. Therefore, $\overline{C}(x)$ will have a local minimum. If this occurs at $x = x_0$, then $\overline{C}(x_0) = \dfrac{C(x_0)}{x_0} = C'(x_0)$. We conclude that *when the average cost is minimized, it equals the marginal cost*. This is a second economic principle.

Minimum Average Cost Principle

When the average cost is minimized, it equals the marginal cost.

Example 4 Suppose the cost function of producing a monitor is

$$2000 + 10\sqrt{x^3} + \frac{x(x - 200)^2}{5}.$$

Find the production level x_0 at which the average cost is minimized, using geometry.

Solution

$$\overline{C}(x) = \frac{C(x)}{x}$$

$$= \frac{2000 + 10\sqrt{x^3} + \left(\dfrac{x(x - 200)^2}{5}\right)}{x}$$

$$= \frac{2000}{x} + 10\sqrt{x} + \frac{(x - 200)^2}{5}$$

The graph of $\overline{C}(x)$ is shown in Figure 6.

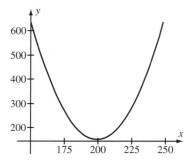

Figure 6

From the graph of $\overline{C}(x)$, the minimum initially appears to be in the neighborhood of 200. By restricting the domain to this neighborhood, we can plot a graph of $\overline{C}(x)$ with more detail. This is shown in Figure 7.

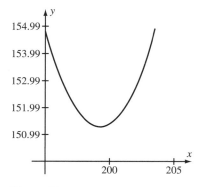

Figure 7

Evaluate $\overline{C}(x)$ for the values of x that appear in the following table.

x	199.239	199.24	199.241	199.24111	199.24112	199.24113	199.24114	199.24115
$\overline{C}(x)$	151.3060652	151.3060648	151.3060649	151.3060649	151.3060649	151.3060649	151.3060649	151.3060650

We conclude that the minimum value of $\overline{C}(x)$ is \$151.31, which occurs when $x = 199.24$. According to the Minimum Average Cost Principle, this number equals the lowest cost. Obviously, the manufacturing company can't produce partial monitors, so they would have to produce either 199 monitors or 200 monitors. $\overline{C}(199) = \$151.32$ and $\overline{C}(200) = \$151.42$; therefore, if they produce 199 monitors, the lowest average cost is achieved. ∎

Elasticity of Demand

Recall the demand equation $x = \mathcal{D}(p)$. If x_0 is the demand of a commodity sold at price p_0 and Δp represents the change in price, then the *change in demand* Δx is given by $\Delta x = \mathcal{D}(p_0 + \Delta p) - \mathcal{D}(p_0)$.

The *percentage change in demand* is $100\dfrac{\Delta x}{x_0} = \dfrac{100(\mathcal{D}(p_0 + \Delta p) - \mathcal{D}(p_0))}{x_0}$.

The *ratio of the percentage change in demand to the percentage change in price* is given by

$$\frac{100\dfrac{\Delta x}{x_0}}{100\dfrac{\Delta p}{p_0}} = \frac{p_0}{x_0}\frac{(\mathcal{D}(p_0 + \Delta p) - \mathcal{D}(p_0))}{\Delta p}.$$

The *elasticity of demand* $E(p_0)$ of the commodity at price p_0 is defined as

$$E(p_0) = -\lim_{\Delta p \to 0}\frac{100\dfrac{\Delta x}{x_0}}{100\dfrac{\Delta p}{p_0}} = -\frac{p_0}{x_0}\mathcal{D}'(p_0) = -p_0\frac{\mathcal{D}'(p_0)}{\mathcal{D}(p_0)}.$$

If $E(p_0) < 1$, then the demand is said to be *inelastic* at price p_0. If $E(p_0) > 1$, then the demand is said to be *elastic* at price p_0. If $E(p_0) = 1$, then the demand is said to have *unit elasticity*.

An important relationship exists between marginal revenue and elasticity of demand. If the formula for revenue as a function of price, $R = (\mathcal{D}(p)) \cdot p$, is differentiated with respect to price, then $\dfrac{dR}{dp}\bigg|_{p=p_0} = x_0(1 - E(p_0))$. Therefore, $R'(p_0) > 0$ when $E(p_0) < 1$ and $R'(p_0) < 0$ when $E(p_0) > 1$.

The results appear in Figure 8.

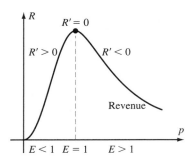

Figure 8

In other words, *revenue is an increasing function of price when the demand is inelastic and a decreasing function of price when the demand is elastic.* Since marginal revenue is zero when elasticity of demand is equal to 1, we have a third business principle.

Maximum Revenue Principle

Revenue is maximized when the demand has unit elasticity.

Example 5 A product is selling at a price at which elasticity of demand is 3. Approximately what effect on sales will a 5% price increase have?

Solution From the definition of elasticity of demand, we have

$$3 = E(p_0) = -\lim_{\Delta p \to 0} \frac{100 \dfrac{\Delta x}{x_0}}{100 \dfrac{\Delta p}{p_0}} \approx -\frac{100 \dfrac{\Delta x}{x_0}}{100\,(.05)},$$

which implies that $\dfrac{\Delta x}{x_0} \approx -.15$—i.e., there will be a *decrease* in demand of approximately 15%. ∎

Self-Test Exercises

1. A product is selling at a price at which elasticity of demand is 1.5. Approximately what effect on sales will a 1% price increase have?
2. Suppose the cost function of producing a certain product is $x^3 + 2x^2 - 5x + 8$. Find, graphically, the production level x_0 at which the average cost is minimized.

The Integral

5.1 The Area Problem

Important Concepts

- Partitions
- Sigma Notation
- Approximation of Area
- Precise Definition of Area

Partitions

For positive integer N, the *uniform partition of order N of the closed interval $[a, b]$* is defined to be the set $\{x_0, x_1, \ldots, x_N\}$ of equally spaced points

$$x_j = a + j \cdot \left(\frac{b - a}{N}\right), 0 \le j \le N,$$

where the expression $j \cdot \left(\dfrac{b - a}{N}\right)$ denotes the product of j and $\dfrac{b - a}{N}$. The actual number of points listed in the partition is $N + 1$.

You should associate the concept of partition with dividing a given object into smaller parts; in this context, the object is the closed interval $[a, b]$, which is a line segment beginning with the number a and ending with the number b. The closed interval $[a, b]$ can be thought of as a line segment that is part of the x-axis.

The *uniform partition of order N* is simply a subdivision of the line segment $[a, b]$ into N smaller line segments, each of which are equal in length. The length of every subinterval segment is represented by $\Delta x = \dfrac{b - a}{N}$. This process is illustrated in Figure 1.

Figure 1

The following examples will further illustrate the process.

Example 1 Describe the uniform partition of order 4 of the closed interval [3, 7].

Solution In this example, order 4 means $N = 4$, and there are four subintervals. The closed interval [3, 7] indicates that the interval begins with the number $a = 3$ and ends with the number $b = 7$. The width of each subinterval is $\Delta x = \dfrac{7 - 3}{4} = \dfrac{4}{4} = 1$. The subintervals from left to right are $I_1 = [3, 4]$, $I_2 = [4, 5]$, $I_3 = [5, 6]$ and $I_4 = [6, 7]$. Therefore, the uniform partition of order 4 of the closed interval [3, 7] is $\{3, 4, 5, 6, 7\}$. ∎

The next example illustrates the changes that occur when the order is changed to 8.

Example 2 Describe the uniform partition of order 8 of the closed interval [3, 7].

Solution In this example, order 8 means $N = 8$, and there are eight subintervals. The closed interval [3, 7] indicates that the interval begins with the number $a = 3$ and ends with the number $b = 7$. The width of each subinterval is $\Delta x = \dfrac{7 - 3}{8} = \dfrac{4}{8} = 0.5$. Therefore, the subintervals from left to right are $I_1 = [3, 3.5]$, $I_2 = [3.5, 4]$, $I_3 = [4, 4.5]$, $I_4 = [4.5, 5]$, $I_5 = [5, 5.5]$, $I_6 = [5.5, 6]$, $I_7 = [6, 6.5]$ and $I_8 = [6.5, 7]$. We conclude that the uniform partition of order 8 of the closed interval [3, 7] is $\{3, 3.5, 4, 4.5, 5, 5.5, 6, 6.5, 7\}$. ∎

What differences occurred when the order increased from 4 in Example 1 to order 8 in Example 2? Basically, the number of subdivisions of the line segment [3, 7] increased from four to eight and the number of points in the partition increased from five to nine.

Self-Test Exercises

1. Describe the uniform partition of order 5 of the closed interval [8, 9].
2. Describe the uniform partition of order 2 of the closed interval [0, 100].

Sigma (Summation) Notation

The expression $\displaystyle\sum_{i=M}^{N} a_i$ represents a sum that is expressed in *sigma notation*. $\displaystyle\sum$ is the uppercase Greek letter *sigma*, which represents addition. The letter i in the sum is called the *index of summation*. It acts as a device that distributes integer values beginning with $i = M$ to the first term a_i followed by $i = M + 1$ to the second term a_i, and it continues in this fashion until the value $i = N$ is assigned to the final term a_i (i.e., $\displaystyle\sum_{i=M}^{N} a_i = a_M + a_{M+1} + a_{M+2} + \cdots + a_N$). The integer M is less than or equal to the integer N. The letter i is considered a *dummy variable* (that is, a letter other than i may be used as the index of summation). Conventionally, the letters i, j, and k are used. Therefore, the expressions $\displaystyle\sum_{i=M}^{N} a_i$ and $\displaystyle\sum_{j=M}^{N} a_j$ represent the same sum.

Summation is linear. Therefore, sigma notation satisfies the following properties:

$$\sum_{i=M}^{N} (a_i \pm b_i) = \sum_{i=M}^{N} a_i \pm \sum_{i=M}^{N} b_i,$$

and if α is a constant, then

$$\sum_{i=M}^{N} \alpha \cdot a_i = \alpha \cdot \sum_{i=M}^{N} a_i.$$

These linearity properties are useful when computing sums. The finite sum of geometric series, the finite sum of integers, and the finite sum of integers squared are illustrated in the following examples.

Finite Geometric Series

If c is a real number that is not equal to zero and r is not equal to one, then

$$c \cdot r^0 + c \cdot r^1 + c \cdot r^2 + \cdots + c \cdot r^{N-1}$$

represents the finite sum of a geometric series. The first term in the sum, $c \cdot r^0$, is actually equal to c because $r^0 = 1$. The second term, $c \cdot r^1$, is obtained by multiplying the first term c by r. The third term, $c \cdot r^2$, is obtained by multiplying the second term, $c \cdot r^1$, by r. This process is repeated until the next to the last term, $c \cdot r^{N-2}$, is multiplied by r to obtain the last term $c \cdot r^{N-1}$. The real number c is called the *first term* of the geometric series, and r is called the *ratio* of the geometric series.

In sigma notion, $c \cdot r^0 + c \cdot r^1 + c \cdot r^2 + \cdots + c \cdot r^{N-1}$ is expressed as $\sum_{k=0}^{N-1} c \cdot r^k$, and its sum is

equal to $c \cdot \left(\dfrac{r^N - 1}{r - 1} \right)$. Can you see why $r \neq 1$? (Hint: Check the denominator.) Please note that the exponent of the last term is $N - 1$ in the geometric series, but the exponent is N in the sum.

Finite Sum of Integers

$1 + 2 + 3 + \cdots + N$ represents the finite sum of integers squared, with 1 as a first term and N as a last term. In sigma notation, $1 + 2 + 3 + \cdots + N$ is expressed as $\sum_{j=1}^{N} j$. There are N terms in the sum; the value 1 appears below sigma and N appears above sigma. The value of $\sum_{j=1}^{N} j$ is

$$\dfrac{N \cdot (N + 1)}{2}.$$

Finite Sum of Integers Squared

$1^2 + 2^2 + 3^2 + \cdots + N^2$ represents the finite sum of integers squared, with 1^2 as a first term and N^2 as a last term. In sigma notation, $1^2 + 2^2 + 3^2 + \cdots + N^2$ is expressed as $\sum_{i=1}^{N} i^2$. There are N terms in the sum; the value 1 appears below sigma and N appears above sigma. The value of $\sum_{i=1}^{N} i^2$ is equal to $\dfrac{N \cdot (N + 1) \cdot (2N + 1)}{6}$.

We can use these special sums and the linearity of summation to solve several examples.

Example 3 Evaluate $\displaystyle\sum_{k=0}^{100} \left(\dfrac{1}{2} \right)^k$.

Solution $\displaystyle\sum_{k=0}^{100} \left(\dfrac{1}{2} \right)^k$ is a finite geometric series $\displaystyle\sum_{k=0}^{N-1} c \cdot r^k$ where $c = 1$ and $r = \dfrac{1}{2}$.

$N - 1 = 100$, so $N = 101$. Therefore,

$$\sum_{k=0}^{100} \left(\frac{1}{2}\right)^k = \left(\frac{1}{2}\right)^0 + \left(\frac{1}{2}\right)^1 + \cdots + \left(\frac{1}{2}\right)^{100}$$

$$= \left(\frac{\left(\frac{1}{2}\right)^{101} - 1}{\left(\frac{1}{2} - 1\right)}\right)$$

$$\approx 2.$$

Example 4 Evaluate $\displaystyle\sum_{k=3}^{10} 5 \cdot (2)^k$.

Solution $\displaystyle\sum_{k=3}^{10} 5 \cdot (2)^k$ is a finite geometric series $\displaystyle\sum_{k=3}^{10} c \cdot r^k$ where $c = 5$ and $r = 2$. There are eight terms in the sum beginning with $5 \cdot (2)^3$, i.e., the $k = 3$ term, and ending with $5 \cdot (2)^{10}$, the $k = 10$ term.

We make a very useful observation:

$$\sum_{k=0}^{10} 5 \cdot (2)^k = 5 \cdot (2)^0 + 5 \cdot (2)^1 + 5 \cdot (2)^2 + 5 \cdot (2)^3 + 5 \cdot (2)^4 + \cdots + 5 \cdot (2)^{10}.$$

If we subtract $5 \cdot (2)^0 + 5 \cdot (2)^1 + 5 \cdot (2)^2$ from $\displaystyle\sum_{k=0}^{10} 5 \cdot (2)^k$, we obtain $5 \cdot (2)^3 + 5 \cdot (2)^4 + \cdots + 5 \cdot (2)^{10}$; in sigma notation, this is equal to $\displaystyle\sum_{k=3}^{10} 5 \cdot (2)^k$. This implies that

$$\sum_{k=3}^{10} 5 \cdot (2)^k = \sum_{k=0}^{10} 5 \cdot (2)^k - \sum_{k=0}^{2} 5 \cdot (2)^k.$$

Since

$$\sum_{k=0}^{10} 5 \cdot (2)^k = \frac{5 \cdot (2^{11} - 1)}{2 - 1} = 5 \cdot (2^{11} - 1)$$

and

$$\sum_{k=0}^{2} 5 \cdot (2)^k = \frac{5 \cdot (2^3 - 1)}{2 - 1} = 5 \cdot (2^3 - 1),$$

we have

$$\sum_{k=3}^{10} 5 \cdot (2)^k = \sum_{k=0}^{10} 5 \cdot (2)^k - \sum_{k=0}^{2} 5 \cdot (2)^k$$

$$= 5 \cdot (2^{11} - 1) - 5 \cdot (2^3 - 1)$$

$$= 10200.$$

Example 5 Evaluate $\displaystyle\sum_{j=1}^{20} j$.

Solution $\displaystyle\sum_{j=1}^{20} j = 1 + 2 + 3 + \cdots + 20$ is the sum of integers from 1 through 20. $N = 20$, so $N + 1 = 21$. Therefore, the sum of the first consecutive twenty positive integers is equal to

$$\frac{N \cdot (N+1)}{2} = \frac{(20)(21)}{2} = 210.$$

■

Example 6 Evaluate $\displaystyle\sum_{i=10}^{15} i^2$.

Solution $\displaystyle\sum_{i=10}^{15} i^2$ is the sum of integers squared beginning with 10^2 and ending with 15^2. Using the same technique as in Example 4, we can determine that $\displaystyle\sum_{i=10}^{15} i^2 = \sum_{i=1}^{15} i^2 - \sum_{i=1}^{9} i^2$. By using the formula

$$\sum_{i=1}^{N} i^2 = \frac{N \cdot (N+1) \cdot (2N+1)}{6},$$

we have

$$\sum_{i=1}^{15} i^2 = \frac{(15) \cdot (16) \cdot (2(15)+1)}{6} = \frac{(15)(16)(31)}{6}$$

and

$$\sum_{i=1}^{9} i^2 = \frac{(9) \cdot (10) \cdot (2(9)+1)}{6} = \frac{(9)(10)(19)}{6}.$$

Therefore,

$$\sum_{i=10}^{15} i^2 = \sum_{i=1}^{15} i^2 - \sum_{i=1}^{9} i^2$$

$$= \frac{(15)(16)(31)}{6} - \frac{(9)(10)(19)}{6}$$

$$= 955.$$

■

Example 7 Evaluate $\displaystyle\sum_{j=1}^{3} (2j+3)$.

Solution Using the linearity of summation, we write $\displaystyle\sum_{j=1}^{3} (2j+3) = 2 \cdot \sum_{j=1}^{3} j + 3 \cdot \sum_{j=1}^{3} 1$. Notice that $\displaystyle\sum_{j=1}^{3} j$ is simply the sum of the integers from one to three, which is equal to

$$\frac{N(N+1)}{2} = \frac{(3)(4)}{2} = 6$$

and

$$\sum_{j=1}^{3} 1 = \sum_{j=1}^{3} j^0 = (1)^0 + (2)^0 + (3)^0 = 1 + 1 + 1 = 3.$$

Therefore,

$$\sum_{j=1}^{3} (2j + 3) = 2 \cdot \sum_{j=1}^{3} j + 3 \cdot \sum_{j=1}^{3} 1 = (2) \cdot (6) + (3) \cdot (3) = 21.$$

Self-Test Exercises

3. Evaluate $\displaystyle\sum_{j=1}^{10} 2 \cdot \left(\frac{1}{4}\right)^j$.

4. Evaluate $\displaystyle\sum_{k=2}^{10} (2k^2 + 3k - 7)$.

Approximation of Area

You can approximate the area of a region bounded above by the graph of a continuous function f, below by the x-axis, on the left by the vertical line $x = a$, and on the right by the vertical line $x = b$ by subdividing the bounded region into adjacent rectangles, each of equal width. If the area of each individual rectangle is calculated and the total area of the rectangles is obtained, then this total is an approximation of the area of the bounded region. Partitions and sigma notation are useful tools.

In general, follow these steps to approximate the area using *right endpoint approximation*:

1. Obtain a uniform partition of order N of the closed interval $[a, b]$.
2. Choose the right endpoint x_j in each subinterval $I_j = [x_{j-1}, x_j]$ for $j = 1, 2, \ldots, N$ created by the partition.
3. Because the area of a rectangle is equal to the vertical height times the horizontal width, each rectangle will have area equal to $f(x_j)\Delta x$. Here, $f(x_j)$ represents the vertical height of rectangle R_j, and $\Delta x = \dfrac{b - a}{N}$ represents the horizontal width of rectangle R_j. (See Figure 2.)

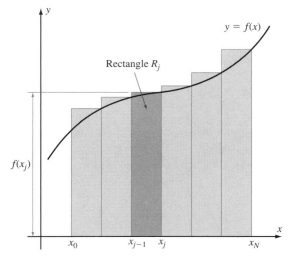

Figure 2

4. The area approximation is equal to $\displaystyle\sum_{j=1}^{N} f(x_j)\Delta x$. As N increases in value, $\displaystyle\sum_{j=1}^{N} f(x_j)\Delta x$ approximates the actual area of the bounded region more closely.

Example 8 Use the right endpoint approximation based on a uniform partition of order 4 to estimate the area of the region that lies under the graph of $f(x) = 5 - x^2$ and above the interval $[0, 2]$. The bounded region appears in Figure 3.

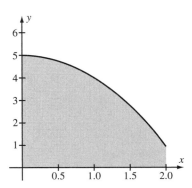

Figure 3

Solution

1. The uniform partition of order 4 of the closed interval $[0, 2]$ is the set $\{0, .5, 1, 1.5, 2\}$; i.e., $x_0 = 0$, $x_1 = .5$, $x_2 = 1$, $x_3 = 1.5$, $x_4 = 2$.
2. Choose the right endpoint of the subintervals $I_j = [x_{j-1}, x_j]$ for $j = 1, 2, 3, 4$; i.e., $x_1 = .5$, $x_2 = 1$, $x_3 = 1.5$, $x_4 = 2$.
3. Rectangle R_1 has height equal to $f(.5) = 5 - (.5)^2 = 4.75$ and width equal to $\Delta x = \dfrac{2 - 0}{4} = 0.5$, so its area A_1 is equal to $(4.75)(.5) = 2.375$. Rectangle R_2 has height equal to $f(1) = 5 - (1)^2 = 4$ and width equal to $\Delta x = 0.5$, so its area A_2 is equal to $(4)(0.5) = 2$. Rectangle R_3 has height equal to $f(1.5) = 5 - (1.5)^2 = 2.75$ and width equal to $\Delta x = 0.5$, so its area A_3 is equal to $(2.75)(.5) = 1.375$. Rectangle R_4 has height equal to $f(2) = 5 - 2^2 = 1$ and width equal to $\Delta x = 0.5$, so its area A_4 is equal to $(1)(.5) = .5$.
4.
$$\sum_{j=1}^{4} f(x_j)\Delta x = A_1 + A_2 + A_3 + A_4$$
$$= 2.375 + 2 + 1.375 + .5$$
$$= 6.25 \text{ square units}$$

This is an approximation of the bounded area. The rectangles R_1, R_2, R_3, and R_4 are represented in Figure 4.

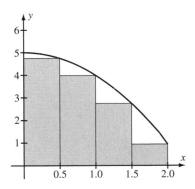

Figure 4

Self-Test Exercise

5. Use the right endpoint approximation based on a uniform partition of order 4 to estimate the area of the region that lies under the graph of $f(x) = 20 - x^2$ and above the interval $[0, 4]$.

Precise Definition of Area

Each rectangle in Figure 4 is physically below the graph of function $f(x) = 5 - x^2$. Therefore, in this particular case, the estimated area is smaller than the actual area. If we subdivide the bounded region into a large number of rectangles, the estimated area will approach the actual area. (See Figure 5.) This is the basis of the following definition.

The *area A of the region* that is bounded above by the graph of f, below by the x-axis, and on the sides by $x = a$ and $x = b$, respectively, is defined as $A = \lim\limits_{N \to \infty} \sum\limits_{j=1}^{N} f(x_j) \Delta x$.

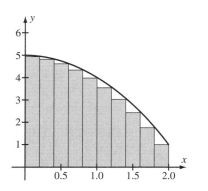
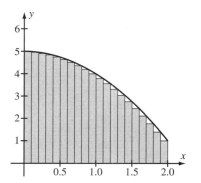

Figure 5

Example 9 Calculate the area bounded by the graph of $f(x) = 5 - x^2$, the vertical line $x = 0$, the vertical line $x = 2$, and the x-axis. The bounded region appears in Figure 3 of this section.

Solution In this case, we have $a = 0$ and $b = 2$. When we divide the interval $[0, 2]$ into N equal subintervals, the length of each subinterval is $\Delta x = \dfrac{2 - 0}{N} = \dfrac{2}{N}$. The uniform partition of order N is $\{x_0, x_1, \ldots, x_j, \ldots, x_N\}$ where

$$x_j = x_0 + j \cdot (\Delta x) = 0 + j \cdot \left(\frac{2}{N}\right) = \frac{2j}{N},$$

i.e., x_j is the right endpoint of interval $I_j = [x_{j-1}, x_j]$.

Therefore,

$$A = \lim_{N \to \infty} \sum_{j=1}^{N} f(x_j) \Delta x$$

$$= \lim_{N \to \infty} \sum_{j=1}^{N} f\left(\frac{2j}{N}\right) \cdot \left(\frac{2}{N}\right)$$

$$= \lim_{N \to \infty} \sum_{j=1}^{N} \left(5 - \left(\frac{2j}{N}\right)^2\right) \cdot \left(\frac{2}{N}\right)$$

$$= \lim_{N \to \infty} \left(\frac{2}{N} \cdot \sum_{j=1}^{N} \left(5 - \frac{4j^2}{N^2}\right)\right)$$

$$= \lim_{N \to \infty} \left(\frac{2}{N} \cdot \sum_{j=1}^{N} 5 \right) - \lim_{N \to \infty} \left(\frac{2}{N} \cdot \sum_{j=1}^{N} \frac{4 j^2}{N^2} \right)$$

$$= \lim_{N \to \infty} \left(\frac{(2)(5)}{N} \sum_{j=1}^{N} 1 \right) - \lim_{N \to \infty} \left(\frac{(2)(4)}{(N)(N^2)} \sum_{j=1}^{N} j^2 \right)$$

$$= \lim_{N \to \infty} \left(\frac{10}{N} \sum_{j=1}^{N} 1 \right) - \lim_{N \to \infty} \left(\frac{8}{N^3} \sum_{j=1}^{N} j^2 \right).$$

Using the formulas $\sum_{j=1}^{N} 1 = N$ and $\sum_{j=1}^{N} j^2 = \dfrac{N(N + 1)(2N + 1)}{6}$, we have

$$A = \lim_{N \to \infty} \left(\frac{10}{N} \cdot N \right) - \lim_{N \to \infty} \left(\frac{8}{N^3} \cdot \frac{N(N + 1)(2N + 1)}{6} \right)$$

$$= \lim_{N \to \infty} 10 - \lim_{N \to \infty} \frac{4(N + 1)(2N + 1)}{3N^2}$$

$$= \lim_{N \to \infty} 10 - \lim_{N \to \infty} \frac{8N^2 + 12N + 4}{3N^2}$$

$$= \lim_{N \to \infty} 10 - \lim_{N \to \infty} \left(\frac{8N^2}{3N^2} + \frac{12N}{3N^2} + \frac{4}{3N^2} \right)$$

$$= \lim_{N \to \infty} 10 - \lim_{N \to \infty} \left(\frac{8}{3} + \frac{4}{N} + \frac{4}{3N^2} \right)$$

$$= 10 - \left(\frac{8}{3} + 0 + 0 \right)$$

$$= 7\frac{1}{3} \text{ square units.}$$

Self-Test Exercise

6. Calculate the area of the region that lies under the graph of $f(x) = 20 - x^2$ and above the interval $[0, 4]$ on the x-axis.

5.2 The Riemann Integral

Important Concepts

- Riemann Sums
- The Riemann Integral
- Calculating Integrals

Riemann Sums

The following concepts were presented in Section 5.1:

- The uniform partition of order N of the closed interval $[a, b]$
- The region bounded above by function f, below by the x-axis, on the left by the vertical line $x = a$, and on the right by the vertical line $x = b$

- The use of sigma notation to represent an approximation of the area of the bounded region

The uniform partition of order N of the closed interval $[a, b]$ is the set $\{x_0, x_1, \ldots, x_N\}$. Depending on the choice we make for s_j in each subinterval I_j, $1 \leq j \leq N$, we obtain a sequence of points $S_N = \{s_1, s_2, \ldots, s_N\}$. This sequence $S_N = \{s_1, s_2, \ldots, s_N\}$ is called *a choice of points associated with the uniform partition of order N*.

Choices for s_j could be the left endpoint, right endpoint, midpoint, or any point of the subinterval I_j. In Example 8, Section 5.1, the right endpoint of the subinterval I_j was used, which resulted in the sequence $S_4 = \{.5, 1, 1.5, 2\}$.

The expression $R(f, S_N) = \sum_{j=1}^{N} f(s_j)\Delta x$, with $\Delta x = \dfrac{b-a}{N}$, is called a *Riemann sum of*
f. Because function f can be *any* function defined on the interval $[a, b]$, *it is possible to have negative Riemann sums*. A negative Riemann sum is illustrated in the following example.

Example 1 Write down two different Riemann sums for the function $f(x) = 2x$ and the interval $[-3, 3]$ using the uniform partition $\{-3, 0, 3\}$.

Solution The subintervals are $I_1 = [-3, 0]$ and $I_2 = [0, 3]$, with $\Delta x = 3$. We can arbitrarily select *any* point s_j in each subinterval. One choice can be: $s_1 = -2 \in I_1$, $s_2 = 3 \in I_2$. The corresponding Riemann sum for this collection of points is

$$R(f, S) = \sum_{j=1}^{2} f(s_j)\Delta x$$
$$= f(-2) \cdot 3 + f(3) \cdot 3$$
$$= (-4)(3) + (6)(3)$$
$$= 6.$$

If we choose s_j to be the left endpoint of each subinterval, that is, $s_1 = -3 \in I_1$, $s_2 = 0 \in I_2$, then the corresponding Riemann sum is

$$R(f, S) = \sum_{j=1}^{2} f(s_j)\Delta x$$
$$= f(-3) \cdot 3 + f(0) \cdot 3$$
$$= (-6)(3) + (0)(3)$$
$$= -18.$$

Of course, a different choice of points for s_1 and s_2 would lead to a different Riemann sum. ∎

Special Cases of Riemann Sums

Case 1
If each s_j is the point where continuous function f achieves its *maximum* in subinterval I_j, then the sequence of points is represented as $U_N = \{u_1, u_2, \ldots, u_N\}$ and the Riemann sum
$R(f, U_N) = \sum_{j=1}^{N} f(u_j)\Delta x$ is called the *upper Riemann sum of order N*.

Case 2
If each s_j is the point where continuous function f achieves its *minimum* in subinterval I_j, then the sequence of points is represented as $L_N = \{\ell_1, \ell_2, \ldots, \ell_N\}$ and the Riemann sum
$R(f, L_N) = \sum_{j=1}^{N} f(\ell_j)\Delta x$ is called the *lower Riemann sum of order N*.

Example 2 If $f(x) = 2x$, find the upper and lower Riemann sums of order 4, $R(f, U_4)$ and $R(f, L_4)$ on closed interval $[0, 8]$.

Solution The four equal-length subintervals are $I_1 = [0, 2]$, $I_2 = [2, 4]$, $I_3 = [4, 6]$, and $I_4 = [6, 8]$. Function $f(x) = 2x$ is an increasing continuous function; that is to say, as x increases, the value of $f(x)$ also increases. Therefore, the candidate for u_j, the maximum point, in subinterval I_j is the right endpoint x_j of the subinterval. We have $\Delta x = \dfrac{8 - 0}{4} = 2$ and $U_4 = \{2, 4, 6, 8\}$. (See Figure 1.) We conclude that

$$
\begin{aligned}
R(f, U_4) &= \sum_{j=1}^{4} f(u_j)\Delta x \\
&= f(2) \cdot 2 + f(4) \cdot 2 + f(6) \cdot 2 + f(8) \cdot 2 \\
&= (f(2) + f(4) + f(6) + f(8))(2) \\
&= (4 + 8 + 12 + 16)(2) \\
&= 80.
\end{aligned}
$$

Similarly, the candidate for ℓ_j, the minimum point, in subinterval I_j is the left endpoint x_{j-1} of the subinterval. We have $\Delta x = \dfrac{8 - 0}{4} = 2$ and $L_4 = \{0, 2, 4, 6\}$. (See Figure 2.) We conclude that

$$
\begin{aligned}
R(f, L_4) &= \sum_{j=1}^{4} f(\ell_j)\Delta x \\
&= f(0) \cdot 2 + f(2) \cdot 2 + f(4) \cdot 2 + f(6) \cdot 2 \\
&= (f(0) + f(2) + f(4) + f(6))(2) \\
&= (0 + 4 + 8 + 12)(2) \\
&= 48.
\end{aligned}
$$

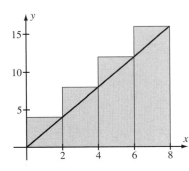

Figure 1

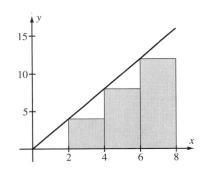

Figure 2   ∎

Theorem 1   Suppose that f is continuous on an interval $[a, b]$.

- A Riemann sum $R(f, S_N)$ is always "pinched" between the lower Riemann sum $R(f, L_N)$ and the upper Riemann sum $R(f, U_N)$. This fact is often expressed as $R(f, L_N) \le R(f, S_N) \le R(f, U_N)$.
- For sufficiently large positive integer N, $R(f, L_N)$ and $R(f, U_N)$ become arbitrarily close to each other. This fact is often expressed as $\lim\limits_{N \to \infty} (R(f, U_N) - R(f, L_N)) = 0$.

Self-Test Exercises

1. If $f(x) = 2x$, find $R(f, U_6)$ and $R(f, L_6)$ on closed interval $[0, 6]$.
2. If $f(x) = 4 - x^2$, find $R(f, U_6)$ and $R(f, L_6)$ on closed interval $[-1, 2]$.

The Riemann Integral

This is the formal definition of the Riemann integral of f on closed interval $[a, b]$: Suppose f is a function defined on the interval $[a, b]$. Let S_N be a choice of points associated with the uniform partition of order N. If $\lim_{N \to \infty} R(f, S_N)$ exists, then we say that f is *integrable* and define the *Riemann integral of f on* $[a, b]$, denoted by $\int_a^b f(x)\, dx$, to be $\lim_{N \to \infty} R(f, S_N)$.

More precisely, this means that if for any $\epsilon > 0$, there exists a positive integer M such that $|R(f, S_N) - \ell| < \epsilon$ for all N greater than or equal to M, then ℓ is the limit of $R(f, S_N)$. If this limit ℓ exits, we say that *f is integrable on* $[a, b]$. Symbolically, $\int_a^b f(x)\, dx = \ell$ is the *Riemann integral of f on* $[a, b]$. The process of finding the Riemann integral of f on $[a, b]$ is called *integration*, or *antidifferentiation*.

The following terms are associated with the symbol $\int_a^b f(x)\, dx$:

- $\int$ is the *integral sign*.
- a is the *lower limit of integration*.
- b is the *upper limit of integration*.
- $f(x)$ is the *integrand*.

Please note that:

- $\int_a^b f(x)\, dx$ can also be referred as a *definite integral* because it has a lower limit of integration and an upper limit of integration. *The value of $\int_a^b f(x)\, dx$ is a number.*
- $\int f(x)\, dx$, the antiderivative notation we used in Section 4.8, is called an *indefinite integral* because it does not have a lower limit of integration or an upper limit of integration. *The value of the indefinite integral $\int f(x)\, dx$ is a family of functions.*

Remember the following fact; it is important.

Theorem 2   If f is continuous on $[a, b]$, then f is integrable on $[a, b]$—i.e., the Riemann integral $\int_a^b f(x)\, dx$ exists.

Example 3 Does the Riemann integral of $f(x) = |x|$ exist on interval $I = [-1, 5]$? Substantiate your answer.

Solution Since $f(x) = |x|$ is continuous on interval $[-1, 5]$, we can conclude that the Riemann integral $\int_{-1}^5 |x|\, dx$ exists by Theorem 2. ∎

Self-Test Exercises

3. Does $\int_{-1}^2 (4 - x^2)\, dx$ exist? Substantiate your answer.
4. Is $f(x) = x^3 + x^2 + x + 1$ integrable on $[-100, 1000]$? Substantiate your answer.

Calculating Integrals

Function F is an *antiderivative* of function f if f is equal to the first derivative of function F (i.e., if $F' = f$).

Theorem 3 One form of the Fundamental Theorem of Calculus states that if f is continuous on $[a, b]$ and if $F' = f$ for $a < x < b$, then $\int_a^b f(x)\,dx = F(b) - F(a)$.

This important theorem can be used to calculate some definite integrals.

Example 4 Evaluate $\int_1^6 2x\,dx$.

Solution Instead of finding the limit of the Riemann sum $\lim\limits_{N \to \infty} R(f, S_N)$ with $f(x) = 2x$ to compute the definite integral, we can apply the Fundamental Theorem of Calculus. Notice that $F(x) = x^2$ is an antiderivative of $2x$. We have

$$\int_1^6 2x\,dx = F(6) - F(1)$$
$$= 6^2 - 1^2$$
$$= 35.$$ ■

In Example 9 of Section 5.1, we calculated the area bounded by the graph $f(x) = 5 - x^2$, the vertical line $x = 0$, the vertical line $x = 2$, and the x-axis using a limit of Riemann sums. In this example, we will calculate the same area using the Fundamental Theorem of Calculus.

Example 5 Use the Fundamental Theorem of Calculus to calculate the area bounded by the graph of $f(x) = 5 - x^2$, the vertical line $x = 0$, the vertical line $x = 2$, and the x-axis.

Solution In Section 5.1, we learned that the area, A, is defined to be a limit of Riemann sums. Since the function $f(x) = 5 - x^2$ is continuous, Theorem 2 tells us that this limit exists. By the definition of the Riemann integral, we have $\int_0^2 (5 - x^2)\,dx = \lim\limits_{N \to \infty} R(f, S_N) = A$. We can apply the Fundamental Theorem of Calculus to evaluate this integral. Since $F(x) = 5x - \dfrac{x^3}{3}$ is an antiderivative of $f(x) = 5 - x^2$, we have

$$A = \int_0^2 (5 - x^2)\,dx$$
$$= F(2) - F(0)$$
$$= \left(5 \cdot 2 - \frac{2^3}{3}\right) - \left(0 \cdot 2 - \frac{0^3}{3}\right)$$
$$= 7\frac{1}{3} \text{ square units.}$$ ■

Compare the solutions for Example 9 in Section 5.1 and Example 5 in Section 5.2. Which is more efficient?

Self-Test Exercises

5. Calculate $\int_0^3 (3 + 4x^2)\,dx$ by finding an antiderivative of the integrand.

6. Calculate $\int_{\pi/2}^{\pi/2} \cos(x)\,dx$ by finding an antiderivative of the integrand.

5.3 Rules for Integration

Important Concept

- Integration Rules

Integration Rules

If functions $|f|$, f, g, and h are integrable functions on closed interval $[a, b]$ and α is a constant, then the following rules hold:

i. $\displaystyle\int_a^b (f(x) \pm g(x))\, dx = \int_a^b f(x)\, dx \pm \int_a^b g(x)\, dx$

ii. $\displaystyle\int_a^b \alpha \cdot f(x)\, dx = \alpha \cdot \int_a^b f(x)\, dx$

iii. $\displaystyle\int_a^b \alpha\, dx = \alpha \cdot (b - a)$

iv. If $a \le c \le b$, then $\displaystyle\int_a^b f(x)\, dx = \int_a^c f(x)\, dx + \int_c^b f(x)\, dx$.

v. $\displaystyle\int_a^b f(x)\, dx = -\int_b^a f(x)\, dx$

vi. If $f(x) \le g(x) \le h(x)$ for $a \le x \le b$, then $\displaystyle\int_a^b f(x)\, dx \le \int_a^b g(x)\, dx \le \int_a^b h(x)\, dx$.

vii. If $f(x) \ge 0$, then $\displaystyle\int_a^b f(x)\, dx \ge 0$.

viii. $\left| \displaystyle\int_a^b f(x)\, dx \right| \le \int_a^b |f(x)|\, dx$

ix. If $f(x)$ is continuous on $[a, b]$, then there is a value c, $a < c < b$, such that $f(c) = \dfrac{1}{(b-a)} \displaystyle\int_a^b f(x)\, dx$.

Rules (i) through (iv) help simplify the process of calculating integrals. Rule (v) deals with interchanging the order of the lower and upper limits of integration. Rules (vi), (vii), and (viii) deal with order properties of integrals. Rule (ix), the *Mean Value Theorem for Integrals*, states that a continuous function assumes its average value.

Example 1 If $a \le c \le b$, $\displaystyle\int_a^b f(x)\, dx = 10$, and $\displaystyle\int_a^c f(x)\, dx = 2$, calculate the value of $\displaystyle\int_c^b f(x)\, dx$.

Solution The rule "if $a \le c \le b$, then $\displaystyle\int_a^b f(x)\, dx = \int_a^c f(x)\, dx + \int_b^b f(x)\, dx$" implies that $\displaystyle\int_c^b f(x)\, dx = \int_a^b f(x)\, dx - \int_a^c f(x)\, dx$, so $\displaystyle\int_c^b f(x)\, dx = 10 - 2 = 8$. ■

Example 2 If $\int\limits_1^5 f(x)\,dx = 7$ and $\int\limits_1^5 g(x)\,dx = 3$, find the values of $\int\limits_5^1 f(x)\,dx$, $\int\limits_1^5 (2f(x) - 8g(x))\,dx$, and $\int\limits_1^5 (f(x) + 6)\,dx$.

Solution The rule "$\int\limits_a^b f(x)\,dx = -\int\limits_b^a f(x)\,dx$" implies that $\int\limits_5^1 f(x)\,dx = -\int\limits_1^5 f(x)\,dx = -7$. We can use the rules "$\int\limits_a^b (f(x) \pm g(x))\,dx = \int\limits_a^b f(x)\,dx \pm \int\limits_a^b g(x)\,dx$" and "$\int\limits_a^b \alpha \cdot f(x)\,dx = \alpha \cdot \int\limits_a^b f(x)\,dx$" to find

$$\int\limits_1^5 (2f(x) - 8g(x))\,dx = \int\limits_1^5 2f(x)\,dx - \int\limits_1^5 8g(x)\,dx$$

$$= 2 \cdot \int\limits_1^5 f(x)\,dx - 8 \cdot \int\limits_1^5 g(x)\,dx$$

$$= 2 \cdot (7) - 8 \cdot (3)$$

$$= -10.$$

We can use the rules "$\int\limits_a^b (f(x) \pm g(x))\,dx = \int\limits_a^b f(x)\,dx \pm \int\limits_a^b g(x)\,dx$" and "$\int\limits_a^b \alpha\,dx = \alpha \cdot (b-a)$" to find

$$\int\limits_1^5 (f(x) + 6)\,dx = \int\limits_1^5 f(x)\,dx + \int\limits_1^5 6\,dx$$

$$= 7 + 6 \cdot (5 - 1)$$

$$= 31. \qquad \blacksquare$$

Example 3 If $g(x)$ is a function that satisfies $5 \leq g(x) \leq 13$ for all x in closed interval $[0, 2]$, determine a lower bound and an upper bound for $\int\limits_0^2 g(x)\,dx$.

Solution The following rules are used to solve this problem: "$\int\limits_a^b \alpha \cdot dx = \alpha \cdot (b - a)$" and "if $f(x) \leq g(x) \leq h(x)$ for $a \leq x \leq b$, then $\int\limits_a^b f(x)\,dx \leq \int\limits_a^b g(x)\,dx \leq \int\limits_a^b h(x)\,dx$."

$5 \leq g(x) \leq 13$ for all x in $[0, 2]$ implies that $\int\limits_0^2 5\,dx \leq \int\limits_0^2 g(x)\,dx \leq \int\limits_0^2 13\,dx$; therefore, $5 \cdot (2 - 0) \leq \int\limits_0^2 g(x)\,dx \leq 13 \cdot (2 - 0)$—i.e., $\int\limits_0^2 g(x)\,dx$ is bounded below by 10 and above by 26. $\qquad \blacksquare$

Self-Test Exercises

1. If $\int_1^{10} f(x)\,dx = 7$ and $\int_5^{10} f(x)\,dx = 3$, find the values of $\int_1^5 f(x)\,dx$ and $\int_5^1 f(x)\,dx$.

2. If $\int_2^6 f(x)\,dx = 5$ and $\int_2^6 g(x)\,dx = -3$, find the values of $\int_2^6 (3f(x) - 2g(x))\,dx$ and $\int_2^6 (g(x) + 7)\,dx$.

3. If $1 \le h(x) \le (x^2 + 2)$ for x in the closed interval $[0, 3]$, determine a lower bound and an upper bound for $\int_0^3 h(x)\,dx$.

5.4 The Fundamental Theorem of Calculus

Important Concept

- Two Versions of the Fundamental Theorem

Two Versions of the Fundamental Theorem

The Fundamental Theorem of Calculus is stated in two ways:

- The most common expression states that if f is continuous on $[a, b]$ and if $F' = f$ for $a < x < b$, then $\int_a^b f(x)\,dx = F(b) - F(a)$.
- An equivalent expression states that if function $F(x)$ is defined as the integral $F(x) = \int_a^x f(t)\,dt, a \le x \le b$, then F is differentiable on the open interval (a, b) and F is an antiderivative of f. In *Leibniz notation,* this form of the Fundamental Theorem of Calculus is expressed as $\dfrac{d}{dx} \int_a^x f(t)\,dt = f(x)$.

Example 1 Use the first part of the Fundamental Theorem of Calculus to evaluate $\int_0^{\pi/4} \sec^2 (x)\,dx$.

Solution $F(x) = \tan(x)$ is an antiderivative for $\sec^2(x)$ because $\dfrac{d}{dx}\tan(x) = \sec^2(x)$. Using the first part of the Fundamental Theorem of Calculus, we see that

$$\int_0^{\pi/4} \sec^2 (x)\,dx = F(x)\Big|_{x=0}^{x=\pi/4}$$
$$= F\left(\frac{\pi}{4}\right) - F(0)$$
$$= \tan\left(\frac{\pi}{4}\right) - \tan(0)$$
$$= 1 - 0$$
$$= 1.$$

Example 2 If $F(x) = \int\limits_{3}^{x} (2t + 5)\, dt,\ 3 \le x \le 10$, calculate $F'(x)$.

Solution Using the second form of the Fundamental Theorem of Calculus, we see that $F'(x) = f(x) = 2x + 5$. ■

Self-Test Exercises

1. Show that $F(x) = 2e^{3x}$ is an antiderivative for $f(x) = 6e^{3x}$. Use this result and the first part of the Fundamental Theorem of Calculus to evaluate $\int\limits_{0}^{5} \dfrac{2}{3} e^{3x}\, dx$.

2. If $F(x) = \int\limits_{2}^{x} (2\cos^2(t) + 5t + 7)\, dt,\ 2 \le x \le 10$, calculate $F'(x)$.

5.5 Integration by Substitution

Important Concept

- Integration by Substitution

Steps to Integrate by Substitution

Integration by substitution deals with functions that are compositions of other functions. This method is based on the Chain Rule and is sometimes called a *change of variable* method. Use the following steps to integrate by substitution.

1. Find an expression $\phi(x)$ in the integrand that has a derivative $\phi'(x)$ that also appears in the integrand. Identifying a proper choice for $\phi(x)$ is a critical step in the method of substitution. Although the choices will be different case by case, usually $\phi(x)$ is the "inside function" of a composition function.
2. Substitute u for $\phi(x)$ and du for $\phi'(x)\, dx$. If du in the problem is not exactly what is needed, but a *constant multiple* is needed, then simply multiply and divide by the needed constant value, place (1/constant value) in front of the integral sign, and place the constant value with the du term. (See Examples 2 and 3 in this section for details.)
3. *Do not proceed unless the entire integrand is expressed in terms of the new variable u.* (No x's can remain.)
4. Evaluate the integral to obtain an answer expressed in terms of u.
5. Substitute the expression $u = \phi(x)$ in your integration results, and obtain an answer as a function of x.
6. Steps 1 to 5 are sufficient if the original problem is an indefinite integral. If the original problem is a definite integral, be sure to convert the lower and upper limits of integration to u-values in Steps 2 and 3. Then follow steps 4 and 5; your results will be a numerical value.

Example 1 Evaluate $\int (\sin(x))^3 \cos(x)\, dx$.

Solution

1. Recognize that $(\sin(x))^3$ is a composition of functions with the inside function $\phi(x) = \sin(x)$. The derivative, $\phi'(x) = \cos(x)$, appears in the remaining part of the integrand.

2. The critical step is to define $u = \sin(x)$, which implies $du = \left(\dfrac{d}{dx}\sin(x)\right)dx = \cos(x)\,dx$. Make the substitutions u for $\sin(x)$ and du for $\cos(x)\,dx$ to obtain

$$\int (\sin(x))^3 \cos(x)\,dx = \int u^3 du.$$

3. The entire integrand is expressed in terms of u only.

4. $\int u^3 du = \dfrac{u^4}{4} + C$. ($C$ is the constant of integration, which must be part of the answer, because $\int (\sin(x))^3 \cos(x)\,dx$ is an indefinite integral.)

5. Substitute $u = \sin(x)$ into the integration answer $\dfrac{u^4}{4} + C$ to obtain the final answer, $\dfrac{(\sin(x))^4}{4} + C$.

■

Example 2 Evaluate $\displaystyle\int \dfrac{x^2}{(1+x^3)^5}\,dx$.

Solution

1. Recognize that $(1 + x^3)^5$ is a composition of functions with the inside function $\phi(x) = 1 + x^3$. The derivative, $\phi'(x) = 3x^2$, appears in the remaining part of the integrand.

2. The critical step is to define $u = 1 + x^3$, which implies $du = \left(\dfrac{d}{dx}(1 + x^3)\right)dx = 3x^2\,dx$. Notice that our original expression of the integral does not have the required constant multiple of 3. Therefore, we multiply by 3 and divide by 3, then make the substitutions u for $(1 + x^3)$ and du for $3x^2\,dx$ to obtain

$$\int \dfrac{x^2}{(1+x^3)^5}\,dx = \dfrac{1}{3}\cdot\int \dfrac{3x^2}{(1+x^3)^5}\,dx$$
$$= \dfrac{1}{3}\cdot\int \dfrac{1}{(1+x^3)^5}\cdot 3x^2\,dx$$
$$= \dfrac{1}{3}\cdot\int \dfrac{1}{u^5}\,du.$$

3. The entire integrand is expressed in terms of u only.

4. We have

$$\dfrac{1}{3}\cdot\int \dfrac{1}{u^5}\,du = \dfrac{1}{3}\cdot\int u^{-5}du$$
$$= \dfrac{1}{3}\cdot\dfrac{u^{-4}}{(-4)} + C$$
$$= \dfrac{-u^{-4}}{12} + C.$$

5. Substitute $u = 1 + x^3$ into the integration answer $\dfrac{-u^{-4}}{12} + C$ to obtain the final answer, $\dfrac{-(1+x^3)^{-4}}{12} + C$, or $\dfrac{-1}{12(1+x^3)^4} + C$.

■

Example 3 Calculate $\int_0^{10} (4x^2 + 4x + 1)\,dx$ using the method of Integration by Substitution.

Solution Because the method of Integration by Substitution is required, we *cannot* integrate term by term. Rather, we express $4x^2 + 4x + 1$ as $(2x + 1)^2$. The function $f(x) = (2x + 1)^2$ is a composition of functions with inside function $\phi(x) = (2x + 1)$. Note that $\phi'(x) = 2\,dx$, so our original expression does not have the required constant multiple of 2. Therefore, we multiply by 2 and divide by 2, then substitute u for $(2x + 1)$ and du for $2\,dx$. The result of these actions appears to be

$$\int_0^{10} (4x^2 + 4x + 1)\,dx = \int_0^{10} (2x + 1)^2\,dx$$

$$= \left(\frac{1}{2}\right) \int_0^{10} (2x + 1)^2 \cdot 2\,dx$$

$$= \left(\frac{1}{2}\right) \int_0^{10} (u)^2\,du.$$

But the format of the statement $\left(\frac{1}{2}\right) \int_0^{10} (u)^2\,du$ is incorrect! What's wrong with it? We have an expression in the variable u, but the values for the lower limit 0 and the upper limit 10 are x-values! We need to convert them to u-values. If $x = 0$ and $u = (2x + 1)$, then $u = 1$. Similarly, if $x = 10$ and $u = (2x + 1)$, then $u = 21$. *These are our new u-values for the lower limit and the upper limit, respectively.* Now we can proceed using the correct format.

$$\int_0^{10} (4x^2 + 4x + 1)\,dx = \left(\frac{1}{2}\right) \int_0^{10} (2x + 1)^2 \cdot 2\,dx$$

$$= \left(\frac{1}{2}\right) \int_1^{21} u^2\,du$$

$$= \left(\frac{1}{2}\right) \cdot \frac{u^3}{3}\Big|_{u=1}^{u=21}$$

$$= \frac{(21)^3}{6} - \frac{(1)^3}{6}$$

$$= \frac{4630}{3} \qquad\blacksquare$$

Self-Test Exercises

1. Evaluate $\int (\cos(x))^6 \sin(x)\,dx$.

2. Evaluate $\int_2^7 (\sqrt{x^3 + 2})(x^2)\,dx$.

3. Evaluate $\int_{20}^{10} \left(\dfrac{1}{e^{3x}}\right) dx$.

5.6 Calculating Area

Important Concepts

- Area of a Bounded Region That Is Above the x-Axis
- Area of Bounded Regions That Are Above and Below the x-Axis
- Area Between Two Curves

Area of a Bounded Region That Is Above the x-Axis

In Sections 5.1 and 5.2, we explained that if $f(x) \geq 0$ for $x \in [a, b]$, then the area of the region under the graph of f, above the x-axis, and between $x = a$ and $x = b$ is given by $\int_a^b f(x)\, dx$. We will use the Fundamental Theorem of Calculus where possible to compute this definite integral.

Example 1 Calculate the area bounded by the function $f(x) = x^2 + 3$, $x = 2$, $x = 5$, and the x-axis.

Solution Figure 1 illustrates the bounded region.

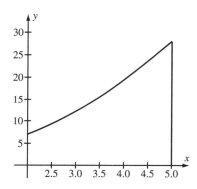

Figure 1

Since $f(x) = x^2 + 3 \geq 0$ for $x \in [2, 5]$, the area of the region is given by $\int_2^5 (x^2 + 3)\, dx$.

An antiderivative of $x^2 + 3$ is $F(x) = \dfrac{x^3}{3} + 3x$, because $F'(x) = x^2 + 3 = f(x)$. We have

$$\text{Area} = \int_2^5 (x^2 + 3)\, dx$$
$$= F(5) - F(2)$$
$$= \left(\frac{5^3}{3} + (3)(5) \right) - \left(\frac{2^3}{3} + (3)(2) \right)$$
$$= 48 \text{ square units.}$$

Area of Bounded Regions That Are Above and Below the x-Axis

If $f(x) \le 0$ for $x \in [a, b]$, then the region bounded between the graph of f, the x-axis, and between $x = a$ and $x = b$ will be *below* the x-axis. In this case, the area of this region is given by the *negative* of $\int_a^b f(x)\,dx$.

In general, if we are asked to find the total area bounded by the graph of $f(x)$ and the x-axis, we need to break down the region into components that are above the x-axis and that are below the x-axis, compute the area of each component separately, and combine the results to obtain the total area. The following example will illustrate this idea.

Example 2 Calculate the area bounded by the graph of $f(x) = \cos(x) - \dfrac{1}{2}$ and the x-axis between the interval $[0, 3]$.

Solution Figure 2 illustrates the bounded regions.

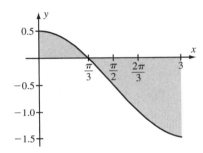

Figure 2

We first notice that for $x \in [0, 3]$, $f(x) = 0$ implies that $\cos(x) - \frac{1}{2} = 0$, which means $\cos(x) = \frac{1}{2}$ and hence $x = \dfrac{\pi}{3}$. Therefore, $f(x) \ge 0$ for $x \in \left[0, \dfrac{\pi}{3}\right]$ and $f(x) \le 0$ for $x \in \left[\dfrac{\pi}{3}, 3\right]$.

The area of the region lying *above* $\left[0, \dfrac{\pi}{3}\right]$ is

$$
\begin{aligned}
A_1 &= \int_0^{\pi/3} f(x)\,dx \\
&= \int_0^{\pi/3} \left(\cos(x) - \frac{1}{2}\right) dx \\
&= \left(\sin(x) - \frac{1}{2}x\right)\Big|_0^{\pi/3} \\
&= \left(\sin\left(\frac{\pi}{3}\right) - \frac{1}{2} \cdot \frac{\pi}{3}\right) - (\sin(0) - 0) \\
&= \frac{\sqrt{3}}{2} - \frac{\pi}{6} \\
&\approx 0.3424 \text{ square units.}
\end{aligned}
$$

The area of the region lying *below* $\left[\frac{\pi}{3}, 3\right]$ is

$$A_2 = -\int_{\pi/3}^{3} f(x)\, dx$$

$$= -\int_{\pi/3}^{3} \left(\cos(x) - \frac{1}{2}\right) dx$$

$$= -\left(\sin(x) - \frac{1}{2}x\right)\Big|_{\pi/3}^{3}$$

$$= -\left(\sin(3) - \frac{3}{2}\right) + \left(\sin\left(\frac{\pi}{3}\right) - \frac{1}{2} \cdot \frac{\pi}{3}\right)$$

$$= -\sin(3) + \frac{3}{2} + \frac{\sqrt{3}}{2} - \frac{\pi}{6}$$

$$\approx 1.7013 \text{ square units.}$$

Notice that areas are always of positive values. The total area is, the sum of the two component areas, that is, $A_1 + A_2 \approx 0.3424 + 1.7013 = 2.0437$ square units. ∎

Calculation of the Area of a Region Bounded by Two Curves

Theorem 1
Let f and g be continuous functions on the interval $[a, b]$ and suppose that $f(x) \geq g(x)$ in this interval. The area of the region bounded between the graphs $y = f(x)$ and $y = g(x)$ on the interval $[a, b]$ is given by $\int_{a}^{b} (f(x) - g(x))\, dx$. In other words, the area is given by

$$\int_{a}^{b} ((\text{top function}) - (\text{bottom function}))\, dx.$$

Example 3 Calculate the area bounded by the functions $f(x) = 2\sqrt{x}$ and $g(x) = 2x^2$.

Solution Figure 3 illustrates the bounded region.

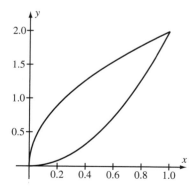

Figure 3

By solving the two equations $y = 2\sqrt{x}$ and $y = 2x^2$, we find that the two curves intersect at $(0, 0)$ and $(1, 2)$. Since $2\sqrt{x} \geq 2x^2$ for x in the interval $[0, 1]$, the top function is $2\sqrt{x}$ and the bottom function is $2x^2$. Therefore, the area of the region bounded by the two graphs is

$$\int_0^1 (2\sqrt{x} - 2x^2)\, dx = \int_0^1 (2x^{1/2} - 2x^2)\, dx$$

$$= \left(\frac{2x^{3/2}}{3/2} - \frac{2x^3}{3} \right)\Big|_0^1$$

$$= \frac{4}{3} - \frac{2}{3}$$

$$= \frac{2}{3} \text{ square units.} \qquad \blacksquare$$

By interchanging the roles of the x-axis and y-axis in Theorem 1, we are able to find the area of the region bounded between the curves $x = f_1(y)$ and $x = g_1(y)$. Namely, suppose that $f_1(y) \geq g_1(y)$ for $y \in [c, d]$; then the area of the region bounded between the "right-side curve" $x = f_1(y)$ and the "left-side curve" $x = g_1(y)$ for $y \in [c, d]$ is given by $\int_c^d (f_1(y) - g_1(y))\, dy$.

In other words, the area is given by $\int_c^d ((\text{right-side curve}) - (\text{left-side curve}))\, dy$.

Example 4 Calculate the area bounded by the functions $f(x) = 2\sqrt{x}$ and $g(x) = 2x^2$, using the "(right-side) – (left-side)" approach.

Solution Figure 3 (shown earlier in this section) illustrates the bounded region.

In the "(right-side) – (left-side)" approach, functions of y are used. The curve on the right-side of the bounded region is $y = 2x^2$, and the curve on the left-side of the bounded region $y = 2\sqrt{x}$. *The curves must be expressed as functions of y.* The left-side curve is $y = 2\sqrt{x}$, which can be rewritten as $\left(\frac{y}{2}\right)^2 = x$. Similarly, the right-side curve $y = 2x^2$ implies $\left(\frac{y}{2}\right)^{1/2} = x$. The area of the region bounded between the right-side curve $x = \left(\frac{y}{2}\right)^{1/2}$ and the left-side curve $x = \left(\frac{y}{2}\right)^2$ for $y \in [0, 2]$ is

$$\int_0^2 \left(\frac{y}{2}\right)^{1/2} - \left(\frac{y}{2}\right)^2 dy = \left(\frac{1}{2^{1/2}} \cdot \frac{y^{3/2}}{3/2} - \frac{y^3}{2^2 \cdot 3} \right)\Big|_{y=0}^{y=2}$$

$$= \frac{1}{2^{1/2}} \cdot \frac{2^{3/2}}{3/2} - \frac{2^3}{2^2 \cdot 3} - 0$$

$$= \frac{4}{3} - \frac{2}{3}$$

$$= \frac{2}{3}. \qquad \blacksquare$$

Self-Test Exercises

1. Calculate the area bounded by the graph of $f(x) = \sin(x) - \frac{1}{2}$ and the x-axis between the interval $[0, 2]$.
2. Calculate the area bounded by the functions $f(x) = 3x$ and $g(x) = 3x^2$.
3. Calculate the area bounded by the functions $f(x) = 3x$ and $g(x) = 3x^2$, using the "(right-side) – (left-side)" approach.

5.7 Numerical Methods of Integration

Important Concepts

- The Midpoint Rule
- The Trapezoidal Rule
- Simpson's Rule

Some definite integrals cannot be calculated directly, so the best we can do is to use numerical methods to obtain an approximation. The three numerical methods discussed are the Midpoint Rule, the Trapezoidal Rule, and Simpson's Rule.

The Midpoint Rule

Recall the uniform partition of order N of the closed interval $[a, b]$. Consider the choice for s_j in subinterval I_j to be the midpoint; that is to say,

$$\bar{x}_j = \frac{(x_{j-1} + x_j)}{2} = a + \left(j - \frac{1}{2} \right) \cdot \Delta x.$$

The Riemann sum $M_N = \Delta x \cdot (f(\bar{x}_1) + f(\bar{x}_2) + \cdots + f(\bar{x}_N))$ is called the *midpoint approximation* of order N. (Figure 1 shows the midpoint approximation on the interval I_j.)

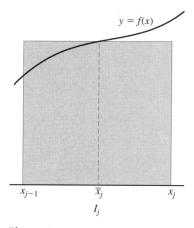

$y = f(x)$

x_{j-1} $\bar{x}_j$ x_j

I_j

Figure 1

The following theorem gives an error estimate for the midpoint approximation.

Theorem 1   If f is a continuous function on the closed interval $[a, b]$ and C is a constant such that $|f''(x)| \leq C$ for $a \leq x \leq b$, then $\left| \int_a^b f(x)\,dx - M_N \right| \leq \left(\frac{C}{24} \right) \cdot \left(\frac{(b-a)^3}{N^2} \right).$

Example 1 Use the midpoint approximation with $N = 4$ to estimate the area bounded by the graph of $f(x) = x^2$, the vertical line $x = 0$, the vertical line $x = 2$, and the x-axis.

Solution $\Delta x = \dfrac{2 - 0}{4} = \dfrac{1}{2}$. The uniform partition of order 4 of closed interval $[0, 2]$ is $\{0, .5, 1, 1.5, 2\}$.

In I_j for $1 \leq j \leq 4$,

$$\bar{x}_j = \frac{(x_{j-1} + x_j)}{2} = a + \left(j - \frac{1}{2}\right) \cdot \Delta x.$$

Therefore, the midpoints are $\bar{x}_1 = .25$, $\bar{x}_2 = .75$, $\bar{x}_3 = 1.25$, $\bar{x}_4 = 1.75$. The Riemann sum using the midpoint approximation is $M_N = \Delta x \cdot (f(\bar{x}_1) + f(\bar{x}_2) + \cdots + f(\bar{x}_N))$. Our approximation is

$$\begin{aligned} M_4 &= (.5) \cdot (f(.25) + f(.75) + f(1.25) + f(1.75)) \\ &= (.5)(.25^2 + .75^2 + 1.25^2 + 1.75^2) \\ &= 2.625 \text{ square units.} \end{aligned}$$ ■

Example 2 Calculate the error estimate for Example 1.

Solution $f(x) = x^2$ implies that $f''(x) = 2$, so $|f''(x)| = 2$. Therefore, we choose $C = 2$ in the error estimate formula.

$$\left| \int_0^2 x^2 dx - 2.625 \right| \leq \left(\frac{2}{24}\right) \cdot \left(\frac{(2-0)^3}{4^2}\right) = \frac{1}{24} \approx 0.0417.$$

The error estimate is no greater than .0417. ■

Self-Test Exercises

1. For $f(x)$ defined in Example 1, calculate the exact value of $\int_0^2 f(x)\, dx$ and verify that M_4 satisfies the error estimate obtained in Example 2.
2. For $f(x)$ defined in Example 1 and the approximation M_4, determine if M_4 is an overestimate or underestimate. Substantiate your answer.

The Trapezoidal Rule

Suppose function f is positive over the subinterval $I_j = [x_{j-1}, x_j]$ created by a uniform partition of order N of the closed interval $[a, b]$. In the Trapezoidal Rule, the area under the graph of f over each subinterval I_j is approximated by a trapezoid. Figure 2 shows the trapezoid used on the subinterval I_j. Notice that this trapezoid has width $\Delta x = x_j - x_{j-1}$ and two edges of heights $f(x_{j-1})$ and $f(x_j)$. Recall that the area of a trapezoid *in this particular form* is equal to one-half the width times the sum of the two edge lengths; i.e.,

$$\begin{aligned} A_j &= \frac{\Delta x}{2} \cdot \left(f(x_{j-1}) + f(x_j)\right) \\ &= \left(\frac{f(x_{j-1}) + f(x_j)}{2}\right) \cdot \Delta x. \end{aligned}$$

The order N *trapezoidal approximation* T_N is equal to

$$\begin{aligned} T_N &= A_1 + A_2 + \cdots + A_N \\ &= \frac{\Delta x}{2}(f(x_0) + f(x_1)) + \frac{\Delta x}{2}(f(x_1) + f(x_2)) + \cdots + \frac{\Delta x}{2}(f(x_{N-1}) + f(x_N)). \end{aligned}$$

After combining the terms, we have the formula

$$T_N = \frac{\Delta x}{2} \cdot (f(x_0) + 2f(x_1) + 2f(x_2) + \cdots + 2f(x_{N-1}) + f(x_N)).$$

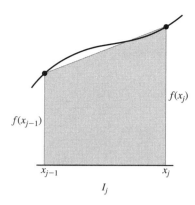

Figure 2

An error estimate for trapezoidal approximation T_N is given by Theorem 2.

Theorem 2

If f is a continuous function on $[a, b]$ and $|f''(x)| \leq C$ for all x in $[a, b]$, then an error estimate for T_N is $\left| \int_a^b f(x)\, dx - T_N \right| \leq \left(\dfrac{C}{12} \right) \cdot \left(\dfrac{(b-a)^3}{N^2} \right).$

Example 3 Calculate T_4 of $f(x) = 2 - x^2$ in $[0, 1]$.

Solution $\Delta x = \dfrac{1 - 0}{4} = .25$; the uniform partition of order 4 of $[0, 1]$ is $\{0, .25, .5, .75, 1\}$.

$$T_4 = \left(\frac{0.25}{2} \right)(f(0) + 2f(.25) + 2f(.5) + 2f(.75) + f(1))$$

$$= \left(\frac{1}{8} \right)(2 + 3.875 + 3.5 + 2.875 + 1) = 1.65625 \qquad \blacksquare$$

Example 4 Calculate the error estimate of T_4 in Example 3.

Solution $f''(x) = -2$ so $|f''(x)| = 2$. For $C = 2$,

$$\left| \int_0^1 f(x)\, dx - T_4 \right| \leq \left(\frac{2}{12} \right) \cdot \left(\frac{(1)^3}{4^2} \right) = \frac{1}{96} \approx 0.0104.$$

Therefore, the error estimate of T_4 is approximately within 1 percent of the actual value. $\qquad \blacksquare$

Self-Test Exercises

3. For $f(x)$ defined in Example 3, calculate the exact value of $\int_0^1 f(x)\, dx$ and verify that T_4 satisfies the error estimate obtained in Example 4.

4. For $f(x)$ defined in Example 3, calculate T_4 to approximate $\int_0^2 f(x)\, dx$. Is T_4 an overestimate or underestimate? Substantiate your answer.

Simpson's Rule

Suppose function f is positive over a pair of adjacent subintervals $I_j = [x_{j-1}, x_j]$ and $I_{j+1} = [x_j, x_{j+1}]$, created by a uniform partition of order N of the closed interval $[a, b]$. In Simpson's Rule, the graph of f over $I_j \cup I_{j+1}$ is approximated by a parabola.

The area under an arc of a parabola is calculated by the rule stated in Theorem 3.

Theorem 3   If the equation of the parabola is given by $P(x) = Ax^2 + Bx + C$ and γ is the midpoint of interval $I = [\alpha, \beta] = [\alpha, \gamma] \cup [\gamma, \beta]$, then the area under the arc of the parabola is equal to

$$\int_\alpha^\beta P(x)\,dx = \left(\frac{\beta - \alpha}{6}\right)(P(\alpha) + 4P(\gamma) + P(\beta)).$$

This formula is used to derive Simpson's Rule.

For a partition of $[a, b]$ with an even number, $N = 2\ell$, of subintervals of equal length Δx, the *order* $N = 2\ell$ *Simpson's Rule approximation* S_N is defined to be

$$S_N = \left(\frac{\Delta x}{3}\right)(f(x_0) + 4f(x_1) + 2f(x_2) + 4f(x_3) + \cdots + 2f(x_{N-2}) + 4f(x_{N-1}) + f(x_N))$$

$$= \left(\frac{\Delta x}{3}\right) \cdot \sum_{i=1}^{\ell} (f(x_{2 \cdot i - 2}) + 4f(x_{2 \cdot i - 1}) + f(x_{2 \cdot i})).$$

An error estimate for the Simpson's Rule approximation S_N, $N = 2\ell$, of continuous function f on $[a, b]$ for which $|f^{(4)}(x)| \le C$ for all x in $[a, b]$ is given by

$$\left| \int_a^b f(x)\,dx - S_N \right| \le \left(\frac{C}{180}\right) \cdot \left(\frac{(b-a)^5}{N^4}\right).$$

Example 5   Calculate S_2 of $f(x) = x^3$ in $[0, 1]$.

Solution $\Delta x = \dfrac{1 - 0}{2} = .5$; the uniform partition of order 2 of $[0, 1]$ is $\{0, .5, 1\}$.

$$S_2 = \left(\frac{0.5}{3}\right)(f(0) + 4f(0.5) + f(1))$$

$$= \frac{1}{6}(0 + (4)(0.125) + 1)$$

$$= \frac{1.5}{6}$$

$$= .25 \qquad \blacksquare$$

Example 6   Calculate the error estimate of S_2 in Example 5.

Solution $f^{(4)}(x) = 0$ so $|f^{(4)}(x)| = 0$. For $C = 0$, $\left| \int_0^1 f(x)\,dx - S_2 \right| \le \left(\dfrac{0}{180}\right) \cdot \left(\dfrac{(1)^5}{2^4}\right) =$ 0. Therefore, the error estimate of S_2 is equal to zero; i.e., S_2 is equal to the actual value. $\blacksquare$

Self-Test Exercises

5. For $f(x)$ defined in Example 5, calculate the exact value of $\int_0^1 f(x)\,dx$ and verify that S_2 satisfies the error estimate obtained in Example 6.

6. For $f(x)$ defined in Example 5, calculate S_4 to approximate $\int_0^2 f(x)\,dx$. Is S_4 an over-estimate or underestimate? Substantiate your answer.

Differential Equations and Transcendental Functions

6.1 First Order Differential Equations

Important Concepts

- Terminology
- Method: Separable Equations
- Applications

Introduction

Differential equations are frequently used to model events that occur in real life. In some models, differential equation solutions can be expressed as formulas; however, this is not always the case. In either case, graphical methods provide valuable information and insight.

A *differential equation* is an equation that involves one or more derivatives of an unknown function. Its *order* is equal to the highest derivative that appears in the differential equation. Solutions to differential equations are either *general* or *particular*. A *general solution* of a differential equation is *a family of functions* that satisfies the given differential equation for all x in some open interval containing x. A *particular solution* of a differential equation is *a unique function* that satisfies the given differential equation and additional given information. The graph of the solution is called a *solution curve* of the differential equation.

Example 1 Identify the order of the differential equation $\dfrac{d^2 y}{dx^2} = 9e^{3x}$.

Solution The second derivative is the highest derivative that appears in the given equation, so $\dfrac{d^2 y}{dx^2} = 9e^{3x}$ is a *second-order* differential equation. ∎

Example 2 Verify that $y(x) = e^{3x} + C$ is a general solution of $\dfrac{d^2 y}{dx^2} = 9e^{3x}$.

Solution $y(x) = e^{3x} + C$ implies that

$$
\begin{aligned}
\frac{dy}{dx} &= \frac{d}{dx}(e^{3x} + C) \\
&= 3e^{3x} + 0 \\
&= 3e^{3x}.
\end{aligned}
$$

Therefore,

$$\frac{d^2 y}{dx^2} = \frac{d}{dx}(3e^{3x})$$
$$= (3)(3e^{3x})$$
$$= 9e^{3x}.$$

Because $y(x) = e^{3x} + C$ satisfies the differential equation $\dfrac{d^2 y}{dx^2} = 9e^{3x}$ for each value of C, it is a general solution. ∎

Example 3 Graphically represent the general solution of $\dfrac{d^2 y}{dx^2} = 9e^{3x}$.

Solution Example 2 determined that $y(x) = e^{3x} + C$ is a general solution of $\dfrac{d^2 y}{dx^2} = 9e^{3x}$. The constant C represents any real number, an infinite number of values. Because an infinite number of possibilities cannot be represented in a single graph, the following graph will illustrate the solution curves $y(x) = e^{3x} + C$ for values $C = 0, 1$. In Figure 1, the solid line curve represents the graph of $y(x) = e^{3x}$, and the dashed line curve represents the graph of $y(x) = e^{3x} + 1$.

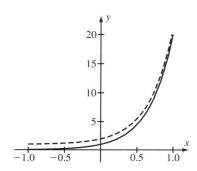

Figure 1 ∎

If the first derivative is the highest order derivative of the unknown function that appears in the equation, the equation is called a *first order differential equation*. In this section, the focus is on first order differential equations; that is, equations that are of the form $\dfrac{dy}{dx} = F(x, y)$. The pair of equations $\dfrac{dy}{dx} = F(x, y)$, $y(x_0) = y_0$ is called an *initial value problem* (IVP). Function y is a *solution of an IVP* if $\dfrac{dy}{dx} = F(x, y)$ and $y(x_0) = y_0$ for all x in some interval containing x_0. The equation $y(x_0) = y_0$ is called an *initial condition*. It can be shown that under mild restrictions on $F(x, y)$, an IVP has a *unique* solution.

Example 4 Verify that $y(x) = x^2 + 5$ is the solution of the IVP $\dfrac{dy}{dx} = 2x$, $y(0) = 5$.

Solution If $y(x) = x^2 + 5$, then

$$\frac{dy}{dx} = \frac{d}{dx}(x^2 + 5)$$
$$= 2x + 0$$
$$= 2x.$$

If $x = 0$ and $y(x) = x^2 + 5$, then $y(0) = 0^2 + 5 = 5$. Because $y(x) = x^2 + 5$ satisfies both conditions, $\dfrac{dy}{dx} = 2x$, $y(0) = 5$, it is the unique solution of the given IVP. ∎

Self-Test Exercises

1. Determine the order of the following differential equation: $\dfrac{dy}{dx} + \dfrac{d^4 y}{dx^4} = \dfrac{d^3 y}{dx^3}$.

2. Verify that $y(x) = \ln|3x| + C$ is a general solution of the differential equation $\dfrac{dy}{dx} = \dfrac{1}{x}, x > 0$.

Method: Separable Equations

The first order differentiable equation of the form

$$\frac{dy}{dx} = F(x, y) = g(x)h(y)$$

is said to be *separable* because $F(x, y)$ is defined as the product of a function of x, $g(x)$, and a function of y, $h(y)$. These functions can be *separated* using arithmetic operations so that they appear on opposite sides of the equation.

To solve a first order, separable differential equation, use arithmetic operations to separate the functions $g(x)$ and $h(y)$ so that they are grouped with their respective differentials dx and dy. Each side will be integrated with respect to either x or y. If no initial conditions are given, then a single constant, C, of integration appears on one side of the result. This procedure is known as the *Method of Separation of Variables*.

Example 5 Solve the first order differential equation $\dfrac{dy}{dx} = \dfrac{x+5}{y-4}, y \neq 4$ and describe the solution curves.

Solution $\dfrac{dy}{dx} = \dfrac{x+5}{y-4}$ can be expressed as

$$\frac{dy}{dx} = (x+5)\left(\frac{1}{y-4}\right),$$

where $g(x) = x+5$ and $h(y) = \dfrac{1}{y-4}$. This differential equation is separable. Multiplying each side of the equation by the expression $y-4$ and expressing the result in differential form, we can rewrite the differential equation as $(y-4)dy = (x+5)\,dx$. Now integrate the left side with respect to y, and integrate the right side with respect to x. The equation $\int (y-4)\,dy = \int (x+5)\,dx$ becomes

$$\frac{y^2}{2} - 4y = \frac{x^2}{2} + 5x + C,$$

which is equivalent to $y^2 - 8y = x^2 + 10x + C_1$ where $C_1 = 2C$. Thus, the general solution of the differential equation is $y^2 - 8y = x^2 + 10x + C_1$.

Now, in order to describe the solution curves, we need to apply the *Method of Completing the Square*. We obtain

$$y^2 - 8y + 16 = x^2 + 10x + 25 + (C_1 - 25 + 16).$$

This equation is equivalent to $(y-4)^2 = (x+5)^2 + C_2$, where $C_2 = (C_1 - 9)$ is a constant. Therefore, the solution curves comprise a family of hyperbolas centered at $(-5, 4)$. ∎

If $\dfrac{dy}{dx} = g(x)$, that is, $h(y) = 1$, then the solution, y, is an antiderivative of x. The following theorem cites conditions that ensure the existence of a unique solution.

Theorem 1   If g is a continuous function on an open interval containing a, then the IVP $\dfrac{dy}{dx} = g(x)$, $y(a) = b$ has the *unique solution* $y(x) = b + \int_a^x g(u)\,du$.

Self-Test Exercises

3. Describe the solution curves of the first order differential equation $\dfrac{dy}{dx} = 1$.

4. Solve the IVP $\dfrac{dy}{dx} = \dfrac{y}{x}$, $x \neq 0$, $y(1) = 10$.

Applications

Differential equations are often used to describe relationships in the physical sciences.

Example 6   The half-life of a certain radioactive substance is 100 years. How much remains of a 15 mg sample of this radioactive substance after eight years? The differential equation that describes the rate of decay of $y(t)$, the amount of the substance, and t, the number of years that have passed, is $\dfrac{dy}{dt} = k \cdot y(t)$, $k < 0$.

Solution   Using the Method of Separation of Variables, the differential equation $\dfrac{dy}{dt} = k \cdot y$ becomes $\dfrac{1}{y} \cdot dy = k \cdot dt$. Integrate both sides to obtain $\ln|y| = k \cdot t + C$. Apply the exponential function to both sides to obtain $e^{\ln|y|} = e^{kt+C} = (e^{kt}) \cdot (e^C)$, which implies that $|y(t)| = C_1 e^{kt}$, where $C_1 = e^C$. The amount of the radioactive substance, $y(t)$, is a positive number; therefore, we have $y(t) = |y(t)| = C_1 e^{kt}$. At the beginning, there are 15 mg of substance. This means that if $t = 0$, then $y = 15$ mg. Using this information, we obtain $15 = y(0) = C_1 e^{k \cdot 0} = C_1$, because $e^0 = 1$. Therefore, we have $y(t) = 15e^{kt}$. It is given that the half-life of the substance is 100 years, which means that if $t = 100$, then $y = 7.5$ mg. Using this information, we obtain $7.5 = y(100) = 15e^{k \cdot 100}$. This implies $\left(\dfrac{7.5}{15}\right) = e^{k \cdot 100}$. Because the exponential function and the natural logarithm function are mutually inverse functions, we take the natural logarithm of both sides to obtain $\ln\left(\dfrac{7.5}{15}\right) = \ln(e^{k \cdot 100}) = 100k$. Solving for k, we obtain

$$k = \left(\dfrac{1}{100}\right)\ln\left(\dfrac{7.5}{15}\right) \approx -0.0069.$$

Therefore, the solution is given by $y(t) = 15e^{-0.0069t}$.

We are now in a position to answer the question: How much remains of a 15 mg sample of this radioactive substance after eight years? If $t = 8$, then $y(8) = 15e^{(-0.0069)(8)} \approx 14.2$; that is, after eight years, approximately 14.2 mg of the radioactive substance remains.   ∎

Self-Test Exercise

5. The formula for continuously compounded interest is $A = Pe^{rt}$, where A represents the accumulated amount, P represents the principal, r is the annual rate of interest, and t represents time in years. If $1000 is deposited into an account with an interest rate of 5%, which is compounded continuously, how long does it take to accumulate $25,000?

6.2 A Calculus Approach to the Logarithm

Important Concepts

- The Natural Logarithm: Definition, Properties, Visualization, Differentiation, and Integration
- Integration of Trigonometric Functions

The Natural Logarithm

In Section 3.6, we defined $\ln(x)$, the natural logarithm, to be the inverse function of e^x. However, this definition is not rigorous because we do not know if the number e, which is defined as $\lim_{n \to \infty} \left(1 + \dfrac{1}{n}\right)^n$, exists. A precise way to define the logarithm function is the following. The *natural logarithm function*, $x \mapsto \ln(x)$, is the unique solution of the IVP $\dfrac{dy}{dx} = \dfrac{1}{x}, x > 0$, $y(1) = 0$. As an integral, the natural logarithm is defined as $\ln(x) = \displaystyle\int_1^x \dfrac{1}{t} dt, x > 0$.

For $x > 1$, you can think of $\ln(x)$ as the area below the graph of $y = \dfrac{1}{t}$ and above the interval $[1, x]$. This concept is represented in Figure 1.

For $0 < x < 1$, the value of $\ln(x)$ is the *negative* of the area between the graph and the x-axis because

$$\ln(x) = \int_1^x \frac{1}{t} dt$$

$$= -\int_x^1 \frac{1}{t} dt$$

for $0 < x < 1$. This concept is represented in Figure 2.

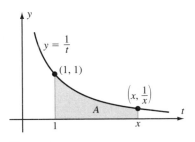

Figure 1

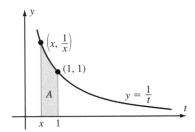

Figure 2

Properties

The following theorems identify properties of the natural logarithm.

Theorem 1   The natural logarithm has the following properties:

a.   If $x > 1$, then $\ln(x) > 0$.
b.   If $x = 1$, then $\ln(x) = 0$.
c.   If $0 < x < 1$, then $\ln(x) < 0$.
d.   The natural logarithm has a continuous derivative: $\dfrac{d}{dx}(\ln(x)) = \dfrac{1}{x}, x > 0$.
e.   The natural logarithm is an increasing function: if $0 < x_1 < x_2$ then $\ln(x_1) < \ln(x_2)$.

Theorem 2   The natural logarithm satisfies $\ln(w \cdot x) = \ln(w) + \ln(x)$ for $w, x > 0$.

Theorem 3

a.   If x is positive, then $\ln(x^{-1}) = -\ln(x)$.
b.   If y is also positive, then $\ln\left(\dfrac{x}{y}\right) = \ln(x) - \ln(y)$.
c.   If p is any number, then $\ln(x^p) = p \cdot (\ln(x))$.

Example 1   Calculate the value: $\ln\left(\dfrac{(2^{10}) \cdot (5)}{(3) \cdot (\sqrt{7})}\right)$.

Solution

$$\ln\left(\frac{(2^{10}) \cdot (5)}{(3) \cdot (\sqrt{7})}\right) = \ln((2^{10}) \cdot (5)) - \ln((3) \cdot (7^{1/2}))$$

$$= \ln(2^{10}) + \ln(5) - \ln(3) - \left(\frac{1}{2}\right) \cdot \ln(7)$$

$$= 10 \cdot \ln(2) + \ln(5) - \ln(3) - \frac{1}{2} \cdot \ln(7)$$

$$\approx 6.47$$ ∎

Visualization

Theorem 4   The natural logarithm is an increasing function with domain equal to the set of positive real numbers. Its range is the set of all real numbers. The equation $\ln(x) = \gamma$ has a *unique* solution, $x \in \mathbb{R}^+$, for every $\gamma \in \mathbb{R}$. The graph (Figure 3) of the natural logarithm function is concave down. The y-axis is a vertical asymptote for the graph.

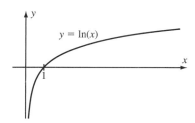

Figure 3

Differentiation and Integration

The following two rules are important:

$$\frac{d}{dx}(\ln|u(x)|) = \frac{1}{u(x)}\frac{du}{dx} \quad \text{and} \quad \int \frac{1}{u}\,du = \ln|u| + C.$$

If $u(x) = x$, and $x \neq 0$, then these two rules reduce to $\frac{d}{dx}(\ln|x|) = \frac{1}{x}dx$ and $\int \frac{1}{x}dx = \ln|x| + C$.

Example 2 Calculate the first derivative of $y = \ln|3x|$.

Solution Let $u(x) = 3x$. Then $\frac{du}{dx} = 3$. Implement the differentiation rule

$$\frac{d}{dx}(\ln|u(x)|) = \frac{1}{u(x)}\frac{du}{dx}$$

to obtain

$$\frac{d}{dx}(\ln|3x|) = \left(\frac{1}{3x}\right)(3) \quad (3)$$

$$= \frac{1}{x}.$$

■

Example 3 Calculate $\int \frac{1}{3x+2}\,dx$.

Solution Using the Method of Substitution, we let $u(x) = 3x+2$; then $du = 3\,dx$. Substituting these values back into the integral, we have

$$\int \frac{1}{3x+2}\,dx = \frac{1}{3}\int \frac{1}{3x+2}\cdot 3\,dx$$

$$= \frac{1}{3}\int \frac{1}{u}\,du$$

$$= \frac{1}{3}\ln|u| + C$$

$$= \frac{1}{3}\ln|3x+2| + C.$$

■

Self-Test Exercises

1. Calculate $\frac{d}{dx}\ln|5x+3|$.

2. Calculate $\int \frac{1}{5x+3}\,dx$.

Integration of Trigonometric Functions

The six trigonometric functions were first introduced in Chapter 1 and the antiderivatives of the sine function and the cosine function were introduced in Chapter 4. This section introduces the antiderivatives of the remaining trigonometric functions. For your convenience, all of them are listed here:

$$\int \sin(x)\,dx = -\cos(x) + C \qquad \int \csc(x)\,dx = -\ln|\csc(x) + \cot(x)| + C$$
$$\int \cos(x)\,dx = \sin(x) + C \qquad \int \sec(x)\,dx = \ln|\sec(x) + \tan(x)| + C$$
$$\int \tan(x)\,dx = \ln|\sec(x)| + C \qquad \int \cot(x)\,dx = -\ln|\csc(x)| + C$$

Example 4 Calculate $\int \csc(x)\tan(x)\,dx$.

Solution

$$\int \csc(x) \cdot \tan(x)\,dx = \int \frac{1}{\sin(x)} \cdot \frac{\sin(x)}{\cos(x)}\,dx$$

$$= \int \frac{1}{\cos(x)}\,dx$$

$$= \int \sec(x)\,dx = \ln|\sec(x) + \tan(x)| + C$$

Self-Test Exercise

3. Calculate $\int_{\pi/4}^{\pi/2} \sec(x)\cot(x)\,dx$.

6.3 The Exponential Function

Important Concepts

- The Exponential Function: Definition, Properties, Differentiation, and Integration
- The Number e

The Exponential Function

The inverse of the natural logarithm function is called the *(natural) exponential function*. It is defined by $x \mapsto \exp(x)$. As such, $\ln(\exp(a)) = a$ for all $a \in \mathbb{R}$, and $\exp(\ln(b)) = b$ for all $b > 0$. The domain of the exponential function is all real numbers, and the image of the exponential function is all positive real numbers.

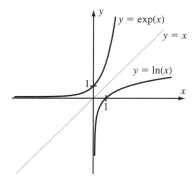

Figure 4

The relationship between the functions $y = \exp(x)$ and $y = \ln(x)$ is illustrated in Figure 1. The graph of the exponential function is the reflection about the line $y = x$ of the graph of the logarithm function and vice versa.

Properties

Some properties of the exponential function and its relationship to the logarithm function are listed here. The symbol −1 denotes inverse.

- $\exp(x) = \ln^{-1}(x)$
- $\ln(x) = \exp^{-1}(x)$
- $\exp(0) = 1$
- $\exp(s + t) = (\exp(s)) \cdot (\exp(t))$ for real numbers s, t
- $(\exp(s))^t = (\exp(s \cdot t))$ for real numbers s, t

The exponential function has *no* x-intercept.

Example 1 Find the value of $\sqrt{\exp(2\ln(5))}$ without using a calculator.

Solution
$$\sqrt{\exp(2\ln(5))} = (\exp(2\ln(5)))^{1/2}$$
$$= \exp\left(\left(\tfrac{1}{2}\right)(2\ln(5))\right)$$
$$= \exp(\ln(5))$$
$$= 5 \qquad\blacksquare$$

Differentiation and Integration

Theorem 1

The exponential function satisfies $\dfrac{d}{dx}(\exp(x)) = \exp(x)$ and $\int (\exp(x))\,dx = \exp(x) + C$. Generally, if u is a function of x [that is, $u \equiv u(x)$], then

$$\frac{d}{dx}(\exp(u)) = \exp(u)\frac{du}{dx} \quad \text{and} \quad \int (\exp(u))\,du = \exp(u) + C.$$

Example 2 Calculate $\dfrac{d}{dx}((x) \cdot (\exp(3x)))$.

Solution Because the function $x \mapsto (x) \cdot (\exp(3x))$ is the product of two functions of x, we use the differentiation rule for products:

$$\frac{d}{dx}((x) \cdot (\exp(3x))) = (x) \cdot \frac{d}{dx}(\exp(3x)) + \exp(3x) \cdot \frac{d}{dx}(x)$$
$$= (x) \cdot (3\exp(3x)) + \exp(3x) \cdot (1)$$
$$= (\exp(3x)) \cdot (3x + 1). \qquad\blacksquare$$

Example 3 Calculate $\int \exp(10)\,dx$.

Solution Because $\exp(10)$ is a real number, $\int \exp(10)\,dx = (\exp(10))x + C$. $\qquad\blacksquare$

Example 4 Calculate $\int \exp(10x)\,dx$.

Solution We will use the Method of Substitution. Let $u = 10x$, then $du = 10\,dx$. We have

$$
\begin{aligned}
\int \exp(10x)\,dx &= \frac{1}{10}\int \exp(10x)\cdot 10\,dx \\
&= \frac{1}{10}\int \exp(u)\,du \\
&= \frac{1}{10}\exp(u) + C \\
&= \frac{1}{10}\exp(10x) + C.
\end{aligned}
$$

Self-Test Exercises

1. Calculate $\dfrac{d}{dx}(x^5\exp(5x))$.

2. Show that if k is a constant not equal to zero, then $\int \exp(k\cdot x)\,dx = \dfrac{1}{k}\exp(k\cdot x) + C$.

The Number e

In Section 2.6, the number e was defined as $e = \lim\limits_{n\to\infty}\left(1 + \dfrac{1}{n}\right)^n$. With the help of the Fundamental Theorem of Calculus and the properties of the exponential function, we can finally show that $e = \exp(1)$. Utilizing the inverse property between the exponential function and the natural logarithm function, we have $\ln(e) = 1$.

Figures 2, 3, and 4 illustrate $e = \exp(1)$, $\ln(e) = 1$, and $\displaystyle\int_1^e \frac{1}{x}\,dx = 1$, respectively.

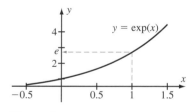

Figure 1

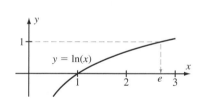

Figure 2

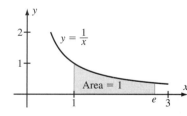

Figure 3

From the result $e = \exp(1)$ and the properties of the exponential function, we can show that

$$
\exp(x) = \exp(1\cdot x) = (\exp(1))^x = e^x.
$$

This means that e^x, the exponential function we discussed in Section 2.6, is the same as $\exp(x)$, the natural exponential function we defined in this section.

Self-Test Exercises

3. Calculate $\displaystyle\int_1^{e^2}\frac{1}{x}\,dx$.

4. Calculate $\dfrac{d}{dx}\left(\dfrac{\exp(x)}{1 + \ln(x)}\right)$.

6.4 Logarithms and Powers with Arbitrary Bases

Important Concepts

- Exponential Function with Arbitrary Bases
- Logarithm Functions with Arbitrary Bases
- Elementary Properties of Exponential Functions and Logarithm Functions with Arbitrary Bases
- Differentiation and Integration of Exponential Functions and Logarithm Functions with Arbitrary Bases

Exponential Functions with Arbitrary Bases

If a is any positive number, we define a^x by $a^x = \exp(x \ln(a))$, $x \in \mathbb{R}$. The function $x \mapsto a^x$ is called the *exponential function a^x with base a*. Therefore,

$$\ln(a^x) = \ln(\exp(x \ln(a))) = x \ln(a).$$

Example 1 Simplify the expression $\ln(2^x)$, given that $\ln(2) \approx 0.69$.

Solution
$$\ln(2^x) = \ln(\exp(x \ln(2)))$$
$$= x \ln(2) \approx .69x$$ ■

Self-Test Exercise

1. Simplify the expression $\dfrac{\ln(5^x)}{\ln(5)}$.

Logarithm Functions with Arbitrary Bases

If $a \neq 1$, $a \in \mathbb{R}^+$, we define the logarithm function with base a, denoted as $\log_a$, by

$$\log_a(x) = \frac{\ln(x)}{\ln(a)}, \quad x \in \mathbb{R}^+.$$

The following theorem summarizes information about exponential functions with arbitrary bases and logarithm functions with arbitrary bases.

Theorem 1   For any fixed positive $a \neq 1$, function $x \mapsto a^x$ has domain $\mathbb{R}$ and image $\mathbb{R}^+$.

The function $x \mapsto \log_a(x)$ has domain $\mathbb{R}^+$ and image $\mathbb{R}$. These two functions satisfy

$$a^{\log_a(x)} = x, \quad x \in \mathbb{R}^+$$

and

$$\log_a(a^x) = x, \quad x \in \mathbb{R}.$$

That is, $x \mapsto a^x$ and $x \mapsto \log_a(x)$ are *mutually inverse* functions.

Figures 1, 2, and 3 visualize the statements found in Theorem 1.

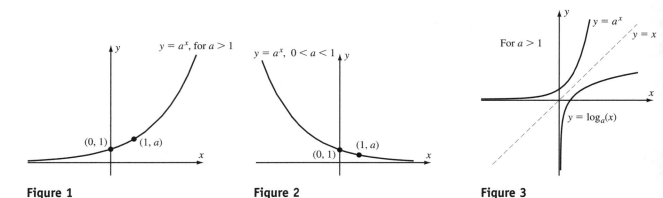

Figure 1   **Figure 2**   **Figure 3**

Example 2 Sketch the graphs $y = \left(\dfrac{1}{2}\right)^x$ and $y = \ln_{1/2}(x)$ using the same set of axes. What do you notice?

Solution In Figure 4, the solid line curve represents the graph of $y = f(x) = \left(\dfrac{1}{2}\right)^x$ and the dashed line curve represents the graph of

$$y = g(x) = \ln_{1/2}(x) = \frac{\ln(x)}{\ln \frac{1}{2}}.$$

We see that both $f(x) = \left(\dfrac{1}{2}\right)^x$ and $g(x) = \ln_{1/2}(x)$ are decreasing functions. Also, we notice that each graph is a reflection of the other along the line $y = x$. This is because the functions, $f(x) = \left(\dfrac{1}{2}\right)^x$ and $g(x) = \ln_{1/2}(x)$, are inverse to each other.

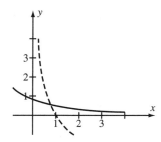

Figure 4

Self-Test Exercises

2. Verify graphically that $f(x) = 0.8^x$ and $g(x) = \log_{0.8}(x)$ are inverse functions to each other.
3. Verify that $y = \left(\frac{1}{2}\right)^x$ is a decreasing function, using algebraic methods.

Elementary Properties of Exponential Functions and Logarithm Functions with Arbitrary Bases

Theorem 2   ***Laws of Exponents***   If $a, b > 0$ and $x, y \in \mathbb{R}$, then

    a. $a^0 = 1$;

    b. $a^1 = a$;

    c. $a^{x+y} = a^x \cdot a^y$;

    d. $a^{x-y} = \dfrac{a^x}{a^y}$;

    e. $(a^x)^y = a^{x \cdot y}$;

    f. $a^x = b$ if and only if $b^{1/x} = a$, $x \neq 0$; and

    g. $(a \cdot b)^x = a^x \cdot b^x$.

Example 3   Simplify $\left(a^{(\log_a 9)/2}\right)^{12}$.

Solution
$$
\begin{aligned}
\left(a^{(\log_a 9)/2}\right)^{12} &= a^{((\log_a 9)/2)\cdot(12)} \\
&= a^{6 \cdot \log_a 9} \\
&= a^{\log_a 9^6} \\
&= 9^6
\end{aligned}
$$

    ■

Theorem 3   Let a and b be positive bases. If $x > 0$ and $y > 0$, then

    a. $\log_a(1) = 0$;

    b. $\log_a(a) = 1$;

    c. $\log_a(x \cdot y) = \log_a(x) + \log_a(y)$;

    d. $\log_a\left(\dfrac{x}{y}\right) = \log_a(x) - \log_a(y)$;

    e. for any exponent p, $\log_a(x^p) = p \cdot (\log_a(x))$;

    f. $\log_a(x) = \dfrac{\log_b(x)}{\log_b(a)}$; and

    g. $\log_a(b) = \dfrac{1}{\log_b(a)}$.

Example 4   Solve for x if $\dfrac{5^x}{3} = \dfrac{7^{2x}}{4^{3x}}$.

Solution   We take the natural logarithm of both sides of the equation: $\ln\left(\dfrac{5^x}{3}\right) = \ln\left(\dfrac{7^{2x}}{4^{3x}}\right)$. Applying the rules for logarithms mentioned above, we obtain

$$\ln(5^x) - \ln(3) = \ln(7^{2x}) - \ln(4^{3x}),$$

or equivalently,

$$x \cdot \ln(5) - \ln(3) = 2x \cdot \ln(7) - 3x \cdot \ln(4).$$

Gathering all the terms involving x yields

$$x \cdot \ln(5) - 2x \cdot \ln(7) + 3x \cdot \ln(4) = \ln(3)$$
$$x \cdot (\ln(5) - 2 \cdot \ln(7) + 3 \cdot \ln(4)) = \ln(3).$$

Solving for x results in $x = \dfrac{\ln(3)}{(\ln(5) - 2 \cdot \ln(7) + 3 \cdot \ln(4))} \approx 0.59.$

    ■

Self-Test Exercises

4. Simplify $\left(e^{\log_{(2a)} 9}\right)^{(\ln(a)+\ln 2)}$.

5. Solve for x if $2^x = \left(\frac{1}{4}\right)^{5x}$.

Differentiation and Integration of Exponential Functions and Logarithm Functions with Arbitrary Bases

For fixed positive constant $a \neq 1$, we have $\dfrac{d}{dx}(a^x) = (\ln(a)) \cdot a^x$ and

$$\frac{d}{dx}(\log_a(x)) = \frac{1}{x \cdot (\ln(a))}.$$

Example 5 Calculate $\dfrac{d}{dx}\left(3^x \cdot \log_{10}(x)\right)$.

Solution We need to use the Product Rule in the differentiation.

$$\frac{d}{dx}\left(3^x \cdot \log_{10}(x)\right) = \left(\frac{d}{dx} 3^x\right) \cdot (\log_{10}(x)) + 3^x \cdot \left(\frac{d}{dx} \log_{10}(x)\right)$$

$$= (\ln(3) \cdot 3^x) \cdot (\log_{10}(x)) + 3^x \cdot \left(\frac{1}{x(\ln(10))}\right) \qquad \blacksquare$$

Because of the relationship between differentiation and integration, the rule $\dfrac{d}{dx}(a^x) = (\ln(a)) \cdot a^x$ readily implies that $\int a^x dx = \dfrac{a^x}{\ln(a)} + C$.

Example 6 Evaluate $\int 2^{\sin(x)} \cos(x)\,dx$.

Solution Let $u = \sin(x)$. Then $du = \cos(x)\,dx$. Therefore,

$$\int 2^{\sin(x)} \cos(x)\,dx = \int 2^u\,du$$

$$= \frac{2^u}{\ln(2)} + C$$

$$= \frac{2^{\sin(x)}}{\ln(2)} + C. \qquad \blacksquare$$

Logarithmic Differentiation

The derivative of the logarithm of f is called the *logarithmic derivative of f*. The process of differentiating $\ln(f(x))$ is called *logarithmic differentiation*.

Example 7 Use logarithmic differentiation to calculate the derivative of $f(x) = 5x^{2x} \cdot e^{3x}$.

Solution First we take the natural logarithm of both sides and apply Theorem 3:

$$\begin{aligned}
\ln(f(x)) &= \ln(5x^{2x} \cdot e^{3x}) \\
&= \ln(5x^{2x}) + \ln(e^{3x}) \\
&= \ln(5) + 2x \cdot \ln(x) + 3x.
\end{aligned}$$

Then we differentiate each side:

$$\frac{d}{dx} \ln(f(x)) = \frac{d}{dx}(\ln(5) + 2x \cdot \ln(x) + 3x).$$

This becomes:

$$\begin{aligned}
\frac{f'(x)}{f(x)} &= \frac{d}{dx}(\ln(5)) + \frac{d}{dx}(2x \cdot \ln(x)) + \frac{d}{dx}(3x) \\
&= 0 + \left(2 \cdot \ln(x) + 2x \cdot \frac{1}{x}\right) + 3 \\
&= 2\ln(x) + 5.
\end{aligned}$$

Finally, we solve for $f'(x)$ to obtain

$$\begin{aligned}
f'(x) &= f(x) \cdot (2\ln(x) + 5) \\
&= (5x^{2x} \cdot e^{3x}) \cdot (2\ln(x) + 5) \\
&= 10x^{2x} \cdot e^{3x} \cdot \ln(x) + 25x^{2x} \cdot e^{3x}.
\end{aligned}$$ ∎

Self-Test Exercises

6. Calculate the derivative of $f(x) = \dfrac{1 - x^3}{2^x}$.

7. Evaluate the integral $\int 2^{\cos(x)} \sin(x)\, dx$.

8. Evaluate the integral $\int 2^{3x-1}\, dx$.

9. Use logarithmic differentiation to calculate the derivative of $f(x) = x^x \cdot e^{\cos(x)}$.

6.5 Applications of the Exponential Function

Important Concepts

- Growth
- Decay

Growth

The exponential function has many applications, including the calculation of growth.

Theorem 1 If the rate of change of a function y is proportional to the value of y [that is, if $y'(t) = k \cdot y(t)$, $k \in \mathbb{R}$], then $y(t) = Ae^{kt}$, where $A = y(0)$.

Example 1 If interest is compounded continuously at a rate of 4.25%, how long will it take an initial deposit to double?

Solution Interest compounded at a rate of 4.25% implies $y'(t) = .0425 \cdot y(t)$. By Theorem 1, the solution to this differential equation is $y(t) = Ae^{.0425t}$. The problem to be solved is "If $y(t) = 2A$ and $k = .0425$, what is the value of t?" $2A = Ae^{.0425t}$ becomes $2 = e^{.0425t}$, which implies that $\ln(2) = .0425t$. This means that $t = \dfrac{\ln(2)}{.0425} \approx 16.3$ years. Therefore, under the given conditions, an initial deposit will double in slightly more than 16 years. ∎

Self-Test Exercise

1. A culture contained an initial number of 5000 bacteria. Three hours later, the culture contained 7800 bacteria. Determine the number of bacteria present in the culture five hours after the initial observation. You may assume that the growth is exponential.

Law of Radioactive Decay

The Law of Radioactive Decay states that $\dfrac{dR}{dt} = -\lambda R$, where $R(t)$ is the amount at time t of a particular radioactive isotope which disintegrates or decays. $\lambda > 0$ is called the *decay* or *disintegration constant*. Using the Method of Separable Equations,

$$\frac{dR}{dt} = -\lambda R \Rightarrow R(t) = R(0)e^{-\lambda t}.$$

You can think of this differential equation as being similar to the exponential growth differential equation, but $k = -\lambda$ in the decay situation.

Example 2 Determine the value of the decay constant if 50 g (grams) of a radioactive substance reduces to 35 g over a period of 15 years.

Solution The problem states that at time $t = 0$, $R(0) = 50$ g and at time $t = 15$, $R(15) = 35$ g. Therefore, from the first condition we have $R(t) = 50e^{-\lambda t}$. The second condition implies that $35 = R(15) = 50e^{-\lambda(15)}$, which means $\dfrac{35}{50} = e^{-15\lambda}$. Hence $\ln\left(\dfrac{35}{50}\right) = -15\lambda$, and we have

$$\lambda = \left(\frac{-1}{15}\right)\left(\ln\left(\frac{35}{50}\right)\right) \approx .0238. \qquad ∎$$

Self-Test Exercise

2. If the decay constant is .0478, determine the amount of a radioactive substance that remains after five years if the initial amount is 75 g.

6.6 Inverse Trigonometric Functions

Important Concepts

- Inverse Trigonometric Functions
- Derivatives of Inverse Trigonometric Functions
- Integrals of Inverse Trigonometric Functions

Inverse Trigonometric Functions

Trigonometric functions are periodic and not one-to-one. To discuss inverses for them, you must restrict the domains of the trigonometric functions on an individual basis.

Definitions and Restricted Domains

Define $x \mapsto \text{Sin}(x)$ to be the sine function with restricted domain $\left[\dfrac{-\pi}{2}, \dfrac{\pi}{2}\right]$. Its inverse function is $x \mapsto \arcsin(x)$. The graphs of $y = \text{Sin}(x)$ and $y = \arcsin(x)$ appear in Figure 1.

Define $x \mapsto \text{Cos}(x)$ to be the cosine function with restricted domain $[0, \pi]$. Its inverse function is $x \mapsto \arccos(x)$. The graphs of $y = \text{Cos}(x)$ and $y = \arccos(x)$ appear in Figure 2.

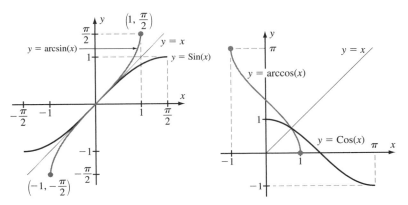

Figure 1   **Figure 2**

Define $x \mapsto \text{Tan}(x)$ to be the tangent function with restricted domain $\left(\dfrac{-\pi}{2}, \dfrac{\pi}{2}\right)$. Its inverse function is $x \mapsto \arctan(x)$. The graphs of $y = \text{Tan}(x)$ and $y = \arctan(x) = \text{Tan}^{-1}(x)$ appear in Figures 3 and 4, respectively.

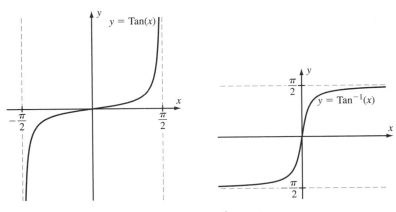

Figure 3   **Figure 4**

Define $x \mapsto \text{Cot}(x)$ to be the cotangent function with restricted domain $(0, \pi)$. Its inverse function is $x \mapsto \text{Cot}^{-1}(x)$. The graphs of $y = \text{Cot}(x)$ and $y = \text{Cot}^{-1}(x)$ appear in Figures 5 and 6, respectively.

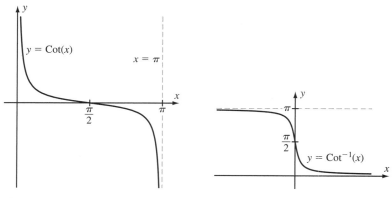

Figure 5   **Figure 6**

Define $x \mapsto \mathrm{Csc}(x)$ to be the cosecant function with restricted domain $\left[\dfrac{-\pi}{2}, 0\right) \cup \left(0, \dfrac{\pi}{2}\right]$.
Its inverse function is $x \mapsto \mathrm{Csc}^{-1}(x)$. The graphs of $y = \mathrm{Csc}(x)$ and $y = \mathrm{Csc}^{-1}(x)$ appear in Figures 7 and 8, respectively.

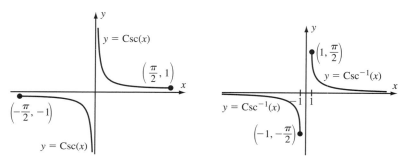

Figure 7   **Figure 8**

Define $x \mapsto \mathrm{Sec}(x)$ to be the secant function with restricted domain $\left[0, \dfrac{\pi}{2}\right) \cup \left(\dfrac{\pi}{2}, \pi\right]$.
Its inverse function is $x \mapsto \mathrm{Sec}^{-1}(x)$. The graphs of $y = \mathrm{Sec}(x)$ and $y = \mathrm{Sec}^{-1}(x)$ appear in Figures 9 and 10, respectively.

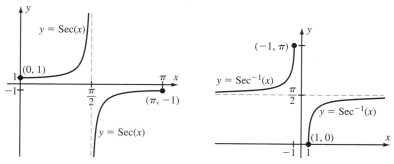

Figure 9   **Figure 10**

Example 1 Calculate the value of the $\vartheta = \arcsin\left(\dfrac{\sqrt{3}}{2}\right)$, without using a calculator.

Solution $\vartheta = \arcsin\left(\dfrac{\sqrt{3}}{2}\right) \Leftrightarrow \mathrm{Sin}(\vartheta) = \dfrac{\sqrt{3}}{2} \Rightarrow \vartheta = \dfrac{\pi}{3}$ ∎

Self-Test Exercise

1. Calculate the value of $\vartheta = \arctan(1)$ without using a calculator.

Derivatives of Inverse Trigonometric Functions

Theorem 1

The functions $t \mapsto \arcsin(t)$ and $t \mapsto \arccos(t)$ are differentiable on the open interval $(-1, 1)$ and $\dfrac{d}{dt}(\arcsin(t)) = \dfrac{1}{\sqrt{1 - t^2}}$, $\dfrac{d}{dt}(\arccos(t)) = \dfrac{-1}{\sqrt{1 - t^2}}$.

Theorem 2

The function $t \mapsto \arctan(x)$ [or equivalently, $\mathrm{Tan}^{-1}(x)$] is differentiable for each t and $\dfrac{d}{dt}(\arctan(t)) = \dfrac{1}{1 + t^2}, -\infty < t < \infty.$

Similarly, the functions $t \mapsto \mathrm{Cot}^{-1}(t), t \mapsto \mathrm{Sec}^{-1}(t)$, and $t \mapsto \mathrm{Csc}^{-1}(t)$ are differentiable and $\dfrac{d}{dt}(\mathrm{Cot}^{-1}(t)) = \dfrac{-1}{1 + t^2}, -\infty < t < \infty, \dfrac{d}{dt}(\mathrm{Sec}^{-1}(t)) = \dfrac{1}{|t|\sqrt{t^2 - 1}}, |t| > 1$, and $\dfrac{d}{dt}(\mathrm{Csc}^{-1}(t)) = \dfrac{-1}{|t|\sqrt{t^2 - 1}}, |t| > 1.$

Example 2 Calculate $\dfrac{d}{dt}(\mathrm{Cot}^{-1}(2t)).$

Solution Let $u = 2t$. Then $\dfrac{du}{dt} = 2$. Using the Chain Rule, we have,

$$
\begin{aligned}
\frac{d}{dt}(\mathrm{Cot}^{-1}(2t)) &= \frac{d}{du}(\mathrm{Cot}^{-1}(u)) \cdot \frac{du}{dt} \\
&= \frac{-1}{1 + u^2} \cdot \frac{du}{dt} \\
&= \frac{-1}{1 + (2t)^2} \cdot (2) \\
&= \frac{-2}{1 + 4t^2}. \quad\blacksquare
\end{aligned}
$$

Self-Test Exercise

2. Calculate $(\arctan(-t))'$.

Integrals of Inverse Trigonometric Functions

The relationship between differentiation and integration suggests the following integration formulas.

$$
\int \frac{du}{\sqrt{1 - u^2}} = \arcsin(u) + C \text{ and } \int \frac{du}{1 + u^2} = \arctan(u) + C
$$

Example 3 Calculate $\displaystyle\int \frac{du}{\sqrt{4 - 4u^2}}.$

Solution

$$\int \frac{du}{\sqrt{4 - 4u^2}} = \int \frac{du}{\sqrt{4(1 - u^2)}}$$

$$= \int \frac{du}{(\sqrt{4})\sqrt{1 - u^2}}$$

$$= \frac{1}{2} \int \frac{du}{\sqrt{1 - u^2}}$$

$$= \frac{1}{2}(\arcsin(u)) + C$$

∎

Self-Test Exercise

3. Calculate $\displaystyle\int \frac{du}{1 + \left(\dfrac{u}{a}\right)^2}$, where $a > 0$ is a constant.

6.7 The Hyperbolic Functions

Important Concepts

- Hyperbolic Functions
- Derivatives
- Inverse Hyperbolic Functions

Hyperbolic Functions

For all real t, define the *hyperbolic sine function* and the *hyperbolic cosine function* by $\sinh(t) = \dfrac{e^t - e^{-t}}{2}$ and $\cosh(t) = \dfrac{e^t + e^{-t}}{2}$, respectively. Their graphs appear in Figures 1 and 2.

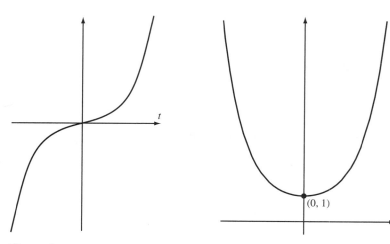

Figure 1   **Figure 2**

The remaining hyperbolic functions can be defined in terms of $\sinh(t)$ and $\cosh(t)$. That is, $\tanh(t) = \dfrac{\sinh(t)}{\cosh(t)}, t \in \mathbb{R}$; $\coth(t) = \dfrac{\cosh(t)}{\sinh(t)}, t \neq 0$; $\mathrm{sech}(t) = \dfrac{1}{\cosh(t)}, t \in \mathbb{R}$; and $\mathrm{csch}(t) = \dfrac{1}{\sinh(t)}, t \neq 0$.

The graphs of these hyperbolic functions appear in Figures 3, 4, 5, and 6.

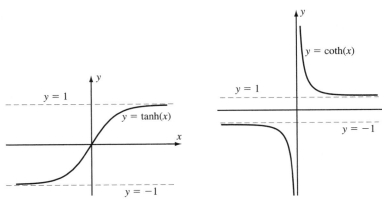

Figure 3

Figure 4

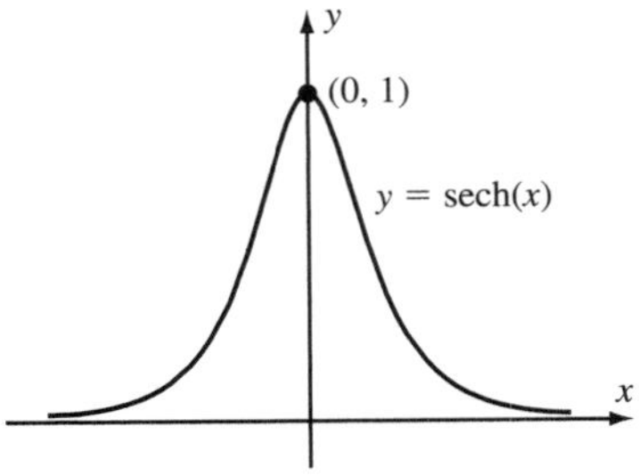

Figure 5

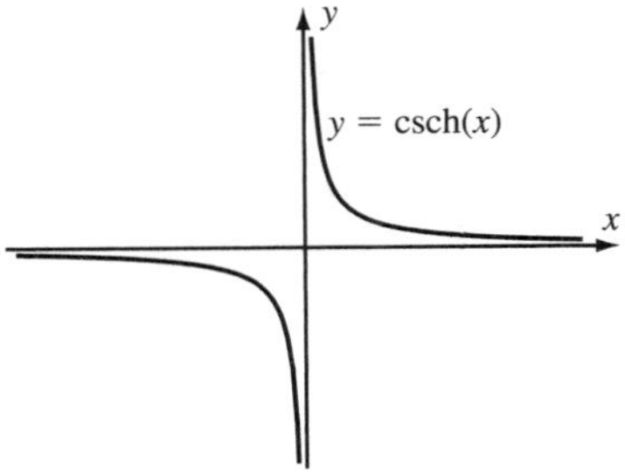

Figure 6

The differentiation formulas for the hyperbolic trigonometric functions are:

$$\frac{d}{dt}\sinh(t) = \cosh(t) \qquad \frac{d}{dt}\coth(t) = -\operatorname{csch}^2(t)$$

$$\frac{d}{dt}\cosh(t) = \sinh(t) \qquad \frac{d}{dt}\operatorname{sech}(t) = -\operatorname{sech}(t) \cdot \tanh(t)$$

$$\frac{d}{dt}\tanh(t) = \operatorname{sech}^2(t) \qquad \frac{d}{dt}\operatorname{csch}(t) = -\operatorname{csch}(t) \cdot \coth(t)$$

Example 1 Calculate $\dfrac{d}{dt}(\cosh(2t) \cdot \sinh(3t))$.

Solution

$$\frac{d}{dt}(\cosh(2t)\sinh(3t)) = \left(\frac{d}{dt}\cosh(2t)\right) \cdot \sinh(3t) + \cosh(2t) \cdot \left(\frac{d}{dt}\sinh(3t)\right)$$

$$= (2\sinh(2t)) \cdot (\sinh(3t)) + (\cosh(2t)) \cdot (3\cosh(3t)) \qquad \blacksquare$$

Self-Test Exercises

1. Calculate $\dfrac{d}{dt}(\sinh(5t) \cdot \cosh(9t))$.

2. Calculate $\dfrac{d}{dt}(\operatorname{csch}(5t))^2$.

Inverse Hyperbolic Functions

The formulas for the inverses of the hyperbolic functions can be explicitly expressed in terms of functions that we already understand.

$$\sinh^{-1}(x) = \ln\left(x + \sqrt{x^2 + 1}\right), x \in \mathbb{R}; \qquad \coth^{-1}(x) = \frac{1}{2}\ln\left(\frac{x+1}{x-1}\right), |x| > 1;$$

$$\cosh^{-1}(x) = \ln\left(x + \sqrt{x^2 - 1}\right), x \geq 1; \qquad \text{sech}^{-1}(x) = \ln\left(\frac{1 + \sqrt{1 - x^2}}{x}\right), 0 < x \leq 1;$$

$$\tanh^{-1}(x) = \frac{1}{2}\ln\left(\frac{1+x}{1-x}\right), -1 < x < 1; \qquad \text{csch}^{-1}(x) = \ln\left(\frac{1}{x} + \frac{\sqrt{1 + x^2}}{|x|}\right), x \neq 0.$$

We can use these explicit formulas for the inverse hyperbolic functions to find their derivatives. As a result, we have:

$$\frac{d}{dx}(\sinh^{-1}(x)) = \frac{1}{\sqrt{1 + x^2}}, x \in \mathbb{R}; \qquad \frac{d}{dx}(\cosh^{-1}(x)) = \frac{1}{\sqrt{x^2 - 1}}, x > 1;$$

$$\frac{d}{dx}(\tanh^{-1}(x)) = \frac{1}{1 - x^2}, -1 < x < 1; \qquad \frac{d}{dx}(\text{sech}^{-1}(x)) = \frac{-1}{x\sqrt{1 - x^2}}, 0 < x < 1.$$

By expressing these differentiation formulas in antiderivative form, we have the following:

$$\int \frac{dx}{\sqrt{1 + x^2}} = \sinh^{-1}(x) + C; \qquad \int \frac{dx}{\sqrt{x^2 - 1}} = \cosh^{-1}(x) + C$$

$$\int \frac{dx}{x\sqrt{1 - x^2}} = -\text{sech}^{-1}(x) + C; \qquad \int \frac{dx}{1 - x^2} = \tanh^{-1}(x) + C.$$

Example 2 Calculate $\int \frac{dx}{1 - 4x^2}$.

Solution Let $u = 2x$, then $du = 2\,dx$.

$$\int \frac{dx}{1 - 4x^2} = \int \frac{dx}{1 - (2x)^2}$$

$$= \frac{1}{2}\int \frac{2dx}{1 - (2x)^2}$$

$$= \frac{1}{2}\int \frac{du}{1 - (u)^2}$$

$$= \frac{1}{2}\tanh^{-1}(u) + C$$

$$= \frac{1}{2}\tanh^{-1}(2x) + C$$

Self-Test Exercises

3. Calculate $\frac{d}{dx}(\tanh^{-1}(2x))$.

4. Calculate $\int \frac{dx}{\sqrt{1 + 9x^2}}$.

Techniques of Integration

7.1 Integration by Parts

Important Concepts

- Integration by Parts Formula
- Steps to Integration by Parts

Integration by Parts (IBP) can be thought of as "the differentiation product rule in reverse." When the function to be integrated is expressed as a product of two functions, you may use this integration method.

Integration by Parts Formula

In the following, we assume that the function to be integrated is equal to the product of two functions of independent variable x; they are $u \equiv u(x)$ and $\dfrac{dv}{dx} \equiv \dfrac{d}{dx}(v(x))$.

The IBP formula is

$$\int u \cdot \frac{dv}{dx}dx = u \cdot v - \int v \cdot \frac{du}{dx}dx.$$

It is often expressed in *differential notation* as $\int u \, dv = uv - \int v \, du$. The definite integral version of the IBP formula is

$$\int_{x=a}^{x=b} u \, dv = u \cdot v \Big|_{x=a}^{x=b} - \int_{x=a}^{x=b} v \, du.$$

Steps to Integration by Parts

1. The *integrand* is expressed as a product of two functions of x. Choose one of the functions to represent u and the other to represent dv in the IBP formula.
2. Differentiate u once with respect to x to obtain du.
3. Integrate dv once with respect to x to obtain v. Note that, by abuse of notation, we do not include the constant of integration, C, in the expression for v. (For indefinite integrals, the constant of integration will appear in the expression for the final answer.)
4. Substitute the information obtained from steps 1 through 3 into the IBP formula.

5. Finish the integration by computing uv and evaluating the integral $\int v\,du$ in the IBP formula. However, if the expression for $\int v\,du$ obtained in step 4 is more complicated than the original (e.g., the terms have higher exponents than in the original expression), your selections for u and dv were *incorrect*. Switch your original choices for u and dv, then start again.

Please note that because of the relationship between differentiation and integration, you can check to see if the answer obtained is correct. Simply differentiate your answer, and if it is equal to the original integrand, then there are no errors in your application of the IBP formula.

Example 1 Calculate $\int xe^{3x}dx$.

Solution

Step 1: Choose $u = x$ and $dv = e^{3x}dx$.

Step 2: $u = x$ implies $du = dx$.

Step 3: $dv = e^{3x}dx$ implies that $v = \int e^{3x}dx = \dfrac{1}{3}e^{3x}$. Do *not* include the constant of integration, C, in the expression for v.

Step 4: Substitute these expressions into the IBP formula $\int u\,dv = uv - \int v\,du$ to obtain

$$\int xe^{3x}dx = (x)\cdot\left(\frac{1}{3}e^{3x}\right) - \int \frac{1}{3}e^{3x}dx.$$

The expression for $\int v\,du$ is equal to $\int \dfrac{1}{3}e^{3x}dx$. This expression is *not* more complicated than the original expression, $\int xe^{3x}dx$, because it is not a product of two functions of x and also we can find its antiderivative easily.

Step 5: Since $\int \dfrac{1}{3}e^{3x}dx = \dfrac{1}{9}e^{3x} + C$, we have

$$\int xe^{3x}\,dx = (x)\cdot\left(\frac{1}{3}e^{3x}\right) - \int \frac{1}{3}e^{3x}\,dx$$

$$= \frac{1}{3}xe^{3x} - \frac{1}{9}e^{3x} + C. \qquad\blacksquare$$

Example 2 Calculate $\displaystyle\int_{1}^{e} \frac{\ln(x)}{\sqrt{x}}\,dx$.

Solution The definite integral can be rewritten as

$$\int_{1}^{e} \frac{\ln(x)}{\sqrt{x}}dx = \int_{1}^{e} x^{-1/2}\cdot\ln(x)\,dx = \int_{1}^{e}\ln(x)\cdot x^{-1/2}\,dx.$$

If we choose $dv = \ln(x)\,dx$, $u = x^{-1/2}$ and we use the IBP formula, the resulting expression is more complicated than in the original expression. So instead we take $u = \ln(x)$ and $dv = x^{-1/2}\,dx$. We then have

$$u = \ln(x), \quad dv = x^{-1/2}\,dx,$$

$$du = \frac{1}{x}\,dx, \quad v = \int x^{-1/2}\,dx = \frac{x^{-(1/2)+1}}{\frac{-1}{2}+1} = \frac{x^{1/2}}{\frac{1}{2}} = 2x^{1/2}.$$

Substitute this information into the formula $\int_{x=a}^{x=b} u\,dv = u \cdot v \Big|_{x=a}^{x=b} - \int_{x=a}^{x=b} v\,du$ to obtain

$$\int_1^e \ln(x) \cdot x^{-1/2}\,dx = \ln(x) \cdot 2x^{1/2} \Big|_{x=1}^{x=e} - \int_1^e 2x^{1/2} \cdot \frac{1}{x}\,dx$$

$$= \ln(x) \cdot 2x^{1/2} \Big|_{x=1}^{x=e} - \int_1^e 2x^{-1/2}\,dx.$$

We have converted the original difficult integral into a new integral that is easier to integrate. Since $\int 2x^{-1/2}dx = 4x^{1/2}$, we can now finish the computation:

$$\int_1^e \ln(x) \cdot x^{-1/2}\,dx = \ln(x) \cdot 2x^{1/2} \Big|_{x=1}^{x=e} - 4x^{1/2} \Big|_{x=1}^{x=e}$$

$$= (\ln(e) \cdot 2e^{1/2} - \ln(1) \cdot 2) - (4e^{1/2} - 4)$$

$$= 4 - 2e^{1/2}. \qquad \blacksquare$$

Sometimes it is necessary to implement the IBP formula more than once in the same example, as the next example illustrates.

Example 3 Calculate $\int_{\frac{\pi}{4}}^{\frac{\pi}{2}} x^2(\sin(x))\,dx.$

Solution Choose $u = x^2$ and $dv = \sin(x)\,dx$. We then have

$$u = x^2, \qquad dv = \sin(x)\,dx,$$

$$du = 2x\,dx, \qquad v = \int \sin(x)\,dx = -\cos(x).$$

Substitute this information into the formula $\int_{x=a}^{x=b} u\,dv = u \cdot v \Big|_{x=a}^{x=b} - \int_{x=a}^{x=b} v\,du$ to obtain

$$\int_{\frac{\pi}{4}}^{\frac{\pi}{2}} x^2(\sin(x))\,dx = (x^2)(-\cos(x)) \Big|_{\frac{\pi}{4}}^{\frac{\pi}{2}} - \int_{\frac{\pi}{4}}^{\frac{\pi}{2}} (-\cos(x))(2x)\,dx$$

$$= (-x^2 \cos(x)) \Big|_{\frac{\pi}{4}}^{\frac{\pi}{2}} + 2\int_{\frac{\pi}{4}}^{\frac{\pi}{2}} (x\cos(x))\,dx.$$

The choices we made for u and dv were correct because the degree of the x-term in the expression $2\int_{\frac{\pi}{4}}^{\frac{\pi}{2}} (x\cos(x))\,dx$ is less than the degree of the x-term in the original expression $\int_{\frac{\pi}{4}}^{\frac{\pi}{2}} (x^2\sin(x))\,dx$. Now we apply the IBP formula to the expression $2\int_{\frac{\pi}{4}}^{\frac{\pi}{2}} (x\cos(x))\,dx$. In the first application of the IBP formula, the function x^2 was selected for u, so we choose x for u in the second application of the IBP formula. *When applying the IBP formula more than once in the same example, be consistent in your choices.* Doing so is important.

Substitute these values:

$$u = x, \quad dv = \cos(x)\,dx,$$
$$du = dx, \quad v = \int \cos(x)\,dx = \sin(x)$$

into the formula $\int_{x=a}^{x=b} u\,dv = u \cdot v \Big|_{x=a}^{x=b} - \int_{x=a}^{x=b} v\,du$ to obtain

$$2\int_{\frac{\pi}{4}}^{\frac{\pi}{2}} (x\cos(x))\,dx = 2\left[x\sin(x)\Big|_{\frac{\pi}{4}}^{\frac{\pi}{2}} - \int_{\frac{\pi}{4}}^{\frac{\pi}{2}} \sin(x)\,dx \right]$$

$$= 2\left[x\sin(x)\Big|_{\frac{\pi}{4}}^{\frac{\pi}{2}} + \cos(x)\Big|_{\frac{\pi}{4}}^{\frac{\pi}{2}} \right].$$

Combine the results of the first and second applications of the IBP formula to obtain

$$\int_{\frac{\pi}{4}}^{\frac{\pi}{2}} x^2(\sin(x))dx = (-x^2\cos(x))\Big|_{\frac{\pi}{4}}^{\frac{\pi}{2}} + 2\int_{\frac{\pi}{4}}^{\frac{\pi}{2}} (x\cos(x))\,dx$$

$$= (-x^2\cos(x))\Big|_{\frac{\pi}{4}}^{\frac{\pi}{2}} + (2x\sin(x) + 2\cos(x))\Big|_{\frac{\pi}{4}}^{\frac{\pi}{2}}$$

$$= \left(\left(-\frac{\pi^2}{4} \right)(0) + \left(\frac{\pi^2}{16} \right)\left(\frac{\sqrt{2}}{2} \right) \right)$$

$$+ \left((2)\left(\frac{\pi}{2} \right)(1) + (2)(0) - (2)\left(\frac{\pi}{4} \right)\left(\frac{\sqrt{2}}{2} \right) - (2)\left(\frac{\sqrt{2}}{2} \right) \right)$$

$$= 0 + \frac{\sqrt{2}\pi^2}{32} + \pi - \frac{\sqrt{2}\pi}{4} - \sqrt{2} \approx 1.053. \qquad \blacksquare$$

Self-Test Exercises

1. Use Integration by Parts twice to calculate $\int x^2 e^{2x}\,dx$.
2. Calculate $\int_1^e \sqrt{x} \cdot \ln(2x)\,dx$.

7.2 Partial Fractions—Linear Factors

Important Concepts

- The Method of Partial Fractions for Distinct Linear Factors
- The Method of Partial Fractions for Repeated Linear Factors

A *rational function* is a function expressed as a fraction where both the numerator and denominator are polynomial functions. The *Method of Partial Fractions* is applied only to rational functions; it allows the rational functions to be expressed as the sum of functions that are easier to handle.

Simple Linear Building Blocks

A *linear function* is a degree 1 polynomial. A *simple linear building block* is defined as $\dfrac{A}{(x-a)}$, where both A and a represent constants. The integral of this building block is $\displaystyle\int \dfrac{A}{(x-a)}\,dx = A \cdot (\ln|x-a|) + C$. By abuse of notation, the constant of integration, C, is not included until the final expression of the solution.

The Method of Partial Fractions for Distinct Linear Factors

To integrate a rational function of the form $\dfrac{p(x)}{(x-a_1)(x-a_2)\cdots(x-a_k)}$, where $p(x)$ is a polynomial and each a_j is a distinct real number, perform the following steps:

1. Make sure that the degree of polynomial $p(x)$ in the numerator of the rational function is less than the degree of polynomial $q(x) = (x-a_1)(x-a_2)\cdots(x-a_k)$ in the denominator. If it is not, divide $p(x)$ by $q(x)$ and apply step 2 to the remainder function of this division.
2. Decompose the original rational function (or the remainder function as mentioned in step 1) into the form $\dfrac{A_1}{(x-a_1)} + \dfrac{A_2}{(x-a_2)} + \cdots + \dfrac{A_k}{(x-a_k)}$.
3. Solve for $A_1, A_2, \ldots, A_k$.

Example 1 Calculate $\displaystyle\int \dfrac{5x+2}{x^2-5x+6}\,dx$.

Solution

Step 1: The denominator, $x^2 - 5x + 6$, of the rational function $\dfrac{5x+2}{x^2-5x+6}$ is of degree 2 and can be factored into the product of two linear factors, $(x-2)$ and $(x-3)$. We notice that the degree of the numerator is 1, which is less than the degree of the denominator.

Step 2: The basic simple linear building blocks are $\dfrac{A}{(x-2)}$ and $\dfrac{B}{(x-3)}$. Note that

$$
\begin{aligned}
\frac{A}{(x-2)} + \frac{B}{(x-3)} &= \frac{A(x-3) + B(x-2)}{(x-2)(x-3)} \\
&= \frac{Ax - 3A + Bx - 2B}{(x-2)(x-3)} \\
&= \frac{(A+B)x + (-3A - 2B)}{(x-2)(x-3)}.
\end{aligned}
$$

Step 3: Now equate the original rational function and this last result to obtain

$$
\frac{5x+2}{x^2-5x+6} = \frac{(A+B)x + (-3A-2B)}{(x-2)(x-3)}.
$$

Because the two denominators are equal, the two fractions can be equal only if the two numerators are equal. After equating the coefficients of the x-terms and the constants, we conclude that $5 = A + B$ and $2 = -3A - 2B$. Either use the Elimination (Addition) Method or the

Substitution Method to solve these two equations simultaneously; we obtain $A = -12$ and $B = 17$. Therefore, $\dfrac{5x + 2}{x^2 - 5x + 6}$ becomes $\dfrac{-12}{(x - 2)} + \dfrac{17}{(x - 3)}$.

Now you can compute the original integral.

$$\int \frac{5x + 2}{x^2 - 5x + 6}\, dx = \int \frac{-12}{(x - 2)}\, dx + \int \frac{17}{(x - 3)}\, dx$$

$$= -12 \ln|x - 2| + 17 \ln|x - 3| + C$$

$$= \ln\left|\frac{(x - 3)^{17}}{(x - 2)^{12}}\right| + C$$ ∎

Self-Test Exercise

1. Calculate $\displaystyle\int \frac{3}{x^2 - x - 12}\, dx$.

Repeated Linear Building Blocks

A *repeated linear building block* is defined as $\dfrac{A}{(x - a)^m}$, where both A and a represent constants and m is an integer greater than 1. The integral of such a building block is

$$\int \frac{A}{(x - a)^m}\, dx = \frac{-A}{m - 1} \cdot \frac{1}{(x - a)^{m-1}} + C.$$

By abuse of notation, the constant of integration, C, is not included until the final expression of the solution.

The Method of Partial Fractions for Repeated Linear Factors

To integrate a rational function of the form

$$\frac{p(x)}{(x - a_1)^{m_1}(x - a_2)^{m_2} \cdots (x - a_k)^{m_k}}$$

where $p(x)$ is a polynomial, each a_j is a distinct real number, and each m_j is a positive integer greater than or equal to 1, perform the following steps:

1. Make sure that the degree of polynomial $p(x)$ in the numerator of the rational function is less than the degree of polynomial $q(x) = (x - a_1)^{m_1}(x - a_2)^{m_2} \cdots (x - a_k)^{m_k}$ in the denominator. If it is not, divide $p(x)$ by $q(x)$ and apply step 2 to the remainder term of this division.
2. For each of the factors $(x - a_j)^{m_j}$ in the denominator $q(x)$ where $m_j = 1$, the partial fraction decomposition has the form $\dfrac{A_1}{(x - a_j)^1}$, as in the case of a distinct linear factor. For each of the factors $(x - a_j)^{m_j}$ in the denominator $q(x)$ where $m_j \geq 2$, the partial fraction decomposition has the form $\dfrac{A_1}{(x - a_j)^1} + \dfrac{A_2}{(x - a_j)^2} + \cdots + \dfrac{A_{m_j}}{(x - a_j)^{m_j}}$.
3. Solve for each A_j that appears in the numerators of the partial fraction decomposition.

Example 2 Calculate $\displaystyle\int \frac{x^3 + 5x + 1}{(x - 1)(x + 1)^2}\, dx$.

Solution

Step 1: $p(x) = x^3 + 5x + 1$ and $q(x) = (x - 1)(x + 1)^2 = x^3 + x^2 - x - 1$. Because degree $p(x) = 3$ and degree $q(x) = 3$, divide the numerator $p(x)$ by the denominator $q(x)$ to obtain

$$\frac{p(x)}{q(x)} = 1 + \frac{(-x^2 + 6x + 2)}{(x - 1)(x + 1)^2}.$$

Step 2: Apply the Method of Partial Fractions for distinct linear factors and repeated linear factors to the remainder term $\dfrac{(-x^2 + 6x + 2)}{(x - 1)(x + 1)^2}$.

$$\frac{(-x^2 + 6x + 2)}{(x - 1)(x + 1)^2} = \frac{A_1}{(x - 1)} + \frac{B_1}{(x + 1)^1} + \frac{B_2}{(x + 1)^2}$$

$$= \frac{A_1(x + 1)^2 + B_1(x - 1)(x + 1) + B_2(x - 1)}{(x - 1)(x + 1)^2}$$

$$= \frac{(A_1 + B_1)x^2 + (2A_1 + B_2)x + (A_1 - B_1 - B_2)}{(x - 1)(x + 1)^2}.$$

By equating coefficients of like terms, we obtain the equations $\begin{cases} -1 = A_1 + B_1 \\ 6 = 2A_1 + B_2 \\ 2 = A_1 - B_1 - B_2 \end{cases}$. By

solving these equations simultaneously using either the Method of Elimination (Addition) or Substitution, we obtain the values $A_1 = \dfrac{7}{4}$, $B_1 = \dfrac{-11}{4}$, $B_2 = \dfrac{5}{2}$.

Therefore,

$$\int \frac{x^3 + 5x + 1}{(x - 1)(x + 1)^2}\, dx = \int 1\, dx + \int \frac{(-x^2 + 6x + 2)}{(x - 1)(x + 1)^2}\, dx$$

$$= \int 1\, dx + \int \frac{\frac{7}{4}}{(x - 1)}\, dx + \int \frac{\frac{-11}{4}}{(x + 1)}\, dx + \int \frac{\frac{5}{2}}{(x + 1)^2}\, dx.$$

We conclude that $\displaystyle\int \frac{x^3 + 5x + 1}{(x - 1)(x + 1)^2}\, dx = x + \frac{7}{4}\ln|x - 1| - \frac{11}{4}\ln|x + 1| - \frac{5}{2}\frac{1}{(x + 1)} + C.$ ■

Self-Test Exercises

2. Calculate $\displaystyle\int \frac{1}{x^2 - 9}\, dx.$

3. Calculate $\displaystyle\int \frac{x^4}{(x + 2)(x^2 + 2x + 1)}\, dx.$

7.3 Powers and Products of Trigonometric Functions

Important Concepts

- Squares of the Sine or Cosine Functions
- Reduction Formulas
- Odd Powers of Sine or Cosine
- Product of Sine and Cosine

Squares of the Sine or Cosine Functions

To integrate the functions $\sin^2(\theta)$ or $\cos^2(\theta)$, use the half-angle formulas

$$\sin^2(\theta) = \frac{1 - \cos(2\theta)}{2} \text{ and } \cos^2(\theta) = \frac{1 + \cos(2\theta)}{2}.$$

Example 1 Calculate $\int_0^{\frac{\pi}{6}} \cos^2(\theta)\, d\theta$.

Solution

$$\int_0^{\frac{\pi}{6}} \cos^2(\theta)\, d\theta = \int_0^{\frac{\pi}{6}} \frac{1 + \cos(2\theta)}{2}\, d\theta$$

$$= \frac{1}{2}\left(\theta + \frac{1}{2}\sin(2\theta)\right)\Big|_{\theta=0}^{\theta=\frac{\pi}{6}}$$

$$= \frac{1}{2}\left(\left(\frac{\pi}{6} + \frac{1}{2}\sin\left(\frac{\pi}{3}\right)\right) - \left(0 + \frac{1}{2}\sin(0)\right)\right)$$

$$= \frac{\pi}{12} + \frac{1}{4}\left(\frac{\sqrt{3}}{2}\right)$$

$$\approx .48$$

Self-Test Exercise

1. Calculate $\int_0^{\frac{\pi}{6}} \sin^2(\theta)\, d\theta$.

Reduction Formulas

When integrating trigonometric functions that are raised to powers greater than 2, sometimes it is useful to use *reduction formulas*. It is sometimes necessary to implement a reduction formula more than once in the same example. The reduction formula is usually used repeatedly until the power of the trigonometric function is reduced to 1 or 0. Here are several reduction formulas involving trigonometric functions.

$$\int \sin^n(x)\, dx = \frac{-1}{n}\sin^{n-1}(x)\cos(x) + \frac{n-1}{n}\int \sin^{n-2}(x)\, dx, \quad n \neq 0$$

$$\int \cos^n(x)\, dx = \frac{1}{n}\sin(x)\cos^{n-1}(x) + \frac{n-1}{n}\int \cos^{n-2}(x)\, dx, \quad n \neq 0$$

$$\int \sec^n(x)\, dx = \frac{1}{n-1}\sec^{n-2}(x)\tan(x) + \frac{n-2}{n-1}\int \sec^{n-2}(x), \quad n \neq 1$$

$$\int \tan^n(x)\, dx = \frac{1}{n-1}\tan^{n-1}(x) - \int \tan^{n-2}(x)\, dx, \quad n \neq 1$$

Example 2 Calculate $\int \tan^4(x)\,dx$.

Solution If we apply the reduction formula

$$\int \tan^n(x)\,dx = \frac{1}{n-1}\tan^{n-1}(x) - \int \tan^{n-2}(x)\,dx,$$

with $n = 4$, to $\int \tan^4(x)\,dx$, we have $\int \tan^4(x)\,dx = \frac{1}{3}\tan^3(x) - \int \tan^2(x)\,dx$. Apply the reduction formula once again to $\int \tan^2(x)\,dx$.

$$\int \tan^2(x)\,dx = \tan(x) - \int \tan^0(x)\,dx$$

$$= \tan(x) - \int 1\,dx$$

$$= \tan(x) - x$$

By combining the results of the two applications of the reduction formula, we obtain the solution:

$$\int \tan^4(x)\,dx = \frac{1}{3}\tan^3(x) - \int \tan^2(x)\,dx = \frac{1}{3}\tan^3(x) - \tan(x) + x + C. \qquad \blacksquare$$

Self-Test Exercise

2. Calculate $\int \tan^5(x)\,dx$. Note that $\int \tan(x)\,dx = -\ln|\cos(x)| + C$.

Odd Powers of the Sine and Cosine Functions

Although we can use reduction formulas to compute $\int \sin^n(x)\,dx$ and $\int \cos^n(x)\,dx$, in general, there is a quicker way to evaluate these integrals when the exponent n is an odd number. The identities $\sin^2(x) = 1 - \cos^2(x)$ and $\cos^2(x) = 1 - \sin^2(x)$ are used in the following technique. To integrate an odd power $\sin^{2k+1}(x)$ of $\sin(x)$, we first rewrite

$$\int \sin^{2k+1}(x)\,dx = \int \sin^{2k}(x)\sin(x)\,dx$$

$$= \int (\sin^2(x))^k \sin(x)\,dx$$

$$= \int (1 - \cos^2(x))^k \sin(x)\,dx,$$

then use the Method of Substitution with $u = \cos(x)$ and $du = -\sin(x)\,dx$.

Similarly, to integrate an odd power $\cos^{2k+1}(x)$ of $\cos(x)$, we first rewrite

$$\int \cos^{2k+1}(x)\,dx = \int \cos^{2k}(x)\cos(x)\,dx$$

$$= \int (\cos^2(x))^k \cos(x)\,dx$$

$$= \int (1 - \sin^2(x))^k \cos(x)\,dx,$$

then use the Method of Substitution with $u = \sin(x)$ and $du = \cos(x)\,dx$.

Example 3 Calculate $\int \sin^3(x)\, dx$.

Solution

$$\int \sin^3(x)\, dx = \int \sin^2(x) \sin(x)\, dx$$

$$= \int (1 - \cos^2(x)) \sin(x)\, dx$$

We make the substitution $u = \cos(x)$ and $du = -\sin(x)\, dx$ to obtain

$$\int \sin^3(x)\, dx = -\int (1 - u^2)\, du$$

$$= -u + \frac{u^3}{3} + C$$

$$= -\cos(x) + \frac{1}{3} \cos^3(x) + C.$$ ∎

Self-Test Exercise

3. Calculate $\int \cos^3(x)\, dx$.

Integrals Involving Both Sine and Cosine Functions

To solve $\int \sin^m(x) \cos^n(x)\, dx$ where m and n are positive integers, consider the following two cases.

- If *at least one* of m or n is odd, apply the techniques for odd powers of the sine or cosine function. If *both m and n are odd*, apply that technique to the term with *smaller* power.
- If *neither m* nor *n* is odd, then both are even. In this case, use the identity $\sin^2(x) + \cos^2(x) = 1$ to convert the integrand to a sum of even powers of sine or cosine. Then apply the appropriate reduction formula.

Example 4 Calculate $\int \cos^2(x) \sin^4(x)\, dx$.

Solution In this example, both the powers of the sine and cosine functions are even. We need to convert the integrand to all sines or all cosines. In this case, we'll convert to all even powers of sine by substituting $\cos^2(x) = 1 - \sin^2(x)$.

$$\int \cos^2(x) \sin^4(x)\, dx = \int (1 - \sin^2(x)) \cdot \sin^4(x)\, dx$$

$$= \int \sin^4(x)\, dx - \int \sin^6(x)\, dx$$

If we apply the reduction formula to $\int \sin^6(x)\, dx$, we have $\int \sin^6(x)\, dx = \dfrac{-1}{6} \sin^5(x) \cos(x) + \dfrac{5}{6} \int \sin^4(x)\, dx$. Thus

$$\int \cos^2(x) \sin^4(x)\, dx = \int \sin^4(x)\, dx - \left(\frac{-1}{6} \sin^5(x) \cos(x) + \frac{5}{6} \int \sin^4(x)\, dx \right)$$

$$= \frac{1}{6} \sin^5(x) \cos(x) + \frac{1}{6} \int \sin^4(x)\, dx.$$

We apply the reduction formula again, first to $\int \sin^4(x)\,dx$, then to $\int \sin^2(x)\,dx$:

$$
\begin{aligned}
\int \cos^2(x)\sin^4(x)\,dx &= \frac{1}{6}\sin^5(x)\cos(x) + \frac{1}{6}\left(\frac{-1}{4}\sin^3(x)\cos(x) + \frac{3}{4}\int \sin^2(x)\,dx\right) \\
&= \frac{1}{6}\sin^5(x)\cos(x) + \frac{-1}{24}\sin^3(x)\cos(x) + \frac{3}{24}\int \sin^2(x)\,dx \\
&= \frac{1}{6}\sin^5(x)\cos(x) + \frac{-1}{24}\sin^3(x)\cos(x) \\
&\quad + \frac{3}{24}\left(\frac{-1}{2}\sin(x)\cos(x) + \frac{1}{2}\int 1\,dx\right) \\
&= \frac{1}{6}\sin^5(x)\cos(x) + \frac{-1}{24}\sin^3(x)\cos(x) + \frac{-3}{48}\sin(x)\cos(x) + \frac{3}{48}x + C \\
&= \frac{1}{6}\sin^5(x)\cos(x) + \frac{-1}{24}\sin^3(x)\cos(x) \\
&\quad + \frac{-1}{16}\sin(x)\cos(x) + \frac{1}{16}x + C.
\end{aligned}
$$
∎

Example 5 Calculate $\int \cos^3(x)\sin^3(x)\,dx$.

Solution In this example, both the powers of the sine and cosine functions are odd and have the same degree. We can apply the technique "Odd Powers of the Sine and Cosine Functions" to either the sine or cosine term.

$$
\begin{aligned}
\int \cos^3(x)\sin^3(x)\,dx &= \int \cos^2(x)\cdot\cos(x)\cdot\sin^3(x)\,dx \\
&= \int (1 - \sin^2(x))\cdot\cos(x)\cdot\sin^3(x)\,dx
\end{aligned}
$$

Let $u = \sin(x)$, and $du = \cos(x)\,dx$. The integral becomes

$$
\begin{aligned}
\int \cos^3(x)\sin^3(x)\,dx &= \int (1 - u^2)\cdot u^3\,du \\
&= \int (u^3 - u^5)\,du \\
&= \frac{u^4}{4} - \frac{u^6}{6} + C \\
&= \frac{\sin^4(x)}{4} - \frac{\sin^6(x)}{6} + C.
\end{aligned}
$$
∎

Self-Test Exercises

4. Calculate $\int \sin^2(x)\cos^4(x)\,dx$.

5. Calculate $\int \sin^4(x)\cos^3(x)\,dx$.

7.4 Integrals Involving Quadratic Expressions

Important Concepts

- Trigonometric Substitution
- Quadratic Substitution Under a Radical
- Quadratic Substitution not Under a Radical

Trigonometric Substitution

The Method of Inverse (or Indirect) Substitution applies to the following quadratic expressions: $a^2 - x^2$, $a^2 + x^2$, and $x^2 - a^2$. The following table summarizes the trigonometric substitutions that are to be used.

Expression	Substitution	Result of Substitution	After Simplification
$a^2 - x^2$	$x = a \cdot \sin(\theta)$, $dx = a \cdot \cos(\theta)d\theta$	$a^2 \cdot (1 - \sin^2(\theta))$	$a^2 \cdot \cos^2(\theta)$
$a^2 + x^2$	$x = a \cdot \tan(\theta)$, $dx = a \cdot \sec^2(\theta)d\theta$	$a^2 \cdot (1 + \tan^2(\theta))$	$a^2 \cdot \sec^2(\theta)$
$x^2 - a^2$	$x = a \cdot \sec(\theta)$, $dx = a \cdot \sec(\theta)\tan(\theta)d\theta$	$a^2 \cdot (\sec^2(\theta) - 1)$	$a^2 \cdot \tan^2(\theta)$

Example 1 Calculate $\displaystyle\int \frac{x^2}{\sqrt{16 + 16x^2}}\, dx$.

Solution

$$\int \frac{x^2}{\sqrt{16 + 16x^2}}\, dx = \int \frac{x^2}{\sqrt{16(1 + x^2)}}\, dx$$

$$= \int \frac{x^2}{\sqrt{16}\sqrt{1 + x^2}}\, dx$$

$$= \frac{1}{4} \int \frac{x^2}{\sqrt{1 + x^2}}\, dx$$

According to the table, you should use the substitutions $x = \tan(\theta)$, $dx = \sec^2(\theta)\, d\theta$ to obtain

$$\frac{1}{4} \int \frac{x^2}{\sqrt{1 + x^2}}\, dx = \frac{1}{4} \int \frac{\tan^2(\theta)}{\sqrt{1 + \tan^2(\theta)}} \sec^2(\theta)\, d\theta$$

$$= \frac{1}{4} \int \frac{\tan^2(\theta)}{|\sec(\theta)|} \sec^2(\theta)\, d\theta.$$

Since $x = \tan(\theta)$ assumes every value in $(-\infty, \infty)$ as θ ranges from $-\dfrac{\pi}{2}$ to $\dfrac{\pi}{2}$, we can assume that θ lies inside the interval $\left(-\dfrac{\pi}{2}, \dfrac{\pi}{2}\right)$, and hence $|\sec(\theta)| = \sec(\theta)$. Now we have

$$
\begin{aligned}
\frac{1}{4} \int \frac{x^2}{\sqrt{1+x^2}}\, dx &= \frac{1}{4} \int \frac{\tan^2(\theta)}{\sec(\theta)} \sec^2(\theta)\, d\theta \\
&= \frac{1}{4} \int \tan^2(\theta) \sec(\theta)\, d\theta \\
&= \frac{1}{4} \int \frac{\sin^2(\theta)}{\cos^2(\theta)} \cdot \frac{1}{\cos(\theta)}\, d\theta \\
&= \frac{1}{4} \int \frac{1 - \cos^2(\theta)}{\cos^3(\theta)}\, d\theta \\
&= \frac{1}{4} \int \frac{1}{\cos^3(\theta)} - \frac{1}{\cos(\theta)}\, d\theta \\
&= \frac{1}{4} \int \sec^3(\theta)\, d\theta - \frac{1}{4} \int \sec(\theta)\, d\theta.
\end{aligned}
$$

Using the reduction formula for $\int \sec^3(\theta)\, d\theta$, the integral becomes

$$
\begin{aligned}
\frac{1}{4} \int \frac{x^2}{\sqrt{1+x^2}}\, dx &= \frac{1}{4}\left(\frac{1}{2}\sec(\theta)\tan(\theta) + \frac{1}{2}\int \sec(\theta)\, d\theta\right) - \frac{1}{4}\int \sec(\theta)\, d\theta \\
&= \frac{1}{8}\sec(\theta)\tan(\theta) - \frac{1}{8}\int \sec(\theta)\, d\theta \\
&= \frac{1}{8}\sec(\theta)\tan(\theta) - \frac{1}{8}(\ln|\sec(\theta) + \tan(\theta)|) + C.
\end{aligned}
$$

We complete the integration by resubstituting. Since $x = \tan(\theta)$, we have $\sec(\theta) = \sqrt{1+x^2}$. Therefore,

$$
\frac{1}{4} \int \frac{x^2}{\sqrt{1+x^2}}\, dx = \frac{1}{8}(\sqrt{1+x^2}) \cdot x - \frac{1}{8}(\ln|\sqrt{1+x^2} + x|) + C. \qquad \blacksquare
$$

Self-Test Exercises

1. Calculate $\displaystyle\int \sqrt{4 + 4x^2}\, dx$.

2. Calculate $\displaystyle\int \frac{dx}{\sqrt{9 - 9x^2}}$.

General Quadratic Expressions Under a Radical

If required, use the Method of Completing the Square to convert the quadratic expression to a sum or difference of squares to which trigonometric substitution can then be applied.

Example 2 Calculate $\displaystyle\int \frac{dt}{\sqrt{t^2 + 2t - 3}}$.

Solution

$$
\begin{aligned}
\int \frac{dt}{\sqrt{t^2 + 2t - 3}} &= \int \frac{dt}{\sqrt{t^2 + 2t + 1 - 4}} \\
&= \int \frac{dt}{\sqrt{(t+1)^2 - 2^2}}
\end{aligned}
$$

Use the substitution $x = t + 1$, $dx = dt$ to obtain $\displaystyle\int \frac{dt}{\sqrt{(t+1)^2 - 2^2}} = \int \frac{dx}{\sqrt{x^2 - 2^2}}$. Now, utilize the trigonometric substitution $x = 2\sec(\theta)$, $dx = 2\sec(\theta)\tan(\theta)\,d\theta$ to obtain

$$\int \frac{dx}{\sqrt{x^2 - 2^2}} = \int \frac{2\sec(\theta)\tan(\theta)\,d\theta}{\sqrt{4\sec^2(\theta) - 4}}$$
$$= \int \frac{2\sec(\theta)\tan(\theta)\,d\theta}{\sqrt{4}\sqrt{\sec^2(\theta) - 1}}$$
$$= \int \frac{2\sec(\theta)\tan(\theta)\,d\theta}{2|\tan(\theta)|}.$$

Because of the integrand $\dfrac{1}{\sqrt{x^2 - 2^2}}$, the value of $x = 2\sec(\theta)$ lies on the open interval $(2, \infty)$, and we can assume that θ lies on the interval $\left(0, \dfrac{\pi}{2}\right)$. As a result, $|\tan(\theta)| = \tan(\theta)$ and the integral becomes

$$\int \frac{dx}{\sqrt{x^2 - 2^2}} = \int \frac{2\sec(\theta)\tan(\theta)\,d\theta}{2\tan(\theta)}$$
$$= \int \sec(\theta)\,d\theta$$
$$= \ln|\sec(\theta) + \tan(\theta)| + C.$$

However, $x = 2\sec(\theta) \Rightarrow \dfrac{x}{2} = \sec(\theta) \Rightarrow \dfrac{t+1}{2} = \sec(\theta)$. Because $\dfrac{t+1}{2} = \sec(\theta) = \dfrac{hypotenuse}{adjacent}$ implies that in the associated right triangle, the hypotenuse is $t+1$, the adjacent side is 2, and using the Pythagorean Theorem, the opposite side is $\sqrt{(t+1)^2 - 2^2} = \sqrt{t^2 + 2t - 3}$. (See Figure 1.) Therefore, $\tan(\theta) = \dfrac{opposite}{adjacent} = \dfrac{\sqrt{t^2 + 2t - 3}}{2}$. Using this information, we obtain

$$\int \frac{dt}{\sqrt{t^2 + 2t - 3}} = \ln(|\sec(\theta) + \tan(\theta)|) + C$$
$$= \ln\left(\left|\frac{t+1}{2} + \frac{\sqrt{t^2 + 2t - 3}}{2}\right|\right) + C.$$

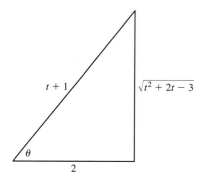

Figure 1

Self-Test Exercise

3. Calculate $\int \sqrt{t^2 + 4t + 3}\, dt$.

Quadratic Expressions Not Under a Radical Sign

Suppose $P(x)$ is a polynomial. To calculate $\int \dfrac{P(x)}{Ax^2 + Bx + C}\, dx$, do *not* use trigonometric substitution immediately. Instead, follow the suggested methodology, which is listed in these steps.

1. If the degree of the polynomial $P(x)$ is less than 2, which is the degree of the polynomial in the denominator, then proceed to step 2. Otherwise, divide the denominator into the numerator to obtain a quotient, $Q(x)$, and a remainder term, $\dfrac{R(x)}{Ax^2 + Bx + C}$ (i.e., degree of $R(x) < 2$). Apply step 2 to the remainder term.
2. If the degree of $P(x)$ is 0 (i.e., $P(x)$ is a constant) or if you have divided and the degree of $R(x)$ is 0 (i.e., $R(x)$ is a constant), proceed to step 3. If $P(x)$ or $R(x)$ has degree one, then separate the numerator into a multiple of $(2Ax + B)$ plus a constant K. Then integrate the expression $\dfrac{2Ax + B}{Ax^2 + Bx + C}$ by making the substitution $u = Ax^2 + Bx + C$, $du = (2Ax + B)\, dx$, and we have

$$\int \frac{2Ax + B}{Ax^2 + Bx + C}\, dx = \ln|Ax^2 + Bx + C| + \text{Constant}.$$

3. Integrate the expression $\dfrac{K}{Ax^2 + Bx + C}$ by completing the square in the denominator.

Example 3 Calculate $\displaystyle\int \frac{2x - 6}{x^2 - 4x + 6}\, dx$.

Solution

$$\int \frac{2x - 6}{x^2 - 4x + 6}\, dx = \int \frac{(2x - 4) - 2}{x^2 - 4x + 6}\, dx$$

$$= \int \frac{2x - 4}{x^2 - 4x + 6}\, dx - \int \frac{2}{x^2 - 4x + 6}\, dx$$

$$= \ln(|x^2 - 4x + 6|) - 2\int \frac{dx}{(x^2 - 4x + 4) + (\sqrt{2})^2}$$

$$= \ln(|x^2 - 4x + 6|) - 2\int \frac{dx}{(x - 2)^2 + (\sqrt{2})^2}$$

Let $u = x - 2 \Rightarrow du = dx$. Then

$$-2\int \frac{dx}{(x - 2)^2 + (\sqrt{2})^2} = -2\int \frac{du}{u^2 + (\sqrt{2})^2}.$$

Using the Method of Inverse Substitution, let $u = \sqrt{2}\tan(\theta)$, $du = \sqrt{2}\sec^2(\theta)\,d\theta$. Then

$$
\begin{aligned}
-2\int \frac{du}{u^2 + (\sqrt{2})^2} &= -2\int \frac{\sqrt{2}\sec^2(\theta)\,d\theta}{2\tan^2(\theta) + 2} \\
&= -2\int \frac{\sqrt{2}\sec^2(\theta)\,d\theta}{2\sec^2(\theta)} \\
&= -\int \sqrt{2}\,d\theta \\
&= -\sqrt{2}\theta + C \\
&= -\sqrt{2}\left(\arctan\left(\frac{u}{\sqrt{2}}\right)\right) + C \\
&= -\sqrt{2}\left(\arctan\left(\frac{x-2}{\sqrt{2}}\right)\right) + C.
\end{aligned}
$$

Therefore,

$$
\begin{aligned}
\int \frac{2x-6}{x^2 - 4x + 6}\,dx &= \ln(|x^2 - 4x + 6|) - 2\int \frac{dx}{(x-2)^2 + (\sqrt{2})^2} \\
&= \ln(|x^2 - 4x + 6|) - \sqrt{2}\left(\arctan\left(\frac{x-2}{\sqrt{2}}\right)\right) + C. \quad \blacksquare
\end{aligned}
$$

Self-Test Exercise

4. Calculate $\displaystyle\int \frac{2}{2x^2 + 6x + 32}\,dx$.

7.5 Irreducible Quadratic Factors

Important Concepts

- Quadratic Building Blocks
- Method of Partial Fractions

Simple Quadratic Building Blocks

Simple quadratic building blocks have the form $\dfrac{Bx + C}{x^2 + bx + c}$, where we assume that the denominator cannot be factored (i.e., it is irreducible). If $(b^2 - 4ac) < 0$, then the quadratic $ax^2 + bx + c$ is irreducible.

Repeated Quadratic Building Blocks

Repeated quadratic building blocks have the form $\dfrac{Bx + C}{(x^2 + bx + c)^n}$, where n is an integer greater than one and $(x^2 + bx + c)$ in the denominator cannot be factored.

The Method of Partial Fractions

To integrate a rational function of the form

$$\frac{p(x)}{(x-a_1)^{m_1}\cdots(x-a_k)^{m_k}(x^2+b_1x+c_1)^{n_1}\cdots(x^2+b_Lx+c_L)^{n_L}},$$

perform the following steps.

1. Be sure that the degree of numerator $p(x)$ is less than the degree of the denominator; if it is not, divide the denominator into the numerator.
2. Factor the denominator into linear and irreducible quadratic factors. Make sure that the quadratic factors $x^2+b_jx+c_j$ cannot be factored into linear factors with real coefficients.
3. For each of the factors $(x-a_j)^{m_j}$ in the denominator of the rational function being considered, the partial fractions decomposition must contain terms of the form

$$\frac{A_1}{(x-a_j)^1}+\frac{A_2}{(x-a_j)^2}+\cdots+\frac{A_{m_j}}{(x-a_j)^{m_j}}.$$

4. For each of the factors $\left(x^2+b_jx+c_j\right)^{n_j}$ in the denominator of the integrand being considered, the partial fractions decomposition must contain terms of the form

$$\frac{B_1x+C_1}{(x^2+b_jx+c_j)}+\frac{B_2x+C_2}{(x^2+b_jx+c_j)^2}+\cdots+\frac{B_{n_j}x+C_{n_j}}{(x^2+b_jx+c_j)^{n_j}}.$$

Example 1 Calculate $\displaystyle\int\frac{2x^2+5}{(x-1)(x^2+1)}dx$.

Solution The degree of the polynomial in the numerator is 2, which is less than 3, the degree of the polynomial in the denominator. Furthermore, the denominator is equal to the product of a linear factor, $(x-1)$, and an irreducible quadratic factor, (x^2+1). Note that $a=1$, $b=0$, and $c=1$, for the quadratic ax^2+bx+c, demonstrates that (x^2+1) is irreducible because $b^2-4ac=(0)^2-4(1)(1)=-4<0$. Therefore,

$$\int\frac{2x^2+5}{(x-1)(x^2+1)}dx=\int\frac{A}{(x-1)}dx+\int\frac{Bx+C}{(x^2+1)}dx$$

$$=\int\frac{A(x^2+1)+(Bx+C)(x-1)}{(x-1)(x^2+1)}dx$$

$$=\int\frac{(A+B)x^2+(-B+C)x+(A-C)}{(x-1)(x^2+1)}dx.$$

Equating coefficients of like terms, we have the equations $\begin{cases} A + B = 2 \\ -B + C = 0 \\ A - C = 5 \end{cases}$. By solving these

simultaneous equations, we find $A = \dfrac{7}{2}$, $B = \dfrac{-3}{2}$, $C = \dfrac{-3}{2}$.

$$\int \frac{2x^2 + 5}{(x-1)(x^2+1)}\,dx = \int \frac{\dfrac{7}{2}}{(x-1)}\,dx + \int \frac{\dfrac{-3}{2}x - \dfrac{3}{2}}{(x^2+1)}\,dx$$

$$= \frac{7}{2}\ln(|x-1|) - \frac{3}{4}\int \frac{2x}{(x^2+1)}\,dx - \frac{3}{2}\int \frac{dx}{(x^2+1)}$$

$$= \frac{7}{2}\ln(|x-1|) - \frac{3}{4}\ln(|x^2+1|) - \frac{3}{2}\arctan(x) + C$$

Self-Test Exercises

1. Calculate $\displaystyle\int \frac{3x^2 + 4x}{(x+1)(x^2+1)}\,dx$.

2. Calculate $\displaystyle\int \frac{2x^2 - 4x + 2}{x^2 - x + 1}\,dx$.

Applications of the Integral

8.1 Volumes

Important Concepts

- Disk Method
- Method of Washers
- Method of Cylindrical Shells

Volumes by Slicing

In the following material, several instances of calculating volume by the Method of Slicing will be presented. In each case, the following methodology will be used.

1. Identify the shape of each slice.
2. Identify the independent variable that gives the position of each slice.
3. Write an expression, in terms of the independent variable, that describes the cross-sectional area of each slice.
4. Identify the interval $[a, b]$ over which the independent variable ranges.
5. With respect to the independent variable from step 2, integrate the expression for the cross-sectional area from step 3 over the interval $[a, b]$ from step 4.

The Disk Method

Consider a cross section of a solid created by the intersection of a solid and a plane perpendicular to the x-axis. If $A(x_j)$ represents the area of the jth cross section and Δx represents its thickness, then the volume V obtained by this Method of Slicing over the interval $I = [a, b]$ is given by

$$V = \lim_{N \to \infty} \sum_{j=1}^{N} A(x_j)\Delta x = \int_{x=a}^{x=b} A(x)\, dx.$$

Please note that we will often use the notation

$$\int_{x=a}^{x=b} f(x)\, dx \text{ or } \int_{y=a}^{y=b} g(y)\, dy$$

to emphasize which variable the limit of integration is.

Example 1 Using the Method of Slicing, calculate the volume of a cube with 3 m sides (Figure 1).

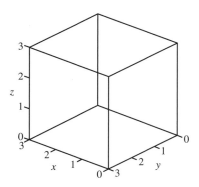

Figure 1

Solution The jth cross section representing the intersection of a plane perpendicular to the x-axis and a cube with 3 m sides is a square with 3 m sides. Therefore, its area $A(x_j)$ is 9 square meters. As such, its volume is

$$V = \lim_{N \to \infty} \sum_{j=1}^{N} A(x_j) \Delta x$$

$$= \int_{x=0}^{x=3} 9 \, dx$$

$$= 9x \Big|_{x=0}^{x=3}$$

$$= 9(3 - 0)$$

$$= 27 \text{ cubic meters.}$$

This answer agrees with the geometric analysis of volume = length × width × height = $3 \times 3 \times 3 = 27$ cubic meters. ∎

Solids of Revolution

A solid that is obtained by rotating a region R in the xy-plane about a line in the xy-plane is called a *solid of revolution*. Generic representations of region R and the solid of revolution that results from rotating region R about the x-axis appear in Figures 2 and 3, respectively.

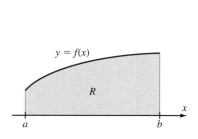

Figure 2

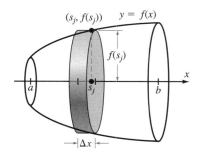

Figure 3

The jth cross section in Figure 3 is a circle; therefore, its area is equal to πr^2, where $r = f(x)$. As such, the volume of the solid of revolution is given by $V = \int_{x=a}^{x=b} \pi(f(x))^2 dx$. This is formalized in Theorem 1.

Theorem 1

Method of Disks (Rotation about the x-axis)   Suppose that f *is* a nonnegative, continuous function on the interval $[a, b]$. Let R denote the region of the xy-plane that is bounded by the graph of f, the x-axis, and the vertical lines $x = a$ and $x = b$. The volume V of the solid, which is obtained by rotating R about the x-axis, is given by $V = \pi \int_{x=a}^{x=b} (f(x))^2 dx$.

Example 2 Calculate the volume of the solid of revolution generated by rotating about the x-axis the region of the xy-plane that is bounded by $y = 9 - x^2$, $y = 0$, $x = 0$, and $x = 3$.

Solution Region R, the area shaded in gray, appears in Figure 4.

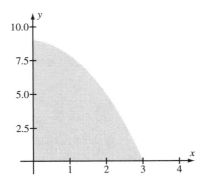

Figure 4

The radius of the solid of revolution is given by $y = 9 - x^2$ and the interval $[a, b] = [0, 3]$. Implementing the formula given in Theorem 1 results in

$$V = \pi \int_{x=a}^{x=b} (f(x))^2 dx$$

$$= \pi \int_{x=0}^{x=3} (9 - x^2)^2 dx$$

$$= \pi \int_{x=0}^{x=3} (81 - 18x^2 + x^4)\, dx$$

$$= \pi \cdot \left(81x - \frac{18x^3}{3} + \frac{x^5}{5} \right) \Bigg|_{x=0}^{x=3}$$

$$= 129.6\pi$$

$$\approx 407.15 \text{ cubic units.} \qquad \blacksquare$$

Sometimes, a planar region R is rotated about the y-axis. A generic representation appears in Figure 5, and the formula to calculate volume is given in Theorem 2.

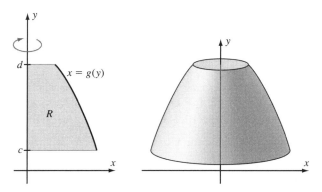

Figure 5

Theorem 2

Method of Disks (Rotating about the y-axis)   Suppose that $x = g(y)$ is a nonnegative, continuous function on the interval $c \leq y \leq d$. Let R denote the region of the xy-plane bounded by the graph of $x = g(y)$, the y-axis, and the horizontal lines $y = c$ and $y = d$. The volume V of the solid generated by rotating R about the y-axis is given by $V = \pi \int_{y=c}^{y=d} (g(y))^2 \, dy$.

Example 3 Calculate the volume of the solid of revolution generated by rotating about the y-axis the region of the xy-plane that is bounded by $y = 9 - x^2$, $y = 0$, $x = 0$, and $x = 3$.

Solution Region R, the area shaded in gray, appears in Figure 6.

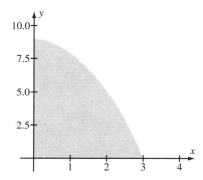

Figure 6

Note that for the graph $y = 9 - x^2$, we need to express x as a function of y:

$$x = g(y) = \sqrt{9 - y}.$$

Notice that $x = 0$ corresponds to $y = 9$ and $x = 3$ corresponds to $y = 0$.
Therefore, utilizing the formula given in Theorem 2, we obtain:

$$V = \pi \int_{y=c}^{y=d} (g(y))^2 \, dy$$

$$= \pi \int_{y=0}^{y=9} \left(\sqrt{9 - y}\right)^2 \, dy$$

$$= \pi \int_{y=0}^{y=9} (9 - y)\, dy$$

$$= \pi \cdot \left(9y - \frac{y^2}{2}\right)\Bigg|_{y=0}^{y=9}$$

$$= \frac{81\pi}{2}$$

$$\approx 127.24 \text{ cubic units.}$$

The volume of the solid of revolution obtained by rotating the region about the x-axis is ≈ 56.55 cubic units, and the volume of the solid of revolution obtained by rotating the *same* region about the y-axis is ≈ 127.24 cubic units.

Self-Test Exercise

1. Region R is bounded above by $y = \dfrac{1}{x}$, below by $y = 0$, and on the sides by $x = \dfrac{1}{2}$ and $x = 2$. Calculate the volume of the solid of revolution that results when region R is revolved about the x-axis.

The Method of Washers

Sometimes we want to generate a solid of revolution by rotating the region *between* two graphs. This method is formalized by Theorem 3.

Theorem 3 Suppose that U and L are nonnegative, continuous functions on the interval $[a, b]$ with $L(x) \le U(x)$ for each x in this interval. Let R denote the region of the xy-plane that is bounded above by the graph of U, below by the graph of L, and on the sides by the vertical lines $x = a$ and $x = b$. The volume V of the solid obtained by rotating R about the x-axis is given by

$$V = \pi \int_{x=a}^{x=b} \left((U(x))^2 - (L(x))^2\right) dx.$$

Example 4 Let R be the region of the xy-plane that is bounded above by $y = \sqrt{x}$, below by $y = x^2$, on the left by the vertical line $x = 0$, and on the right by the vertical line $x = 1$. Calculate the volume of the solid of revolution that is generated when R is rotated about the x-axis.

Solution Region R is the region bounded by the graph of $y = \sqrt{x}$ (which appears as a dotted curve) and the graph of $y = x^2$ (which appears as a solid curve) in the graph in Figure 7.

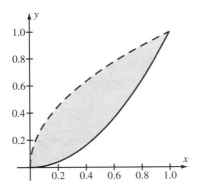

Figure 7

We see from Figure 7 that $U(x) = \sqrt{x}$ and $L(x) = x^2$. The two curves intersect at $(0, 0)$ and $(1, 1)$, therefore $a = 0, b = 1$. The graph of region R demonstrates that the conditions stated in Theorem 3 are satisfied. Therefore, the volume of the solid of revolution obtained by rotating region R about the x-axis is given by

$$V = \pi \int_{x=a}^{x=b} ((U(x))^2 - (L(x))^2)\,dx$$

$$= \pi \int_{x=0}^{x=1} ((\sqrt{x})^2 - (x^2)^2)\,dx$$

$$= \pi \int_{x=0}^{x=1} (x - x^4)\,dx$$

$$= \pi \left(\frac{x^2}{2} - \frac{x^5}{5} \right) \Big|_{x=0}^{x=1}$$

$$= \frac{3\pi}{10}$$

$$\approx .94 \text{ cubic units.} \qquad \blacksquare$$

Please note that the following equation:

$$\pi \int_{x=0}^{x=1} ((\sqrt{x})^2 - (x^2)^2)\,dx = \pi \int_{x=0}^{x=1} x\,dx - \pi \int_{x=0}^{x=1} x^4\,dx$$

can be interpreted using these steps.

1. Let R_1 be the region bounded by the graph of $y = \sqrt{x}$, $x = 0$, $x = 1$, $y = 0$, and let V_1 represent the volume of the solid of revolution S_1 obtained by rotating R_1 about the x-axis. (The region R_1 in the xy-plane and the solid S_1 are shown in Figures 8 and 9, respectively.)

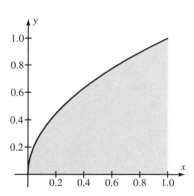

Figure 8

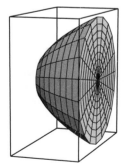

Figure 9

2. Let R_2 be the region bounded by the graph of $y = x^2$, $x = 0$, $x = 1$, $y = 0$, and let V_2 represent the volume of the solid of revolution S_2 obtained by rotating R_2 about the x-axis. (The region R_2 in the xy-plane and the solid S_2 are shown in Figures 10 and 11, respectively.)

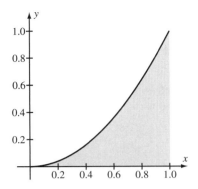

Figure 10

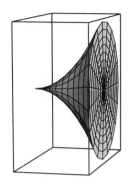

Figure 11

3. Volume $V_3 = V_2 - V_1$ represents the volume of the solid of revolution S_3, which results when S_2 is detached from S_1.
4. Volume V_3 is equal to volume V, which was calculated in Example 4.

Rotation About a Line That Is Not a Coordinate Axis

You can calculate the volume of a solid that results from rotating a region R about a line that is not a coordinate axis, as Example 5 illustrates.

Example 5 Calculate the volume of the solid of revolution generated by rotating about the line $y = 9$ the region of the xy-plane that is bounded by $y = 9 - x^2$, $y = 0$, $x = 0$, and $x = 3$.

Solution Region R (the area shaded in gray) and the graph of $y = 9$ (the thick horizontal line) appear in Figures 12 and 13.

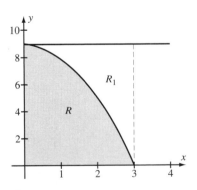

Figure 12

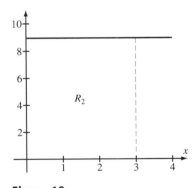

Figure 13

Let R_1 be the region bounded by $y = 9$, $y = 9 - x^2$, $x = 0$, and $x = 3$. (See Figure 12.) Also, let R_2 be the region bounded by $y = 9$, $y = 0$, $x = 0$, and $x = 3$. (See Figure 13.) Denote the solids of revolution obtained by revolving the regions R_1 and R_2 about the line $y = 9$ as S_1 and S_2, respectively. If you analyze the geometry of this example, you can see that the solid of revolution obtained by revolving region R about the line $y = 9$ is equivalent to the solid obtained by detaching the solid of revolution S_1 from the solid of revolution S_2.

We note the following:

1. In region R_2, the radius r_2 is equal to 9 and the solid of revolution S_2 is a cylinder. Let V_2 be the volume of S_2.
2. In region R_1, the radius r_1 is equal to $9 - y$ where $y = 9 - x^2$. S_1 is the solid of revolution, which results when R_1 is rotated about the line $y = 9$. S_1 resembles a megaphone with curved sides. Let V_1 be the volume of S_1.
3. In both cases, $a = 0$ and $b = 3$.
4.

$$V_2 = \pi \int_{x=0}^{x=3} ((r_2(x))^2)\, dx$$

$$= \pi \int_{x=0}^{x=3} (9)^2\, dx$$

$$= 81\pi \left(x \Big|_{x=0}^{x=3} \right)$$

$$= 243\pi$$

$$\approx 763.41 \text{ cubic units}$$

$$V_1 = \pi \int_{x=0}^{x=3} ((r_1(x))^2)\, dx$$

$$= \pi \int_{x=0}^{x=3} (9 - (9 - x^2))^2\, dx$$

$$= \pi \int_{x=0}^{x=3} (x^2)^2\, dx$$

$$= \pi \int_{x=0}^{x=3} x^4\, dx$$

$$= \pi \left(\frac{x^5}{5} \right) \Big|_{x=0}^{x=3}$$

$$= \frac{243\pi}{5} \approx 152.68 \text{ cubic units}$$

5.
$$V = V_2 - V_1$$

$$= 243\pi - \frac{243\pi}{5}$$

$$= \frac{4}{5}(243\pi)$$

$$\approx 610.73 \text{ cubic units}$$

■

Self-Test Exercise

2. Region R is bounded above by $y = 2x$, below by $y = 0$, and on the sides by $x = 1$ and $x = 4$. Calculate the volume of the solid of revolution, which results when region R is revolved about the line $x = 0$.

The Method of Cylindrical Shells

This method depends on observation of the geometry of the figures. Theorems 4 and 5 formalize this method.

Theorem 4

Method of Cylindrical Shells (Rotation about the y-axis) Let f be a nonnegative, continuous function on an interval $[a, b]$ of nonnegative numbers. If V denotes the volume of the solid generated when the region below the graph of f and above the interval $[a, b]$ is rotated about the y-axis, then $V = 2\pi \int_{x=a}^{x=b} x \cdot f(x)\, dx$.

Theorem 5

Method of Cylindrical Shells (Rotation about the x-axis)   Let g be a nonnegative, continuous function on an interval $[c, d]$ of nonnegative numbers. The volume of the solid generated when the region bounded by $x = g(y)$, the y-axis, $y = c$, and $y = d$ is rotated about the x-axis is

$$V = 2\pi \int_{y=c}^{y=d} y \cdot g(y)\, dy.$$

Example 6 Calculate the volume of the solid obtained by rotating the region R bounded by $y = x^5$, $x = 0$, $x = 1$, and $y = 0$ about the y-axis.

Solution Region R is represented by the gray area in Figure 14.

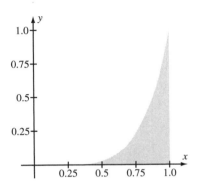

Figure 14

We see from Figure 14 that $y = x^5 = f(x)$, $a = 0$, and $b = 1$. The graph establishes that the conditions of Theorem 4 are met; therefore,

$$V = 2\pi \int_{x=a}^{x=b} x \cdot f(x)\, dx$$

$$= 2\pi \int_{x=0}^{x=1} x \cdot x^5\, dx$$

$$= 2\pi \int_{x=0}^{x=1} x^6\, dx$$

$$= 2\pi \left. \frac{x^7}{7} \right|_{x=0}^{x=1}$$

$$= \frac{2\pi}{7}$$

$$\approx .90 \text{ cubic units.}$$

Self-Test Exercises

3. Let R be the region of the xy-plane that is bounded above by $y = \sqrt{x}$, below by $y = x^2$, on the left by the vertical line $x = 0$, and on the right by the vertical line $x = 1$. Calculate the volume of the solid of revolution that is generated when R is rotated about the y-axis.

4. Calculate the volume of the solid obtained by rotating the region R bounded by $y = x^5$, $x = 0$, $x = 1$, and $y = 0$ about the x-axis, using the method of cylindrical shells.

8.2 Arc Length and Surface Area

Important Concepts

- Arc Length
- Surface Area

Arc Length

If f has a continuous derivative on an interval containing $[a, b]$, then the *arc length L of the graph of f over the interval $[a, b]$* is given by $L = \int_{x=a}^{x=b} \sqrt{1 + (f'(x))^2}\,dx$.

If g' is continuous, then the *arc length L of the graph of $x = g(y)$ for $c \le y \le d$* is given by $L = \int_{y=c}^{y=d} \sqrt{1 + (g'(y))^2}\,dy$.

Example 1 Calculate the length of the curve $y = 5x$ from $x = 1$ to $x = 9$.

Solution To calculate the arc length L of the curve $y = 5x = f(x)$ from $x = 1$ to $x = 9$, as shown in Figure 1, we use the formula

$$L = \int_{x=a}^{x=b} \sqrt{1 + (f'(x))^2}\,dx$$

$$= \int_{x=1}^{x=9} \sqrt{1 + (5)^2}\,dx$$

$$= \int_{x=1}^{x=9} \sqrt{26}\,dx$$

$$= \sqrt{26}x \Big|_{x=1}^{x=9}$$

$$= \sqrt{26}\,(9 - 1)$$

$$= 8\sqrt{26}$$

$$\approx 40.79 \text{ units.}$$

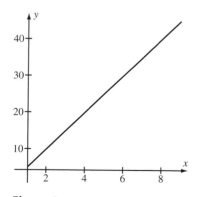

Figure 1

This result agrees with the length of the line segment from point $A = (1, 5)$ to point $B = (9, 45)$, which is calculated using the distance formula:

$$d = \sqrt{(x_1 - x_2)^2 + (y_1 - y_2)^2}$$
$$= \sqrt{(1 - 9)^2 + (5 - 45)^2}$$
$$= \sqrt{64 + 1600}$$
$$= \sqrt{1664}$$
$$\approx 40.79 \text{ units.}$$ ∎

Example 2 Calculate the arc length of the curve $y = 1 + x^{2/3}$ from $x = 1$ to $x = 8$.

Solution To calculate the arc length of the curve $y = 1 + x^{2/3}$ from $x = 1$ to $x = 8$, which is shown in Figure 2, we use the formula $L = \int_{y=c}^{y=d} \sqrt{1 + (g'(y))^2}\, dy$. The curve $y = 1 + x^{2/3}$ can be rewritten as:

$$y = 1 + x^{2/3} \Leftrightarrow (y - 1)^{3/2} = x = g(y).$$

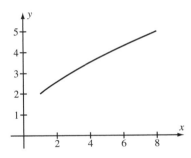

Figure 2

This implies that $g'(y) = \dfrac{3}{2}(y - 1)^{1/2}$ so

$$1 + (g'(y))^2 = 1 + \frac{9}{4}(y - 1) = \frac{9}{4}y - \frac{5}{4} = \frac{1}{4}(9y - 5).$$

$y = 1 + x^{2/3}$ and $x = 1$ imply that $y = 1 + (1)^{2/3} = 1 + 1 = 2.$ $y = 1 + x^{2/3}$ and $x = 8$ imply that $y = 1 + 8^{2/3} = 1 + 4 = 5.$ Therefore,

$$L = \int_{y=c}^{y=d} \sqrt{1 + (g'(y))^2}\, dy$$
$$= \int_{y=2}^{y=5} \sqrt{\frac{1}{4}(9y - 5)}\, dy$$
$$= \frac{1}{2} \int_{y=2}^{y=5} \sqrt{9y - 5}\, dy.$$

We implement the Method of Integration by Substitution with $u = 9y - 5$ and $du = 9\,dy$. Therefore,

$$\frac{1}{2} \int_{y=2}^{y=5} \sqrt{9y - 5}\, dy = \left(\frac{1}{2}\right) \cdot \left(\frac{1}{9}\right) \int_{u=13}^{u=40} u^{1/2}\,du$$

$$= \left(\frac{1}{2}\right)\left(\frac{1}{9}\right)\left(\frac{2}{3}\right) u^{3/2}\Big|_{u=13}^{u=40}$$

$$= \frac{1}{27}(40^{3/2} - 13^{3/2})$$

$$\approx 7.63 \text{ units.} \qquad\blacksquare$$

Self-Test Exercise

1. Calculate the arc length of the curve $y^2 = x^3$ from $x = 4$ to $x = 8$.

Surface Area

Let function f be nonnegative (i.e., $f \geq 0$) with a continuous derivative on interval $[a, b]$. If the graph of f is rotated about the x-axis, then a surface of revolution is generated (see Figure 3).

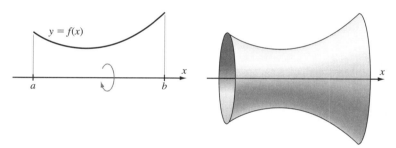

Figure 3

Surface Area Formulas

If a nonnegative function f has a continuous derivative on an interval containing $[a, b]$, then the *surface area* of the surface of revolution obtained when the graph of f over $[a, b]$ is rotated about the x-axis is given by $2\pi \int_{x=a}^{x=b} f(x) \cdot \sqrt{1 + (f'(x))^2}\, dx$.

If a nonnegative function g has a continuous derivative on an interval containing $[c, d]$, then the *surface area* of the surface of revolution obtained when the graph of $x = g(y)$ for $c \leq y \leq d$ is rotated about the y-axis is given by $2\pi \int_{y=c}^{y=d} g(y) \cdot \sqrt{1 + (g'(y))^2}\, dy$.

Example 3 Calculate the surface area of the solid of revolution obtained when the curve $y = x^{1/3}$ from $x = 1$ to $x = 8$ is rotated about the y-axis.

Solution
$$y = x^{1/3} \Leftrightarrow y^3 = x = g(y).$$
$$g'(y) = 3y^2 \Rightarrow 1 + (g'(y))^2 = 1 + 9y^4.$$

If $y = x^{1/3}$ and $x = 1$, then $y = 1$. If $y = x^{1/3}$ and $x = 8$, then $y = 2$. Using the formula $2\pi \int_{y=c}^{y=d} g(y) \cdot \sqrt{1 + (g'(y))^2}\, dy$, we see that the surface area is equal to $2\pi \int_{y=1}^{y=2} y^3 \cdot \sqrt{1 + 9y^4}\, dy$. Using the Method of Integration by Substitution with $u = 1 + 9y^4$ and $du = 36y^3 dy$, we obtain

$$2\pi \int_{y=1}^{y=2} y^3 \cdot \sqrt{1 + 9y^4}\, dy = \frac{2\pi}{36} \int_{u=10}^{u=145} u^{1/2} du$$

$$= \left(\frac{\pi}{18}\right)\left(\frac{2}{3}\right) u^{3/2}\Big|_{u=10}^{u=145}$$

$$= \frac{\pi}{27}(145^{3/2} - 10^{3/2})$$

$$\approx 199.48 \text{ square units.} \qquad\blacksquare$$

Self-Test Exercise

2. Set up, but do not evaluate, the integral for finding the surface area of the surface of revolution obtained when the curve $y = x^2$ from $x = 2$ to $x = 4$ is rotated about the x-axis.

8.3 The Average Value of a Function

Important Concepts

- Basic Technique
- Probability Theory Application

Basic Technique

The underlying concept to determine the average value of a function involves Riemann Sums. If f is a continuous function with a domain that contains the interval $[a, b]$, we take a uniform partition of the interval with $\Delta x = \dfrac{b - a}{N}$. If we add up the values of f at the right endpoints of the partition (see Figure 1) and divide by the number of points, then the resulting expression is an approximation to the average value f_{ave} of f:

$$f_{ave} \approx \frac{f(x_1) + f(x_2) + \cdots + f(x_{N-1}) + f(x_N)}{N}$$

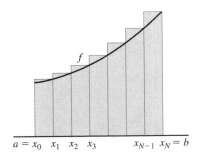

Figure 1

This expression can be rewritten as:

$$f_{ave} \approx \frac{(f(x_1) + f(x_2) + \cdots + f(x_{N-1}) + f(x_N)) \cdot \Delta x}{N \cdot \Delta x} = \frac{\sum\limits_{j=1}^{N} f(x_j) \Delta x}{b - a}.$$

The accuracy of f_{ave} is directly related to the value of N. As $N \to \infty$, the Riemann Sum $\sum\limits_{j=1}^{N} f(x_j) \Delta x \to \int\limits_{x=a}^{x=b} f(x)\,dx$; this implies $\frac{1}{b-a} \sum\limits_{j=1}^{N} f(x_j) \Delta x \to \frac{1}{b-a} \int\limits_{x=a}^{x=b} f(x)\,dx$. Our findings are summarized in the following definition.

Suppose that f is a Riemann integral function on the interval $[a, b]$. The *average value of function f on the interval* $[a, b]$ is the number $f_{ave} = \frac{1}{b-a} \int\limits_{x=a}^{x=b} f(x)\,dx$.

Example 1 Calculate the average value of the function $f(x) = x^5$ on the interval $[6, 8]$.

Solution Recall that $f_{ave} = \frac{1}{b-a} \int\limits_{x=a}^{x=b} f(x)\,dx$. Therefore, the average value of function f on the interval $[6, 8]$ is

$$\frac{1}{8-6} \int\limits_{x=6}^{x=8} x^5\,dx = \frac{1}{2} \left(\frac{1}{6}\right) x^6 \Big|_{x=6}^{x=8}$$

$$= \frac{1}{12} (8^6 - 6^6)$$

$$\approx 17957.33.$$

Self-Test Exercise

1. Calculate the average value of the function $f(x) = \sin(x)$ on the interval $\left[0, \frac{\pi}{4}\right]$.

Probability Theory Application

Suppose that X is a random variable with values that all are in an interval I. If there is a function f such that $P(\alpha \leq X \leq \beta) = \int\limits_{x=\alpha}^{x=\beta} f(x)\,dx$ for every subinterval $[\alpha, \beta]$ of I, then we say that f is a *probability density function* (p.d.f.) *of X*. $P(\alpha \leq X \leq \beta)$ represents the probability that X obtains a value between α and β.

A continuous p.d.f. on $I = [a, b]$ is nonnegative (i.e., $f \geq 0$ for all x in I). Also, $\int\limits_{x=a}^{x=b} f(x)\,dx = 1$. Conversely, any continuous function that satisfies both these properties is a p.d.f. of some random variable.

Example 2 If $f(x) = \frac{3}{26} x^2$ is a p.d.f. of a random variable X on the interval $I = [1, 3]$, calculate $P(1 \leq X \leq 2)$.

Solution Recall that $P(\alpha \le X \le \beta) = \int\limits_{x=\alpha}^{x=\beta} f(x)\,dx$. Therefore,

$$
\begin{aligned}
P(1 \le X \le 2) &= \int\limits_{x=1}^{x=2} \frac{3}{26}x^2\,dx \\
&= \left(\frac{3}{26}\right)\left(\frac{1}{3}\right)x^3 \Big|_{x=1}^{x=2} \\
&= \frac{1}{26}\left(2^3 - 1^3\right) \\
&= \frac{7}{26} \\
&\approx 0.27.
\end{aligned}
$$

∎

Average Values in Probability Theory

If f is the probability density function of a random variable X that takes values in an interval $I = [a, b]$ and if $x \cdot f(x)$ is Riemann integrable over I, then the *average value (or mean)* μ of X is defined as $\mu = \int\limits_{x=a}^{x=b} x \cdot f(x)\,dx$. This value is also said to be the *expectation of X*, which is denoted as $\overline{X}$ or equivalently as $E(X)$.

Example 3 If $f(x) = \frac{3}{26}x^2$ is a p.d.f. of a random variable X on the interval $I = [1, 3]$, calculate $E(X)$.

Solution

$$
\begin{aligned}
E(X) &= \int\limits_{x=1}^{x=3} x \cdot \frac{3}{26}x^2\,dx \\
&= \frac{3}{26} \int\limits_{x=1}^{x=3} x^3\,dx \\
&= \frac{3}{26}\left(\frac{1}{4}x^4 \Big|_{x=1}^{x=3}\right) \\
&= \frac{3}{104}\left(3^4 - 1^4\right) \\
&= \frac{240}{104} \\
&\approx 2.31.
\end{aligned}
$$

∎

Self-Test Exercise

2. Calculate the mean of the random variable with p.d.f. $f(x) = 6x^5$ on $I = [0, 1]$.

8.4 Center of Mass

Important Concept

- Center of Mass

The Balancing Point

Suppose that two point masses m_1 and m_2 are situated at endpoints of an interval $[x_1, x_2]$ on the x-axis, as illustrated in Figure 1. If a fulcrum could be positioned underneath the axis and if the interval could pivot about the fulcrum, the *lever law of physics* states that the masses are in balance if and only if the distances d_1 and d_2 satisfy the equation $m_1 d_1 = m_2 d_2$.

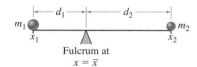

Figure 1

This principle of physics can be used in conjunction with calculus to determine the balancing point of a planar region, as shown in Figure 2.

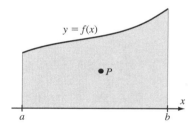

Figure 2

Moments of Two-Point Systems

In Figure 1, the point $x = \bar{x}$ represents the *center of mass* of the system. If we rewrite the equation $m_1 d_1 = m_2 d_2$ in terms of the center of mass, we obtain $m_1 \cdot (x_1 - \bar{x}) + m_2 \cdot (x_2 - \bar{x}) = 0$. The left side of this equation is called the *moment about the axis* $x = \bar{x}$. In general, if we position the fulcrum at any point $x = c$, then we define the *moment about the axis* $x = c$ by $M_{x=c} = m_1 \cdot (x - c) + m_2 \cdot (x - c)$. The *moment about the y-axis* is defined by $M_{x=0} = m_1 \cdot x + m_2 \cdot x$.

If we expand the equation $m_1 \cdot (x_1 - \bar{x}) + m_2 \cdot (x_2 - \bar{x}) = 0$ and solve for $\bar{x}$, we obtain $\bar{x} = \dfrac{m_1 x_1 + m_2 x_2}{m_1 + m_2}$. If we let $M = m_1 + m_2$, then the equation for $\bar{x}$ can be expressed as $\bar{x} = \dfrac{M_{x=0}}{M}$.

Moments of a Region

The definition of moment for a two-point system can also be generalized to the case of a region. Let R be the region illustrated in Figure 3, where it is assumed that R has uniform density δ.

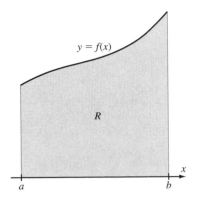

$y = f(x)$

R

a b x

Figure 3

Let c be any real number. Suppose that f is continuous and nonnegative on the interval $[a, b]$. R denotes the planar region bounded above by the graph of $y = f(x)$, below by the x-axis, and on the sides by the lines $x = a$ and $x = b$. If R has a uniform density δ, then the *moment $M_{x=c}$ of R about the axis $x = c$* is defined by the equation:

$$M_{x=c} = \int_{x=a}^{x=b} (x - c)\delta f(x)\, dx = \delta \int_{x=a}^{x=b} (x - c)f(x)\, dx.$$

Notice that $M_{x=c}$ is defined even if c is not between a and b.

Example 1 Let R be the region bounded by $y = 16 - x^2$, $y = 0$, $x = 1$, and $x = 2$. If R has uniform mass density $\delta = 3$, calculate the moment about the axis $x = 6$.

Solution Region R appears in Figure 4.

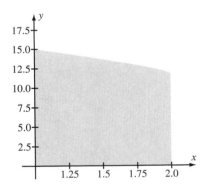

Figure 4

$$M_{x=6} = 3 \int_{x=1}^{x=2} (x - 6)(16 - x^2)\, dx$$

$$= 3 \int_{x=1}^{x=2} (-x^3 + 6x^2 + 16x - 96)\, dx$$

$$= 3 \left(-\frac{1}{4}x^4 + 2x^3 + 8x^2 - 96x \Big|_{x=1}^{x=2} \right)$$

$$= \frac{-741}{4}$$

$$= -185.25$$

Since the whole region lies to the left of the axis $x = 6$, we can expect the moment about this axis to be negative. ∎

Center of Mass

The center of mass $(\bar{x}, \bar{y})$ of a region R is the point at which the region balances (as shown in Figure 2 of this section). Let R be a region as shown in Figure 3 of this section. The x-coordinate $\bar{x}$ of the center of mass of R is the real number which satisfies $M_{x=\bar{x}} = 0$.

Theorem 1 Let f be a continuous, nonnegative function on the interval $[a, b]$. Let R denote the region bounded above by the graph of $y = f(x)$, below by the x-axis, and on the sides by the lines $x = a$ and $x = b$. Let M denote the mass of R. If R has a uniform mass density δ, then the x-coordinate of the center of mass of R is given by

$$\bar{x} = \frac{M_{x=0}}{M} = \frac{\int\limits_{x=a}^{x=b} x \cdot f(x)\,dx}{\int\limits_{x=a}^{x=b} f(x)\,dx}.$$

The y-coordinate $\bar{y}$ of the center of mass is given by

$$\bar{y} = \frac{M_{y=0}}{M} = \frac{\dfrac{1}{2}\int\limits_{x=a}^{x=b} (f(x))^2\,dx}{\int\limits_{x=a}^{x=b} f(x)\,dx}.$$

Example 2 Let R be the region bounded by $y = 16 - x^2$, $y = 0$, $x = 1$, and $x = 2$. If R has uniform mass density $\delta = 3$, calculate the center of mass of R.

Solution Region R appears in Figure 5.

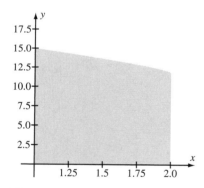

Figure 5

$$\bar{x} = \frac{M_{x=0}}{M}$$

$$= \frac{\displaystyle\int_{x=1}^{x=2} x(16-x^2)\,dx}{\displaystyle\int_{x=1}^{x=2} (16-x^2)\,dx}$$

$$= \frac{\displaystyle\int_{x=1}^{x=2} (16x-x^3)\,dx}{\displaystyle\int_{x=1}^{x=2} (16-x^2)\,dx}$$

$$= \frac{\left(8x^2 - \frac{1}{4}x^4\right)\Big|_{x=1}^{x=2}}{\left(16x - \frac{1}{3}x^3\right)\Big|_{x=1}^{x=2}}$$

$$= \frac{\dfrac{81}{4}}{\dfrac{41}{3}}$$

$$= \frac{243}{164}$$

$$\approx 1.48$$

$$\bar{y} = \frac{M_{y=0}}{M}$$

$$= \frac{\dfrac{1}{2}\displaystyle\int_{x=1}^{x=2} (16-x^2)^2\,dx}{\displaystyle\int_{x=1}^{x=2} (16-x^2)\,dx}$$

$$= \frac{\dfrac{1}{2}\left(\dfrac{1}{5}x^5 - \dfrac{32}{3}x^3 + 256x\right)\Big|_{x=1}^{x=2}}{\dfrac{41}{3}}$$

$$= \frac{\dfrac{1}{2}\left(\dfrac{2813}{15}\right)}{\dfrac{41}{3}}$$

$$= \frac{2813}{410}$$

$$\approx 6.86$$

Therefore, the center of mass $(\bar{x}, \bar{y})$ of region R is $(\bar{x}, \bar{y}) \approx (1.48, 6.86)$. ■

Self-Test Exercise

1. Calculate the center of mass of region R, which is bounded above by $y = x^2$, below by the x-axis, and on the sides by the lines $x = 2$ and $x = 4$. R has uniform mass density $\delta = 2$.

8.5 Work

Important Concepts

*   Work Done by Varying Amounts of Force

Work Done by Constant Force

Figure 1 illustrates that a body is moved a distance d along a straight line while being acted on by a force of *constant* magnitude F *in the direction of motion*. By definition, the *work performed in this movement is force times distance*.

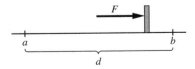

Figure 1

Example 1 How much work is done lifting a 1000 lb steel beam upward 20 ft?

Solution The constant force is 1000 lb and the distance is 20 ft. Therefore, the work done lifting this steel beam upward 20 ft is $W = (1000)(20) = 20000$ ft-lb. ∎

Work Done by Varying Force

Suppose a body is moved linearly from $x = a$ to $x = b$ by a force in the direction of motion. If the magnitude of the force at each point $x \in [a, b]$ is $F(x)$, then the work performed is defined to be $\int_{x=a}^{x=b} F(x)\, dx$. In other words, when the magnitude of the force is not constant, we will use an integral to calculate work.

Examples with Varying Weights

Example 2 How much work is done moving a block of concrete along a five-foot path if at each point x the amount of force is x^2 pounds?

Solution Work W is defined by

$$
\begin{aligned}
W &= \int_{x=0}^{x=5} x^2\, dx \\
&= \frac{1}{3}x^3 \Big|_{x=0}^{x=5} \\
&= \frac{1}{3}(5^3 - 0^3) \\
&= \frac{125}{3} \\
&\approx 41.67 \text{ ft-lbs.}
\end{aligned}
$$

∎

A Spring Example

Figures 2, 3, and 4 illustrate a spring at rest, a stretched spring, and a compressed spring, respectively.

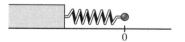

Figure 2

Figure 3

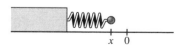

Figure 4

To stretch or compress a spring, a force must be exerted that is equal in magnitude to the restorative spring force, but opposite in direction. Suppose that $0 \le a \le b$. The work done in stretching a spring from a to b is $W = \int\limits_{x=a}^{x=b} (kx)\,dx$, where k is the *spring constant* and x represents the amount of the extension.

Example 3 Calculate the work performed to stretch a spring 3 in. beyond its equilibrium position. The spring constant is 2 lb/in.

Solution

$$W = \int\limits_{x=a}^{x=b} (kx)\,dx$$

$$= \int\limits_{x=0}^{x=3} (kx)\,dx$$

$$= k\frac{x^2}{2}\bigg|_{x=0}^{x=3}$$

$$= \frac{k}{2}(9-0)$$

$$= \frac{9k}{2}\ \text{in.-lb.}$$

If we substitute the value $k = 2$, the work performed is 9 in.-lb.   ■

Self-Test Exercise

1. If 4 J of work is done in extending a spring 0.1 m beyond its equilibrium position, then how much extra work is required to extend it an extra 0.3 m?

8.6 Improper Integrals and Unbounded Integrands

Important Concept

- Integrals with Infinite Integrands

Integrals with Infinite Integrands

Figure 1 illustrates an unbounded integrand.

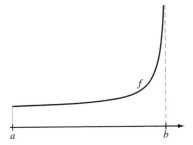

Figure 1

If $\int_a^b f(x)\,dx$ is an improper integral with infinite integrand at b, then the value of the integral is defined to be $\lim\limits_{\epsilon \to 0^+} \int_a^{b-\epsilon} f(x)\,dx$, provided that this limit exists and is finite. In this case, the integral is said to *converge*. Otherwise, the integral is said to *diverge*.

Figure 2 illustrates the method by which we evaluate this type of improper integral.

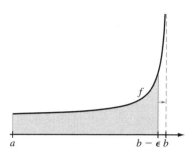

Figure 2

Example 1 Evaluate the integral $\int_0^2 \dfrac{1}{(x-2)^2}\,dx$.

Solution The given integral is improper with infinite integrand at 2. We calculate

$$\lim_{\epsilon \to 0^+} \int_0^{2-\epsilon} (x-2)^{-2}dx = \lim_{\epsilon \to 0^+} \left(-(x-2)^{-1}\right)\Big|_{x=0}^{x=2-\epsilon}$$

$$= -\left(\lim_{\epsilon \to 0^+} ((2-\epsilon-2)^{-1}) - (0-2)^{-1}\right)$$

$$= \infty.$$

This integral diverges. ∎

If $f(x)$ is continuous on $(a, b]$ and unbounded as $x \to a^+$, the value of the improper integral $\int_a^b f(x)\,dx$ is defined to be $\lim\limits_{\epsilon \to 0^+} \int_{a+\epsilon}^b f(x)\,dx$, provided that this limit exists and is finite. In this case, the integral is said to *converge*. Otherwise, the integral is said to *diverge*. Figure 3 illustrates this technique.

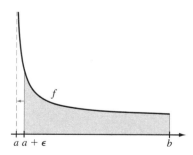

Figure 3

Example 2 Evaluate the improper integral $\displaystyle\int_1^2 \frac{1}{\sqrt{x-1}}\, dx$.

Solution This is an improper integral with infinite integrand at 1.

$$\int_1^2 \frac{1}{\sqrt{x-1}}\, dx = \lim_{\epsilon\to 0^+} \int_{1+\epsilon}^2 \frac{1}{\sqrt{x-1}}\, dx$$

$$= 2\lim_{\epsilon\to 0^+} \sqrt{x-1}\,\Big|_{x=1+\epsilon}^{x=2}$$

$$= 2\left(\lim_{\epsilon\to 0^+}(\sqrt{2-1} - \sqrt{1+\epsilon-1})\right)$$

$$= 2(1-0)$$

$$= 2.$$

The improper integral converges.   ■

When an integrand f is continuous on an open interval (a, b) and unbounded at both endpoints a and b, choose any interior point c and investigate the two improper integrals $\int_a^c f(x)\, dx$ and $\int_c^b f(x)\, dx$. If *both* integrals converge, the improper integral is said to converge. Otherwise, the improper integral is said to diverge.

Example 3 Calculate $\displaystyle\int_0^2 \frac{1}{x(x-2)}\, dx$.

Solution This is an improper integral with infinite integrand at both $x=0$ and $x=2$. Therefore, choose $c=1$ and express

$$\int_0^2 \frac{1}{x(x-2)}\, dx = \lim_{\epsilon\to 0^+}\int_{0+\epsilon}^1 \frac{1}{x(x-2)}\, dx + \lim_{\epsilon\to 0^+}\int_1^{2-\epsilon} \frac{1}{x(x-2)}\, dx.$$

Using the Method of Partial Fractions,

$$\int_{0+\epsilon}^1 \frac{1}{x(x-2)}\, dx = \lim_{\epsilon\to 0^+}\left(\frac{-1}{2}\ln(|x|) + \frac{1}{2}\ln(|x-2|)\right)\Big|_{x=0+\epsilon}^{x=1}$$

$$= \lim_{\epsilon\to 0^+}\left(\left(\frac{-1}{2}\ln(|1|) + \frac{1}{2}\ln(|1-2|)\right) + \frac{1}{2}\ln(|\epsilon|) - \frac{1}{2}\ln(|\epsilon-2|)\right)$$

$$= \infty.$$

Because $\displaystyle\lim_{\epsilon\to 0^+}\int_{0+\epsilon}^1 \frac{1}{x(x-2)}\, dx$ diverges, the improper integral $\displaystyle\int_0^2 \frac{1}{x(x-2)}\, dx$ diverges.   ■

Self-Test Exercise

1. Calculate $\displaystyle\int_0^2 \frac{1}{\sqrt{x}}\, dx$.

8.7 Improper Integrals and Unbounded Intervals

Important Concepts

- The Integral on an Infinite Interval
- Applications

The Integral on an Infinite Interval

Let f be a continuous function on the interval $[A, \infty)$. The value of the improper integral $\int_A^\infty f(x)\,dx$ is defined to be $\lim\limits_{N \to \infty} \int_A^N f(x)\,dx$, provided that the limit exists and is finite (see Figure 1). When the limit exists, the integral is said to *converge*. Otherwise, it is said to *diverge*.

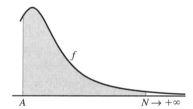

Figure 1

Similarly, if g is a continuous function on the interval $(-\infty, B]$, the value of the improper integral $\int_{-\infty}^B g(x)\,dx$ is defined to be $\lim\limits_{M \to -\infty} \int_M^B g(x)\,dx$, provided that the limit exists and is finite (see Figure 2). When the limit exists, the integral is said to *converge*. Otherwise, it is said to *diverge*.

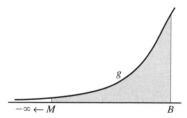

Figure 2

Example 1 Calculate $\displaystyle\int_2^\infty \frac{1}{x}\,dx$.

Solution

$$\int_2^\infty \frac{1}{x}\,dx = \lim_{N \to \infty} \int_2^N \frac{1}{x}\,dx$$

$$= \lim_{N \to \infty} \left. (\ln |x|) \right|_{x=2}^{x=N}$$

$$= \lim_{N \to \infty} (\ln |N| - \ln |2|)$$

$$= \infty$$

The improper integral diverges.

Self-Test Exercise

1. Calculate $\displaystyle\int_{-\infty}^{-5} \frac{1}{(x)^2}\,dx.$

Financial Applications

If an amount A of money is allowed to compound continuously at an annual rate r, then it will grow to Ae^{rt}. The amount Ae^{-rt} invested now will return A in t years. Ae^{-rt} is the *present value* of an amount A that is to be received t years from now.

Now, consider a future *income stream* rather than a single payment A. If r is the current annual interest rate, then $\int_{T_1}^{T_2} f(t)e^{-rt}dt$ is the *present value of the income stream* $f(t)$ for $T_1 \leq t \leq T_2$.

If an income stream continues in perpetuity, then we let $T_2 \to \infty$—i.e., $\int_{T_1}^{\infty} f(t)e^{-rt}dt$ is the *present value of an income stream that begins T_1 years in the future and continues in perpetuity.*

Example 2 Suppose a trust is established that pays t dollars per year for every year in perpetuity where t is time measured in years. Note that $t = 0$ corresponds to the present. Assume a constant interest rate of 5%. What is the total value, in today's dollars, of all the money that will ever be earned by this trust account?

Solution The present value of the income stream is $\int_0^{\infty} te^{-0.05t}dt = \lim_{N\to\infty}\int_0^N te^{-0.05t}dt$. Use the Method of Integration by Parts, with $u = t$ and $dv = e^{-0.05t}dt$, to obtain

$$\int_0^N te^{-0.05t}dt = \frac{-1}{0.05}te^{-0.05t}\Big|_0^N - \int_0^N \left(\frac{-1}{0.05}\right)e^{-0.05t}dt$$

$$= \frac{-1}{0.05}Ne^{-0.05N} - \left(\frac{-1}{0.05}\right)^2 (e^{-0.05N} - 1).$$

Therefore,

$$\lim_{N\to\infty}\int_0^N te^{-0.05t}dt = \left(\frac{-1}{0.05}\right)\lim_{N\to\infty}(Ne^{-0.05N}) - \frac{1}{(.0025)}\lim_{N\to\infty}(e^{-0.05N}) + \left(\frac{1}{.0025}\right).$$

Apply l'Hôpital's Rule to $\left(\dfrac{-1}{0.05}\right)\lim_{N\to\infty}(Ne^{-0.05N})$ to obtain the value 0. The value of $\dfrac{1}{(0.0025)}\lim_{N\to\infty}(e^{-0.05N})$ is also 0. Therefore, the present value of the income stream is equal to $\dfrac{1}{.0025} = 400$ dollars. ∎

Random Variables

Suppose that X is a random variable that takes values in an interval I, where I can be of the form $[a, b]$, $[a, \infty)$, $(-\infty, b]$ or $(-\infty, \infty)$. Let f be a probability density function of X. Then the mean (or expectation) μ of X is given by $\mu = \int_I x \cdot f(x)\, dx$, where I is the interval of integration. The interval I can be finite or infinite. We discussed the case when the interval I was finite in Section 8.3.

An especially important probability density function is called a *normal* probability density function. It is defined to be

$$f(x) = \frac{1}{\sqrt{2\pi}\sigma} \exp\left(\frac{-1}{2}\left(\frac{x-\mu}{\sigma}\right)^2\right), \quad -\infty < x < \infty,$$

where μ is any real number and σ is any positive number. μ is the mean and σ is the standard deviation of the density function.

Example 3 Suppose the life of a certain light bulb is normally distributed with a mean of 100 hours (h) and a standard deviation of 10 hours. What is the probability that the light bulb will last at least 100 hours?

Solution

$$P(100 \le X \le \infty) = \int_{100}^{\infty} \left(\frac{1}{10\sqrt{2\pi}}\right) e^{\frac{-(x-100)^2}{2(10)^2}}\, dx$$

$$= \left(\frac{1}{10\sqrt{2\pi}}\right) \int_{100}^{\infty} e^{\frac{-(x-100)^2}{2(10)^2}}\, dx$$

$$= \frac{1}{2}$$

A computer algebra system was used to compute the improper integral. ∎

Self-Test Exercise

2. Calculate the mean of a random variable X with p.d.f. $f(x) = 3e^{-3x}$ for $x \in [0, \infty)$.

Infinite Series

9.1 Series

Important Concepts

- Definitions
- Some Special Series
- Basic Properties

Definitions

An *infinite sequence,* denoted as $\{a_n\}_{n=1}^{\infty}$ or simply as $\{a_n\}$, is a list of terms $a_1, a_2, a_3, \ldots$ indexed by the positive integers.

If $a_1, a_2, a_3, \ldots$ is an infinite sequence, the sum of the terms $a_1 + a_2 + a_3 + \cdots$ is called an *infinite series.* The series $a_1 + a_2 + a_3 + \cdots$ is often denoted in *sigma notation* as

$$a_1 + a_2 + a_3 + \cdots = \sum_{j=1}^{\infty} a_j.$$

Both $a_1 + a_2 + a_3 + \cdots = \sum_{j=1}^{\infty} a_j$ and $a_1 + a_2 + a_3 + \cdots = \sum_{n=1}^{\infty} a_n$ are valid representations of an infinite series because the indexes j and n are "dummy variables" (i.e., any letter can represent an index of summation). By convention, letters that appear later in the alphabet are used.

If $a_1 + a_2 + a_3 + \cdots$ is an infinite series, the Nth Partial Sum of the infinite series, denoted by S_N, is defined as the sum of the first N terms (i.e., $S_N = \sum_{j=1}^{N} a_j = a_1 + a_2 + \cdots + a_{N-1} + a_N$).

The sequence $\{S_N\}_{N=1}^{\infty}$ is called the *Sequence of Partial Sums* of the infinite series $\sum_{n=1}^{\infty} a_n$. If $\lim_{N \to \infty} S_N = \ell$, we say that the *infinite series* $\sum_{n=1}^{\infty} a_n$ *converges to* ℓ. When we refer to $\sum_{n=1}^{\infty} a_n$ in this case, we mean limit ℓ and we write $\sum_{n=1}^{\infty} a_n = \ell$. We call ℓ *the sum of the infinite series.* If $\lim_{N \to \infty} S_N$ does not exist, i.e. $\{S_N\}_{N=1}^{\infty}$ does *not* converge, we say the infinite series, $\sum_{n=1}^{\infty} a_n$ *diverges.*

Example 1 If $\{a_n\} = \{2, 4, 6, 8, \ldots\}$, define the jth term a_j.

Solution $\{a_n\} = \{2, 4, 6, 8, \ldots\} = \{2(1), 2(2), 2(3), 2(4), \ldots\}$; therefore, a_j is defined by $a_j = 2j$, where j is a positive integer. ∎

Example 2 Consider the infinite series $\sum\limits_{j=1}^{\infty} a_j$, where $a_j = 2j$. Define S_N and determine whether the infinite series converges or diverges.

Solution
$$S_N = \sum_{j=1}^{N} a_j$$
$$= a_1 + a_2 + \cdots + a_{N-1} + a_N$$
$$= \sum_{j=1}^{N} 2j$$
$$= 2\sum_{j=1}^{N} j$$
$$= 2\left(\frac{N(N+1)}{2}\right)$$
$$= N(N+1)$$

Therefore, $\lim\limits_{N\to\infty} S_N = \lim\limits_{N\to\infty} (N(N+1)) = \infty$—i.e., there is no finite limit ℓ, so the infinite series $\sum\limits_{j=1}^{\infty} a_j$ diverges. ∎

A series that can be written in the form $\sum\limits_{n=1}^{\infty} (f(n) - f(n+1))$ is called a *collapsing* or *telescoping series* because each partial sum collapses into two summands.

Example 3 Show that $\dfrac{2n+1}{n^2(n+1)^2} = \dfrac{1}{n^2} - \dfrac{1}{(n+1)^2}$. Discuss the convergence of the series $\sum\limits_{n=1}^{\infty} \dfrac{2n+1}{n^2(n+1)^2}$.

Solution By taking the common denominator, we have
$$\frac{1}{n^2} - \frac{1}{(n+1)^2} = \frac{(n+1)^2 - n^2}{n^2(n+1)^2}$$
$$= \frac{n^2 + 2n + 1 - n^2}{n^2(n+1)^2}$$
$$= \frac{2n+1}{n^2(n+1)^2}.$$

Therefore $\sum\limits_{n=1}^{\infty} \dfrac{2n+1}{n^2(n+1)^2} = \sum\limits_{n=1}^{\infty} \left(\dfrac{1}{n^2} - \dfrac{1}{(n+1)^2}\right)$ is a telescoping series.

$$S_N = \sum_{n=1}^{N} \frac{2n+1}{n^2(n+1)^2}$$

$$= \sum_{n=1}^{N} \left(\frac{1}{n^2} - \frac{1}{(n+1)^2} \right)$$

$$= \left(1 - \frac{1}{2^2} \right) + \left(\frac{1}{2^2} - \frac{1}{3^2} \right) + \left(\frac{1}{3^2} - \frac{1}{4^2} \right) + \cdots$$

$$+ \left(\frac{1}{(N-1)^2} - \frac{1}{N^2} \right) + \left(\frac{1}{N^2} - \frac{1}{(N+1)^2} \right)$$

$$= 1 - \frac{1}{(N+1)^2}$$

Clearly,

$$\lim_{N \to \infty} S_N = \lim_{N \to \infty} \left(1 - \frac{1}{(N+1)^2} \right)$$

$$= \lim_{N \to \infty} 1 - \lim_{N \to \infty} \frac{1}{(N+1)^2}$$

$$= 1 - 0$$

$$= 1.$$

Therefore, the series $\sum_{n=1}^{\infty} \frac{2n+1}{n^2(n+1)^2}$ converges to 1. ∎

Self-Test Exercise

1. Determine whether the infinite series $a_1 + a_2 + a_3 + \cdots = \sum_{j=1}^{\infty} a_j$, $\{a_n\} = \{-1, 1, -1, 1, -1, 1, \ldots\}$, converges or diverges.

Some Special Series

Some series arise sufficiently often in mathematical models that they are identified by name. In the following sections, we will present some of these named series.

Harmonic Series

The *harmonic series* is defined by $\sum_{n=1}^{\infty} \frac{1}{n}$. Theorem 1 establishes that the harmonic series is divergent.

Theorem 1   The harmonic series $\sum_{n=1}^{\infty} \frac{1}{n}$ is divergent.

The harmonic series is a *p*-series. The *p-series* is defined by $\sum_{n=1}^{\infty} \frac{1}{n^p}$, where exponent p is a fixed number. In Section 3, it will be shown that a *p*-series converges for fixed $p > 1$ and diverges for fixed $p \leq 1$.

Example 4 Verify numerically that the series $\sum_{n=1}^{\infty} \frac{1}{n^4}$ is convergent.

Solution This is a p-series with $p = 4$. Utilizing a computer algebra system or a calculator, the following partial sums of the series $\sum_{n=1}^{\infty} \dfrac{1}{n^4}$ were computed:

$$\sum_{n=1}^{100} \frac{1}{n^4} = 1.082322906\ldots, \quad \sum_{n=1}^{500} \frac{1}{n^4} = 1.082323232\ldots, \quad \sum_{n=1}^{1000} \frac{1}{n^4} = 1.082323233996\ldots,$$

$$\text{and } \sum_{n=1}^{5000} \frac{1}{n^4} = 1.08232323432615\ldots.$$

These results suggest that $\sum_{n=1}^{\infty} \dfrac{1}{n^4}$ converges to $1.08232323\ldots$. Actually, using advanced mathematics, one can show that $\sum_{n=1}^{\infty} \dfrac{1}{n^4} = \dfrac{\pi^4}{90} = 1.08232323371113\ldots.$ ∎

Geometric Series

The *geometric series* is defined by $\sum_{n=0}^{\infty} r^n = 1 + r + r^2 + r^3 + \cdots$ with terms that are nonnegative powers of a fixed number r. Theorem 2 establishes conditions for r under which the geometric series either converges or diverges.

Theorem 2   If $|r| < 1$, the (geometric) series $\sum_{n=0}^{\infty} r^n$ converges to $\dfrac{1}{1-r}$ (i.e., $\sum_{n=0}^{\infty} r^n = \dfrac{1}{1-r}, |r| < 1$). If $|r| \geq 1$, the (geometric) series $\sum_{n=0}^{\infty} r^n$ diverges.

Example 5   Discuss the convergence or divergence of the series $\dfrac{1}{3} + \dfrac{1}{9} + \dfrac{1}{27} + \dfrac{1}{81} + \cdots$.

Solution

$$\frac{1}{3} + \frac{1}{9} + \frac{1}{27} + \frac{1}{81} + \cdots = \frac{1}{3}\left(1 + \frac{1}{3} + \frac{1}{9} + \frac{1}{27} + \cdots\right) = \frac{1}{3}\left(\sum_{n=0}^{\infty}\left(\frac{1}{3}\right)^n\right)$$

Because $|r| = \left|\dfrac{1}{3}\right| < 1$, we know that the geometric series $\sum_{n=0}^{\infty}\left(\dfrac{1}{3}\right)^n$ converges to $\dfrac{1}{1-\dfrac{1}{3}}$.

Therefore,

$$\frac{1}{3}\left(\sum_{n=0}^{\infty}\left(\frac{1}{3}\right)^n\right) = \frac{1}{3}\left(\frac{1}{1-\dfrac{1}{3}}\right)$$

$$= \frac{1}{3}\left(\frac{1}{\dfrac{2}{3}}\right)$$

$$= \frac{1}{3}\left(\frac{3}{2}\right)$$

$$= \frac{1}{2}.$$

We conclude that the series $\frac{1}{3} + \frac{1}{9} + \frac{1}{27} + \frac{1}{81} + \cdots = \frac{1}{2}$ (i.e., the series $\frac{1}{3} + \frac{1}{9} + \frac{1}{27} + \frac{1}{81} + \cdots$ converges to $\frac{1}{2}$). ∎

Self-Test Exercise

2. Discuss the convergence of the series $\sum_{j=1}^{\infty} 5j$.

Basic Properties

Theorems 3 and 4 present some of the basic properties of series. These properties are quite similar to the properties of sequences discussed in Chapter 2.

Theorem 3 Suppose that $\sum_{n=1}^{\infty} a_n$ converges to A and $\sum_{n=1}^{\infty} b_n$ converges to B; then

a. $\sum_{n=1}^{\infty} (a_n + b_n)$ converges to $A + B$ and $\sum_{n=1}^{\infty} (a_n - b_n)$ converges to $A - B$; and

b. $\sum_{n=1}^{\infty} (\lambda a_n)$ converges to λA for any real constant λ.

Theorem 4 If $\sum_{n=1}^{\infty} a_n$ diverges and $\lambda \neq 0$, then $\sum_{n=1}^{\infty} (\lambda a_n)$ also diverges.

Example 6 Discuss the convergence or divergence of $\sum_{n=1}^{\infty} \frac{3^n}{110}$.

Solution $\sum_{n=1}^{\infty} \frac{3^n}{110} = \sum_{n=1}^{\infty} \frac{1}{110}(3^n) = \frac{1}{110} \sum_{n=1}^{\infty} (3^n)$. If $\sum_{n=1}^{\infty} (3^n)$ diverges, then by Theorem 4 the given series diverges. If $\sum_{n=1}^{\infty} (3^n)$ converges, then by Theorem 3 the given series converges.

$$\sum_{n=1}^{\infty} (3^n) = 3 + 3^2 + 3^3 + 3^4 + \cdots = 3(1 + 3 + 3^2 + 3^3 + \cdots) = 3 \sum_{n=0}^{\infty} 3^n$$

$\sum_{n=0}^{\infty} 3^n$ is a geometric series where $r = 3$. Because $|r| = 3 \geq 1$, Theorem 2 establishes that $\sum_{n=0}^{\infty} 3^n$ diverges. $\sum_{n=1}^{\infty} \frac{3^n}{110} = \sum_{n=1}^{\infty} (\lambda a_n)$ where $\lambda = \frac{1}{110}$ and $a_n = 3^n$. Since $\sum_{n=0}^{\infty} 3^n$ diverges, by Theorem 4, $\sum_{n=1}^{\infty} \frac{3^n}{110}$ also diverges. ∎

Self-Test Exercise

3. Discuss the convergence or divergence of the series $\sum_{k=1}^{\infty} \frac{5^k}{6^{k+1}}$.

9.2 Determining Convergence

Important Concepts

- The Divergence Test
- Series with Nonnegative Terms
- The Tail End of a Series

The Divergence Test

Theorem 1 is known as the *Divergence Test*. This test gives *no* information about convergence.

Theorem 1   ***Divergence Test***   If the summands a_n of an infinite series $\sum_{n=1}^{\infty} a_n$ do not tend to zero, then the series $\sum_{n=1}^{\infty} a_n$ diverges.

Example 1   Use the Divergence Test to establish that the series $\sum_{n=1}^{\infty} (3^n)$ diverges.

Solution   The nth summand of the series $\sum_{n=1}^{\infty} (3^n)$ is $a_n = 3^n$. We have $\lim_{n \to \infty} a_n = \lim_{n \to \infty} 3^n = \infty$. Therefore, by Theorem 1, the series $\sum_{n=1}^{\infty} (3^n)$ diverges.   ∎

Symbolically, the Divergence Test can be stated $\lim_{n \to \infty} a_n \neq 0 \Rightarrow \sum_{n=1}^{\infty} a_n$ diverges. However, the statement $\lim_{n \to \infty} a_n = 0 \Rightarrow \sum_{n=1}^{\infty} a_n$ converges is *false*. A counterexample to this (false) statement is given by the harmonic series. The nth summand a_n of the harmonic series is $a_n = \frac{1}{n}$. Clearly, $\lim_{n \to \infty} a_n = \lim_{n \to \infty} \frac{1}{n} = 0$, but $\sum_{n=1}^{\infty} a_n = \sum_{n=1}^{\infty} \frac{1}{n}$ is the *divergent harmonic series*!

Self-Test Exercise

1. Establish that the series $\sum_{n=1}^{\infty} \frac{6^n}{5^{n+1}}$ diverges using the Divergence Test.

Series with Nonnegative Terms

If the series consists of terms each of which is nonnegative, then identifying convergence or divergence of the series is more straightforward. Theorem 2 identifies conditions under which a series with nonnegative terms converges.

Theorem 2   Suppose that $a_n \geq 0$ for each n. Let $S_N = \sum_{n=1}^{N} a_n = a_1 + a_2 + \cdots + a_{N-1} + a_N$. If a real number U exists such that $S_N \leq U$ for all N, the infinite series $\sum_{n=1}^{\infty} a_n$ converges.

Example 2 Use Theorem 2 to establish the convergence of the series $\sum\limits_{n=1}^{\infty} 10^{-n}$.

Solution $a_n = \dfrac{1}{10^n} \geq 0$ for each n. The formula for the sum of a *finite* geometric series is

$$\sum_{n=0}^{N-1} r^n = 1 + r + r^2 + \cdots + r^{N-1} = \frac{1 - r^N}{1 - r}.$$ We can use this formula to obtain a sum for S_N

of the given series.

$$S_N = \sum_{n=1}^{N} a_n$$

$$= \frac{1}{10} + \frac{1}{10^2} + \cdots + \frac{1}{10^N}$$

$$= \frac{1}{10} \left(\frac{1 - \dfrac{1}{10^N}}{1 - \dfrac{1}{10}} \right)$$

$$= \frac{1}{10} \cdot \frac{10}{9} \left(1 - \frac{1}{10^N} \right)$$

$$= \frac{1}{9} \left(1 - \frac{1}{10^N} \right)$$

$$= \frac{1}{9} - \frac{1}{9(10^N)}$$

Therefore, $S_N \leq \dfrac{1}{9}$ for all N (i.e., $U = \dfrac{1}{9}$). Because all the conditions in the premise of

Theorem 2 are satisfied, we conclude that the given series $\sum\limits_{n=1}^{\infty} 10^{-n}$ converges.   ■

Self-Test Exercises

2. Discuss the convergence or divergence of the series $\sum\limits_{n=1}^{\infty} 5$.

3. Use Theorem 2 to show the convergence of the series $\sum\limits_{n=0}^{\infty} 3^{-n}$.

The Tail End of a Series

If M is any positive integer, then $\sum\limits_{n=1}^{\infty} a_n$ *is convergent if and only if* $\sum\limits_{n=M+1}^{\infty} a_n$ *is convergent.*
This statement stresses that convergence is determined by the behavior of the terms toward the "tail end" of the series and *not* the beginning.

Example 3 Does the series $\sum\limits_{n=150}^{\infty} (-.7)^n$ converge?

Solution $\sum\limits_{n=150}^{\infty} (-.7)^n$ converges if and only if $\sum\limits_{n=1}^{\infty} (-.7)^n$ converges. However, $\sum\limits_{n=1}^{\infty} (-.7)^n =$

$\sum\limits_{n=0}^{\infty} (-.7)^n - 1$ and $\sum\limits_{n=0}^{\infty} (-.7)^n$ is a geometric series with $|r| = |-.7| = .7 < 1$. Therefore,

$$\sum_{n=0}^{\infty}(-.7)^n \text{ converges to some number } A, \text{ which establishes that } \sum_{n=1}^{\infty}(-.7)^n \text{ converges to some}$$

number $A-1$. Because $\sum_{n=1}^{\infty}(-.7)^n$ converges, this establishes that $\sum_{n=150}^{\infty}(-.7)^n$ converges. ∎

Self-Test Exercise

4. Discuss the convergence or divergence of the series $\sum_{n=85}^{\infty}\dfrac{1}{\sqrt[3]{n^3}}$.

9.3 Series with Nonnegative Terms—The Integral Test

Important Concepts

- The Integral Test
- p-Series

The Integral Test

Theorem 1 is known as the Integral Test. It provides yet another test for convergence for infinite series of positive terms.

Theorem 1

The Integral Test Let f be a positive, continuous, decreasing function on the interval $[1, \infty)$. The infinite series $\sum_{n=1}^{\infty}f(n)$ converges if and only if the improper integral $\int_{1}^{\infty}f(x)\,dx$ converges. In this case,

$$\int_{1}^{\infty}f(x)\,dx \le \sum_{n=1}^{\infty}f(n) \le f(1) + \int_{1}^{\infty}f(x)\,dx.$$

Example 1 Discuss the convergence or divergence of the series $\sum_{n=1}^{\infty}\dfrac{n}{3^n}$.

Solution The function $f(x) = \dfrac{x}{3^x}$ is positive, continuous and decreasing for $x \ge 1$ as the graph in Figure 1 indicates.

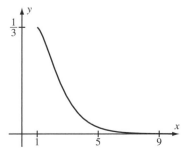

Figure 1

We can apply the Method of Integration by Parts and l'Hôpital's Rule to the improper integral $\int_1^\infty f(x)\,dx$ to obtain:

$$\int_{x=1}^\infty f(x)\,dx = \lim_{N\to\infty} \int_1^N f(x)\,dx$$

$$= \lim_{N\to\infty} \int_1^N (x(3^{-x}))\,dx$$

$$= \lim_{N\to\infty} \left(-x\left(\frac{3^{-x}}{\ln(3)}\right)\right)\Big|_{x=1}^{x=N} - \lim_{N\to\infty} \int_1^N \frac{3^{-x}}{-\ln(3)}\,dx$$

$$= \lim_{N\to\infty} \left(-x\left(\frac{3^{-x}}{\ln(3)}\right)\right)\Big|_{x=1}^{x=N} - \lim_{N\to\infty} \left(\frac{3^{-x}}{(\ln(3))^2}\right)\Big|_{x=1}^{x=N}$$

$$= \lim_{N\to\infty} \left(-N\left(\frac{3^{-N}}{\ln(3)}\right) + \frac{3^{-1}}{\ln(3)}\right) - \lim_{N\to\infty}\left(\frac{3^{-N}}{(\ln(3))^2} - \frac{3^{-1}}{(\ln(3))^2}\right)$$

$$= \left(0 + \frac{3^{-1}}{\ln(3)}\right) - \left(0 - \frac{3^{-1}}{(\ln(3))^2}\right)$$

$$= \frac{3^{-1}}{\ln(3)}\left(1 + \frac{1}{\ln(3)}\right)$$

$$\approx .58.$$

By applying the Integral Test, we can conclude that the given infinite series also converges because the improper integral converges. ■

Self-Test Exercise

1. Use the Integral Test to establish that the harmonic series, $\sum_{n=1}^\infty \frac{1}{n}$, diverges.

p-Series

If p is a fixed number, the infinite series $\sum_{n=1}^\infty \frac{1}{n^p}$ is called a p-series. Theorem 2 describes how the convergence of the p-series depends on the value of p.

Theorem 2   Fix a real number p. The series $\sum_{n=1}^\infty \frac{1}{n^p}$ converges if $p > 1$; it diverges if $p \le 1$.

Example 2 Determine whether the series $\sum_{n=1}^\infty \frac{(n+2)^2}{\sqrt{n^7}}$ converges or diverges.

Solution

$$\sum_{n=1}^\infty \frac{(n+2)^2}{\sqrt{n^7}} = \sum_{n=1}^\infty \frac{(n^2 + 4n + 4)}{n^{7/2}}$$

$$= \sum_{n=1}^\infty \frac{n^2}{n^{7/2}} + \sum_{n=1}^\infty \frac{4n}{n^{7/2}} + \sum_{n=1}^\infty \frac{4}{n^{7/2}}$$

$$= \sum_{n=1}^\infty \frac{1}{n^{3/2}} + 4\sum_{n=1}^\infty \frac{1}{n^{5/2}} + 4\sum_{n=1}^\infty \frac{1}{n^{7/2}}$$

Because each of the series $\sum\limits_{n=1}^{\infty} \dfrac{1}{n^{3/2}}$, $4\sum\limits_{n=1}^{\infty} \dfrac{1}{n^{5/2}}$, and $4\sum\limits_{n=1}^{\infty} \dfrac{1}{n^{7/2}}$ is convergent by Theorem 2, the

given series $\sum\limits_{n=1}^{\infty} \dfrac{(n+2)^2}{\sqrt{n^7}}$ is convergent.

Self-Test Exercise

2. Determine whether the series $\sum\limits_{n=1}^{\infty} \dfrac{(n+2)^2}{\sqrt{n^5}}$ is convergent or divergent.

9.4 Series with Nonnegative Terms—The Comparison Test

Important Concepts

- The Comparison Test for Convergence
- The Comparison Test for Divergence
- The Limit Comparison Test

The Comparison Test for Convergence

Theorem 1

The Comparison Test for Convergence　Let $0 \le a_n \le b_n$ for every n. If the series $\sum\limits_{n=1}^{\infty} b_n$ converges, then the series $\sum\limits_{n=1}^{\infty} a_n$ also converges.

Example 1　Show that the series $\sum\limits_{n=1}^{\infty} \dfrac{(n-1)}{n^3 - 1}$ converges.

Solution　For $n \ge 1$, $n^2 + n + 1 > n^2 > 0$; this implies that $0 < \dfrac{1}{n^2 + n + 1} < \dfrac{1}{n^2}$.

$$\sum_{n=1}^{\infty} \frac{(n-1)}{n^3 - 1} = \sum_{n=1}^{\infty} \frac{(n-1)}{(n-1)(n^2 + n + 1)} = \sum_{n=1}^{\infty} \frac{1}{(n^2 + n + 1)} < \sum_{n=1}^{\infty} \frac{1}{n^2}.$$

$\sum\limits_{n=1}^{\infty} \dfrac{1}{n^2}$ is a p-series with $p = 2 > 1$, so it converges. By the Comparison Test for Convergence, the given series converges.

Self-Test Exercise

3. Determine if the series $\sum\limits_{n=1}^{\infty} \dfrac{1}{3n + 3^n}$ converges.

The Comparison Test for Divergence

Theorem 2

The Comparison Test for Divergence Let $0 \le b_n \le a_n$ for all sufficiently large n. If the series $\sum_{n=1}^{\infty} b_n$ diverges, then the series $\sum_{n=1}^{\infty} a_n$ also diverges.

Example 2 Show that the series $\sum_{n=1}^{\infty} \frac{1}{\sqrt{6n-5}}$ diverges.

Solution For $n \ge 1$, $6n - 5 < 6n$. This implies that $0 < \frac{1}{6n} < \frac{1}{6n-5}$, which in turn implies that $0 < \sqrt{\frac{1}{6n}} < \sqrt{\frac{1}{6n-5}}$. However, $\sqrt{\frac{1}{6n}} = \frac{1}{\sqrt{6} \cdot \sqrt{n}}$ and $\sqrt{\frac{1}{6n-5}} = \frac{1}{\sqrt{6n-5}}$, so we have $0 < \frac{1}{\sqrt{6} \cdot \sqrt{n}} < \frac{1}{\sqrt{6n-5}}$ for $n \ge 1$. The series $\sum_{n=1}^{\infty} \frac{1}{\sqrt{n}} = \sum_{n=1}^{\infty} \frac{1}{n^{1/2}}$ is a p-series with $p = \frac{1}{2} < 1$, so the series $\sum_{n=1}^{\infty} \frac{1}{\sqrt{6} \cdot \sqrt{n}}$ diverges. By the Comparison Test for Divergence, the series $\sum_{n=1}^{\infty} \frac{1}{\sqrt{6n-5}}$ also diverges. ■

Self-Test Exercise

4. Show that the series $\sum_{n=1}^{\infty} \frac{1}{\sqrt{\sqrt{n}+1}}$ diverges.

The Limit Comparison Test

Theorem 3

The Limit Comparison Test Let $\sum_{n=1}^{\infty} a_n$ and $\sum_{n=1}^{\infty} b_n$ be a series of positive terms. If $\lim_{n \to \infty} \frac{a_n}{b_n}$ exists and is a finite positive number, then $\sum_{n=1}^{\infty} a_n$ converges if and only if $\sum_{n=1}^{\infty} b_n$ converges.

Example 3 Determine whether the series $\sum_{n=1}^{\infty} \frac{n^2}{(n^2+1)^2}$ converges.

Solution $(n^2+1)^2 = n^4 + 2n^2 + 1$, so the nth summand a_n is about $\frac{n^2}{n^4} = \frac{1}{n^2}$ for large values of n. Apply the Limit Comparison Test with the series $\sum_{n=1}^{\infty} b_n = \sum_{n=1}^{\infty} \frac{1}{n^2}$.

$$
\begin{aligned}
\lim_{n \to \infty} \frac{a_n}{b_n} &= \lim_{n \to \infty} \frac{\dfrac{n^2}{(n^2+1)^2}}{\dfrac{1}{n^2}} \\
&= \lim_{n \to \infty} \frac{n^4}{n^4 + 2n^2 + 1} \\
&= \lim_{n \to \infty} \frac{1}{1 + \dfrac{2}{n^2} + \dfrac{1}{n^4}} \\
&= 1 < \infty
\end{aligned}
$$

The limit is 1; therefore, a finite positive number and $\sum_{n=1}^{\infty} b_n = \sum_{n=1}^{\infty} \frac{1}{n^2}$ converges because it is a p-series with $p = 2 > 1$. By applying the Limit Comparison Test, we can determine that the given series $\sum_{n=1}^{\infty} \frac{n^2}{(n^2+1)^2}$ also converges. ∎

Example 4 Show that the series $\sum_{n=1}^{\infty} \frac{1}{\sqrt{6n-5}}$ diverges using the Limit Comparison Test.

Solution Let $a_n = \frac{1}{\sqrt{6n-5}}$. Apply the Limit Comparison Test using the series $\sum_{n=1}^{\infty} b_n = \sum_{n=1}^{\infty} \frac{1}{\sqrt{n}}$ for comparison:

$$
\begin{aligned}
\lim_{n\to\infty} \frac{a_n}{b_n} &= \lim_{n\to\infty} \frac{\dfrac{1}{\sqrt{6n-5}}}{\dfrac{1}{\sqrt{n}}} \\
&= \lim_{n\to\infty} \frac{\sqrt{n}}{\sqrt{6n-5}} \\
&= \lim_{n\to\infty} \sqrt{\frac{n}{6n-5}} \\
&= \lim_{n\to\infty} \sqrt{\frac{1}{6-\dfrac{5}{n}}} \\
&= \sqrt{\frac{1}{6}}.
\end{aligned}
$$

Since this limit is finite and positive and since $\sum_{n=1}^{\infty} \frac{1}{\sqrt{n}}$ diverges, we conclude that $\sum_{n=1}^{\infty} \frac{1}{\sqrt{6n-5}}$ diverges. ∎

Please note that it is easier to use the Limit Comparison Test than the Comparison Test because we do not have to do a term-by-term comparison in the Limit Comparison Test. In Example 4, if we change the series from $\sum_{n=1}^{\infty} \frac{1}{\sqrt{6n-5}}$ to $\sum_{n=1}^{\infty} \frac{1}{\sqrt{6n+5}}$, then a similar argument that we used in there, namely $\lim_{n\to\infty} \frac{\dfrac{1}{\sqrt{6n+5}}}{\dfrac{1}{\sqrt{n}}} = \sqrt{\frac{1}{6}}$, can still be used to show that $\sum_{n=1}^{\infty} \frac{1}{\sqrt{6n+5}}$ diverges. However, we cannot apply a similar argument that we used in Example 2 to this new series $\sum_{n=1}^{\infty} \frac{1}{\sqrt{6n+5}}$, because the term-by-term comparison $\frac{1}{\sqrt{6}\cdot\sqrt{n}} < \frac{1}{\sqrt{6n+5}}$ is no longer true.

Self-Test Exercise

5. Use the Limit Comparison Test to determine whether the series $\displaystyle\sum_{n=1}^{\infty} \frac{n^2}{(n^2 + 1)}$ converges or diverges.

9.5 Alternating Series

Important Concepts

- Alternating Series Test
- Absolute Convergence
- Conditional Convergence

Alternating Series Test

Theorem 1

Alternating Series Test   If $\{a_n\}$ is a sequence of nonnegative numbers that satisfies

    a. $a_1 \geq a_2 \geq a_3 \geq \cdots$, and

    b. $\displaystyle\lim_{n\to\infty} a_n = 0$,

then the series $\displaystyle\sum_{n=1}^{\infty} (-1)^{n+1} a_n$ converges. Furthermore, $\ell = \displaystyle\sum_{n=1}^{\infty} (-1)^{n+1} a_n$ lies between each consecutive pair of consecutive partial sums $S_N = \displaystyle\sum_{n=1}^{N} (-1)^{n+1} a_n$ and $S_{N+1} = S_N + (-1)^N a_{N+1}$. In particular, the limit ℓ satisfies the inequality $|\ell - S_N| \leq a_{N+1}$.

Example 1 Apply the Alternating Series Test to the series $\displaystyle\sum_{n=1}^{\infty} \frac{(-1)^{n+1}}{2^n}$.

Solution Before we can apply the Alternating Series Test, we must verify that the conditions in the hypothesis are met. Choose $a_n = \dfrac{1}{2^n}$. Because $2 \leq 2^2 \leq 2^3 \leq \cdots$ implies that $\dfrac{1}{2} \geq \dfrac{1}{2^2} \geq \dfrac{1}{2^3} \geq \cdots$, we have established that $a_1 \geq a_2 \geq a_3 \geq \cdots$ and that $a_n \to 0$. Therefore, the series has the form of an alternating series, as in Theorem 1. The Alternating Series Test applies and the series $\displaystyle\sum_{n=1}^{\infty} \frac{(-1)^{n+1}}{2^n}$ converges. ∎

Self-Test Exercise

1. Analyze the series $\displaystyle\sum_{n=1}^{\infty} \frac{(-1)^{n+1}}{\sqrt{2n - 1}}$.

Absolute Convergence

Let $\sum\limits_{n=1}^{\infty} a_n$ be a series, possibly containing both positive and negative terms. If the series $\sum\limits_{n=1}^{\infty} |a_n|$ of absolute values converges, we say that the series $\sum\limits_{n=1}^{\infty} a_n$ *converges absolutely.*

 Theorem 2 If the series $\sum\limits_{n=1}^{\infty} |a_n|$ converges, then the series $\sum\limits_{n=1}^{\infty} a_n$ converges.

In other words, Theorem 2 states that absolute convergence implies convergence.

Example 2 Determine whether the series $\sum\limits_{n=1}^{\infty} \left(\dfrac{-1}{n}\right)^5$ converges absolutely.

Solution $\sum\limits_{n=1}^{\infty} \left(\dfrac{-1}{n}\right)^5 = \sum\limits_{n=1}^{\infty} \dfrac{(-1)^5}{n^5}$ and $\sum\limits_{n=1}^{\infty} \left|\left(\dfrac{-1}{n}\right)^5\right| = \sum\limits_{n=1}^{\infty} \left|\dfrac{(-1)^5}{n^5}\right| = \sum\limits_{n=1}^{\infty} \dfrac{1}{n^5}$

$\sum\limits_{n=1}^{\infty} \dfrac{1}{n^5}$ converges because it is a p-series with $p = 5 > 1$. Therefore, the given series $\sum\limits_{n=1}^{\infty} \left(\dfrac{-1}{n}\right)^5$ converges absolutely. ∎

Theorem 3 Let $\sum\limits_{n=0}^{\infty} a_n$ and $\sum\limits_{n=0}^{\infty} b_n$ be absolutely convergent series.

   a. $\sum\limits_{n=0}^{\infty} (a_n + b_n)$ and $\sum\limits_{n=0}^{\infty} (a_n - b_n)$ are absolutely convergent.

   b. $\sum\limits_{n=0}^{\infty} \lambda a_n$ converges absolutely for any real constant λ.

Self-Test Exercise

2. If $\sum\limits_{n=0}^{\infty} a_n = \sum\limits_{n=0}^{\infty} \dfrac{1}{n^3 + 5}$ and $\sum\limits_{n=0}^{\infty} b_n = \sum\limits_{n=0}^{\infty} \dfrac{(-1)^n}{\sqrt{n^3 + 5}}$, determine if $\sum\limits_{n=0}^{\infty} (a_n + b_n)$ is absolutely convergent.

Conditional Convergence

If a series $\sum\limits_{n=1}^{\infty} a_n$ converges but does not converge absolutely, then we say that the series *converges conditionally.*

Example 3 Verify that the series $\sum\limits_{n=1}^{\infty} \dfrac{(-1)^n}{\sqrt[6]{n}}$ is conditionally convergent.

Solution $\displaystyle\sum_{n=1}^{\infty} \frac{(-1)^n}{\sqrt[6]{n}} = \sum_{n=1}^{\infty} \frac{(-1)^n}{n^{1/6}}$

$\displaystyle\sum_{n=1}^{\infty} \left| \frac{(-1)^n}{n^{1/6}} \right| = \sum_{n=1}^{\infty} \frac{1}{n^{1/6}}$ is not convergent because it is a p-series with $p = \dfrac{1}{6} < 1$.

Therefore, the given series does not converge absolutely. However, $1 < 2 < 3 < \cdots$ implies

that $1^{1/6} < 2^{1/6} < 3^{1/6} < \cdots$, which in turn implies that $\dfrac{1}{1^{1/6}} > \dfrac{1}{2^{1/6}} > \dfrac{1}{3^{1/6}} > \cdots$. This

establishes that $a_1 \geq a_2 \geq a_3 \geq \cdots$. In addition, $\displaystyle\lim_{n\to\infty} a_n = \lim_{n\to\infty} \frac{1}{n^{1/6}} = 0$. Therefore, by

the Alternating Series Test, the given series $\displaystyle\sum_{n=1}^{\infty} \frac{(-1)^n}{\sqrt[6]{n}}$ is convergent. Because the given series

$\displaystyle\sum_{n=1}^{\infty} \frac{(-1)^n}{\sqrt[6]{n}}$ is not absolutely convergent but is convergent, it is conditionally convergent. ■

Self-Test Exercise

3. Determine if the series $\displaystyle\sum_{n=1}^{\infty} \frac{(-1)^{n-1}}{e^{3n}}$ is absolutely convergent, convergent, or conditionally convergent.

9.6 The Ratio and Root Tests

Important Concepts

- The Ratio Test
- The Root Test

The Ratio Test

Theorem 1

The Ratio Test   Let $\displaystyle\sum_{n=1}^{\infty} a_n$ be a series. Suppose $\displaystyle\lim_{n\to\infty} \left| \frac{a_{n+1}}{a_n} \right| = L$.

a.   If $L < 1$, then the series converges absolutely.
b.   If $L > 1$, then the series diverges.
c.   If $L = 1$, then the test gives no information.

Example 1 Use the Ratio Test to analyze the series $\displaystyle\sum_{n=1}^{\infty} \frac{n!}{3^n}$.

Solution

$$\lim_{n\to\infty} \left| \frac{a_{n+1}}{a_n} \right| = \lim_{n\to\infty} \left| \frac{\dfrac{(n+1)!}{3^{n+1}}}{\dfrac{n!}{3^n}} \right|$$

$$= \lim_{n\to\infty} \frac{(n+1)!}{3^{n+1}} \frac{3^n}{n!}$$

$$= \lim_{n\to\infty} \frac{n+1}{3}$$

$$= \infty$$

$L = \infty > 1$, so the given series $\displaystyle\sum_{n=1}^{\infty} \frac{n!}{3^n}$ diverges. ■

Self-Test Exercise

1. Use the Ratio Test to analyze the series $\sum_{n=1}^{\infty} \dfrac{3^n}{n!}$.

The Root Test

The Root Test   Let $\sum_{n=1}^{\infty} a_n$ be a series. Suppose $\lim_{n\to\infty} |a_n|^{1/n} = L$.

   a.   If $L < 1$, then the series converges absolutely.
   b.   If $L > 1$, then the series diverges.
   c.   If $L = 1$, then the test gives no information.

Example 2   Apply the Root Test to the series $\sum_{n=1}^{\infty} \dfrac{3^n}{n}$.

Solution $a_n = \dfrac{3^n}{n}$. Therefore,

$$\lim_{n\to\infty} |a_n|^{1/n} = \lim_{n\to\infty} \left| \frac{3^n}{n} \right|^{1/n}$$
$$= \lim_{n\to\infty} \frac{3}{n^{1/n}}$$
$$= \frac{3}{1}$$
$$= 3 > 1.$$

Therefore, the given series $\sum_{n=1}^{\infty} \dfrac{3^n}{n}$ diverges.

Self-Test Exercise

2. Apply the Root Test to the series $\sum_{n=1}^{\infty} \dfrac{3^n}{n^3}$.

Taylor Series

10.1 Introduction to Power Series

Important Concepts

- Definition
- Radius and Interval of Convergence
- Power Series About an Arbitrary Base Point

Definition

An expression of the form $a_0 + a_1 x + a_2 x^2 + a_3 x^3 + \cdots$, where the coefficients a_n are constants, is called a *power series in x*. In sigma notation, $a_0 + a_1 x + a_2 x^2 + a_3 x^3 + \cdots = \sum_{n=0}^{\infty} a_n x^n$.

Note that $\sum_{n=0}^{\infty} a_n x^n$ converges for $x = 0$ and can be regarded as a function of x with domain consisting of all x for which the series converges.

Example 1 Determine if the power series $\sum_{n=0}^{\infty} (3x)^n$ converges at the following points: $x = 0$, 1, and $-\dfrac{1}{5}$.

Solution At $x = 0$, the power series becomes the series $\sum_{n=0}^{\infty} 0^n = \sum_{n=0}^{\infty} 0$, which is, of course, a convergent series.

At $x = 1$, the power series becomes the geometric series $\sum_{n=0}^{\infty} 3^n$. Recall that a geometric series $\sum_{n=0}^{\infty} u^n = 1 + u + u^2 + u^3 + \cdots$ diverges when $|u| > 1$; therefore, the power series $\sum_{n=0}^{\infty} (3x)^n$ diverges at $x = 1$.

At $x = -\dfrac{1}{5}$, the power series becomes the geometric series $\sum_{n=0}^{\infty} \left(\dfrac{-3}{5}\right)^n$. Since a geometric series $\sum_{n=0}^{\infty} u^n = 1 + u + u^2 + u^3 + \cdots$ converges when $|u| < 1$, we conclude that the power series $\sum_{n=0}^{\infty} (3x)^n$ converges at $x = -\dfrac{1}{5}$. ∎

Radius and Interval of Convergence

Theorem 1 If we let $\displaystyle\sum_{n=0}^{\infty} a_n x^n$ be a power series, then precisely one of the following statements holds:

    a. The series converges absolutely for every real x.
    b. There is a positive number R such that the series converges absolutely for $|x| < R$ and diverges for $|x| > R$.
    c. The series converges only at $x = 0$.

In case a of Theorem 1, $R = \infty$ and in case c of Theorem 1, $R = 0$. In all cases, R is called the *radius of convergence* of the power series.

Theorem 2 Suppose that the limit $\ell = \displaystyle\lim_{n \to \infty} |a_n|^{1/n}$ exists (as a nonnegative real number or ∞). Let R denote the radius of convergence of the power series $\displaystyle\sum_{n=0}^{\infty} a_n x^n$. If $0 < \ell < \infty$, then $R = \dfrac{1}{\ell}$. If $\ell = 0$, then $R = \infty$. If $\ell = \infty$, then $R = 0$.

The set of points at which a power series $\displaystyle\sum_{n=0}^{\infty} a_n x^n$ converges is called the *interval of convergence*. Various possibilities that can arise for the intervals of convergence are illustrated in Figure 1.

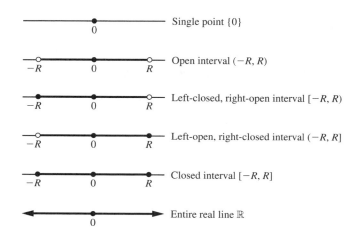

Figure 1

Example 2 Identify the interval of convergence for the power series $\displaystyle\sum_{n=0}^{\infty} (3x)^n$.

Solution $\displaystyle\sum_{n=0}^{\infty} u^n = 1 + u + u^2 + u^3 + \cdots$ is both a power series in u and a geometric series. As a geometric series, $\displaystyle\sum_{n=0}^{\infty} u^n = 1 + u + u^2 + u^3 + \cdots$ converges when $|u| < 1$, i.e., when $u \in (-1, 1)$. If we substitute $u = 3x$, we obtain $\displaystyle\sum_{n=0}^{\infty} (3x)^n = 1 + (3x) + (3x)^2 + (3x)^3 + \cdots$. It converges when $|(3x)| < 1$. However, $|(3x)| = |(3)||(x)| = 3|(x)|$, so $|(3x)| < 1 \Rightarrow 3|(x)| < 1 \Rightarrow |(x)| < \dfrac{1}{3}$. Therefore, the interval of convergence I for the power series $\displaystyle\sum_{n=0}^{\infty} (3x)^n$ is $I = \left(\dfrac{-1}{3}, \dfrac{1}{3} \right)$. ∎

Example 3 Use Theorem 2 to determine the interval of convergence for the power series $\sum_{n=0}^{\infty} (3x)^n$.

Solution

$$\sum_{n=0}^{\infty} (3x)^n = \sum_{n=0}^{\infty} 3^n x^n, \text{ so } a_n = 3^n.$$

$$\ell = \lim_{n \to \infty} |a_n|^{1/n}$$
$$= \lim_{n \to \infty} |3^n|^{1/n}$$
$$= \lim_{n \to \infty} 3$$
$$= 3$$

is a finite positive number. Therefore, $R = \dfrac{1}{\ell} = \dfrac{1}{3}$. This tells us that the power series converges absolutely when $|x| < \dfrac{1}{3}$ and diverges when $|x| > \dfrac{1}{3}$. Now let us check the two endpoints. When $x = \dfrac{1}{3}$, the power series becomes $\sum_{n=0}^{\infty} 1^n = \sum_{n=0}^{\infty} 1$, which diverges. When $x = \dfrac{-1}{3}$, the power series becomes $\sum_{n=0}^{\infty} (-1)^n$, which also diverges. The interval of convergence is therefore

$$I = (-R, R) = \left(\dfrac{-1}{3}, \dfrac{1}{3} \right).$$ ■

Self-Test Exercise

1. Determine the interval of convergence for the power series $\sum_{n=0}^{\infty} \left(\dfrac{x}{5} \right)^n$.

Power Series About an Arbitrary Base Point

An expression of the form $S = a_0 + a_1 (x - c) + a_2 (x - c)^2 + a_3 (x - c)^3 + \cdots$, where center c and the coefficients a_n are constants, is called a *power series in x with base point (or center) c*.

In sigma notation, $S = a_0 + a_1 (x - c) + a_2 (x - c)^2 + a_3 (x - c)^3 + \cdots = \sum_{n=0}^{\infty} a_n (x - c)^n$.

Theorem 4 Let $\sum_{n=0}^{\infty} a_n (x - c)^n$ be a power series for which the limit $\ell = \lim_{n \to \infty} |a_n|^{1/n}$ exists as a nonnegative real number or ∞. Set $R = \dfrac{1}{\ell}$ if ℓ is a positive real number. If $\ell = 0$, then set $R = \infty$. If $\ell = \infty$, then set $R = 0$.

a. If the limit $\lim_{n \to \infty} \left| \dfrac{a_{n+1}}{a_n} \right|$ also exists as a nonnegative real number or ∞, then $\ell = \lim_{n \to \infty} \left| \dfrac{a_{n+1}}{a_n} \right|$.

b. The series converges absolutely for $|x - c| < R$ and diverges for $|x - c| > R$. In particular, if $R = \infty$, then the series converges absolutely for every real x. If $R = 0$, then the series converges only for $x = c$.

As in the case $c = 0$, R is called the *radius of convergence* of the power series. The set on which the series converges is called the *interval of convergence*. If $R = \infty$, then the interval of convergence is the entire real line. If $R = 0$, then the interval of convergence is the single point $\{c\}$. When $0 < R < \infty$, the series may or may not converge at the endpoints $x = c + R$ and $x = c - R$. To determine the interval of convergence, we must test each endpoint separately by substituting the endpoints $x = c + R$ and $x = c - R$ into the series. Therefore, when R is positive and finite, the interval of convergence will have the form $[c - R, c + R]$ or $(c - R, c + R]$ or $[c - R, c + R)$ or $(c - R, c + R)$. Figure 2 illustrates the possibilities.

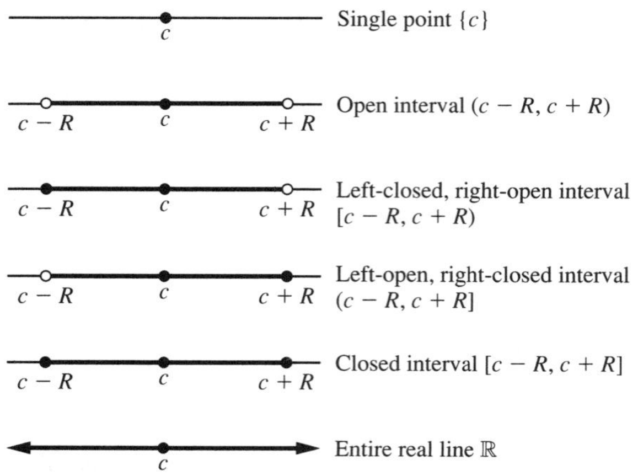

Figure 2

Example 4 Calculate the interval of convergence for the power series $\sum_{n=0}^{\infty} (5x - 5)^n$.

Solution
$$\sum_{n=0}^{\infty} (5x - 5)^n = \sum_{n=0}^{\infty} (5(x - 1))^n = \sum_{n=0}^{\infty} 5^n (x - 1)^n$$

It is a power series with $a_n = 5^n$ and center $c = 1$.

$$\lim_{n\to\infty} \left| \frac{a_{n+1}}{a_n} \right| = \lim_{n\to\infty} \left| \frac{5^{n+1}}{5^n} \right|$$
$$= \lim_{n\to\infty} 5$$
$$= 5$$

is a positive real number. Therefore, $\ell = 5$ and $R = \frac{1}{\ell} = \frac{1}{5}$.

We need to test the endpoints $x = c + R = 1 + \frac{1}{5} = \frac{6}{5}$ and $x = c - R = 1 - \frac{1}{5} = \frac{4}{5}$. When $x = \frac{4}{5}$, the power series becomes

$$\sum_{n=0}^{\infty} 5^n \left(\frac{4}{5} - 1 \right)^n = \sum_{n=0}^{\infty} 5^n \left(\frac{-1}{5} \right)^n$$
$$= \sum_{n=0}^{\infty} 5^n \frac{(-1)^n}{5^n}$$
$$= \sum_{n=0}^{\infty} (-1)^n,$$

which diverges. When $x = \frac{6}{5}$, the power series becomes

$$\sum_{n=0}^{\infty} 5^n \left(\frac{6}{5} - 1 \right)^n = \sum_{n=0}^{\infty} 5^n \left(\frac{1}{5} \right)^n$$
$$= \sum_{n=0}^{\infty} 5^n \frac{(1)^n}{5^n}$$
$$= \sum_{n=0}^{\infty} (1)^n,$$

which also diverges. Therefore, the interval of convergence is $(c - R, c + R) = \left(\frac{4}{5}, \frac{6}{5} \right)$. ∎

Self-Test Exercise

2. Calculate the interval of convergence for the power series $\displaystyle\sum_{n=0}^{\infty} \frac{(x-6)^n}{3^n}$.

10.2 Operations on Power Series

Important Concepts

- Addition and Scalar Multiplication
- Differentiation and Antidifferentiation of Power Series
- Power Series Expansions of Some Standard Functions
- A Uniqueness Theorem with Applications to Differential Equations
- The Relationship between the Coefficients and Derivatives of Power Series

Addition and Scalar Multiplication

Theorem 1 deals with adding or subtracting two power series *with the same base point* and multiplying a power series by a constant.

Theorem 1

Let λ be any real number, and let k be a positive integer. Suppose that $\displaystyle\sum_{n=0}^{\infty} a_n (x-c)^n$ and $\displaystyle\sum_{n=0}^{\infty} b_n (x-c)^n$ are power series that converge absolutely on an interval I.

a. The power series $\displaystyle\sum_{n=0}^{\infty} (a_n + b_n)(x-c)^n$ and $\displaystyle\sum_{n=0}^{\infty} (a_n - b_n)(x-c)^n$ also converge absolutely on I; moreover,

$$\sum_{n=0}^{\infty} a_n (x-c)^n + \sum_{n=0}^{\infty} b_n (x-c)^n = \sum_{n=0}^{\infty} (a_n + b_n)(x-c)^n$$

and

$$\sum_{n=0}^{\infty} a_n (x-c)^n - \sum_{n=0}^{\infty} b_n (x-c)^n = \sum_{n=0}^{\infty} (a_n - b_n)(x-c)^n.$$

b. The power series $\displaystyle\sum_{n=0}^{\infty} \lambda a_n (x-c)^n$ converges absolutely on I and

$$\lambda \sum_{n=0}^{\infty} a_n (x-c)^n = \sum_{n=0}^{\infty} \lambda a_n (x-c)^n.$$

c. The power series $\displaystyle\sum_{n=0}^{\infty} a_n (x-c)^{n+k}$ converges absolutely on I and

$$\sum_{n=0}^{\infty} a_n (x-c)^{n+k} = (x-c)^k \sum_{n=0}^{\infty} a_n (x-c)^n.$$

Example 1 Express $\dfrac{x^4}{(8 - 4x^2)}$ as a power series with base point 0.

Solution

$$\frac{x^4}{(8 - 4x^2)} = \frac{x^4}{8\left(1 - \dfrac{1}{2}x^2\right)}$$

$$= \frac{x^4}{8\left(1 - \left(\dfrac{1}{\sqrt{2}}x\right)^2\right)}$$

$$= \frac{1}{8}\left(\frac{x^4}{1}\right)\left(\frac{1}{\left(1 - \left(\dfrac{1}{\sqrt{2}}x\right)^2\right)}\right)$$

Recall that $\dfrac{1}{1 - u} = \displaystyle\sum_{n=0}^{\infty} u^n$. By substituting $u = \left(\dfrac{1}{\sqrt{2}}x\right)^2$, we obtain

$$\frac{1}{\left(1 - \left(\dfrac{1}{\sqrt{2}}x\right)^2\right)} = \sum_{n=0}^{\infty}\left(\left(\frac{1}{\sqrt{2}}x\right)^2\right)^n$$

$$= \sum_{n=0}^{\infty}\left(\frac{1}{2}\right)^n (x)^{2n}.$$

$\displaystyle\sum_{n=0}^{\infty}\left(\left(\dfrac{1}{\sqrt{2}}x\right)^2\right)^n$ converges for $\left|\dfrac{x^2}{2}\right| < 1$, i.e., for $|x| < \sqrt{2}$.

$$\frac{1}{8}\left(\frac{x^4}{1}\right)\left(\frac{1}{\left(1 - \left(\dfrac{1}{\sqrt{2}}x\right)^2\right)}\right) = \frac{1}{8}\left(\frac{x^4}{1}\right)\sum_{n=0}^{\infty}\left(\frac{1}{2}\right)^n (x)^{2n}$$

$$= \sum_{n=0}^{\infty}\left(\frac{1}{2}\right)^{n+3} (x)^{2n+4} \text{ for } |x| < \sqrt{2}$$

We conclude that $\dfrac{x^4}{(8 - 4x^2)} = \displaystyle\sum_{n=0}^{\infty}\left(\frac{1}{2}\right)^{n+3} (x)^{2n+4}$ for $|x| < \sqrt{2}$. ∎

Self-Test Exercise

1. Express $\dfrac{x^5}{(2 + 3x^2)}$ as a power series with base point 0.

Differentiation and Antidifferentiation of Power Series

Theorem 2   Let I denote the interval $(c - R, c + R)$, where R is the radius of convergence of $f(x) = \displaystyle\sum_{n=0}^{\infty} a_n (x - c)^n$.

a. The function f is infinitely differentiable on I and

$$f'(x) = \sum_{n=1}^{\infty} n a_n (x - c)^{n-1}.$$

The power series for f' also converges absolutely on the interval I.

b. The power series

$$F(x) = \sum_{n=0}^{\infty} \frac{a_n}{n+1} (x - c)^{n+1}$$

converges absolutely on the interval I. The function F satisfies the equation $F(x) = \int_c^x f(t)\, dt$ for all x in I. In particular, $F'(x) = f(x)$ on I. The indefinite integral of $f(x)$ is given by $\int f(x)\, dx = F(x) + C$, where C is an arbitrary constant.

Example 2 Calculate the derivative of the power series $f(x) = \sum_{n=0}^{\infty} 5^{-n} (x - 1)^n$ for x inside the interval of convergence $I = (-4, 6)$.

Solution By implementing Theorem 2, we obtain the derivative $f'(x) = \sum_{n=0}^{\infty} 5^{-n} \left(n (x - 1)^{n-1} \right)$, for x inside the interval of convergence $I = (-4, 6)$. ∎

Example 3 Calculate $\int \dfrac{1}{1 + x^3}\, dx$ as a power series and identify the interval of convergence.

Solution

$$\frac{1}{1 + x^3} = \frac{1}{1 - (-x^3)}$$

$$\frac{1}{1 - u} = 1 + u + u^2 + u^3 + \cdots = \sum_{n=0}^{\infty} u^n \text{ converges when } |u| < 1, \text{ i.e., when } u \in (-1, 1). \text{ If}$$

we substitute $u = -x^3$, we obtain

$$\frac{1}{1 - (-x^3)} = \sum_{n=0}^{\infty} (-x^3)^n = 1 + (-x^3) + (-x^3)^2 + (-x^3)^3 + \cdots,$$

which converges when $|(-x^3)| < 1$. However, $|(-x^3)| = |(-1)||(x^3)| = 1|(x^3)|$, so $|(-x^3)| < 1$ $\Rightarrow |(x)|^3 < 1 \Rightarrow |(x)| < 1$. Therefore, the interval of convergence I for the power series $\sum_{n=0}^{\infty} (-x^3)^n$ is $I = (-1, 1)$. By Theorem 2,

$$\int \frac{1}{1 + x^3}\, dx = \int \sum_{n=0}^{\infty} (-x^3)^n\, dx$$

$$= \int \sum_{n=0}^{\infty} (-1)^n x^{3n}\, dx$$

$$= \sum_{n=0}^{\infty} (-1)^n \int x^{3n}\, dx$$

$$= \sum_{n=0}^{\infty} \frac{(-1)^n}{3n + 1} x^{3n+1} + C.$$

Therefore, $\displaystyle\int \frac{1}{1 + x^3}\, dx = \sum_{n=0}^{\infty} \frac{(-1)^n}{3n + 1} x^{3n+1} + C$ converges in $(-1, 1)$. We must check if convergence occurs at the endpoint. If $x = 1$, then $\sum_{n=0}^{\infty} \frac{(-1)^n}{3n+1} x^{3n+1}$ becomes $\sum_{n=0}^{\infty} \frac{(-1)^n}{3n+1}$. We

apply the Alternating Series Test to the series: $a_n = \frac{1}{3n+1}$.

$1 < 4 < 7 < 10 < \cdots \Rightarrow 1 > \frac{1}{4} > \frac{1}{7} > \frac{1}{10} > \cdots$ establishes that we have a decreasing

sequence; $\lim_{n\to\infty} a_n = \frac{1}{3n+1} = 0$. Therefore, the series $\sum_{n=0}^{\infty} \frac{(-1)^n}{3n+1} x^{3n+1}$ converges at $x = 1$. If

$x = -1$, then $\sum_{n=0}^{\infty} \frac{(-1)^n}{3n+1} x^{3n+1}$ becomes

$$\sum_{n=0}^{\infty} \frac{(-1)^n}{3n+1} (-1)^{3n+1} = \sum_{n=0}^{\infty} \frac{(-1)^{4n+1}}{3n+1}$$

$$= \sum_{n=0}^{\infty} \frac{(-1)}{3n+1}$$

$$= -\sum_{n=0}^{\infty} \frac{1}{3n+1}.$$

Comparing this series with $\sum_{n=1}^{\infty} \frac{1}{n}$ using the Limit Comparing Test, we conclude that it diverges.

Therefore, the series $\sum_{n=0}^{\infty} \frac{(-1)^n}{3n+1} x^{3n+1}$ diverges at $x = -1$.

Therefore, the interval of convergence for the series $\sum_{n=0}^{\infty} \frac{(-1)^n}{3n+1} x^{3n+1}$ is $I = (-1, 1]$. ∎

Self-Test Exercise

2. Identify the interval of convergence for the power series $f(x) = \sum_{n=0}^{\infty} 5^{-n} (x-1)^n$ and its derivative.

Power Series Expansions of Some Standard Functions

We can express some standard functions as power series through the use of differentiation and integration.

Example 4 Find a power series expansion for $F(x) = \arctan(3x)$, and determine its interval of convergence.

Solution Recall that $\frac{d}{du}(\arctan(u)) = \frac{1}{1+u^2}$. Therefore,

$$F'(x) = \frac{d}{dx} \arctan(3x)$$

$$= \frac{1}{1+(3x)^2} \cdot \frac{d}{dx} 3x$$

$$= \frac{3}{1+9x^2}.$$

We can express $F'(x)$ in a power series,

$$F'(x) = \frac{3}{1 + 9x^2}$$

$$= 3\left(\frac{1}{1 - (-9x^2)}\right)$$

$$= 3\sum_{n=0}^{\infty} (-9x^2)^n$$

$$= \sum_{n=0}^{\infty} (-1)^n (3)^{2n+1} x^{2n},$$

which converges when $\left|-9x^2\right| < 1$, or equivalently, when $\frac{-1}{3} < x < \frac{1}{3}$. Therefore,

$$F(x) = \int F'(x)\,dx$$

$$= \int \left(\sum_{n=0}^{\infty} (-1)^n (3)^{2n+1} x^{2n}\right) dx$$

$$= \sum_{n=0}^{\infty} (-1)^n (3)^{2n+1} \int x^{2n}\,dx$$

$$= \sum_{n=0}^{\infty} \frac{(-1)^n (3)^{2n+1}}{2n + 1} x^{2n+1} + C.$$

Because $F(0) = \arctan(0) = 0$, we have $C = 0$, and $F(x) = \arctan(3x) = \sum_{n=0}^{\infty} \frac{(-1)^n (3)^{2n+1}}{2n + 1} x^{2n+1}$.

Now we need to determine the interval of convergence for this power series. This series converges for $\frac{-1}{3} < x < \frac{1}{3}$ because the power series for $F'(x)$ converges in that interval. We need to check for convergence at the endpoints, $x = \frac{-1}{3}$ and $x = \frac{1}{3}$. If $x = \frac{-1}{3}$ and $\sum_{n=0}^{\infty} \frac{(-1)^n (3)^{2n+1}}{2n + 1} x^{2n+1}$, then $\sum_{n=0}^{\infty} \frac{(-1)^{3n+1}}{2n + 1}$. If $x = \frac{1}{3}$ and $\sum_{n=0}^{\infty} \frac{(-1)^n (3)^{2n+1}}{2n + 1} x^{2n+1}$, then $\sum_{n=0}^{\infty} \frac{(-1)^n}{2n + 1}$. Apply the Alternating Series Test with $a_n = \frac{1}{2n + 1}$ (for both endpoints). $1 > \frac{1}{3} > \frac{1}{5} > \cdots$; therefore, the series is decreasing. Furthermore, $\lim_{n \to \infty} \frac{1}{2n + 1} = 0$. Because the conditions of the hypothesis of the Alternating Series Test are satisfied, we conclude that the series converges at the endpoints, $x = \frac{-1}{3}$ and $x = \frac{1}{3}$. Therefore, the series $F(x) = \sum_{n=0}^{\infty} \frac{(-1)^n (3)^{2n+1}}{2n + 1} x^{2n+1}$ converges for $\frac{-1}{3} \le x \le \frac{1}{3}$.   ■

Self-Test Exercise

3. Find a power series representation for $\ln(1 + 2x)$.

A Uniqueness Theorem with Applications to Differential Equations

Theorem 3 If $\displaystyle\sum_{n=0}^{\infty} a_n (x - c)^n = \sum_{n=0}^{\infty} b_n (x - c)^n$ for all x in an open interval containing the base point c, then $a_n = b_n$ for every n. In other words, if we fix a base point, the power series expansion of a function is unique.

Example 5 Find a power series solution of the initial value problem $\dfrac{dy}{dx} = ky$, $y(0) = 1$.

Solution Suppose that $y(x) = \displaystyle\sum_{n=0}^{\infty} a_n x^n$. Using Theorem 2, we obtain $\dfrac{dy}{dx} = \displaystyle\sum_{n=0}^{\infty} a_n n x^{n-1} = \displaystyle\sum_{n=1}^{\infty} a_n n x^{n-1}$. For y to satisfy the given differential equation, we must have $\displaystyle\sum_{n=1}^{\infty} a_n n x^{n-1} = k \displaystyle\sum_{n=0}^{\infty} a_n x^n$. If we make a change of summation index on the left side, we obtain $\displaystyle\sum_{n=0}^{\infty} a_{n+1} (n + 1) x^n = k \displaystyle\sum_{n=0}^{\infty} a_n x^n$. Expanding the series on both sides, we obtain

$$a_1 + 2a_2 x + 3a_3 x^2 + \sum_{n=3}^{\infty} a_{n+1} (n + 1) x^n = a_0 k + a_1 k x + a_2 k x^2 + \sum_{n=3}^{\infty} a_n k x^n.$$

Use the Uniqueness Theorem (Theorem 3) to deduce $a_1 = k a_0$, $2a_2 = k a_1$, $3a_3 = k a_2$, and $(n + 1)a_{n+1} = k a_n$ for $n \geq 3$. The initial condition $y(0) = 1$ tells us that $a_0 = 1$. Therefore, $a_1 = k$, $a_2 = \dfrac{k^2}{2}$, $a_3 = \dfrac{k^3}{6}$, and $a_n = \dfrac{k^n}{n!}$, for $n \geq 4$. It follows that on its interval of convergence,

$$y(x) = 1 + kx + \frac{k^2}{2}x^2 + \frac{k^3}{6}x^3 + \sum_{n=4}^{\infty} \frac{k^n}{n!} x^n = 1 + kx + \frac{(kx)^2}{2} + \frac{(kx)^3}{6} + \sum_{n=4}^{\infty} \frac{(kx)^n}{n!}. \quad\blacksquare$$

Self-Test Exercise

4. Use the Uniqueness Theorem to find a power series solution of the initial value problem $\dfrac{dy}{dx} = 2xy$, $y(0) = 3$.

The Relationship Between the Coefficients and Derivatives of a Power Series

Theorem 4 If $f(x) = \displaystyle\sum_{n=0}^{\infty} a_n (x - c)^n$ converges on an open interval centered at c, then

$$a_n = \frac{f^{(n)}(c)}{n!}, \; n = 0, 1, 2, 3, \ldots.$$

For $n \geq 1$, $f^{(n)}(c)$ represents the nth derivative of f at c and $f^{(0)}(c) = f(c)$.

10.3 Taylor Polynomials

Important Concepts

- Generalizing the Tangent Line Approximation
- Higher-Order Approximating Polynomials
- Taylor's Theorem
- Concluding Remarks about Convergence and the Size of the Error Term

Generalizing the Tangent Line Approximation

The equation of the tangent line to a curve, $y = f(x)$, at a point of tangency, $(c, f(c))$, is given by $y = f'(c)(x - c) + f(c)$. In this case, y is a degree 1 polynomial, so we will refer to this y as $y = P_1(x)$ where $P_1(c) = f(c)$ and $P_1'(c) = f'(c)$. Refer to Figure 1.

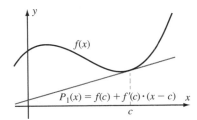

Figure 1

To better approximate the graph of f near c, we will need a curve that has the same concavity as the graph of f. We will need a degree 2 polynomial with derivatives at c that agree with those of f up to order 2, i.e., $P_2(c) = f(c), P_2'(c) = f'(c)$, and $P_2''(c) = f''(c)$. We may write $P_2(x) = f(c) + f'(c)(x - c) + \alpha(x - c)^2$. The polynomial $P_2(x)$ satisfies the hypothesis of Theorem 4 of Section 10.2, so $\alpha = a_2 = \dfrac{f^{(2)}(c)}{2!} = \dfrac{f''(c)}{2}$. Therefore, $P_2(x) = f(c) + f'(c)(x - c) + \dfrac{f''(c)}{2}(x - c)^2$ is the required degree 2 polynomial.

Example 1 Let $f(x) = e^{2x}$. Find a degree 2 polynomial $P_2(x)$ that approximates the graph of f near $(2, e^4)$.

Solution Utilize the equation $P_2(x) = f(c) + f'(c)(x - c) + \dfrac{f''(c)}{2}(x - c)^2$ with $c = 2$ and $f(x) = e^{2x}$.

$$f(x) = e^{2x} \Rightarrow \quad f(2) = e^4$$
$$f'(x) = 2e^{2x} \Rightarrow \quad f'(2) = 2e^4$$
$$f''(x) = 4e^{2x} \Rightarrow \quad f''(2) = 4e^4$$

Therefore,

$$P_2(x) = e^4 + 2e^4(x - 2) + \frac{4e^4}{2}(x - 2)^2$$
$$\approx 54.6 + 109.2(x - 2) + 109.2(x - 2)^2$$
$$= 273 - 327.6x + 109.2x^2.$$

Figure 2 exhibits the graph of $f(x) = e^{2x}$, which appears as a solid curve, and $P_2(x) \approx 273 - 327.6x + 109.2x^2$, which appears as a dotted curve, on the same set of axes in a neighborhood of the point of tangency $(2, e^4) \approx (2, 54.6)$.

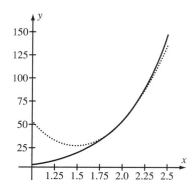

Figure 2

Self-Test Exercise

1. Find a degree 2 polynomial $P_2(x)$ that approximates the graph of $f(x) = \ln(2x)$ near $(2, \ln(4))$.

Higher-Order Approximating Polynomials

Theorem 1   Let N be a nonnegative integer. If f is an N times continuously differentiable function, then the polynomial $P_N(x) = \sum_{n=0}^{N} \dfrac{f^{(n)}(c)}{n!} (x - c)^n$ is the unique degree N polynomial such that $P_N(c) = f(c)$, $P_N'(c) = f'(c)$, $P_N''(c) = f''(c), \ldots, P_N^{(N)}(c) = f^{(N)}(c)$.

We call the polynomial P_N of Theorem 1 the *Taylor polynomial of degree N and base point c for the function f.*

Example 2   Compute the Taylor polynomials of degree 1, 2, and 3 for the function $f(x) = x^4$ expanded about the point $x = 1$.

Solution

$$f(1) = 1, \; f'(1) = 4x^3\big|_{x=1} = 4, \; f''(1) = 12x^2\big|_{x=1} = 12, \text{ and } f'''(1) = 24x\big|_{x=1} = 24$$

Therefore, according to Theorem 1,

$$P_1(x) = \frac{f(1)}{0!}(x-1)^0 + \frac{f'(1)}{1!}(x-1) = 1 + 4(x-1) = 4x - 3,$$

$$P_2(x) = \frac{f(1)}{0!}(x-1)^0 + \frac{f'(1)}{1!}(x-1) + \frac{f''(1)}{2!}(x-1)^2$$
$$= 1 + 4(x-1) + 6(x-1)^2 = 6x^2 - 8x + 3,$$

$$P_3(x) = \frac{f(1)}{0!}(x-1)^0 + \frac{f'(1)}{1!}(x-1) + \frac{f''(1)}{2!}(x-1)^2 + \frac{f'''(1)}{3!}(x-1)^3$$
$$= 1 + 4(x-1) + 6(x-1)^2 + 4(x-1)^3 = 4x^3 - 6x^2 + 4x - 1.$$

Figure 3 exhibits the graphs of $f(x) = x^4$, $P_1(x)$, $P_2(x)$, and $P_3(x)$.

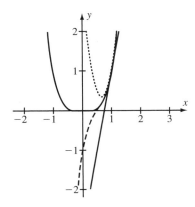

Figure 3

■

Self-Test Exercise

2. Compute the Taylor polynomials of degree 1, 2, and 3 for the function $f(x) = x^{-4}$ expanded about the point $x = 1$.

Taylor's Theorem

Theorem 2 Suppose that f is $N + 1$ times continuously differentiable on an open interval I centered at c. If $x_0 \in I$, then there is a number ϵ between c and x_0 such that

$$f(x_0) = P_N(x_0) + R_N(x_0)$$

where

$$R_N(x_0) = \frac{f^{(N+1)}(\epsilon)}{(N+1)!}(x_0 - c)^{N+1}.$$

Example 3 Apply Taylor's Theorem with $N = 3$ to the function $f(x) = \cos(x)$ about the point $c = \dfrac{\pi}{2}$.

Solution Fix x_0. Calculate $f\left(\dfrac{\pi}{2}\right)$, $f'\left(\dfrac{\pi}{2}\right)$, $f''\left(\dfrac{\pi}{2}\right)$, $f'''\left(\dfrac{\pi}{2}\right)$, and $f^{(4)}(x)$.

$$f\left(\frac{\pi}{2}\right) = \cos\left(\frac{\pi}{2}\right) = 0, \qquad f'\left(\frac{\pi}{2}\right) = -\sin\left(\frac{\pi}{2}\right) = -1,$$

$$f''\left(\frac{\pi}{2}\right) = -\cos\left(\frac{\pi}{2}\right) = 0, \qquad f'''\left(\frac{\pi}{2}\right) = \sin\left(\frac{\pi}{2}\right) = 1, \text{ and}$$

$$f^{(4)}(x) = \cos(x).$$

Therefore,

$$P_3(x_0) = \frac{0}{0!}\left(x_0 - \frac{\pi}{2}\right)^0 - \frac{1}{1!}\left(x_0 - \frac{\pi}{2}\right)^1 + \frac{0}{2!}\left(x_0 - \frac{\pi}{2}\right)^2 + \frac{1}{3!}\left(x_0 - \frac{\pi}{2}\right)^3$$

$$= -\left(x_0 - \frac{\pi}{2}\right)^1 + \frac{1}{6}\left(x_0 - \frac{\pi}{2}\right)^3.$$

The remainder term is given by

$$R_3(x_0) = \frac{f^{(4)}(\epsilon)}{4!} \left(x_0 - \frac{\pi}{2} \right)^4 = \frac{\cos(\epsilon)}{4!} \left(x_0 - \frac{\pi}{2} \right)^4,$$

where ϵ is an undetermined value between $\frac{\pi}{2}$ and x_0.

Therefore,

$$\cos(x_0) = -\left(x_0 - \frac{\pi}{2} \right) + \frac{1}{6} \left(x_0 - \frac{\pi}{2} \right)^3 + \frac{\cos(\epsilon)}{4!} \left(x_0 - \frac{\pi}{2} \right)^4.$$

Self-Test Exercise

3. Apply Taylor's Theorem with $N = 2$ to the function $f(x) = \tan(x)$ about the point $x = 0$.

Convergence and the Size of the Error Term

The formal infinite Taylor expansion for a function f actually converges to $f(x)$ at a point x if and only if the error term $R_N(x)$ tends to zero as N tends to infinity.

10.4 The Rate of Convergence of Taylor's Expansion

Important Concepts

• Estimating the Error Term
• Achieving a Desired Degree of Accuracy

Estimating the Error Term

Theorem 1
Let f be a function that is $N + 1$ times continuously differentiable on an open interval I centered at c. For each x_0 in I, let J denote the closed interval with endpoints x_0 and c. Therefore, $J = [c, x_0]$ if $c \le x_0$ and $J = [x_0, c]$ if $x_0 \le c$. The error term $R_N(x_0)$ in the Nth order Taylor expansion of f about the point c satisfies

$$|R_N(x_0)| \le M_{N+1} \frac{|x_0 - c|^{N+1}}{(N+1)!} \quad (x_0 \in I)$$

where

$$M_{N+1} = \max_{x \in J} \left| f^{(N+1)}(x) \right|.$$

Example 1 Use the degree 3 Taylor polynomial expanded about $\frac{\pi}{2}$ to approximate $\cos(1.6)$ and estimate the (absolute) error from the error term.

Solution In Example 3 of Section 10.3, we found

$$\cos(x_0) = -\left(x_0 - \frac{\pi}{2} \right) + \frac{1}{6} \left(x_0 - \frac{\pi}{2} \right)^3 + \frac{\cos(\epsilon)}{4!} \left(x_0 - \frac{\pi}{2} \right)^4.$$

Therefore,

$$\cos(1.6) \approx P_3(1.6) = -\left(1.6 - \frac{\pi}{2}\right) + \frac{1}{6}\left(1.6 - \frac{\pi}{2}\right)^3 = -0.02919952212\ldots.$$

By Theorem 1, Section 10.4, we see that the error term satisfies $|R_N(x_0)| \leq M_{N+1} \dfrac{|x_0 - c|^{N+1}}{(N+1)!}$, where $M_{N+1} = \max\limits_{x \in J}\left|f^{(N+1)}(x)\right|$, $x_0 = 1.6$, and $N = 3$.

$$M_4 = \max_{x \in J}\left|f^{(4)}(x)\right| = \max_{\frac{\pi}{2} \leq x \leq 1.6}|\cos(x)|$$

Because the value of the cosine function lies between -1 and 1, we conclude that $M_4 = \max\limits_{\frac{\pi}{2} \leq x \leq 1.6}|\cos(x)| \leq 1$.

$$|R_3(1.6)| \leq M_4 \frac{\left|1.6 - \dfrac{\pi}{2}\right|^4}{(4)!} \leq \frac{\left|1.6 - \dfrac{\pi}{2}\right|^4}{(4)!} \approx 3.03 \times 10^{-8}$$

This means that if we use $P_3(1.6) = -0.02919952212\ldots$ to approximate $\cos(1.6)$, the absolute error will be less than 3.03×10^{-8}. We remark that $\cos(1.6) = -0.029199522301288\ldots$, which confirms our analysis. ∎

Degrees of Accuracy

Example 2 Compute $\sin(1.2)$ to an accuracy of two decimal places.

Solution Let $f(x) = \sin(x)$, $c = 0$, and $x_0 = 1.2$; we have $f'(x) = \cos(x)$, $f''(x) = -\sin(x)$, $f^{(3)}(x) = -\cos(x)$, ...

Thus, $f(0) = 0$, $f'(0) = 1$, $f''(0) = 0$, $f^{(3)}(0) = -1$, ... When N is odd, i.e., $N = 2n+1$, the degree N Taylor polynomial of $f(x) = \sin(x)$ with base point 0 is:

$$P_N(x) = P_{2n+1}(x) = f(0) + \frac{f'(0)}{1}x + \frac{f''(0)}{2!}x^2 + \frac{f^{(3)}(0)}{3!}x^3 + \cdots + \frac{f^{(N)}(0)}{N!}x^N$$

$$= x - \frac{x^3}{3!} + \frac{x^5}{5!} - \cdots + (-1)^n\frac{x^N}{N!}.$$

To estimate the error, we note that $f^{(N+1)}(x)$ is either equal to $\sin(x)$, $-\sin(x)$, $\cos(x)$, or $-\cos(x)$; therefore, $\max|f^{(N+1)}(x)| \leq 1$. Thus,

$$|R_N(1.2)| \leq 1 \cdot \frac{|1.2 - 0|^{N+1}}{(N+1)!} = \frac{(1.2)^{N+1}}{(N+1)!}.$$

We require two decimal places of accuracy, which means we want the error term to be less than 5×10^{-3}. We need to find the smallest N such that $\dfrac{(1.2)^{N+1}}{(N+1)!} < 5 \times 10^{-3}$. Substituting in a few values for N, we see that $\dfrac{1.2^{(2+1)}}{(2+1)!} \approx 0.288$, $\dfrac{1.2^{(3+1)}}{(3+1)!} \approx 0.086$, $\dfrac{1.2^{(4+1)}}{(4+1)!} \approx 0.02$, and $\dfrac{1.2^{(5+1)}}{(5+1)!} \approx 0.0041$. Therefore, the inequality will first be true when $N = 5$. We will use $P_5(1.2)$ in the estimation, and the inequality $|R_5(1.2)| < 5 \times 10^{-3}$ assures that $P_5(1.2)$ and $\sin(1.2)$ will agree to two decimal places. Our estimation is therefore $\sin(1.2) \approx P_5(1.2) = 1.2 - \dfrac{1.2^3}{3!} + \dfrac{1.2^5}{5!} = 0.93273\ldots$. ∎

Self-Test Exercise

1. Estimate the error term for an estimate of $e^{-0.5}$ to an accuracy of four decimal places.

10.5 Taylor Series

Important Concepts

- Taylor Series; Maclaurin Series
- Power Series Expansions of the Common Transcendental Functions
- The Binomial Series
- Using Taylor Series to Approximate

Taylor Series and Maclaurin Series

Suppose that for every nonnegative integer N the function f is N times continuously differentiable on an open interval I centered at c. The power series

$$T(x) = \sum_{n=0}^{\infty} \frac{f^{(n)}(c)}{n!} (x - c^n)$$

is called the *Taylor series of f with base point c*. We say that the Taylor series is *expanded about c* (or centered at c).

A Taylor series with base point 0 is also called a *Maclaurin series*.

Theorem 1   Suppose that for every nonnegative integer N, the function f is N times continuously differentiable on an open interval I centered at c. Let

$$R_N(x) = f(x) - \sum_{n=0}^{N} \frac{f^{(n)}(c)}{n!} (x - c)^n$$

be the remainder in the Nth order Taylor expansion about c. If

$$\lim_{N \to \infty} R_N(x) = 0,$$

then $f(x)$ equals the Taylor series of f expanded about c:

$$f(x) = \sum_{n=0}^{\infty} \frac{f^{(n)}(c)}{n!} (x - c)^n .$$

Power Series Expansions of the Common Transcendental Functions

Theorem 2   If b is a fixed positive constant, then $\lim_{k \to \infty} \dfrac{b^k}{k!} = 0$.

Example 1 Show that the Maclaurin series of $f(x) = \arctan(x)$ is $\displaystyle\sum_{n=0}^{\infty} (-1)^n \frac{x^{2n+1}}{(2n + 1)}$ for $-1 \le x \le 1$.

Solution

$$f^0(x) = \arctan(x) \Rightarrow f^0(0) = 0$$

$$f'(x) = \frac{1}{1+x^2} \Rightarrow f'(0) = 1$$

$$f''(x) = \frac{-2x}{(1+x^2)^2} \Rightarrow f''(0) = 0$$

$$f^{(3)}(x) = \frac{8x^2}{(1+x^2)^3} - \frac{2}{(1+x^2)^2} \Rightarrow f^{(3)}(0) = -2$$

$$f^{(4)}(x) = \frac{-48x^3}{(1+x^2)^4} + \frac{24x}{(1+x^2)^3} \Rightarrow f^{(4)}(x) = 0$$

$$f^{(5)}(x) = \frac{384x^4}{(1+x^2)^5} - \frac{288x^2}{(1+x^2)^4} + \frac{24}{(1+x^2)^3} \Rightarrow f^{(5)}(0) = 24$$

$$f^{(6)}(x) = \frac{-3840x^5}{(1+x^2)^6} + \frac{3840x^3}{(1+x^2)^5} - \frac{720x}{(1+x^2)^4} \Rightarrow f^{(6)}(x) = 0$$

$$f^{(7)}(x) = \frac{46080x^6}{(1+x^2)^7} - \frac{57600x^4}{(1+x^2)^6} + \frac{17280x^2}{(1+x^2)^5} - \frac{720}{(1+x^2)^4} \Rightarrow f^{(7)}(0) = -720$$

The Maclaurin series of $\arctan(x)$ is

$$\frac{0}{0!}x + \frac{1}{1!}x + \frac{0}{2!}x^2 - \frac{2}{3!}x^3 + \frac{0}{4!}x^4 + \frac{24}{5!}x^5 + \frac{0}{6!}x^6 - \frac{720}{7!}x^7 + \cdots$$

$$= x - \frac{x^3}{3} + \frac{x^5}{5} - \frac{x^7}{7} + \cdots + (-1)^n \frac{x^{2n+1}}{2n+1} + \cdots$$

$$= \sum_{n=0}^{\infty} (-1)^n \frac{x^{2n+1}}{2n+1}.$$

Since $\lim\limits_{n\to\infty} \left| \frac{a_{n+1}}{a_n} \right| = \lim\limits_{n\to\infty} \left| \frac{\frac{1}{2n+1}}{\frac{1}{2n+3}} \right| = 1$, the radius of convergence of this Maclaurin series is 1 by Theorem 4 of Section 10.1. At the endpoint $x = -1$ or $x = 1$, this power series is an alternating series with $a_n = \frac{1}{2n+1}$, which converges according to the Alternating Series Test (Theorem 1 of Section 9.5). Hence the Maclaurin series of $\arctan(x)$ converges for $-1 \le x \le 1$. ∎

Here are a few examples of convergent power series for familiar functions:

$$\sin(x) = \sum_{n=0}^{\infty} \frac{(-1)^n}{(2n+1)!} x^{2n+1}, \quad -\infty < x < \infty$$

$$\cos(x) = \sum_{n=0}^{\infty} \frac{(-1)^n}{(2n)!} x^{2n}, \quad -\infty < x < \infty$$

$$\ln(1+x) = \sum_{n=1}^{\infty} \frac{(-1)^{n-1}}{n} x^n, \quad -1 < x \le 1$$

$$e^x = \sum_{n=0}^{\infty} \frac{x^n}{n!}, \quad -\infty < x < \infty$$

The Binomial Series

The binomial series is a power series representation for the function $f(x) = (1 + x)^\alpha$.

$$(1 + x)^\alpha = \sum_{n=0}^{\infty} \binom{\alpha}{n} x^n,$$

where $\binom{\alpha}{n} = \dfrac{\alpha(\alpha - 1) \cdots (\alpha - n + 1)}{n!}$ and $|x| < 1$. The binomial series converges to $f(x) = (1 + x)^\alpha$ in its interval of convergence for any value of α.

Example 2 Calculate the binomial series for $\left(\dfrac{1}{2}\right)^2$.

Solution If $x = \dfrac{-1}{2}$ and $\alpha = 2$, then $\left(\dfrac{1}{2}\right)^2 = \left(1 + \left(\dfrac{-1}{2}\right)\right)^2 = (1 + x)^\alpha$. The power series representation of $f(x) = (1 + x)^\alpha$ is given by $(1 + x)^\alpha = \sum_{n=0}^{\infty} \binom{\alpha}{n} x^n$. Therefore, the binomial series representation of $\left(\dfrac{1}{2}\right)^2 = \left(1 + \left(\dfrac{-1}{2}\right)\right)^2 = \sum_{n=0}^{\infty} \binom{2}{n} \left(\dfrac{-1}{2}\right)^n$.   ∎

Using a Taylor Series to Approximate

Example 3 Given that $f(x) = \arctan(x)$ is represented by its Maclaurin series for $-1 \le x \le 1$, approximate the value of $\arctan(.7)$ to five decimal places.

Solution In Example 1, we demonstrated that the Maclaurin series of $\arctan(x)$ is $\sum_{n=0}^{\infty} (-1)^n \dfrac{x^{2n+1}}{(2n + 1)}$. Given that $f(x) = \arctan(x)$ is represented by its Maclaurin series, we have

$$\arctan(x) = \sum_{n=0}^{\infty} (-1)^n \frac{x^{2n+1}}{(2n + 1)} = x - \frac{x^3}{3} + \frac{x^5}{5} - \frac{x^7}{7} + \cdots + (-1)^n \frac{x^{2n+1}}{(2n + 1)} + \cdots,$$

for $-1 \le x \le 1$. Therefore,

$$\arctan(.7) = .7 - \frac{(.7)^3}{3} + \frac{(.7)^5}{5} - \frac{(.7)^7}{7} + \cdots + (-1)^n \frac{(.7)^{2n+1}}{(2n + 1)} + \cdots.$$

This is an alternating series, so the truncation error is less than the first term omitted (Theorem 1 of Section 9.5). The maximum allowable error is 5×10^{-6}. The first term that is less than 5×10^{-6} is $\dfrac{(.7)^{27}}{27} \approx 0.000002433$. Therefore, we should use $\sum_{n=0}^{12} (-1)^n \left(\dfrac{(.7)^{2n+1}}{2n + 1}\right) = 0.6107276365$. The actual value of $\arctan(.7)$ is 0.6107259644. We see that

$$|0.6107259644 - 0.6107276365| = 0.0000016721 < 0.000005,$$

which confirms our analysis.   ∎

Self-Test Exercise

1. Estimate $\sin(.32)$ to three decimal places.

Section II
Textbook Solutions and Answers

Solutions to Odd Exercises

Answers to Quick Quiz Exercises

Basics

Number Systems

Problems for Practice

1. If $x = 2.13$ then $100x = 213$ or, equivalently, $x = 213/100$.

3. If $x = 0.232323\ldots$ then $100x = 23.2323\ldots$ and $(100 - 1)x = 23.2323\ldots - 0.232323\ldots$, or $x = 23/99$.

5. If $x = 5.001001001\ldots$ then $1000x = 5001.001001\ldots$ and $(1000 - 1)x = 5001.001001\ldots - 5.001001001\ldots$, or $x = 4996/999$.

7. 0.025

9. $1.\overline{6}$ (The overline identifies the repeating block.)

11. 0.72

13. $[1, 3]$

15. $(1, \infty)$

17. $[-14, 6]$

19.

21.

23.

25. $\left\{ x : |x - 1| \le 2 \right\}$

27. $\left\{ x : |x - 1| < \pi + 1 \right\}$

Further Theory and Practice

29. Let x and y be two rational numbers; $x = a/b$ and $y = c/d$, where a, b, c, and d are integers and b and d are not zero. The sum is $x + y = a/b + c/d = (ad + bc)/bd = s/t$, a rational number because $s = ad + bc$ is an integer and $t = bd$ is an integer and not zero. Similarly, $x \cdot y = ac/bd = s/t$ is always rational because $t = ac$ is an integer and $t = bd$ is an integer and not zero.

31. Both approximations are rational numbers, but π is not. An irrational number cannot be exactly equal to a rational number, however, the error can be made as small as necessary.

33. To avoid a penalty, the interval should be $[34.155, 34.845]$; $\{x: |x - 34.5| \le 0.345\}$.

35.   **37.**

39.   **41.**

43.   **45.** $\{-3\}$

47. $\left\{x : 1 < x \le 2\sqrt{2}\right\}$

49. We calculate as follows:

$$|s - 4| > |2s + 9| \Leftrightarrow \begin{bmatrix} \begin{cases} s < -9/2 \\ -s + 4 > -2s - 9 \end{cases} \\ \begin{cases} -9/2 \le s \le 4 \\ -s + 4 > 2s + 9 \end{cases} \\ \begin{cases} s > 4 \\ s - 4 > 2s + 9 \end{cases} \end{bmatrix} \Leftrightarrow \begin{bmatrix} \begin{cases} s < -9/2 \\ s > -13 \end{cases} \\ \begin{cases} -9/2 \le s \le 4 \\ s < -5/3 \end{cases} \\ \begin{cases} s > 4 \\ s < -13 \end{cases} \end{bmatrix} \Leftrightarrow \begin{bmatrix} -13 < s < -9/2 \\ -9/2 \le s < -5/3 \\ \varnothing \end{bmatrix} \Leftrightarrow -13 < s < -5/3$$

The answer is $\{s : -13 < x < -5/3\}$.

51. We calculate as follows: $\dfrac{w}{w + 1} < 3w + 2 \Leftrightarrow \dfrac{3w^2 + 4w + 2}{w + 1} > 0$. As $3w^2 + 4w + 2 > 0$ for all w, we obtain $w + 1 > 0$, or $w > -1$. Using the absolute value sign we derive $\left\{w : \dfrac{|w + 1|}{w + 1} = 1\right\}$.

53. Whereas a and b are nonpositive, $|a + b| = -(a + b)$ and $|a| + |b| = -a + (-b) = -(a + b)$. Therefore, $|a| + |b| = |a + b|$, and equality holds in the Triangle Inequality.

55. Supposing that the smallest positive number S exists, notice that, because $S/2$ is also a positive number, the inequality $S < S/2$ must hold. From this inequality it follows that $2S < S$, or, equivalently, $S < 0$. This contradicts the assumption that S is positive.

57. **a.** Suppose that $x = a/b$ where a and b are integers with no common factors.

  b. Then, $2 = x^2 = a^2/b^2$.

  c. Therefore, $a^2 = 2b^2$.

  d. Then, a^2 is even and a is even as well: $a = 2\alpha$, where α is an integer.

  e. According to part **b**: $2 = 4\alpha^2/b^2$, hence $b^2 = 2\alpha^2$.

  f. Then, b^2 is even and b is even as well: $b = 2u$, where u is an integer.

  g. We conclude that both a and b have 2 as a common factor. This contradicts our assumption in part **a**. Therefore, 2 cannot be a square of a rational number and x is irrational.

Calculator/Computer Exercises

59. $[0.4485, 0.4495]$

61. $[-0.00049001, 0.00050999]$

63. 4.001

65. The following table records the calculation of $\alpha_0 \cdot \beta_0$ using $n = 12$, 14, 16, 18 and 20 digits.

n	$\alpha_0 \cdot \beta_0$
12	.999999999996
14	.99999999999995
16	.9999999999999996
18	.99999999999999993
20	.99999999999999999990

67.

x	Relative Error	x	Relative Error
10^{10}	0.5	-10^{12}	0.0101
10^{12}	9.901×10^{-3}	-10^{14}	1.0001×10^{-4}
10^{14}	9.999×10^{-5}	-10^{16}	1.000001×10^{-6}
10^{16}	9.99999×10^{-7}	-10^{18}	$1.00000001 \times 10^{-8}$
10^{18}	9.9999999×10^{-9}	-10^{20}	$1.000000000 \times 10^{-10}$

69.

x	Relative Error
10^3	1.031
10^6	0.0345
10^9	0.00003
10^{12}	0.3×10^{-7}
10^{15}	0.3×10^{-10}

x	Relative Error
-10^3	0.9709
-10^6	0.0322
-10^9	0.00003
-10^{12}	0.3×10^{-7}
-10^{15}	0.3×10^{-10}

Section 1.2   Planar Coordinates and Graphing in the Plane

Problems for Practice

1.

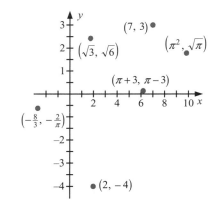

3.

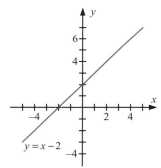

5.   Using the Pythagorean Theorem, we calculate the distances as follows:

$$\left|\overline{AB}\right| = \sqrt{\left((-4)-2\right)^2 + \left(7-3\right)^2} = \sqrt{36+16} = 2\sqrt{13} \,;$$

$$\left|\overline{AC}\right| = \sqrt{\left((-5)-2\right)^2 + \left((-6)-3\right)^2} = \sqrt{49+81} = \sqrt{130} \,;$$

$$\left|\overline{BC}\right| = \sqrt{\left((-5)-(-4)\right)^2 + \left((-6)-7\right)^2} = \sqrt{1+169} = \sqrt{170} \,.$$

7.   Center: (1, 3); radius: 3

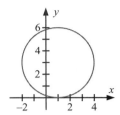

9.   We complete the square as follows:
$$4\left(x^2 + \tfrac{16}{4}x\right) + 4\left(y^2 - \tfrac{8}{4}y\right) + 2 = 0,$$

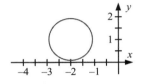

$$4\left(x^2 + 4x + \left(\tfrac{4}{2}\right)^2\right) + 4\left(y^2 - 2y + \left(-\tfrac{2}{2}\right)^2\right)$$
$$+ 2 - 4\cdot\left(\tfrac{4}{2}\right)^2 - 4\cdot\left(-\tfrac{2}{2}\right)^2 = 0,$$

$4(x+2)^2 + 4(y-1)^2 - 18 = 0$, $(x+2)^2 + (y-1)^2 = \tfrac{9}{2}$. The equation describes a circle with center: (–2, 1) and radius: $\tfrac{\sqrt{3}}{2}$.

11.   Center: $(0, -5)$; radius: $\sqrt{2}$

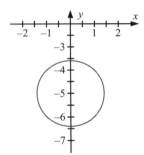

13.   We complete the square as follows:
$$3\left(x^2 + 4x\right) + 3\left(y^2 - 2y\right) = 2,$$

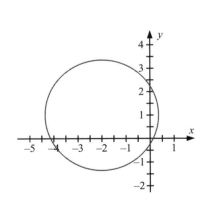

$$3\left(x^2 + 4x + \left(\tfrac{4}{2}\right)^2\right) + 3\left(y^2 - 2y + \left(-\tfrac{2}{2}\right)^2\right)$$
$$= 2 + 3\cdot\left(\tfrac{4}{2}\right)^2 + 3\cdot\left(\tfrac{2}{2}\right)^2,$$

$3(x+2)^2 + 3(y-1)^2 = 17$, $(x-(-2))^2 + (y-1)^2 = \tfrac{17}{3}$.
This equation describes the circle with center: (–2, 1)
and radius: $\sqrt{\tfrac{17}{3}}$.

15. We complete the square as follows:

$$x^2 + \left(y^2 - y + \left(-\tfrac{1}{2}\right)^2\right) = \left(\tfrac{1}{2}\right)^2, \quad x^2 + \left(y - \tfrac{1}{2}\right)^2 = \left(\tfrac{1}{2}\right)^2.$$

This equation describes the circle with center: $\left(0, \tfrac{1}{2}\right)$ and radius: $\tfrac{1}{2}$.

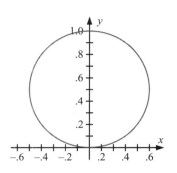

17. Using formula (1.1) of the textbook the equation of the circle is: $(x+3)^2 + (y-5)^2 = 6^2$, or $(x+3)^2 + (y-5)^2 = 36$.

19. Using formula (1.1) of the textbook the equation of the circle is: $(x+4)^2 + (y-\pi)^2 = 5^2$, or $(x+4)^2 + (y-\pi)^2 = 25$.

21. Vertex: $(0, -3)$; axis of symmetry: $x = 0$.

23. We complete the square to find: $y = 1 - (x-1)^2$. Vertex: $(1, 1)$; axis of symmetry: $x = 1$.

25.

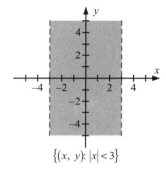

$\{(x, y) : |x| < 3\}$

27.

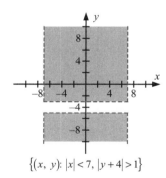

$\{(x, y) : |x| < 7, |y+4| > 1\}$

29.

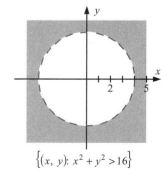

$\{(x, y) : x^2 + y^2 > 16\}$

31.

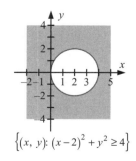

$\{(x, y) : (x-2)^2 + y^2 \geq 4\}$

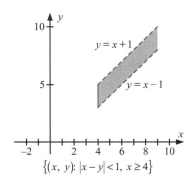

33.

$$\{(x, y): |x-y| < 1, \ x \geq 4\}$$

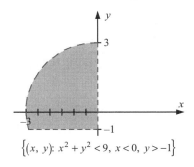

35.

$$\{(x, y): x^2 + y^2 < 9, \ x < 0, \ y > -1\}$$

Further Theory and Practice

37. Let $M(x, y)$ be the point that is equidistant from points $A(2, 3)$, $B(8, 2)$, and $C(7, 9)$. Using the Pythagorean Theorem we see that $\left|\overline{AM}\right| = \sqrt{(x-2)^2 + (y-3)^2}$,

$$\left|\overline{BM}\right| = \sqrt{(x-8)^2 + (y-2)^2} \ ,$$

and $\left|\overline{CM}\right| = \sqrt{(x-7)^2 + (y-9)^2}$. Point M is equidistant from A, B, and C; therefore

$$\left|\overline{AM}\right| = \left|\overline{BM}\right| = \left|\overline{CM}\right|,$$

and we have the following system of equations:

$$\begin{cases} \sqrt{(x-2)^2 + (y-3)^2} = \sqrt{(x-8)^2 + (y-2)^2} \\ \sqrt{(x-8)^2 + (y-2)^2} = \sqrt{(x-7)^2 + (y-9)^2} \end{cases},$$

or $\begin{cases} x^2 - 4x + 4 + y^2 - 6y + 9 = x^2 - 16x + 64 + y^2 - 4y + 4 \\ x^2 - 16x + 64 + y^2 - 4y + 4 = x^2 - 14x + 49 + y^2 - 18y + 81 \end{cases}$, or $\begin{cases} 12x - 2y = 55 \\ -x + 7y = 31 \end{cases}$. From the

second equation we see that $x = 7y - 31$. We now substitute x in the first equation so that $12(7y - 31) - 2y = 55$ and $y = \frac{427}{82}$, $x = \frac{447}{82}$. We, therefore, find that point $\left(\frac{447}{82}, \frac{427}{82}\right)$ is equidistant from the three given points.

39.   Let $M(x, y)$ be the point that is equidistant from points $A(1, 0)$, $B(0, 1)$, and $C(0, -1)$. Using the Pythagorean Theorem and equidistance condition we have the following system of equations: $\begin{cases} \sqrt{(x-1)^2 + y^2} = \sqrt{x^2 + (y-1)^2} \\ \sqrt{x^2 + (y-1)^2} = \sqrt{x^2 + (y+1)^2} \end{cases}$. We solve to find $x = 0$ and $y = 0$. The origin, $(0, 0)$, is the only point that is equidistant from the three given points.

41.   We first move all the terms to the left-hand side of the equation: $5x^2 - 2x + 5y^2 + 3y - 6 = 0$. We then complete the square as follows:

$$5\left(x^2 - \tfrac{2}{5}x + \left(\tfrac{1}{5}\right)^2\right) + 5\left(y^2 + \tfrac{3}{5}y + \left(\tfrac{3}{10}\right)^2\right) - 6 - 5 \cdot \left(\tfrac{1}{5}\right)^2 - 5 \cdot \left(\tfrac{3}{10}\right)^2 = 0,$$

$5\left(x - \tfrac{1}{5}\right)^2 + 5\left(y + \tfrac{3}{10}\right)^2 = \tfrac{133}{20}$, $\left(x - \tfrac{1}{5}\right)^2 + \left(y - \left(-\tfrac{3}{10}\right)\right)^2 = \tfrac{133}{100}$. This is the equation of a circle with center: $\left(\tfrac{1}{5}, -\tfrac{3}{10}\right)$ and radius: $\tfrac{\sqrt{133}}{10}$.

43.   We move all the terms to the left-hand side of the equation: $2x^2 - x - 2y^2 + 5y + 7 = 0$. We then complete the square as follows:

$$2\left(x^2 - \tfrac{1}{2}x + \left(-\tfrac{1}{4}\right)^2\right) - 2\left(y^2 - \tfrac{5}{2}y + \left(-\tfrac{5}{4}\right)^2\right) + 7 - 2 \cdot \left(\tfrac{1}{4}\right)^2 + 2 \cdot \left(\tfrac{5}{4}\right)^2 = 0,$$

$\left(x - \tfrac{1}{4}\right)^2 - \left(y - \tfrac{5}{4}\right)^2 = -5$. This equation differs from equation (1.1) of the textbook as evidenced by the minus sign that precedes the term y^2 and the minus sign on the right-hand side. This equation does not describe a circle; it is the equation of a hyperbola.

45.

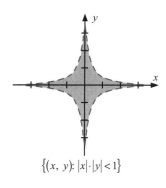

$\{(x, y): |x| \cdot |y| < 1\}$

47.

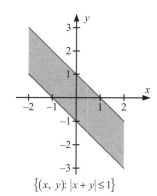

$\{(x, y): |x + y| \leq 1\}$

49.

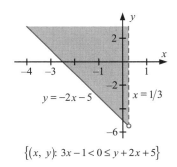

$$\{(x, y): 3x-1 < 0 \le y+2x+5\}$$

51.

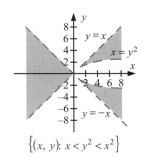

$$\{(x, y): x < y^2 < x^2\}$$

53.

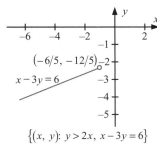

$$\{(x, y): y > 2x, \, x-3y=6\}$$

55. Let $(u, \, 2u)$ be the point on the line. If the point is also to lie on the circle

$$(x-h)^2 + (y-k)^2 = r^2$$

then u must satisfy the quadratic equation $(u-h)^2 + (2u-k)^2 = r^2$. This quadratic equation could have zero, one, or two real solutions.

57. The intersecting points of two circles are the solutions of the system:

$$\begin{cases} (x-h_1)^2 + (y-k_1)^2 = r_1^2 \\ (x-h_2)^2 + (y-k_2)^2 = r_2^2 \end{cases}, \text{ or } \begin{cases} x^2 - 2h_1x + h_1^2 + y^2 - 2k_1y + k_1^2 = r_1^2 \\ x^2 - 2h_2x + h_2^2 + y^2 - 2k_2y + k_2^2 = r_2^2 \end{cases}.$$

We subtract the second line of the equation from the first line to yield:

$$2x(h_2 - h_1) + 2y(k_2 - k_1) = (r_1^2 - r_2^2) + (h_2^2 - h_1^2) + (k_2^2 - k_1^2).$$

Let $a = 2(h_2 - h_1)$, $b = 2(k_2 - k_1)$, and $c = (r_1^2 - r_2^2) + (h_2^2 - h_1^2) + (k_2^2 - k_1^2)$ to obtain $ax + by = c$. If $a = 0$, $b = 0$, and $c \ne 0$, then there can be no solution and no intersecting points. If $a = 0$, $b = 0$, and $c = 0$, the circles are identical. If $a \ne 0$, then substituting $x = \dfrac{c-by}{a}$ into the first equation yields a quadratic equation in y that may have zero, one, or two solutions that correspond to zero, one, or two intersecting points of the two circles, respectively. If $a = 0$ and $b \ne 0$, the same argument applies after substituting $y = \dfrac{c-ax}{b}$.

59.   For the point $P(x, y)$, the distance from $(0, 1)$ is

$d = \sqrt{(x-0)^2 + (y-1)^2}$. The condition $d = y$ leads to

the equation $y = \sqrt{(x-0)^2 + (y-1)^2}$, or $y = \dfrac{1}{2} + \dfrac{x^2}{2}$.

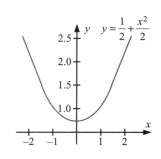

Calculator/Computer Exercises

61.   a.

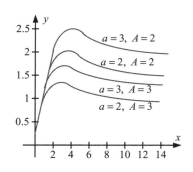   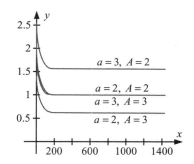

The function approaches $\dfrac{a}{A}$ as x becomes infinite.

b.   We divide each term of the following fraction by x^2 to obtain

$$\frac{ax^2 + 7x + 2}{Ax^2 - 3x + 12} = \frac{a + \frac{7}{x} + \frac{2}{x^2}}{A - \frac{3}{x} + \frac{12}{x^2}} .$$

As x tends to infinity, the only nonvanishing terms are a in the numerator and A in the

denominator. The resulting fraction is $\dfrac{a}{A}$.

63.

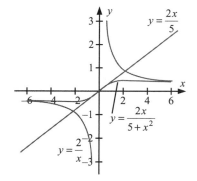   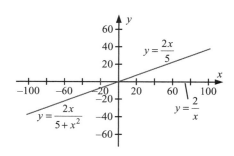

For small x, the serpentine behaves as $y = \dfrac{ax}{b}$; for large x, it is similar to the hyperbola $y = \dfrac{a}{x}$.

Lines and Their Slopes

Problems for Practice

1.

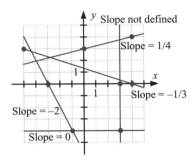

3.

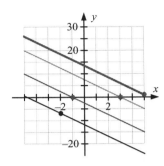

5. $y = 5(x+3)+7$

7. $y = -4x+9$

9. $\dfrac{7x}{9} + \dfrac{5y}{18} = 1$

11. $y = \left(\dfrac{147}{220}\right)(x-18)+14$

13. $y = 3(x+4)$

15. $y = \left(-\dfrac{1}{3}\right)x - 5$

17. $-2x+4y+8=0$

19. $y = \left(-\dfrac{1}{3}\right)x - \dfrac{1}{8}$

21. The first line has slope $5/3$. It is *not* parallel to the second line, which has slope -3.

It is *not* perpendicular to the third line, which has slope 8.

23. The first line has slope 3. It is *not* parallel to the second line, which has slope $22/7$.

It *is* perpendicular to the third line, which has slope $-1/3$.

25.

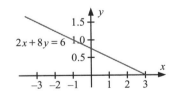

27.

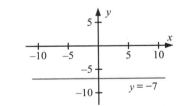

29.

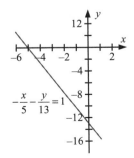

31.

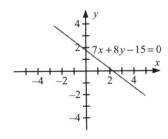

Further Theory and Practice

33.   Simplify the point-slope form for a line: $y = m(x - x_0) + y_0 = mx + (-mx_0 + y_0) = mx + b$. This is the slope-intercept form for a line.

35.   All points $(a, \, 2a - 17)$ with $a \neq 5$.

37.   All points $(a, \, b)$ with $a \neq -2$ and $b = 7 - 3(a + 2)$; for example, $(1, \, -2)$.

39.   Example: $(a, \, b) = (1, \, 0)$ and $(c, \, d) = (3, \, 4)$.

41.   Points on the line can be represented as $\left(x, \, \dfrac{3x - 4}{8} \right)$, so that their distance from $(2, \, 8)$ is

$$d(x) = \sqrt{(x - 2)^2 + \left(\frac{3x - 4}{8} - 8 \right)^2}, \text{ or } d(x) = \sqrt{\frac{73}{64} x^2 - \frac{83}{8} x + \frac{305}{4}}. \text{ The minimum distance}$$

corresponds to the vertex of this parabola, that is, to $x = \dfrac{332}{73}$. We find that $y = \dfrac{88}{73}$ and the point is $\left(\dfrac{332}{73}, \, \dfrac{88}{73} \right)$.

43.   We solve the system $\begin{cases} x + 2y = 4 \\ 2x + y = 5 \end{cases}$ to find the point of intersection at $(2, \, 1)$. If there are two lines with slopes k' and k'' that have an angle of intersection ϕ, we obtain

$$\tan(\phi) = \frac{k' - k''}{1 + k'k''}.$$

For the first line, the slope is $k_1 = -\dfrac{1}{2}$, and for the second line it is $k_2 = -2$. Given that the slope of the line we are seeking is k, we find that $\dfrac{k - k_1}{1 + kk_1} = \dfrac{k_2 - k}{1 + kk_2}$, which leads to $k = 1$ and $k = -1$. The slope-intercept form of the line through the point of intersection $(2, \, 1)$ is $y - 1 = k(x - 2)$. Therefore, we obtain two lines $y = x - 1$ and $y = -x + 3$.

45.   The slope of the line through points $(x_1, \, y_1)$ and $(x_2, \, y_2)$ is

$$\frac{y_2 - y_1}{x_2 - x_1}.$$

The equation of the line is

$$y = \frac{(y_2 - y_1)(x - x_1)}{x_2 - x_1} + y_1.$$

We transform this equation to:

$$y - y_1 = \frac{(y_2 - y_1)(x - x_1)}{x_2 - x_1}, \text{ or } \frac{y - y_1}{x - x_1} = \frac{y_2 - y_1}{x_2 - x_1}.$$

47. Because the y-coordinates are the same, the distance is only the difference in x-values. Solve

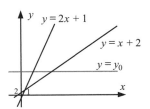

$$y_0 = 2x + 1 \Rightarrow x = \frac{y_0 - 1}{2}$$

$$y_0 = x + 2 \Rightarrow x = y_0 - 2.$$

Solve the equation $\frac{y_0 - 1}{2} - (y_0 - 2) = 1{,}000{,}000$ to get $\frac{y_0}{2} = -999998\frac{1}{2}$, so $y_0 = -199997$

and $y_0 - 2 - \frac{y_0 - 1}{2} = 1{,}000{,}000$ to get $\frac{y_0}{2} = 100001\frac{1}{2}$, so $y_0 = 2000003$.

49. The slope of the regression line is about $m = 0.00194$.

A car with 100,000 mi would emit about 0.452 g of hydrocarbons per mile.

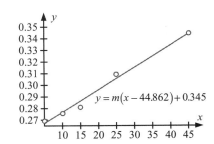

51. The total budget C is the sum of the amounts spent on X and Y. The amount spent on X is $p_X \cdot x$, and the amount spent on Y is $p_Y \cdot y$. Hence, the equation for the budget line is

$C = p_X \cdot x + p_Y \cdot y$. The x-intercept is $\dfrac{C}{p_X}$, the y-intercept is $\dfrac{C}{p_Y}$, and the slope is $-\dfrac{p_X}{p_Y}$.

The new budget line is $C' = p_X \cdot x + p_Y \cdot y$. The only change is in the constant term, so the new line is parallel to the old one, but shifted vertically. It is shifted up if $C' > C$ and down if $C' < C$.

53. Three points are collinear if they lie on the same straight line. Let the three points have coordinates $(x_1,\ y_1)$, $(x_2,\ y_2)$, and $(x_3,\ y_3)$. The three lines determined by these points are

$$f_1(x) = \frac{(y_2 - y_1)(x - x_1)}{x_2 - x_1} + y_1$$

$$f_2(x) = \frac{(y_3 - y_1)(x - x_1)}{x_3 - x_1} + y_1$$

$$f_3(x) = \frac{(y_3 - y_2)(x - x_2)}{x_3 - x_2} + y_2.$$

If these points are collinear, i.e., lie on the same line, then $f_1(x) = f_2(x) = f_3(x)$ and the slopes of the lines are the same. Therefore, three points are collinear only if the slope determined by any pair of these points is the same as the others. Next, if the slopes of these lines are the same, then

$$f_1(x) = (y_2 - y_1)\frac{x - x_1}{x_2 - x_1} + y_1 = (y_3 - y_1)\frac{x - x_1}{x_3 - x_1} + y_1,$$

and $f_1(x_3) = (y_3 - y_1)(x_3 - x_1)/(x_3 - x_1) + y_1 = y_3$. Therefore, point (x_3, y_3) lies on the line determined by points (x_1, y_1), and (x_2, y_2); hence all points lie on the same line.

55.   Let L be any line passing through at least two of the given points and let Q be one of the given points that is not on that line. We note that there is some positive distance d from the point Q to the line L. Given that there are only a finite number of possibilities, one can select an L and a Q that minimize the number d. The line L is the one we seek. It passes through only two of the given points. One verifies this by checking cases;

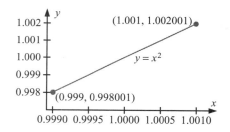

Suppose that 3 points P_1, P_2, P_3 are on the line L. Compute (using Problem 1.3.54) the distance from Q to L, from P_1 to $\overline{P_2Q}$ and to $\overline{P_3Q}$, from P_2 to $\overline{P_1Q}$ and to $\overline{P_3Q}$, and from P_3 to $\overline{P_1Q}$ and to $\overline{P_2Q}$. Note that one of these distances is less than the distance from Q to L.

Calculator/Computer Exercises

57.   Example of zoom on window:
$[0.999, 1.001] \times [0.999^2, 1.001^2]$.

The slope of the line segment is $\dfrac{1.002001 - 0.998001}{1.001 - 0.999}$,

or 2.0.

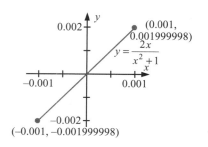

59.   Example of zoom on window:
$[-0.001, 0.001] \times [-0.001999998, 0.001999998]$.
The slope of the line segment is
$\dfrac{0.001999998 - (-0.001999998)}{0.001 - (-0.001)}$, or 1.999998.

61.   Example of zoom on window:
$\left[1-10^{-8},\ 1+10^{-8}\right]\times[0,\ 2]$.
The slope of the line segment is 0.

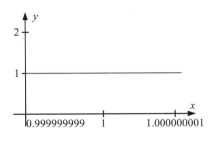

63.   The coordinates of the point Q are $Q=\left(\dfrac{3}{4},\ \dfrac{\sqrt{3}}{2}\right)$.

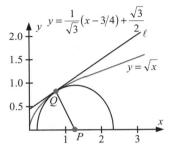

The equation of the line $\overline{PQ}$ is $y=-\sqrt{3}\cdot x+\dfrac{5\sqrt{3}}{4}$.

The equation of the line ℓ perpendicular to $\overline{PQ}$ is

$$y=\left(\dfrac{1}{\sqrt{3}}\right)\left(x-\dfrac{3}{4}\right)+\dfrac{\sqrt{3}}{2}\ .$$

65.   The slope $m\approx 4.76923$ minimizes the sum of the squared errors. The slope $m\approx 4.6666667$ minimizes the sum of the absolute errors.

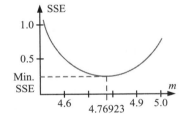

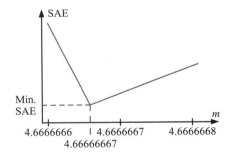

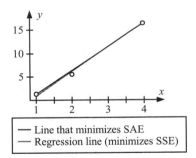

Section 1.4   **Functions and Their Graphs**

Problems for Practice

1. $\{x\in\mathbb{R}:x\neq-1\}$

3. $\{x\in\mathbb{R}:x\leq-\sqrt{2}\ \text{or}\ x\geq\sqrt{2}\}$ (The two inequalities can also be expressed as $|x|\geq\sqrt{2}$).

5. $\{x \in \mathbb{R} : x \neq -1 \text{ and } x \neq 1\}$

7. $\mathbb{R}$ (all real numbers)

9.

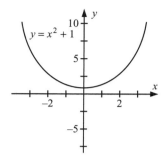

11.

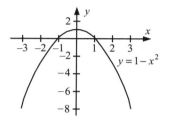

13.

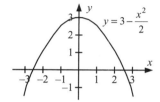

15. Let $F(x) = x^{-2}$. The domain of $F(x)$ is $\{x \in \mathbb{R} : x \neq 0\}$. The coordinates for several points on the graph of $F(x)$ are as follows:

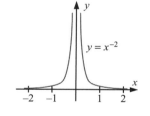

x	-2	-1	-0.75	-0.5	0.5	0.75	1	2
$F(x)$	0.25	1	$16/9$	4	4	$16/9$	1	0.25

17. The domain of $F(x) = \sqrt{2x+4}$ is $\{x \in \mathbb{R} : x \geq -2\}$. The coordinates for several points on the graph of $F(x)$ are as follows:

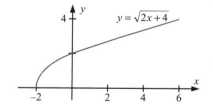

x	-2	1	0	1	2	2.5
$F(x)$	0	$\sqrt{2}$	2	$\sqrt{6}$	$2\sqrt{2}$	3

19. The domain is $\{x \in \mathbb{R} : x \geq 4\}$. The coordinates for several points on the graph of $F(x)$ are as follows:

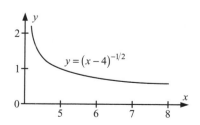

x	4.25	4.5	5	8	13
$F(x)$	2	$\sqrt{2}$	1	$1/2$	$1/3$

21. The domain is $\{x \in \mathbb{R} : x > -1\}$. The coordinates for several points on the graph of $F(x)$ are as follows:

x	-0.99	-0.75	0	3	8
$F(x)$	10	2	1	1/2	1/3

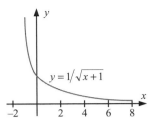

23. The domain is $\mathbb{R}$. The coordinates for several points on the graph of $F(x)$ are as follows:

x	-2	-1.5	-1	-0.5	0	0.5	1	1.5	2
$F(x)$	4	2.25	1	0.25	0	0.25	1	-0.25	-2

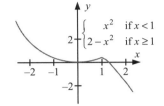

Further Theory and Practice

25. For $\alpha = 360$ the arc length of the arc of the circle of radius r subtended by this angle is $s(\alpha) = 2\pi r$, hence $2\pi r = k \cdot 360$ and $k = \pi r / 180$. For an arbitrary $\alpha \leq 360$, the length of the arc is $s(\alpha) = (\pi/180) \cdot r\alpha$.

27. For $\alpha = 360$ the sector becomes a disk with the area of πr^2. Therefore, $\pi r^2 = k \cdot r^2 \cdot 360$ and $k = \pi/360$. The area of a sector with an angle α is $A(r, \alpha) = (\pi/360) \cdot r^2 \cdot \alpha$.

29. $A = -(r+s)$; $B = rs$.

31. For $0 \leq x \leq 1$ the straight line contains points $(0, 1)$ and $(1, 3)$ and its equation is $y = 2x + 1$. Similarly, for $1 \leq x \leq 3$ its equation is $y = -1.5x + 4.5$, and for $3 \leq x \leq 4$ the equation is $y = x - 3$. We, therefore, solve to find

$$f(x) = \begin{cases} 1 + 2x & \text{if } 0 \leq x \leq 1 \\ 1.5(3 - x) & \text{if } 1 < x \leq 3 \\ x - 3 & \text{if } 3 < x \leq 4 \\ 1 & \text{if } 4 < x \leq 5 \end{cases}$$

33.   We complete the template as follows: $m(x) = \begin{cases} 0.15 & \text{if } 0 < x \le 23350 \\ 0.28 & \text{if } 23350 < x \le 56550 \\ 0.31 & \text{if } 56550 < x < 117950 \end{cases}$.

For $0 < x \le 23350$ the area under the graph of $m(x)$ is the area of the rectangle based on $[0, x]$ with height 0.15, i.e. $A(x) = 0.15x = T(x)$. For $23350 < x \le 56550$ the area under $m(x)$ includes two parts; a rectangle with sides 23350 and 0.15 (the area is 3202.5) and a rectangle with sides $x - 23350$ and 0.28 $\big($the area is $0.28(x - 23350)\big)$. We conclude that $A(x) = T(x)$. For $56550 < x \le 117950$,

$$A(x) = 23350 \cdot 0.15 + (56550 - 23350) \cdot 0.28 + (x - 56550) \cdot 0.31 = T(x).$$

Therefore, $A(x) = T(x)$ everywhere.

35.   The terms of the sequence are $f_1 = 2$, $f_n = 2^n = 2^{n-1} \cdot 2 = f_{n-1} \cdot 2$. A recursive definition is $f_1 = 2$, $f_n = 2 \cdot f_{n-1}$ for $n \ge 2$.

37.   The terms of the sequence are $f_1 = 1$, $f_n = n! = 1 \cdot 2 \cdot 3 \cdot \ldots \cdot (n-1) \cdot n = f_{n-1} \cdot n$. A recursive definition is $f_1 = 1$, $f_n = n \cdot f_{n-1}$ for $n \ge 2$.

39.   The terms of the sequence are $f_1 = 1$, $f_n = \dfrac{n(n+1)}{2} = \dfrac{n(n-1+2)}{2} = \dfrac{n(n-1)}{2} + n = f_{n-1} + n$. A recursive definition is $f_1 = 1$, $f_n = n + f_{n-1}$ for $n \ge 2$ (see Problem 1.4.36).

41.   For $x \ge 0$ we see that $\lfloor x \rfloor = \text{Int}(x)$. For $x < 0$, the function is $\text{Int}(x) = \lfloor x \rfloor + 1$, hence

$$\lfloor x \rfloor = \begin{cases} \text{Int}(x) & \text{if } x \ge 0 \\ \text{Int}(x) - 1 & \text{if } x < 0 \end{cases}$$

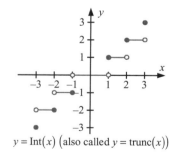

$y = \text{Int}(x)$ $\big($also called $y = \text{trunc}(x)\big)$

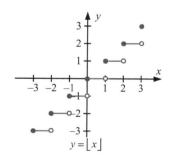

$y = \lfloor x \rfloor$

43.   Annual payment is $12 \cdot m$ dollars; total payment is $12 \cdot m \cdot n$ dollars. Therefore, $I(P, m, n) = 12 \cdot m \cdot n - P$ dollars.

45.   The terms of the sequence are 1, 1, 2, 5, 14, 42, 132, 429, 1430. Using the first recursive definition, we see that c_0 and $c_1 = c_0 c_0 = 1$ are integers. It follows that $c_2 = c_0 c_1 + c_1 c_0$ is an integer. In general, if $c_0, c_1, \cdots c_n$ are integers, then $c_{n+1} = c_0 c_n + c_1 c_{n-1} + \cdots + c_{n-1} c_1 + c_n c_0$ is also an integer.

47. $H(x) = (x \geq 0)$

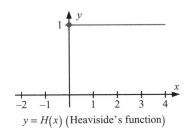

$y = H(x)$ (Heaviside's function)

49. $b(x) = x(0 < x)(x \leq 1) + (2-x)(1 < x)(x \leq 2)$

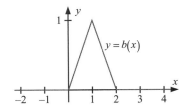

$y = b(x)$

Calculator/Computer Exercises

51.   If the incidence of a particular disease is small, then the probability that a positive result is a true positive is small. Routine screening (without regard to the presence of risk factors) is not advisable.

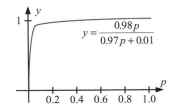

$y = \dfrac{0.98p}{0.97p + 0.01}$

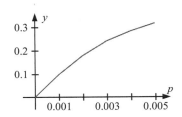

53.   The maximum of $f(x)$ occurs when $x = 4$. Set $x_0 = 4$. Let $x_1 = 3$. (The choice of a different value in the interval $(2, x_0)$ will lead to similar results.) The following table compares the values of $F(h, x_1)$ and $F(h, x_0)$ for $h = 10^{-4}$, $h = 10^{-5}$, $h = 10^{-6}$ and $h = 10^{-8}$.

$f(x) = 5 - 24x^2 + 20x^3 - 3x^4$

	$h = 10^{-4}$	$h = 10^{-5}$	$h = 10^{-6}$	$h = 10^{-8}$
$F(h, x_1)$	0.72e − 2	0.72e − 3	0.72e − 4	0.72e − 6
$F(h, x_0)$	0.72e − 6	0.72e − 8	0.72e − 10	0.72e − 12

At the point x_0, the function f attains a maximum value. The backward differences are several orders of magnitude smaller than the backward differences at other points x_1.

55.   **a.** $\left| x_{n+1} - \sqrt{\eta} \right| = \left| \dfrac{1}{2}\left(x_n + \dfrac{\eta}{x_n} \right) - \sqrt{\eta} \right| = \left| \dfrac{x_n^2 - 2x_n\sqrt{\eta} + \eta}{2x_n} \right| = \dfrac{\left(x_n - \sqrt{\eta} \right)^2}{2x_n}$.

   b.   If $\left| x_n - \sqrt{\eta} \right| \leq 10^{-k}$, then $\left| x_{n+1} - \sqrt{\eta} \right| \leq \dfrac{10^{-2k}}{2x_n} < 10^{-2k}$ for $\eta \in (1, 4)$.

c.   Using the algorithm from part **b** we compute the terms: $x_0 = \dfrac{3}{2}$,

$$x_1 = \frac{1}{2}\left(\frac{3}{2} + \frac{3.75}{\frac{3}{2}}\right) = 2,\; x_2 = \frac{1}{2}\left(2 + \frac{3.75}{2}\right) = 1.9375,$$

$$x_3 = \frac{1}{2}\left(1.9375 + \frac{3.75}{1.9375}\right) = 1.9364919235,$$

$$x_4 = \frac{1}{2}\left(1.9364919235 + \frac{3.75}{1.9364919235}\right) = 1.9364916731.$$

Comparing with $\sqrt{3.75} \approx 1.9364916731$, x_n for $n \geq 3$ approximates $\sqrt{3.75}$ up to six decimal places.

57. $p_1 = 2.828427124$, $p_2 = 3.061467458$, $p_3 = 3.121445152$, $p_4 = 3.136548492$, $p_5 = 3.140331158$, $p_6 = 3.141277254$, $p_7 = 3.141513803$, $p_8 = 3.141572945$

Section 1.5   Combining Functions

Problems for Practice

1. $x^2 + 5 + \left(\dfrac{x+1}{x-1}\right) = \dfrac{x^3 - x^2 + 6x - 4}{x-1}$

3. $\dfrac{(2x-5)+1}{(2x-5)-1} = \dfrac{x-2}{x-3}$

5. $(2x-5)\left(\dfrac{x+1}{x-1}\right)$

7. $\left(\dfrac{(2x-5)+1}{(2x-5)-1}\right)^2 + 5 = \dfrac{(x-2)^2}{(x-3)^2 + 5}$

9. $\left(x^2 + 5\right)\left(\left(\dfrac{x+1}{x-1}\right) + 2x - 5\right) = 2\left(x^2 + 5\right)\dfrac{x^2 - 3x + 3}{x-1}$

11. $\dfrac{\left(x^2 + 5 + \left(\frac{x+1}{x-1}\right)\right)}{(2x-5)^2} = \dfrac{x^3 - x^2 + 6x -}{(x-1)(2x-5}$

13.   Function $h(x)$ can be written as a composition function $(g \circ f)(x)$ where $f(x) = x - 2$ and $g(x) = x^2$.

15.   Function $h(x)$ can be written as a composition function $(g \circ f)(x)$ where $f(x) = x^3 + 3x$ and $g(x) = x^4$.

17.   Function $h(x)$ can be written as a composition function $(g \circ f)(x)$ where $f(x) = x^2 + 4x$ and $g(x) = x + 4$.

19.   Calculate: $g(x) = 3x^2 + 1$, since $3(x+1)^2 + 1 = 3x^2 + 6x + 4$.

21. Calculate: $g(x) = \dfrac{x}{x^2 + 2}$, since $\dfrac{x^2 + 1}{\left(x^2 + 1\right)^2 + 2} = \dfrac{x^2 + 1}{x^4 + 2x^2 + 3}$.

23. We calculate $g\left(\dfrac{1}{8}\right) = \left(\dfrac{1}{8}\right)^{-1/3} = 2$ and $f(2) = \sqrt{2 \cdot 2 + 5} = 3$, hence $(f \circ g)\left(\dfrac{1}{8}\right) = 3$.

25. We calculate $f(11) = \sqrt{2 \cdot 11 + 5} = \sqrt{27}$ and $g(54) = 54^{-1/3}$, hence $f^2(11) \cdot g^3(54) = \dfrac{27}{54} = \dfrac{1}{2}$.

27. $(x+5)(x-1) = x^2 + 4x - 5$

29. $(x-2)(x+2)\left(x^2 + 2x + 2\right) = x^4 + 2x^3 - 2x^2 - 8x - 8$

31. $g(x) = f(x+2)$

The plot of g is obtained by shifting the plot of f by 2 units to the left.

33. $g(x) = f(x-3)$

The plot of g is obtained by shifting the plot of f by 3 units to the right.

35. $g(x) = f(x+1) + 4$

The plot of g is obtained by shifting the plot of f by 1 unit to the left and 4 units up.

37. $g(x) = f(x-1) + 1$

The plot of g is obtained by shifting the plot of f by 1 unit to the right and 1 unit up.

39. $y = 3x - 7$

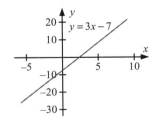

41.   The curve that has the parametric equation

$$\begin{cases} x = \dfrac{1}{t} \\ y = 3 \quad 0 < t \le 1 \end{cases}$$

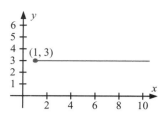

is the horizontal ray that has $(1, 3)$ as a left endpoint.

43. $y = \dfrac{1}{x},\ 0 < x \le 1$

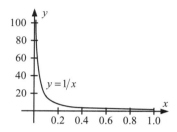

Further Theory and Practice

45.   Suppose that $(p \circ p)(x) = x$ for all x. Because $\deg\big(p(x)\big)^2 = \deg\big((p \circ p)(x)\big) = 1$, we see that $\deg\big(p(x)\big) = 1$. It follows that $p(x) = mx + k$ for some constants m and k with $m \ne 0$. Thus, $x = (p \circ p)(x) = m(mx + k) + k = m^2 x + (m + 1)k$, or $(m + 1)\big((m - 1)x + k\big) = 0$. If $m = -1$ then this last equation is satisfied for all k. Otherwise $(m - 1)x + k = 0$ for all x. For this to happen, we must have $m = 1$ and $k = 0$.

47.   Assume that a rational function is written in the form $f(x) + p(x)/h(x)$ where the degree n of $p(x)$ is equal or larger than the degree m of $h(x)$. We can always write $p(x) = ax^{n-m} h(x) + q(x)$, where $q(x)$ has a degree smaller than n. Our rational function can now be written as $\big(f(x) + ax^{n-m}\big) + q(x)/h(x)$. If the degree of $q(x)$ is still not smaller than m, we can repeat the same procedure until the degree of $q(x)$ becomes $m - 1$. At that point the expression for the rational function becomes $\tilde{f}(x) + q(x)/h(x)$ and the degree of $q(x)$ is strictly smaller than m. Similarly,

$$\frac{3x^5 + 2x^4 - x^2 + 6}{x^3 - x + 3} = \big(3x^2 + 2x + 3\big) - \frac{8x^2 + 3x + 3}{x^3 - x + 3}.$$

49.   As $h(x) = x^2 + 2x + 3 = (x + 1)^2 + 2$, we calculate $g(x) = x^2 + 2$.

51.   As $h(x) = 2x^2 = 2(x^2 - 9) + 18$, we calculate $g(x) = 2x + 18$.

53.   As $h(x) = g(x - 4)$, we calculate $f(x) = x - 4$.

55.   As $h(x) = (x^2 - 1) + 1 = \left((x^2 - 1)^{1/3} \right)^3 + 1 = g\left((x^2 - 1)^{1/3} \right)$, we calculate $f(x) = (x^2 - 1)^{1/3}$.

57.   Replacing x by $x - 3$, we obtain $f(x) = (x - 3)^2$.

59.   Replacing x by $-x$, we obtain $f(x) = \dfrac{1 - x^3}{x^2 + 1}$.

61.   Note that $f \circ f$ is the result of applying f to $f(x)$, but $f \cdot f$ is the square of $f(x)$. We compute the combined function and the product: $(f \circ f)(x) = x^{p^2}$ and $(f \cdot f)(x) = x^{2p}$. Then $f \circ f = f \cdot f$ when $p^2 = 2p$. Solving this equation we obtain $p = 0$ and $p = 2$.

63.   One function is $g(u) = \pi$ for all u. One function is $h(u) = u^{1/5}$. One function is $k(u) = u^{1/5} + \pi$.

65.

67.   Verification: $G \circ F = G(F(x))$, then $H \circ (G \circ F) = H(G(F(x)))$; $H \circ G = H(G(x))$, and $H(G(x)) \circ F = H(G(F(x)))$.

69. $x = y^{2/3}$ **71.** $y = 3 \cdot 2^x$

73.   The approximating least square line is
$y = 3.12x + 1.91$.

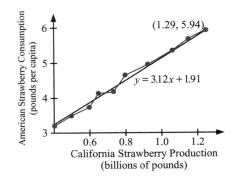

75.   If a point (x_0, y_0) is on the curve C then there is t_0 in I such that $f(t_0) = x_0$ and $g(t_0) = y_0$.
Given that $x_0 = f(t_0) = t_0$, we get $y_0 = g(t_0) = \varphi(t_0) = \varphi(x_0)$. Therefore, a point (x_0, y_0) is
on the graph $\mathcal{G}$. Now we show that if a point (x_0, y_0) is on the graph $\mathcal{G}$ then it is on the curve
C. If a point (x_0, y_0) is on the graph $\mathcal{G}$ then $y_0 = \varphi(x_0)$. We can always find t_0 in I such that
$t_0 = f(t_0) = x_0$ for a t_0 in I. Then $y_0 = \varphi(x_0) = \varphi(t_0) = g(t_0)$. As $x_0 = f(t_0)$ and
$y_0 = g(t_0)$, (x_0, y_0) is on the curve C. Therefore, a point (x_0, y_0) is on the graph $\mathcal{G}$ only if it
is on the curve C.

77.   We can write any point in the plane as $(x_0 \pm d, y)$ where $d \geq 0$. Then

$$R(x, y) = R(x_0 \pm d, y) = (2x_0 - (x_0 \pm d), y) = (x_0 \mp d, y),$$

and $(x_0 \pm d, y) \mapsto (x_0 \mp d, y)$ under R, which shows that R is a reflection about the line $x = x_0$.
Given that $f(2x_0 - x) = R(x, f(x))$ the graph of $f(2x_0 - x)$ is a reflection about the line
$x = x_0$ of the graph of $f(x)$. Therefore, if $f(2x_0 - x) = f(x)$, the graph of $f(x)$ is
symmetric about the line $x = x_0$.

Calculator/Computer Exercises

79. C' is the graph of $y = (x - 3)^2$.

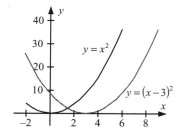

81. C' is obtained by replacing x with $-x$ in the equation that defines C.

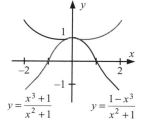

83.

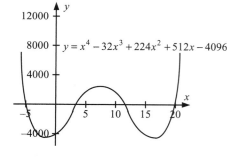

a. $x_0 = 8$

b.

Two functions coincide. Therefore,
function $y(x)$ is symmetric about the line
$x = 8$.

c.   The function $p(x) = f(x-8)$ is indicated by the solid line, and the function

$$p(-x) = f(8-x)$$

is shown with points. The graphs are the same, hence $p(x) = p(-x)$. Therefore $p(x)$ is an even function. The explicit form of the polynomial $p(x)$ is found by calculating $f(x-8)$: $p(x) = x^4 - 160x^2 - 2048$.

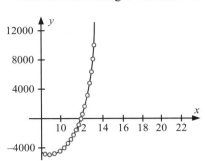

The roots of $p(x)$ are: $\left(-4\sqrt{5-\sqrt{17}},\, -4\sqrt{5+\sqrt{17}},\, 4\sqrt{5-\sqrt{17}},\, 4\sqrt{5+\sqrt{17}}\right)$.

The roots of $f(x)$ are: $\left(8-4\sqrt{5-\sqrt{17}},\, 8-4\sqrt{5+\sqrt{17}},\, 8+4\sqrt{5-\sqrt{17}},\, 8+4\sqrt{5+\sqrt{17}}\right)$.

85. $r(3) = 0.288675$; The line is $y = 0.288675(x-3) + \sqrt{3}$.

87. $r(2) = 1.64645$; The line is $y = 1.64645(x-2) + \left(4-\sqrt{2}\right)$.

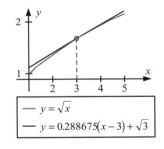

$$\cdots\cdots\ y = \sqrt{x}$$
$$\text{——}\ y = 0.288675(x-3) + \sqrt{3}$$

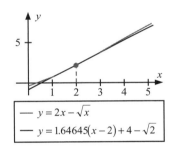

$$\cdots\cdots\ y = 2x - \sqrt{x}$$
$$\text{——}\ y = 1.64645(x-2) + 4 - \sqrt{2}$$

89.

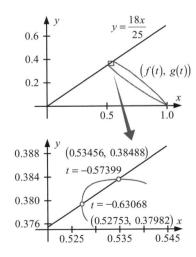

Section 1.6   Trigonometry

Problems for Practice

1. $\cos\left(\dfrac{\pi}{6}\right)=\dfrac{\sqrt{3}}{2}$, $\sin\left(\dfrac{\pi}{6}\right)=\dfrac{1}{2}$, $\tan\left(\dfrac{\pi}{6}\right)=\dfrac{\sqrt{3}}{3}$, $\cot\left(\dfrac{\pi}{6}\right)=\sqrt{3}$, $\sec\left(\dfrac{\pi}{6}\right)=\dfrac{2\sqrt{3}}{3}$, $\csc\left(\dfrac{\pi}{6}\right)=2$

3. $\cos\left(\dfrac{2\pi}{3}\right)=-\dfrac{1}{2}$, $\sin\left(\dfrac{2\pi}{3}\right)=\dfrac{\sqrt{3}}{2}$, $\tan\left(\dfrac{2\pi}{3}\right)=-\sqrt{3}$, $\cot\left(\dfrac{2\pi}{3}\right)=-\dfrac{\sqrt{3}}{3}$, $\sec\left(\dfrac{2\pi}{3}\right)=-2$,

$\csc\left(\dfrac{2\pi}{3}\right)=\dfrac{2\sqrt{3}}{3}$

5. $\sin\left(\dfrac{\pi}{3}\right)\sin\left(\dfrac{\pi}{6}\right)=\left(\dfrac{\sqrt{3}}{2}\right)\left(\dfrac{1}{2}\right)=\dfrac{\sqrt{3}}{4}$ **7.** $\cos\left(\dfrac{\pi}{6}\right)+\cos\left(\dfrac{\pi}{3}\right)=\left(\dfrac{\sqrt{3}}{2}\right)+\dfrac{1}{2}=\dfrac{\sqrt{3}+1}{2}$

9. $\dfrac{\tan\left(\pi/3\right)}{\tan\left(\pi/6\right)}=\dfrac{\sqrt{3}}{\sqrt{3}/3}=3$ **11.** $\sin\left(\pi\cdot\sin\left(\dfrac{\pi}{6}\right)\right)=\sin\left(\pi\cdot\left(\dfrac{1}{2}\right)\right)=1$

13. $\sin\left(\dfrac{19\pi}{2}\right)^{\cos(33\pi)}=\sin\left(8\pi+\dfrac{3\pi}{2}\right)^{\cos(32\pi+\pi)}=\sin\left(\dfrac{3\pi}{2}\right)^{\cos(\pi)}=(-1)^{-1}=-1$

15. $\cos(\theta)=\sqrt{1-\sin^2(\theta)}=\sqrt{1-\dfrac{1}{9}}=\dfrac{2\sqrt{2}}{3}$

17. $\sin(2\theta)=2\sin(\theta)\cos(\theta)=2\sqrt{1-\cos^2(\theta)}\,\cos(\theta)=2\sqrt{1-\dfrac{16}{25}}\left(\dfrac{4}{5}\right)=2\left(\dfrac{3}{5}\right)\left(\dfrac{4}{5}\right)=\dfrac{24}{25}$

19. $\cos\left(\dfrac{\theta}{2}\right)=\sqrt{\dfrac{1+\cos(\theta)}{2}}=\sqrt{\dfrac{1+\sqrt{1-\sin^2(\theta)}}{2}}=\sqrt{\dfrac{1+\sqrt{1-25/144}}{2}}=\sqrt{\dfrac{12+\sqrt{119}}{24}}$

$=\dfrac{1}{2}\sqrt{\dfrac{12+\sqrt{119}}{6}}$

21.   All six   **23.**   Tangent, cotangent

25.   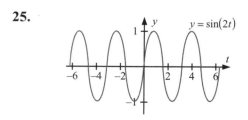   **27.**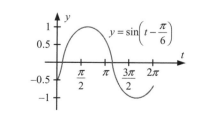

29. $x=\cos(t)$, $y=\sin(t)$ **31.** $x=\sin(t)$, $y=\cos(t)$

Further Theory and Practice

33. The proportionality means that $s(r, \theta) = kr\theta$. For $\theta = 2\pi$, the arc length is $2\pi r$. Therefore, $2\pi r = kr \cdot 2\pi$ and $k = 1$. Finally, $s(r, \theta) = r\theta$.

35. The first equation holds for some values of its variable. The solutions are $n\pi/2$ where n is an integer. The second equation is an *identity*. It holds for all values of its variable.

37. $\tan(\theta + \varphi) = \dfrac{\sin(\theta + \varphi)}{\cos(\theta + \varphi)} = \dfrac{\sin(\theta)\cos(\varphi) + \cos(\theta)\sin(\varphi)}{\cos(\theta)\cos(\varphi) - \sin(\theta)\cos(\varphi)} = \dfrac{\sin(\theta) + \cos(\theta)\tan(\varphi)}{\cos(\varphi) - \sin(\theta)\tan(\varphi)}$

$= \dfrac{\tan(\theta) + \tan(\varphi)}{1 - \tan(\theta)\tan(\varphi)}$

39. Adding the expressions for $\cos(\theta - \varphi)$ and $\cos(\theta + \varphi)$ we find:

$$\cos(\theta - \varphi) + \cos(\theta + \varphi) = 2\cos(\theta)\cos(\varphi)$$

and so $\cos(\theta)\cos(\varphi) = \frac{1}{2}\left[\cos(\theta - \varphi) + \cos(\theta + \varphi)\right]$.

41. $\sin(\theta) = \sin\left((\theta - \pi/2) + \pi/2\right) = \sin(\theta - \pi/2)\cos(\pi/2) + \cos(\theta - \pi/2)\sin(\pi/2)$
$= \sin(\theta - \pi/2) \cdot 0 + \cos(\theta - \pi/2) \cdot 1 = \cos(\theta - \pi/2)$

43. $\cos(\theta + \pi) = \cos(\theta)\cos(\pi) - \sin(\theta)\sin(\pi) = \cos(\theta) \cdot (-1) - \sin(\theta) \cdot 0 = -\cos(\theta)$

45. $\cos\left(\dfrac{7\pi}{12}\right) = \cos\left(\dfrac{\pi}{3} + \dfrac{\pi}{4}\right) = \cos\left(\dfrac{\pi}{3}\right)\cos\left(\dfrac{\pi}{4}\right) - \sin\left(\dfrac{\pi}{3}\right)\sin\left(\dfrac{\pi}{4}\right) = \left(\dfrac{1}{2}\right)\left(\dfrac{\sqrt{2}}{2}\right) - \left(\dfrac{\sqrt{3}}{2}\right)\left(\dfrac{\sqrt{2}}{2}\right)$

$= \sqrt{2}\,\dfrac{1 - \sqrt{3}}{4}$

$\sin\left(\dfrac{7\pi}{12}\right) = \sin\left(\dfrac{\pi}{3} + \dfrac{\pi}{4}\right) = \sin\left(\dfrac{\pi}{3}\right)\cos\left(\dfrac{\pi}{4}\right) + \cos\left(\dfrac{\pi}{3}\right)\sin\left(\dfrac{\pi}{4}\right) = \left(\dfrac{\sqrt{3}}{2}\right)\left(\dfrac{\sqrt{2}}{2}\right) + \left(\dfrac{1}{2}\right)\left(\dfrac{\sqrt{2}}{2}\right)$

$= \sqrt{2}\,\dfrac{1 + \sqrt{3}}{4}$

$\tan\left(\dfrac{7\pi}{12}\right) = \dfrac{\sin(7\pi/12)}{\cos(7\pi/12)} = \dfrac{1 + \sqrt{3}}{1 - \sqrt{3}} = \dfrac{\left(1 + \sqrt{3}\right)^2}{1 - 3} = -2 - \sqrt{3}$

$\cot\left(\dfrac{7\pi}{12}\right) = \dfrac{1}{\tan(7\pi/12)} = \dfrac{-1}{2 + \sqrt{3}} = -\dfrac{2 - \sqrt{3}}{4 - 3} = -2 + \sqrt{3}$

$\sec\left(\dfrac{7\pi}{12}\right) = \dfrac{1}{\cos(7\pi/12)} = \dfrac{2\sqrt{2}}{1 - \sqrt{3}} = -\sqrt{2}\left(1 + \sqrt{3}\right)$

$\csc\left(\dfrac{7\pi}{12}\right) = \dfrac{1}{\sin(7\pi/12)} = \dfrac{2\sqrt{2}}{1 + \sqrt{3}} = \sqrt{2}\left(\sqrt{3} - 1\right)$

47.   Let $\theta = \alpha + \beta$ and $\varphi = \alpha - \beta$. Then $\alpha = \dfrac{\theta + \varphi}{2}$, $\beta = \dfrac{\theta - \varphi}{2}$, and

$$\sin(\alpha)\cos(\beta) = \tfrac{1}{2}\big(\sin(\theta) + \sin(\varphi)\big), \text{ or } \sin(\theta) + \sin(\varphi) = 2\sin\left(\frac{\theta + \varphi}{2}\right)\cos\left(\frac{\theta - \varphi}{2}\right).$$

49.   For $\theta = \alpha + \beta$ and $\varphi = \alpha - \beta$: $\alpha = \dfrac{\theta + \varphi}{2}$, $\beta = \dfrac{\theta - \varphi}{2}$, and

$$\cos(\alpha)\cos(\beta) = \tfrac{1}{2}\big(\cos(\theta) + \cos(\varphi)\big),$$

or $\cos(\theta) + \cos(\varphi) = 2\cos\left(\dfrac{\theta + \varphi}{2}\right)\cos\left(\dfrac{\theta - \varphi}{2}\right)$.

51.   Let ϕ be the phase shift. Then

$$A\cos(\theta) + B\sin(\theta) = C\sin(\theta + \phi) = C\big(\sin(\phi)\cos(\theta) + \cos(\theta)\sin(\phi)\big),$$

where $C = \sqrt{A^2 + B^2}$. This equation has to be satisfied for any θ; therefore with this value it is possible to simultaneously set $\sin(\phi) = \dfrac{A}{C}$ and $\cos(\phi) = \dfrac{B}{C}$.

53.   Let A and C be the two points on the earth's surface, and B be the projection of the top of the mountain onto the line $\overline{AC}$. Then $\ell = |AB| = |AC| - |BC| = \big|h\cot(\theta) - h\cot(\phi)\big|$ and

$$h = \frac{\ell}{\big|\cot(\theta) - \cot(\phi)\big|}.$$

55.   We first note that $\dfrac{x}{a} + \dfrac{y}{b} = 1$. Therefore, the equation of the curve is $y = b - \dfrac{b}{a}\cdot x$. It represents a line segment connecting $(a,\ 0)$ to $(0,\ b)$; from $(0,\ b)$, it doubles back to $(a,\ 0)$ and then repeats, ending at $(a,\ 0)$.

57.   Given that $\left(\dfrac{x}{a}\right)\left(\dfrac{y}{b}\right) = 1$, we conclude that $y = \dfrac{ab}{x}$. For $\theta \in [0,\ \pi/2)$ this is part of a hyperbola $(x > 0)$.

59. $p = 2\pi$

61.   According to the definition of the tangent function,

$$\tan(x + p) - \tan(x) = \frac{\sin(x + p)}{\cos(x + p)} - \frac{\sin(x)}{\cos(x)} = \frac{2\sin(p)}{\cos(2x + p) + \cos(p)}.$$

In order for p to be the period, this difference has to be equal to zero which is true if $p = \pi$.

63. The period of $\tan(2x)$ is $\dfrac{\pi}{2}$; the period of $\sin(3x)$ is $\dfrac{2\pi}{3}$. The period of their sum is $p = 2\pi$.

65. In the model, $T(t) = 27.5 + 23.5\sin\left(\dfrac{\pi t}{6} - \dfrac{\pi}{2}\right)$, we see that the low occurs at the end of

December $(t = 12)$, and the high occurs at the beginning of July $(t = 6)$.

67. **a.** This n-gon can be decomposed into n equal isosceles triangles with equal sides r and

an angle $\dfrac{2\pi}{n}$ between them. The third side of this triangle is $2r\sin\left(\dfrac{\pi}{n}\right)$ and

$$p(n,\, r) = 2rn\sin\left(\dfrac{\pi}{n}\right).$$

   b. The area of one triangle is $r^2\dfrac{\sin(2\pi/n)}{2}$. The entire area is $A(n,\, r) = nr^2\dfrac{\sin(2\pi/n)}{2}$.

Calculator/Computer Exercises

69.

π	$n \cdot \sin(1/n)$
10	0.9983342
10^2	0.9999833
10^3	0.9999998
10^4	1.0000000
10^5	1.0000000
10^6	1.0000000

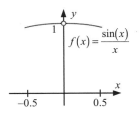

We compute $\lim_{n\to\infty} n \cdot \sin\left(\dfrac{1}{n}\right) = 1$; $\lim_{x\to 0} f(x) = 1$. As $a_n = f\left(\dfrac{1}{n}\right)$, the limiting behavior

of a_n at large n can be determined from the graph by locating the value $f(x = 0)$ which is
equal to 1.

71. The function $f(x) = 90x + \cos(x)$ appears to coincide with $y = 90x$ when $-10 \le x \le 10$.
The function $f(x) = 90x + \cos(x)$ appears to coincide with $y = 1$ when $-0.0001 \le x \le 0.0001$.

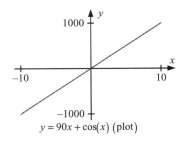

$y = 90x + \cos(x)$ (plot)

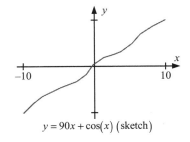

$y = 90x + \cos(x)$ (sketch)

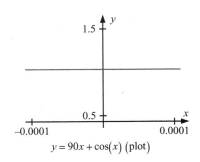

$y = 90x + \cos(x)$ (plot)

73. $T(t) = 59 + 19\sin\left(\dfrac{\pi t}{6} - \dfrac{7\pi}{12}\right)$

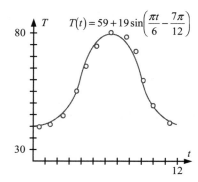

75.   Using the trigonometric relation for cosine we obtain
$\cos(v_1 \cdot t) + \cos(v_2 \cdot t) = A(t) \cdot \cos(\omega \cdot t)$ where

$A(t) = 2\cos\left(\dfrac{(v_1 - v_2)t}{2}\right)$ and $\omega = \dfrac{v_1 + v_2}{2}$. For $v_1 = 8$

and $v_2 = 6$, $\omega = 7$.

The frequency of the modulated amplitude is 1.

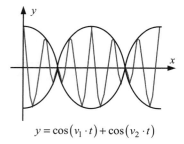

77.

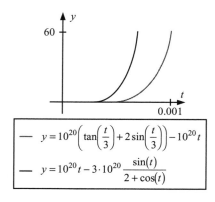

Limits

The Concept of Limit

Problems for Practice

1. $\lim\limits_{x \to 2}(x+3) = 2+3 = 5$

3. $\lim\limits_{h \to 4}\left(3h^2 + 2h + 1\right) = 3 \cdot 4^2 + 2 \cdot 4 + 1 = 57$

5. $\lim\limits_{h \to 1}\dfrac{h-3}{h+1} = \dfrac{1-3}{1+1} = -1$

7. If we choose any $\epsilon > 0$, then the values of $g(x)$ for x *outside* the interval $(2-\epsilon,\ 2+\epsilon)$ cannot influence the value of $\lim\limits_{x \to 2} g(x)$. For example, if we were to use $\epsilon = 0.1$, we see that $g(x)$ is identically -4 for x in the interval $(1.9,\ 2.1)$. Therefore, we obtain $\lim\limits_{x \to 2} g(x) = g(2) = -4$.

9. The limit as x approaches 5 exists as follows:
$$\lim\limits_{x \to 5}\frac{x^2 - 25}{x - 5} = \lim\limits_{x \to 5}\frac{(x-5)(x+5)}{x-5} = \lim\limits_{x \to 5}(x+5) = 10.$$

11. The limit does exist as shown: $\lim\limits_{t \to -7}\dfrac{t+7}{t^2 - 49} = \lim\limits_{t \to -7}\dfrac{t+7}{(t+7)(t-7)} = \lim\limits_{t \to -7}\dfrac{1}{t-7} = -\dfrac{1}{14}.$

13. The limit does not exist. We factor as follows: $\dfrac{x^2 + 6x - 8}{x^2 - 2x - 24} = \dfrac{x^2 + 6x - 8}{x - 6} \cdot \dfrac{1}{x+4}$. When x approaches -4, the ratio $\left(\dfrac{x^2 + 6x - 8}{x - 6}\right)$ approaches $\dfrac{(-4)^2 + 6(-4) - 8}{-4 - 6} = \dfrac{8}{5}$. However, $\dfrac{1}{x+4}$ does not have a limit as x approaches -4. For example, if we let $x = -4 + \dfrac{1}{n}$, then, as n increases without bound, x gets closer to -4, yet $\dfrac{1}{x+4} = \dfrac{1}{1/n} = n$ increases without bound.

15.   As we approach the limit of $x = 2$, we see that

$$\lim_{x \to 2^-} f(x) = \lim_{x \to 2^-} \left(x^2 - 3x \right) = 4 - 3(2) = -2 .$$

From the positive direction we find $\lim_{x \to 2^+} f(x) = \lim_{x \to 2^+} \dfrac{-x}{x-1} = -\dfrac{2}{2-1} = -2$. As both the

limits $\lim_{x \to 2^-} f(x)$ and $\lim_{x \to 2^+} f(x)$ exist and equal -2, we conclude that $\lim_{x \to 2} f(x)$ exists and equals -2.

17.   As $\dfrac{x^2 - 3x - 10}{x^2 - 9x + 20} = \dfrac{(x+2)(x-5)}{(x-4)(x-5)} = \dfrac{x+2}{x-4}$, we have $\lim_{x \to 5^+} f(x) = \lim_{x \to 5^+} \dfrac{x+2}{x-4} = \dfrac{5+2}{5-4} = 7$. We

can also see that $\lim_{x \to 5^-} f(x) = \lim_{x \to 5^-} \dfrac{x^2 - 4}{3} = \dfrac{25 - 4}{3} = 7$. In conclusion, the $\lim_{x \to 5} f(x)$ exists and

equals 7, because $\lim_{x \to 5^-} f(x)$ and $\lim_{x \to 5^+} f(x)$ both exist and equal 7.

19.   Given that $\left| (x+1) - 3 \right| < 0.01$ the range can be simplified to $\left| x - 2 \right| < 0.01$ or
$-0.01 < x - 2 < 0.01$, and finally $1.99 < x < 2.01$. Therefore, if x is within $\delta = 0.01$ of $c = 2$,
$f(x) = x + 1$ is within $\epsilon = 0.01$ of $\ell = 3$.

21.   Given that $\left| (5x + 2) - 17 \right| < 0.01$ the range can be simplified to $\left| 5x - 15 \right| < 0.01$ or
$-0.01 < 5x - 15 < 0.01$ and finally $2.998 < x < 3.002$. Therefore, if x is within $\delta = 0.002$ of
$c = 3$, $f(x) = 5x + 2$ is within $\epsilon = 0.01$ of $\ell = 17$.

23.   From the definition of $f(x)$: $\lim_{x \to 2^-} f(x) = 5$ and $\lim_{x \to 2^+} f(x) = 3$ we can see that for $\ell \geq 4$ and
$x = 2 + 1/n$: $\ell - f(x) \geq 1 > 0.01$. Hence there are no candidates for the limit inside $[4, \infty)$. For
$\ell < 4$ and $x = 2 - 1/n$: $f(x) - \ell > 1 > 0.01$ we find again, that there are no candidates for the
limit inside $(-\infty, 4)$.

Further Theory and Practice

25.   **a.**   The domains of g, f, h, and k are, respectively, $\{ x \in \mathbb{R} : x \neq 1 \}$, $\mathbb{R}$, $\mathbb{R}$, and $\mathbb{R}$.

   b.   The functions f and h are the same because they have the same domain and
   $f(x) = h(x)$ for each x in their domain.

   The function g is not defined at $x = 1$, whereas function k has a different value at
   $x = 1$: $f(1) = h(1) = 2$, $k(1) = 1$.

   c.   When $x \neq 1$, we have $\dfrac{x^2 - 1}{x - 1} = \dfrac{(x+1)(x-1)}{x-1} = x + 1$. Therefore,
   $g(x) = f(x) = h(x) = k(x)$ for $x \neq 1$. It follows that
   $\lim_{x \to 1} g(x) = \lim_{x \to 1} f(x) = \lim_{x \to 1} h(x) = \lim_{x \to 1} k(x) = 2$.

27. **a.** The domains of f, g, k, and h are $\mathbb{R}$, $\mathbb{R}$, $\mathbb{R}$, and $\{x \in \mathbb{R} : x \neq 0\}$, respectively.

b. The functions f and k are the same because they have the same domain and $f(x) = k(x)$ for each x in their domain (including $x = 0$)

The function h differs from f and k because it has a different domain; g differs from f and k because $1 = g(0) \neq f(0) = k(0) = 0$.

c. When $x \neq 0$, we have $g(x) = f(x) = h(x) = k(x)$. It follows that

$$\lim_{x \to 0} g(x) = \lim_{x \to 0} f(x) = \lim_{x \to 0} h(x) = \lim_{x \to 0} k(x) = 0.$$

29. We compute $\lim_{x \to 0^+} H(x) = 1$, $\lim_{x \to 0^-} H(x) = 0$. As $\lim_{x \to 0^+} H(x) \neq \lim_{x \to 0^-} H(x)$, $\lim_{x \to 0} H(x)$ does not exist.

31. Let x be the length in meters and y be the same length in feet: $x = 0.3048 y$. The accuracy of measurement in feet is 0.001, $|y - y_0| < 0.001$, so we find that $\left| \dfrac{x - x_0}{0.3048} \right| < 0.001$, or

$$|x - x_0| < 0.3048 \cdot 0.001 = 3.048 \cdot 10^{-4} \text{ m.}$$

33. **a.** The $\lim_{x \to 0} \lfloor x \rfloor$ does not exist because $\lim_{x \to 0^+} \lfloor x \rfloor = 0$, and $\lim_{x \to 0^-} \lfloor x \rfloor = -1$.

b. The limit exists as follows: $\lim_{x \to 1/2} \lfloor x \rfloor = 0$.

c. Given that $\lfloor x \rfloor = 0$ for $x \in [0, 1)$, the expression $\dfrac{1}{\lfloor x \rfloor}$ is not defined on an interval just to the left of $c = 1$. Therefore, we do not discuss the (two-sided) limit $\lim_{x \to 1^-} \dfrac{1}{\lfloor x \rfloor}$.

d. Given that $\lfloor x \rfloor = -1$ for $x \in [-1, 0)$, we find $\lim_{x \to -1/2} \dfrac{1}{\lfloor x \rfloor} = \dfrac{1}{-1} = -1$.

35. See the figure for the answer to Problem 1.4.41. The graph shows that

$$\lim_{x \to n^+} \lfloor x \rfloor = \left(\lim_{x \to n^-} \lfloor x \rfloor \right) + 1;$$

hence, $\lim_{x \to n^+} \lfloor x \rfloor \neq \lim_{x \to n^-} \lfloor x \rfloor$.

37. The distance traveled by the particle over time interval t is $d(t) = t^2$. The average velocity over time interval $[t_1, t_2]$ is $v = \dfrac{|d(t_2) - d(t_1)|}{t_2 - t_1}$.

a. $d(2) = 4$, $d(1) = 1$, $v = \dfrac{4 - 1}{2 - 1} = 3$

b. $d(1.5) = 2.25$, $d(1) = 1$, $v = \dfrac{2.25 - 1}{1.5 - 1} = 2.5$

c. $d(1.1)=1.21$, $d(1)=1$, $v=\dfrac{1.21-1}{1.1-1}=2.1$

d. $d(1+h)=1+2h+h^2$, $d(1)=1$, $v=\dfrac{1+2h+h^2-1}{h}=2+h$

e. $\lim\limits_{h\to0}(2+h)=2$

39.

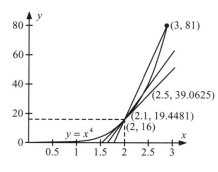

a. $f(2)=2^4=16$, $f(3)=3^4=81$,

   $m=\dfrac{f(3)-f(2)}{3-2}=65$

b. $f(2)=16$, $f(2.5)=2.5^4=39.0625$,

   $m=\dfrac{f(2.5)-f(2)}{2.5-2}=46.125$

c. $f(2)=16$, $f(2.1)=2.1^4=19.4481$,

   $m=\dfrac{f(2.1)-f(2)}{2.1-2}=34.481$

d. $f(2)=16$, $f(2+h)=(2+h)^4=16+32h+24h^2+8h^3+h^4$,

   $m=\dfrac{f(2+h)-f(2)}{h}=32+24h+8h^2+h^3$

e. $\lim\limits_{h\to0}\left(32+24h+8h^2+h^3\right)=32$

41.   **a.**   Let $\dfrac{x^{5/3}}{|x|}$ as $\dfrac{(x)}{|x|}x^{2/3}$. For $x>0$ we have $\left(\dfrac{x}{|x|}\right)x^{2/3}=x^{2/3}$ and for $x<0$ we have

$\left(\dfrac{x}{|x|}\right)x^{2/3}=-x^{2/3}$. In either case, the expression tends to 0 as x tends to 0. Therefore,

$\lim\limits_{x\to0}\dfrac{x^{5/3}}{|x|}=0$.

b.   We write $\sqrt{\left|x^2-1\right|}=\sqrt{\left|(x-1)(x+1)\right|}=\sqrt{|x-1|}\sqrt{|x+1|}$. Therefore,

$\lim\limits_{x\to1}\dfrac{\sqrt{|x-1|}}{\sqrt{\left|x^2-1\right|}}=\lim\limits_{x\to1}\dfrac{1}{\sqrt{|x+1|}}=\dfrac{1}{\sqrt{2}}$.

Calculator/Computer Exercises

43. $|x-2| < 0.0095$

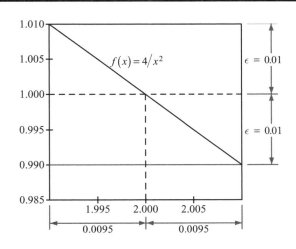

45. $|x-10| < 0.0046$

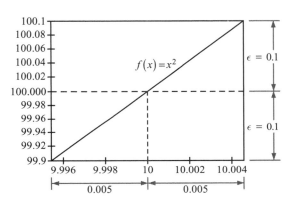

47. $|x-5| < 0.2381$

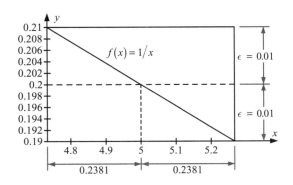

49. In order to produce work of the desired tolerance, $1.2|m| + 100m^2 < 10^{-3}$, we must solve for m: $|m| < 7.82 \times 10^{-4}$ mm.

51.   a.

x	$f(x)$
$\pi/2 - 1/10$	-0.99833
$\pi/2 - 1/100$	-0.99998
$\pi/2 - 1/1000$	-0.99999

d.

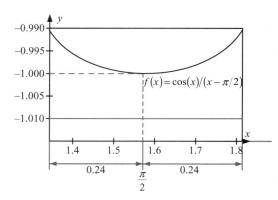

b. -1

c. $\delta = 0.24$

53.   a.

x	$f(x)$
$1/10$	0.16658
$1/100$	0.16666
$1/1000$	0.16666

d.

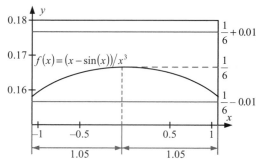

b. $\lim_{x \to 0} (x - \sin(x))/x^3 = 1/6$

c. $\delta = 1.05$

55.   Average velocity between t_1 and t_2 is $v = \dfrac{\left| p(t_2) - p(t_1) \right|}{t_2 - t_1}$.

a.   Plugging $t = 2$ into the function $p(t)$ we find that $p(2) = 6$, and

$p(2+h) = (2+h)^2 = 6 + 4h + h^2$, giving an average velocity

$v = \dfrac{6 + 4h + h^2 - 6}{h} = 4 + h$. Therefore $|h| < 0.1$ because $3.9 < 4 + h < 4.1$.

b.   From $3.99 < 4 + h < 4.01$ we get $|h| < 0.01$.

c.   The range $4 - \epsilon < 4 + h < 4 + \epsilon$ leads to $|h| < \epsilon$.

d.   The instantaneous velocity of the body at time $t = 2$ is 4.

57.   a. $\overline{v}(t) = \dfrac{\sin(t + 0.0001) - \sin(t - 0.0001)}{0.0002}$

See figure.

b.   The function $\overline{v}(t)$ is the greatest near

$t = 0$ and 2π; therefore, $\overline{v}(t)$ is the most

negative near $t = \pi$.

c.   The function $\overline{v}(t)$ is zero near $t = \dfrac{\pi}{2}$

and $\dfrac{3\pi}{2}$.

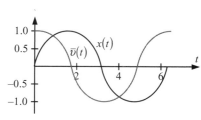

d.   The function $\bar{v} < 0$ for t in $\left[\dfrac{\pi}{2}, \dfrac{3\pi}{2}\right]$. As t increases from $\dfrac{\pi}{2}$ to $\dfrac{3\pi}{2}$, the value of $x(t)$ decreases from 1 to -1. If $\bar{v}(t) < 0$, then $x(t+0.0001) < x(t-0.0001)$. This suggests that x is decreasing.

e.   The function $\bar{v}$ appears to approximate the function $\cos(t)$.

Section 2.2   Limit Theorems

Problems for Practice

1.   The limit is 6. To verify this using the rigorous definition of the limit, we find δ such that $0 < |x-3| < \delta$ which implies that $0 < |2x-6| < \epsilon$. Reducing, we find $|2x-6| = 2|x-3|$, or $|x-3| < \delta$. This in turn gives $|2x-6| < \epsilon$ if $\delta = \dfrac{\epsilon}{2}$. Hence, the limit is 6.

3.   The limit is 0. To verify this using the rigorous definition of the limit, we find δ such that $0 < |x+2| < \delta$ which implies that $0 < \left|\dfrac{x^2-4}{x-2}\right| < \epsilon$. Reducing, we find

$$\frac{x^2-4}{x-2} = \frac{(x-2)(x+2)}{(x-2)} = (x+2),$$

or $\left|\dfrac{x^2-4}{x-2}\right| = |x+2|$, $|x+2| < \delta$. This in turn gives $\left|\dfrac{x^2-4}{x-2}\right| < \epsilon$ if $\delta = \epsilon$. Hence, the limit is 0.

5.   The limit from the right is 4 and the limit from the left is 15. Consider the limit from the right. If $5 < x < 5+\delta$, then for any $\epsilon > 0$, $|f(x)-4| = |4-4| = 0 < \epsilon$. Therefore, $\lim\limits_{x\to 5^+} f(x) = 4$. For the limit from the left, $5-\delta < x < 5$, and for any $\epsilon > 0$, $|f(x)-15| = |3x-15| = 3|x-5| < \delta$ if $\delta = \dfrac{\epsilon}{3}$. Therefore, $\lim\limits_{x\to 5^-} f(x) = 15$. As the limit from the right does not equal the limit from the left, $\lim\limits_{x\to 5} f(x)$ does not exist.

7.   The limit from the right is -4 and the limit from the left is 0. Consider the limit from the right. If $1 < x < 1+\delta$, then for any $\epsilon > 0$, $|H(x)-(-4)| = |-4+4| = 0 < \epsilon$. Therefore, $\lim\limits_{x\to 1^+} H(x) = -4$. For the limit from the left, $1-\delta < x < 1$, and for any $\epsilon > 0$, $|H(x)-0| = |0-0| = 0 < \epsilon$. Therefore, $\lim\limits_{x\to 1^-} H(x) = 0$. As the limit from the right does not equal the limit from the left, $\lim\limits_{x\to 1} H(x)$ does not exist.

9. The limit from the right is –7 and the limit from the left is 17. Consider the limit from the right. If $4 < x < 4 + \delta$, then for any $\epsilon > 0$, $|f(x) - (-7)| = |-x - 3 + 7| = |x - 4| < \epsilon$ if $\delta = \epsilon$. Therefore, $\lim_{x \to 4^+} f(x) = -7$. For the limit from the left, $4 - \delta < x < 4$, and for any $\epsilon > 0$,

$|f(x) - 17| = |x^2 + 1 - 17| = |(x-4)(x+4)| = 8 \cdot |x - 4| < \epsilon$ if $\delta = \dfrac{\epsilon}{8}$. Therefore, $\lim_{x \to 4^-} f(x) = 17$.

As the limit from the right does not equal the limit from the left, $\lim_{x \to 4} f(x)$ does not exist.

11. The limit from the right is 64 and the limit from the left is 64. Consider the limit from the right. If $4 < x < 4 + \delta$, then for any $\epsilon > 0$,

$$|f(x) - 64| = |4x^2 - 64| = 4|(x-4)(x+4)| = 32|x - 4| < \epsilon$$

if $\delta = \dfrac{\epsilon}{32}$. Therefore, $\lim_{x \to 4^+} f(x) = 64$. For the limit from the left, $4 - \delta < x < 4$, and for any

$\epsilon > 0$, $|f(x) - 64| = |x^3 - 64| = |(x-4)(x^2 + 4x + 16)| = 48|x - 4| < \epsilon$ if $\delta = \dfrac{\epsilon}{48}$.

$$|f(x) - 17| = |x^2 + 1 - 17| = |(x-4)(x+4)| = 8 \cdot |x - 4| < \epsilon$$

if $\delta = \dfrac{\epsilon}{8}$. Therefore, $\lim_{x \to 4^-} f(x) = 64$. As the limit from the right equals the limit from the left,

and $f(4) = \lim_{x \to 4^-} f(x) = \lim_{x \to 4^+} f(x) = 64$, $\lim_{x \to 4} f(x)$ exists and is equal to 64.

13. $\lim_{x \to 4} (2x + 6) = \lim_{x \to 4} 2x + \lim_{x \to 4} 6 = 8 + 6 = 14$

15. $\lim_{x \to 1} (x^2 - 6) = \lim_{x \to 1} x^2 - \lim_{x \to 1} 6 = 1 - 6 = -5$

17. $\lim_{x \to 0} \dfrac{x^2 + 2}{x + 1} = \dfrac{\lim_{x \to 0} x^2 + \lim_{x \to 0} 2}{\lim_{x \to 0} x + \lim_{x \to 0} 1} = \dfrac{0 + 2}{0 + 1} = 2$

19. $\lim_{x \to 4} (x - 5) \dfrac{x}{x+1} = \left(\lim_{x \to 4} x - \lim_{x \to 4} 5 \right) \cdot \dfrac{\lim_{x \to 4} x}{\lim_{x \to 4} x + \lim_{x \to 4} 1} = (4 - 5) \cdot \dfrac{4}{4 + 1} = -\dfrac{4}{5}$

21. Let $p(x) = (x - 2)^2$, $q(x) = x + 1$, and $q(1) \neq 0$. Utilizing Theorem 3 we find that the limit

$\lim_{x \to 1} \dfrac{p(x)}{q(x)}$ exists and $\lim_{x \to 1} \dfrac{p(x)}{q(x)} = \dfrac{p(1)}{q(1)} = \dfrac{1}{2}$.

23. Let $f(x) = x^2 - 9$, $g(x) = x - 3$, $\dfrac{f(x)}{g(x)} = \dfrac{x^2 - 9}{x - 3} = (x - 3)\dfrac{x + 3}{x - 3} = x + 3$. Utilizing Theorem 3,

we find that the limit $\lim_{x \to 3} \dfrac{f(x)}{g(x)}$ exists and $\lim_{x \to 3} \dfrac{f(x)}{g(x)} = \lim_{x \to 3} (x + 3) = 6$.

25. Let $f(x) = x^2 - 1$, $g(x) = x + 1$, and $\dfrac{f(x)}{g(x)} = \dfrac{x^2 - 1}{x + 1} = x - 1$. Utilizing Theorem 3, we find that

the limit $\displaystyle\lim_{x \to 3} \dfrac{f(x)}{g(x)}$ exists and $\displaystyle\lim_{x \to -1} \dfrac{f(x)}{g(x)} = \lim_{x \to -1}(x - 1) = -2$.

27. Let $f(x) = x - \sqrt[3]{x}$, $g(x) = \sqrt{x} - 3$, and $g(\pi) \neq 0$. Utilizing Theorem 3 we find that the limit

$\displaystyle\lim_{x \to \pi} \dfrac{f(x)}{g(x)}$ exists and $\displaystyle\lim_{x \to \pi} \dfrac{x - \sqrt[3]{x}}{\sqrt{x} - 3} \overset{\text{Theorem 2c}}{=} \dfrac{\displaystyle\lim_{x \to \pi}\left(x - \sqrt[3]{x}\right)}{\displaystyle\lim_{x \to \pi}\left(\sqrt{x} - 3\right)} \overset{\text{Theorem 2a}}{=} \dfrac{\displaystyle\lim_{x \to \pi} x - \lim_{x \to \pi} \sqrt[3]{x}}{\displaystyle\lim_{x \to \pi} \sqrt{x} - 3}$. From

Theorem 4, we can demonstrate that $\dfrac{\displaystyle\lim_{x \to \pi} x - \lim_{x \to \pi} \sqrt[3]{x}}{\displaystyle\lim_{x \to \pi} \sqrt{x} - 3} = \dfrac{\pi - \sqrt[3]{\pi}}{\sqrt{\pi} - 3}$.

29. Let $-\left|x^3 \cos\left(\dfrac{1}{x}\right)\right| \leq x^3 \cos\left(\dfrac{1}{x}\right) \leq \left|x^3 \cos\left(\dfrac{1}{x}\right)\right|$. Given that $\left|\cos\left(\dfrac{1}{x}\right)\right| \leq 1$ and $\displaystyle\lim_{x \to 0}\left|x^3\right| = 0$ and

from the Pinching Theorem, we find that $\displaystyle\lim_{x \to 0} x^3 \cos\left(\dfrac{1}{x}\right) = 0$.

31. Let $-\left|(x-5)^2 \sin(\csc(\pi x))\right| \leq (x-5)^2 \sin(\csc(\pi x)) \leq \left|(x-5)^2 \sin(\csc(\pi x))\right|$. Given that

$\left|\sin(\csc(\pi x))\right| \leq 1$ and $\displaystyle\lim_{x \to 5}\left|(x-5)^2\right| = 0$, $\displaystyle\lim_{x \to 5}\left((x-5)^2 \sin(\csc(\pi x))\right) = 0$ and from the

Pinching Theorem, we find that $\displaystyle\lim_{x \to 5}\left(1 + (x-5)^2 \sin(\csc(\pi x))\right) = 1$.

33. Let $-\left|(x-2)^2 \sin\left(\sec\left(\dfrac{\pi}{x}\right)\right)\right| \leq (x-2)^2 \sin\left(\sec\left(\dfrac{\pi}{x}\right)\right) \leq \left|(x-2)^2 \sin\left(\sec\left(\dfrac{\pi}{x}\right)\right)\right|$. Given that

$\left|\sin\left(\sec\left(\dfrac{\pi}{x}\right)\right)\right| \leq 1$, $\displaystyle\lim_{x \to 2}\left|(x-2)^2\right| = 0$, and $\displaystyle\lim_{x \to 2}(x+1) = 3$, $\displaystyle\lim_{x \to 2}\left((x-2)^2 \sin\left(\sec\left(\dfrac{\pi}{x}\right)\right)\right) = 0$

and from the Pinching Theorem, we find that $\displaystyle\lim_{x \to 2}\left((x+1) + (x-2)^2 \sin\left(\sec\left(\dfrac{\pi}{x}\right)\right)\right) = 3$.

35. Utilizing the Pinching Theorem, $\displaystyle\lim_{x \to 2} f(x) = 2$.

37. There is an insufficient amount of necessary information for the Pinching Theorem to be used.

39. The domain of definition for the function is $1 < x < 2$. Therefore the one-sided limits are given as follows: $\displaystyle\lim_{x \to 1^+} f(x) = 0$ and $\displaystyle\lim_{x \to 2^-} f(x) = 0$.

41. The $\displaystyle\lim_{x \to -2^+} f(x) = 2$ and $\displaystyle\lim_{x \to 2^-} f(x) = 0$.

Further Theory and Practice

43. The limit is 1. To verify this we use the rigorous definition of the limit. Finding δ such that $|x-4|<\delta$ implies $\left|\dfrac{|x|}{x-1}\right|<\epsilon$. For $x>0$, $\dfrac{|x|}{x}=1$ and $\left|\dfrac{|x|}{x-1}\right|=0<\epsilon$ for any $\epsilon>0$. Hence the limit is 1.

45. The limit is -14. To verify this we use the rigorous definition of the limit. Finding δ such that $|x-7|<\delta$ implies $0<\big||x|-3x+14\big|<\epsilon$. For $x>0$, $\big||x|-3x+14\big|=|x-3x+14|=2|x-7|$, and $|x-7|<\delta$ leads to $\big||x|-3x+14\big|<\epsilon$ if $\delta=\epsilon/2$. Hence, the limit is -14.

47. The limit is 36. Use the rigorous definition of the limit. Finding δ such that $|x+3|<\delta$ implies that $\left|\big(x-|x|\big)^2-36\right|<\epsilon$. For $x<0$

$$\left|\big(x-|x|\big)^2-36\right|=\left|(x+x)^2-36\right|=\left|4x^2-36\right|=4\left|x^2-9\right|=4|x+3|\cdot|x-3|.$$

Then, using $|x+3|<\delta$, $|f(x)|=24|x+3|$ and $|f(x)|<\epsilon$ if $\delta=\epsilon/24$. Therefore, the limit is 36.

49. The limit is 0. Consider the limit from the right. If $0<x<0+\delta$, then

$$|H(x)-0|=\left|x^2\right|=|x|^2<\epsilon$$

if $\delta=\sqrt{\epsilon}$. This in turn gives us $\lim\limits_{x\to0^+}H(x)=0$. For the limit from the left: $0-\delta<x<0$, $|H(x)-0|=|x|<\epsilon$ if $\delta=\epsilon$. Therefore, $\lim\limits_{x\to0^-}H(x)=0$. Whereas the limits from the right and the left of $x=0$ are equal, the limit exists and is 0.

51. The limit from the right is 4 and the limit from the left is 6. Consider the limit from the right. If $5<x<5+\delta$, then $|g(x)-4|=|x-1-4|=|x-5|<\epsilon$ if $\delta=\epsilon$. Therefore, $\lim\limits_{x\to5^+}g(x)=4$. For the limit from the left: $5-\delta<x<5$, $|g(x)-6|=|x+1-6|=|x-5|<\epsilon$ if $\delta=\epsilon$. This in turn gives us $\lim\limits_{x\to5^-}g(x)=6$. Whereas the limits from the right and the left of $x=5$ are different, $\lim\limits_{x\to5}g(x)$ does not exist.

53. Let $u=3x$. Note that $u\to0$ as $x\to0$. We now find that

$$\lim_{x\to0}\left(\frac{\sin(3x)}{x}\right)=3\lim_{x\to0}\left(\frac{\sin(3x)}{3x}\right)=3\lim_{u\to0}\left(\frac{\sin(u)}{u}\right)=3.$$

55. Let $u = 2x$. Note that $u \to 0$ as $x \to 0$. We now find that

$$\lim_{x \to 0}\left(\frac{\sin(2x)}{\sin(x)}\right) = 2\lim_{x \to 0}\left(\frac{\sin(2x)/(2x)}{\sin(x)/(x)}\right) = 2\frac{\lim\limits_{u \to 0}\left(\sin(u)/u\right)}{\lim\limits_{x \to 0}\left(\sin(x)/x\right)} = 2 \cdot \frac{1}{1} = 2 .$$

This answer can also be obtained by using the identity $\sin(2x) = 2\sin(x)\cos(x)$.

57. $\displaystyle\lim_{x \to 0}\frac{1 - \cos(x)}{\tan(x)} = \lim_{x \to 0}\left(\frac{\frac{1 - \cos(x)}{x}}{\frac{\tan(x)}{x}}\right) = \frac{\lim\limits_{x \to 0}\frac{1 - \cos(x)}{x}}{\left(\lim\limits_{x \to 0}\frac{1}{\cos(x)}\right)\left(\lim\limits_{x \to 0}\frac{\sin(x)}{x}\right)} = \frac{0}{1 \cdot 1} = 0$

59. $\displaystyle\lim_{x \to 0}\frac{\sin(x°)}{x} = \lim_{x \to 0}\frac{\sin(\pi x/180)}{x} = \lim_{x \to 0}\frac{\sin(\pi x/180) \cdot (\pi/180)}{x \cdot (\pi/180)} = (\pi/180) \cdot \lim_{x \to 0}\frac{\sin(\pi x/180)}{\pi x/180} = \frac{\pi}{180}$

61. Let $u = 2x$ and $v = 3x$. Note that u and v tend to 0 as x tends to 0. We now find that

$$\lim_{x \to 0}\frac{\sin(2x)\tan(3x)}{x^2} = 2\lim_{x \to 0}\frac{\sin(2x)}{2x} \cdot 3\lim_{x \to 0}\frac{\sin(3x)}{3x}\lim_{x \to 0}\frac{1}{\cos(3x)}$$

$$= 2\lim_{u \to 0}\frac{\sin(u)}{u} \cdot 3\lim_{v \to 0}\frac{\sin(v)}{v}\lim_{v \to 0}\frac{1}{\cos(v)} = 2 \cdot 1 \cdot 3 \cdot 1 \cdot 1 = 6 .$$

63. $\displaystyle\lim_{x \to 1}\left(\frac{x^2 - 1}{\sqrt{x} - 1}\right) = \lim_{x \to 1}\left(\frac{(x + 1)(\sqrt{x} + 1)(\sqrt{x} - 1)}{\sqrt{x} - 1}\right) = \lim_{x \to 1}(x + 1) \cdot \lim_{x \to 1}(\sqrt{x} + 1) = 2 \cdot 2 = 4$

65. $\displaystyle\lim_{h \to 0}\frac{h}{\sqrt{1 + 2h} - 1} = \lim_{h \to 0}\frac{h(\sqrt{1 + 2h} + 1)}{(\sqrt{1 + 2h} - 1)(\sqrt{1 + 2h} + 1)} = \lim_{h \to 0}\frac{h(\sqrt{1 + 2h} + 1)}{(1 + 2h - 1)} = \frac{1}{2} \cdot \lim_{h \to 0}(\sqrt{1 + 2h} + 1) = 1$

67. $\displaystyle\lim_{h \to 0}\left(\frac{2}{h\sqrt{4 + h}} - \frac{1}{h}\right) = \lim_{h \to 0}\frac{2 - \sqrt{4 + h}}{h\sqrt{4 + h}} = \lim_{h \to 0}\frac{(2 - \sqrt{4 + h})(2 + \sqrt{4 + h})}{h(\sqrt{4 + h})(2 + \sqrt{4 + h})}$

$$= \lim_{h \to 0}\frac{4 - (4 + h)}{h(\sqrt{4 + h})(2 + \sqrt{4 + h})} = -\lim_{h \to 0}\frac{1}{(\sqrt{4 + h})(2 + \sqrt{4 + h})} = -\frac{1}{2 \cdot 4} = -\frac{1}{8}$$

69. $\displaystyle\lim_{x \to 4}\left(\frac{\sqrt{x} - 2}{\sqrt{x + 5} - 3}\right) = \lim_{x \to 4}\left(\frac{(\sqrt{x} - 2)(\sqrt{x} + 2)(\sqrt{x + 5} + 3)}{(\sqrt{x + 5} - 3)(\sqrt{x + 5} + 3)(\sqrt{x} + 2)}\right)$

$$= \lim_{x \to 4}\left(\frac{(x - 4)(\sqrt{x + 5} + 3)}{(x + 5 - 9)(\sqrt{x} + 2)}\right) = \lim_{x \to 4}\left(\frac{\sqrt{x + 5} + 3}{\sqrt{x} + 2}\right) = \frac{\lim\limits_{x \to 4}(\sqrt{x + 5} + 3)}{\lim\limits_{x \to 4}(\sqrt{x} + 2)} = \frac{6}{4} = \frac{3}{2}$$

71.   The notation $x \to 3^-$ means that x approaches 3 through values less than 3. For such values of x, we have $x - 3 < 0$ and $|x - 3| = -(x - 3)$. Therefore,

$$\lim_{x \to 3^-} \frac{(x-3)}{|x-3|} = \lim_{x \to 3^-} \frac{(x-3)}{-(x-3)} = \lim_{x \to 3^-} (-1) = -1.$$

73.   Given that both $f(x) = \sqrt{16 - 3x}$ and $g(x) = \sqrt{16 - 4x}$ as $x \to 4^-$ exist (because

$16 - 4x > 0$), we find $\displaystyle \lim_{x \to 4^-} \frac{x + \sqrt{16 - 3x}}{x + \sqrt{16 - x^2}} = \frac{\displaystyle \lim_{x \to 4^-} \left(x + \sqrt{16 - 3x} \right)}{\displaystyle \lim_{x \to 4^-} \left(x + \sqrt{16 - x^2} \right)} = \frac{4 + 2}{4} = \frac{3}{2}.$

75.   The notation $x \to 0^+$ means that x approaches 0 through values greater than 0. For such values of x, the expression $u = \sqrt{x}$ is defined. Inasmuch as u tends to 0 as x tends to 0, we have

$$\lim_{x \to 0^+} \frac{\sin\left(\sqrt{x}\right)}{\sqrt{x}} = \lim_{u \to 0^+} \frac{\sin(u)}{u} = 1.$$

77.   Let $\epsilon > 0$. Then $\epsilon_1 = \dfrac{\epsilon}{10^6}$ is also positive. Given that $\lim_{x \to 0} g(x) = 0$, it follows that there is a

$\delta > 0$ such that $|g(x) - 0| < \epsilon_1$ whenever $0 < |x - 0| < \delta$. Consequently, for these values of x,

$|f(x) \cdot g(x) - 0| = |f(x)| \cdot |g(x)| < 10^6 \cdot \epsilon_1 = \epsilon$. According to the rigorous definition of limits,

$\lim_{x \to 0} f(x) \cdot g(x)$ exists and equals 0.

Notice that the choice of δ depends on the number 10^6. The value 10^6 therefore influences how close to 0 we must take x in order for $f(x)g(x)$ to be within ϵ of the limiting value 0.

However, the same proof can be used when 10^6 is replaced by any fixed positive number M. In other words, if M is a fixed positive number, if $|f(x)| \leq M$ for all x, and if $\lim_{x \to 0} g(x) = 0$,

then $\lim_{x \to 0} (f(x) \cdot g(x)) = 0$.

Although the value of M does not affect the limit, some hypothesis is necessary. If $g(x) = x$

and $f(x) = 0$ for $x \leq 0$ and $f(x) = \dfrac{1}{|x|}$ for $x \neq 0$, then $f(x)g(x) = \text{signum}(x)$, which does

not have a limit as x tends to 0.

79.   For $x \neq 0$ we multiply by $\dfrac{1}{|x|}$. Given that $\dfrac{1}{|x|}$ is positive, multiplication by $\dfrac{1}{|x|}$ preserves

inequalities. Therefore, for $x \neq 0$, we have $0 \leq \dfrac{f(x)}{|x|} \leq \dfrac{x^2}{|x|} = \dfrac{|x|^2}{|x|} = |x|$. Notice that, because

$f(x)$ is nonnegative, $\dfrac{f(x)}{|x|}$ is equal to $\left|\dfrac{f(x)}{x}\right|$. Therefore, $\left|\dfrac{f(x)}{x}\right| \leq |x|$, or $-|x| \leq \dfrac{f(x)}{x} \leq |x|$.

Setting $g(x) = -|x|$ and $h(x) = |x|$, we have $g(x) \leq \dfrac{f(x)}{x} \leq h(x)$. Knowing that

$$\lim_{x \to 0} g(x) = \lim_{x \to 0} h(x) = 0,$$

the Pinching Theorem tells us that $\displaystyle\lim_{x \to 0} \dfrac{f(x)}{x} = 0$.

81.   For any $f(x)$, $-|f(x)| \leq f(x) \leq |f(x)|$. As $\displaystyle\lim_{x \to 0} |f(x)| = \lim_{x \to 0}(-|f(x)|) = 0$, the Pinching

Theorem yields $\displaystyle\lim_{x \to c} f(x) = 0$.

83.   Let $g(u) = f(1/u)$ for $u > 0$. Given that $g(u)$ is defined for all $u > 0$, it is meaningful to

consider the one-sided limit $\displaystyle\lim_{u \to 0^+} g(u)$. If we let $x = 1/u$, then x grows ever larger as u

approaches 0 from the right. Knowing that $g(u) = f(x)$, it makes sense to define

$\displaystyle\lim_{x \to \infty} f(x) = \lim_{u \to 0^+} g(u)$, if the limit on the right exists.

Computer/Calculator Exercises

85.   0.00062

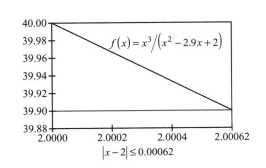

87.   0.046

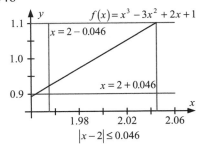

89.

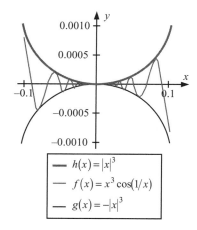

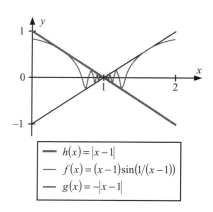

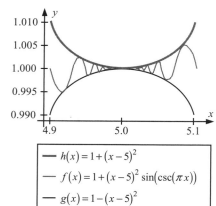

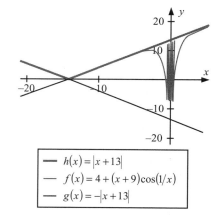

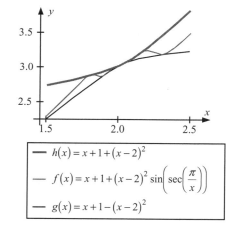

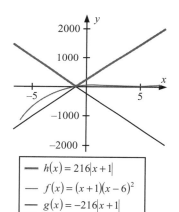

91.

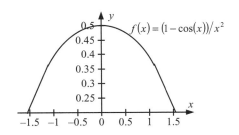
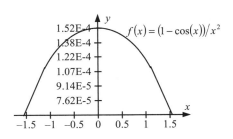

The limits of f and g at 0 are $\displaystyle\lim_{x\to 0} f(x) = \frac{1}{2}$ and $\displaystyle\lim_{x\to 0} g(x) = \frac{\pi^2}{16200} \approx 1.523 \cdot 10^{-4}$.

We note that $x = \left(\dfrac{\pi}{180}\right) x^{\circ}$, or $x^{\circ} = \left(\dfrac{180}{\pi}\right) x$. Therefore,

$$f(x) = \frac{1 - \cos(x)}{x^2} = \left(\frac{180}{\pi}\right)^2 \frac{1 - \cos(180x/\pi)^{\circ}}{(180x/\pi)^2} = \left(\frac{180}{\pi}\right)^2 g\left(\frac{180x}{\pi}\right).$$

This yields $\alpha = \left(\dfrac{180}{\pi}\right)^2$.

$$\lim_{x\to 0} g(x) = \frac{1}{\alpha} \cdot \lim_{x\to 0} f(x) = \left(\frac{180}{\pi}\right)^{-2} \cdot \frac{1}{2} = \frac{\pi^2}{16200}$$

93.   **a.** $[1.99,\ 2.01] \times [2.9999,\ 3.0001]$

b.

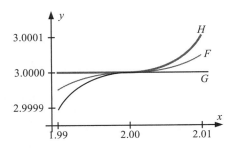

c.   The Pinching functions are
$G(x) \le F(x) \le H(x)$, $G(x)$
and $H(x)$.

d. $\displaystyle\lim_{x\to c} G(x) = \lim_{x\to c} H(x)$
$\qquad = \lim_{x\to c} F(x)$

$c = 2$

e. $\displaystyle\lim_{x\to 2} F(x) = 3$

95.   **a.** $[2.95,\ 3.05] \times [6.5,\ 8.5]$

b.

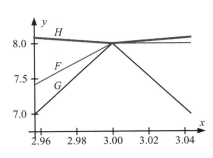

c.   The Pinching functions are
$G(x)$ and $H(x)$,
$G(x) \le F(x) \le H(x)$.

d. $\displaystyle\lim_{x\to c} G(x) = \lim_{x\to c} H(x)$
$\qquad = \lim_{x\to c} F(x)$

$c = 3$

e. $\displaystyle\lim_{x\to c} F(x) = 8$

Section 2.3   Continuity

Problems for Practice

The functions of Exercises 1–7 are rational functions. A rational function is continuous at each point of its domain. The domain of a rational function consists of all real values except those that are roots of the denominator.

1.   The function f is continuous at all values of its domain, which is $\mathbb{R}$.

3.   The function g is continuous at all values of its domain, which is $\{x \in \mathbb{R} : x \neq -1\}$.

5.   The function f is continuous at all values of its domain, which is $\mathbb{R}$.

7.   The function H is continuous at all values of its domain, which is $\{x \in \mathbb{R} : x \neq 3, \; x \neq -4\}$.

Note: The denominator and numerator of $H(x)$ each have the factor $x - 3$:

$$H(x) = \frac{(x-2)(x-3)}{(x+4)(x-3)}.$$ For $x \neq 3$, this common factor does cancel. However, we may not

evaluate the denominator at $x = 3$ so this value is excluded from the domain of H.

In Exercises 8–14, each piece of the given function is defined by a rational function. Therefore each piece is continuous on its domain. Continuity at the point of transition between the two pieces must be investigated.

9.   Here $\lim_{x \to 3^-} g(x) = \lim_{x \to 3^-} (x+1) = 4$, $\lim_{x \to 3^+} g(x) = \lim_{x \to 3^+} 6 = 6$, and $g(3) = (3+1) = 4$. Given that

these three value are not equal, g is *not* continuous at 3. Thus, g is continuous on $\{x \in \mathbb{R} : x \neq 3\}$.

11.   Here $\lim_{x \to -4^-} H(x) = \lim_{x \to -4^-} (x+1)^4 = 81$, $\lim_{x \to -4^+} H(x) = \lim_{x \to -4^+} (-20x+1) = 81$, and

$H(-4) = -20(-4) + 1 = 81$. Given that these three values are not all equal, H is continuous at -4. Thus, H is continuous on $\mathbb{R}$.

13.   Here $\lim_{x \to 7^-} g(x) = \lim_{x \to 7^-} (2x-19) = -5$, $\lim_{x \to 7^+} g(x) = \lim_{x \to 7^+} (2-x) = -5$, and $g(7) = -5$. Given

that these values are not all equal, g *is* continuous at 7. Thus, g is continuous on $\mathbb{R}$.

15.   As x approaches 2 from the left we have $\lim_{x \to 2^-} f(x) = \lim_{x \to 2^-} (x^2+7) = 11$ and from the right,

$\lim_{x \to 2^+} f(x) = \lim_{x \to 2^+} (x^3+3) = 11$. We now set $F(2) = 11$ to obtain a continuous extension of f.

17.   As x approaches 2 from the left we have $\lim_{x \to 2^-} f(x) = \lim_{x \to 2^-} \left(\frac{x^6}{18-x} \right) = 4$ and from the right,

$\lim_{x \to 2^+} f(x) = \lim_{x \to 2^+} \left(\frac{x^4}{x+2} \right) = 4$. We now set $F(2) = 4$ to obtain a continuous extension of f.

In Exercises 19–22, each piece of the given function is defined by a polynomial. Polynomials are left and right continuous at all points. Left and right continuity at the point of transition between the two pieces must be investigated.

19.   We have $\lim\limits_{x\to 5^-} f(x) = \lim\limits_{x\to 5^-} 2 = 2 = f(5)$ and $\lim\limits_{x\to 5^+} f(x) = \lim\limits_{x\to 5^+} 3 = 3 \neq f(5)$. Therefore f is left continuous but not right continuous at 5. In conclusion, f is left continuous on $\mathbb{R}$, right continuous on $\{x \in \mathbb{R} : x \neq 5\}$, and continuous on $\{x \in \mathbb{R} : x \neq 5\}$.

21.   We have $\lim\limits_{x\to 3^-} f(x) = \lim\limits_{x\to 5^-} (x+1) = 4 \neq f(3)$ and $\lim\limits_{x\to 3^+} f(x) = \lim\limits_{x\to 3^+} 2 = 2 = f(3)$. Therefore, f is right continuous but not left continuous at $x = 3$. In conclusion, f is left continuous on $\{x \in \mathbb{R} : x \neq 3\}$, right continuous on $\mathbb{R}$, and continuous on $\{x \in \mathbb{R} : x \neq 3\}$

23.

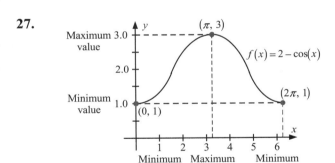

25.

27.

Further Theory and Practice

29.   For example (many solutions are possible here)

a.   The limits as x approaches 1 from the left and 3 from the right are

$\lim\limits_{x\to 1^-} f(x) = 1^2 + 16 = 17$, and $\lim\limits_{x\to 3^+} f(x) = 3 - 4 = -1$ respectively. The equation for a

line passing through the points $(1, 17)$ and $(3, -1)$ is: $\dfrac{x-1}{3-1} = \dfrac{y-17}{-1-17}$, i.e.,

$\dfrac{x-1}{2} = \dfrac{y-17}{-18}$ or $y = 26 - 9x$. Therefore, $f(x) = 26 - 9x$ for $x \in [1, 3]$.

b.   The limits as x approaches 1 from the left and 3 from the right are

$$\lim_{x\to1^-} f(x) = \frac{1+1}{1+2} = \frac{2}{3}, \text{ and } \lim_{x\to3^+} f(x) = 3^2 = 9 \text{ respectively. The equation for a line}$$

passing through the points $\left(1, \frac{2}{3}\right)$ and $(3, 9)$ is $\dfrac{x-1}{3-1} = \dfrac{y-2/3}{9-2/3}$, i.e., $\dfrac{x-1}{2} = \dfrac{3y-2}{25}$

or $y = \dfrac{25x-21}{6}$. Therefore, $f(x) = \dfrac{25x-21}{6}$ for $x \in [1, 3]$.

c.   The limits as x approaches 1 from the left and 3 from the right are $\lim_{x\to1^-} f(x) = 1$,

$\lim_{x\to3^+} f(x) = -6$ respectively. The equation for a line passing through the points $(1, 1)$

and $(3, -6)$ is $\dfrac{x-1}{3-1} = \dfrac{y-1}{-6-1}$ or $y = \dfrac{9-7x}{2}$ and $f(x) = \dfrac{9-7x}{2}$ for $x \in [1, 3]$.

31.   Given that $\lim_{x\to1} f(x)$, $\lim_{x\to1} g(x)$, $\lim_{x\to1} h(x)$, and $\lim_{x\to1} k(x)$ all exist and equal 2, g and h are

continuous at 1, and k is discontinuous at 1. The functions g and h are identical.

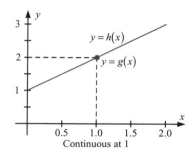

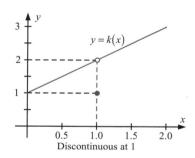

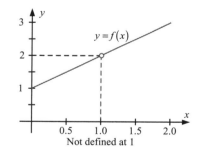

33. $T(x) = \begin{cases} 0.10x & \text{if } 0 < x \le 23350 \\ 2335 + 0.15(x - 23350) & \text{if } 23350 < x \le 56550 \\ 7315 + 0.25(x - 56550) & \text{if } 56550 < x \le 117950 \\ 22665 + 0.30(x - 117950) & \text{if } 117950 < x \le 256500 \\ 64230 + 0.35(x - 256500) & \text{if } 256500 < x \end{cases}$

For $x \notin \{23350, 56550, 117950, 256500\}$ the function $T(x)$ is linear in the vicinity of x, and therefore is continuous at x. For the four special points:

$$\lim_{x \to 23350^-} T(x) = \lim_{x \to 23350^+} T(x) = T(23350) = 2335$$

$$\lim_{x \to 56550^-} T(x) = \lim_{x \to 56550^+} T(x) = T(56550) = 7315$$

$$\lim_{x \to 117950^-} T(x) = \lim_{x \to 117950^+} T(x) = T(117950) = 22665$$

$$\lim_{x \to 256500^-} T(x) = \lim_{x \to 256500^+} T(x) = T(256500) = 64230$$

Therefore, $T(x)$ is continuous on $\{x \in \mathbb{R} : x > 0\}$.

35. **a.** See Exercise 41 of Section 1.4 for the plot of $\lfloor x \rfloor$.

b. For all $x \in [2, 3)$: $\lfloor x \rfloor = 2$ and $\lim_{x \to 3^-} \lfloor x \rfloor = 2$.

c. For all $x \in (2, 2.5]$: $-1 \leq 4 - 2x < 0$ and $\lim_{x \to 2^+} \lfloor 4 - 2x \rfloor = -1$.

d. For x in the interval $\left(-\dfrac{1}{n}, -\dfrac{1}{n+1} \right]$, we can write $x = -\dfrac{1}{n} + \delta$ where

$$0 \leq \delta \leq \left(-\frac{1}{n+1} \right) - \left(-\frac{1}{n} \right) = \frac{1}{(n+1)n}.$$

For these values of x, we calculate

$$x \cdot \left\lfloor -\frac{1}{x} \right\rfloor = \left(-\frac{1}{n} + \delta \right)(-n - 1) = 1 + \frac{1}{n} - \delta - \delta n = 1 - x - \delta n.$$

Given that $0 < \delta n \leq \dfrac{1}{n+1}$, we see $1 - x - \dfrac{1}{n+1} \leq x \left\lfloor -\dfrac{1}{x} \right\rfloor < 1 - x$ for

$$x \in \left(-\frac{1}{n}, -\frac{1}{n+1} \right].$$

On this interval, we have $x \leq -\dfrac{1}{n+1}$ which in turn gives us

$$1 - x - \frac{1}{n+1} \geq 1 + \frac{1}{n+1} - \frac{1}{n+1} = 1.$$

Thus, $1 \leq x \left\lfloor -\dfrac{1}{x} \right\rfloor < 1 - x$ for $x \in \left(-\dfrac{1}{n}, -\dfrac{1}{n+1} \right]$. By the Pinching Theorem, it follows

that $\lim_{x \to 0^-} x \left\lfloor -\dfrac{1}{x} \right\rfloor = 1$.

e. For all $x \in (-1, 0)$: $\lfloor -x \rfloor = 0$, $\lim_{x \to 0^-} \lfloor -x \rfloor = 0$.

37.   At the moment of departure and the stopping point, the train velocity is 0; average velocity is 80 km/h. $v_{min} = 0$ and at some time $t_0 \in (0, 1)$, $v_{max} = v(t_0) > 80$. If $v(t)$ is continuous, by the Intermediate Value Theorem there exists $t_1 \in (0, t_0)$ and $t_2 \in (t_0, 1)$ such that $v(t_1) = v(t_2) = 60$.

39.   Let $f(t) = 10t^3 - 7t^2 + 20t - 14$. As $f(0) = -14$, $f(1) = 9$ by the Intermediate Value Theorem there exists $t_0 \in (0, 1)$ such that $f(t_0) = 0$. For $f(\sin(x)) = 0$ this means that $\sin(x) = t_0$. This equation has infinite number of solutions. For example, $\sin(0) = 0$ and $\sin(\pi/2) = 1$. The Intermediate Value Theorem tells us that, given $t_0 \in (0, 1)$, the equation $\sin(x) = t_0$ has a solution x in $(0, \pi/2)$. The same argument can be used to obtain a solution in every interval of the form $(2\pi n, 2\pi n + \pi/2)$ where n is an integer.

41.   The function p has a root in the interval $(-1, 0)$ because $p(1) = -1$ and $p(0) = 1$.

43.   The function p has a root in the intervals $(1, 2)$ and $(2, 3)$ because $p(1) = 14$, $p(2) = -10$, and $p(3) = 90$.

45.   The statement is false as demonstrated by the following graph.

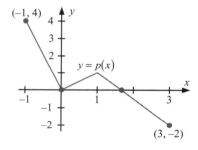

47.   The statement is false. (The figure for Problem 2.3.45 illustrates this exercise as well.)

49.   The statement is false as demonstrated by the following graphs.

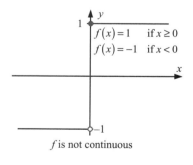

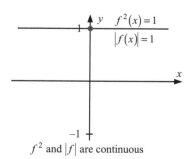

51.   The statement is false. (The figure for the Problem 2.3.49 illustrates this exercise as well.)

53.   The statement is true; otherwise, $f(x)$ would satisfy the Extreme Value Theorem and would therefore have a maximum.

55.   We compute $\ell = \lim\limits_{x \to 0} f(x) = 0$, $L = \lim\limits_{y \to 0} g(y) = 0$, $\lim\limits_{x \to 0} g(f(x)) = \lim\limits_{x \to 0} g(0) = g(0) = 1$. This looks like the result of Theorem 2, but Theorem 2 does not actually apply since the function $g(x)$ is not continuous.

57.   The limits $\ell = \lim\limits_{x \to 0} f(x) = 0$, and $L = \lim\limits_{y \to 0} g(y) = 0$. If x_n are rational, then

$$\lim\limits_{x_n \to 0} g(f(x_n)) = \lim\limits_{x_n \to 0} g(0) = g(0) = 1.$$

If x_n are irrational, then $\lim\limits_{x_n \to 0} g(f(x_n)) = \lim\limits_{x_n \to 0} g(x_n) = 0$. Therefore, $\lim\limits_{x \to 0} g(f(x))$ does not exist. This does not violate Theorem 2 since the functions $f(x)$ and $g(x)$ are not continuous at $x = 0$.

59.   Let $f(x) = h_1(x) - h_2(x)$, where $h_1(x)$ $(h_2(x))$ is the hiker's elevation on the first (second) day after the x hours of the climb. If L is the height of the mountain, then $f(0) = -L$, $f(10) = L$. Here we see that h_1 and h_2 are continuous functions. Therefore, by the Intermediate Value Theorem there exists $t \in (0, 10)$ such that $f(t) = 0$, or $h_1(t) = h_2(t)$.

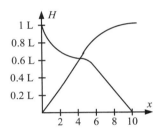

61. $\lim\limits_{x \to 0} f(x) = f(0) = -\dfrac{1}{1000000}$

Function	Reason for Continuity
$F(x) = \sqrt[1001]{x}$	Theorem 4, Section 2.2
$G(x) = (\pi/2) \cdot F(x)$	Theorem 1a, Section 2.3
$H(x) = \cos(x)$	Theorem 7, Section 2.2
$J(x) = H(G(x))$	Theorem 2, Section 2.3
$K(x) = x^3$	Theorem 3, Section 2.2
$L(x) = K(J(x))$	Theorem 2, Section 2.3
$M(x) = L(x)/1000000$	Theorem 1a, Section 2.3
$N(x) = x^2$	Theorem 3, Section 2.2
$f(x) = N(x) - M(x)$	Theorem 1a, Section 2.3

Calculator/Computer Exercises

63.   Let $f(x) = p(\tan(x))$ for $x \in \left(-\dfrac{\pi}{2}, \dfrac{\pi}{2}\right)$. Whereas f is the composition of continuous

functions, it is also continuous. According to Exercise 38, $\displaystyle\lim_{u \to \infty} p(u) = \infty$ and

$\displaystyle\lim_{u \to -\infty} p(u) = -\infty$ if the leading coefficient of p is 1. The same limits hold if the leading

coefficient of p is positive. The signs are reversed if the leading coefficient of p is negative.

Thus, as $x \to \left(\dfrac{\pi}{2}\right)^{-}$, we have $\tan(x) \to \infty$ and $p(\tan(x)) \to \pm\infty$. As $x \to \left(-\dfrac{\pi}{2}\right)^{+}$, we have

$\tan(x) \to -\infty$ and $p(\tan(x)) \to \mp\infty$. Therefore, given any real number c, we may find a and

b in the interval $\left(-\dfrac{\pi}{2}, \dfrac{\pi}{2}\right)$ such that c lies between $f(a)$ and $f(b)$. Applying the

Intermediate Value Theorem to f on the closed interval $[a, b]$, we see that there is a

$\gamma \in (a, b)$ such that $f(\gamma) = c$.

When $p(x) = x^3 + 3x$ and $c = 19$, we find that $\gamma \approx 4.301695685$.

65.   The function does not have a
continuous extension at point c.

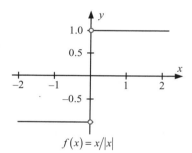

$f(x) = x/|x|$

67.   A continuous extension f is obtained
by setting $f(2) = -8$.

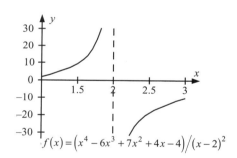

$f(x) = \left(x^4 - 6x^3 + 7x^2 + 4x - 4\right)/(x-2)^2$

69.   Let $x^4 - 10x^3 + 38x^2 - 64x + 40 = \left(x^2 - 6x + 10\right)(x - 2)^2$. Completing the square of the first

factor on the right, we obtain $x^4 - 10x^3 + 38x^2 - 64x + 40 = \left((x-3)^2 + 1\right)(x - 2)^2$, which is

nonnegative. Therefore, we find that

$$p(x) = x^4 - 10x^3 + 38x^2 - 64x + 50 = \left((x-3)^2 + 1\right)(x-2)^2 + 10 \geq 10.$$

Given that $p(2) = 10$, we see that 10 is the minimum value of p. On the other hand, $p(x)$

becomes arbitrarily large as x tends to infinity. (This statement is clear for the monomial x^4.
In the next section we deal with such limits in greater detail. For now, it is a good idea to

convince yourself that $p(x)$ and x^4 have comparable values for large x by plotting both in

the same window over an interval such as $\left[10^4, 2 \times 10^4\right]$. The plots will be barely distinguishable.) It follows that for any $\gamma > 10$, there is a $b > 2$ with $p(b) > \gamma$. Applying the Intermediate Value Theorem to p over the closed interval $[2, b]$, we see that we cannot solve the equation. Yet, $p(x) = \gamma$ can be solved for any $\gamma \geq 10$. Of course, the inequality $p(x) \geq 10$ means that we cannot solve the equation $p(x) = \gamma$ for $\gamma < 10$. By successive zooming, we find that $p(x) = 20$ for $x \approx 0.7267$ and $x \approx 4.1132$.

Section 2.4   Infinite Limits and Asymptotes

Problems for Practice

1.   The limit is $+\infty$ because the function becomes arbitrarily large and positive as $x \to 6$. The limit is ∞.

3.   To find the limit, we must use Table 1 or divide each term by x^5:

$$\lim_{x \to +\infty} \frac{x^4 + 3x - 92}{x^5 - 7x^2 + 44} = \lim_{x \to +\infty} \frac{1/x + 3/x^4 - 92/x^5}{1 - 7/x^3 + 44/x^5} = \frac{0 + 0 - 0}{1 - 0 + 0} = 0.$$

5.   The limit exists and is equal to 1 because $\dfrac{x + \sqrt{x}}{x - \sqrt{x}} = \dfrac{1 + 1/\sqrt{x}}{1 - 1/\sqrt{x}}$ and $\dfrac{1}{\sqrt{x}} \to 0$ as $x \to +\infty$.

7.   The limit exists and is equal to 1 because $\dfrac{x + \cos(x)}{x - \sin(x)} = \dfrac{1 + \cos(x)/x}{1 - \sin(x)/x}$, $-\dfrac{1}{|x|} \leq \dfrac{\cos(x)}{x} \leq \dfrac{1}{|x|}$,

$-\dfrac{1}{|x|} \leq \dfrac{\sin(x)}{x} \leq \dfrac{1}{|x|}$ and $\dfrac{1}{|x|} \to 0$ as $x \to +\infty$.

9.   The limit is $+\infty$ because the function becomes arbitrarily large and positive as $x \to 1^+$.

11.   The limit exists and is equal to 1/3. To calculate it, use Table 1 or divide each term by x^2 as shown: $\lim_{x \to +\infty} \dfrac{x^2 - 4x + 9}{3x^2 - 8x + 18} = \lim_{x \to +\infty} \dfrac{1 - 4/x + 9/x^2}{3 - 8/x + 18/x^2} = \dfrac{1 - 0 + 0}{3 - 0 + 0} = \dfrac{1}{3}$.

13.   The limit exists and is equal to ∞, because the function becomes arbitrarily large and positive as $x \to 0^+$.

15.   The limit exists and is equal to ∞, because the function becomes arbitrarily large and positive as $x \to 0$.

17.   The limit is $-\infty$ because the function becomes arbitrarily large and negative at $x \to 0^-$. The limit is $-\infty$.

19.   The limit exists and is equal to ∞, because the function becomes arbitrarily large and positive as $x \to 2^-$.

21.   **Vertical asymptote:** $x = 7$

   Calculations:
   The limit as x approaches 7 from the left is
   $\lim\limits_{x \to 7^-} f(x) = -\infty$, and from the right is
   $\lim\limits_{x \to 7^+} f(x) = \infty$. Therefore, $x = 7$ is a vertical
   asymptote of f.

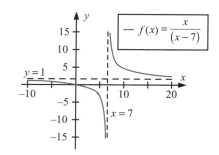

   Horizontal asymptote: $y = 1$

   Calculations:
   From the limits: $\lim\limits_{x \to \pm\infty} f(x) = \lim\limits_{x \to \pm\infty} \dfrac{1}{1 - 7/x} = 1$. We see that $y = 1$ is a horizontal

   asymptote of f.

23.   **Vertical asymptote:** $x = 0$

   Calculations:
   As x approaches 0 from the right and left we find
   the limits to $+$ and $-\infty$ respectively as shown:

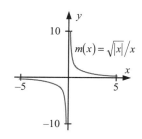

$$\lim_{x \to 0^+} m(x) = \lim_{x \to 0^+} \frac{\sqrt{|x|}}{x} = \lim_{x \to 0^+} \frac{\sqrt{x}}{x} = \lim_{x \to 0^+} \frac{1}{\sqrt{x}}$$
$$= +\infty$$

$$\lim_{x \to 0^-} m(x) = -\lim_{x \to 0^-} \frac{1}{\sqrt{-x}} = -\infty$$

   Therefore, $x = 0$ is a vertical asymptote.

   Horizontal asymptote: $y = 0$

   Calculations:
   The limits as x approaches ∞ from the right and left are equal and 0 as given:

$$\lim_{x \to \infty} m(x) = \lim_{x \to \infty} \frac{1}{\sqrt{x}} = 0$$

$$\lim_{x \to -\infty} m(x) = -\lim_{x \to -\infty} \frac{1}{\sqrt{-x}} = 0$$

   Therefore $y = 0$ is a horizontal asymptote.

25.   **Vertical asymptote:** There is no vertical asymptote.

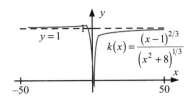

Horizontal asymptote:
   Calculations:
   The limits as x approaches $\pm\infty$ are calculated:

$$\lim_{x\to\pm\infty} k(x) = \lim_{x\to\pm\infty} \frac{(x-1)^{2/3}}{(x^2+8)^{1/3}} = \lim_{x\to\pm\infty} \frac{(1-1/x)^{2/3}}{(1+8/x^2)^{1/3}} = 1.$$

   Therefore, $y=1$ is a horizontal asymptote.

27.   **Vertical asymptote:** There is no vertical asymptote.

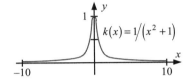

Horizontal asymptote: $y=0$
   Calculations:
   The limits as x approaches $\pm\infty$ are calculated:
$$\lim_{x\to\pm\infty} k(x) = \lim_{x\to\pm\infty} \frac{1}{x^2+1} = 0.$$ Therefore, $y=0$
   is a horizontal asymptote.

29.   **Vertical asymptote:** $x=0$
   Calculations:
   The limits as x approaches $\pm\infty$ are calculated:

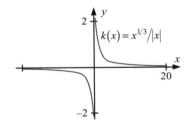

$$\lim_{x\to 0^+} k(x) = \lim_{x\to 0^+} \frac{x^{1/3}}{x} = \lim_{x\to 0^+} \frac{1}{x^{2/3}} = \infty$$

$$\lim_{x\to 0^-} k(x) = -\lim_{x\to 0^-} \frac{x^{1/3}}{x} = -\lim_{x\to 0^-} \frac{1}{x^{2/3}} = -\infty$$

Horizontal asymptote: $y=0$
   Calculations:
   The limits as x approaches 0 from the right and left respectively are shown below:

$$\lim_{x\to\infty} k(x) = \lim_{x\to\infty} \frac{1}{x^{2/3}} = 0$$

$$\lim_{x\to-\infty} k(x) = -\lim_{x\to-\infty} \frac{1}{x^{2/3}} = 0$$

Therefore $y=0$ is a horizontal asymptote.

31. **Vertical asymptote:** $x = 0$, $x = -1$, and $x = -4$
Calculations:

Given the function $h(x) = \dfrac{1}{x(x+1)\sqrt[3]{x+4}}$, we

have

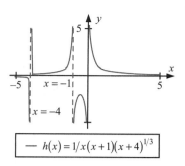

$$\lim_{x \to 0^+} h(x) = \lim_{x \to 0^+} \frac{1}{x(x+1)\sqrt[3]{x+4}} = \infty$$

$$\lim_{x \to 0^-} h(x) = \lim_{x \to 0^-} \frac{1}{x(x+1)\sqrt[3]{x+4}} = -\infty$$

Therefore $x = 0$ is a vertical asymptote. The limits as x approaches -1 from the right and left are shown below:

$$\lim_{x \to -1^+} h(x) = \lim_{x \to -1^+} \frac{1}{x(x+1)\sqrt[3]{x+4}} = -\infty$$

$$\lim_{x \to -1^-} h(x) = \lim_{x \to -1^-} \frac{1}{x(x+1)\sqrt[3]{x+4}} = \infty$$

Therefore, $x = -1$ is a vertical asymptote. In addition, the limits as x approaches -4 from the right and left

$$\lim_{x \to -4^+} h(x) = \lim_{x \to -4^+} \frac{1}{x(x+1)\sqrt[3]{x+4}} = \infty$$

$$\lim_{x \to -4^-} h(x) = \lim_{x \to -4^-} \frac{1}{x(x+1)\sqrt[3]{x+4}} = -\infty$$

give us a vertical asymptote at $x = -4$.

Horizontal asymptote: $y = 0$
Calculations:
Finally, $\lim\limits_{x \to \pm\infty} h(x) = 0$, which in turn yields a horizontal asymptote at $y = 0$.

33. **Vertical asymptote:** $x = 0$
Calculations:
The limits as x approaches 0 from the right and left are given:

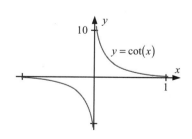

$$\lim_{x \to 0^+} k(x) = \frac{\lim\limits_{x \to 0^+} \cos(x)}{\lim\limits_{x \to 0^+} \sin(x)} = \frac{1}{\lim\limits_{x \to 0^+} \sin(x)} = \infty$$

$$\lim_{x \to 0^-} k(x) = \frac{\lim\limits_{x \to 0^-} \cos(x)}{\lim\limits_{x \to 0^-} \sin(x)} = \frac{1}{\lim\limits_{x \to 0^-} \sin(x)} = -\infty$$

Therefore, $x = 0$ is a vertical asymptote.

Horizontal asymptote: There is no horizontal asymptote.

Further Theory and Practice

35. A function $f(x)$ with $x \in \mathbb{R}$ cannot have two distinct limits at $x \to +\infty$ or $x \to -\infty$.

Proof: Suppose that α_1 and α_2 are distinct limits of function $f(x)$ as $x \to +\infty$:

$$\lim_{x \to +\infty} f(x) = \alpha_1, \ \lim_{x \to +\infty} f(x) = \alpha_2, \ \alpha_1 < \alpha_2. \text{ Letting } \epsilon = \frac{\alpha_2 - \alpha_1}{3}, \text{ we can take } x$$

sufficiently large, so that $\alpha_1 - \epsilon < f(x) < \alpha_1 + \epsilon < \alpha_2 - \epsilon < f(x) < \alpha_2 + \epsilon$, i.e.,
$f(x) < f(x)$, which is impossible. Therefore $\alpha_1 = \alpha_2$. The same arguments can be repeated for $x \to -\infty$.

37. Suppose f, g, and h are functions with domains that each contain $S = (a, +\infty)$. Moreover assume that $g(x) \le f(x) \le h(x)$ for all $x \in S$. If $\lim_{x \to +\infty} g(x) = \ell$ and $\lim_{x \to +\infty} h(x) = \ell$, then

$\lim_{x \to +\infty} f(x) = \ell$. Suppose f, g, and h are functions with domains that each contain $S = (-\infty, a)$.
Moreover assume that $g(x) \le f(x) \le h(x)$ for all $x \in S$. If $\lim_{x \to -\infty} g(x) = \ell$ and

$\lim_{x \to -\infty} h(x) = \ell$, then $\lim_{x \to -\infty} f(x) = \ell$.

Proof For any $\epsilon > 0$ and sufficiently large x we have $\ell - \epsilon < g(x) \le f(x) \le h(x) < \ell + \epsilon$,
therefore $\ell - \epsilon < f(x) < \ell + \epsilon$ and $\lim_{x \to \infty} f(x) = \ell$. The second statement can be proved the same way.

39. For any $A > 0$ there is $\delta > 0$ such that $G(x) > A - \ell + 1$ and $|F(x) - \ell| < 1$ for $|x - c| < \delta$.
Then for this x, $F(x) + G(x) > A$, i.e., $\lim_{x \to c} (F(x) + G(x)) = \infty$.

For any $B < 0$ there is $\delta > 0$ such that $H(x) < B - \ell - 1$ and $|F(x) - \ell| < 1$ for $|x - c| < \delta$.

41. **Vertical asymptotes:** $x = n\pi \, (n \in \mathbb{R}, \, n \ne 0)$
Calculations:
This function is not defined at $x = \pi n$, $n \in \mathbb{Z}$.
Inasmuch as $k \in \mathbb{Z}$,

$$\lim_{x \to 0} \frac{\sin(x)}{x} = 1$$

$$\lim_{x \to 0} g(x) = 1$$

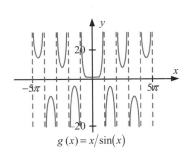

$g(x) = x/\sin(x)$

and the function $g(x)$ has no vertical asymptote at $x = 0$. For $c = 2k\pi$, $k \neq 0$ we have

$$\lim_{x \to c^+} g(x) = \lim_{x \to c^+} \frac{x}{\sin(x)} = \infty$$

$$\lim_{x \to c^-} g(x) = \lim_{x \to c^-} \frac{x}{\sin(x)} = -\infty$$

Therefore $x = c$ is a vertical asymptote. For $c = (2k+1)\pi$, $k \in \mathbb{Z}$ we have

$$\lim_{x \to c^+} g(x) = \lim_{x \to c^+} \frac{x}{\sin(x)} = -\infty$$

$$\lim_{x \to c^-} g(x) = \lim_{x \to c^-} \frac{x}{\sin(x)} = \infty$$

In other words, $x = c$ is a vertical asymptote too.

Horizontal asymptote: There are no horizontal asymptotes.
    Calculations:
    Given that $g(x) = \dfrac{x}{\sin(x)} = \dfrac{1}{\sin(x)/x}$, $\displaystyle\lim_{x \to \pm\infty} \dfrac{\sin(x)}{x} = 0$ (see Problem 2.4.40) and

$\sin(x)$ changes its sign each half-period, then $\displaystyle\lim_{x \to \pm\infty} g(x)$ does not exist and $g(x)$ has no horizontal asymptotes.

43.   **Vertical asymptote:** $x = 0$
    Calculations:
    Given that $\displaystyle\lim_{x \to 0^+} \cos^2 x = 1$, we have

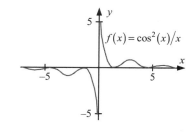

$$\lim_{x \to 0^+} \frac{\cos^2(x)}{x} = \infty$$

$$\lim_{x \to 0^-} \frac{\cos^2(x)}{x} = -\infty$$

Therefore, $x = 0$ is a vertical asymptote.

Horizontal asymptote: $y = 0$
    Calculations:
    Within the range of $-\dfrac{1}{|x|} \leq \dfrac{\cos^2(x)}{x} \leq \dfrac{1}{|x|}$, $\displaystyle\lim_{x \to \pm\infty} \dfrac{1}{|x|} = 0$. Therefore $\displaystyle\lim_{x \to \pm\infty} f(x) = 0$ and $y = 0$ is a horizontal asymptote.

45. **Vertical asymptote:** $x = n\pi \, (n \in \mathbb{R})$

Calculations:

The domain of $g(x)$ is $\{x \in \mathbb{R} : x \neq \pi n, \ n \in \mathbb{Z}\}$, because

$$\lim_{x \to \pi n} \left| \sin(x) \right| = 0$$

$$\lim_{x \to \pi n} g(x) = \lim_{x \to \pi n} \frac{1}{\left| \sin(x) \right|} = \infty$$

so that $x = \pi n$ are upward vertical asymptotes.

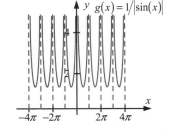

Horizontal asymptote: There are no horizontal asymptotes.

Calculations:

Inasmuch as $\lim\limits_{x \to \pm\infty} \left| \sin(x) \right|$ does not exist, $g(x)$ has no horizontal asymptotes.

47. The graph shows that $x = 0$ is not a vertical asymptote,

since $\lim\limits_{x \to 0^+} \dfrac{\sin\left(\frac{1}{x}\right)}{x}$ and $\lim\limits_{x \to 0^-} \dfrac{\sin\left(\frac{1}{x}\right)}{x}$ do not exist.

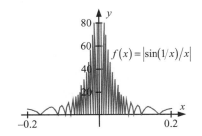

49. Although $f\left(10^{80}\right) = 10^{-20}$, which is many orders of magnitude less than $g\left(10^{80}\right) = 10^{20}$, we have $\lim\limits_{x \to \infty} f(x) = \infty$ and $\lim\limits_{x \to \infty} g(x) = 0$. The point is that the behavior of a function on a bounded interval $(0, \ M)$ does not give information about its behavior at infinity, no matter how great M is. In particular, a comparison of functions f and g on the interval $\left(0, \ 10^{80}\right)$ has no significance for their behaviors at infinity.

51. The volume V of a quantity of gas must be greater than the volume b occupied by the gas molecules themselves. Therefore, the domain of the pressure P as a function of the volume V is the interval $(b, \ \infty)$ and the limit is

$$\lim_{V \to b^+} P(V) = \lim_{V \to b^+} \left(\frac{RT}{V - b} - \frac{a}{V^2} \right) = \infty . \text{ Therefore, } V = b$$

is a vertical asymptote.

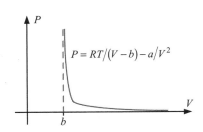

53.   The continuity of the function f means that for any $c \in (a, b)$ there is

$$\lim_{x \to c^-} f(x) = \lim_{x \to c^+} f(x) = \lim_{x \to c} f(x) = f(c) < \infty.$$

Therefore c is not a vertical asymptote.

Given that $\lim\limits_{x \to a^+} f(x) = f(a) < \infty$, and $\lim\limits_{x \to b^-} f(x) = f(b) < \infty$, there are no asymptotes at $x = a$ and $x = b$.

55.   **a.**   If $m = 0$, then $\lim\limits_{x \to +\infty} \left(f(x) - (mx + b) \right) = \lim\limits_{x \to +\infty} \left(f(x) - b \right) = 0$, or $\lim\limits_{x \to +\infty} f(x) = b$. This is the definition of horizontal asymptote $y = b$ of function f.

   b.   Suppose that $f(x) = \dfrac{p(x)}{q(x)}$ is a quotient of polynomials. Then

$$\lim_{x \to \infty} \left(f(x) - (mx + b) \right) = \lim_{x \to \infty} \left(\frac{p(x) - (mx + b)q(x)}{q(x)} \right).$$

In order for this limit to be 0, the degree of the numerator must be less than the degree N of the denominator. (See Table 1.) Notice that, if $m \neq 0$, the degree of $(mx + b)q(x)$ is $N + 1$. If $p(x)$ had degree greater than $N + 1$ then the numerator would have no cancellation to prevent $p(x) - (mx + b)q(x)$ from having degree greater than $N + 1$. Therefore, the degree of $p(x)$ is no greater than $N + 1$. If the degree of $p(x)$ were less than $N + 1$, then $p(x) - (mx + b)q(x)$ would have no cancellation to prevent its degree from being the degree of $(mx + b)q(x)$, namely $N + 1$. In conclusion, if $f(x) = \dfrac{p(x)}{q(x)}$ is the quotient of polynomials $p(x)$ and $q(x)$, if $m \neq 0$, and if $\lim\limits_{x \to \infty} \left(f(x) - (mx + b) \right) = 0$, then the degree of $p(x)$ must be exactly one greater than the degree of $q(x)$. This is a necessary condition: it must be satisfied in order for there to be a skew-asymptote $y = mx + b$ with $m \neq 0$. (Going beyond what the exercise calls for, we note that the degree of $p(x)$ being exactly one greater than the degree of $q(x)$ is also a sufficient condition: if $p(x)$ has a degree one greater than the degree of $q(x)$, then $\dfrac{p(x)}{q(x)}$ has a skew-asymptote.)

   c.   If the degree of $p(x)$ is 2 greater than the degree of $q(x)$, then the division $\dfrac{p(x)}{q(x)}$ results in a degree 2 polynomial $ax^2 + bx + c$ and a remainder $r(x)$ that has degree less than the degree of $q(x)$. In other words, $p(x) = \left(ax^2 + bx + c \right) q(x) + r(x)$, or where $r(x)$ has degree less than $q(x)$. Using the bottom entry of Table 1, we have

$\lim_{x\to\infty}\left(\dfrac{p(x)}{q(x)}-\left(ax^2+bx+c\right)\right)=\lim_{x\to\infty}\dfrac{r(x)}{q(x)}=0$. We say that $f(x)$ has a parabolic

asymptote if there are constants $a\neq 0$, b, and c for which

$\lim_{x\to\infty}\left(f(x)-\left(ax^2+bx+c\right)\right)=0$ or $\lim_{x\to-\infty}\left(f(x)-\left(ax^2+bx+c\right)\right)=0$. With this

definition, we have shown that if p is a polynomial of degree 2 greater than the degree

of q, then the rational function $\dfrac{p(x)}{q(x)}$ has a parabolic asymptote.

57.   If $y=ax^2+bx+c$ is a parabolic asymptote for the

graph of $f(x)$ as $x\to\infty$, then

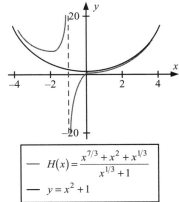

$$a=\lim_{x\to\infty}\frac{f(x)}{x^2}$$

$$b=\lim_{x\to\infty}\frac{f(x)-ax^2}{x}$$

$$c=\lim_{x\to\infty}\left(f(x)-\left(ax^2+bx\right)\right)$$

$$\lim_{x\to\infty}\left(f(x)-\left(ax^2+bx+c\right)\right)=0$$

$$\cdots\quad H(x)=\frac{x^{7/3}+x^2+x^{1/3}}{x^{1/3}+1}$$

$$\text{—}\quad y=x^2+1$$

The same computations can be done as $x\to-\infty$ as shown here:

$$\lim_{x\to-\infty}\frac{F(x)}{x^2}=\lim_{x\to-\infty}\frac{x}{x+3}=\lim_{x\to-\infty}\frac{1}{1+3/x}=1$$

$$\lim_{x\to\infty}\frac{F(x)-x^2}{x}=\lim_{x\to\infty}\left(\frac{x^2}{x+3}-x\right)=\lim_{x\to\infty}\left(\frac{-3x}{x+3}\right)=\lim_{x\to\infty}\left(\frac{-3}{1+3/x}\right)=-3$$

$$\lim_{x\to\infty}\frac{F(x)-\left(x^2-3x\right)}{x}=\lim_{x\to\infty}\frac{9x}{x+3}=9\lim_{x\to\infty}\frac{1}{1+3/x}=9$$

$$\lim_{x\to\infty}\frac{F(x)-\left(x^2-3x+9\right)}{x}=-\lim_{x\to\infty}\frac{27}{x+3}=0$$

Nothing changes as $x\to-\infty$. Therefore, $y=x^2-3x+9$ is a parabolic asymptote of F.

A parabolic asymptote of G is calculated below:

$$\lim_{x\to\infty}\frac{G(x)}{x^2}=\lim_{x\to\infty}\frac{x^2}{x^2+5x}=\lim_{x\to\infty}\frac{1}{1+5/x}=1$$

$$\lim_{x\to\infty}\frac{G(x)-x^2}{x}=\lim_{x\to\infty}\left(\frac{-5x^2}{x^2+5x}\right)=-5\lim_{x\to\infty}\frac{1}{1+5/x}=-5$$

$$\lim_{x\to\infty}\left(G(x)-\left(x^2-5x\right)\right)=\lim_{x\to\infty}\frac{25x}{x+5}=25\lim_{x\to\infty}\frac{1}{1+5/x}=25$$

$$\lim_{x\to\infty}\left(G(x)-\left(x^2-5x+25\right)\right)=-\lim_{x\to\infty}\frac{125}{x+5}=0$$

Nothing will change as $x \to -\infty$. Therefore, $y = x^2 - 5x + 25$ is a parabolic asymptote of G.

A parabolic asymptote of H is calculated as follows:

$$\lim_{x \to \infty} \frac{H(x)}{x^2} = \lim_{x \to \infty} \frac{x^{7/3} + x^2 + x^{1/3}}{x^{7/3} + x^2} = \lim_{x \to \infty} \frac{1 + x^{-1/3} + x^{-2}}{1 + x^{-1/3}} = 1$$

$$\lim_{x \to \infty} \frac{H(x) - x^2}{x} = \lim_{x \to \infty} \frac{1}{x + x^{2/3}} = 0$$

$$\lim_{x \to \infty} \left(H(x) - x^2 \right) = \lim_{x \to \infty} \frac{x^{1/3}}{x^{1/3} + 1} = \lim_{x \to \infty} \frac{1}{1 + x^{-1/3}} = 1$$

$$\lim_{x \to \infty} \left(H(x) - \left(x^2 + 1 \right) \right) = -\lim_{x \to \infty} \frac{1}{x^{1/3} + 1} = 0$$

Nothing will change as $x \to -\infty$. Therefore, $y = x^2 + 1$ is a parabolic asymptote of H.

In summary, $x^2 - 3x + 9$ is a parabolic asymptote of $F(x)$; $x^2 - 5x + 25$ is a parabolic asymptote of $G(x)$; and $x^2 + 1$ is a parabolic asymptote of $H(x)$.

Calculator/Computer Exercises

59.   We must solve the inequality $\dfrac{\sqrt{x}}{x+1} < 0.001$, or $\dfrac{0.001x - \sqrt{x} + 0.001}{x+1} < 0$. The result is

$x \in \left[0, \left(500 - \sqrt{249999} \right)^2 \right) \cup \left(\left(500 + \sqrt{249999} \right)^2, \infty \right)$, which is approximately

$x \in \left[0, 10^{-5} \right) \cup (999997.999999, \infty)$. Therefore, $x > \left(500 + \sqrt{249999} \right)^2 \approx 999997.999999$.

61.   There is a horizontal asymptote at $y = 3$.

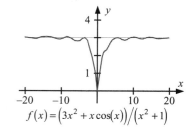

$f(x) = \left(3x^2 + x \cos(x) \right) / \left(x^2 + 1 \right)$

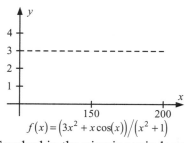

$f(x) = \left(3x^2 + x \cos(x) \right) / \left(x^2 + 1 \right)$

Graphed in the viewing window
$[100, 200] \times [0, 4]$

63.   There is a horizontal asymptote at $y = 1$.

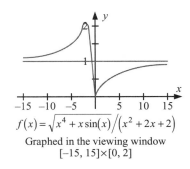

$f(x) = \sqrt{x^4 + x\sin(x)} \big/ (x^2 + 2x + 2)$

Graphed in the viewing window
$[-15, 15] \times [0, 2]$

$f(x) = \sqrt{x^4 + x\sin(x)} \big/ (x^2 + 2x + 2)$

Graphed in the viewing window
$[500, 600] \times [0.9, 1.1]$

65.   The numerator of this fraction is not less than 3, and the denominator on the segment $[0, 3]$ has only one root $x = \dfrac{\pi}{2}$, which is a vertical asymptote.

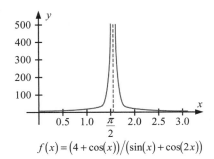

$f(x) = (4 + \cos(x)) \big/ (\sin(x) + \cos(2x))$

Section 2.5   **Limits of Sequences**

Problems for Practice

---

1.   Given that $0 < a_n < \dfrac{1}{n}$ and $\dfrac{1}{n} \to 0$ as $n \to \infty$, we conclude that $a_n \to 0$.

3.   Given that $a_n = 1 - \dfrac{1}{n+1}$; $\dfrac{1}{n+1} \to 0$ as $n \to \infty$, we find $a_n \to 1$ as $n \to \infty$.

5.   Given that $0 < a_n < \dfrac{1}{n}$ and $\dfrac{1}{n} \to 0$ as $n \to \infty$, we conclude that $a_n \to 0$.

7.   The limit does not exist and $\{a_n\}$ is not convergent because $a_{2k} = 1$, and $a_{2k+1} = -1$.

9.   Given that $3^{-n} \to 0$, $2^{-n} \to 0$ as $n \to \infty$, we find $a_n \to 0$ as $n \to \infty$.

11.   We compute $a_n = \dfrac{3n - 5/n^2}{4 + 5/n^2}$; $5/n^2 \to 0$ as $n \to \infty$; then $4 + \dfrac{5}{n^2} \to 4$ as $n \to \infty$, but

$3n - \dfrac{5}{n^2} \to \infty$ and $a_n \to \infty$ as $n \to \infty$.

13. Given that $\dfrac{1}{n^2+1} \to 0$, we find $a_n \to 4$ as $n \to \infty$.

15. By Theorem 1c, we see that $a_n \to 1$ as $n \to \infty$.

17. $\lim\limits_{j\to\infty}\left(2^{-j}+3\right) \overset{\text{Th. 2a}}{=} \lim\limits_{j\to\infty} 2^{-j} + \lim\limits_{j\to\infty} 3 \overset{\text{Th. 2e}}{=} \lim\limits_{j\to\infty} 2^{-j} + 3 \overset{\text{Th. 1d}}{=} 0+3 = 3$

19. $\lim\limits_{m\to\infty}\left(1+m^{-1}\right)\cdot\left(2-6m^{-2}\right) \overset{\text{Th. 2b}}{=} \left(\lim\limits_{m\to\infty}\left(1+m^{-1}\right)\right)\cdot\left(\lim\limits_{m\to\infty}\left(2-6m^{-2}\right)\right)$

$\overset{\text{Th. 2a}}{=} \left(\lim\limits_{m\to\infty} 1 + \lim\limits_{m\to\infty} m^{-1}\right)\cdot\left(\lim\limits_{m\to\infty} 2 - \lim\limits_{m\to\infty}\left(6m^{-2}\right)\right)$

$\overset{\text{Th. 2d}}{=} \left(\lim\limits_{m\to\infty} 1 + \lim\limits_{m\to\infty} m^{-1}\right)\cdot\left(\lim\limits_{m\to\infty} 2 - 6\lim\limits_{m\to\infty} m^{-2}\right)$

$\overset{\text{Th. 1b}}{=} \left(\lim\limits_{m\to\infty} 1 + 0\right)\cdot\left(\lim\limits_{m\to\infty} 2 - 6\cdot 0\right) \overset{\text{Th. 2e}}{=} 2$

21. $\lim\limits_{n\to\infty}\dfrac{2^n-3^n}{3^n+4^n} = \lim\limits_{n\to\infty}\dfrac{(1/2)^n-(3/4)^n}{(3/4)^n+1} \overset{\text{Th. 2c}}{=} \dfrac{\lim\limits_{n\to\infty}\left((1/2)^n-(3/4)^n\right)}{\lim\limits_{n\to\infty}\left((3/4)^n+1\right)} \overset{\text{Th. 2a}}{=} \dfrac{\lim\limits_{n\to\infty}(1/2)^n - \lim\limits_{n\to\infty}(3/4)^n}{\lim\limits_{n\to\infty}(3/4)^n + \lim\limits_{n\to\infty} 1}$

$\overset{\text{Th. 1d}}{=} \dfrac{0-0}{0+\lim\limits_{n\to\infty} 1} \overset{\text{Th. 2e}}{=} \dfrac{0}{1} = 0$

23. Given the range $-1 \le \cos(2k) \le 1$, $-2^{-k} \le \dfrac{\cos(2k)}{2^k} \le 2^{-k}$, by Theorem 1d, $\lim\limits_{k\to\infty} 2^{-k} = 0$ and

$\lim\limits_{k\to\infty}\dfrac{\cos(2k)}{2^k} = 0$.

25. Given the range $m^{-2} \le \dfrac{2+\cos(m)}{m^2} \le 3m^{-2}$, $\lim\limits_{m\to\infty} m^{-2} = 0$ (Theorem 1b). Therefore,

$\lim\limits_{m\to\infty}\dfrac{2+\cos(m)}{m^2} = 0$.

27. $\lim\limits_{j\to\infty}\cos\left(\dfrac{1}{j}\right) \overset{\text{Th. 4}}{=} \cos\left(\lim\limits_{j\to\infty}\left(\dfrac{1}{j}\right)\right) \overset{\text{Th. 1b}}{=} \cos(0) = 1$

29. $\lim\limits_{n\to\infty}\left(2^{-n}\right)\dfrac{n}{n+4} = \lim\limits_{n\to\infty}\dfrac{2^{-n}}{1+4n^{-1}} \overset{\text{Th. 2c}}{=} \dfrac{\lim\limits_{n\to\infty} 2^{-n}}{\lim\limits_{n\to\infty}\left(1+4n^{-1}\right)} \overset{\text{Th. 2a}}{=} \dfrac{\lim\limits_{n\to\infty} 2^{-n}}{\lim\limits_{n\to\infty} 1 + \lim\limits_{n\to\infty} 4n^{-1}}$

$\overset{\text{Th. 2d}}{=} \dfrac{\lim\limits_{n\to\infty} 2^{-n}}{\lim\limits_{n\to\infty} 1 + 4\lim\limits_{n\to\infty} n^{-1}} \overset{\text{Th. 1b}}{=} \dfrac{\lim\limits_{n\to\infty} 2^{-n}}{\lim\limits_{n\to\infty} 1 + 4\cdot 0} \overset{\text{Th. 2e}}{=} \dfrac{\lim\limits_{n\to\infty} 2^{-n}}{1} \overset{\text{Th. 1d}}{=} \dfrac{0}{1} = 0$

31. $\lim\limits_{j\to\infty}\dfrac{\sqrt{4j+1}}{\sqrt{j}}=\lim\limits_{j\to\infty}\sqrt{4+j^{-1}}\overset{\text{Th. 4}}{=}\sqrt{\lim\limits_{j\to\infty}\left(4+j^{-1}\right)}\overset{\text{Th. 2a}}{=}\sqrt{\lim\limits_{j\to\infty}4+\lim\limits_{j\to\infty}j^{-1}}\overset{\text{Th. 2b}}{=}\sqrt{\lim\limits_{j\to\infty}4+0}$

$\overset{\text{Th. 2e}}{=}\sqrt{4}=2$

33. $\lim\limits_{m\to\infty}\cos\left(\pi\sin\left(\dfrac{\pi m}{2m+2}\right)\right)\overset{\text{Th. 4}}{=}\cos\left(\pi\sin\left(\pi\lim\limits_{m\to\infty}\left(\dfrac{m}{2m+2}\right)\right)\right)$

$=\cos\left(\pi\sin\left(\pi\lim\limits_{m\to\infty}\left(\dfrac{1}{2+2m^{-1}}\right)\right)\right)$

$\overset{\text{Th. 2c}}{=}\cos\left(\pi\sin\left(\pi\left(\dfrac{\lim\limits_{m\to\infty}1}{\lim\limits_{m\to\infty}\left(2+2m^{-1}\right)}\right)\right)\right)$

$\overset{\text{Th. 2a}}{=}\cos\left(\pi\sin\left(\pi\left(\dfrac{\lim\limits_{m\to\infty}1}{\lim\limits_{m\to\infty}2+\lim\limits_{m\to\infty}2m^{-1}}\right)\right)\right)$

$\overset{\text{Th. 2d}}{=}\cos\left(\pi\sin\left(\pi\left(\dfrac{\lim\limits_{m\to\infty}1}{\lim\limits_{m\to\infty}2+2\lim\limits_{m\to\infty}m^{-1}}\right)\right)\right)$

$\overset{\text{Th. 1b}}{=}\cos\left(\pi\sin\left(\pi\left(\dfrac{\lim\limits_{m\to\infty}1}{\lim\limits_{m\to\infty}2+2\cdot 0}\right)\right)\right)\overset{\text{Th. 2e}}{=}\cos\left(\pi\sin\left(\pi\left(\dfrac{1}{2}\right)\right)\right)$

$=\cos\left(\pi\right)=-1$

35. $\lim\limits_{j\to\infty}\dfrac{4j}{\sqrt{j^{2}+5j+2}}=\lim\limits_{j\to\infty}\dfrac{4}{\sqrt{1+5j^{-1}+2j^{-2}}}\overset{\text{Th. 4}}{=}\dfrac{4}{\sqrt{\lim\limits_{j\to\infty}\left(1+5j^{-1}+2j^{-2}\right)}}$

$\overset{\text{Th. 2a}}{=}\dfrac{4}{\sqrt{\lim\limits_{j\to\infty}1+\lim\limits_{j\to\infty}5j^{-1}+\lim\limits_{j\to\infty}2j^{-2}}}\overset{\text{Th. 2d}}{=}\dfrac{4}{\sqrt{\lim\limits_{j\to\infty}1+5\lim\limits_{j\to\infty}j^{-1}+2\lim\limits_{j\to\infty}j^{-2}}}$

$\overset{\text{Th. 1b}}{=}\dfrac{4}{\sqrt{\lim\limits_{j\to\infty}1+5\cdot 0+2\cdot 0}}\overset{\text{Th. 2e}}{=}\dfrac{4}{\sqrt{1}}=4$

37. The limit of the sequence as N approaches ∞ is given as follows:

$$1-\frac{1}{3}+\frac{1}{9}-\frac{1}{27}+...=\lim\limits_{N\to\infty}\left(1-\frac{1}{3}+\frac{1}{9}-\frac{1}{27}+...+\left(\frac{1}{3}\right)^{N}\right)=\frac{1}{1+1/3}=\frac{3}{4}.$$

We have $r=-\dfrac{1}{3}$, $\left|r\right|<1$.

39.   The limit of the sequence as N approaches ∞ is given as follows:

$$1.11111111... = 1 + \frac{1}{10} + \frac{1}{100} + \frac{1}{1000} + ... = \lim_{N \to \infty} \left(1 + \frac{1}{10} + \frac{1}{100} + \frac{1}{1000} + ... \left(\frac{1}{10}\right)^N \right)$$

$$= \frac{1}{1 - 1/10} = \frac{10}{9}$$

We have $r = \frac{1}{10}$, $|r| < 1$.

Further Theory and Practice

41.   As $\ell = 0$, $\left| \frac{1}{n+7} - 0 \right| < \epsilon$ leads to $n + 7 > \frac{1}{\epsilon}$ and $n > \frac{1}{\epsilon} - 7$. Let $N = \max \left\{ \left[\frac{1}{\epsilon} - 7 \right] + 1, \, 1 \right\}$. For

each $\epsilon > 0$ and $n \geq N$ we have $\left| \frac{1}{n+7} \right| < \epsilon$ and $\lim_{n \to \infty} \frac{1}{n+7} = 0$.

43.   As $\ell = 2$, $\left| \frac{2m+3}{m+5} - 2 \right| < \epsilon$ leads to $\frac{7}{m+5} < \epsilon$ and $m > \frac{7}{\epsilon} - 5$. Let $N = \max \left\{ \left[\frac{7}{\epsilon} - 5 \right] + 1, \, 1 \right\}$.

For each $\epsilon > 0$ and $m \geq N$ we have $\left| \frac{2m+3}{m+5} - 2 \right| < \epsilon$ and $\lim_{m \to \infty} \frac{2m+3}{m+5} = 2$.

45.   The limit of the sequence as N approaches ∞ is given as follows:

$$\frac{1}{4} + \frac{1}{8} + \frac{1}{16} + \frac{1}{32} + ... = \left(\frac{1}{4}\right) \lim_{N \to \infty} \left(1 + \frac{1}{2} + \frac{1}{4} + \frac{1}{8} + ... \left(\frac{1}{2}\right)^N \right) = \left(\frac{1}{4}\right) \left(\frac{1}{1 - 1/2}\right) = \frac{1}{2}. \text{ We have } r = \frac{1}{2},$$

$|r| < 1$.

47.   The limit of the sequence as N approaches ∞ is given as follows:

$$\frac{16}{3} + \frac{32}{9} + \frac{64}{27} + \frac{128}{81} + ... = \left(\frac{16}{3}\right) \lim_{N \to \infty} \left(1 + \frac{2}{3} + \frac{4}{9} + \frac{8}{27} + ... + \left(\frac{2}{3}\right)^N \right) = \left(\frac{16}{3}\right) \left(\frac{1}{1 - 2/3}\right) = 16.$$

We have $r = \frac{2}{3}$, $|r| < 1$.

49.   The limit of the sequence as N approaches ∞ is given as follows:

$$123.01232323... = 123.01 + \frac{23}{10000} + \frac{23}{1000000} + ...$$

$$= 123.01 + \left(\frac{23}{10000}\right) \lim_{N \to \infty} \left(1 + \frac{1}{100} + \frac{1}{10000} + ... + \left(\frac{1}{100}\right)^N\right)$$

$$= 123.01 + \left(\frac{23}{10000}\right) \cdot \left(\frac{1}{1 - 1/100}\right) = 123.01 + \left(\frac{23}{10000}\right) \cdot \left(\frac{100}{99}\right)$$

$$= 123.01 + \frac{23}{9900} = \frac{608911}{4590}$$

We have $r = \dfrac{1}{100}$, $|r| < 1$.

51.   The sequence $a_n = (-1)^n$ diverges, but $a_n^2 \equiv 1$ converges.

53. $\lim_{j \to \infty} (a_j - a_{j+1}) \stackrel{\text{Th. 2a}}{=} \lim_{j \to \infty} a_j - \lim_{j \to \infty} a_{j+1} = \ell - \ell = 0$

55.   By Theorem 2c we have $\lim_{n \to \infty} \left(\dfrac{a_n}{a_{n+1}}\right) = \dfrac{\lim_{n \to \infty} a_n}{\lim_{n \to \infty} a_{n+1}} = \dfrac{\ell}{\ell} = 1$.

57.   Let $p(x) = p_m x^m + p_{m-1} x^{m-1} + ... + p_1 x + p_0$ be a polynomial of degree m, $p_m \neq 0$, and $q(x) = q_n x^n + q_{n-1} x^{n-1} + ... + q_1 x + q_0$ be a polynomial of degree n, $q_n \neq 0$. Then

$$\frac{p(j)}{q(j)} = \frac{p_m j^m + p_{m-1} j^{m-1} + ... + p_1 j + p_0}{q_n j^n + q_{n-1} j^{n-1} + ... + q_1 j + q_0} = \frac{p_m j^{m-n} + p_{m-1} j^{m-n-1} + ... + p_1 j^{1-n} + p_0 j^{-n}}{q_n + q_{n-1} j^{-1} + ... + q_1 j^{1-n} + q_0 j^{-n}}$$

and $\lim_{j \to \infty} \left(q_n + q_{n-1} j^{-1} + ... + q_1 j^{1-n} + q_0 j^{-n}\right) = q_n$.

a.   If $m < n$, then $m - n < 0$ and $\lim_{j \to \infty} \left(p_m j^{m-n} + p_{m-1} j^{m-n-1} + ... + p_1 j^{1-n} + p_0 j^{-n}\right) = 0$.

Therefore, $\lim_{j \to \infty} \dfrac{p(j)}{q(j)} = \dfrac{0}{q_n} = 0$.

b.   If $m = n$, then $\lim_{j \to \infty} \left(p_m j^{m-n} + p_{m-1} j^{m-n-1} + ... + p_1 j^{1-n} + p_0 j^{-n}\right) = p_m$ and

$\lim_{j \to \infty} \dfrac{p(j)}{q(j)} = \dfrac{p_m}{q_n}$.

c.   If $m > n$, $m - n > 0$ and $\lim_{j \to \infty} \left(p_m j^{m-n} + p_{m-1} j^{m-n-1} + ... + p_1 j^{1-n} + p_0 j^{-n}\right) = \infty$.

Therefore, $\lim_{j \to \infty} \dfrac{p(j)}{q(j)} = \dfrac{\infty}{q_n} = \infty$.

59. We first compute $\displaystyle\lim_{n\to\infty}\lim_{m\to\infty} a_{n,m} = \lim_{n\to\infty}\lim_{m\to\infty}\frac{n/m}{n/m+1} = \lim_{n\to\infty} 0 = 0$. However,

$$\lim_{m\to\infty}\lim_{n\to\infty} a_{n,m} = \lim_{m\to\infty}\lim_{n\to\infty}\frac{1}{1+m/n} = \lim_{m\to\infty} 1 = 1 .$$

Calculator/Computer Exercises

61.

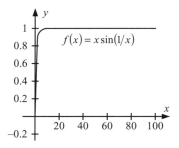

Zooming:

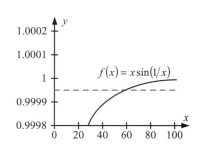

From the above graph, we see that the limit exists: $\displaystyle\lim_{j\to\infty} j\cdot\sin\left(\frac{1}{j}\right)=1$.

As $57\sin\left(\dfrac{1}{57}\right)\approx 0.999949$, $58\sin\left(\dfrac{1}{58}\right)\approx 0.99995$, $j=58$ results in an error smaller than 5×10^{-5}.

63.

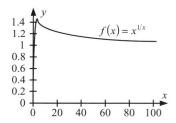

Zooming:

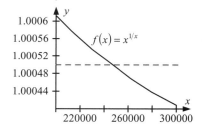

From the above graph we see that the limit exists: $\displaystyle\lim_{j\to\infty} j^{1/j}=1$.

As $250000^{1/250000}\approx 1.00005$, $j=250000$ results in an error smaller than 5×10^{-5}.

65.

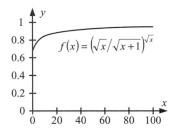

Zooming:

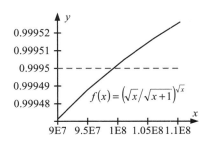

From the above graph, we can see that the

limit exists: $\lim\limits_{j \to \infty} \left(\dfrac{\sqrt{j}}{\sqrt{j+1}} \right)^{\sqrt{j}} = 1$.

As $\left(\dfrac{\sqrt{10^8}}{\sqrt{10^8+1}} \right)^{\sqrt{10^8}} \approx 0.99995$, $j = 10^8$ results in an error smaller than 5×10^{-5}.

67.

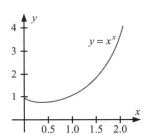

$$\lim\limits_{x \to 0^+} x^x = 1$$

$$\lim\limits_{j \to \infty} j^{1/j} = \lim\limits_{j \to \infty} \dfrac{1}{(1/j)^{1/j}} = \dfrac{1}{\lim\limits_{j \to \infty} (1/j)^{1/j}}$$

$$= \dfrac{1}{1} = 1$$

Section 2.6 Exponential Functions

Problems for Practice

1. $\sqrt{2}^{\sqrt{3}} \cdot \sqrt{2}^{\sqrt{3}} = \sqrt{2}^{\left(\sqrt{3}+\sqrt{3}\right)} = \sqrt{2}^{2\sqrt{3}} = \left(\sqrt{2}^2\right)^{\sqrt{3}} = 2^{\sqrt{3}}$

3. $(1/8)^{-\pi/3} = 8^{\pi/3} = \left(2^3\right)^{\pi/3} = 2^{\pi}$

5. $\left(\sqrt{11}^{\sqrt{2}}\right)^{\sqrt{2}} = \sqrt{11}^{\sqrt{2}\sqrt{2}} = \sqrt{11}^2 = 11$

7.

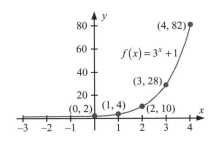

9.

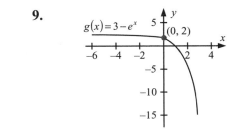

11.

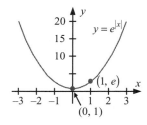

13.   By dividing each term in the numerator and denominator by e^{2x}, we have

$$\lim_{x\to\infty}\frac{e^{2x}-e^{-2x}}{e^{2x}+e^{-2x}}=\lim_{x\to\infty}\frac{1-e^{-4x}}{1+e^{-4x}}=\frac{1-0}{1+0}=1.$$

15.   We let $v=1/(x-e)$ and note that, as x tends to e through values greater than e, the expression $x-e$ tends to 0 through positive values, and the reciprocal $v=1/(x-e)$ tends to positive infinity. We have $\lim_{x\to e^{+}}\pi^{1/(x-e)}=\lim_{v\to\infty}\pi^{v}=\infty$.

17.   **a.**   The denominator and numerator are continuous at every x. The denominator is 0 only for $x=0$. Therefore, the only possible vertical asymptote is $x=0$. Since the numerator is not also 0 when $x=0$, the line $x=0$ is, indeed, a vertical asymptote. Horizontal asymptotes are investigated using the method of Exercise 2.6.13: dividing numerator and denominator by e^{7x}, we have

$$\lim_{x\to\infty}\frac{e^{7x}+e^{-7x}}{e^{7x}-e^{-7x}}=\lim_{x\to\infty}\frac{1+e^{-17x}}{1-e^{-14x}}=\frac{1+0}{1-0}=1$$

and, dividing numerator and denominator by e^{-7x}, we have

$$\lim_{x\to-\infty}\frac{e^{7x}+e^{-7x}}{e^{7x}-e^{-7x}}=\lim_{x\to\infty}\frac{e^{14x}+1}{e^{14x}-1}=\frac{0+1}{0-1}=-1.$$

Therefore $y=-1$ and $y=1$ are horizontal asymptotes.

b.   The function $x\mapsto\exp(-x^{2})$ is continuous for every x. Therefore, its graph has no vertical asymptote. Since $\lim_{x\to\infty}\exp(-x^{2})=\lim_{x\to\infty}1/\exp(x^{2})=0$ we see that $y=0$ is a horizontal asymptote. This line is also a horizontal asymptote as x tends to negative infinity since $\lim_{x\to-\infty}\exp(-x^{2})=\lim_{x\to\infty}1/\exp(x^{2})=0$.

c. The reasoning of part (a) shows that $x = e$ is a vertical asymptote. Since $\lim_{x \to \infty} 3x/(x-e) = \lim_{x \to \infty} 3/(1-e/x) = 3$, we see that $y = 3$ is a horizontal asymptote. This line is also a horizontal asymptote as x tends to negative infinity since $\lim_{x \to -\infty} 3x/(x-e) = \lim_{x \to -\infty} 3/(1-e/x) = 3$.

d. The point $x = 0$ is not in the domain of the given expression. Since the denominator is continuous and nonzero for all other values of x, we conclude that $x = 0$ is the only possible vertical asymptote. As $x \to 0$, the denominator tends to 0. Since we have $\lim_{x \to 0^-} \left(3^{x+1/x} + 2^x\right) = 0 + 1$ and $\lim_{x \to 0^+} \left(3^{x+1/x} + 2^x\right) = \infty$, we see that $x = 0$ is a vertical asymptote as x tends to 0 from both the left and the right. As x tends to positive infinity, we have, by dividing each term by 3^x,

$$\lim_{x \to \infty} \frac{3^{x+1/x} + 2^x}{3^x - 3} = \lim_{x \to \infty} \frac{3^{1/x} + (2/3)^x}{1 - 3/3^x} = \frac{3^0 + 0}{1 - 0} = 1.$$

As x tends to negative infinity, we have, letting $u = -x$,

$$\lim_{x \to -\infty} \frac{3^{x+1/x} + 2^x}{3^x - 3} = \lim_{u \to \infty} \frac{(1/3)^{u+1/u} + (1/2)^u}{(1/3)^u - 3} = \frac{0 + 0}{0 - 3} = 0.$$

We conclude that $y = 1$ and $y = 0$ are horizontal asymptotes.

19.   **a.** $1000(1 + 0.07)^5 = 1402.55$

   b. $1000(1 + 0.07/2)^{10} = 1410.60$

   c. $1000(1 + 0.07/4)^{20} = 1414.78$

   d. $1000(1 + 0.07/365)^{365 \cdot 5} = 1419.02$

   e. $1000\left(e^{0.07}\right)^5 = 1419.07$

21.   Using Theorem 4 with $u = 2$, we have $\lim_{n \to \infty} \left(1 + \frac{2}{n}\right)^n = e^2$.

23.   Using Theorem 4 with $u = -1$, we have $\lim\limits_{n \to \infty} \left(1 + \frac{1}{n}\right)^n = e^{-1}$.

25.   Let $a_n \to \ell$ as $n \to \infty$. Then $\ell = \sqrt{2 + 3\ell}$. Therefore, $\ell^2 = 2 + 3\ell$, and ℓ must be positive. We solve to find $\dfrac{3 + \sqrt{17}}{2}$.

27.   From Example 2, we see that $\ell = \dfrac{\ell + 6}{\ell + 2}$. Given that $a_1 = 1$, we note that all terms of the sequence are positive, so the limit must also be positive. We, therefore, find that $\ell = 2$.

Further Theory and Practice

29.   Let $P(t + \tau) = Ae^{k(t+\tau)} = 2P(t) = 2Ae^{kt}$; therefore, $e^{k\tau} = 2$.

31.   Let $V(t)$ be the volume. Because $V(t)$ grows exponentially, we have $V(t) = V(0)e^{kt}$ for some constant $k > 0$. Since $V(1) = V(0)e^k$, the information $V(1) = 1.035 \cdot V(0)$ tells us that $e^k = 1.035$. Thus, $V(t) = V(0)e^{kt} = V(0)\left(e^k\right)^t = V(0) \cdot 1.035^t$. The percentage increase that is expected in ten years is $100\left(1.035^{10} - 1\right) \approx 41$. To find the doubling time τ, we solve $1.035^\tau \approx 20.15$.

33.   Given that $\lim\limits_{k \to \infty} \left(1 + \dfrac{3}{k}\right)^k = e^3$, we derive $\left(\lim\limits_{k \to \infty} \left(1 + \dfrac{3}{k}\right)^k\right)^2 = \left(e^3\right)^2 = e^6$.

35.   According to the definition of limit, there is an N such that $|a_j - \ell| < 1$ for all $j > N$. For these values of N we have $\ell - 1 < a_j < \ell + 1$. These inequalities tell us that at most N terms, namely $a_1, a_2, \ldots, a_N$ can be greater than $\ell + 1$ or less than $\ell - 1$. If we set M equal to the maximum value of $|a_1|, |a_2|, \ldots |a_N|, |\ell + 1|, |\ell - 1|$, then we have $|a_j| \le M$ for all j.

37. Using mathematical induction, we will prove the assertion $x_{n-1} < x_n < \alpha$ for all $n \geq 1$. When $n = 1$, this assertion becomes $1 < \dfrac{1+\alpha}{2} < \alpha$. Both inequalities of this assertion follow from the hypothesis $\alpha > 1$. Suppose now that $x_{m-1} < x_m < \alpha$ holds for a certain value m. Then the equation $x_{m+1} = \dfrac{\alpha + x_m}{2}$ tells us that $x_m = \dfrac{\alpha + x_{m-1}}{2} < \dfrac{\alpha + x_m}{2} = x_{m+1}$ and $x_{m+1} < \dfrac{\alpha + \alpha}{2} = \alpha$.

In other words, our assertion $x_{n-1} < x_n < \alpha$ is true when $n = 1$ and once it is true for any particular value n it is also true for the next greatest integer $n+1$. We conclude that $x_{n-1} < x_n < \alpha$ is true for all $n \geq 1$. The inequalities we have just established tell us that $\{x_n\}$ is an increasing sequence that is bounded above by α.

From the Monotone Convergence Property of the real numbers we see that $\ell = \lim\limits_{n \to \infty} a_n$ exists.

This limit must satisfy the equation $\ell = \dfrac{\alpha + \ell}{2}$, from which we calculate $\ell = \alpha$.

39. Solving the equation $\ell = 1 + \dfrac{1}{\ell}$, we obtain $\ell = \dfrac{1 + \sqrt{5}}{2}$ after choosing the positive solution.

Instructor Note: Ratios of successive Fibonnaci numbers generate a sequence $\{r_k\}$ that satisfies the given equation. Let $f_0 = 0$, $f_1 = 1$ and $f_{n+2} = f_n + f_{n+1}$ for $n \geq 0$. This is the Fibonnaci sequence. Set $r_1 = 1$ and, for $k \geq 2$, set $r_k = \dfrac{f_k}{f_{k-1}}$. Then

$$r_{k+1} = \frac{f_{k+1}}{f_k} = \frac{f_{k-1} + f_k}{f_k} = 1 + \frac{f_{k-1}}{f_k} = 1 + \frac{1}{r_k}.$$

41. From $|\sin(n)| \leq 1$, it follows that $\dfrac{\sin^2(1)}{2} \leq \dfrac{1}{2}$, $\dfrac{\sin^2(2)}{4} \leq \dfrac{1}{4}$, $\dfrac{\sin^2(3)}{8} \leq \dfrac{1}{8}$, $\ldots$, $\dfrac{\sin^2(n)}{2^n} \leq \dfrac{1}{2^n}$.

Therefore: $s_N \leq \dfrac{1}{2} + \dfrac{1}{4} + \dfrac{1}{8} + \ldots + \dfrac{1}{2^n}$. From Example 9 of Section 2.5, $|s_N| \leq 1$, therefore, $\{s_N\}$ is bounded by 1. Inasmuch as it is an increasing bounded sequence, $\{s_N\}$ has a limit and is, therefore, convergent.

43.

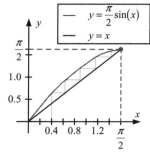

$$\lim_{n \to \infty} x_n = \frac{\pi}{2}$$

45. The terminal velocity is $v_\infty = kg$.

The horizontal asymptote is $v = v_\infty$.

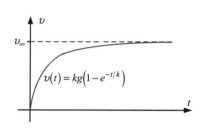

47. The concentration of the solute is $\lim_{t \to \infty} c(t) = C$.

The horizontal asymptote is $y = C$.

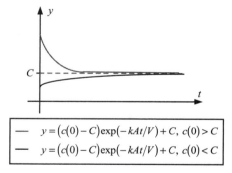

49. **a.** Letting $m = n^2$, we have $\displaystyle\lim_{n \to \infty} \left(1 + \frac{1}{n^2}\right)^{n^2} = \lim_{m \to \infty} \left(1 + \frac{1}{m}\right)^m = e$.

b. Letting $m = n - 2$, we have

$$\lim_{n \to \infty} \left(\frac{n}{n-2}\right)^{2n} = \lim_{m \to \infty} \left(\frac{m+2}{m}\right)^{2m+4} = \left(\lim_{m \to \infty} \left(1 + \frac{2}{m}\right)^m\right)^2 \lim_{m \to \infty} \left(1 + \frac{2}{m}\right)^4 = \left(e^2\right)^2 (1)^4 = e^4.$$

Calculator/Computer Exercises

51. **a.** $v(t) = e^{-0.045t}$

b.

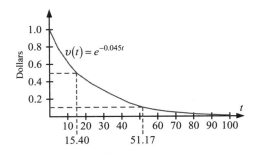

c. The value will be \$0.50 in 15.40 years.

The value will be \$0.10 in 51.17 years.

53. Measuring t in months, the number $N(t)$ of transistors that can be pressed onto a fixed area is given by $N(t) = N(0)e^{kt}$ where $18k \approx 0.6931471806$, or $k \approx 0.0385081767$. Solving the equation $e^{0.0385081767t} = 100$ for t, we find $t \approx 119.59$ months.

55. $m_1 = \dfrac{1}{2}$, $m_2 = \dfrac{3}{4}$, $m_3 = \dfrac{5}{8}$, $m_4 = \dfrac{11}{16}$, $m_5 = \dfrac{23}{32}$, $m_6 = \dfrac{45}{64}$, $m_7 = \dfrac{89}{128}$, $m_8 = \dfrac{177}{256}$, $m_9 = \dfrac{355}{512}$,

$m_{10} = \dfrac{709}{1024} \approx 0.6924$

The length of I_n is $\dfrac{1}{2^n}$.

Using $\left| \gamma - m_n \right| < \dfrac{1}{2^n}$, gives us maximum error $\left| \gamma - \dfrac{709}{1024} \right| < \dfrac{1}{2^{10}} \approx 0.00098$.

57. Let $m(t)$ be the mass of a ^{226}Ra sample. Since ^{226}Ra decays exponentially, there is a $k < 0$ such that $m(t) = m(0)\exp(kt)$. Since the amount of ^{226}Ra is halved every 1620 years, we have $m(0)/2 = m(0)\exp(1620k)$, which we solve to find $k = -4.2786863 \times 10^{-4}$. Thus, $m(t) = m(0)\exp\left(-4.2786863 \times 10^{-4} t\right)$. Let s be the maximum safe amount of ^{226}Ra. The initial amount of ^{226}Ra is given to be $5s$, so that $m(t) = 5s\exp\left(-4.2786863 \times 10^{-4} t\right)$. We solve $s = 5s\exp\left(-4.2786863 \times 10^{-4} t\right)$ to find $t \approx 3761.52$ years.

59.   We calculate $r(4000) = r(0) \cdot e^{-(4000)1.212856 \times 10^{-4}} \approx r(0) \cdot 0.616$. The amount of ^{14}C in the Siberian mammoth fossils was about 61.6% of $r(0)$.

61.   We solve $0.0879 \cdot r(0) = r(0) \cdot e^{-(t)1.212856 \times 10^{-4}}$ to find $t \approx 20048$ years old.

63.   We calculate $r(9400) = r(0) \cdot e^{-(9400)1.212856 \times 10^{-4}} \approx 0.3198 r(0)$. The fraction of $r(0)$ found was about 0.32.

The Derivative

Section 3.1 **Rates of Change and Tangent Lines**

Problems for Practice

In the odd-numbered solutions to Exercises 1–43 that follow, each instantaneous rate of change is calculated from its definition as a limit of a quotient of increments. As an alternative, by using equation (3.4), the *Insight* at the top of page 160, and Theorem 1, the following generalization of formula (3.4) results: If

$$f(x) = \frac{A}{x} + B + Cx + Dx^2 + Ex^3$$

then

$$\lim_{\Delta x \to 0} \frac{f(x + \Delta x) - f(x)}{\Delta x} = -\frac{A}{x^2} + C + 2Dx + 3Ex^2.$$

Each instantaneous rate of change required in Exercises 1–44 of Section 3.1 may be obtained from this formula by appropriate specifications of the constants A, B, C, D, and E.

1. $$\lim_{\Delta t \to 0} \frac{p(c + \Delta t) - p(c)}{\Delta t} = \lim_{\Delta t \to 0} \frac{5 - 5}{\Delta t} = 0$$

3. $$\lim_{\Delta t \to 0} \frac{p(c + \Delta t) - p(c)}{\Delta t} = \lim_{\Delta t \to 0} \frac{-5(c + \Delta t)^2 + 2 + 5c^2 - 2}{\Delta t} = \lim_{\Delta t \to 0} \frac{-10c\Delta t - 5\Delta t^2}{\Delta t} = -10c = -10$$

5. $$\lim_{\Delta t \to 0} \frac{p(c + \Delta t) - p(c)}{\Delta t} = \lim_{\Delta t \to 0} \frac{\dfrac{1}{c + \Delta t} - \dfrac{1}{c}}{\Delta t} = \lim_{\Delta t \to 0} \frac{-\Delta t}{\Delta t \, c(c + \Delta t)} = -\frac{1}{c^2} = -4$$

7. The velocity of the body at $t = 4$ is:

$$\lim_{\Delta t \to 0} \frac{p(c + \Delta t) - p(c)}{\Delta t} = \lim_{\Delta t \to 0} \frac{6(c + \Delta t) + 3 - 6c - 3}{\Delta t} = \lim_{\Delta t \to 0} \frac{6\Delta t}{\Delta t} = 6 > 0.$$

The body is moving forward.

9. The velocity of the body at $t = 4$ is:

$$\lim_{\Delta t \to 0} \frac{p(c + \Delta t) - p(c)}{\Delta t} = \lim_{\Delta t \to 0} \frac{-(c + \Delta t)^3 + 5(c + \Delta t)^2 + c^3 - 5c^2}{\Delta t}$$

$$= -3c^2 + 10c = -8 < 0$$

The body is moving backward.

11. $m = \lim_{\Delta x \to 0} \frac{f(x + \Delta x) - f(x)}{\Delta x} = \lim_{\Delta x \to 0} \frac{(x + \Delta x)^2 - x^2}{\Delta x} = \lim_{\Delta x \to 0} \frac{2x\Delta x + \Delta x^2}{\Delta x} = 2x = 6$

13. $m = \lim_{\Delta x \to 0} \frac{H(x + \Delta x) - H(x)}{\Delta x} = \lim_{\Delta x \to 0} \frac{3(x + \Delta x)^2 - 6 - 3x^2 + 6}{\Delta x} = 6x = -6$

15. $m = \lim_{\Delta x \to 0} \frac{f(x + \Delta x) - f(x)}{\Delta x} = 4x = 20$ 17. $m = \lim_{\Delta x \to 0} \frac{f(x + \Delta x) - f(x)}{\Delta x} = -6x = 12$

$y = 20(x - 5) + 50$ $y = 12(x + 2) - 7$

19. $m = \lim_{\Delta x \to 0} \frac{f(x + \Delta x) - f(x)}{\Delta x} = 4x = 20$ 21. $m = \lim_{\Delta x \to 0} \frac{f(x + \Delta x) - f(x)}{\Delta x} = -6x = 12$

$y = -\frac{1}{20}(x - 5) + 50$ $y = -\frac{1}{12}(x + 2) - 7$

23. The rate of change of the population of the colony of bacteria is

$$\lim_{\Delta t \to 0} \frac{B(t + \Delta t) - B(t)}{\Delta t} = \lim_{\Delta t \to 0} \frac{6(t + \Delta t)^3 - 6t^3}{\Delta t} = 18t^2.$$

After 2 hours it is equal to 72.

25. The death rate for the reindeer is determined as $\lim_{\Delta t \to 0} \frac{r(t + \Delta t) - r(t)}{\Delta t} = -800 - 80t - 3t^2$. After 11 months, it is equal to –2043; the reindeer are dying out at a rate of 2,043 per month.

27. The surface area of a sphere of radius r is $S(r) = 4\pi r^2$. The rate of change is

$$\lim_{\Delta r \to 0} \frac{S(r + \Delta r) - s(r)}{\Delta r} = 8\pi r.$$

It is 64π in^2 per in when the radius is 8 in.

Further Theory and Practice

29. The velocity of the body is $\lim\limits_{\Delta t \to 0} \dfrac{p(t + \Delta t) - p(t)}{\Delta t} = 2t + 1$. The velocity is positive when

$t > -\dfrac{1}{2}$, and is negative when $t < -\dfrac{1}{2}$.

31. The rate of change of a function f at the value x is the slope of the tangent line to the graph of $y = f(x)$ at the point $(x, f(x))$. Although we cannot, by inspection, determine the values of these slopes at the indicated points, we *can* order these points by increasing slope, from most negative to most positive: D, C, E, B, A, F. By ordering the six given slopes in the same way, we obtain the following correspondences: A: 10; B: 3; C: −1; D: −3; E: 0; F: 20.

33.   **a.** $H(0) = 18$ feet

     b.   The rate of the ball rising as a function of time may be calculated as:

$$\lim\limits_{\Delta t \to 0} \dfrac{H(t + \Delta t) - H(t)}{\Delta t} = 13.8 - 32t$$

This is not the velocity of the ball because the motion of the ball is not vertical.

     c.   The ball reaches its maximum height when $13.8 - 32t = 0$ or $t = 0.43125$ seconds.

The maximum height is $H(0.43125) = 20.9756$ feet.

     d.   The average rate of change is $\dfrac{H(0.43125) - H(0)}{0.43125} = \dfrac{2.9756}{0.43125} = 6.9$ feet/sec.

     e.   The average rate of change is equal to zero when $H(t) = H(0) = 18$ feet or

$13.8t - 16t^2 = 0$ or $t = 0.8625$ seconds so the interval is $[0, \ 0.8625]$ s.

     f.   The ball is on the ground when $H(t) = 0$ feet or $t = 1.57623$ seconds, so the ball is in the air for 1.57623 seconds.

     g.   The rate of change is $13.8 - 32 \cdot 1.57623 = -36.6393$ feet/sec.

     h.   The average rate of change is $\dfrac{H(1.57623) - H(0)}{1.57623} = \dfrac{0 - 18}{1.57623} = -11.4197$ feet/sec.

     i.   The rate of change is equal to the overall average when $13.8 - 32t = -11.4197$ or $t = 0.788114$ seconds.

35. Let $y = ax^2 + bx + c$ be the equation of the polynomial. Then the equation of the tangent line to the graph at $x = x_0$ is $y = (2ax_0 + b)(x - x_0) + ax_0^2 + bx_0 + c = 2axx_0 - ax_0^2 + bx + c$. The equation $ax^2 + bx + c = 2axx_0 - ax_0^2 + bx + c$ or $ax^2 - 2axx_0 + ax_0^2 = 0$ has the only root $x = x_0$, so the tangent line can only intersect f at one point, $x = x_0$.

37. Substituting $x = 1$ and $y = -9$ into the equation of the tangent line, we get $-9 = 6x_0 - 3x_0^2$.
The solutions are $x_0 = -1$ and $x_0 = 3$. Thus, the line $y = -6(x+1) + 3$, which is tangent to the
graph of $y = 3x^2$ at the point $(1, -3)$, and the line $y = 18(x-3) + 27$, which is tangent to the
graph of $y = 3x^2$ at the point $(3, 27)$, pass through the point $(1, -9)$.

39. Velocity is $v(t) = \lim_{\Delta t \to 0} \dfrac{p(t + \Delta t) - p(t)}{\Delta t} = 3t^2$ and acceleration at time $t = 1$ is

$$\lim_{\Delta t \to 0} \frac{v(t + \Delta t) - v(t)}{\Delta t} = 6t.$$

It is equal to 6 when $x = 1$.

41. Given that the slope of the tangent line to the graph of $f(x)$ is constant, $f(x)$ is a linear function
with the form $f(x) = ax + b$ where a is the slope of the tangent line. So the function is:

$$f(x) = \begin{cases} -2x + b_1 & \text{if} \quad -\infty < x < 1 \\ x + b_2 & \text{if} \quad 1 < x < 3 \\ -x + b_3 & \text{if} \quad 3 < x < \infty \end{cases}$$

As $f(2) = 5$, then $b_2 = 3$. Moreover, because $f(x)$ is continuous at $x = 1$, $-2 + b_1 = 1 + b_2$,
and $b_1 = b_2 + 3 = 6$. Similarly, because $f(x)$ is continuous at $x = 3$, $3 + b_2 = -3 + b_3$, or
$b_3 = b_2 + 6 = 9$. Finally, we have

$$f(x) = \begin{cases} -2x + 6 & \text{if} \quad -\infty < x < 1 \\ x + 3 & \text{if} \quad 1 < x < 3 \\ -x + 9 & \text{if} \quad 3 < x < \infty \end{cases}.$$

43. It may be noted that the existence of both one-sided limits ℓ_L and ℓ_R implies that f is
continuous at c.

a. Except for their common initial point $(c, f(c))$, the half-lines $y = T_R(x)$ and
$y = T_L(x)$, lie in the half-planes $\{(x, y): x > c\}$ and $\{(x, y): x < c\}$ respectively. It
follows that $\alpha_f(c)$ must lie in the open interval $(0, 2\pi)$. The graph of f has a corner
at $(c, f(c))$ if and only if $\alpha_f(c) \neq \pi$.

b. The graph of f has a tangent line at $(c, f(c))$ precisely when $\alpha_f(c) = \pi$. It is
precisely for this value of $\alpha_f(c)$ that a straight line is formed from the union of the
half-line $y = T_L(x)$, which is tangent to that part of the graph of f that lies in the half-
plane $\{(x, y): x \leq c\}$, and the half-line $y = T_R(x)$, which is tangent to that part of the
graph of f that lies in the half-plane $\{(x, y): x \geq c\}$.

c.   Under the given hypotheses, namely the existence of the one-sided limits ℓ_L and ℓ_R, there is no value of $\alpha_f(c)$ that corresponds to a cusp. When a cusp like that shown in Figure 13 occurs, the angle between the right tangent half-line and the left tangent half-line is 0. When the graph of Figure 13 is turned upside down, we obtain a cusp at which the angle between the right tangent half-line and the left tangent half-line is 2π. However, under the stated hypotheses, $\alpha_f(c)$ cannot attain either of these values.

d.   If the graph of f has a vertical tangent line at $(c, \ f\ (c))$, then $\alpha_f(c)$ is not defined. The analytic definition of $\alpha_f(c)$ requires both one-sided limits ℓ_L and ℓ_R to exist as real numbers.

Calculator/Computer Exercises

45.   **a.**   **i.** $\dfrac{H(3)-H(2)}{3-2} = \dfrac{76-84}{1} = -8 \ \text{ft/s}$

　　　　ii. $\dfrac{H(2.1)-H(2)}{2.1-2} = \dfrac{84.64-84}{0.1} = 6.4 \ \text{ft/s}$

　　　　iii. $\dfrac{H(2.01)-H(2)}{2.01-2} = \dfrac{84.0784-84}{0.01} = 7.84 \ \text{ft/s}$

　　　　iv. $\dfrac{H(2.001)-H(2)}{2.001-2} = \dfrac{84.008-84}{0.001} = 8 \ \text{ft/s}$

　　b. $v(t) = \lim\limits_{\Delta t \to 0} \dfrac{H(t+\Delta t)-H(t)}{\Delta t} = 72 - 32t$

　　　　$v(2) = 8 \ \text{ft/s}$

47.

n	0	1	2	3	4
Ave. Velocity over $\left[0,\ 10^{-n}\right]$	0.079401	1.6209	1.7219	1.7310	1.7320

Estimate for instantaneous velocity is $v = \sqrt{3}$.

The graph supports the conjecture.

Its slope is close to $\sqrt{3}$ at $t = 0$.

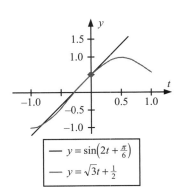

$$— \ y = \sin\left(2t + \tfrac{\pi}{6}\right)$$
$$— \ y = \sqrt{3}t + \tfrac{1}{2}$$

49.

n	1	2	3	4
Ave. Velocity over $\left[2,\ 2+10^{(-n)}\right]$	7.7711	7.4261	7.3928	7.3894
Ave. Velocity over $\left[3,\ 3+10^{(-n)}\right]$	21.124	20.186	20.096	20.087

n	1	2	3	4
Position at $2+10^{(-n)}$	8.16616	7.4633	7.3964	7.3898
Position at $3+10^{(-n)}$	22.1980	20.2874	20.1056	20.0875

Notice that the values

$$p(2)=(2.718281828)^2 = 7.389\ldots$$

and $p(3)=(2.718281828)^3 = 20.085\ldots$ are very nearly equal to the values of the velocity function at $t = 2$ and $t = 3$. Extrapolating, it appears that, for the given position function, velocity equals position.

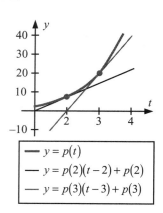

51.   Zooming in to the viewing window

$$\left[\frac{\pi}{4}-h,\ \frac{\pi}{4}+h\right]\times\left[\tan\left(\frac{\pi}{4}-h\right),\ \tan\left(\frac{\pi}{4}+h\right)\right]$$

results in a slope of

$$\frac{\tan\left(\dfrac{\pi}{4}+h\right)-\tan\left(\dfrac{\pi}{4}-h\right)}{2h} \approx 2.0006\ .$$

For $Q(\Delta x)=\dfrac{\tan\left(\frac{\pi}{4}+\Delta x\right)-\tan\left(\frac{\pi}{4}\right)}{\Delta x}$,

Δx	0.1	0.01	0.001	0.0001
$Q(\Delta x)$	2.23049	2.02027	2.0020	2.0002

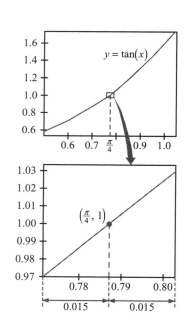

It compares well with the visual estimate of the slope also being approximately 2.

Section 3.2   **The Definition of the Derivative**

Problems for Practice

1. $f'(x) = 2x$, $f'(-3) = -6$

3. $\dfrac{dF}{dt} = 5$, $\dfrac{dF}{dt}(8) = 5$

5. $\dot{g}(t) = 24t^2 - 8$, $\dot{g}(5) = 592$

7. $g'(x) = -\pi$

9. $g'(x) = 12x^2 + 12x$

11.   From $f'(x) = 2x$, we obtain $2x = 6$, so $x = 3$.

13.   From $f'(x) = 3x^2$ we obtain $3x^2 = 6$, so $x = -\sqrt{2}$ and $x = \sqrt{2}$.

15. $f'(x) = -6x^2$

17. $f'(x) = \begin{cases} 2x - 1 & \text{if } x < 0 \\ 2x + 1 & \text{if } x > 0 \end{cases}$

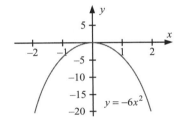

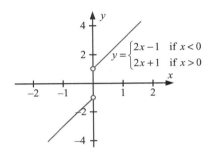

Further Theory and Practice

19.   To find the derivative of the function at the point $x = 0$, we first calculate

$$\varphi(x) = \frac{f(x) - f(0)}{x - 0} = \frac{|x|^{3/2}}{x},$$

so that $\varphi(x) = \begin{cases} -|x|^{1/2} & \text{if } x < 0 \\ x^{1/2} & \text{if } x > 0 \end{cases}$. Then $\lim\limits_{x \to 0} \varphi(x) = 0$ and $f'(0) = 0$.

21.   To find the derivative of the function at the point $x = 0$, we first calculate

$$\varphi(x) = \frac{f(x) - f(9)}{x - 9} = \frac{\sqrt{x} - 3}{x - 9} = \frac{1}{\sqrt{x} + 3}.$$

Then $\lim\limits_{x \to 9} \varphi(x) = \dfrac{1}{6}$ and $f'(9) = \dfrac{1}{6}$.

23.   From the graph we see that there are positive constants a and b such that $H(x) = ax$, $G(x) = b$, and $F(x) = 0$ for all x. From this we infer that $H' = a$, $G' = 0$, and $F' = 0$. It follows that K, which is not a constant function, is not the derivative of any of the three other functions. From the given hypothesis, we deduce that F, G, and H are the three graphed functions that are derivatives. It follows that $a = b$, $G = H'$, $F = G'$, and, by a process of elimination, $H = K'$. (We can further deduce that $K(x) = ax^2/2$. We also note that $F = F'$.)

25.   Let $p(x) = ax + b$. Given that $p(2) = 2a + b = 6$ and $p'(4) = a = -5$, we see that $a = -5$, $b = 16$, and $p(x) = -5x + 16$.

27.   Let $\varphi(x) = \dfrac{f(x) - f(0)}{x - 0} = \dfrac{f(x)}{x}$, then $\varphi(x) = \begin{cases} x & \text{if } x < 0 \\ 1 & \text{if } x > 0 \end{cases}$. Whereas $\lim\limits_{x \to 0^-} \varphi(x) = 0$ and $\lim\limits_{x \to 0^+} \varphi(x) = 1$ are different, the function $f(x)$ is not differentiable at $x = 0$.

29.   Let $\varphi(x) = \dfrac{f(x) - f(0)}{x - 0} = \dfrac{f(x)}{x}$, then $\varphi(x) = \begin{cases} -1 & \text{if } x < 0 \\ x + 1 & \text{if } x > 0 \end{cases}$. Given that $\lim\limits_{x \to 0^-} \varphi(x) = -1$ and $\lim\limits_{x \to 0^+} \varphi(x) = 1$ are different, the function $f(x)$ is not differentiable at $x = 0$.

31.   We can take $f(x) = \sqrt{x}$, $c = 4$; or $f(x) = \sqrt{4 + x}$, $c = 0$.

33.   As $\dfrac{1}{\sqrt{3 + \Delta x} + \sqrt{3}} = \dfrac{\sqrt{3 + \Delta x} - \sqrt{3}}{\left(\sqrt{3 + \Delta x} + \sqrt{3}\right)\left(\sqrt{3 + \Delta x} - \sqrt{3}\right)} = \dfrac{\sqrt{3 + \Delta x} - \sqrt{3}}{\Delta x}$, we can take: $f(x) = \sqrt{x}$, $c = 3$ and $f(x) = \sqrt{3 + x}$, $c = 0$.

35.   The tangent line of the graph at $(a,\ f(a))$ slopes upward; the tangent line of the graph at $(b,\ f(b))$ slopes downward.

37.   Given that $f(x) > 0$, we have $f'(x) > 0$, so $f(x)$ increases when x increases. Furthermore, $f(1) > f(0)$ and the slope of the tangent line at the point $(1,\ f(1))$ is greater than the slope at $(0,\ f(0))$. The slope at point $(2,\ f(2))$ is greater than the slope at $(1,\ f(1))$ and so on. The slope of the tangent line at point $(-1,\ f(-1))$ is smaller than the slope at $(0,\ f(0))$; the slope at the point $(-2,\ f(-2))$ is smaller than the slope at $(-1,\ f(-1))$ and so on.

39.   According to Problem 3.2.38, f' is an odd function, so $f'(x) = -f'(x)$ and $f'(0) = -f'(0)$. Therefore, $f'(0) = 0$.

41.   Let $\varphi(x) = \dfrac{f(x) - f(c)}{x - c} = \dfrac{\frac{x}{1+x^2} - \frac{c}{1+c^2}}{x - c} = -\dfrac{xc - 1}{(1+x^2)(1+c^2)}$. Then

$$f'(c) = \lim_{x \to c} \varphi(x) = \lim_{x \to c} \left[-\dfrac{xc - 1}{(1+x^2)(1+c^2)} \right] = \dfrac{1 - c^2}{(1+c^2)^2}.$$

43. $g'(c - k) = \lim_{\Delta x \to 0} \dfrac{g(c - k + \Delta x) - g(c - k)}{\Delta x} = \lim_{\Delta x \to 0} \dfrac{f(c + \Delta x) - f(c)}{\Delta x} = f'(c)$

Since this limit exists, g is differentiable at $c - k$.

45.   We compute:

$$E(p) = -\lim_{\Delta p \to 0} \dfrac{(q(p + \Delta p) - q(p)) \cdot p}{\Delta p \cdot q(p)} = \dfrac{-q'(p)}{q(p)} \cdot p.$$

If the price increases from \$1.00 to \$1.20, then

$$1.89 \approx -\dfrac{q(1.20) - q(1.00)}{q(1.00)} \cdot \dfrac{1.00}{0.20} = \dfrac{5(q(1.20) - q(1.00))}{q(1.00)}.$$

So, the percent change is approximately $-\dfrac{1.89}{5} = -37.8\%$.

47.   **a.**   Fixed costs equal C_0 (point P).

   b.   From $\dfrac{x}{x_1} = \dfrac{y - C_0}{C_1 - C_0}$ deduce $y = C_0 + \dfrac{C_1 - C_0}{x_1} x$, so $C'(x) \approx \dfrac{C_1 - C_0}{x_1}$.

   c.   From $\dfrac{x - x_2}{x_3 - x_2} = \dfrac{y - C_2}{C_3 - C_2}$ deduce $y = C_2 + \dfrac{C_3 - C_2}{x_3 - x_2}(x - x_2)$ so $C'(x) \approx \dfrac{C_3 - C_2}{x_3 - x_2}$.

   d.   From $\dfrac{x - x_4}{x_5 - x_4} = \dfrac{y - C_4}{C_5 - C_4}$ deduce $y = C_4 + \dfrac{C_5 - C_4}{x_5 - x_4}(x - x_4)$, so $C'(x) = \dfrac{C_5 - C_4}{x_5 - x_4}$.

   e.   As $C_3 - C_2 = C_2 - C_1$, but $x_3 - x_2 > x_2 - x_1$, the arc of $y = C(x)$ between R and S corresponds to the producer's economies of scale.

   This corresponds to $C'(x)$ decreasing on this interval.

   f.   The arc of $y = C(x)$ between T and U corresponds to $C'(x)$ increasing on this interval.

49.   Let us write x for $P(t)$. By completing the square, we have

$$P'(t) = kx(M-x) = k\left(Mx - x^2\right) = k\left(\frac{M^2}{4} - \left(\left(\frac{M}{2}\right)^2 - Mx + x^2\right)\right)$$

$$= k\left(\frac{M^2}{4} - \left(\left(\frac{M}{2}\right) - x\right)^2 + \frac{M}{2}x\right)$$

We see that $P'(t)$ is greatest when $x = \dfrac{M}{2}$.

51.   Based on the definition of $f(x)$, we obtain $0 \le \lim_{x\to 0}\left|f(x)\right| = \lim_{x\to 0}\left|x\sin\left(\dfrac{1}{x}\right)\right| \le \lim_{x\to 0}|x| = 0$.

Therefore $\lim_{x\to 0}\left|f(x)\right| = 0$ and f is continuous at $x = 0$. Moreover,

$$\lim_{x\to 0}\frac{f(x)-f(0)}{x-0} = \lim_{x\to 0}\sin\left(\frac{1}{x}\right)$$

does not exist, so f is not differentiable at $x = 0$.

53.   Based on the definition of $f(x)$, we obtain

$$\lim_{x\to c}\frac{f(x)-(ax+b)}{x-c} = \lim_{x\to c}\frac{\left(f(x)-f(c)\right)+\left(f(c)-(ax+b)\right)}{x-c}$$

$$= \lim_{x\to c}\frac{f(x)-f(c)}{x-c} - \lim_{x\to c}\frac{(ax+b)-f(c)}{x-c}$$

$$= f'(c) - \lim_{x\to c}\frac{a(x-c)+\left(ac+b-f(c)\right)}{x-c} = f'(c) - a = 0$$

because $a = f'(c)$ and $ac+b = f(c)$.

Calculator/Computer Exercises

55.

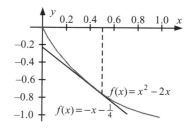

57.

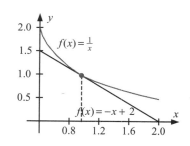

59. The slope of the secant line through $\left(\zeta,\, f\left(\zeta\right)\right)$ gives the approximation $f'\left(c\right) \approx 0.2272727275$.

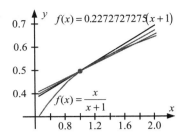

61. The slope of the secant line through $\left(\zeta,\, f\left(\zeta\right)\right)$ gives the approximation $f'\left(c\right) \approx 0.7019192169$.

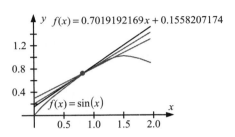

63. $f'\left(c\right) \approx \dfrac{y_R - y_Q}{x_R - x_Q}$, $y_R = 1.00034$, $y_Q = 0.99966$,

$x_R = 1.001$, $x_Q = 0.999$, $f'\left(c\right) \approx 0.34$

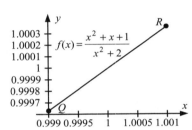

65. $f'\left(c\right) \approx \dfrac{y_R - y_Q}{x_R - x_Q}$, $y_R = 321.5$, $y_Q = 315.2$,

$x_R = 0.99901$, $x_Q = 0.99899$, $f'\left(c\right) \approx 320000$

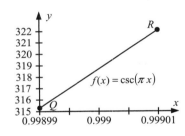

67. The function $\varphi\left(x\right)$ does not have a continuous extension at $x = 1$, therefore $f'\left(1\right)$ does not exist.

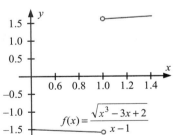

69. The function $\varphi\left(x\right)$ does not have a continuous extension at $x = \dfrac{\pi}{2}$, therefore $f'\left(\dfrac{\pi}{2}\right)$ does not exist.

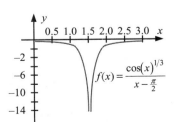

71.   The function $\varphi(x)$ is continuous at $x=0$, therefore $f'(0)$ exists and $f'(0)=0$.

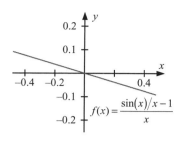

73.

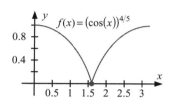

a. $f'\left(\dfrac{\pi}{2}\right)$ does not exist.

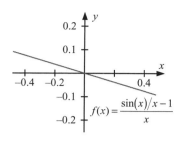

b.   The graph of f does not have a tangent line at $x=\dfrac{\pi}{2}$, therefore f is not differentiable at this point.

75.

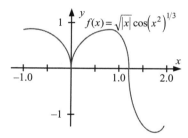

a. $f'(0)$ and $f'\left(\sqrt{\dfrac{\pi}{2}}\right)$ do not exist.

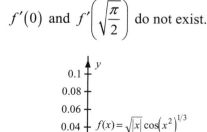

b.   The graph of f has vertical tangent lines at $x=0$ and $x=\sqrt{\dfrac{\pi}{2}}$, therefore f is not differentiable at these points.

77.

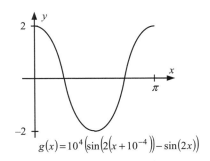

$$g(x)=10^4\left(\sin\left(2\left(x+10^{-4}\right)\right)-\sin(2x)\right)$$

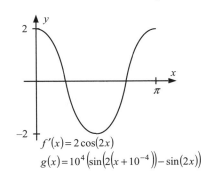

$$f'(x)=2\cos(2x)$$
$$g(x)=10^4\left(\sin\left(2\left(x+10^{-4}\right)\right)-\sin(2x)\right)$$

Function $g(x)$ is a good approximation of $f'(x)=2\cos(2x)$ (these functions are indistinguishable on the previous figure).

Section 3.3   Rules for Differentiation

Problems for Practice

1. $\frac{d}{dx}\left(4x^3+3x^2\right)=4\frac{d}{dx}x^3+3\frac{d}{dx}x^2=4\cdot3x^2+3\cdot2x=12x^2+6x$

3. $\frac{d}{dx}\left(\left(x^2-5x\right)\cdot\left(4x^3+x^2\right)\right)=\left(\frac{d}{dx}\left(x^2-5x\right)\right)\cdot\left(4x^3+x^2\right)+\left(x^2-5x\right)\frac{d}{dx}\left(4x^3+x^2\right)$
$$=(2x-5)\left(4x^3+x^2\right)+\left(x^2-5x\right)\left(12x^2+2x\right)=20x^4-76x^3-15x^2$$

5. $\frac{d}{dx}\left(5\sin(x)-6x\cos(x)\right)=5\left(\frac{d}{dx}\sin(x)\right)-6\frac{d}{dx}\left(x\cos(x)\right)$
$$=5\cos(x)-6\left(\left(\frac{d}{dx}x\right)\cos(x)+x\frac{d}{dx}\cos(x)\right)$$
$$=5\cos(x)-6\left(\cos(x)-x\sin(x)\right)=6x\sin(x)-\cos(x)$$

7. $\frac{d}{dx}\left(\frac{x}{\sin(x)}\right)=\frac{\sin(x)\frac{d}{dx}x-x\frac{d}{dx}\sin(x)}{\sin^2(x)}=\frac{\sin(x)-x\cos(x)}{\sin^2(x)}$

9. $\frac{d}{dx}\left(\frac{1}{x^3+x^2+1}\right)=-\frac{\frac{d}{dx}\left(x^3+x^2+1\right)}{\left(x^3+x^2+1\right)^2}=-\frac{3x^2+2x}{\left(x^3+x^2+1\right)^2}$

11. $\frac{d}{dx}\left(\frac{x^2+1}{x^2+2}\right)=\frac{\left(x^2+2\right)\frac{d}{dx}\left(x^2+1\right)-\left(x^2+1\right)\frac{d}{dx}\left(x^2+2\right)}{\left(x^2+2\right)^2}=\frac{\left(x^2+2\right)2x-\left(x^2+1\right)2x}{\left(x^2+2\right)^2}=\frac{2x}{\left(x^2+2\right)^2}$

13. $\frac{d}{dx}\left(\frac{3x}{x^2+1}\right)=\frac{\left(x^2+1\right)\frac{d}{dx}(3x)-3x\frac{d}{dx}\left(x^2+1\right)}{\left(x^2+1\right)^2}=\frac{3\left(x^2+1\right)-(3x)(2x)}{\left(x^2+1\right)^2}=3\frac{\left(1-x^2\right)}{\left(x^2+1\right)^2}$

15. $\frac{d}{dx}\left(\frac{x\sin(x)}{x+\sin(x)}\right) = \frac{(x+\sin(x))\frac{d}{dx}(x\sin(x)) - x\sin(x)\frac{d}{dx}(x+\sin(x))}{(x+\sin(x))^2}$

$= \frac{(x+\sin(x))(\sin(x)+x\cos(x)) - x\sin(x)(1+\cos x)}{(x+\sin(x))^2} = \frac{x^2\cos(x)+\sin^2(x)}{(x+\sin(x))^2}$

17. $\frac{d}{dx}\left(\frac{(x^2+4)(x-5)}{(x-1)}\right) = \frac{(x-1)\frac{d}{dx}((x^2+4)(x-5)) - (x^2+4)(x-5)\frac{d}{dx}(x-1)}{(x-1)^2}$

$= \frac{(x-1)\left(\left(\frac{d}{dx}(x^2+4)\right)(x-5) + (x^2+4)\frac{d}{dx}(x-5)\right) - (x^2+4)(x-5)}{(x-1)^2}$

$= \frac{(x-1)\left((2x)(x-5) + (x^2+4)\right) - (x^2+4)(x-5)}{(x-1)^2} = 2\frac{(x^3-4x^2+5x+8)}{(x-1)^2}$

19. $\frac{d}{dx}\left(x(x^2+1)(x^3+2x)\right) = \frac{d}{dx}\left(x(x^2+1)\right)(x^3+2x) + x(x^2+1)\frac{d}{dx}(x^3+2x)$

$= \left(\left(\frac{d}{dx}x\right)(x^2+1) + x\frac{d}{dx}(x^2+1)\right)(x^3+2x) + x(x^2+1)(3x^2+2)$

$= \left((x^2+1) + x(2x)\right)(x^3+2x) + x(x^2+1)(3x^2+2) = 6x^5 + 12x^3 + 4x$

21. $\frac{d}{dx}\left(\sin^2(x)\cdot\sec(x)\right) = \frac{d}{dx}\left(\frac{\sin^2(x)}{\cos(x)}\right) = \frac{\cos(x)\frac{d}{dx}\sin^2(x) - \sin^2(x)\frac{d}{dx}\cos(x)}{\cos^2(x)}$

$= \frac{\cos(x)\left(\left(\frac{d}{dx}\sin(x)\right)\sin(x) + \sin(x)\left(\frac{d}{dx}\sin(x)\right)\right) + \sin^3(x)}{\cos^2(x)}$

$= \frac{\left(2\cos^2(x)\sin(x)\right) + \sin^3(x)}{\cos^2(x)} = \sin(x)\left(\frac{2\cos^2(x)}{\cos^2(x)} + \frac{\sin^2(x)}{\cos^2(x)}\right)$

$= \sin(x)\left(2 + \tan^2(x)\right)$

23. We calculate the derivative: $f'(x) = \frac{(x+4)\frac{d}{dx}(3x^2-5) - (3x^2-5)\frac{d}{dx}(x+4)}{(x+4)^2} = \frac{3x^2+24x+5}{(x+4)^2}$.

Its value *at* $x=1$ is $f'(1) = \frac{32}{25}$, and the tangent line to the graph is $y = f'(1)(x-1) + f(1)$ or

$y = \left(\frac{32}{25}\right)(x-1) - \frac{2}{5}$.

25.   We calculate the derivative:

$$f'(x) = \frac{(x+1)\frac{d}{dx}\sin(x) - \sin(x)\frac{d}{dx}(x+1)}{(x+1)^2} = \frac{x\cos(x) + \cos(x) - \sin(x)}{(x+1)^2}.$$

Its value at $x = 0$ is $f'(0) = 1$ and the tangent line to the graph is $y = f'(0)(x-0) + f(0)$ or $y = x$.

27.   Using a forward difference quotient, we find that $f'(4) \approx \dfrac{f(4.1) - f(4)}{4.1 - 4} = \dfrac{6.2 - 5.7}{0.1} = 5.0$.

29.   Using a forward difference quotient, we find that $f'(\pi) \approx \dfrac{f(\pi + 0.01) - f(\pi)}{(\pi + 0.01) - \pi} = \dfrac{0.2}{0.01} = 20.0$

Further Theory and Practice

31.   From the Product Rule, we obtain

$$f'_{n+1}(x) = \frac{d}{dx}(xf_n(x)) = f_n(x) \cdot \frac{d}{dx}x + x\frac{d}{dx}f_n(x) = f_n(x) + xf'_n(x).$$

Let us assume that (3.16) is true for some nonnegative integer n. Then the expression $x\frac{d}{dx}x^n = x(nx^{n-1})$, or $xf'_n(x) = nx^n$ is also true. From equation (3.15) we see that

$$xf'_n(x) = f'_{n+1}(x) - f_n(x), \text{ or } f'_{n+1}(x) - f_n(x) = nx^n.$$

From $f_n(x) = x^n$ we calculate $f'_{n+1}(x) = x^n + nx^n = (n+1)x^n$.

Equation (3.16) is true for $n+1$ as well.

Using this formula we obtain: $\frac{d}{dx}x^{10} = 10 \cdot x^{10-1} = 10x^9$

33. $\frac{d}{dx}\tan(x) = \frac{d}{dx}\left(\dfrac{\sin(x)}{\cos(x)}\right) = \dfrac{\cos(x)\frac{d}{dx}\sin(x) - \sin(x)\frac{d}{dx}\cos(x)}{\cos^2(x)} = \dfrac{\cos^2(x) + \sin^2(x)}{\cos^2(x)}$

$$= \frac{1}{\cos^2(x)} = \sec^2(x)$$

35. $\frac{d}{dx}\sin(2x) = \frac{d}{dx}(2\sin(x)\cos(x)) = 2\left(\left(\frac{d}{dx}\sin(x)\right)\cos(x) + \sin(x)\frac{d}{dx}\cos(x)\right)$

$$= 2(\cos^2(x) - \sin^2(x)) = 2\cos(2x)$$

37. $\left(\dfrac{1}{f}\right)'(3) = \dfrac{-f'(3)}{f(3)^2} = \dfrac{-6}{(-4)^2} = -\dfrac{3}{8}$

$(6f)'(3) = 6 \cdot f'(3) = 36$

39.   Using the product rule, we calculate:

$$\left(f \cdot g^2\right)'(4) = f'(4)g^2(4) + f(4)\left(g^2\right)'(4) = (-5)(1)^2 + 2\left(g^2\right)'(4).$$

Given that $\left(g^2\right)'(4) = g'(4)g(4) + g(4)g'(4) = (-9)(1) + (1)(-9) = -18$ we have

$\left(f \cdot g^2\right)'(4) = -5 + 2(-18) = -41$.

Similarly $\left(f^2\right)'(4) = f'(4)f(4) + f(4)f'(4) = (-5)(2) + (2)(-5) = -20$ and

$\left(g \cdot f^2\right)'(4) = g'(4)f^2(4) + g(4)\left(f^2\right)'(4) = (-9)(2)^2 + (1)(-20) = -56$.

Using $\left(\dfrac{1}{f^2}\right)'(4) = -\left(f^2\right)' \dfrac{4}{\left(f^2(4)\right)^2} = \dfrac{20}{2^4} = \dfrac{5}{4}$, we finally obtain

$$\left(\dfrac{g}{f}\right)'(4) = \dfrac{f(4)g'(4) - g(4)f'(4)}{f(4)^2} = \dfrac{2 \cdot (-9) - (1)(-5)}{2^2} = -\dfrac{13}{4}.$$

41.   We first calculate the derivatives of $f(x)$ and $g(x)$ respectively: $f'(x) = 2x - 7$ and

$g'(x) = \dfrac{-9}{x^2}$. Given that their slopes are supposed to be equal, we obtain

$$2x - 7 = -\dfrac{9}{x^2}$$

or $2x^3 - 7x^2 + 9 = 0$. Upon examination, we see that $x = -1$ is a root. Dividing $2x^3 - 7x^2 + 9$

by $x + 1$ we obtain $2x^2 - 9x + 9 = (2x - 3)(x - 3)$; therefore, $x = 3$ and $x = \dfrac{3}{2}$ are two other

roots. The three values of c are $c = 3$, $c = \dfrac{3}{2}$, and $c = -1$.

43. **a.** The domain of f is $\{x \neq -1\}$.

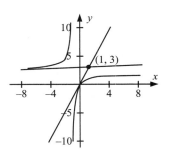

There appear to be two points, x_1 and x_2, with $x_1 < -1 < x_2$ that have the required property. We estimate $x_1 \approx 7$, $x_2 \approx 0$.

b. As $f'(x) = \dfrac{2}{(x+1)^2}$, the slope of the tangent line to the graph of f at $(c, f(c))$ is

$$\dfrac{2}{(c+1)^2}.$$

The common property of the tangents to the graph of f is that they rise; in other words, all tangents have positive slope. This property is consistent with the formula

$$f'(c) = \dfrac{2}{(c+1)^2}.$$

c. The equation of the tangent line to the graph of f at $(c, f(c))$ is

$$y = \dfrac{2}{(c+1)^2}(x-c) + \dfrac{2c}{c+1}.$$

d. The property of the tangent line passing through $(1, 3)$ requires that c must satisfy

$$3 = \left(\dfrac{2}{(c+1)^2}\right)(1-c) + \dfrac{2c}{c+1} \text{ or } 3(c+1)^2 = 2(1-c) + 2c(c+1) \text{ or } c^2 + 6c + 1 = 0.$$

There are two solutions: $c_1 = -3 - 2\sqrt{2}$ and $c_2 = -3 + 2\sqrt{2}$.

45. There is a polynomial $q(x)$ such that $p(x) = (x-\alpha)^3 q(x)$ and $(x-\alpha)$ is not a factor of $q(x)$. By the Product Rule: $p'(x) = 3(x-\alpha)^2 q(x) + (x-\alpha)^3 q'(x) = (x-\alpha)^2 Q(x)$ where $Q(x) = (x-\alpha)q'(x) + 3q(x)$. As $(x-\alpha)$ is not a factor of $q(x)$, it is not a factor of $Q(x)$ either. We conclude that $(x-\alpha)^2$ is a factor of $p'(x)$ but that $(x-\alpha)^3$ is not.

47. For $\alpha \leq 0$ the point 0 is not in the domain of f^2 and therefore $f^2(x)$ is not differentiable at 0. (Note: 0^0 is not defined.) For $0 < \alpha < 1$ we see that $f^2(x) = x^{2\alpha}$, which is differentiable provided that $2\alpha \geq 1$ or $\alpha \geq \dfrac{1}{2}$.

49. The rule $\dfrac{d}{dx}\{f_1(x)+\cdots+f_n(x)\}=\dfrac{d}{dx}f_1(x)+\cdots+\dfrac{d}{dx}f_n(x)$ is only valid when n is a *fixed* positive integer. In fact, $x=\underbrace{1+\cdots+1}_{x \text{ summands}}$ is only meaningful when x is a positive integer.

Therefore, we cannot differentiate the function $\mathbb{R} \ni x \to \underbrace{1+\cdots+1}_{x \text{ summands}}$ with respect to x.

51. $D_0 f(c,\,h)=\dfrac{f\left(c+\frac{h}{2}\right)-f\left(c-\frac{h}{2}\right)}{h}=\dfrac{f\left(c+\frac{h}{2}\right)-f(c)}{2\frac{h}{2}}+\dfrac{f(c)-f\left(c-\frac{h}{2}\right)}{2\frac{h}{2}}$

$=\dfrac{D_+ f\left(c,\,\frac{h}{2}\right)+D_- f\left(c,\,\frac{h}{2}\right)}{2}$

53. If $h(x)\cdot g(x)=1$ for all x, then using the Product Rule we see that

$$h'(x)g(x)+h(x)g'(x)=0\,.$$

Thus, $h'(x)=-h(x)\dfrac{g'(x)}{g(x)}=-\dfrac{g'(x)}{g(x)^2}\,.$

55. **a.** By definition,

$$E(p)=-q'(p)\cdot\dfrac{p}{q(p)}=-\lim_{\Delta p\to 0}\dfrac{q(p+\Delta p)-q(p)}{\Delta p}\cdot\dfrac{p}{q(p)}=-\lim_{\Delta p\to 0}\dfrac{\Delta q/q(p)}{\Delta p/p}\,,$$

and the asserted approximation results when a small Δp is chosen.

b. If p is the standard supermarket price, then $0.31\approx-\dfrac{\Delta q/q(p)}{\Delta p/p}=-\dfrac{\Delta q/q(p)}{0.02}$ or

$\dfrac{\Delta q}{q(p)}=-(0.31)(0.02)=-0.0062$. Potato consumption will therefore decrease by about 0.62%.

c. $E(70)\approx-\dfrac{\left(q(80)-q(70)\right)/300}{10/70}=-\dfrac{\left(200-300\right)/300}{10/70}=2.33$

$E(90)\approx-\dfrac{\left(q(80)-q(90)\right)/96}{(-10)/90}=-\dfrac{\left(200-96\right)/96}{(-10)/90}=9.75$

$E(80)\approx-\dfrac{\left(q(90)-q(70)\right)/200}{20/80}=-\dfrac{\left(96-300\right)/200}{20/80}=4.08$

Calculator/Computer Exercises

57.

n	Forward	Backward	Central
1	0.48943	−0.48943	0.00000
2	0.04934	−0.04934	0.00000
3	0.00493	−0.00493	0.00000
4	0.00049	−0.00049	0.00000
5	0.00005	−0.00005	0.00000

59.

n	Forward	Backward	Central
1	0.24280	0.24401	0.24343
2	0.24337	0.24349	0.24343
3	0.24343	0.24344	0.24343
4	0.24344	0.24343	0.24343
5	0.24350	0.24340	0.24350

61. $f'\left(\dfrac{\pi}{6}\right) = \cos\left(\dfrac{\pi}{6}\right) = \dfrac{\sqrt{3}}{2}$

   a.   0.86603

   b. $h = \dfrac{1}{1000}$ suffices

63. $f'\left(\dfrac{1}{\sqrt[4]{2}}\right) = -\sqrt{2}$

   a.   −1.41421

   b. $h = \dfrac{1}{1000}$ suffices

65. $f'(x) = -\sin(x)$

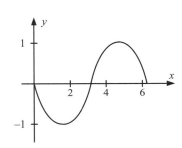

67. $L(2.70) \approx 0.99322$ and

$L(2.73) \approx 1.00435$; $a_0 \approx 2.178$ and

$\dfrac{d}{dx} a_0^x\Big|_{x=0} \approx L(2.718) = 1$

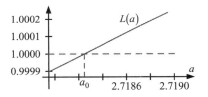

Section 3.4   **Differentiation of Some Basic Functions**

Problems for Practice

1. $f'(x) = 8\frac{d}{dx}\left(x^{10}\right) - 6\frac{d}{dx}\left(x^{-5}\right) = 80x^9 + 30x^{-6}$

3. $H'(x) = \cot(x)\frac{d}{dx}\left(\csc(x)\right) + \csc(x)\frac{d}{dx}\left(\cot(x)\right) = -\csc(x)\cot^2(x) - \csc^3(x)$

$= -\csc(x)\left(\cot^2(x) + \csc^2(x)\right) = -\csc(x)\left(2\cot^2(x) + 1\right)$

5. $g'(x) = \dfrac{\left(\sqrt{x}+1\right)\frac{d}{dx}\left(\sec(x)\right) - \sec(x)\frac{d}{dx}\left(\sqrt{x}+1\right)}{\left(\sqrt{x}+1\right)^2} = \dfrac{\left(\sqrt{x}+1\right)\sec(x)\tan(x) - \sec(x)\left(1/2\sqrt{x}\right)}{\left(\sqrt{x}+1\right)^2}$

$= \dfrac{\sec(x)\left(\left(2x + 2\sqrt{x}\right)\tan(x) - 1\right)}{2\sqrt{x}\left(\sqrt{x}+1\right)^2}$

7. $f'(x) = \sec(x)\tan(x) - \sec^2(x)$

9. $H'(x) = \cot(x)\frac{d}{dx}(x) + x\frac{d}{dx}(\cot(x)) - \frac{d}{dx}(\csc(x)) = \cot(x) - x\csc^2(x) + \csc(x)\cot(x)$

11. $g'(x) = \sec(x)\frac{d}{dx}(\tan(x)) + \tan(x)\frac{d}{dx}(\sec(x)) = \sec^3(x) + \sec(x)\tan^2(x)$

13. $f'(x) = e^x\frac{d}{dx}(x^{-5}) + x^{-5}\frac{d}{dx}(e^x) = e^x(-5x^{-6}) + x^{-5}e^x = x^{-6}e^x(x-5)$

15. $H'(x) = 2\tan(x)\frac{d}{dx}(\tan(x)) = 2\tan(x)\sec^2(x)$

17. $g'(x) = \dfrac{(2+\tan(x))\frac{d}{dx}(x) - x\frac{d}{dx}(2+\tan(x))}{(2+\tan(x))^2} = \dfrac{2+\tan(x) - x\sec^2(x)}{(2+\tan(x))^2}$

19. $f'(x) = \dfrac{(e^x+\sin(x))\frac{d}{dx}(x^2) - x^2\frac{d}{dx}(e^x+\sin(x))}{(e^x+\sin(x))^2} = \dfrac{(e^x+\sin(x))(2x) - x^2(e^x+\cos(x))}{(e^x+\sin(x))^2}$

$$= \dfrac{x((2-x)e^x + 2\sin(x) - x\cos(x))}{(e^x+\sin(x))^2}$$

21.   The slope is $\frac{d}{dx}(x^{-3}\sin(x)) = -3x^{-4}\sin(x) + x^{-3}\cos(x)$. At $x = \pi$ we find the slope to be

$-3\pi^{-4}\sin(\pi) - \pi^{-3}\cos(\pi) = \pi^{-3}$. As $f(\pi) = 0$, the point-slope form of the tangent line is

$y = -\dfrac{x-\pi}{\pi^3}$.

23.   The slope is $\frac{d}{dx}(3\tan(x)\sec(x)) = 3\sec^3(x) + 3\sec(x)\tan^2(x)$. At $x = \dfrac{\pi}{3}$ this slope equals 42.

Given that $f\left(\dfrac{\pi}{3}\right) = 6\sqrt{3}$, the tangent line is $y = 42\left(x - \dfrac{\pi}{3}\right) + 6\sqrt{3}$.

25.   The slope is $\frac{d}{dx}(x\tan(x)) = \tan(x) + x\sec^2(x)$. At $x = \pi$ it equals π. Given that $f(\pi) = 0$,

the tangent line is $y = \pi(x - \pi)$.

27. $f'(x) = 2 + e^x$

29. $f'(x) = \cos(x)\frac{d}{dx}(e^{-x}) + e^{-x}\frac{d}{dx}(\cos(x)) = -e^{-x}(\cos(x) + \sin(x))$

Further Theory and Practice

31. As $\frac{d}{dx}\left(x^7\right)=7x^6$, $\frac{d}{dx}\left(x^2\right)=2x$, $\frac{d}{dx}(x)=1$, $\frac{d}{dx}(C)=0$, the desired polynomial is

$$p(x)=x^7-4\left(\frac{x^2}{2}\right)+6x+C,\text{ or } p(x)=x^7-2x^2+6x+C \ (C \text{ can be any constant}).$$

33. As $\frac{d}{dx}\left(x^9\right)=9x^8$, $\frac{d}{dx}\left(x^6\right)=6x^5$, $\frac{d}{dx}\left(x^2\right)=2x$, $\frac{d}{dx}(C)=0$, the desired polynomial is

$$p(x)=\frac{x^9}{9}+x^6-\frac{3x^2}{2}+C \ (C \text{ can be any constant}).$$

35. As $\frac{d}{dx}(\sin(x))=\cos(x)$ and $\frac{d}{dx}(C)=0$, the desired function is $\sin(x)+C$ (C can be any constant).

37. As $\frac{d}{dx}(\cot(x))=-\csc^2(x)$, $\frac{d}{dx}(C)=0$, the desired function is $8\cot(x)+C$ (C can be any constant).

39. Let $f(x)=a\cos(x)+b\sin(x)$. Then $f'(x)=-a\sin(x)+b\cos(x)$ and

$$f(x)+f'(x)=(a+b)\cos(x)+(b-a)\sin(x).$$

Hence $a+b=3$, $b-a=-6$, from which we obtain: $a=\frac{9}{2}$, $b=-\frac{3}{2}$ and

$$f(x)=\left(\frac{9}{2}\right)\cos(x)+\left(-\frac{3}{2}\right)\sin(x).$$

41. As $\frac{dy}{dx}(t)=e^t$, the equation of the tangent line is $y=e^t(x-t)+e^t$.

The x-intercept is $t-1$ and the y-intercept is $e^t(1-t)$.

43. As $\frac{dy}{dx}=C\frac{d}{dx}\left(x^p\right)=Cpx^{p-1}=\frac{p}{x}\left(Cx^p\right)$, we have $\frac{dy}{dx}=p\frac{y}{x}$.

45. $f'(x)=\dfrac{(\cos(x)+\sin(x))\frac{d}{dx}(\cos(x)-\sin(x))-(\cos(x)-\sin(x))\frac{d}{dx}(\cos(x)+\sin(x))}{(\cos(x)+\sin(x))^2}$

$=\dfrac{(\cos(x)+\sin(x))(-\sin(x)-\cos(x))-(\cos(x)-\sin(x))(-\sin(x)+\cos(x))}{(\cos(x)+\sin(x))^2}$

$=\dfrac{-2}{(\cos(x)+\sin(x))^2}$

$f'\left(\frac{\pi}{4}\right)=-1$

47. Let $h(x)=x^5+2x^4-5x^3+x^2+x-3$, $g(x)=2x^3-4x^2+7x-2$, then $f(x)=\dfrac{h(x)}{g(x)}$ and

$f'(x)=\dfrac{g(x)\cdot h'(x)-h(x)\cdot g'(x)}{g^2(x)}$. As $h'(x)=5x^4+8x^3-15x^2+2x+1$ and

$g'(x)=6x^2-8x+7$, we have $f'(1)=\dfrac{g(1)\cdot h'(1)-h(1)\cdot g'(1)}{g^2(1)}=\dfrac{3\cdot1-(-3)\cdot5}{9}=2$.

49. $\dfrac{d}{dx}f_n(x)=e^x\dfrac{d}{dx}(x^n)+x^n\dfrac{d}{dx}(e^x)=ne^xx^{n-1}+x^ne^x=nf_{n-1}(x)+f_n(x)$

$f_n'(x)=nf_{n-1}(x)+f_n(x)$.

51. As $y-\tan(t)=f'(t)(x-t)$, we have $x=\left(\dfrac{1}{f'(t)}\right)(y-\tan(t))+t$ and

$$M(t)=\dfrac{1}{f'(t)}=\cos^2(t);$$

$x=\lim_{t\to(\pi/2)^-}\left(M(t)(y-\tan(t))+t\right)=\dfrac{\pi}{2}$, which is an asymptote.

53. $\lim_{h\to0}\dfrac{e^{h/2}-e^{-h/2}}{h}=\dfrac{1}{2}\lim_{h\to0}\dfrac{\left(e^{h/2}-1\right)+\left(1-e^{-h/2}\right)}{h/2}=\dfrac{1}{2}\lim_{h\to0}\dfrac{e^{h/2}-1}{h/2}+\dfrac{1}{2}\lim_{h\to0}\dfrac{e^{-h/2}-1}{-h/2}$

$=\dfrac{1}{2}\dfrac{de^x}{dx}\bigg|_{x=0}+\dfrac{1}{2}\dfrac{de^x}{dx}\bigg|_{x=0}=1$

55. For $x>0$: $|x|^p=x^p$, $\dfrac{d}{dx}|x|^p=px^{p-1}=\dfrac{px^p}{x}=\dfrac{p|x|^p}{x}$, for $x<0$: $|x|^p=(-x)^p$,

$\dfrac{d}{dx}|x|^p=-p(-x)^{p-1}=\dfrac{p(-x)^p}{x}=\dfrac{p|x|^p}{x}$.

57. For the product function: $\deg(p\cdot q)=n+m$, hence $\deg\left((p\cdot q)'\right)=n+m-1$.

For the product function: $\deg(p\circ q)=nm$, therefore, $\deg\left((p\circ q)'\right)=nm-1$.

The degree of the numerator of $\left(\dfrac{p}{q}\right)'$ is $n+m-1$; the degree of the denominator is $2m$.

59. $g'(x) = \lim\limits_{h \to 0} \dfrac{g(x+h) - g(x)}{h} = \lim\limits_{h \to 0} \dfrac{e^{kx+kh} - e^{kx}}{h} = e^{kx} \lim\limits_{h \to 0} \dfrac{e^{kh} - 1}{h} = e^{kx} g'(0) = g(x) g'(0)$

61. The derivatives $f'(t)$ and $g'(t)$ are the rates of change of mass and distance respectively with respect to time; therefore $f'(t)g(t)$ and $f(t)g'(t)$ have the dimension of mass times distance over time. The derivative $(fg)'(t)$ has the same dimension. On the contrary, $f'(t)g'(t)$ has the dimension of mass times distance over time squared and cannot enter the expression for $(fg)'(t)$.

63. $\dfrac{d}{dx}\left(\cot(x)\right) = \dfrac{d}{dx}\left(\dfrac{\cos(x)}{\sin(x)}\right) = \dfrac{\sin(x)\frac{d}{dx}\left(\cos(x)\right) - \cos(x)\frac{d}{dx}\left(\sin(x)\right)}{\sin^2(x)} = \dfrac{-\sin^2(x) - \cos^2(x)}{\sin^2(x)}$

$\qquad = -\csc^2(x)$

Calculator/Computer Exercises

65.

h	$D_0f(c,h)$
0.1	−1.000416715
0.01	−1.000004180
0.001	−1.000000000
0.0001	−1.000000000
0.00001	−1.000000000

Hence $\dfrac{d}{dx} e^{-x}\Big|_{x=0} = -1$.

67. The set on which f increases is the same as the set on which $f' > 0$. The set on which f decreases is the same as the set on which $f' < 0$. The points at which f changes from increasing to decreasing, or from decreasing to increasing, are the points at which $f' = 0$.

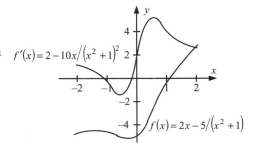

Interval where f increases	Interval where f decreases	Point(s) at which f has a horizontal tangent
(−2, −1.207), (−0.2198, 2)	(−1.207, −0.2198)	{−1.207, −0.2198}
Interval where $f' > 0$	Interval where $f' < 0$	Point(s) at which $f' = 0$
(−2, −1.207), (−0.2198, 2)	(−1.207, −0.2198)	{−1.207, −0.2198}

69. The inferences are the same as those stated in Exercise 67.

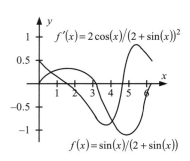

Interval where f increases	Interval where f decreases	Point(s) at which f has a horizontal tangent
$(0, \pi/2)$, $(3\pi/2, 2\pi)$	$(\pi/2, 3\pi/2)$	$\{\pi/2, 3\pi/2\}$
Interval where $f' > 0$	Interval where $f' < 0$	Point(s) at which $f' = 0$
$(0, \pi/2)$, $(3\pi/2, 2\pi)$	$(\pi/2, 3\pi/2)$	$\{\pi/2, 3\pi/2\}$

71. $g_k'(0) = k$

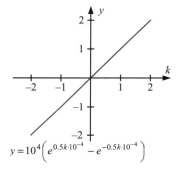

$$y = 10^4 \left(e^{0.5k \cdot 10^{-4}} - e^{-0.5k \cdot 10^{-4}} \right)$$

73. The function e^x grows faster than x^{20} on $(88.083\ldots, \infty)$

It catches up when $e^x = x^{20}$ for $x \approx 89.9951$.

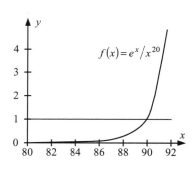

75.   **a.**   As $\tau\left(p + p_0\mu^2\right) = p\mu^2\tau_0 + p_0\tau_0 - 2\mu^2\Delta$, we have $\tau = \dfrac{p\mu^2\tau_0 + p_0\tau_0 - 2\mu^2\Delta}{p + p_0\mu^2}$ and

$$\frac{d\tau}{dp} = \frac{\left(p + p_0\mu^2\right)\frac{d}{dp}\left(p\mu^2\tau_0 + p_0\tau_0 - 2\mu^2\Delta\right) - \left(p\mu^2\tau_0 + p_0\tau_0 - 2\mu^2\Delta\right)\frac{d}{dp}\left(p + p_0\mu^2\right)}{\left(p + p_0\mu^2\right)^2}$$

$$= \frac{\left(p + p_0\mu^2\right)\left(\mu^2\tau_0\right) - \left(p\mu^2\tau_0 + p_0\tau_0 - 2\mu^2\Delta\right)}{\left(p + p_0\mu^2\right)^2} = \frac{\mu^4 p_0\tau_0 - p_0\tau_0 + 2\mu^2\Delta}{\left(p + p_0\mu^2\right)^2}.$$

b.   We calculate: $\tau_1 = \dfrac{p_1\mu^2\tau_0 + p_0\tau_0 - 2\mu^2\Delta}{p_1 + p_0\mu^2}$, $\tau'(p_1) = \dfrac{\mu^4 p_0\tau_0 - p_0\tau_0 + 2\mu^2\Delta}{\left(p_1 + p_0\mu^2\right)^2}$. The

equation of the tangent line is $\tau = \tau_1 + \tau'(p_1)(p - p_1)$.

c. $\tau(p_0) = \dfrac{\tau_0\left(p_0 + p_0\mu^2\right) - 2\mu^2\Delta}{p_0 + p_0\mu^2} = \tau_0 - \dfrac{2\mu^2\Delta}{p_0 + p_0\mu^2} > \tau_0$ for $\Delta < 0$.

d.   From $\tau_0 = \tau_1 + \tau'(p_1)(p_0 - p_1)$ we have a quadratic equation:

$$p_1^2\left(\mu^2\tau_0 - \tau_0\right) + p_1\left(2p_0\tau_0 - 4\mu^2\Delta - 2p_0\tau_0\mu^2\right)$$
$$+ p_0^2\tau_0\mu^2 - 2p_0\mu^4\Delta - p_0^2\tau_0 + 2p_0\mu^2\Delta = 0.$$

The roots $Q = (p_1, \tau_1)$ correspond to the coordinates of the Chapman-Jouguet points.

e.   For the given parameters the Hugoniot curve is

defined by the equation $\tau = \dfrac{0.425p + 4.57}{p + 0.525}$.

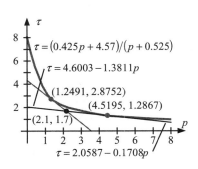

The equation for the Chapman-Jouguet points
is $1.257p_1^2 - 7.355p_1 + 7.19775 = 0$. The first
root is $p_1 = 1.2491$, then $\tau_1 = 2.8752$, and the
tangent line is $\tau = 4.6003 - 1.3811p$. The
second root is $p_1 = 4.5195$, then $\tau_1 = 1.2867$
and the tangent line is $\tau = 2.0587 - 0.1708p$. The Chapman-Jouguet points are
$(1.2491, 2.8752)$ and $(4.5195, 1.2867)$ and the tangent lines at these points are
Rayleigh lines.

Section 3.5   The Chain Rule

Problems for Practice

1. $f'(x) = \frac{d}{du} u^{10}\Big|_{u=x^2+3x} \cdot \frac{d}{dx}\left(x^2+3x\right) = 10\left(x^2+3x\right)^9 (2x+3)$

3. $H'(x) = \frac{d}{du} \tan(u)\Big|_{u=\sin(x)} \cdot \frac{d}{dx}\sin(x) = \sec^2\left(\sin(x)\right)\cos(x)$

5. $g'(x) = \frac{d}{du} \sec(u)\Big|_{u=\cos(x)} \cdot \frac{d}{dx}\cos(x) = -\sec\left(\cos(x)\right)\tan\left(\cos(x)\right)\sin(x)$

7. $f'(x) = \frac{d}{du} \sin(u)\Big|_{u=x^8+x^2} \cdot \frac{d}{dx}\left(x^8+x^2\right) = \left(8x^7+2x\right)\cos\left(x^8+x^2\right)$

9. $H'(x) = \frac{d}{du} u^6\Big|_{u=\sin(x)+x^2} \cdot \frac{d}{dx}\left(\sin(x)+x^2\right) = 6\left(\sin(x)+x^2\right)^5 \left(\cos(x)+2x\right)$

11. $g'(x) = \frac{d}{du} \tan(u)\Big|_{u=x^3} \cdot \frac{d}{dx} x^3 = 3x^2 \sec^2\left(x^3\right)$

13. $f'(x) = \frac{d}{du} \exp(u)\Big|_{u=x^2+1} \cdot \frac{d}{dx}\left(x^2+1\right) = 2x\exp\left(x^2+1\right)$

15. $H'(x) = \frac{d}{du} \cos(u)\Big|_{u=\sqrt{x}} \cdot \frac{d}{dx}\sqrt{x} = -\frac{1}{2\sqrt{x}}\sin\left(\sqrt{x}\right)$

17. $g'(x) = \frac{d}{du} \tan(u)\Big|_{u=1-7x} \cdot \frac{d}{dx}\left(1-7x\right) = -7\sec^2\left(1-7x\right)$

19. $(f \circ g)' = \frac{d}{du} f(u)\Big|_{u=g(x)} \cdot \frac{d}{dx}g(x) = \left(4u^3+14u\right)\Big|_{u=\sqrt{x}} \cdot \frac{d}{dx}\sqrt{x} = 2x+7$

   $(g \circ f)' = \frac{d}{du} g(u)\Big|_{u=f(x)} \cdot \frac{d}{dx}f(x) = \left(\frac{1}{2\sqrt{u}}\right)\Big|_{u=x^4+7x^2} \cdot \frac{d}{dx}\left(x^4+7x^2\right) = \frac{2x^3+7x}{\sqrt{x^4+7x^2}}$

21. $(f \circ g)' = \frac{d}{du} f(u)\Big|_{u=g(x)} \cdot \frac{d}{dx}g(x) = \frac{1}{(u+1)^2}\Big|_{u=x^3} \cdot \frac{d}{dx}x^3 = \frac{3x^2}{\left(x^3+1\right)^2}$

   $(g \circ f)' = \frac{d}{du} g(u)\Big|_{u=f(x)} \cdot \frac{d}{dx}f(x) = 3u^2\Big|_{u=x/(x+1)} \cdot \frac{d}{dx}\left(\frac{x}{x+1}\right) = \frac{3x^2}{(x+1)^4}$

23. $(g \circ f)'(2) = g'(f(2)) \cdot f'(2) = 8 \cdot g'(3) = 8 \cdot 2 = 16$

25. $(g \circ f)'(6) = g'(f(6)) \cdot f'(6) = 0 \cdot g'\left(\dfrac{1}{2}\right) = 0 \cdot \left(\dfrac{1}{3}\right) = 0$

Further Theory and Practice

27. $F'(u) = 1 + 2u$, $F'(G \circ H(x)) = 1 + 2\sin\left(\dfrac{x}{x^2+3}\right)$, $F'(G \circ H(2)) = 1 + 2\sin\left(\dfrac{2}{7}\right)$,

$G'(s) = \cos(s)$, $G'(H(x)) = \cos\left(\dfrac{x}{x^2+3}\right)$, $G'(H(2)) = \cos\left(\dfrac{2}{7}\right)$, $H'(x) = \dfrac{3 - x^2}{\left(x^2+3\right)^2}$,

$H'(2) = -\dfrac{1}{49}$,

$$(F \circ G \circ H)'(2) = \dfrac{d}{ds}F(s)\Big|_{s=G(2) \circ H(2)} \cdot \dfrac{d}{du}G(u)\Big|_{u=H(2)} \cdot \dfrac{d}{dx}H(x)\Big|_{x=2}$$

$$= \left(-\dfrac{1}{49}\right)\left(1 + 2\sin\left(\dfrac{2}{7}\right)\right)\cos\left(\dfrac{2}{7}\right)$$

29. $(F \circ F \circ F)' = F'(F \circ F) \cdot (F \circ F)' = F'(F \circ F)F'(F)F'$, $(F \circ F)(x) = \left(x + x^2\right)\left(1 + x + x^2\right)$,

$F'(F \circ F)(x) = 1 + 2(F \circ F)(x) = 1 + 2\left(x + x^2\right)\left(1 + x + x^2\right)$, $F'(F \circ F)(6) = 3613$,

$F'(F(x)) = 1 + 2F(x) = 1 + 2\left(x + x^2\right)$, $F'(F(6)) = 85$, $F'(x) = 1 + 2x$, $F'(6) = 13$,

$(F \circ F \circ F)'(6) = 3613 \cdot 85 \cdot 13 = 3992365$

31.   Whereas $x(t)$ is measured in meters and $\sin(\omega t)$ is unitless, we assume that A also carries the units of meters. The argument of $\sin(\omega t)$ has to be unitless, so ωt is unitless. We, therefore measure ω as 1/second(s).

Then $x'(t) = \omega A \cos(\omega t)$ is measured in meters/seconds as it should be because $x'(t)$ is a velocity.

The role of the Chain Rule is to introduce a factor ω, resulting in the right-hand side becoming the *rate of change of the distance with respect to time*.

33. Arguments of exponential and trigonometric functions are unitless, so both kt and ωt are unitless. The units of the positive constants k and ω, have the dimension of $1/\text{second}$. Given that the trigonometric and exponential functions are unitless and $x(t)$ is measured in meters, A is also measured in meters. Also, the derivative is $x'(t) = Ae^{-kt}(-k\sin(\omega t) + \omega\cos(\omega t))$.

The dimension of the expression in parentheses is $1/\text{second}$, therefore the dimension of the derivative is $\text{meters}/\text{second}$, which is dimensionally consistent with $x'(t)$ being an instantaneous velocity.

35. The rate at which the mass of ^{14}C is changing is $m'(t) = -0.121286 \cdot e^{-0.00121286t}$.

Then $m'(t) = -0.00121286 \cdot m(t)$. We see that $\lim_{t \to \infty} m(t) = \lim_{t \to \infty} m'(t) = 0$.

Due to the latter limit, the decay process slows down as the mass decreases and in theory takes infinitely long time.

37. Let a be the length of leg of the isosceles right triangle and L be the length of its hypotenuse. Then $L = \sqrt{2}a$, and the area of the triangle is $A = \dfrac{a^2}{2} = \dfrac{L^2}{4}$. Then

$$\frac{dA}{dt} = \frac{dA}{dL}\cdot\frac{dL}{da}\cdot\frac{da}{dt} = \frac{L}{2}\cdot\sqrt{2}\cdot\frac{da}{dt}.$$

Given that $L = 8$ in. and $\dfrac{da}{dt} = 2$ in./min, the rate of the change of the area is

$$\frac{dA}{dt} = \frac{8 \text{ in.}}{2}\cdot\sqrt{2}\cdot 2 \text{ in./min} = 8\sqrt{2} \text{ in.}^2/\text{min}.$$

39. The difference between $f'(t^2)$ and $g'(t)$ is the order in which the operations of differentiation and evaluation are performed. In calculating $f'(t^2)$, we first differentiate f and then evaluate the derived function at t^2. In calculating $g'(t)$, we first evaluate f at t^2 and then differentiate the resulting composed function. Analytically, the difference shows up in the factor $2t$ in the formula $g'(t) = 2t\,f'(t^2)$. This formula gives us $g'(\sqrt{t}) = 2\sqrt{t}f'(t)$, so that $g'(\sqrt{t})$ and $f'(t)$ also differ. We can explain this difference directly from the Chain Rule. Notice that, for positive t, we have $f(t) = g(\sqrt{t})$. The rate of change of f with respect to t depends not only on the rate of change of g at $\sqrt{t}$ but also on the rate of change of $\sqrt{t}$.

41. $f'(x) = \dfrac{d}{du}u^5\Big|_{u=\sec(7x)} \cdot \dfrac{d}{du}\sec(v)\Big|_{v=7x} \cdot \dfrac{d}{dx}(7x) = 35\sec^5(7x)\tan(7x)$

43. $f'(x) = \frac{d}{du}e^u\big|_{u=\cos(2x)} \cdot \frac{d}{dv}\cos(v)\big|_{v=2x} \cdot \frac{d}{dx}(2x) = -2\sin(2x)\exp(\cos(2x))$

45. $f'(x) = \frac{d}{du}\cos(u)\big|_{u=\sqrt{2x^2+3}} \cdot \frac{d}{dv}\sqrt{v}\big|_{v=2x^2+3} \cdot \frac{d}{dx}(2x^2+3) = -\frac{2x}{\sqrt{2x^2+3}}\sin\left(\sqrt{2x^2+3}\right)$

47. $f'(x) = \frac{d}{du}\sqrt{u}\big|_{u=\sin^2(3x)+\cos^2(3x)}$

$\cdot\left(\frac{d}{dv}v^2\big|_{v=\sin(3x)} \cdot \frac{d}{dw}\sin(w)\big|_{w=3x} \cdot \frac{d}{dx}(3x) + \frac{d}{dv}v^2\big|_{v=\cos(3x)} \cdot \frac{d}{dw}\cos(w)\big|_{w=3x} \cdot \frac{d}{dx}(3x)\right)$

$= \frac{2\sin(3x)\cos(3x)\cdot 3 - 2\cos(3x)\sin(3x)\cdot 3}{2\sqrt{\sin^2(3x)+\cos^2(3x)}} = 0$

Alternatively, we can obtain $f'(x) = 0$ immediately from the observation that $f(x) = 1$ for all x.

49. $f'(x) = \frac{d}{du}\sqrt{u}\big|_{u=\tan(5x)+\sqrt{2x+1}} \cdot \left(\frac{d}{dv}\tan(v)\big|_{v=5x} \cdot \frac{d}{dx}(5x) + \frac{d}{dv}\sqrt{v}\big|_{v=2x+1} \cdot \frac{d}{dx}(2x+1)\right)$

$= \frac{5\sec^2(5x)+1/\sqrt{2x+1}}{2\sqrt{\tan(5x)+\sqrt{2x+1}}} = \frac{5\sqrt{2x+1}\cdot\sec^2(5x)+1}{2\sqrt{2x+1}\sqrt{\tan(5x)+\sqrt{2x+1}}}$

51. Observe that, for any value of c, we have $f(0) = (c\cdot 0+1)^3 = 1$. The derivative $f'(0)$ exists if and only if the two-sided limit $\lim_{h\to 0}\frac{f(0+h)-f(0)}{h}$ exists. We calculate each one-sided limit

$$\lim_{h\to 0^+}\frac{f(0+h)-f(0)}{h} = \lim_{h\to 0^+}\frac{f(h)-1}{h} = \lim_{h\to 0^+}\frac{h+1-1}{h} = 1$$

and

$$\lim_{h\to 0^-}\frac{f(0+h)-f(0)}{h} = \lim_{h\to 0^-}\frac{f(h)-1}{h} = \lim_{h\to 0^-}\frac{(ch+1)^3-1}{h} = \lim_{h\to 0^-}(c^3h^2+3c^2h+3c) = 3c.$$

We conclude that the two-sided limit $\lim_{h\to 0}\frac{f(0+h)-f(0)}{h}$ exists if and only if $3c = 1$. By setting $c = \frac{1}{3}$ we make $f(x)$ differentiable at $x=1$ and this is the only specification of c that results in differentiability.

53. **a.** The initial population size is P_0.

The limiting size is $\lim\limits_{t \to \infty} P(t) = M$.

b. $P'(t) = \frac{d}{du}\left(\frac{P_0 M}{u}\right)\Big|_{u=P_0+(M-P_0)e^{-kMt}} \cdot \frac{d}{dt}\left(P_0 + (M-P_0)e^{-kMt}\right)$

$= P_0 M^2 (M - P_0)\dfrac{ke^{-kMt}}{\left(P_0 + (M-P_0)e^{-kMt}\right)^2} = k \cdot P(t) \cdot M(M - P_0)\dfrac{ke^{-kMt}}{P_0 + (M-P_0)e^{-kMt}}$

$M - P(t) = M - \dfrac{P_0 M}{P_0 + (M-P_0)e^{-kMt}} = \dfrac{M\left(P_0 + (M-P_0)e^{-kMt}\right) - P_0 M}{P_0 + (M-P_0)e^{-kMt}}$

$= M(M - P_0)\dfrac{ke^{-kMt}}{P_0 + (M-P_0)e^{-kMt}}$

$\therefore\ P'(t) = k \cdot P(t) \cdot (M - P(t))$

c. We calculate the limit: $\lim\limits_{t \to \infty} P'(t) = k \cdot \lim\limits_{t \to \infty} P(t) \cdot \lim\limits_{t \to \infty}(M - P(t)) = k \cdot M \cdot 0 = 0$. The

population growth rate has a finite limit at $t \to \infty$.

55. The terminal velocity is $v_\infty = \sqrt{\dfrac{g}{\kappa}}$.

The acceleration of the object is

$$v'(t) = \frac{d}{du}\left(\sqrt{\frac{g}{k}}\left(\frac{u-1}{u+1}\right)\right)\Bigg|_{u=\exp\left(2t\sqrt{g\kappa}\right)} \cdot \frac{d}{dv}e^{v}\Big|_{v=2t\sqrt{g\kappa}} \cdot \frac{d}{dt}\left(2t\sqrt{g\kappa}\right) = 4g\frac{\exp\left(2t\sqrt{g\kappa}\right)}{\left(\exp\left(2t\sqrt{g\kappa}\right)+1\right)^2}.$$

The limit of the acceleration is $\lim\limits_{t \to \infty} v'(t) = 0$.

$$g - k \cdot v^2 = g - k \cdot \left(\frac{g}{k}\right)\frac{\left(\exp\left(2t\sqrt{g\kappa}\right)-1\right)^2}{\left(\exp\left(2t\sqrt{g\kappa}\right)+1\right)^2} = g\frac{\left(\exp\left(2t\sqrt{g\kappa}\right)+1\right)^2 - \left(\exp\left(2t\sqrt{g\kappa}\right)-1\right)^2}{\left(\exp\left(2t\sqrt{g\kappa}\right)+1\right)^2}$$

$$= 4g\frac{\exp\left(2t\sqrt{g\kappa}\right)}{\left(\exp\left(2t\sqrt{g\kappa}\right)+1\right)^2} = v'$$

Calculator/Computer Exercises

57. We calculate $f'(x_0) = \left(3x_0^2 + 1\right)\sec^2\left(x_0^3 + x_0 + 1\right)$ and solve $f'(x_0) = 20$ to find $x_0 = 0.64466$ to five decimal places.

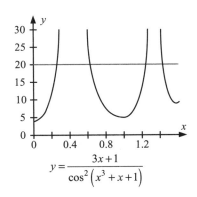

$$y = \frac{3x+1}{\cos^2\left(x^3 + x + 1\right)}$$

59. Using the Chain Rule, we calculate

$$f'(x) = -2\pi\cos(x)\sin(x) = -\pi\sin(2x).$$

We see that this function attains its maximum at $x = \dfrac{3\pi}{4} = 2.35619\ldots$. Similarly, we see that $g'(x)$

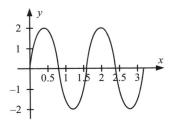

attains its maximum at $x = 0$. So f and g increase most rapidly at $\dfrac{3\pi}{4}$ and 0 respectively. By

plotting $(g \circ f)'(x)$ and zooming in to its maximum values, we see that $g \circ f$ increases most rapidly at $0.445738\ldots$ and $2.01653\ldots$. As discussed in the preceding exercise,

$$(g \circ f)'(x) = g'(f(x))f'(x)$$

may attain its maximum at a point where neither of its factors $g'(f(x))$ and $f'(x)$ is maximized.

61.   Given that $m = e^{-k \cdot t}$, we obtain the equation $0.5 = e^{-k \cdot 4.51 \cdot 10^9}$. Solving it, we have:

$$-0.6931 = -4.51 \cdot 10^9 \cdot k,$$

or $k = 1.537 \times 10^{-10}$ 1/yr. The rate of change of this mass of ^{238}U is

$$m' = -1.537 \cdot 10^{-10} \cdot e^{-1.537 \cdot 10^{-10} t}.$$

When $m = 0.125$ g, $t = -10^{10} \cdot \log(0.125)/1.537 = 1.353 \times 10^{10}$ yr, and

$$m' = -1.537 \cdot 10^{-10} \cdot e^{-1.537 \cdot 10^{-10} \cdot 1.353 \cdot 10^{10}} \text{ g/yr}.$$

Therefore, $m' = -1.921 \times 10^{-11}$ g/yr.

63.   **a.**   **b.**

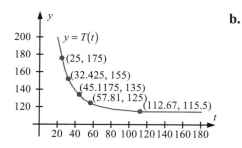

c.

T °F	175	155	135	125	115.5
$T'(t)$	−3.277	−2.184	−1.092	−0.5461	−0.0273

65.

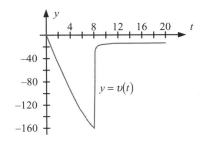

The function v is everywhere continuous, and differentiable at all points except $t = 8$.

The function $v'(t)$ is continuous and differentiable in its domain, which does not include the point $t = 8$ as illustrated by the figure.

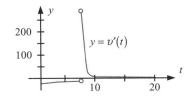

Section 3.6   Inverse Functions and the Natural Logarithm

Problems for Practice

1. The function is invertible: $f^{-1} : [1, \infty) \to [0, \infty)$; $f^{-1}(t) = \sqrt{t-1}$.

3. The function is invertible: $f^{-1} : [0, 2] \to [0, 1]$; $f^{-1}(t) = \dfrac{\sqrt{1 + 4t} - 1}{2}$.

5. The function is invertible: $f^{-1} : [1, 1000) \to [1, 10)$; $f^{-1}(t) = t^{1/3}$.

7. The function is invertible: $h^{-1} : \left(1, \dfrac{16}{15}\right) \to (4, \infty)$; $h^{-1}(t) = \sqrt{\dfrac{t}{t-1}}$.

9. The function is invertible: $h^{-1} : (2, 3) \to (1, 6)$; $h^{-1}(t) = t^2 - 3$.

11. The function is not invertible, because g is not onto; its image is $\left(0, \dfrac{1}{2}\right]$.

13.   **a.**   The function is invertible.   **b.**   The function is not invertible.
                                                                c.   The function is not invertible.
                                                                d.   The function is not invertible.

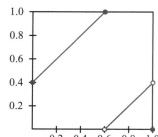

e.   The function is invertible.   **f.**   The function is invertible.

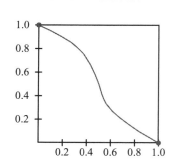   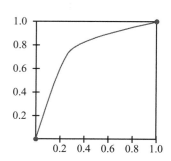

15. $\left(f^{-1}\right)'(4) = \dfrac{1}{f'\left(f^{-1}(4)\right)} = \dfrac{1}{f'(1)} = \dfrac{1}{2}$ **17.** $\left(f^{-1}\right)'(4) = \dfrac{1}{f'\left(f^{-1}(4)\right)} = \dfrac{1}{f'(-1)} = \dfrac{1}{3}$

19. $f^{-1} : (2,245) \to (0,3)$; $f^{-1}(t) = (t-2)^{1/5}$; $\left(f^{-1}\right)'(t) = \dfrac{1}{5(t-2)^{4/5}}$

21. $f^{-1} : \left[\dfrac{1}{64},1\right] \to [1,8]$; $f^{-1}(t) = \dfrac{1}{\sqrt{t}}$; $\left(f^{-1}\right)(t) = -\dfrac{1}{2t^{3/2}}$

23. $e^{\ln(3)} = 3$

25. $\exp(-3\ln(2)) = \left(\exp\left(\ln(2)\right)\right)^{-3} = 2^{-3} = \dfrac{1}{8}$

27. $\left(\ln\left(1+\exp(x)\right)\right)' = \left(\dfrac{1}{1+\exp(x)}\right) \cdot \left(\exp(x)\right)' = \dfrac{\exp(x)}{1+\exp(x)} = \dfrac{e^x}{1+e^x}$

29. $\left(2^{\ln(x)}\right)' = 2^{\ln(x)} \cdot \ln(2) \cdot \left(\ln(x)\right)' = 2^{\ln(x)} \dfrac{\ln(2)}{x}$

31. $\left(\sqrt{\ln(x)}\right)' = \left(\dfrac{1}{2}\right)\left(\dfrac{1}{\sqrt{\ln(x)}}\right) \cdot \left(\ln(x)\right)' = \dfrac{1}{2x\sqrt{\ln(x)}}$

33. $\left(\cos\left(\ln(x)\right)\right)' = -\sin\left(\ln(x)\right) \cdot \left(\ln(x)\right)' = \dfrac{-\sin\left(\ln(x)\right)}{x}$

35. $\left(\sin\left(2\cdot3^x\right)\right)' = \cos\left(2\cdot3^x\right)\cdot\left(2\cdot3^x\right)' = 2\cdot3^x\ln(3)\cos\left(2\cdot3^x\right)$

37. $\left(2^{\left(\pi^x\right)}\right)' = 2^{\pi^x}\ln(2)\cdot\left(\pi^x\right)' = 2^{\pi^x}\ln(2)\pi^x\ln(\pi)$

39. 1) $f'(x) = \left(\dfrac{1}{x^2}\right)\cdot\left(x^2\right)' = \dfrac{2x}{x^2} = \dfrac{2}{x}$

   2) $f(x) = 2\ln(x)$, $f'(x) = \dfrac{2}{x}$

Further Theory and Practice

41. The domain of $f + c$ is the domain of f. If the range of f is $[a,b]$, the range of $f + c$ is $[a+c, b+c]$.

The function $f + c$ is one-to-one and onto, hence $f + c$ is invertible.

The relationship is: $(f+c)^{-1}(t) = f^{-1}(t-c)$.

43. The function $g \circ f : (a,b) \to (\alpha, \beta)$ is one-to-one and onto, hence it is invertible.

The function $(g\circ f)(s) = g(f(s))$, $s = f^{-1}\left(g^{-1}(t)\right) = \left(f^{-1}\circ g^{-1}\right)(t)$ and therefore

$$(g\circ f)^{-1} = f^{-1}\circ g^{-1}.$$

Then $\left((g\circ f)^{-1}\right)'(t) = \dfrac{1}{f'\left(f^{-1}\left(g^{-1}(t)\right)\right)}\cdot\dfrac{1}{g'\left(g^{-1}(t)\right)}.$

45. Let $s^5 + 2s^3 + 2s + 3 = 3$. This leads to $s = 0$, $f^{-1}(3) = 0$; $f'(s) = 5s^4 + 6s^2 + 2$,

$$\left(f^{-1}\right)'(3) = \dfrac{1}{f'(0)} = \dfrac{1}{2}.$$

47. Let $\pi s - \cos\left(\dfrac{\pi s}{2}\right) = \pi$. This leads to $s = 1$ (see figure), $f^{-1}(\pi) = 1$; $f'(s) = \pi + \left(\dfrac{\pi}{2}\right)\sin\left(\dfrac{\pi s}{2}\right)$,

$$\left(f^{-1}\right)'(\pi) = \dfrac{1}{f'(1)} = \dfrac{2}{3\pi}.$$

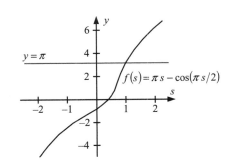

49. As $2Ae^{kt} = Ae^{k(t+\tau)}$, hence $k\tau = \ln(2)$ and $k = \left(\dfrac{1}{\tau}\right)\ln(2)$.

51. By definition of the doubling time $2^{k(\tau+t)} = 2 \cdot 2^{kt}$, hence $k\tau = 1$ and $\tau = \dfrac{1}{k}$.

$$P'(t) = P(0)2^{kt}\ln(2)k = P(t)\ln(2)k = 2k\ln(2)P(0)$$

53. Let $\ln(f(x)) = v(x)\ln(u(x))$.

Hence $f(x) = \exp\big(v(x)\ln(u(x))\big)$ and

$$f'(x) = \exp\big(v(x)\ln(u(x))\big) \cdot \big(v(x)\ln(u(x))\big)'$$
$$= \exp\big(v(x)\ln(u(x))\big) \cdot \left(v'(x)\ln(u(x)) + v(x)u'\dfrac{x}{u(x)}\right)$$

or $f'(x) = v'(x) \cdot u(x)^{v(x)} \cdot \ln(u(x)) + v(x) \cdot u(x)^{v(x)-1} \cdot u'(x)$.

55. Because $\frac{d}{dx}e^x = e^x$ and $\frac{d}{dx}\ln(x) = \dfrac{1}{x}$, the tangent lines to the graphs of $y = e^x$ and $y = \ln(x)$ at $\left(c,\ e^c\right)$ and $\left(c,\ \ln(c)\right)$ have slopes e^c and $\dfrac{1}{c}$. The condition for the tangent lines to be perpendicular is that the product of the slopes be -1. That is, $e^c \cdot \dfrac{1}{c} = -1$. But c must be positive to be in the domain of the natural logarithm. It follows that $e^c \cdot \dfrac{1}{c} > 0$. Since the equation $e^c \cdot \dfrac{1}{c} = -1$ has no positive root, we conclude that the tangent lines cannot be perpendicular.

57. Let $f(x) = a^x$, then $f'(0) = \lim\limits_{x \to 0} \dfrac{f(x) - f(0)}{x} = \lim\limits_{x \to 0} \dfrac{a^x - 1}{x} = \ln(a)$.

59. For the function $P(r) = P_0 \exp\left(\dfrac{2\tau v}{rkT_0}\right)$: $(0,\infty) \to (P_0,\infty)$ is one-to-one and onto; hence it is

invertible. The inverse function is $r(P) = \dfrac{(2\tau v)/(kT_0)}{\ln(P/P_0)}$. We calculate the two derivatives as

follows:

$$\frac{dP}{dr} = P_0 \exp\left(\frac{2\tau v}{rkT_0}\right)\left(-\frac{2\tau v}{r^2 kT_0}\right) = -\frac{2\tau v P_0}{r^2 kT_0}\cdot \exp\left(\frac{2\tau v}{rkT_0}\right)$$

$$\frac{dr}{dP} = \left(\frac{2\tau v}{kT_0}\right)\left(\frac{-1}{\ln^2\left(\frac{P}{P_0}\right)}\right)\left(\frac{P_0}{P}\right)\left(\frac{1}{P_0}\right) = -\frac{2\tau v}{kT_0 P \ln^2(P/P_0)}$$

$$= -\frac{2\tau v}{kT_0 P_0 \exp((2\tau v)/(rkT_0))\ln^2\left(\exp((2\tau v)/(rkT_0))\right)} = -\frac{r^2 kT_0}{2\tau v P_0 \exp((2\tau v)/(rkT_0))}$$

We, therefore, conclude that $\frac{dP}{dr}\cdot\frac{dr}{dP} = 1$.

61. $v(t) = y'(t) = \left(\sqrt{\dfrac{g}{\kappa}}\right)\dfrac{1 - \exp\left(2t\sqrt{g\kappa}\right)}{1 + \exp\left(2t\sqrt{g\kappa}\right)}$

63. If $y = Ax^m$, we have $\ln(y) = \ln(A) + m\ln(x)$, implying that $\ln(y)$ is a linear function of $\ln(x)$ with slope m and y-intercept $\ln(A)$, $A > 0$.

Calculator/Computer Exercises

65. The slopes of the tangent lines are $\dfrac{3}{4}$ and $\dfrac{4}{3}$ having a product equal to 1, which explains the Inverse Function Derivative Rule.

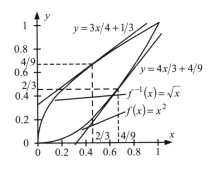

67. The equation for the mass is: $m(t) = m(0)e^{-t\ln(2)/5.3}$, or $m(t) = m(0)e^{-0.13078t}$. For

$$m(0) = 10$$

we see that $m(t) = 10e^{-0.13078t}$ and $m'(t) = -1.3078e^{-0.13078t}$. The rate at which the cobalt is decaying is -1.3078 g/yr.

To calculate the time when the rate becomes 1 g/yr we find t such that $|m'(t)| = 1$. It follows that the instantaneous rate is $t = \dfrac{\ln(1.3078)}{0.13078}$ or $t = 2.05189$ yr.

69. The equation $s^3 + 2s + \sin(s) = 1$ has a unique root $s_0 \approx 0.3239$. For the derivative we have:

$$f'(s) = 3s^2 + 2 + \cos(s), \ \left(f^{-1}\right)'(c) = \frac{1}{f'}\left(f^{-1}(c)\right) = \frac{1}{f'}(s_0) \approx 0.3065.$$

71. The equation $s^5 + 4s - 2 = 15$ has a unique root $s_0 \approx 1.6031$. For the derivative we have:

$$f'(s) = 5s^4 + 4, \ \left(f^{-1}\right)'(c) = \frac{1}{f'}\left(f^{-1}(c)\right) = \frac{1}{f'}(s_0) \approx 0.02701.$$

73. Two tangent lines are parallel if and only if $e^c = \dfrac{1}{c}$, implying that $c = 0.56714$.

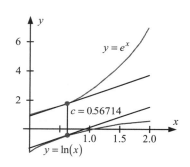

75. The functions $f(x) = 10^5 \dfrac{\ln\left(x + 10^{-5}\right) - \ln\left(x - 10^{-5}\right)}{2}$

and $y(x) = \dfrac{1}{x}$ coincide in the viewing window; hence

$\dfrac{d}{dx}\ln(x) = \dfrac{1}{x}$. This is an illustration of the natural

logarithm differentiation rule.

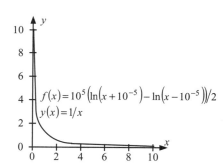

77.   Macaques: $y = e^{-2.056} \cdot x^{1.0164}$, i.e.

$$y = 0.128 \cdot x^{1.0164}$$

$$y' = 0.1301 \cdot x^{0.0164}$$

$$y'(500) = 0.1440$$

Humans: $y = e^{-2.211} \cdot x^{1.0160}$, i.e.

$$y = 0.1096 \cdot x^{1.0160}$$

$$y' = 0.1114 \cdot x^{0.0160}$$

$$y'(500) = 0.1230$$

79.

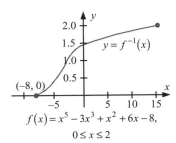

$$f(x) = x^5 - 3x^3 + x^2 + 6x - 8,$$
$$0 \le x \le 2$$

81.

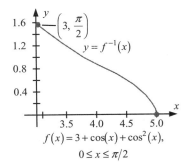

$$f(x) = 3 + \cos(x) + \cos^2(x),$$
$$0 \le x \le \pi/2$$

Section 3.7   Higher Derivatives

Problems for Practice

1. $f'(x) = \frac{d}{dx}\left(4x^3 - 7x^{-5} + 2x^{5/2}\right) = 12x^2 + 35x^{-6} + 5x^{3/2}$

$f''(x) = \left(f'(x)\right)' = \frac{d}{dx}\left(12x^2 + 35x^{-6} + 5x^{3/2}\right) = 24x - 210x^{-7} + \left(\frac{15}{2}\right)\sqrt{x}$

$f'''(x) = \left(f''(x)\right)' = \frac{d}{dx}\left(24x - 210x^{-7} + \left(\frac{15}{2}\right)\sqrt{x}\right) = 24 + 1470x^{-8} + \frac{15}{4\sqrt{x}}$

3. $g'(t) = \frac{d}{dt}\tan(t) = \sec^2(t)$;

$g''(t) = \left(g'(t)\right)' = \frac{d}{dt}\sec^2(t) = 2\sec(t)\frac{d}{dt}\sec(t) = 2\sec^2(t)\tan(t)$;

$g'''(t) = \left(g''(t)\right)' = \frac{d}{dt}\left(2\sec^2(t)\tan(t)\right) = 2\tan(t)\frac{d}{dt}\sec^2(t) + 2\sec^2(t)\frac{d}{dt}\tan(t)$

$\qquad = 2\tan(t) \cdot 2\sec^2(t)\tan(t) + 2\sec^2(t)\sec^2(t) = 2\sec^2(t)\left(2\tan^2(t) + \sec^2(t)\right)$

5. $k'(x) = \frac{d}{dx}\left(\frac{x+1}{x-1}\right) = \frac{(x-1)-(x+1)}{(x-1)^2} = -\frac{2}{(x-1)^2}$

$k''(x) = (k'(x))' = \frac{d}{dx}\left(-\frac{2}{(x-1)^2}\right) = \frac{4}{(x-1)^3}$

$k'''(x) = (k''(x))' = -\frac{12}{(x-1)^4}$

7. $k'(x) = \frac{d}{dx}\sin(x^3 - 3x) = \cos(x^3 - 3x)\frac{d}{dx}(x^3 - 3x) = (3x^2 - 3)\cos(x^3 - 3x)$

$k''(x) = (k'(x))' = \frac{d}{dx}\left((3x^2 - 3)\cos(x^3 - 3x)\right)$

$= \cos(x^3 - 3x)\frac{d}{dx}(3x^2 - 3) + (3x^2 - 3)\frac{d}{dx}\cos(x^3 - 3x)$

$= 6x \cdot \cos(x^3 - 3x) - (3x^2 - 3)^2 \sin(x^3 - 3x)$

$k'''(x) = (k''(x))' = 6x\frac{d}{dx}\cos(x^3 - 3x) + \cos(x^3 - 3x)\frac{d}{dx}(6x) - (3x^2 - 3)^2\frac{d}{dx}\sin(x^3 - 3x)$

$\qquad - \sin(x^3 - 3x)\frac{d}{dx}(3x^2 - 3)^2$

$= -6x(3x^2 - 3)\sin(x^3 - 3x) + 6\cos(x^3 - 3x)$

$\qquad - (3x^2 - 3)^3 \cos(x^3 - 3x) - 2(3x^2 - 3) \cdot 6x \cdot \sin(x^3 - 3x)$

$= \cos(x^3 - 3x)\left(6 - (3x^2 - 3)^3\right) - 18x(3x^2 - 3)\sin(x^3 - 3x)$

9. $H'(t) = \frac{d}{dt}(t\tan(t)) = t\frac{d}{dt}\tan t + \tan(t)\frac{d}{dt}t = \tan(t) + t\sec^2(t)$

$H''(t) = (H'(t))' = \frac{d}{dt}(\tan(t) + t\sec^2(t)) = \sec^2(t) + \sec^2(t) + t\frac{d}{dt}\sec^2(t)$

$= 2\sec^2(t) + 2t\sec^2(t)\tan(t) = 2\sec^2(t)(1 + t \cdot \tan(t))$

$H'''(t) = (H''(t))' = \frac{d}{dt}\left(2\sec^2(t)(1 + t \cdot \tan(t))\right)$

$= 2(1 + t\tan(t))\frac{d}{dt}\sec^2(t) + 2\sec^2(t)\frac{d}{dt}(1 + t\tan(t))$

$= 2(1 + t\tan(t)) \cdot 2\sec^2(t)\tan(t) + 2\sec^2(t)\left(\tan(t) + t\sec^2(t)\right)$

$= 2\sec^2(t)\left(2\tan(t) + 2t\tan^2(t) + \tan(t) + t\sec^2(t)\right)$

$= 2\sec^2(t)\left(3\tan(t) + 2t\tan^2(t) + t(1 + \tan^2(t))\right)$

$= 2\sec^2(t)\left(3\tan(t) + 3t\tan^2(t) + t\right)$

11. $f^{(1)}(x) = \cos(x)$; $f^{(2)}(x) = -\sin(x)$; $f^{(3)}(x) = -\cos(x)$; $f^{(4)}(x) = \sin(x)$;

$f^{(5)}(x) = \cos(x)$

13. $\dot{g}(t) = -6\left(3t^2 - 5t\right)^{-7}(6t - 5)$

$\ddot{g}(t) = 42\left(3t^2 - 5t\right)^{-8}(6t - 5)^2 - 36\left(3t^2 - 5t\right)^{-7} = 6\left(7(6t - 5)^2 - 6\left(3t^2 - 5t\right)\right)\left(3t^2 - 5t\right)^{-8}$

$\qquad = 6\left(234t^2 - 390t + 175\right)\left(3t^2 - 5t\right)^{-8}$

15. $H^{(1)}(x) = -48x^5 + 35x^4 - 18x$

$H^{(2)}(x) = -240x^4 + 140x^3 - 18$

$H^{(3)}(x) = -960x^3 + 420x^2$

$H^{(4)}(x) = -2880x^2 + 840x$

17. $f'(t) = -4\sin(4t + 3);\ f''(t) = -16\cos(4t + 3)$

19. $g'(x) = -\sin(\sin(x)) \cdot \cos(x)$

$g''(x) = -\cos'(x)\sin(\sin(x)) - \cos(x)\sin'(\sin(x))$

$\qquad = \sin(x)\sin(\sin(x)) - \cos(x)\cos(\sin(x)) \cdot \cos(x)$

$\qquad = \sin(x)\sin(\sin(x)) - \cos^2(x)\cos(\sin(x))$

21. $a(t) = \frac{d}{dt}\upsilon = \frac{d}{dt}\left(t^2 - 5t\right) = 2t - 5$

The acceleration at $t = 3$ is $a(3) = \frac{d}{dt}\upsilon\Big|_3 = 2\cdot 3 - 5 = 1\ \text{ft/s}^2$.

The acceleration at $t = 6$ is $a(6) = \frac{d}{dt}\upsilon\Big|_6 = 2\cdot 6 - 5 = 7\ \text{ft/s}^2$.

23. $a(t) = \frac{d^2}{dt^2}p = \frac{d}{dt}\left(\frac{d}{dt}p\right) = \frac{d}{dt}\left(2t^3 + t^2 + 6t\right) = \frac{d}{dt}\left(6t^2 + 2t + 6\right) = 12t + 2$

The acceleration at time $t = 3$ is $a(3) = \frac{d^2}{dt^2}p\Big|_3 = 12\cdot 3 + 2 = 38\ \text{m/min}^2$.

The acceleration at time $t = 6$ is $a(6) = \frac{d^2}{dt^2}p\Big|_3 = 12\cdot 6 + 2 = 74\ \text{m/min}^2$.

Further Theory and Practice

25.   a.   We calculate

$$\lim_{\Delta x \to 0^+} \frac{f(0 + \Delta x) - f(0)}{\Delta x} = \lim_{\Delta x \to 0^+} \frac{(\Delta x)^3 - 0}{\Delta x} = \lim_{\Delta x \to 0^+} (\Delta x)^2 = 0$$

and

$$\lim_{\Delta x \to 0^-} \frac{f(0 + \Delta x) - f(0)}{\Delta x} = \lim_{\Delta x \to 0^-} \frac{-(\Delta x)^3 - 0}{\Delta x} = \lim_{\Delta x \to 0^-} -(\Delta x)^2 = 0.$$

Since these limits exist and are both equal to 0, we see that

$$f'(0) = \lim_{x \to 0} \frac{f(0 + \Delta x) - f(0)}{\Delta x}$$

exists and equals 0.

b.   We calculate

$$\lim_{\Delta x \to 0^+} \frac{f'(0 + \Delta x) - f'(0)}{\Delta x} = \lim_{\Delta x \to 0^+} \frac{3(\Delta x)^2 - 0}{\Delta x} = \lim_{\Delta x \to 0^+} 3\Delta x = 0$$

and

$$\lim_{\Delta x \to 0^-} \frac{f'(0 + \Delta x) - f'(0)}{\Delta x} = \lim_{\Delta x \to 0^-} \frac{-3(\Delta x)^2 - 0}{\Delta x} = \lim_{\Delta x \to 0^-} -3\Delta x = 0.$$

It follows that $f''(0)$ exists and equals 0.

c.   We calculate

$$\lim_{\Delta x \to 0^+} \frac{f''(0 + \Delta x) - f''(0)}{\Delta x} = \lim_{\Delta x \to 0^+} \frac{6(\Delta x) - 0}{\Delta x} = \lim_{\Delta x \to 0^+} 6 = 6$$

and

$$\lim_{\Delta x \to 0^-} \frac{f''(0 + \Delta x) - f''(0)}{\Delta x} = \lim_{\Delta x \to 0^-} \frac{-6(\Delta x) - 0}{\Delta x} = \lim_{\Delta x \to 0^-} -6 = -6.$$

We conclude that $f'''(0)$ does not exist.

27.   Let k be any fixed nonnegative integer. Let j and n be any nonnegative integers. Observe that

$$\frac{d^j}{dx^j}x^n = \begin{cases} 0 & \text{if } n < j \\ n! & \text{if } n = j \\ n(n-1)\cdots(n-j+1)x^{n-j} & \text{if } n > j \end{cases}.$$

It follows that

$$\left.\frac{d^j}{dx^j}x^n\right|_{x=0} = \begin{cases} 0 & \text{if } n < j \\ n! & \text{if } n = j \\ 0 & \text{if } n > j \end{cases}.$$

Supposing $a_k \neq 0$ and $p(x) = a_k x^k + a_{k-1}x^{k-1} + \ldots + a_1 x + a_0$, we conclude that $p^{(k+1)}(x) = 0$ for all x and $p^{(k)}(0) = k!\, a_k \neq 0$. Thus, if $\deg(p) = k$, then $p^{(k+1)}(x) = 0$ for all $p^{(k)}(x_0) \neq 0$ for some x_0. In the other direction, we suppose that p has degree $d \neq k$ and show that either $p^{(k+1)}(x) \neq 0$ for some x or $p^{(k)}(x) = 0$ for every x. There are two cases. If $d < k$ then $k \geq d + 1$. Since $p^{(d+1)}(x) = 0$ for all x (by the direction already proved), we infer that $p^{(k)}(x) = 0$ for every x. If $d > k$, then $p^{(k)}(x)$ is a polynomial of degree $d - k > 0$. As such, $p^{(k)}(x)$ can have at most $d - k$ roots. In particular, $p^{(k+1)}(x) \neq 0$ for some x.

29. $p(x) = ax^3 + bx^2 + cx + d$

$p'(x) = 3ax^2 + 2bx + c$

$p''(x) = 6ax + 2b$

$p'''(x) = 6a$

$p'''(5) = 6a = 6, \ a = 1$

$p''(-1) = 6 \cdot (-1) + 2b = -14, \ b = -4$

$p'(2) = 3 \cdot 2^2 - 8 \cdot 2 + c = -2, \ c = 2$

$p(1) = 1 \cdot 1^3 - 4 \cdot 1^2 + 2 \cdot 1 + d = 1, \ d = 2$

$p(x) = x^3 - 4x^2 + 2x + 2$

31. In Problem 3.7.27 we show that for a polynomial of degree n, the k-th derivative is

$$p^{(k)}(x) = n(n-1)\ldots(n-k+1)x^{n-k} + (n-1)\ldots(n-k)x^{n-k-1} + \ldots$$

if $n \geq k$. Therefore, $p''(x) = n(n-1)x^{n-2} + (n-1)(n-2)x^{n-3} + \ldots$. For $n > 2$, $p''(x)$ is unbounded and the condition $|p''(x)| \leq 1$ cannot be met for all x. For $n < 2$, $p''(x) = 0$ for all x. Therefore, $p(x)$ is a polynomial of degree 2, $p(x) = ax^2 + bx + c$. Its second derivative is $p''(x) = (p'(x))' = (2ax + b)' = 2a$. From the condition $|p''(x)| \leq 1$, we see that $|a| \leq 1/2$.

33. For the Product Rule $n = 1$:

$$(f \cdot g)^{(1)} = \binom{1}{0}f^{(1)}g^{(0)} + \binom{1}{1}f^{(0)}g^{(1)} = f'g + fg'$$

For Leibniz's Rule $n = 2$:

$$(f \cdot g)^{(2)} = \binom{2}{0}f^{(2)}g^{(0)} + \binom{2}{1}f^{(1)}g^{(1)} + \binom{2}{2}f^{(0)}g^{(2)} = f^{(2)}g^{(0)} + \frac{2}{1}f^{(1)}g^{(1)} + f^{(0)}g^{(2)}$$

$$= f''g + 2f'g' + fg''$$

$$(f \cdot g)^{(2)} = (f'g + fg')' = f''g + f'g' + f'g' + fg'' = f''g + 2f'g' + fg''$$

For Leibniz's Rule $n = 3$:

$$(f \cdot g)^{(3)} = \binom{3}{0}f^{(3)}g^{(0)} + \binom{3}{1}f^{(2)}g^{(1)} + \binom{3}{2}f^{(1)}g^{(2)} + \binom{3}{3}f^{(0)}g^{(3)}$$

$$= f^{(3)}g^{(0)} + \frac{3}{1}f^{(2)}g^{(1)} + \frac{3 \cdot 2}{1 \cdot 2}f^{(1)}g^{(2)} + f^{(0)}g^{(3)} = f'''g + 3f''g' + 3f'g'' + fg'''$$

$$(f \cdot g)^{(3)} = (f''g + 2f'g' + fg'')' = f'''g + f''g' + 2f''g' + 2f'g'' + f'g'' + fg'''$$

$$= f'''g + 3f''g' + 3f'g'' + fg'''$$

35.   Let $f(x) = x^2$, $g(x) = \sqrt{x^2+1}$. Then $f'(x) = 2x$, $f''(x) = 2$, $g'(x) = \dfrac{x}{\sqrt{x^2+1}}$, and

$$g''(x) = \frac{\sqrt{x^2+1} - x\left(\sqrt{x^2+1}\right)'}{x^2+1} = \frac{\sqrt{x^2+1} - \left(x^2/\sqrt{x^2+1}\right)}{x^2+1} = \frac{1}{\left(x^2+1\right)^{3/2}}.$$

So,

$$\frac{d^2}{dx^2}\left(x^2\sqrt{x^2+1}\right) = \frac{d^2}{dx^2}(f \cdot g) = 2\sqrt{x^2+1} + 2 \cdot 2x \cdot \frac{x}{\sqrt{x^2+1}} + \frac{x^2}{\left(x^2+1\right)^{3/2}}$$

$$= 2\sqrt{x^2+1} + \frac{4x^2}{\sqrt{x^2+1}} + \frac{x^2}{\left(x^2+1\right)^{3/2}} = \frac{6x^4 + 9x^2 + 2}{\left(x^2+1\right)^{3/2}}.$$

37. $f'(x) = 3x^2$, $f''(x) = 6x$

$g'(x) = -\sin(x)$, $g''(x) = -\cos(x)$

$D^2(f \cdot g)(x) = x^3 \cdot (-\cos(x)) + 2 \cdot 3x^2 \cdot (-\sin(x)) + 6x \cdot \cos(x)$

$\qquad = -x^3\cos(x) - 6x^2\sin(x) + 6\cos(x)$

39. $P_1(x) = \left(\dfrac{1}{2!}\right)\dfrac{d}{dx}(x^2-1) = \left(\dfrac{1}{2}\right)\cdot 2x = x$

$P_2(x) = \left(\dfrac{1}{2^2\cdot 2!}\right)\dfrac{d^2}{dx^2}(x^2-1)^2$

$\quad = \left(\dfrac{1}{8}\right)\dfrac{d}{dx}\left(\dfrac{d}{dx}(x^2-1)^2\right)$

$\quad = \left(\dfrac{1}{8}\right)\dfrac{d}{dx}\left(2(x^2-1)\cdot 2x\right)$

$\quad = \left(\dfrac{1}{2}\right)\left((x^2-1)x' + x(x^2-1)'\right) = \left(\dfrac{1}{2}\right)(x^2-1+2x^2)$

$\quad = \dfrac{3x^2-1}{2}$

$P_3(x) = \left(\dfrac{1}{2^3\cdot 3!}\right)\dfrac{d^3}{dx^3}(x^2-1)^3$

$\quad = \left(\dfrac{1}{8\cdot 6}\right)\dfrac{d^3}{dx^3}\left(x^6-3x^4+3x^2-1\right)\dfrac{d}{dx}\left(\dfrac{d}{dx}\left(\dfrac{d}{dx}\left(x^6-3x^4+3x^2-1\right)\right)\right)$

$\quad = \dfrac{d}{dx}\left(\dfrac{d}{dx}\left(6x^5-12x^3+6x\right)\right)$

$\quad = \dfrac{d}{dx}\left(30x^4-36x^2+6\right)$

$\quad = 120x^3-72x$

$P_3(x) = \dfrac{5x^3-3x}{2}$

$P_4(x) = \left(\dfrac{1}{2^4\cdot 4!}\right)\dfrac{d^4}{dx^4}(x^2-1)^4$

$\quad = \left(\dfrac{1}{16\cdot 24}\right)\dfrac{d^4}{dx^4}\left(x^8-4x^6+6x^4-4x^2+1\right)\dfrac{d}{dx}\left(\dfrac{d}{dx}\left(\dfrac{d}{dx}\left(\dfrac{d}{dx}\left(x^8-4x^6+6x^4-4x^2+1\right)\right)\right)\right)$

$\quad = \dfrac{d}{dx}\left(\dfrac{d}{dx}\left(\dfrac{d}{dx}\left(8x^7-24x^5+24x^3-8x\right)\right)\right)$

$\quad = \dfrac{d}{dx}\left(\dfrac{d}{dx}\left(56x^6-120x^4+72x^2-8\right)\right)$

$\quad = \dfrac{d}{dx}\left(336x^5-480x^3+144x\right) = 1680x^4-1440x^2+144$

$P_4(x) = \dfrac{35x^4-30x^2+3}{8}$

41.   Let $p(x) = a_k x^k + a_{k-1} x^{k-1} + \ldots + a_1 x + a_0$. Using the solution of Problem 3.7.27, we see that $m \leq k$:

$$p^{(m)}(x) = a_k \left(x^k\right)^{(m)} + a_{k-1} \left(x^{k-1}\right)^{(m)} + \ldots + a_m \left(x^m\right)^{(m)} + \ldots + a_1 \left(x^1\right)^{(m)} + \left(a_0\right)^{(m)}$$

$$= a_k k(k-1)\ldots(k-m+1)x^{k-m} + a_{k-1}(k-1)(k-2)\ldots(k-m)x^{k-1-m} + a_m \cdot m!$$

Therefore, $p^{(m)}(0) = a_m m!$ and $a_m = \dfrac{p^{(m)}(0)}{m!}$ for any $m \leq k$. So

$$p(x) = p(0) + p'(0)x + \ldots + \left(\frac{p^{(k)}(0)}{k!}\right)x^k .$$

43.   The function $f(x)$ is a polynomial with a non-zero coefficient at x^0 and is the next lowest power of x equal to 93. Therefore, based on the solution provided for Problem 3.7.27, we see that $f^{(83)}(0) = 0$.

45.   From the solution to Problem 3.6.61 we see that y satisfies the differential equation as follows:

$$y'(t) = \left(\sqrt{\frac{g}{k}}\right) \frac{1 - e^{2t\sqrt{gk}}}{1 + e^{2t\sqrt{gk}}} .$$

Then, by differentiating the first derivative, we obtain:

$$y''(t) = \left(\sqrt{\frac{g}{k}}\right) \frac{\left(1 - e^{2t\sqrt{gk}}\right)'\left(1 + e^{2t\sqrt{gk}}\right) - \left(1 - e^{2t\sqrt{gk}}\right)\left(1 + e^{2t\sqrt{gk}}\right)'}{\left(1 + e^{2t\sqrt{gk}}\right)^2}$$

$$= \left(\sqrt{\frac{g}{k}}\right) \frac{\left(-2e^{2t\sqrt{gk}}\sqrt{gk}\right)\left(1 + e^{2t\sqrt{gk}}\right) - \left(2e^{2t\sqrt{gk}}\sqrt{gk}\right)\left(1 - e^{2t\sqrt{gk}}\right)}{\left(1 + e^{2t\sqrt{gk}}\right)^2} = \frac{-4ge^{2t\sqrt{gk}}}{\left(1 + e^{2t\sqrt{gk}}\right)^2}$$

On the other hand, substituting the first derivative in the formula for the second derivative yields

$$-g + k\left(y'(t)\right)^2 = -g + k\left(\left(\sqrt{\frac{g}{k}}\right)\frac{1 - e^{2t\sqrt{gk}}}{1 + e^{2t\sqrt{gk}}}\right)^2 = -g + g\frac{\left(1 - e^{2t\sqrt{gk}}\right)^2}{\left(1 + e^{2t\sqrt{gk}}\right)^2}$$

$$= -g\frac{\left(1 + e^{2t\sqrt{gk}}\right)^2 - \left(1 - e^{2t\sqrt{gk}}\right)^2}{\left(1 + e^{2t\sqrt{gk}}\right)^2} = -4g\frac{e^{2t\sqrt{gk}}}{\left(1 + e^{2t\sqrt{gk}}\right)^2} = y''(t)$$

Calculator/Computer Exercises

47.

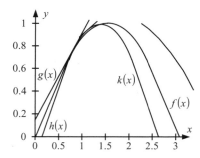

x	$f(x)$	$g(x)$	$h(x)$	$k(x)$
0.67	0.620985	0.625507	0.620799	0.621055
0.68	0.628793	0.632579	0.628651	0.628846
0.69	0.636537	0.639650	0.636432	0.636577
0.70	0.644217	0.646721	0.644142	0.644246
0.71	0.651833	0.653792	0.651782	0.651853
0.72	0.659384	0.660863	0.659351	0.659397

The best approximation of $f(x)$ near c is $k(x)$, then $h(x)$, and finally $g(x)$.

49.

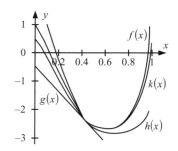

x	$f(x)$	$g(x)$	$h(x)$	$k(x)$
0.35	−1.67995	−1.90000	−1.63000	−1.68400
0.40	−1.99520	−2.10000	−1.98000	−1.99600
0.45	−2.27195	−2.30000	−2.27000	−2.27200
0.55	−2.66795	−2.70000	−2.67000	−2.66800
0.60	−2.76320	−2.90000	−2.78000	−2.76400
0.65	−2.77195	−3.10000	−2.83000	−2.77600

The best approximation of $f(x)$ near c is $k(x)$, then $h(x)$, and finally $g(x)$.

50–53   With $f_1(x) = \sqrt{2x}$, $c_1 = 1$, $f_2(x) = \sin(x)$, $c_2 = \dfrac{\pi}{6}$, $f_3(x) = \sin(\pi x)$, $c_3 = \dfrac{1}{6}$, and

$f_4(x) = \sqrt{\dfrac{x+1}{x}}$, $c_4 = 1$, the Maple code of Figure 1, page 219 of the text produces the

following values:

Increment $h = 10^{-n}$	D2$(f[1], c[1], h)$	D2$(f[2], c[2], h)$	D2$(f[3], c[3], h)$	D2$(f[4], c[4], h)$
$n = 1$	−.3546	−.4995	−4.8943	.7133
$n = 2$	−.3535	−.4999	−4.9343	.7071
$n = 3$	−.3535	−.4999	−4.9347	.7071
$n = 4$	−.3535	−.5000	−4.9348	.7071
$n = 5$	−.3536	−.5000	−4.9348	.7070
$n = 6$	−.3600	−.5000	−4.9340	.7000

Watching for deterioration of accuracy, we are led to the following approximations:

$f_1^{(2)}(c_1) \approx -0.3536$, $f_2^{(2)}(c_2) \approx -0.5000$, $f_3^{(2)}(c_3) \approx -4.9348$, and $f_4^{(2)}(c_4) \approx -0.7071$.

55. $(f \circ g)^{(1)} = f'(g)g'$

$(f \circ g)^{(2)} = f''(g)(g')^2 + f'(g)g''$

$(f \circ g)^{(3)} = f'''(g)(g')^3 + 3f''(g)g''g' + f'(g)g'''$

$(f \circ g)^{(4)} = f^{(4)}(g)(g')^4 + 6f'''(g)g''(g')^2 + 4f''(g)g'''g' + 3f''(g)(g'')^2 + f'(g)g^{(4)}$

The general formula is

$$D^k(f \circ g) = \sum_{j_1, j_2, \ldots, j_k} \frac{k!}{j_1! j_2! \ldots j_k!} \left(D^{j_1, j_2, \ldots, j_k}(f) \circ g\right) \cdot \left(\frac{D(g)}{1!}\right)^{j_1} \cdot \left(\frac{D^2(g)}{2!}\right)^{j_2} \cdots \left(\frac{D^k(g)}{k!}\right)^{j_k}$$

where the sum is taken over all k-tuples $(j_1, j_2, \ldots, j_k)$ of nonnegative integers such that $1 \cdot j_1 + 2 \cdot j_2 + \cdots + k \cdot j_k = k$.

Section 3.8   Implicit Differentiation and Related Rates

Problems for Practice

1. $y^2 + 2xy \frac{dy}{dx} + x^2 \frac{dy}{dx} + 2xy = 0$, $\left(2xy + x^2\right)\frac{dy}{dx} = -2xy - y^2$, $\frac{dy}{dx} = -\frac{2xy + y^2}{2xy + x^2}$, $\frac{dy}{dx}\Big|_P = -\frac{8}{5}$

3. $\left(\frac{3}{5}\right)x^{-2/5} + 4 \cdot \left(\frac{3}{5}\right)y^{-2/5}\frac{dy}{dx} = 0$, $\frac{dy}{dx} = -\left(\frac{1}{4}\right)\left(\frac{y}{x}\right)^{2/5}$, $\frac{dy}{dx}\Big|_P = -\frac{1}{16}$

5. $4x^3 - 4y^3\frac{dy}{dx} = 0$, $\frac{dy}{dx} = \left(\frac{x}{y}\right)^3$, $\frac{dy}{dx}\Big|_P = \frac{1}{8}$

7.   Differentiating the equation with respect to x, we obtain: $\frac{y}{4} + \frac{x}{4}\frac{dy}{dx} = \frac{1}{2\sqrt{x}} + \frac{1}{2\sqrt{y}}\frac{dy}{dx}$,

$\frac{dy}{dx} = \frac{\frac{1}{2\sqrt{x}} - \frac{y}{4}}{\frac{x}{4} - \frac{1}{2\sqrt{y}}}$, $\frac{dy}{dx}\Big|_P = -1$. The tangent line to the curve is $y - 4 = (-1)(x - 4)$, or $y = 8 - x$.

9.   Differentiating the equation with respect to x, we obtain:

$$2\sin(x)\cos(x) + 2\cos(y)(-\sin(y))\frac{dy}{dx} = 0,$$

$\frac{dy}{dx} = \frac{\sin(2x)}{\sin(2y)}$, $\frac{dy}{dx}\Big|_P = 1$. The tangent line to the curve is $y - \frac{\pi}{4} = x - \frac{\pi}{4}$, or $y = x$.

11. Differentiating the equation with respect to x, we obtain:

$$y^4 + 4xy^3\frac{dy}{dx} - 6x^2y^2 - 4x^3y\frac{dy}{dx} + 4y + 4x\frac{dy}{dx} = 0,$$

$\left(4xy^3 - 4x^3y + 4x\right)\frac{dy}{dx} = 6x^2y^2 - y^4 - 4y$, $\frac{dy}{dx} = \dfrac{6x^2y^2 - y^4 - 4y}{4xy^3 - 4x^3y + 4x}$, $\left.\frac{dy}{dx}\right|_P = 9$. The slope of the

normal line to the curve is $\left(-\dfrac{1}{9}\right)$ and its equation is $y - 2 = \left(-\dfrac{1}{9}\right)(x-2)$, or $y = \dfrac{20-x}{9}$.

13. Differentiating the equation with respect to x, we obtain: $\sin\left(\dfrac{y}{2}\right) + \left(\dfrac{x}{2}\right)\cos\left(\dfrac{y}{2}\right)\frac{dy}{dx} - \frac{dy}{dx} = 0$,

$\dfrac{dy}{dx} = \dfrac{\sin(y/2)}{1 - (x/2)\cos(y/2)}$, $\left.\dfrac{dy}{dx}\right|_P = 1$. The slope of the normal line to the curve is (-1) and its

equation is $y - \pi = (-1)(x - \pi)$, or $y = 2\pi - x$.

15. Differentiating the equation with respect to x, we obtain: $\left(\dfrac{1}{3}\right)y^{-2/3}\frac{dy}{dx} - \left(\dfrac{1}{3}\right)x^{-2/3} = 0$,

$\left(-\dfrac{2}{3}\right)y^{-5/3}\left(\dfrac{dy}{dx}\right)^2 + y^{-2/3}\dfrac{d^2y}{dx^2} + \left(\dfrac{2}{3}\right)x^{-5/3} = 0$. Substitution of P into the first equation yields

$\left.\dfrac{dy}{dx}\right|_P = \left.\left(\dfrac{y}{x}\right)^{2/3}\right|_P = 4$. Then from the second equation we see that $\left.\dfrac{d^2y}{dx^2}\right|_P = -\dfrac{4}{3}$.

17. Differentiating the equation with respect to x, we obtain: $4x^3 + 8y^3\frac{dy}{dx} = 0$,

$\left.\dfrac{dy}{dx}\right|_P = -\left.\left(\dfrac{x^3}{2y^3}\right)\right|_P = 4$, $12x^2 + 24y^2\left(\dfrac{dy}{dx}\right)^2 + 8y^3\dfrac{d^2y}{dx^2} = 0$. Hence, $\left.\dfrac{d^2y}{dx^2}\right|_P = -54$.

19. Differentiating the equation with respect to x, we obtain: $y + x\dfrac{dy}{dx} - 2xy\dfrac{dy}{dx} - y^2 = 1$,

$\dfrac{dy}{dx} + \dfrac{dy}{dx} + x\dfrac{d^2y}{dx^2} - 2y\dfrac{dy}{dx} - 2x\left(\dfrac{dy}{dx}\right)^2 - 2xy\dfrac{d^2y}{dx^2} - 2y\dfrac{dy}{dx} = 0$, $\left.\dfrac{dy}{dx}\right|_P = \left.\dfrac{y^2 - y + 1}{x - 2xy}\right|_P = -1$. Hence,

$\left.\dfrac{d^2y}{dx^2}\right|_P = 0$.

21. Differentiating the equation with respect to x, we obtain: $e^{xy}\left(y + x\dfrac{dy}{dx}\right) = 4y\dfrac{dy}{dx}$, $-1 = -4\left.\dfrac{dy}{dx}\right|_P$,

$\left.\dfrac{dy}{dx}\right|_P = \dfrac{1}{4}$. The equation of the tangent line to the curve is $y + 1 = \dfrac{x}{4}$, or $y = \dfrac{x}{4} - 1$.

23. Differentiating the equation with respect to x, we obtain: $\dfrac{1}{xy-1}\left(y + x\dfrac{dy}{dx}\right) + 2y\dfrac{dy}{dx} - 2 = 0$,

$2 + \left.\dfrac{dy}{dx}\right|_P + 4\left.\dfrac{dy}{dx}\right|_P - 2 = 0$, $\left.\dfrac{dy}{dx}\right|_P = 0$. The equation of the tangent line to the curve is $y = 2$.

25.   If the edge of the square is x, then the diagonal z is $x\sqrt{2}$. The area A is given by $A = x^2 = \dfrac{z^2}{2}$.

Therefore, $\dfrac{dA}{dt} = \dfrac{z\,dz}{dt} = (5)(2) = 10$ sq. in./sec.

27.   Let z be the distance between the two particles. Then $z^2 = x^2 + y^2$ when A is at $(x,\,0)$ and B is at $(0,\,y)$. When $x = 4$ and $y = 9$, we have $z = \sqrt{16 + 81} = \sqrt{97}$ and

$$2z\frac{dz}{dt} = 2x\frac{dx}{dt} + 2y\frac{dy}{dt} = 2(4)(5) + 2(9)(8) = 184.$$

Therefore, $\dfrac{dz}{dt} = \dfrac{184}{2\sqrt{97}} = \dfrac{92}{\sqrt{97}}$ and the distance between A and B changes with the rate of

$\dfrac{92}{\sqrt{97}}$ units per second.

29.   Let x be the length of a woman's shadow, and y be her distance from the wall. The woman's

shadow changes at a rate of $\dfrac{x}{35/6} = \dfrac{x+y}{10}$ and $x = \dfrac{7y}{5}$, $\dfrac{dx}{dt} = \left(\dfrac{7}{5}\right)\dfrac{dy}{dt} = \dfrac{28}{5}$ ft/s.

31.   Let x be the distance between the lighthouse and the boat, then $\cot(\theta) = \dfrac{x}{100}$ and

$$\left(-\frac{1}{\sin^2(\theta)}\right)\frac{d\theta}{dt} = \left(\frac{1}{100}\right)\frac{dx}{dt}.$$

From $\sin(\theta) = \dfrac{100}{\sqrt{100^2 + 1000^2}} = \dfrac{1}{\sqrt{101}}$, $\dfrac{dx}{dt} = 300$, we see that the acute angle between the

beam of light and the surface of the water changes at a rate of $\dfrac{d\theta}{dt} = -\dfrac{3}{101}$ radians/min.

Further Theory and Practice

33.   For $y = 0$ we have $x = \pm 9$, hence the line crosses the x-axis at points $A(-3,\,0)$ and $B(3,\,0)$.

Then $2x - 4y - 4x\frac{dy}{dx} + 2y\frac{dy}{dx} = 0$, $\dfrac{dy}{dx} = \dfrac{2y - x}{y - 2x}$, $\left.\dfrac{dy}{dx}\right|_A = \left.\dfrac{dy}{dx}\right|_B = \dfrac{1}{2}$. The tangent lines to the curve

at A and B are, therefore, parallel.

35. If $a^2 - ab + b^2 = 4$, we see that $(-a)^2 - (-a)(-b) + (-b)^2 = 4$ and the point $P'(-a, -b)$ is on the ellipse. Differentiating the equation with respect to x, we obtain:

$$2x - y - x\frac{dy}{dx} + 2y\frac{dy}{dx} = 0, \ \frac{dy}{dx} = \frac{y - 2x}{2y - x}, \ \frac{dy}{dx}\Big|_P = \frac{dy}{dx}\Big|_{P'} = \frac{b - 2a}{2b - a}.$$

Hence, the tangent lines at P and P' are parallel.

37. As the equation does not change by replacing (a, b) on (b, a), the first statement is true.

Differentiating the equation with respect to x, we obtain: $3x^2 - 3ny - 3nx\frac{dy}{dx} + 3y^2\frac{dy}{dx} = 0$, so

that $\frac{dy}{dx} = \frac{ny - x^2}{y^2 - nx}$, $\frac{dy}{dx}\Big|_P = \frac{nb - a^2}{b^2 - na}$, $\frac{dy}{dx}\Big|_{P'} = \frac{na - b^2}{a^2 - nb}$. Therefore, the product of the slopes is:

$\left(\frac{dy}{dx}\Big|_P\right) \cdot \left(\frac{dy}{dx}\Big|_{P'}\right) = 1$.

39. Differentiating the equation $y\ln(x) = x\ln(y)$ with respect to x, we obtain,

$$\ln(x)\frac{dy}{dx} + \frac{y}{x} = \ln(y) + \left(\frac{x}{y}\right)\frac{dy}{dx}, \ \frac{dy}{dx} = \frac{\ln(y) - y/x}{\ln(x) - x/y}.$$

For $x = 2$ we have $y = 2$ and $y = 4$, point $P(2, 4)$ is on the curve C, $\frac{dy}{dx}\Big|_P = 4\frac{\ln(2) - 1}{2\ln(2) - 1}$.

41. Differentiating the equation with respect to x, we obtain: $\left(\frac{1}{2}\right)x^{-1/2} + \left(\frac{1}{2}\right)y^{-1/2}\frac{dy}{dx} = 0$,

$\frac{dy}{dx} = -\sqrt{\frac{y}{x}}$. If $P(x_0, y_0)$ is a point on the curve (i.e., $\sqrt{x_0} + \sqrt{y_0} = a$), we have

$$\frac{dy}{dx}\Big|_P = -\frac{a - \sqrt{x_0}}{\sqrt{x_0}}$$

for $x_0 > 0$. The tangent line through point P is

$$y - \left(a - \sqrt{x_0}\right)^2 = -\frac{a - \sqrt{x_0}}{\sqrt{x_0}}(x - x_0),$$

or $y = (x_0 - a)\frac{x}{\sqrt{x_0}} + a^2 - a\sqrt{x_0}$. Then the y-intercept is $a^2 - a\sqrt{x_0}$, the x-intercept is $a\sqrt{x_0}$,

and their sum is $s = a^2$ for every tangent T.

It is in fact independent of T.

43.   Given that $x^{2/3}(1-y) = Ay^2 \exp\left(\dfrac{b}{x}\right)$, we have

$$\left(\frac{3}{2}\right)x^{1/2}(1-y) - x^{3/2}\frac{dy}{dx} = 2Ay\exp\left(\frac{b}{x}\right)\frac{dy}{dx} - bAy^2\frac{\exp(b/x)}{x^2}$$

and $\dfrac{dy}{dx} = \dfrac{(3/2)x^{1/2}(1-y) + bAy^2\exp(b/x)/x^2}{2Ay + x^{3/2}}$, or $\dfrac{dy}{dx} = -Ay^3(2b+3x)\dfrac{\exp(b/x)}{2x^{7/2}(y-2)}$.

45.   **a.**   Differentiating the equation with respect to q, we obtain:

$$1 + \frac{dq}{dp} + 4pq + 2p^2\frac{dq}{dp} + 3q^3 + 9pq^2\frac{dq}{dp} = 0,$$

$\dfrac{dq}{dp} = -\dfrac{3q^3 + 4pq + 1}{9pq^2 + 2p^2 + 1}$. The slope of the demand curve is $\left.\dfrac{dq}{dp}\right|_{(6,3.454)} = -0.2893$.

b.   Elasticity of demand for the product: $E(6) = -q'(6)\cdot\dfrac{6}{q(6)} = 0.5025$.

47.   Let ℓ, w, and A denote the length, width, and area respectively. Then

$$\frac{dA}{dt} = \left(\frac{d\ell}{dt}\right)w + \ell\frac{dw}{dt} = (5)(8) + (6)(-7) = -2.$$

The area of the rectangle decreases at a rate of 2 square inches per minute.

49.   The surface of the water is a circle of radius r. By the Pythagorean Theorem, we see that $(20-h)^2 + r^2 = 20^2$. The area A is given by $A = \pi r^2 = \pi\left(20^2(20-h)^2\right)$. We differentiate it with respect to time t to find the rate of change of the water surface area:

$$\frac{dA}{dt} = \frac{dh}{dh}\frac{dh}{dt} = \pi(40-2h)\frac{dA}{dt} = 20\pi\frac{dh}{dt},$$

when $h = 10$. On the other hand, we see that

$$-5 = \frac{d}{dt}\left(\frac{\pi h^2(60-h)}{3}\right) = \frac{d}{dh}\left(\frac{\pi h^2(60-h)}{3}\right)\frac{dh}{dt} = \pi h(40-h)\frac{dh}{dt} = 300\pi\frac{dh}{dt}$$

when $h = 10$. Therefore, $\dfrac{dh}{dt} = -\dfrac{5}{300\pi} = -\dfrac{1}{60\pi}$. The area of the water on the surface is decreasing at a rate of $\dfrac{dA}{dt} = 20\pi\left(-\dfrac{1}{60\pi}\right) = -\dfrac{1}{3}$.

Calculator/Computer Exercises

51. We substitute the x-coordinate of P to obtain $10^3 - 30y + y^3 = 0$ and $y_P = -10.99700$. We calculate the derivative to deduce: $\frac{dy}{dx} = \frac{x^2 - y}{x - y^2}$, $\frac{dy}{dx}\big|_P = -1.00057$

53. We substitute the x-coordinate of P to obtain $(-1.4649)^3 - (-1.4649)^2 y^2 + y^3 = 0$ and $y_P = 2.60808$. We calculate the derivative to deduce: $\frac{dy}{dx} = \frac{3x^2 - 2xy^2}{2x^2 y - 3y^2}$, $\frac{dy}{dx}\big|_P = -2.86197$.

55. We substitute the x-coordinate of P to obtain $10^3 - 20y + y^3 = 0$ and $y_P = -10.66574$, $\frac{dy}{dx} = \frac{3x^2 - 2y}{2x - 3y^2}$, $\frac{dy}{dx}\big|_P = -1.00018$. The tangent line of the curve e is

$$y + 10.66574 = -1.00018(x - 10.000),$$

or $y = -1.0002x - 0.66397$.

57. We substitute the x-coordinate of P to obtain $(-2.0125)^3 - (-2.0125)^2 y^2 + y^3 = 0$ and $y_P = 4.45994$, $\frac{dy}{dx} = \frac{3x^2 - 2xy^2}{2x^2 y - 3y^2}$, $\frac{dy}{dx}\big|_P = -3.91620$. The tangent line of the curve is

$$y - 4.45994 = -3.91620(x + 2.0125),$$

or $y = -3.91620x - 3.42141$.

59. $\frac{dy}{dx}\big|_P = 0.69558$

The tangent line of the curve is
$y = 0.69558x + 0.58261$.

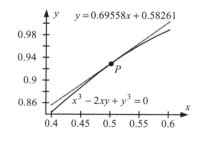

61. $\frac{dy}{dx}\big|_P = -0.38516$

The tangent line of the curve i
$y = -0.38516x + 2.74256$.

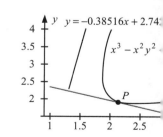

63. From $\dfrac{dy}{dx} = -\dfrac{4x^3 + 1 + 5y^3}{15xy^2}$, there can be

vertical tangent lines for $x = 0$ and $y = 0$.
In the first case there is no solution for
$x^4 + x + 5xy^3 = 1$; in the second case we
have two solutions: $x = -1.22074$ and
$x = 0.724492$.

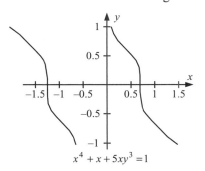

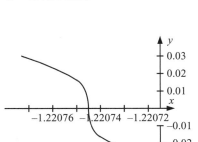

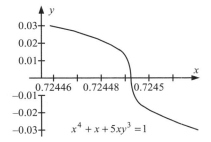

65. $q(4) = 4.152096...$

$q(4.2) = 4.06797...$

$q(3.8) = 4.24084...$

$q'(4) = -0.43170...$

$q'(4) \approx \dfrac{q(4.2) - q(3.8)}{4.2 - 3.8} = -0.43217...$

Section 3.9   Differentials and Approximation of Functions

Problems for Practice

1. $f(c) = 2$, $f'(x) = \dfrac{x^{-1/2}}{2}$, $f'(c) = \dfrac{1}{4}$, $\Delta x = -0.1$, $f(x) \approx 2 + \dfrac{1}{4} \cdot (-0.1) = 2 - 0.025 = 1.975$

3. $f(c) = 0$, $f'(x) = \cos(x) + \sin(x)$, $f'(c) = \sqrt{2}$, $\Delta x = \dfrac{\pi}{12}$,

$f(x) \approx 0 + \sqrt{2} \cdot \dfrac{\pi}{12} = 1.41421 \cdot 0.26179 = 0.37024$

5. $f(c)=1$, $f'(x)=2x\dfrac{\left(x^2+1\right)^{-2/3}}{3}$, $f'(c)=0$, $\Delta x=1$, $f(x)\approx 1+0\cdot 1=1$

7. $f(c)=1$, $f'(x)=\dfrac{1}{\cos^2(x)}$, $f'(c)=2$, $\Delta x=0.8-\dfrac{\pi}{4}$,

$f(x)\approx 1+2\cdot\left(0.8-\dfrac{\pi}{4}\right)=2.6-\dfrac{\pi}{2}=1.0292$

9. $f(c)=\dfrac{1}{\sqrt{3}}$, $f'(x)=-\dfrac{1}{\sin^2(x)}$, $f'(c)=-\dfrac{4}{3}$, $\Delta x=-\dfrac{\pi}{12}$,

$f(x)\approx \dfrac{1}{\sqrt{3}}+\left(-\dfrac{4}{3}\right)\cdot\left(-\dfrac{\pi}{12}\right)=\dfrac{1}{\sqrt{3}}+\dfrac{\pi}{9}=0.926416$

11. $f(c)=\dfrac{1}{8}$, $f'(x)=\dfrac{-3x^{-5/2}}{2}$, $f'(c)=-\dfrac{3}{64}$, $\Delta x=0.21$,

$f(x)\approx \dfrac{1}{8}+\left(-\dfrac{3}{64}\right)\cdot 0.21=0.125-0.00984=0.11516$

13. $f(c)=\dfrac{1}{\sqrt{2}}$, $f'(x)=\sqrt{\pi}\cos\dfrac{\sqrt{\pi x}}{2\sqrt{x}}$, $f'(c)=\sqrt{2}$, $\Delta x=0.2-\dfrac{\pi}{16}$,

$f(x)\approx \dfrac{1}{\sqrt{2}}+\sqrt{2}\cdot\left(0.2-\dfrac{\pi}{16}\right)=0.70711+1.4142\cdot 0.00365=0.7122693098$

15. $f(x)=\sqrt{x}$, $c=25$, $f(c)=5$, $f'(x)=\dfrac{1}{2}\sqrt{x}$, $f'(c)=1/10$, $\Delta x=-1$,

$f(x)\approx 5+\dfrac{1}{10}\cdot(-1)=5-0.1=4.9$

The error is 0.001.

17. $f(x)=\sqrt{1+\sqrt{x}}$, $c=9$, $f(c)=2$, $f'(x)=\dfrac{1}{4}\sqrt{x(1+\sqrt{x})}$, $f'(c)=\dfrac{1}{24}$, $\Delta x=0.1$,

$f(x)\approx 2+\dfrac{1}{24}\cdot 0.1=2+0.00417=2.00417$

The error is 0.00001.

19. $f(x) = \cos(x)$, $c = \dfrac{\pi}{3}$, $f(c) = \dfrac{1}{2}$, $f'(x) = -\sin(x)$, $f'(c) = -\dfrac{\sqrt{3}}{2}$, $\Delta x = -\dfrac{\pi}{180}$,

$f(x) \approx \dfrac{1}{2} + \left(-\dfrac{\sqrt{3}}{2}\right) \cdot \left(-\dfrac{\pi}{180}\right) = 0.51511$

The error is 0.00007.

Further Theory and Practice

21. $df = g'(h(x))h'(x)\,dx$

23. The linearization $L(x) = f(0)$ because $f'(0) = 0$ for any differentiable even function $f(x)$. See Exercise 39, Section 3.2.

25. Calculating the product of the two functions, we obtain:

$$f(x)g(x) = \big(f(c) + f'(c)(x-c)\big)\big(g(c) + g'(c)(x-c)\big)$$
$$= f(c)g(c) + f(c)g'(c)(x-c) + f'(c)g(c)(x-c) + f'(c)g'(c)(x-c)^2$$
$$= f(c)g(c) + \big(f(c)g'(c) + f'(c)g(c)\big)(x-c) + f'(c)g'(c)(x-c)^2$$

The coefficient $\Big(f(c)g'(c) + f(c)' g(c)\Big)$ in front of $x - c$ is $(f \cdot g)'(c)$.

27. $y(x_1) \approx y(x_0) + y'(x_0)\Delta x = 2 + (1+2)(1.2-1) = 2 + 3 \cdot 0.2 = 2.6$

29. $y(x_1) \approx y(x_0) + y'(x_0)\Delta x = 3 + (0 - 2\cdot3)\left(\dfrac{1}{4} - 0\right) = 3 - \dfrac{3}{2} = \dfrac{3}{2}$

31. The slope of the demand curve is $q' = \dfrac{-1 - 4pq - 3q^3}{1 + 2p + 9pq^2} = -0.2612$.

Approximately 3.235 units ($q(6.80) \approx 3.248 - 0.2612 \cdot 0.05 = 3.235$) would be sold if the price is increased to $6.80.

Approximately 3.287 units ($q(6.60) \approx 3.248 - 0.2612 \cdot (-0.15) = 3.287$) would be sold if the price were decreased to $6.60.

33. The slope of the demand curve is $q(10.0) = 40000$; $q' = -2\left(1 + 2\sqrt{q}\right)\dfrac{q}{p}\left(1 + 4\sqrt{q}\right) = -4004.99$.

Approximately 41001.2 units ($q(9.75) \approx 40000 - 4004.99\,(-0.25) = 41001.2$) can be sold at $9.75.

Calculator/Computer Exercises

---

35. For $f(x) = x^{1/3}$, $c = 27$, and $\Delta x = 0.9$, we have $f'(x) = \dfrac{x\big|^{-2/3}}{3}$, $f(c) = 3$, $f'(c) = \dfrac{1}{27}$, and the linear approximation is

$$27.9^{1/3} = f(27 + 0.9) \approx f(27) + f'(27)(27.9 - 27) = 3 + \frac{1}{27}(0.9) = 3.03333$$

to five decimal places. Next, we use the Method of Increments with increment $h = \dfrac{\Delta x}{3} = 0.3$ and three steps. We will carry two extra decimal places until we round to five decimal places for our final answer. Let $x_0 = c = 27$, $x_1 = c + h = 27.3$, $x_2 = c + 2h = 27.6$, and

$$x_3 = c + 3h = c + \Delta x = 27.9.$$

For $k = 1$, 2 and 3, we have $f(x_k) \approx y_k = y_{k-1} + f'(x_{k-1}) \cdot h$. We calculate

$$y_1 = 3 + \frac{(27)^{-2/3}}{3}(0.3) \approx 3.0111111$$

$$y_2 = 3.0111111 + \frac{(27.3)^{-2/3}}{3}(0.3) \approx 3.0221407$$

$$y_3 = 3.0221407 + \frac{(27.6)^{-2/3}}{3}(0.3) \approx 3.03309.$$

The exact value of $f(27.9)$ when rounded to five decimal places is 3.03297. The three step approximation is therefore more accurate than the one step approximation.

37.   For $f(x) = \ln(x)$, $c = e$, and $\Delta x = 3 - 1$, we have $f'(x) = \dfrac{1}{x}$, $f(c) = 1$, $f'(c) = \dfrac{1}{e}$, and the

linear approximation is $1 + \dfrac{1}{e}(3 - e) = 1.10363832351433$,

$$\ln(3) = f\big(e + (3 - e)\big) \approx f(e) + f'(e)(3 - e) = 1 + \dfrac{1}{e}(3 - e) = 1.10364$$

to five decimal places. Next, we use the Method of Increments with increment $h = \dfrac{\Delta x}{2} = \dfrac{3 - e}{2}$
and two steps. We will carry two extra decimal places until we round to five decimal places
for our final answer. Let $x_0 = c = e$, $x_1 = c + h = \dfrac{e}{2} + \dfrac{3}{2}$, and $x_2 = c + 2h = 3$. For $k = 1$ and 2,
we have $f(x_k) \approx y_k = y_{k-1} + f'(x_{k-1}) \cdot h$. We calculate

$$y_1 = 1 + \dfrac{1}{e}\left(\dfrac{3 - e}{2}\right) \approx 1.0518192$$

$$y_2 = 1.0518192 + \dfrac{1}{e/2 + 3/2}\left(\dfrac{3 - e}{2}\right) = 1.10109$$

rounded to five decimal places. The exact value of $f(3)$ when rounded to five decimal places
is 1.09861. The two step approximation is therefore more accurate than the one step
approximation.

39. $y_1 \approx y(x_0) + y'(x_0)\Delta x = 2 + 3 \cdot 0.2 = 2.6$, $m_1 = 3.4$,
$z_1 \approx y(x_0) + m_1\Delta x = 2 + 3.4 \cdot 0.2 = 2.68$

The exact solution is $y(x_1) = 2.68561$; therefore the z_1 approximation is more accurate.

41. $y_1 \approx y(x_0) + y'(x_0)\Delta x = 3 - 6 \cdot 0.25 = 1.5$, $m_1 = -4.46875$,
$z_1 \approx y(x_0) + m_1\Delta x = 3 - 4.46875 \cdot 0.25 = 1.88281$

The exact solution is $y(x_1) = 1.82421$; therefore the z_1 approximation is more accurate.

43. From the demand equation we have: $q(5.10) = 9611.65$; $q' = -4q \dfrac{1 - 3q^{3/4}}{p\left(1 + 2pq^{3/4}\right)}$;

$q'(5.10) = -3769.66$.

The estimated demand at price p_1 is $q(5) \approx 9611.65 + 3769.66 \cdot 0.10 = 9988.62$.

The exact demand is $q(5) = 10000$.

The relative error is 0.11%.

45. From the demand equation we calculate: $q(1.80) = 3950655$; $q' = -2q \dfrac{25 + pq^{4/5}}{p\left(10 + pq^{4/5}\right)}$;

$q'(1.80) = -4389809$.

The estimated demand at the price value p_1 is $q(2) \approx 3950655 + (-4389809) \cdot 0.2 = 3072693$.

The exact demand is $q(2) = 3200000$.

The relative error is 3.98%.

Applications of the Derivative

Problems for Practice

1. The inequality $f(x) = (x-2)^2 + 3 \geq 0 + 3$ has equality at $x = 2$. Thus, the minimum of 3 is at $x = 2$. Local minimum: 2

3. The inequality $f(x) = \dfrac{1}{x^2 + 1} \leq \dfrac{1}{0+1}$ has equality at $x = 0$. Thus, the maximum of 1 is at $x = 0$. Local maximum: 0

5. Because $\sin(x)$ has local maxima at $\left\{(2n+1)\dfrac{\pi}{2} : n \in \mathbb{Z}\right\}$ and minima at $\left\{(2n+3)\dfrac{\pi}{2}\right\}$, our (x) has local maxima: $\left\{(4k+1)\dfrac{\pi}{8} : k \in \mathbb{Z}\right\}$; local minima: $\left\{(4k+3)\dfrac{\pi}{8} : k \in \mathbb{Z}\right\}$.

7. The derivative is $f'(x) = 4x - 24$, so $x = 6$ is the only candidate. We calculate $f(6) = -36$ and $f(6+h) = 2(6+h)^2 - 24(6+h) + 36 = -36 + 2h^2 = f(6) + 2h^2$. As $f(6+h) \geq f(6)$ for every h, f has the only local minimum of -36 at $x = 6$.

9. The derivative is $f'(x) = 4x^3 - 4x = 4x(x-1)(x+1)$, so $x = 0$, $x = 1$, and $x = -1$ are candidates. As $f(x) = (x^4 - 2x^2 + 1) = (x^2 - 1)^2$ we see that f has the only local minimum of 0 at $x = -1$ and $x = 1$. Moreover, $f(x) = 1 - x^2(2 - x^2)$. Given that $(x^2)(2 - x^2)$ is the product of two positive expressions when x is close to 0, we see that $f(x)$ has the only local maximum at $x = 0$.

11. Given that $f'(x) = (x-5)^3 + 3(x-3)(x-5)^2 = 4(x-5)^2(x-1)$, we find the following candidates: $x = 5$, which is not an extremum, and $x = 1$, which is the only local minimum.

13.   Given that $f'(x)=1-1/x=(x-1)/x$, we find the only candidate to be $x=1$, which is the only local maximum.

15.   Given that $f'(x)=e^x-1$, we find the only candidate to be $x=0$, which is the only local minimum.

17.   Since the derivative of $5x$ is 5 and two functions have the same derivatives if and only if they differ by a constant, all the functions with derivative 5 are the functions of form $5x+C$.

19.   Since the derivative of $x^3/3+\pi x$ is $x^2+\pi$ and two functions have the same derivatives if and only if they differ by a constant, all the functions with derivative $x^2+\pi$ are the functions of form $x^3/3+\pi x+C$.

21.   Since the derivative of $\sin(x)$ is $\cos(x)$ and two functions have the same derivatives if and only if they differ by a constant, all the functions with derivative $\cos(x)$ are the functions of form $\sin(x)+C$.

23.   The function is continuous and differentiable on $(0, 1)$. As $f(0)=f(1)=0$, we can assert that $4c^3+21c^2-18c+1=0$ for some $c\in(0, 1)$.

25.   The function is continuous and differentiable on $(1,32)$. As $f(32)-f(1)=1$, we can assert that $(1/5)c^{-4/5}=1/31$ for some $c\in(1, 32)$.

27.   The function is continuous and differentiable on $(0, 2)$. As $f(2)=f(0)=0$, we can assert that $-4(c-1)^3=0$ for some $c\in(0, 2)$.

Further Theory and Practice

29.   The derivative of $g(x)$ is: $g'(x)=f'(x)-\frac{f(b)-f(a)}{b-a}$. In accordance with the Rolle's theorem, there is $c\in(a, b)$ such that $g'(c)=0$ or $f'(c)-\frac{f(b)-f(a)}{b-a}=0$, so that $f'(c)=\frac{f(b)-f(a)}{b-a}$.

31.   On this interval $f'(x)$ is never 0 as the Mean Value Theorem would imply. There is no contradiction because $f(x)$ is not differentiable at $x=1\in I$.

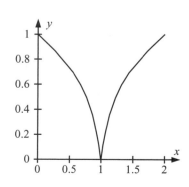

33. On this interval $f'(x)$ is never 0 as the Mean Value Theorem would imply. There is no contradiction because $f(x)$ is not differentiable at $x = 0 \in I$.

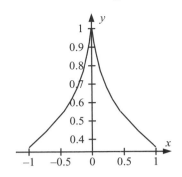

35. $f'(x) = 3x^2 - 1$; $f'\left(\dfrac{1}{\sqrt{3}}\right) = 0$; $c = \dfrac{1}{\sqrt{3}}$

37. $f'(x) = \dfrac{(2x+1)(x^2+1)-(x^2+x)2x}{(x^2+1)^2} = \dfrac{1+2x-x^2}{(x^2+1)^2}$; $f'(1 \pm \sqrt{2}) = 0$; $c = 1 - \sqrt{2}$

39. $f'(x) = 2Ax$; $\dfrac{f(b)-f(a)}{b-a} = A(a+b)$; $f'(c) = A(a+b)$; $c = \dfrac{a+b}{2}$

41. $f'(x) = 1 - \dfrac{1}{x^2}$; $\dfrac{f(b)-f(a)}{b-a} = \dfrac{1}{2}$; $f'(c) = \dfrac{1}{2}$; $c = \sqrt{2}$

43. We calculate the derivatives to be $f'(x) = -2\cos(x)\sin(x)$; $g'(x) = 2\sin(x)\cos(x)$. It follows that $\left(\cos^2(x) + \sin^2(x)\right)' = f'(x) + g'(x) = 0$ for all real x, and that

$$f(x) + g(x) = \text{const}.$$

As $f(0) + g(0) = 1$, we conclude that $f(x) + g(x) = \cos^2(x) + \sin^2(x) = 1$ for all real x.

45. By the Mean Value Theorem there exists $c \in (a,\, a+h)$ such that $f'(c) = \frac{f(a+h)-f(a)}{h}$. We denote $\theta = \frac{c-a}{h}$. It follows that if $c = a + \theta h$ then $f(a+h) = f(a) + h \cdot f'(a+\theta h)$.

47. We denote $p(x) = x^3 - 3x^2 + 4x - 1$. We then see that the derivative, $p'(x) = 3x^2 - 6x + 4$, is everywhere defined and does not have real roots, and consequently $p(x)$ cannot have extrema. Furthermore, $p(0) < 0$ and $p(1) > 0$. Given that the function $p(x)$ is continuous, there exists $c \in (0,\, 1)$ such that $p(c) = 0$. As $p(x)$ does not have extrema, c is the only real root of $p(x)$.

49. We calculate the derivative: $p'(x) = 3x^2 + 2ax$. We see that $p'(x)$ is always defined and can have at most one negative root. If $p(x)$ had three negative roots, then according to Rolle's Theorem $p'(x)$ would have two negative roots.

51. We calculate the derivative: $p'(x) = 3x^2 + a$. We see that $p'(x)$ is always defined and can have at most one negative root. If $p(x)$ had three negative roots, then according to Rolle's Theorem $p'(x)$ would have two negative roots.

53. We denote: $h(x) = f(x) - g(x)$. It follows that $h'(x) > 0$ for $x > a$ and $h(a) \geq 0$. Using the Mean Value Theorem, we find that for every $x > a$ there exists $c \in (a, x)$ such that $h(x) = h(a) + h'(c)(c - a)$. So for every $x > a$, the inequalities $h(x) > 0$, or $f(x) > g(x)$ hold.

55. We denote: $f(x) = 1 + px - (1 + x)^p$. It follows that $f'(x) = p - p(1 + x)^{p-1}$. In accordance with the Mean Value Theorem, we find that for all $x > 0$ there exists $c \in (0, x)$ such that $f(x) = f(0) + f'(c)(x - 0) = x\left(p - p(1 + c)^{p-1}\right)$. So for every $x > 0$, $f(x) > 0$ or $(1 + x)^p < 1 + px$.

57. $x_{N+1} = \Phi(x_N) = \Phi(x_*) = x_*$. Repeating this manipulation for $N + 1$, $N + 2$, ... we show that $x_n = x_*$ for any $n > N$.

59. According to the Mean Value Theorem, for all x there exists c such that $\Phi'(c) = \dfrac{\Phi(x) - \Phi(x_*)}{x - x_*} = \dfrac{\Phi(x) - x_*}{x - x_*}$.

For $x = x_*$ we calculate: $\Phi'(x_*) = \lim\limits_{x \to x_*} \dfrac{\Phi(x) - \Phi(x_*)}{x - x_*}$.

As $|\Phi'(x_*)| < 1$, there exists $\delta > 0$ and $K < 1$ such that $\left|\dfrac{\Phi(x) - x_*}{x - x_*}\right| < K$ for all $x \in (x_* - \delta, x_* + \delta)$.

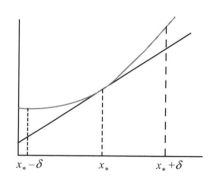

61. As in Problem 4.1.60, $|x_2 - x_*| = |\Phi(x_1) - \Phi(x_*)| = |\Phi'(c)(x_1 - x_*)| \leq K|x_1 - x_*| < K^2|x_0 - x_*|$, and so on. We conclude that $|x_n - x_*| < K^n|x_0 - x_*|$.

63. Let $f(x) = e^{-kx} p(x)$. We calculate the derivative:

$$f'(x) = -ke^{-kx} p(x) + e^{-kx} p'(x) = e^{-kx}(p'(x) - kp(x)).$$

Furthermore, we calculate: $f(a) = f(b) = 0$. Using Rolle's Theorem, we see that there exists $c \in (a, b)$, such that $f'(c) = 0$ or $p'(c) = kp(c)$.

65. We denote $h(x) = f(x)g(x)$. It follows that $h(a) = h(b) = 0$. Using Rolle's Theorem, we see that there exists $c \in (a, b)$, such that $h'(c) = 0$. As $h'(x) = f'(x)g(x) + f(x)g'(x)$, we obtain the formula: $f'(c)g(c) + f(c)g'(c) = 0$.

67.   We calculate the derivative: $f'(x) = \dfrac{1}{2}\sqrt{1+x}$. Using the Mean Value Theorem, we see that

there exists $c \in \left(\dfrac{h}{2}, \, h\right)$, such that $f(1+h) - f\left(1+\dfrac{h}{2}\right) = f'(c) \cdot \dfrac{h}{2} = \dfrac{h}{4}\sqrt{1+c}$. So the limit is

$$\lim_{c \to 0}\left(\frac{\sqrt{1+h} - \sqrt{1+h/2}}{h}\right) = \lim_{c \to 0}\frac{1}{4\sqrt{1+c}} = \frac{1}{4}.$$

69.   We calculate: $g(a) = f\left(\dfrac{a+b}{2}\right)$; $g\left(\dfrac{a+b}{2}\right) = -f\left(\dfrac{a+b}{2}\right)$. Given that $g(x)$ is a continuous

function, there exists $a_1 \in \left[a, \, \dfrac{a+b}{2}\right)$ such that $g(a_1) = 0$.

We calculate: $g(a_1) = f\left(a_1 + \dfrac{b-a}{2}\right) - f(a_1) = f(b_1) - f(a_1) = 0$, so $f(b_1) = f(a_1)$.

Applying the Intermediate Value Theorem to the interval $\left[a_1, \, \dfrac{a_1 + b_1}{2}\right)$, we obtain a pair

$(a_2, \, b_2)$ such that $f(b_2) = f(a_2)$ and $b_2 - a_2 = \dfrac{b_1 - a_1}{2}$, and so on.

Using the Mean Value Theorem, we see that for any pair $(a_n, \, b_n)$ there exists $c_n \in (a_n, \, b_n)$

such that $f'(c_n) = \dfrac{f(b_n) - f(a_n)}{b_n - a_n} = 0$. As $n \to \infty$ we get $c \in (a, \, b)$ such that $f'(c) = 0$.

Calculator/Computer Exercises

71.   Local maximum: $x = -0.68$
Local minimum: $x = 0.97$
Thin line: $f(x) = x^3 - 2x + \cos(x)$
Thick line: $f'(x) = 3x^2 - 2 - \sin(x)$

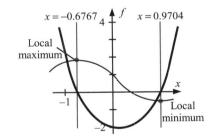

73.   Local maximum: $x = -1.27$
Local minimum: $x = -0.27$
Thin line: $f(x) = x - 2\exp\left(-x^2\right)$
Thick line: $f'(x) = 1 + 4xe^{-x^2}$

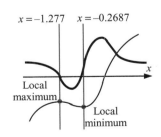

75. $c = 1.3124$

Upper straight line: $y = 20x + 1$

Lower straight line: $y = f'(c)(x - c) + f(c)$

Thick line: $y = x^5 + x^3 + 1$

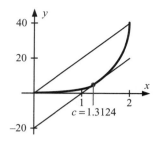

77. $c = 1.826$

Upper straight line: $y = 6x + 25$

Lower straight line: $y = 6(x - c) + f(c)$

Thick line: $y = x^4 - 2x^3 + x^2 - 2x + 13$

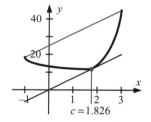

Section 4.2   Maxima and Minima of Functions

Problems for Practice

1. We calculate the derivative: $f'(x) = 3x^2 + 6x - 45 = 3(x^2 + 2x - 15) = 3(x + 5)(x - 3)$.

$f'(x) > 0$ on each of the intervals $(-\infty, -5)$ and $(3, \infty)$; $f'(x) < 0$ on $(-5, 3)$. Therefore, f is increasing on each of the intervals $(-\infty, -5)$ and $(3, \infty)$, and is decreasing on the interval $(-5, 3)$.

3. We calculate the derivative $f'(x) = \dfrac{(x^2 - 5)}{(x^2 + 5)^2} = \dfrac{(x - \sqrt{5})(x + \sqrt{5})}{(x^2 + 5)^2}$, and conclude that

$f'(x) > 0$ on each of the intervals $(-\infty, -\sqrt{5})$ and $(\sqrt{5}, \infty)$; $f'(x) < 0$ on $(-\sqrt{5}, \sqrt{5})$. Hence, f is increasing on each of the intervals $(-\infty, -\sqrt{5})$ and $(\sqrt{5}, \infty)$, and decreasing on the intervals $(-\sqrt{5}, \sqrt{5})$.

5. We calculate the derivative and conclude that $f'(x) = -\dfrac{2}{(x - 1)^2} < 0$ on $(-\infty, 1) \cup (1, \infty)$, hence f is decreasing on each of the intervals $(-\infty, 1)$ and $(1, \infty)$.

7.   We calculate the derivative

$$f'(x) = 2(x+1)(x+2)^2 + 2(x+1)^2(x+2) = 2(x+1)(x+2)(2x+3),$$

and conclude that $f'(x) > 0$ on each of the intervals $\left(-2, -\dfrac{3}{2}\right)$ and $(-1, \infty)$; $f'(x) < 0$ on

each of the intervals $(-\infty, -2)$ and $\left(-\dfrac{3}{2}, -1\right)$. Hence f is increasing on each of the intervals

$\left(-2, -\dfrac{3}{2}\right)$ and $(-1, \infty)$, and decreasing on each of the intervals $(-\infty, -2)$ and $\left(-\dfrac{3}{2}, -1\right)$.

9.   We calculate the derivative and conclude that

$$f'(x) = -4\cos(x)\sin(x) + 3 = -2\sin(2x) + 3 > 0$$

on $\mathbb{R}$, hence f is increasing on the interval $(-\infty, \infty)$.

11.   We calculate the derivative $f'(x) = e^x - 1$, and conclude that $f'(x) > 0$ on $(0, \infty)$ and $f'(x) < 0$ on the interval $(-\infty, 0)$; hence f is increasing on the interval $(0, \infty)$ and decreasing on the interval $(-\infty, 0)$.

13.   For the derivative $f'(x) = 2x + 1$, $c = -\dfrac{1}{2}$ is a critical point. As $f'(x) > 0$ for $x > -\dfrac{1}{2}$, and $f'(x) < 0$ for $x < -\dfrac{1}{2}$, we find that $c = -\dfrac{1}{2}$ is a local minimum.

15.   We calculate the derivative $f'(x) = 5x^4 - 20x^3 = 5x^3(x-4)$. There are two critical points: $c_1 = 0$ and $c_2 = 4$. Given that $f'(x) > 0$ for $x < c_1$ and $x > c_2$, and $f'(x) < 0$ for $c_1 < x < c_2$, the local maximum is at $c_1 = 0$ and the local minimum is at $c_2 = 4$.

17.   We calculate the derivative $f'(x) = -\dfrac{4x^3}{\left(x^4 + 6\right)^2}$. The only critical point is at $c = 0$. Given that $f'(x) > 0$ for $x < c$, and $f'(x) < 0$ for $x > c$, the local maximum is at $c = 0$.

19. $f'(x) = 2xe^{-x} - x^2 e^{-x} = xe^{-x}(2-x)$. There are two critical points: $c_1 = 0$ and $c_2 = 2$. Given that $f'(x) > 0$ for $c_1 < x < c_2$ and $f'(x) < 0$ for $x < c_1$ and $x > c_2$, the local minimum is at $c_1 = 0$ and the local maximum is at $c_2 = 2$.

21.   The function is: $f(x) = \begin{cases} x^2 - x + 7 & \text{if} \quad x \le 7 \\ x^2 + x - 7 & \text{if} \quad x > 7 \end{cases}$. We calculate the derivative

$$f'(x) = \begin{cases} 2x - 1 & \text{if} \quad x < 7 \\ 2x + 1 & \text{if} \quad x > 7 \end{cases}.$$

The critical points are: $c_1 = \dfrac{1}{2}$ and $c_2 = 7$. We find that $f'(x) > 0$ for $x > c_1$, and $f'(x) < 0$ for $x < c_1$. Hence, the local minimum is at $c_1 = \dfrac{1}{2}$ and there are no local extrema at $c_2 = 7$.

23.   The function is: $f(x) = \begin{cases} \sin(x) & \text{if} \quad x \in \bigcup\limits_{k \in \mathbb{Z}} [2\pi k, \ \pi + 2\pi k] \\ -\sin(x) & \text{if} \quad x \in \bigcup\limits_{k \in \mathbb{Z}} (\pi + 2\pi k, \ 2\pi + 2\pi k) \end{cases}$. Its derivative is:

$f'(x) = \begin{cases} \cos(x) & \text{if} \quad x \in \bigcup\limits_{k \in \mathbb{Z}} (2\pi k, \ \pi + 2\pi k) \\ -\cos(x) & \text{if} \quad x \in \bigcup\limits_{k \in \mathbb{Z}} (\pi + 2\pi k, \ 2\pi + 2\pi k) \end{cases}$. The critical points are $c_n = \pi n$ and

$b_n = \dfrac{\pi}{2} + \pi n$, $n \in \mathbb{Z}$. For small positive ϵ we see that: $f'(x) > 0$ for $x = c_n + \epsilon$, and $f'(x) < 0$ for $x = c_n - \epsilon$. Hence $c_n = \pi n$ and $n \in \mathbb{Z}$ are local minima; $f'(x) < 0$ for $x = b_n + \epsilon$, $f'(x) > 0$ for $x = b_n - \epsilon$, hence $b_n = \dfrac{\pi}{2} + \pi n$, $n \in \mathbb{Z}$, are local maxima.

25.   We calculate the derivative

$$f'(x) = \frac{1}{5}(x-4)^{-4/5}(3x+6)^{2/3} + (x-4)^{1/5}\frac{2}{3}(3x+6)^{-1/3}3 = (x-4)^{-4/5}(3x+6)^{-1/3}\frac{13x-34}{5}.$$

There are three critical points: $c_1 = -2$, $c_2 = \dfrac{34}{13}$ and $c_3 = 4$. Given that $f'(x) > 0$ for $x < c_1$, $c_2 < x < c_3$ and $x > c_3$; $f'(x) < 0$ for $c_1 < x < c_2$, there is a local maximum at $c_1 = -2$, a local minimum is at $c_2 = \dfrac{34}{13}$, and there is no local extremum at $c_3 = 4$.

27. The function and its derivative are: $f(x) = \begin{cases} x^{1/3} + x & \text{if} \quad x \le 0 \\ x^{1/3} - x & \text{if} \quad x > 0 \end{cases}$,

$$f'(x) = \begin{cases} \left(\dfrac{1}{3}\right) x^{-2/3} + 1 & \text{if} \quad x < 0 \\ \left(\dfrac{1}{3}\right) x^{-2/3} - 1 & \text{if} \quad x > 0 \end{cases}.$$

There are two critical points: $c_1 = 0$ and $c_2 = \dfrac{\sqrt{3}}{9}$. Given that $f'(x) > 0$ for $x < c_1$ and

$c_1 < x < c_2$, and $f'(x) < 0$ for $x > c_2$, local maximum is at $c_2 = \dfrac{\sqrt{3}}{9}$ and there are no local

extrema at $c_1 = 0$.

29. The function f is increasing on each of the intervals $\left(-\dfrac{3}{2}, -1\right)$, $\left(0, \dfrac{3}{2}\right)$ and $\left(2, \dfrac{5}{2}\right)$, and

decreasing on each of the intervals $(-1, 0)$ and $\left(\dfrac{3}{2}, 2\right)$.

Further Theory and Practice

31. Given that $f'(2) > 0$, the function is increasing to the right of $x = 2$ so that for small positive

ϵ we see that $f(2 + \epsilon) > f(2)$. In our case, $\epsilon = 10^{-5}$.

33. If f is differentiable and $f' > 0$ everywhere, then $(x) > f(y)$ for all $x > y$. Hence, f is one-to-one. If $f' < 0$ everywhere, then $f(x) < f(y)$ for all $x > y$. Hence f is one-to-one. Likewise, if $f' > 0$ except at finitely many points, if $f(x) = f(y)$ then either $f'(x) = 0$ on (x, y), or there is an interval on which f is decreasing.

35. For $\alpha < \beta$, $\alpha, \beta \in I$, the inequality $f(\alpha) < f(\beta)$ holds. Then

$$f^3(\alpha) - f^3(\beta) = (f(\alpha) - f(\beta))\left(f^2(\alpha) + f(\alpha)f(\beta) + f^2(\beta)\right) < 0,$$

because $f^2(\alpha) + f(\alpha)f(\beta) + f^2(\beta) > 0$ and $f(\alpha) - f(\beta) < 0$. Therefore, f^3 is increasing.

37. For $a < b$, the following inequalities hold: $f(a) < f(b)$ and $g(f(a)) < g(f(b))$; therefore, $g \circ f$ is increasing. It follows that $(g \circ f)'(t) = g'(f(t)) \cdot f'(t) > 0$ is a product of two positive values, hence $g \circ f$ is increasing.

39. We calculate the derivative $p'(t) = 12t^2 - 14t + 4 = 12\left(t - \dfrac{1}{2}\right)\left(t - \dfrac{2}{3}\right)$, and conclude that

$p'(t) > 0$ on the intervals $\left(-\infty, \dfrac{1}{2}\right)$ and $\left(\dfrac{2}{3}, \infty\right)$, and $p'(t) < 0$ on the interval $\left(\dfrac{1}{2}, \dfrac{2}{3}\right)$.

Hence the automobile is traveling forward at $t < \dfrac{1}{2}$ and $t > \dfrac{2}{3}$, and is going in reverse at

$\dfrac{1}{2} < t < \dfrac{2}{3}$.

41. The function with roots at a and b can be written as $f(x) = (x-a)(x-b)g(x)$, where g is a polynomial and $g(a) \neq 0$, $g(b) \neq 0$. Then the derivative is:

$$f'(x) = (x-b)g(x) + (x-a)g(x) + (x-a)(x-b)g'(x),$$

so that $f'(a) = (a-b)g(a)$ and $f'(b) = (b-a)g(b)$. We conclude that

$$f'(a)f'(b) = -(a-b)^2 g(a)g(b).$$

Therefore, the sign of $f'(a)g'(b)$ is opposite to the sign of $g(a)g(b)$. For $g(x) = c$ (there are no other roots of f), we see that $f'(a)g'(b) < 0$.

43. The function $f(n)$ is an increasing function—higher level of labor input produces bigger commodity output. Each additional unit of work leads to a smaller increase in product function f. MPL is a decreasing function of n: every additional unit of work results in a smaller additional output.

45. We calculate the derivative $\dfrac{d}{dx}\left(f^k(x)\right) = kf^{k-1}(x)f'(x)$, $f^{k-1}(x) > 0$ for all real x, and

conclude that $\sin\left(\dfrac{d}{dx} f^k(x)\right) = \sin\left(f'(x)\right)$. The two functions have the same critical points and the same character of the sign changes near these critical points. They therefore have the same extrema.

47. The rise of f leads to the rise of $f \cdot g$ and of $\dfrac{f}{g}$, hence the sign in front of $f'(c)g(c)$ is

positive. The rise of g leads to the rise of $f \cdot g$ (plus sign preceding $f(c)g'(c)$) and to the

decrease of $\dfrac{f}{g}$ (minus sign preceding $f(c)g'(c)$).

49. **a.** The function $\dfrac{k}{T}$ is a decreasing function of T, hence $\exp\left(\dfrac{k}{T}\right)$ is decreasing as well. It is multiplied by the decreasing function $AT^{-3/2}$, hence

$$AT^{-3/2}\exp\left(\frac{k}{T}\right)$$

is a decreasing function of T.

b. Let $0 < x_1 < x_2 < 1$, then

$$\frac{1-x_1}{x_1^2} - \frac{1-x_2}{x_2^2} = \frac{x_2^2 - x_1 x_2^2 - x_1^2 + x_2 x_1^2}{x_1^2 x_2^2} = \frac{(x_2 - x_1)(x_1(1-x_2) + x_2)}{x_1^2 x_2^2} > 0.$$

Hence $\dfrac{1-x}{x^2}$ is decreasing for $x \in (0, 1)$.

c. Larger values of T correspond to the smaller right-hand side which, in turn, corresponds to larger x; hence x is an increasing function of T.

d. We can assign a unique value of x to each T; that is, x is a function of T.

e. We differentiate implicitly:

$$\frac{-x'x^2 - 2xx'(1-x)}{x^2} = -\frac{3}{2}AT^{-5/2}\exp\left(\frac{k}{T}\right) + AT^{-3/2}\exp\left(\frac{k}{T}\right)\left(-\frac{k}{T^2}\right),$$

$$x'\left(\frac{2-x}{x}\right) = \frac{3}{2}AT^{-5/2}\exp\left(\frac{k}{T}\right) + AT^{-3/2}\exp\left(\frac{k}{T}\right)\left(\frac{k}{T^2}\right).$$ Then the right-hand side is

positive, $\dfrac{2-x}{x} > 0$, hence $x' > 0$ and x is an increasing function of T.

51. We calculate the derivative $f'(x) = p - px^{p-1} = p\left(1 - x^{p-1}\right)$, to conclude that $f'(x) > 0$ for

$0 < x < 1$, $f'(x) < 0$ for $x > 1$. Hence 1 is a local minimum of f and $f(1) \le f(x)$ for $x > 0$.

As $f(1) = p - 1$, we deduce: $p - 1 \le px - x^p$ and $x^p \le px + 1 - p$. For $x = \sqrt{n}$ and $p = \dfrac{1}{n}$ we

calculate: $n^{1/n} \le 1 + \left(\dfrac{1}{\sqrt{n}}\right) - \left(\dfrac{1}{n}\right)$. It is evident that $n^{1/n} \ge 1^{1/n} = 1$, hence

$$1 \le n^{1/n} \le 1 + \left(\frac{1}{\sqrt{n}}\right) - \left(\frac{1}{n}\right).$$

As $\displaystyle\lim_{n\to\infty}\left(1 + \left(\frac{1}{\sqrt{n}}\right) - \left(\frac{1}{n}\right)\right) = 1$, the limit is $\displaystyle\lim_{n\to\infty} n^{1/n} = 1$.

Calculator/Computer Exercises

53.   The derivative is $f'(x) = 4x^3 - 2x - 7.1$. There is a unique critical point at $c = 1.3479$, and $f'(x) > 0$ for $x > c$; $f'(x) < 0$ for $x < c$. Hence f is increasing on the interval $(1.3479, \infty)$ and decreasing on the interval $(-\infty, 1.3479)$.

55.   The derivative is $f'(x) = 10x^4 - 4.8x^3 - 5.1x^2 + 49x - 53.7$. There are two critical points: $c_1 = -1.90757$ and $c_2 = 1.06832$. Given that the derivative $f'(x) > 0$ for $x < c_1$ and $x > c_2$, and $f'(x) < 0$ for $c_1 < x < c_2$, hence f is increasing on the interval $(-\infty, -1.90757)$ and $(1.06832, \infty)$; it is decreasing on the interval $(-1.90757, 1.06832)$.

57.   The derivative is $f'(x) = \dfrac{-2x^5 - 3x^4 - 4x^3 + 2x + 1}{\left(x^4 + 1\right)^2}$. There is a unique critical point: $c = 0.69725$.

Given that the derivative $f'(x) > 0$ for $x < c$, and $f'(x) < 0$ for $x > c$; hence f is increasing on the interval $(-\infty, 0.69725)$ and decreasing on the interval $(0.69725, \infty)$.

59.   The derivative is $f'(x) = -e^{-x} \ln(x) + \dfrac{e^{-x}}{x}$. There is a unique critical point at $c = 1.76322$.

Given that the derivative $f'(x) > 0$ for $0 < x < c$ and $f'(x) < 0$ for $x > c$, hence f is increasing on the interval $(0, 1.76322)$ and decreasing on the interval $(1.76322, \infty)$.

61.
$$f(x) = \begin{cases} \sqrt{\sin^2(x) - \cos(x)} & \text{if } x_0 \le |x| \le \dfrac{\pi}{2} \\ \sqrt{\cos(x) - \sin^2(x)} & \text{if } |x| < x_0 \end{cases},$$

$$f'(x) = \begin{cases} \dfrac{\sin(x) + \sin(2x)}{2\sqrt{\sin^2(x) - \cos(x)}} & \text{if } x_0 < |x| < \dfrac{\pi}{2} \\ -\dfrac{\sin(x) + \sin(2x)}{2\sqrt{\cos(x) - \sin^2(x)}} & \text{if } |x| < x_0 \end{cases},$$

where $x_0 = \arctan\left(\dfrac{\sqrt{2\sqrt{5} - 2}}{\sqrt{5} - 1}\right) \approx 0.90456$. There are three critical points: $c_1 = -x_0$, $c_2 = 0$

and $c_3 = x_0$. Given that the derivative $f'(x) > 0$ for $c_1 < x < c_2$ and $c_3 < x < \dfrac{\pi}{2}$; $f'(x) < 0$ for

$-\dfrac{\pi}{2} < x < c_1$ and $c_2 < x < c_3$, hence f is increasing on the intervals $(-0.90456, 0)$ and

$\left(0.90456, \dfrac{\pi}{2}\right)$, and decreasing on the intervals $\left(-\dfrac{\pi}{2}, -0.90456\right)$ and $(0, 0.90456)$. There

are two local minima at ± 0.90456 and there is a local maximum at 0.

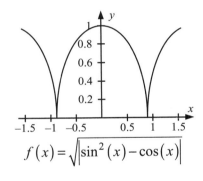

$$f(x) = \sqrt{\left|\sin^2(x) - \cos(x)\right|}$$

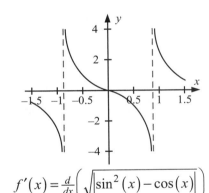

$$f'(x) = \frac{d}{dx}\left(\sqrt{\left|\sin^2(x) - \cos(x)\right|}\right)$$

63. $f'(x) = \exp\left(x^2 - 2x\right) \cdot \left[2x\ln(x)\ln(1-x) - 4x^2\ln(x)\ln(1-x)\right.$

$\left. + 2x^3\ln(x)\ln(1-x) - \ln(1-x) + x\ln(1-x) + x\ln(x)\right]$

There is a unique critical point at $c = 0.27777$. Given that the derivative $f'(x) > 0$ for $0 < x < c$, and $f'(x) < 0$ for $c < x < 1$, hence f increases on the interval $(0,\ 0.27777)$ and decreases on the interval $(0.27777,\ 1)$. Local maximum is at $c = 0.27777$.

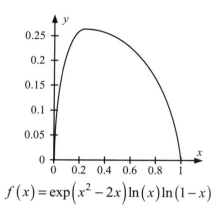

$f(x) = \exp\left(x^2 - 2x\right)\ln(x)\ln(1-x)$

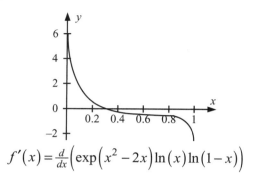

$f'(x) = \frac{d}{dx}\left(\exp\left(x^2 - 2x\right)\ln(x)\ln(1-x)\right)$

65. **a.** The derivative is

$f'(x) = 5x^4 - 48x^3 + 165x^2 - 240x + 124$. It changes from positive $(S = 1)$ to negative $(S = -1)$ at points A and C and from negative to positive at points B and D. The local maxima are at A and C and the local minima are at B and D.

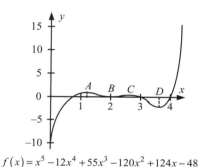

$f(x) = x^5 - 12x^4 + 55x^3 - 120x^2 + 124x - 48$

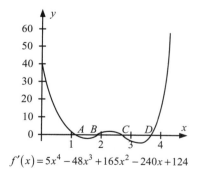

$f'(x) = 5x^4 - 48x^3 + 165x^2 - 240x + 124$

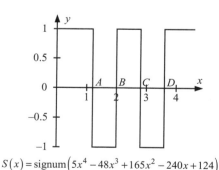

$S(x) = \operatorname{signum}\left(5x^4 - 48x^3 + 165x^2 - 240x + 124\right)$

b.   The derivative is

$$f'(x) = 16x^3 - 72x^2 + 102x - 44.$$

It changes from negative $(S = -1)$ to positive $(S = 1)$ at points A and B, hence these points are the local minima. f' changes from positive to negative at point B, which is, therefore, the local maximum.

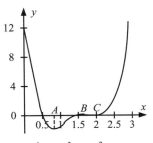

$$f(x) = 4x^4 - 24x^3 + 51x^2 - 44x + 12$$

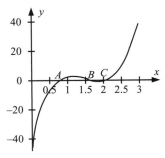

$$f'(x) = 16x^3 - 72x^2 + 102x - 44$$

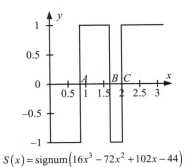

$$S(x) = \mathrm{signum}(16x^3 - 72x^2 + 102x - 44)$$

c.   The derivative is $f'(x) = 4x^3 - 3x^2 - 14x + 1$, It changes from negative $(S = -1)$ to positive $(S = 1)$ at points A and C, hence these are the local minima. f' changes from negative to positive at point B; this is the local maximum.

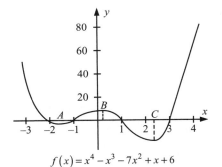

$$f(x) = x^4 - x^3 - 7x^2 + x + 6$$

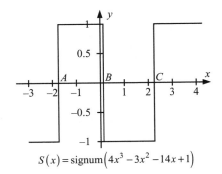

$$S(x) = \mathrm{signum}(4x^3 - 3x^2 - 14x + 1)$$

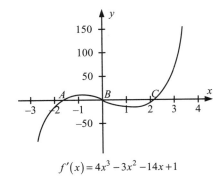

$$f'(x) = 4x^3 - 3x^2 - 14x + 1$$

Section 4.3   Applied Maximum-Minimum Problems

Problems for Practice

1.   Let w be the base and h be the height of the rectangle. The perimeter is $p = 2(w+h)$. We calculate the area $A = w \cdot h = w\left(\dfrac{p}{2} - w\right) = \dfrac{pw}{2} - w^2$ and its derivative: $A' = \dfrac{p}{2} - 2w$. We conclude that $A' = 0$ at $w = \dfrac{p}{4} = \dfrac{100}{4} = 25$, or if each side is 25 ft.

3.   The distance from the point (x, y) to the origin is $S = \sqrt{x^2 + y^2} = \sqrt{x^2 + \dfrac{1}{x}}$, then

$$S' = \frac{2x - 1/x^2}{x^2 + 1/x},$$

$S' = 0$ at $x^3 = \dfrac{1}{2}$, or $x = \left(\dfrac{1}{2}\right)^{1/3}$, and the point is $\left(2^{-1/3}, 2^{1/6}\right)$.

5.   Let x be the side of the squares removed. The volume of the box is:

$$V = (20 - 2x)(20 - 2x)x = 400x - 80x^2 + 4x^3.$$

We calculate the derivative $V' = 400 - 160x + 12x^2$ and find that $V' = 0$ at $x = 10$ and $x = \dfrac{10}{3}$, so that $V(10) = 0$ and $V\left(\dfrac{10}{3}\right) = \dfrac{16000}{27}$. Therefore, the maximum length for each side is $10/3$ in.

7.   The equation of the line is $y = \left(\dfrac{b}{a}\right)x + b$. For the point $(1, 2)$ we find $\dfrac{b}{a} + b = 2$, or $b = \dfrac{2a}{a+1}$, so that $S = ab = \dfrac{2a^2}{a+1}$, and its derivative is $S' = \dfrac{2a(a+2)}{(a+1)^2}$. We find that $S' = 0$ at $a = 0$ and $a = -2$. For $a = -2$ we find $b = 4$ and slope is -2.

9.   Set $F = x - x^3$; $F' = 1 - 3x^2$; $F' = 0$ at $x = \sqrt{\dfrac{1}{3}}$. The number in the set that exceeds its cube the most is $\sqrt{\dfrac{1}{3}}$.

11.   Let x be the width of the type and $\dfrac{80}{x}$ be the height of the type. The width of paper is $x+2$

and its height is $\dfrac{80}{x}+2$. Then $A=\left(\dfrac{80}{x}+2\right)(x+2)=84+2x-\dfrac{160}{x}$; $A'=2-\dfrac{160}{x^2}$; $A'=0$ at

$x=4\sqrt{5}$. The area of the page is minimal if height $=$ width $=2+4\sqrt{5}$ in.

13.   The base of the rectangle is $2x$ and the height is y. The area of the rectangle is:

$$A=2xy=2x\left(4-x^2\right)$$

and $A'=8-6x^2$. Then $A'=0$ at $x=\dfrac{2}{\sqrt{3}}$. The area is maximum when the base $=\dfrac{2}{\sqrt{3}}$ in. and

the height $=\dfrac{8}{3}$ in.

15.   Let x be the side length of the base and let y be the height of the planter. Then $x^2y=3$, or

$$y=\dfrac{3}{x^2}.$$

The material cost is $C=12\cdot4\cdot xy+x^2=\dfrac{144}{x}+x^2$. We find $C'=\dfrac{2\left(x^3-72\right)}{x^2}=0$ when

$$x=72^{1/3}=2\cdot9^{1/3}.$$

Side length of base $=2\cdot9^{1/3}$ m. Height: $3^{-1/3}\cdot2^{-2}$ m.

17.   Let x be the length of the triangle and y be the other side length of the rectangle. Then the

equation for the enclosed area is $A=xy+\dfrac{\sqrt{3}x^2}{4}=100$, or $y=\dfrac{100}{x}-\dfrac{\sqrt{3}x}{4}$. The length L of the

fence is $L=3x+2y=3x+\dfrac{200}{x}-\dfrac{\sqrt{3}x}{2}$. We calculate the derivative: $L'=\dfrac{\left(6-\sqrt{3}\right)x^2-400}{2x^2}$,

which is 0 when $x=\dfrac{20}{\sqrt{6-\sqrt{3}}}$ m (the side length of the triangle) and

$$y=5\sqrt{6-\sqrt{3}}-5\sqrt{\dfrac{6+\sqrt{3}}{11}}\ \text{m}$$

(the other side length of the rectangle).

19.   Let r and h denote the radius and height of the cylinder. Then $\pi r^2 + 2\pi rh + 2\pi r^2 = 5\pi$, or

$h = \dfrac{5 - 3r^2}{2r}$. The volume is $V = \pi r^2 h + \dfrac{2\pi r^3}{3} = \pi r^2 \dfrac{\left(5 - 3r^2\right)}{2r} + \dfrac{2\pi r^3}{3}$. We calculate the

derivative: $V' = \dfrac{\pi\left(5 - 3r^2\right)}{2} - \pi r^2$, which equals 0 when $r = 1$. For $r = 1$ in., $V = \dfrac{5\pi}{3}$ in.^{3}

(radius = 1 in. and cylinder height = 1 in.).

Further Theory and Practice

21.   As $\pi r^2 + \pi r\sqrt{r^2 + h^2} = 2\pi$, we deduce $h = \dfrac{2\sqrt{1 - r^2}}{r}$. Therefore, $V = \dfrac{\pi r^2 h}{3} = \dfrac{2\pi r\sqrt{1 - r^2}}{3}$. We

calculate the derivative: $V' = \dfrac{1 - 2r^2}{\sqrt{1 - r^2}}$, whose root is $r = \dfrac{1}{\sqrt{2}}$. Therefore, $h = \dfrac{2\sqrt{1 - \left(1/\sqrt{2}\right)^2}}{1/\sqrt{2}} = 2$,

and $V = \pi\left(\dfrac{1}{\sqrt{2}}\right)^2 \dfrac{2}{3} = \dfrac{\pi}{3}$.

23.   The derivative is $\dfrac{-\left(a\cos^2(\theta) - a\sin^2(\theta) + 2b\sin(\theta)\sin(\theta)\right)}{\left(a\sin\theta\cos\theta - b\cos^2\theta\right)^2}$. The numerator can be

rewritten as $a\cos(2\theta) + b\sin(2\theta)$, which equals 0 when $\tan(2\theta_0) = -\dfrac{a}{b}$, or, equivalently,

$\tan(\theta_0) = \dfrac{b + \sqrt{a^2 + b^2}}{a}$.

25.   Let x be the length of the piece of wire formed into a square. The length of the piece of wire
formed into a circle is then $L - x$. We write the circumference of the circle as $L - x = 2\pi r$ or

$r = \dfrac{L - x}{2\pi}$ and the length of the square's side as $a = \dfrac{x}{4}$. The area of both pieces is

$$A = a^2 + \pi r^2 = \left(\dfrac{x}{4}\right)^2 + \dfrac{\pi(L - x)^2}{(2\pi)^2} = \dfrac{x^2}{16} + \dfrac{(L - x)^2}{4\pi}.$$

We calculate the derivative, $A' = \dfrac{x}{8} - \dfrac{L - x}{2\pi}$, $A' = 0$ at $x_1 = \dfrac{4\pi L}{\pi + 4}$. We compute the values of

function A at local extremum x_1 and the endpoints of the intervals at $x = 0$ and $x = L$ to

obtain $A(x_1) = \dfrac{x^2}{16} + \dfrac{(L - x)^2}{4\pi} = \dfrac{L^2}{(\pi + 4)^2} + \dfrac{L^2\pi}{4(\pi + 4)^2} = \dfrac{L^2}{4(\pi + 4)}$, $A(0) = \dfrac{L^2}{4\pi}$, $A\left(\dfrac{L}{4}\right) = \dfrac{L^2}{16}$.

The minimum total area is found at the local extremum where $A = \dfrac{L^2}{4(\pi + 4)}$.

27. Let the length of a piece of wire formed into a circle be x, and the circumference of the piece of wire formed into a semicircle be $L-4x$. The radius r_1 of the semicircle is $r_1 = \dfrac{x}{\pi+2}$ and the radius r_2 of the circle is $r_2 = \dfrac{L-x}{2\pi}$. The total area is

$$A = \frac{\pi r_1^2}{2} + \pi r_2^2 = \frac{\pi x^2}{2(\pi+2)^2} + \frac{(L-x)^2}{4\pi}.$$

We calculate its derivative as $A' = \dfrac{\pi x}{(\pi+2)^2} - \dfrac{L-x}{2\pi}$, so that $A'=0$ at $x_1 = \dfrac{L(\pi+2)^2}{3\pi^2+4\pi+4}$.

Compare the values of the function A at the critical point x_1 and at the endpoints of the interval, $x=0$ and $x=L$: $A(x_1) = \dfrac{\pi L^2}{6\pi^2+8\pi+8}$, $A(0) = \dfrac{L^2}{4\pi}$, $A(L) = \dfrac{\pi L^2}{2(\pi+2)^2}$. The largest possible area is $\dfrac{L^2}{4\pi}$ and the smallest possible area is $\dfrac{\pi L^2}{6\pi^2+8\pi+8}$.

29. We calculate the derivative: $F'(v) = \left(\left(a + \dfrac{b}{v^4}\right)v^2\right)' = \left(av^2 + \dfrac{b}{v^2}\right)' = 2av - \dfrac{2b}{v^3}$. We find that $F'(v)=0$ at $v = \left(\dfrac{b}{a}\right)^{1/4}$. Drag is minimized at $v = \left(\dfrac{b}{a}\right)^{1/4}$.

31. We calculate the derivative and solve the equation: $U'(r) = A\left(\dfrac{6}{r^7} - \dfrac{12b}{r^{13}}\right) = 0$. The solution is $r = (2b)^{1/6}$.

33. Let the unequal side be $a = 100 - 2h = 80\sin(\alpha)$. The area of the garden is

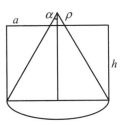

$$A = ah + \left(80^2\alpha - \frac{80^2\sin(2\alpha)}{2}\right)$$

$$= 80\left(50\sin(\alpha) - 40\sin^2(\alpha) + 80\alpha - 40\sin(2\alpha)\right).$$

Then $A' = 80\left(50\cos(\alpha) - 80\sin(2\alpha) - 80\cos(2\alpha) + 80\right)$. The critical point is found from the equation $A'(\alpha) = 0$. The largest possible garden area is the value of function A at the critical point.

35. We calculate the derivative $\Omega' = \dfrac{b}{R^4\sin^2(\theta)} - \dfrac{b\cos(\theta)}{r^4\sin^2(\theta)}$ to find that $\Omega' = 0$ at

$$\cos(\theta) = \left(\frac{r}{R}\right)^{1/4}.$$

37. If we add x m of fence to the straight section of 100 m, the opposite side of the rectangle having the same length will require $(100+x)$ m of the fence. The remainder will go to the shorter sides whose length is $y = \frac{1}{2}\left(200 - (100 + x + x)\right) = 50 - x$ m. The area of the rectangle $A = xy = (100 + x)(50 - x) = 5000 - 50x - x^2$ is a monotonically decreasing function of x. Therefore, the area is largest when $x = 0$, and $A = 50 \cdot 100 = 5000$ m^2.

39. The volume of the cylinder is $V = \pi r^2 h$, or $h = \dfrac{V}{\pi r^2}$. The surface area of the necessary sheet of metal is $A = 2\pi rh + 2 \cdot (2r)^2 = \dfrac{2V}{r} + 8r^2$, so that $A' = -\dfrac{2V}{r^2} + 16r$. Then $A' = 0$ at $r^3 = \dfrac{V}{8}$ and $\dfrac{h}{r} = \dfrac{V}{\pi r^3} = \dfrac{8}{\pi}$. The amount of metal is minimized at $\dfrac{h}{r} = \dfrac{8}{\pi}$.

41. We calculate the derivative:

$$\left(2\pi\sqrt{\frac{2}{3g}}\sqrt{\frac{\ell^2 + 18x^2}{12x + \ell}}\right)' = 4\pi\sqrt{\frac{6}{g}}\,\frac{18x^2 - 3\ell x - \ell^2}{(12x+\ell)^2\sqrt{\left(\ell^2 + 18x^2\right)/(12x + \ell)}}.$$

We find the extremum by solving: $18x^2 - 3\ell x - \ell^2 = 0$, and the period is least at $x = \ell/6$.

43. We find the maximum value of $\dfrac{\tan(\alpha)}{2 + \tan^2(\alpha)}$:

$$\frac{d}{d\tan(\alpha)}\left(\frac{\tan(\alpha)}{2 + \tan^2(\alpha)}\right) = \frac{2 - \tan^2(\alpha)}{\left(\tan^2(\alpha) + 2\right)^2}$$

The derivative equals 0 when $\tan(\alpha) = \sqrt{2}$. Then $\tan(\beta_0) = \dfrac{\sqrt{2}}{2 + \left(\sqrt{2}\right)^2} = \dfrac{1}{2\sqrt{2}}$, or

$\beta_0 = \tan^{-1}\left(\dfrac{1}{2\sqrt{2}}\right)$. Further, if $\sin(\beta_0) = \dfrac{1}{3}$, then $\cos(\beta_0) = \sqrt{1 - \left(\dfrac{1}{3}\right)^2} = \dfrac{3}{2\sqrt{2}}$, and

$\tan(\beta_0) = \dfrac{\sin(\beta_0)}{\cos(\beta_0)} = \dfrac{1}{2\sqrt{2}}$.

Calculator/Computer Exercises

45. By calculating the derivative we find $\theta = 0.78622$.

47.   The expression of heat at point x is $Q = \dfrac{1}{(10-r)^2} + \dfrac{2}{r^2}$. Then $Q' = -\dfrac{4}{r^3} + \dfrac{2}{(10-r)^3}$. The

relative point from the coolest point is $\dfrac{10\left(2 - 2^{2/3} + 2^{1/3}\right)}{3}$ units from the stronger source.

49.   We calculate the derivative to deduce:

$$\frac{d}{dx}\left(\frac{2g}{v}R\right) = \frac{d}{dx}\left(v\sin(2\alpha) + \sqrt{v^2\sin^2(2\alpha) + 4gh + 4gh\cos(2\alpha)}\right)$$

or

$$\frac{2g}{v}R'(\alpha) = 2v\cos(2\alpha) + \frac{2v^2\sin(2\alpha)\cos(2\alpha) - 4gh\sin(2\alpha)}{\sqrt{v^2\sin^2(2\alpha) + 4gh + 4gh\cos(2\alpha)}}.$$

Therefore, $R'(\alpha) = 0$ when

$$4v^2\cos^2(2\alpha)\left(v^2\sin^2(2\alpha) + 4gh + 4gh\cos(2\alpha)\right) = \left(2v^2\sin(2\alpha)\cos(2\alpha) - 4gh\sin(2\alpha)\right)^2.$$

Let $A = \cos(2\alpha)$ and $B = \sin(2\alpha)$. The equation becomes

$$4A^2v^2\left(v^2B^2 + 4gh + 4Agh\right) = \left(2v^2AB - 4ghB\right)^2,$$

or $4A^2v^2\left(v^2B^2 + 4gh + 4Agh\right) - \left(2v^2AB - 4ghB\right)^2 = 0$, or

$$A^2v^2 + A^3v^2 + v^2AB^2 - ghB^2 = 0.$$

As $B^2 \cong 1 - A^2$, so $A^2v^2 + A^3v^2 + \left(v^2A - gh\right)\left(1 - A^2\right) = 0$, or $A^2\left(v^2 + gh\right) + v^2A - gh = 0$.
The solution for this quadratic equation in A is

$$A = \frac{-v^2 \pm \left(v^2 + 2gh\right)}{2\left(v^2 + gh\right)}.$$

The only positive root is $A = \dfrac{gh}{v^2 + gh}$, which means that $\cos(2\alpha) = \dfrac{gh}{v^2 + gh}$ and

$$\alpha = \frac{1}{2}\arccos\left[\frac{gh}{v^2 + gh}\right].$$

51.   Function $\dfrac{Y}{x}$ is monotonically decreasing at large x, therefore the local maximum at $x = 0.512$ is an absolute maximum.

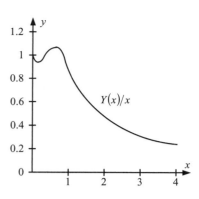

Section 4.4   Concavity

Problems for Practice

1.   We calculate $f'(x) = 3x^2 + 18x - 21 = 3(x+7)(x-1)$ and $f''(x) = 6(x+3)$. Therefore, for f we observe the following.

Concave up on the interval: $(-3, \infty)$

Concave down on the interval: $(-\infty, -3)$

Inflection point: -3

Critical points: $\{-7, 1\}$

Local maximum: -7

Local minimum: 1

3.   We calculate $f'(x) = 6x^2 - 6x - 12 = 6(x+1)(x-2)$ and $f''(x) = 12x - 6$. Therefore, for f we observe the following.

Concave up on the interval: $\left(\dfrac{1}{2}, \infty\right)$

Concave down on the interval: $\left(-\infty, \dfrac{1}{2}\right)$

Inflection point: $\dfrac{1}{2}$

Critical points: $\{-1, 2\}$

Local maximum: -1

Local minimum: 2

5. We calculate $f'(x) = 4x^2\left(x - \dfrac{21}{4}\right)$ and $f''(x) = 12x\left(x - \dfrac{7}{2}\right)$. Therefore, for f we observe the following.

Concave up on the intervals: $(-\infty,\ 0) \cup \left(\dfrac{7}{2},\ \infty\right)$

Concave down on the interval: $\left(0,\ \dfrac{7}{2}\right)$

Inflection points: 0 and $\dfrac{7}{2}$

Critical points: 0 and $\dfrac{21}{4}$

Local maximum: $\dfrac{21}{4}$

Local minima: None

7. We calculate $f'(x) = 10x^2(x+3)^2$ and $f''(x) = 40x(x+3)\left(x + \dfrac{3}{2}\right)$. Therefore, for f we observe the following.

Concave up on the intervals: $\left(-3,\ -\dfrac{3}{2}\right)$ and $(0,\ \infty)$

Concave down on the intervals: $(-\infty, -3)$ and $\left(-\dfrac{3}{2}, 0\right)$

Inflection points: -3, $-\dfrac{3}{2}$, and 0

Critical points: 0 and -3

Local maxima: None

Local minima: None

9. We calculate $f'(x) = \dfrac{1}{\left(x^2 + 1\right)^{3/2}}$ and $f''(x) = \dfrac{-3x}{\left(x^2 + 1\right)^{5/2}}$. Therefore, for f we observe the following.

Concave up on the interval: $(-\infty,\ 0)$

Concave down on the interval: $(0,\ \infty)$

Inflection point: 0

Critical point: None

Local maxima: None

Local minima: None

11.   We calculate $f'(x) = \dfrac{4x-5}{3x^{2/3}}$ and $f''(x) = 2\dfrac{2x+5}{9x^{5/3}}$. Therefore, for f we observe the following.

Concave up on the intervals: $\left(-\infty, -\dfrac{5}{2}\right)$ and $(0, \infty)$

Concave down on the interval: $\left(-\dfrac{5}{2}, 0\right)$

Inflection points: $-\dfrac{5}{2}$ and 0

Critical points: 0 and $\dfrac{5}{4}$

Local maxima: None

Local minimum: $\dfrac{5}{4}$

13.   We calculate $f'(x) = (x+1)\dfrac{(x-1)}{x^2}$ and $f''(x) = \dfrac{2}{x^3}$. Therefore, for f we observe the following.

Concave up on the interval: $(0, \infty)$

Concave down on the interval: $(-\infty, 0)$

Inflection point: None
Critical points: -1 and 1
Local maximum: -1
Local minimum: 1

15.   We calculate $f'(x) = 1 - \sin(x)$ and $f''(x) = -\cos(x)$. Therefore, for f we observe the following.

Concave up on the intervals: $\left(\dfrac{\pi}{2} + 2\pi n, \dfrac{3\pi}{2} + 2\pi n\right)$ for $n \in \mathbb{Z}$

Concave down on the intervals: $\left(-\dfrac{\pi}{2} + 2\pi n, \dfrac{\pi}{2} + 2\pi n\right)$ for $n \in \mathbb{Z}$

Inflection points: $\left\{\dfrac{\pi}{2} + n\pi: n \text{ an integer}\right\}$

Critical points: $\left\{\dfrac{\pi}{2} + 2\pi m: m \text{ an integer}\right\}$

Local maxima: None
Local minima: None

17.   We calculate $f'(x) = \dfrac{1}{x}$ and $f''(x) = -\dfrac{1}{x^2}$. Therefore, for f we observe the following.

Concave up on the interval: None
Concave down on the interval: $(0, \infty)$
Critical point: None
Inflection point: None
No local extrema

19. We calculate $f'(x) = \ln(x) + 1$ and $f''(x) = \frac{1}{x}$. Therefore, for f we observe the following.

Concave up on the interval: $(0, \infty)$

Concave down on the interval: None

Inflection point: None

Critical point: $\frac{1}{e}$

Local maxima: None

Local minimum: $\frac{1}{e}$

Further Theory and Practice

21. Given that $f'(x) > 0$ for $x < 0$ and $f'(x) < 0$ for $0 < x < 1$, a local maximum is at 0.

Given that $f'(x) < 0$ for $0 < x < 1$ and $f'(x) > 0$ for $x > 1$, a local minimum is at 1.

23. Given that $f'(x) > 0$ for $x < -1$ and $f'(x) < 0$ for $-1 < x < 1$, a local maximum is at -1.

Given that $f'(x) < 0$ for $-1 < x < 1$ and $f'(x) > 0$ for $x > 1$, a local minimum is at 1.

25. Given that $f'(x) > 0$ for $x < -1$ and $f'(x) < 0$ for $-1 < x < 1$, a local maximum is at -1.

Given that $f'(x) < 0$ for $-1 < x < 1$ and $f'(x) > 0$ for $x > 1$, a local minimum is at 1.

27. We find that: $f''(x) = 0$ at 0 and 1; $f''(x) > 0$ for $x < 0$ and $x > 1$, $f''(x) < 0$ for $0 < x < 1$. Therefore, there are two inflection points: 0 and 1.

29. We find that: $f''(x) = 0$ at -1 and 1; $f''(x) > 0$ for $x < -1$ and $x > 1$, $f''(x) < 0$ for $-1 < x < 1$. Therefore, there are two inflection points: -1 and 1.

31. We find that: $f''(x) = 0$ at -1 and 1; $f''(x) > 0$ for $x < -1$ and $x > 1$, $f''(x) < 0$ for $-1 < x < 1$. Therefore, there are two inflection points: -1 and 1.

33. We calculate $f'(x) = 3x^2 (2x - 1)(x - 1)^2$ and $f''(x) = 6x(x - 1)(5x^2 - 5x + 1)$. Given that $f'(x)$ changes sign only at $x = \frac{1}{2}$ and $f''\left(\frac{1}{2}\right) = \frac{3}{8} > 0$, a local minimum is at $x = \frac{1}{2}$, and there are no other local extrema.

35. We calculate the derivative $f'(x) = \dfrac{x(4 - x)}{\left(6x^2 - x^3\right)^{2/3}}$, so the critical points are 0 and 4. As $f''(x) = \dfrac{8}{x^{4/3}(x - 6)^{5/3}}$ is negative for $x = 4$, f has a local maximum at 4. Given that $f(0) = 0$ and $f(x) \geq 0$ for $x \leq 6$, a local minimum is at $x = 0$.

37.   Given that $\dfrac{dy}{dx} = -\dfrac{16a^3 x}{\left(4a^2 + x^2\right)^2}$ and $\dfrac{d^2 y}{dx^2} = \dfrac{16a^3\left(3x^2 - 4a^2\right)}{\left(4a^2 + x^2\right)^3}$, we deduce that the curve is

concave up on the intervals $\left(-\infty, -\dfrac{2a}{\sqrt{3}}\right)$ and $\left(\dfrac{2a}{\sqrt{3}}, \infty\right)$ and concave down on the intervals

$\left(-\dfrac{2a}{\sqrt{3}}, \dfrac{2a}{\sqrt{3}}\right)$. The inflection points are $-\dfrac{2a}{\sqrt{3}}$ and $\dfrac{2a}{\sqrt{3}}$.

39.   We calculate $\left(f^2\right)''(x) = 2\left(f(x)f''(x) + f'(x)^2\right) \geq 0$. Therefore, f^2 is concave up.

41.   We calculate $m'(t) = k \cdot m(t)\left(2c - m(t)\right) = k \cdot \left(c^2 - \left(m(ty) - c\right)^2\right)$. Thus, $m'(t)$ increases for

$m(t) < c$ and decreases for $m(t) > c$. For the value of t at which $m(t) = c$, the graph of

$y = m(t)$ has a point of inflection. The graph is concave up on the interval $(0,\ c)$ and concave

down on the interval $(c,\ \infty)$.

43.   **a.**   The second derivative is $T''(t) = -\left(\dfrac{7\alpha}{2}\right)T^{5/2}T'(t) = \left(\dfrac{7\alpha^2}{2}\right)T^6 > 0$, hence the graph

of T is concave up.

b. $L'(t) = L(0)\left(\dfrac{7}{2}\right)\left(\dfrac{T(t)}{T(0)}\right)^{5/2}\left(\dfrac{T'(t)}{T(0)}\right),$

$L''(t) = L(0)\left(\dfrac{35}{4}\right)\left(\dfrac{T(t)}{T(0)}\right)^{3/2}\left(\dfrac{T'(t)}{T(0)}\right)^2 + L(0)\left(\dfrac{7}{2}\right)\left(\dfrac{T(t)}{T(0)}\right)^{5/2}\left(\dfrac{T''(t)}{T(0)}\right) > 0,$

hence L is concave up.

c.   There is no consistency with part **b** because a logarithmic scale is used.

45.   **a.**   The growth of x for a fixed s leads to the reduction of y, hence MRS_{XY} is negative. The indifference curves are decreasing functions.

b.   The absolute value of MRS_{XY} decreases. The indifference curve is concave up.

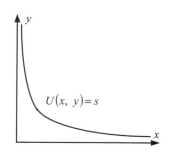

Enrichment Exercises

47.   The derivatives are: $f'(x) = 3x^2 + 2$, $f''(x) = 6x$, hence $\kappa(x) = \dfrac{6|x|}{\left(9x^4 + 12x^2 + 5\right)^{3/2}}$.

49. The derivatives are: $f'(x) = \dfrac{1}{2\sqrt{x}}$, $f''(x) = -\dfrac{1}{4x^{3/2}}$, hence $\kappa(x) = \dfrac{2}{(4x+1)^{3/2}}$.

51. The derivatives are: $f'(x) = \dfrac{1}{x}$, $f''(x) = -\dfrac{1}{x^2}$, hence

$$\kappa(x) = \frac{|x|}{\left(x^2+1\right)^{3/2}} = \frac{x}{\left(x^2+1\right)^{3/2}}$$

on the domain of f. To find its maximum, we consider $\kappa'(x) = \dfrac{1-2x^2}{\left(x^2+1\right)^{5/2}}$. There is a critical

point $\dfrac{\sqrt{2}}{2}$ for κ. As $\kappa'(x) > 0$ for $0 < x < \dfrac{\sqrt{2}}{2}$, and $\kappa'(x) < 0$ for $x > \dfrac{\sqrt{2}}{2}$, there is a local

maximum of $\dfrac{2\sqrt{3}}{9}$ at $\dfrac{\sqrt{2}}{2}$.

53. Given that $\kappa(x) = \dfrac{|\sin(x)|}{\left(1+\cos^2(x)\right)^{3/2}}$ (see Problem 4.4.48),

$$\kappa'(x) = -2\left(\cos^2(x) - 2\right)\frac{\cos(x)}{\left(1+\cos^2(x)\right)^{5/2}}$$

for $x \neq \pi k$, $k \in \mathbb{Z}$. We calculate the critical points to be at $x = \pi n/2$, $n \in \mathbb{Z}$. The sign of κ' coincides with the sign of $\cos(x)$, hence the maximum curvature of 1 occurs at

$$(2n+1)\frac{\pi}{2}, \ n \in \mathbb{Z}.$$

55. By implicit differentiation $2x + 2yy' = 0$, or $y' = -\dfrac{x}{y}$. Then the second derivative is

$$y'' = -\frac{y - xy'}{y^2} = -\frac{y + x^2/y}{y^2} = -\frac{r^2}{y^3}$$

and $\kappa(x) = \dfrac{r^2}{|y|^3\left(1+x^2/y^2\right)^{3/2}} = \dfrac{r^2}{r^3} = \dfrac{1}{r}$.

57. Let $\kappa_1(x)$ be the curvature of the graph of g at $(x, g(x))$ and $\kappa_2(x)$ be the curvature of the graph of f at $(x+a, f(x+a))$. The derivatives are $g'(x) = f'(x+a)$, $g''(x) = f''(x+a)$ and

$$\kappa_1(x) = \frac{|g''(x)|}{\left(1+(g'(x))^2\right)^{3/2}} = \frac{|f''(x+a)|}{\left(1+(f'(x+a))^2\right)^{3/2}} = \kappa_2(x).$$

59.   Let $p(x) = P_n(x)$ be a polynomial of degree $n \geq 2$. Then $p'(x) = P_{n-1}(x)$, $p''(x) = P_{n-2}(x)$

and $\kappa(x) = \dfrac{\left| P_{n-2}(x) \right|}{\left(1 + P_{n-1}^2(x) \right)^{3/2}} = \dfrac{\left| P_{n-2}(x) \right|}{\sqrt{P_{6n-6}(x)}}$. The denominator is the square root of the polynomial

of degree $6n - 6$, i.e. its leading term is proportional to x^{3n-3}, hence $\lim\limits_{x \to \pm\infty} \kappa(x) = 0$. Let N be

some positive integer such that $\kappa(x) < 1$ for all $|x| > N$; given that κ is continuous on $[-N, N]$,

it is bounded: there exists $m < M$ such that $m \leq \kappa(x) \leq M$ for $-N \leq x \leq N$. If $0 \in [m, M]$,

the following inequality holds: $m \leq \kappa(x) \leq M$ for $-\infty < x < \infty$, that is, κ is bounded on $\mathbb{R}$. If

$0 < m$, we have $0 \leq \kappa(x) \leq M$ for $-\infty < x < \infty$, and we see that κ is, again, bounded on $\mathbb{R}$.

Calculator/Computer Exercises

61. $f'(x) = -\dfrac{x \exp(x) - \exp(x) - 1}{\left(1 + \exp(x) \right)^2}$, $f''(x) = \exp(x) \dfrac{(x-2) \exp(x) - x - 2}{\left(1 + \exp(x) \right)^3}$.

Critical point: 1.279 (local maximum)
Inflection points: −2.399 and 2.399

63. $f'(x) = 4x^3 - 2\exp(x) + 3$, $f''(x) = 12x^2 - 2\exp(x)$
Critical point: −0.808 (local minimum)
Inflection points: −0.344 and 0.533

65. $f'(x) = \dfrac{x^2 - 1}{\left(x^2 + x + 1 \right)^2}$, $f''(x) = -2\dfrac{x^3 - 3x - 1}{\left(x^2 + x + 1 \right)^3}$

Critical points: −1 (local maximum) and 1 (local minimum)
Inflection points: −1.532, −0.347, 1.879.

67.   Discontinuity points of C are the points where f''
changes sign; hence they are the points of inflection.

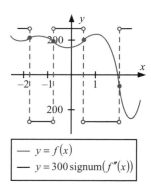

$f(x) = 6x^6 + x^5 - 60x^4 - 35x^3 + 120x^2 + 52x + 160$

Section 4.5   Graphing Functions

Problems for Practice

1.   **a.**   Local maximum: $P = (-1, 12)$

Local minimum: $R = (3, -20)$

b.   Inflection point: $Q = (1, -4)$

c.   Increasing on the intervals: $(-\infty, -1) \cup (3, \infty)$

Decreasing on the interval: $(1, 3)$

d.   Concave up on the interval: $(1, \infty)$

Concave down on the interval: $(-\infty, 1)$

e.   Horizontal asymptote: None
Vertical asymptote: None
Skew asymptote: None

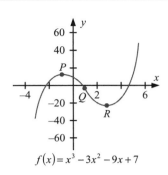

$$f(x) = x^3 - 3x^2 - 9x + 7$$

3.   **a.**   Local maximum: $Q = \left(-\dfrac{27}{8}, \dfrac{4}{27}\right)$

Local minima: None

b.   Inflection point: $P = \left(-\left(\dfrac{9}{5}\right)^3, \dfrac{100}{729}\right)$

c.   Increasing on the interval: $\left(-\infty, \dfrac{\sqrt{-27}}{8}\right)$

Decreasing on the intervals: $\left(-\dfrac{27}{8}, 0\right) \cup (0, 8)$

d.   Concave up on the intervals: $\left(-\infty, -\left(\dfrac{9}{5}\right)^3\right) \cup (0, \infty)$

Concave down on the interval: $\left(-\left(\dfrac{9}{5}\right)^3, 0\right)$

e.   Horizontal asymptote: $y = 0$
Vertical asymptote: $x = 0$
Skew asymptote: None

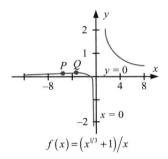

$$f(x) = \left(x^{1/3} + 1\right)/x$$

5.   **a.**   Local maxima: None

Local minimum: $P = \left(\dfrac{1}{2}, -3 \cdot 2^{-4/3} \right)$

b.   Inflection points: $Q = (0, 0)$ and $R = (-1, 3)$

c.   Increasing on the interval: $\left(\dfrac{1}{2}, \infty \right)$

Decreasing on the interval: $\left(-\infty, \dfrac{1}{2} \right)$

d.   Concave up on the intervals:
$(-\infty, -1) \cup (0, \infty)$

Concave down on the interval: $(-1, 0)$

e.   Horizontal asymptote: None
Vertical asymptote: None
Skew asymptote: None

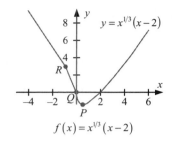

$y = x^{1/3}(x-2)$

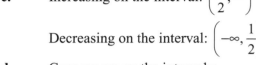

$f(x) = x^{1/3}(x-2)$

7.   **a.**   Local maxima: $P = (0, 4)$

Local minima: $R = (2, 0)$

b.   Inflection point: $Q = (1, 2)$

c.   Increasing on the intervals: $(-\infty, 0) \cup (2, \infty)$

Decreasing on the interval: $(0, 2)$

d.   Concave up on the interval: $(1, \infty)$

Concave down on the interval: $(-\infty, 1)$

e.   Horizontal asymptote: None
Vertical asymptote: None
Skew asymptote: None

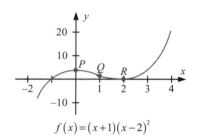

$f(x) = (x+1)(x-2)^2$

9.   **a.**   Local maxima: None
Local minima: None

b.   Inflection points: $\{(n\pi, -n\pi): n \in \mathbb{Z}\}$

c.   Increasing on the interval: None
Decreasing on the interval: $(k\pi, (k+1)\pi)$,

$k \in \mathbb{Z}$

d.   Concave up on the intervals: $((2k-1)\pi, 2k\pi)$

and $k \in \mathbb{Z}$
Concave down on the intervals:
$(2k\pi, (2k+1)\pi)$ and $k \in \mathbb{Z}$

e.   Vertical asymptote: None
Horizontal asymptote: None
Skew asymptote: None

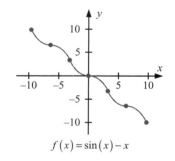

$f(x) = \sin(x) - x$

11. **a.** Local maximum: $s = \left(2, \dfrac{1}{4}\right)$

Local minimum: $Q = \left(-2, -\dfrac{1}{4}\right)$

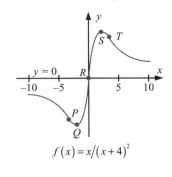

b. Inflection points: $P = \left(-2\sqrt{3}, -\sqrt{\dfrac{3}{8}}\right)$,

$R = (0,0)$, and $T = \left(2\sqrt{3}, \sqrt{\dfrac{3}{8}}\right)$

$f(x) = x/(x+4)^2$

c. Increasing on the interval: $(-2, 2)$

Decreasing on the intervals: $(-\infty, -2) \cup (2, \infty)$

d. Concave up on the intervals: $\left(-2\sqrt{3}, 0\right) \cup \left(2\sqrt{3}, \infty\right)$

Concave down on the intervals: $\left(-\infty, -2\sqrt{3}\right) \cup \left(0, 2\sqrt{3}\right)$

e. Horizontal asymptote: $y = 0$
Vertical asymptote: None
Skew asymptote: None

13. **a.** Local maxima: None
Local minimum: $P = (0, -1)$

b. Inflection point: None

c. Increasing on the intervals:
$(-\infty, -2) \cup (-2, 0)$

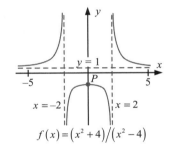

Decreasing on the intervals: $(0, 2) \cup (2, \infty)$

d. Concave up on the intervals: $(-\infty, 2) \cup (2, \infty)$

Concave down on the interval: $(-2, 2)$

$f(x) = (x^2 + 4)/(x^2 - 4)$

e. Horizontal asymptote: $y = 1$
Vertical asymptotes: $x = -2$ and $x = 2$
Skew asymptote: None

15. **a.** Local maximum: $Q = \left(\dfrac{3}{2}, f\left(\dfrac{3}{2} \right) \right)$

Local minima: None

b. Inflection points:

$$P = \left(\dfrac{3}{2} - \dfrac{9\sqrt{5}}{10}, f\left(\dfrac{3}{2} - \dfrac{9\sqrt{5}}{10} \right) \right), \ O = (0,0), \text{ and}$$

$$R = \left(\dfrac{3}{2} + \dfrac{9\sqrt{5}}{10}, f\left(\dfrac{3}{2} + \dfrac{9\sqrt{5}}{10} \right) \right)$$

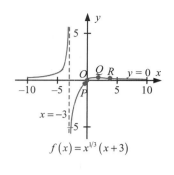

$f(x) = x^{1/3}(x+3)$

c. Increasing on the intervals: $(-\infty, -3) \cup \left(-3, \dfrac{3}{2} \right)$

Decreasing on the interval: $\left(\dfrac{3}{2}, \infty \right)$

d. Concave up on the intervals: $(-\infty, -3) \cup \left(\dfrac{3}{2} - \dfrac{9\sqrt{5}}{10}, 0 \right) \cup \left(\dfrac{3}{2} + \dfrac{9\sqrt{5}}{10}, \infty \right)$

Concave down on the intervals: $\left(-3, \dfrac{3}{2} - 9\sqrt{5} \right) \cup \left(0, \dfrac{3}{2} + 9\sqrt{5} \right)$

e. Horizontal asymptote: $y = 0$
Vertical asymptote $x = -3$
Skew asymptote: None

17. **a.** Local maxima: None

Local minimum: $P = \left(-\dfrac{5}{6}, f\left(-\dfrac{5}{6} \right) \right)$;

b. Inflection points: $Q = \left(-\dfrac{5}{3}, f\left(-\dfrac{5}{3} \right) \right)$ and

$R = (-1, 0)$

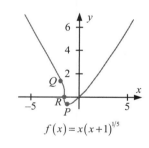

$f(x) = x(x+1)^{1/5}$

c. Increasing on the interval: $\left(-\dfrac{5}{3}, \infty \right)$

Decreasing on the interval: $\left(-\infty, -\dfrac{5}{3} \right)$

d. Concave up on the intervals: $\left(-\infty, -\dfrac{5}{3} \right) \cup (-1, \infty)$

Concave down on the interval: $\left(-\dfrac{5}{3}, -1 \right)$

e. Horizontal asymptote: None
Vertical asymptote: None
Skew asymptote: None

19.   **a.**   Local maxima: None

Local minimum: $P = \left(2^{-4/3}, f\left(2^{-4/3}\right)\right)$

b.   Inflection point: None

c.   Increasing on the interval: $\left(2^{-4/3}, \infty\right)$

Decreasing on the interval: $\left(0, 2^{-4/3}\right)$

d.   Concave up on the interval: $(0, \infty)$

Concave down on the interval: None

e.   Horizontal asymptote: None

Vertical asymptote: None

Skew asymptote: None

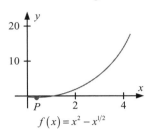

$f(x) = x^2 - x^{1/2}$

Further Theory and Practice

21.   **a.**   Local maximum: $Q = \left(-2, f(-2)\right)$

Local minima: None

b.   Inflection points:

$$P = \left(-2 - \frac{6\sqrt{5}}{5}, f\left(-2 - \frac{6\sqrt{5}}{5}\right)\right), \quad R = (0, 0),$$

$$\text{and } S = \left(-2 + \frac{6\sqrt{5}}{5}, f\left(-2 + \frac{6\sqrt{5}}{5}\right)\right)$$

c.   Increasing on the interval: $(-\infty, -2)$

Decreasing on the intervals: $(-2, 4) \cup (4, \infty)$

d.   Concave up on the intervals: $\left(-\infty, -2 - \frac{6\sqrt{5}}{5}\right) \cup \left(0, -2 + \frac{6\sqrt{5}}{5}\right) \cup (4, \infty)$

Concave down on the intervals: $\left(-2 - \frac{6\sqrt{5}}{5}, 0\right) \cup \left(-2 + \frac{6\sqrt{5}}{5}, 4\right)$

e.   Horizontal asymptote: $y = 0$

Vertical asymptote: $x = 4$

Skew asymptote: None

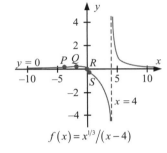

$f(x) = x^{1/3}/(x-4)$

23. **a.** Local maxima: $P = \left(-\sqrt{\dfrac{6}{3}}, 2\sqrt{\dfrac{3}{9}} \right)$ and

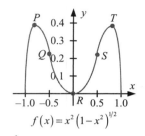

$f(x) = x^2 \left(1 - x^2 \right)^{1/2}$

$T = \left(\dfrac{\sqrt{6}}{3}, 2\sqrt{\dfrac{3}{9}} \right)$

Local minimum: $R = (0, 0)$

b. Inflection points: $Q = \left(-\dfrac{\sqrt{27 - 3\sqrt{33}}}{6}, f\left(-\sqrt{27 - 3\sqrt{33}} \right) \right)$ and

$S = \left(\dfrac{\sqrt{27 - 3\sqrt{33}}}{6}, f\left(\sqrt{27 - 3\sqrt{33}} \right) \right)$

c. Increasing on the intervals: $\left[-1, -\dfrac{\sqrt{6}}{3} \right) \cup \left(0, \dfrac{\sqrt{6}}{3} \right)$

Decreasing on the intervals: $\left(-\dfrac{\sqrt{6}}{3}, 0 \right) \cup \left(\dfrac{\sqrt{6}}{3}, 1 \right]$

d. Concave up on the interval: $\left(-\dfrac{\sqrt{27 - 3\sqrt{33}}}{6}, \dfrac{\sqrt{27 - 3\sqrt{33}}}{6} \right)$

Concave down on the intervals: $\left[-1, -\dfrac{\sqrt{27 - 3\sqrt{33}}}{6} \right) \cup \left(\dfrac{\sqrt{27 - 3\sqrt{33}}}{6}, 1 \right]$

e. Horizontal asymptote: None
Vertical asymptote: None
Skew asymptote: None

25. **a.** Local maxima: None
Local minima: $P = (-2, 0)$ and $Q = (1, 0)$

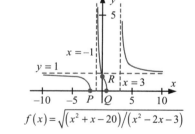

$f(x) = \sqrt{(x^2 + x - 20)/(x^2 - 2x - 3)}$

b. Inflection point: $R = \left(0, \dfrac{\sqrt{6}}{3} \right)$

c. Increasing on the interval: None
Decreasing on the intervals:
$(-\infty, -2] \cup (-1, 1] \cup (3, \infty)$

d. Concave up on the intervals: $(-1, 0) \cup (3, \infty)$

Concave down on the intervals: $(-\infty, -2) \cup (0, 1)$

e. Horizontal asymptote: $y = 1$

Vertical asymptote: $x = -1$ and $x = 3$
Skew asymptote: None

27. **a.** Local maximum: $R = \left(\dfrac{1}{3}, f\left(\dfrac{1}{3}\right)\right)$

Local minimum: $Q = (-2, 0)$

b. Inflection point: $P = \left(-\dfrac{13}{3}, f\left(-\dfrac{13}{3}\right)\right)$

c. Increasing on the intervals:

$(-\infty, -2) \cup \left(\dfrac{1}{3}, \infty\right)$

Decreasing on the interval: $\left(-2, \dfrac{1}{3}\right)$

d. Concave up on the intervals: $\left(-\dfrac{13}{3}, -2\right) \cup (-2, \infty)$

Concave down on the interval: $\left(-\infty, -\dfrac{13}{3}\right)$

e. Horizontal asymptote: None
Vertical asymptote: None
Skew asymptote: None

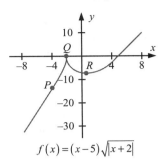

$f(x) = (x-5)\sqrt{|x+2|}$

29. **a.** Local maxima: None

Local minimum: $Q = \left(2^{-1/3}, f\left(2^{-1/3}\right)\right)$

b. Inflection point: $P = (-1, 1)$

c. Increasing on the interval: $\left(2^{-1/3}, \infty\right)$

Decreasing on the intervals:

$(-\infty, 0) \cup \left(0, 2^{-1/3}\right)$

d. Concave up on the intervals:
$(-\infty, -1) \cup (0, \infty)$

Concave down on the interval: $(-1, 0)$

e. Horizontal asymptote: None
Vertical asymptote: $x = 0$
Skew asymptote: None

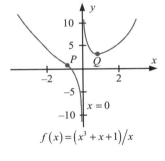

$f(x) = (x^3 + x + 1)/x$

31. **a.** No local extrema

b. Inflection point: None

c. Increasing on the intervals:

$$\left(\left((2k-1)\frac{\pi}{2},\ (2k+1)\frac{\pi}{2}\right),\ k\in\mathbb{Z}\right)$$

Decreasing on the interval: None

d. Concave up on the intervals:

$$\left(\left(4k-1\right)\frac{\pi}{2},\left(4k+1\right)\frac{\pi}{2}\right)(k\in\mathbb{Z})$$

Concave down on the intervals:

$$\left(\left(4k+1\right)\frac{\pi}{2},\left(4k+3\right)\frac{\pi}{2}\right)(k\in\mathbb{Z})$$

e. Horizontal asymptote: None

Vertical asymptotes: lines of the form $x=(2k+1)\dfrac{\pi}{2}(k\in\mathbb{Z})$

Skew asymptote: None

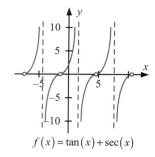

$f(x)=\tan(x)+\sec(x)$

33. **a.** Local maximum: $Q=\left(-\dfrac{1}{3},\dfrac{4}{27}\right)$

Local minima: $P=(-1,0)$ and $R=(0,0)$

b. Inflection points: $S=\left(-\dfrac{2}{3},\dfrac{2}{27}\right)$ and

$R=(0,0)$

c. Increasing on the intervals: $\left(-1,-\dfrac{1}{3}\right)\cup(0,\infty)$

Decreasing on the intervals: $(-\infty,-1)\cup\left(-\dfrac{1}{3},0\right)$

d. Concave up on the intervals: $\left(-\infty,-\dfrac{2}{3}\right)\cup(0,\infty)$

Concave down on the interval: $\left(-\dfrac{2}{3},0\right)$

e. Horizontal asymptote: None
Vertical asymptote: None
Skew asymptote: None

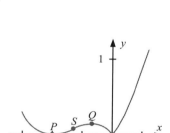

$f(x)=|x|\cdot(x+1)^2$

35.   **a.**   Local maximum:

$$R = \left(\frac{-14}{13} + \frac{3\sqrt{3}}{13}, \; f\left(\frac{-14}{13} + \frac{3\sqrt{3}}{13} \right) \right)$$

Local minimum:

$$Q = \left(\frac{-14}{13} - \frac{3\sqrt{3}}{13}, \; f\left(\frac{-14}{13} - \frac{3\sqrt{3}}{13} \right) \right)$$

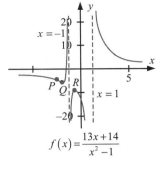

$$f(x) = \frac{13x + 14}{x^2 - 1}$$

b.   Inflection point: $P = (-2, -4)$

c.   Increasing on the intervals:

$$\left(-\frac{14}{13} - \frac{3\sqrt{3}}{13}, \; -1 \right) \cup \left(-1, \; -\frac{14}{13} + \frac{3\sqrt{3}}{13} \right)$$

Decreasing on the intervals: $\left(-\infty, \; -\frac{14}{13} - \frac{3\sqrt{3}}{13} \right) \cup \left(-\frac{14}{13} + \frac{3\sqrt{3}}{13}, \; 1 \right) \cup (1, \; \infty)$

d.   Concave up on the intervals: $(-2, -1) \cup (1, \infty)$

Concave down on the intervals: $(-\infty, -2) \cup (-1, 1)$

e.   Horizontal asymptote: $y = 0$

Vertical asymptotes: $x = -1$ and $x = 1$

Skew asymptote: None

37.   **a.**   Local maxima: None

Local minima: None

b.   Inflection point: None

c.   Increasing on the intervals: $(-\infty, 0) \cup (0, \infty)$

Decreasing on the interval: None

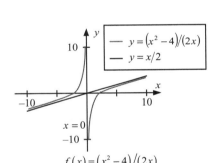

$$f(x) = (x^2 - 4) / (2x)$$

d.   Concave up on the interval: $(-\infty, 0)$

Concave down on the interval: $(0, \infty)$

e.   Horizontal asymptote: None

Vertical asymptote: $x = 0$

Skew asymptote: $y = \dfrac{x}{2}$.

39.   If the graph has a horizontal asymptote then $\lim\limits_{x \to +\infty} f(x) = a < \infty$ or $\lim\limits_{x \to -\infty} f(x) = a < \infty$. It means that either $\lim\limits_{x \to +\infty} f'(x) = 0$ or $\lim\limits_{x \to -\infty} f'(x) = 0$. By the Mean Value Theorem, for any $x > 0$ there exists $c \in (0, x)$ such that $f'(x) = f'(0) + f''(c) \cdot (x - 0) = f'(0) + f''(c) \cdot x$. So $f'(x) \geq f'(0) + x$ for any $x > 0$. Calculating the limit when $x \to +\infty$ we obtain $\lim\limits_{x \to +\infty} f'(x) = +\infty$ the same way one can show that $\lim\limits_{x \to -\infty} f'(x) = -\infty$.

Calculator/Computer Exercises

41.   **a.**   Local maxima: None
Local minima: None
Global maxima: None
Global minimum:

$$P = \left(0.60583,\ f\left(-0.60583\right)\right)$$

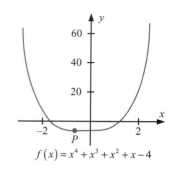

$$f(x) = x^4 + x^3 + x^2 + x - 4$$

b.   Inflection point: None

c.   Increasing on the interval: $\left(-0.60583, \infty\right)$

Decreasing on the interval: $\left(-\infty, -0.60583\right)$

d.   Concave up on the interval: $\left(-\infty, \infty\right)$

Concave down on the interval: None

e.   Vertical asymptote: None
Horizontal asymptote: None
Skew asymptote: None

43.   **a.**   Local maxima: $P = \left(-1.6012,\ f\left(-1.6012\right)\right)$

and $T = \left(0.39915,\ f\left(0.39915\right)\right)$

Local minima: $R = \left(-1.1839,\ f\left(-1.1839\right)\right)$

and $V = \left(1.5859,\ f\left(1.5859\right)\right)$

Global maxima: None
Global minima: None

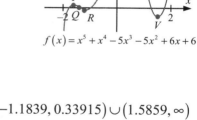

$$f(x) = x^5 + x^4 - 5x^3 - 5x^2 + 6x + 6$$

b.   Inflection points: $Q = \left(-1.4117,\ f\left(1.4117\right)\right)$,

$S = \left(-0.31451,\ f\left(-0.31451\right)\right)$, and

$U = \left(1.1262,\ f\left(1.1262\right)\right)$

c.   Increasing on the intervals: $\left(-\infty, -1.6012\right) \cup \left(-1.1839, 0.33915\right) \cup \left(1.5859, \infty\right)$

Decreasing on the intervals: $\left(-1.6012, -1.1839\right) \cup \left(0.39915, 1.5859\right)$

d.   Concave up on the intervals: $\left(-1.4117, -0.31451\right) \cup \left(1.1262, \infty\right)$

Concave down on the intervals: $\left(-\infty, -1.4117\right) \cup \left(-0.31451, 1.1262\right)$

e.   Vertical asymptote: None
Horizontal asymptote: None
Skew asymptote: None

45.   **a.**   Local maximum: $Q = (0.6071, 1.1425)$

Local minima: None

Global maxima: None

Global minima: $P = (-1, 0)$ and $R = (1, 0)$

b.   Inflection points: $S = (-0.9032, 0.1849)$,

$T = (-0.5385, 0.5548)$, $U = (0.0542, 0.9990)$

and $V = (0.4296, 1.0831)$

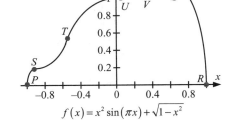

$f(x) = x^2 \sin(\pi x) + \sqrt{1 - x^2}$

c.   Increasing on the interval: $(-1, 0.6071)$

Decreasing on the interval: $(0.6071, 1)$

d.   Concave up on the intervals: $(-0.9032, -0.5385) \cup (0.0542, 0.4296)$

Concave down on the intervals: $(-1, -0.9032) \cup (-0.5385, 0.0542) \cup (0.4296, 1)$

e.   Vertical asymptote: None

Horizontal asymptote: None

Skew asymptote: None

47.   Local maximum: $S = (0, 2)$

Local minima: None

Global maximum: $S = (0, 2)$

Global minima: $P = (-2, 0)$ and $V = (2, 0)$

Inflection points: $Q = (-1.7321, f(-1.7321))$,

$R = (-0.58288, f(-0.58288))$,

$T = (0.58288, f(0.58288))$, and $U = (1.7321, f(1.7321))$

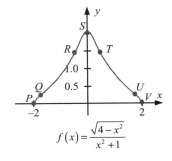

$f(x) = \dfrac{\sqrt{4 - x^2}}{x^2 + 1}$

Increasing on the interval: $(-2, 0)$

Decreasing on the interval: $(0, 2)$

Concave up on the intervals: $(-1.7321, -0.58288) \cup (0.58288, 1.7321)$

Concave down on the intervals: $(-2, -1.7321) \cup (0.58288, 1.7321)$

Vertical asymptote: None

Horizontal asymptote: None

Skew asymptote: None

49.

a.   Local maxima: None
Local minima: None
Global maxima: None
Global minima: None

b.   Inflection point: $P = (-0.0232, 0.5686)$

c.   Increasing on the interval: $(-4, 4)$
Decreasing on the interval: None

d.   Concave up on the interval: $(-0.0232, 4)$
Concave down on the interval: $(-4, -0.0232)$

e.   Vertical asymptote: $x = -4$ and $x = 4$
Horizontal asymptote: None
Skew asymptote: None

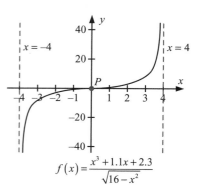

$$f(x) = \frac{x^3 + 1.1x + 2.3}{\sqrt{16 - x^2}}$$

51.   We are given the two functions: $f(x) = \dfrac{x^2 - 1}{x^2 + 1}$ and $g(x) = \dfrac{x^4 - 102x^2 + 100}{x^4 - 99x^2 - 100}$. The expansion

of $g(x)$ leads to: $g(x) \approx \dfrac{(x^2 - 1)(x^2 - 101)}{(x^2 + 1)(x^2 - 100)}$ which is very close to $f(x)$ as long as x is

sufficiently far from $x = 10$ where the denominator vanishes.

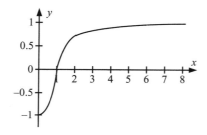

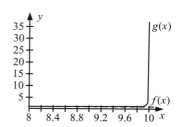

Note that $g(x)$ is not defined at $x = 10$.

53. $f(x) = 12x^5 - 2565x^4 + 146200x^3 + 1$ $f'(x) = 60x^4 - 10260x^3 + 438600x^2$

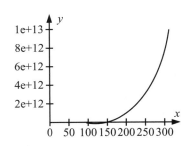

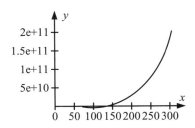

The first derivative has roots at $x = 170$ and $x = 172$.

$f''(x) = 240x^3 - 30780x^2 + 877200x$

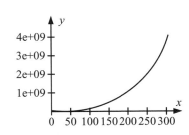

The function f is not increasing on I; f' is negative in (170, 172) and positive elsewhere. The graph of f is not concave up on I: f'' is negative in (85.5, 171) and positive elsewhere in I.

Section 4.6   L'Hôpital's Rule

Problems for Practice

1. $\lim\limits_{x \to 0} \dfrac{1 - e^x}{x} = \lim\limits_{x \to 0} \dfrac{\frac{d}{dx}(1 - e^x)}{\frac{d}{dx}x} = \lim\limits_{x \to 0}(-e^x) = -1$

3. $\lim\limits_{x \to 5} \dfrac{\ln(x/5)}{x - 5} = \lim\limits_{x \to 5} \dfrac{\frac{d}{dx}\ln(x/5)}{\frac{d}{dx}(x - 5)} = \lim\limits_{x \to 5} \dfrac{1}{x} = \dfrac{1}{5}$

5. $\lim\limits_{x \to \pi/2} \dfrac{\ln(\sin(x))}{(\pi - 2x)^2} = \lim\limits_{x \to \pi/2} \dfrac{\frac{d}{dx}(\ln(\sin(x)))}{\frac{d}{dx}((\pi - 2x)^2)} = -\lim\limits_{x \to \pi/2} \dfrac{\cot(x)}{4(\pi - 2x)} = -\dfrac{1}{4}\lim\limits_{x \to \pi/2} \dfrac{\frac{d}{dx}(\cot(x))}{\frac{d}{dx}(\pi - 2x)}$

$\qquad\qquad = -\dfrac{1}{8}\lim\limits_{x \to \pi/2} \csc^2(x) = -\dfrac{1}{8}$

7. $\lim\limits_{x \to -1} \dfrac{\cos(x+1) - 1}{x^3 + x^2 - x - 1} = \lim\limits_{x \to -1} \dfrac{\frac{d}{dx}(\cos(x+1) - 1)}{\frac{d}{dx}(x^3 + x^2 - x - 1)} = \lim\limits_{x \to -1} \dfrac{-\sin(x+1)}{3x^2 + 2x - 1} = -\lim\limits_{x \to -1} \dfrac{\frac{d}{dx}(\sin(x+1))}{\frac{d}{dx}(3x^2 + 2x - 1)}$

$\qquad\qquad = -\lim\limits_{x \to -1} \dfrac{\cos(x+1)}{6x + 2} = \dfrac{1}{4}$

9. We calculate as follows: $\lim\limits_{x\to 0}\dfrac{e^{x}-e^{-x}}{x^{2}}=\lim\limits_{x\to 0}\dfrac{\frac{d}{dx}\left(e^{x}-e^{-x}\right)}{\frac{d}{dx}x^{2}}=\lim\limits_{x\to 0}\dfrac{e^{x}+e^{-x}}{2x}$. This limit does not

exist. It approaches $-\infty$ from the left and ∞ from the right.

11. $\lim\limits_{x\to 1}\dfrac{\ln(x)}{x-\sqrt{x}}=\lim\limits_{x\to 1}\dfrac{\frac{d}{dx}\ln(x)}{\frac{d}{dx}\left(x-\sqrt{x}\right)}=\lim\limits_{x\to 1}\dfrac{1}{x\left(1-1/\left(2\sqrt{x}\right)\right)}=2$

13. $\lim\limits_{x\to+\infty}\dfrac{\sin(3/x)}{\sin(9/x)}=\lim\limits_{x\to+\infty}\dfrac{\frac{d}{dx}\sin(3/x)}{\frac{d}{dx}\sin(9/x)}=\lim\limits_{x\to+\infty}\dfrac{\cos(3/x)\left(-3/x^{2}\right)}{\cos(9/x)\left(-9/x^{2}\right)}=\dfrac{1}{3}\lim\limits_{x\to+\infty}\dfrac{\cos(3/x)}{\cos(9/x)}=\dfrac{1}{3}$

15. $\lim\limits_{x\to-\infty}\dfrac{\ln(1+1/x)}{\sin(1/x)}=\lim\limits_{x\to-\infty}\dfrac{\frac{d}{dx}\ln(1+1/x)}{\frac{d}{dx}\sin(1/x)}=\lim\limits_{x\to-\infty}\dfrac{\left(-1/x^{2}\right)}{(1+1/x)\cos(1/x)\left(-1/x^{2}\right)}$

$$=\lim\limits_{x\to-\infty}\dfrac{1}{(1+1/x)\cos(1/x)}=1$$

17. $\lim\limits_{x\to\infty}\dfrac{x^{2}}{e^{3x}}=\lim\limits_{x\to\infty}\dfrac{\frac{d}{dx}x^{2}}{\frac{d}{dx}e^{3x}}=\lim\limits_{x\to\infty}\dfrac{2x}{3e^{3x}}=\dfrac{2}{3}\lim\limits_{x\to\infty}\dfrac{\frac{d}{dx}x}{\frac{d}{dx}e^{3x}}=\dfrac{2}{3}\lim\limits_{x\to\infty}\dfrac{1}{3e^{3x}}=0$

19. $\lim\limits_{x\to 0}\dfrac{\sin^{2}(x)}{x^{2}}=\lim\limits_{x\to 0}\dfrac{\frac{d}{dx}\sin^{2}(x)}{\frac{d}{dx}x^{2}}=\lim\limits_{x\to 0}\dfrac{2\sin(x)\cos(x)}{2x}=\dfrac{1}{2}\lim\limits_{x\to 0}\dfrac{\sin(2x)}{x}=\dfrac{1}{2}\lim\limits_{x\to 0}\dfrac{\frac{d}{dx}\sin(2x)}{\frac{d}{dx}x}$

$$=\dfrac{1}{2}\lim\limits_{x\to 0}2\cos(2x)=1$$

21. $\lim\limits_{x\to 1}\dfrac{(x-1)^{2}}{\cos^{2}(\pi x/2)}=\lim\limits_{x\to 1}\dfrac{\frac{d}{dx}(x-1)^{2}}{\frac{d}{dx}\cos^{2}(\pi x/2)}=\lim\limits_{x\to 1}\dfrac{2(x-1)}{2\cos(\pi x/2)(-\sin(\pi x/2))(\pi/2)}$

$$=-\dfrac{4}{\pi}\lim\limits_{x\to 1}\dfrac{x-1}{\sin(\pi x)}=-\dfrac{4}{\pi}\lim\limits_{x\to 1}\dfrac{\frac{d}{dx}(x-1)}{\frac{d}{dx}\sin(\pi x)}=-\dfrac{4}{\pi}\lim\limits_{x\to 1}\dfrac{1}{\pi\cos(\pi x)}=\dfrac{4}{\pi^{2}}$$

23. We calculate $\lim\limits_{x\to 0}\dfrac{e^{2x}-e^{3x}}{x^{2}}=\lim\limits_{x\to 0}\dfrac{\frac{d}{dx}\left(e^{2x}-e^{3x}\right)}{\frac{d}{dx}x^{2}}=\lim\limits_{x\to 0}\dfrac{2e^{2x}-3e^{3x}}{2x}$. This limit does not exist.

25. $\lim\limits_{x\to+\infty}x\cdot e^{-2x}=\lim\limits_{x\to+\infty}\dfrac{x}{e^{2x}}=\lim\limits_{x\to+\infty}\dfrac{\frac{d}{dx}x}{\frac{d}{dx}e^{2x}}=\lim\limits_{x\to+\infty}\dfrac{1}{2e^{x}}=0$

27. $\lim\limits_{x\to\infty}e^{-x}\ln(x)=\lim\limits_{x\to\infty}\dfrac{\ln(x)}{e^{x}}=\lim\limits_{x\to\infty}\dfrac{\frac{d}{dx}\ln(x)}{\frac{d}{dx}e^{x}}=\lim\limits_{x\to\infty}\dfrac{1}{xe^{x}}=0$

29. $\lim\limits_{x\to 1^{+}}(x-1)^{-1}\ln(x)=\lim\limits_{x\to 1^{+}}\dfrac{\ln(x)}{x-1}=\lim\limits_{x\to 1^{+}}\dfrac{\frac{d}{dx}\ln(x)}{\frac{d}{dx}(x-1)}=\lim\limits_{x\to 1^{+}}\dfrac{1}{x}=1$

31. $\displaystyle\lim_{x\to-2}(x+2)^2\tan(\pi x/4)=\lim_{x\to-2}\frac{(x+2)^2\sin(\pi x/4)}{\cos(\pi x/4)}=\lim_{x\to-2}\frac{(x+2)^2}{\cos(\pi x/4)}\cdot\lim_{x\to-2}\sin(\pi x/4)$

$$=\lim_{x\to-2}\frac{\frac{d}{dx}(x+2)^2}{\frac{d}{dx}\cos(\pi x/4)}=\lim_{x\to-2}\frac{2(x+2)}{\sin(\pi x/4)\cdot(\pi/4)}=0$$

33.   Denote $f(x)=x^{\sqrt{x}}$, and consider $\ln(f(x))=\sqrt{x}\ln(x)=\ln(x)/(1/\sqrt{x})$, where both numerator and denominator tend to infinity. We calculate the limit as follows:

$$\lim_{x\to0^+}\ln(f(x))=\lim_{x\to0^+}\frac{\ln(x)}{1/\sqrt{x}}=\lim_{x\to0^+}\frac{\frac{d}{dx}\ln(x)}{\frac{d}{dx}(1/\sqrt{x})}=-\lim_{x\to0^+}\frac{2x^{3/2}}{x}=-2\lim_{x\to0^+}x^{1/2}=0$$

and $\displaystyle\lim_{x\to0^+}f(x)=e^0=1$.

35.   Denote $f(x)=\left(\sqrt{x}\right)^{\sqrt{x}}=x^{\sqrt{x}/2}$, and consider $\ln(f(x))=\left(\sqrt{x}/2\right)\ln(x)=(1/2)\ln(x)/(1/\sqrt{x})$, where both numerator and denominator tend to 0. Calculating the limit we deduce:

$$\lim_{x\to0^+}\ln(f(x))=\frac{1}{2}\lim_{x\to0^+}\frac{\ln(x)}{1/\sqrt{x}}=\frac{1}{2}\lim_{x\to0^+}\frac{\frac{d}{dx}\ln(x)}{\frac{d}{dx}(1/\sqrt{x})}=-\frac{1}{2}\lim_{x\to0^+}\frac{2x^{3/2}}{x}=-\lim_{x\to0^+}x^{1/2}=0$$

and $\displaystyle\lim_{x\to0^+}f(x)=e^0=1$.

37.   Denote $f(x)=x^{1/x}$, and consider $\ln(f(x))=\ln(x)/x$, where both numerator and denominator tend to infinity. Calculating the limit we deduce:

$$\lim_{x\to+\infty}\ln(f(x))=\lim_{x\to+\infty}\frac{\ln(x)}{x}=\lim_{x\to+\infty}\frac{\frac{d}{dx}\ln(x)}{\frac{d}{dx}x}=\lim_{x\to+\infty}\frac{1}{x}=0$$

and $\displaystyle\lim_{x\to+\infty}f(x)=e^0=1$.

39.   Denote $f(x)=x^{1/\sqrt{x}}$, and consider $\ln(f(x))=\ln(x)/\sqrt{x}$, where both numerator and denominator tend to infinity. Calculating the limit we deduce:

$$\lim_{x\to+\infty}\ln(f(x))=\lim_{x\to+\infty}\frac{\ln(x)}{\sqrt{x}}=\lim_{x\to+\infty}\frac{\frac{d}{dx}\ln(x)}{\frac{d}{dx}\sqrt{x}}=\lim_{x\to+\infty}\frac{1/x}{-1/(2\sqrt{x})}=-2\lim_{x\to+\infty}\frac{1}{\sqrt{x}}=0$$

and $\displaystyle\lim_{x\to+\infty}f(x)=e^0=1$.

41. $$\lim_{x\to\pi/2}\left(\frac{1}{\cos(x)}-\frac{1}{\cos(3x)}\right)=\lim_{x\to\pi/2}\frac{\cos(3x)-\cos(x)}{\cos(x)\cos(3x)}=\lim_{x\to\pi/2}\frac{\frac{d}{dx}\big(\cos(3x)-\cos(x)\big)}{\frac{d}{dx}\big(\cos(x)\cos(3x)\big)}$$

$$=\lim_{x\to\pi/2}\frac{-3\sin(3x)+\sin(x)}{-\sin(x)\cos(3x)-3\cos(x)\sin(3x)}$$

The last limit does not exist as the denominator vanishes while the numerator does not.

43. $$\lim_{x\to0}\left(\frac{x}{1-\cos(x)}-\frac{2}{x}\right)=\lim_{x\to0}\frac{x^2-2+2\cos(x)}{x(1-\cos(x))}=\lim_{x\to0}\frac{\frac{d}{dx}\big(x^2-2+2\cos(x)\big)}{\frac{d}{dx}\big(x(1-\cos(x))\big)}$$

$$=\lim_{x\to0}\frac{2x-2\sin(x)}{1-\cos(x)+x\sin(x)}=2\lim_{x\to0}\frac{\frac{d}{dx}\big(x-\sin(x)\big)}{\frac{d}{dx}\big(1-\cos(x)+x\sin(x)\big)}$$

$$=2\lim_{x\to0}\frac{1-\cos(x)}{2\sin(x)+x\cos(x)}=2\lim_{x\to0}\frac{\frac{d}{dx}\big(1-\cos(x)\big)}{\frac{d}{dx}\big(2\sin(x)+x\cos(x)\big)}$$

$$=2\lim_{x\to0}\frac{\sin(x)}{3\cos(x)-x\sin(x)}=0$$

45. $$\lim_{x\to0}\left(\frac{1}{\sin(x)}-\frac{1}{\sin(2x)}\right)=\lim_{x\to0}\frac{\sin(2x)-\sin(x)}{\sin(x)\sin(2x)}=\lim_{x\to0}\frac{\frac{d}{dx}\big(\sin(2x)-\sin(x)\big)}{\frac{d}{dx}\big(\sin(x)\sin(2x)\big)}$$

$$=\lim_{x\to0}\frac{2\cos(2x)-\cos(x)}{\cos(x)\sin(2x)+2\sin(x)\cos(2x)}$$

The last limit does not exist as the denominator vanishes while the numerator does not.

47. $$\lim_{x\to+\infty}\left(\sqrt{4x+5}-2\sqrt{x}\right)=\lim_{x\to+\infty}\frac{4x-5-4x}{\sqrt{4x-5}+2\sqrt{x}}=-5\lim_{x\to+\infty}\frac{1}{\sqrt{4x-5}+2\sqrt{x}}=0$$

49. $$\lim_{x\to-\infty}(x+1)\ln\big(x/(x+1)\big)=\lim_{x\to-\infty}\frac{\ln\big(x/(x+1)\big)}{1/(x+1)}=\lim_{x\to-\infty}\frac{\frac{d}{dx}\ln\big(x/(x+1)\big)}{\frac{d}{dx}\big(1/(x+1)\big)}$$

$$=\lim_{x\to-\infty}\frac{\big((x+1)/x\big)\big(1/(x+1)^2\big)}{-1/(x+1)^2}=-\lim_{x\to-\infty}\frac{x+1}{x}=-1$$

51. $$\lim_{x\to+\infty}\left(\big(e^{2x}-e^{x}\big)^{1/2}-e^{x}\right)=\lim_{x\to+\infty}\frac{e^{2x}-e^{x}-e^{2x}}{\sqrt{e^{2x}-e^{x}}+e^{x}}=-\lim_{x\to+\infty}\frac{e^{x}}{\sqrt{e^{2x}-e^{x}}+e^{x}}=-\lim_{x\to+\infty}\frac{1}{\sqrt{1-e^{-x}}+1}$$

$$=-\frac{1}{2}$$

Further Theory and Practice

53. Let $f(h) = (1+h)^{1/h}$, then $\ln\big(f(h)\big) = \dfrac{\ln(1+h)}{h}$ and its limit is

$$\lim_{h\to 0^+} \ln\big(f(h)\big) = \lim_{h\to 0^+} \frac{\ln(1+h)}{h} = \lim_{h\to 0^+} \frac{\frac{d}{dh}\ln(1+h)}{\frac{d}{dh}h} = \lim_{h\to 0^+} \frac{1}{1+h} = 1.$$

We calculate the limit of $f(h)$: $\displaystyle\lim_{h\to 0^+} f(h) = e^1 = e$. As $(1+hx)^{1/h} = \Big((1+hx)^{1/(hx)}\Big)^x$, the limit

of $f(h)$ is: $\displaystyle\lim_{h\to 0^+}(1+hx)^{1/h} = \lim_{h\to 0^+}\Big((1+hx)^{1/(hx)}\Big)^x = \Big(\lim_{h\to 0^+}(1+hx)^{1/hx}\Big)^x = e^x$.

55. Let $f(x) = a\sin(x) - \sin(ax)$ and $g(x) = \tan(bx) - b\tan(x)$. As $x \to 0$, the limits of $f'(x)$, $f''(x)$, $g'(x)$, and $g''(x)$ are all 0. However, $f'''(x) = -a\cos(x) + a^3\cos(ax)$. This tends to $a^3 - a$ as $x \to 0$. Furthermore,

$$g'''(x) = 2b\Big(b^2\sec^4(bx) + 2b^2\tan^2(bx)\sec^2(bx) - \sec^4(x) - 2\tan^2(x)\sec^2(x)\Big).$$

This tends to $2b\big(b^2 - 1\big)$ as $x \to 0$. The given limit is therefore $a\dfrac{a^2-1}{2b\big(b^2-1\big)}$.

57. **a.** $\displaystyle\lim_{x\to\pi/2^-}\frac{\cot(x)}{x-\pi/2} = \lim_{x\to\pi/2^-}\frac{\frac{d}{dx}\cot(x)}{\frac{d}{dx}(x-\pi/2)} = -\lim_{x\to\pi/2^-}\csc^2(x) = -1$

b. $\displaystyle\lim_{x\to\pi^+}\frac{\tan(x/2)}{(x-\pi)^{-1}} = \lim_{x\to\pi^+}\frac{x-\pi}{\cos(x/2)}\cdot\lim_{x\to\pi^+}\sin(x/2) = \lim_{x\to\pi^+}\frac{\frac{d}{dx}(x-\pi)}{\frac{d}{dx}\cos(x/2)} = \lim_{x\to\pi^+}\frac{2}{-\sin(x/2)}$

$= -2$

c. $\displaystyle\lim_{x\to 0^+}\frac{\ln(1+x)}{\ln(1+3x)} = \lim_{x\to 0^+}\frac{\frac{d}{dx}\ln(1+x)}{\frac{d}{dx}\ln(1+3x)} = \lim_{x\to 0^+}\frac{1+3x}{3(1+x)} = \frac{1}{3}$

d. $\displaystyle\lim_{x\to 0^+}\frac{\sin(x)}{\ln(1+x)} = \lim_{x\to 0^+}\frac{\frac{d}{dx}\sin(x)}{\frac{d}{dx}\ln(1+x)} = \lim_{x\to 0^+}\big(\cos(x)(1+x)\big) = 1$

e. $\displaystyle\lim_{x\to 0^+}\frac{\ln(x)}{1/\sqrt{x}} = \lim_{x\to 0^+}\frac{\frac{d}{dx}\ln(x)}{\frac{d}{dx}(1/\sqrt{x})} = \lim_{x\to 0^+}\frac{-2x^{3/2}}{x} = -2\lim_{x\to 0^+}x^{1/2} = 0$

f. $\displaystyle\lim_{x\to 0^+}\frac{\ln(x)}{1/\sin(x)} = \lim_{x\to 0^+}\frac{\frac{d}{dx}\ln(x)}{\frac{d}{dx}(1/\sin(x))} = \lim_{x\to 0^+}\frac{-\sin^2(x)}{x\cos(x)} = -\lim_{x\to 0^+}\frac{\sin^2(x)}{x}\lim_{x\to 0^+}\frac{1}{\cos(x)}$

$= -\lim_{x\to 0^+}\frac{\frac{d}{dx}\sin^2(x)}{\frac{d}{dx}x} = -\lim_{x\to 0^+}2\sin(x)\cos(x) = 0$

59. In Example 7 it was shown that $\lim\limits_{x\to\infty} x^{1/x} = 1$, hence we need to evaluate

$$\lim_{x\to\infty} \frac{x}{\sqrt{1+x^2}} = \lim_{x\to\infty} \frac{1}{\sqrt{x^{-2}+1}} = 1.$$

Finally, $\lim\limits_{x\to\infty} \frac{x^{(x+1)/x}}{\sqrt{1+x^2}} = \lim\limits_{x\to\infty} x^{1/x} \cdot \lim\limits_{x\to\infty} \frac{x}{\sqrt{1+x^2}} = 1\cdot 1 = 1$.

61. Let $A\big(g(a), f(a)\big)$ and $B\big(g(b), f(b)\big)$ be two points. The equation of a straight line through these points is $y - f(a) = \frac{f(b)-f(a)}{g(b)-g(a)}\big(x - g(a)\big)$. For $x = g(t)$ and $y = f(t)$ we

see that $\frac{f(b)-f(a)}{g(b)-g(a)}\big(g(t)-g(a)\big) - \big(f(t)-f(a)\big) = 0$, or $r(t) = 0$, where $r(t)$ is the

distance from $\big(f(t), g(t)\big)$ to the line with slope $g(b) - g(a)$. Cauchy's Mean Value Theorem statement implies that there is a point $\xi \in (a, b)$, where the tangent line to $r(t)$ is horizontal.

Calculator/Computer Exercises

63. The numerically obtained limit is $1/2$.

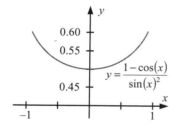

65. The numerically obtained limit is 1.

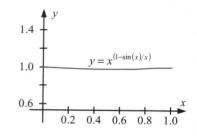

67.

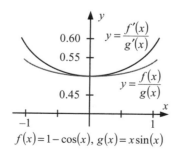

$f(x)=1-\cos(x),\ g(x)=x\sin(x)$

69.

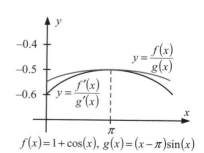

$f(x)=1+\cos(x),\ g(x)=(x-\pi)\sin(x)$

| **Section 4.7** | **The Newton-Raphson Method** |

Problems for Practice

1. Calculator check: $x = \sqrt{6} \approx 2.44949$

$x_1 = 2.0$

$$x_2 = x_1 - \frac{x_1^2 - 6}{2x_1} = 2.0 - \frac{(2)^2 - 6}{2 \cdot 2} = 2.500$$

$$x_3 = x_2 - \frac{x_2^2 - 6}{2x_2} = 2.5 - \frac{(2.5)^2 - 6}{2 \cdot 2.5} = 2.450$$

$$x_4 = x_3 - \frac{x_3^2 - 6}{2x_3} = 2.45 - \frac{(2.45)^2 - 6}{2 \cdot 2.45} = 2.449$$

Four iterations were enough.

3. Calculator check: $x = \dfrac{5 + \sqrt{25-8}}{2} \approx 4.56155$

$x_1 = 4.0$

$$x_2 = x_1 - \frac{x_1^2 - 5x_1 + 2}{2x_1 - 5} = 4.0 - \frac{(4.0)^2 - 5 \cdot 4.0 + 2}{2 \cdot 4.0 - 5} = 4.667$$

$$x_3 = x_2 - \frac{x_2^2 - 5x_2 + 2}{2x_2 - 5} = 4.0 - \frac{(4.667)^2 - 5 \cdot 4.667 + 2}{2 \cdot 4.667 - 5} = 4.5616$$

Four iterations were enough.

5. Calculator check: $x = \sqrt[5]{31} \approx 1.98734$

$x_1 = 2.0$

$$x_2 = x_1 - \frac{x_1^5 - 31}{5x_1^4} = 2.0 - \frac{(2)^5 - 31}{5 \cdot (2)^4} = 1.9875$$

Two iterations were enough.

7. $f(x) = x^3 - 4$; $f'(x) = 3x^2$; $x_1 = 1.500$; $x_2 = 1.5926$; $x_3 = 1.5874$; $x_4 = 1.5874$; $4^{1/3} \approx 1.59$

9. $f(x) = x^2 - 6$; $f'(x) = 2x$; $x_1 = 2.500$; $x_2 = 2.4500$; $x_3 = 2.4495$; $6^{1/2} \approx 2.45$

11. $f(x) = x^2 - 10$; $f'(x) = 2x$; $x_1 = 3.000$; $x_2 = 3.1667$; $x_3 = 3.1623$; $\sqrt{10} \approx 3.16$

Further Theory and Practice

13.   If $f(x) = ax + b$, then $f'(x) = a$, and for any x_1 we deduce

$$x_2 = x_1 - \frac{f(x_1)}{f'(x_1)} = x_1 - \frac{ax_1 + b}{a} = \frac{-b}{a}.$$

On the other hand, the root of $f(x)$ is $x = -\dfrac{b}{a}$. In this case the root of $f(x)$ is found after the first iteration and does not depend on the initial guess.

15.   **a.**   We calculate the derivatives $f'(x) = \cos(x)$, $f''(x) = -\sin(x)$ to obtain

$$\left| \frac{f(x)f''(x)}{(f'(x))^2} \right| = \left| \frac{-\sin^2(x)}{\cos^2(x)} \right| = \tan^2(x) < 1,$$

for $x \in I$.

b. $f'(x) = 2x - 1$, $f''(x) = 2$, $\left| \dfrac{f(x)f''(x)}{(f'(x))^2} \right| = \left| 2\dfrac{x^2 - x}{(2x-1)^2} \right|.$

Whereas the denominator is equal to zero at $x = \dfrac{1}{2}$, $2\dfrac{x^2 - x}{2x-1} \gg 1$ as $x \to \dfrac{1}{2}^+$, the condition is not satisfied for every $x \in I$.

c. $f'(x) = 3x^2$, $f''(x) = 6x$, $\left| \dfrac{f(x)f''(x)}{(f'(x))^2} \right| = \left| 6x\dfrac{x^3 + 8}{9x^4} \right| = \left| \dfrac{2}{3} + \dfrac{16}{9x^3} \right|.$ When $x \to -1^-$,

$\left| \dfrac{2}{3} + \dfrac{16}{9x^3} \right| > 1$, therefore the condition is not satisfied for every $x \in I$.

d.   We calculate the derivatives $f'(x) = 6x$, $f''(x) = 6$ to obtain

$$\left| \frac{f(x)f''(x)}{(f'(x))^2} \right| = \left| 6\frac{3x^2 - 12}{36x^2} \right| = \left| \frac{1}{2} - \frac{2}{x^2} \right| < 1,$$

if $x \in I$.

17.   If $f(x) = x^2 - c$, then $f'(x) = 2x$ and $x_{j+1} = x_j - \dfrac{f(x_j)}{f'(x_j)} = x_j - \dfrac{x_j^2 - c}{2x_j} = \left(\dfrac{1}{2}\right)\left(x_j + \dfrac{c}{x_j}\right)$. If

x_j is exactly equal to $\sqrt{c}$, then so is x_{j+1}. Otherwise, one of the two estimates x_j and $\dfrac{c}{x_j}$ undershoots $\sqrt{c}$ and the other overshoots $\sqrt{c}$; the oppositely-signed errors tend to cancel when the average x_{j+1} of these two estimates is formed.

19. Since f' has no root, its sign does not change. Thus, neither f' nor f'' changes sign. Let us assume that both are positive. (The other cases are handled in a similar manner.) Let c be the root of f: there is only one since f is increasing. If $x_1 > c$ then $f(x_1) > 0$, $f'(x_1) > 0$, and

$x_2 = x_1 - \dfrac{f(x_1)}{f'(x_1)} < x_1$. Also, because the graph of f is concave up, it lies above its tangent

lines. Since x_2 is the x-intercept of the tangent line to the graph of f at $(x_1, f(x_1))$, it follows that $f(x_2) \geq 0$. Thus, $x_2 \geq c$. If we repeat this argument, then we obtain a decreasing sequence $\{x_n\}$ that is bounded below by c. It follows that the sequence of Newton-Raphson iterates converges to a number $\ell \geq c$. Suppose that $f(\ell) > 0$. Let M be the maximum value of

f' on $[c, x_1]$. Then, for all j, we have $f(x_j) > f(\ell)$ and $x_j - x_{j+1} = \dfrac{f(x_j)}{f'(x_j)} > \dfrac{f(\ell)}{M}$. Such a

sequence tends to $-\infty$, which is a contradiction. In this case $x_1 < c$, we have $x_2 > c$ and we are back to the first case.

21. As indicated in the previous problem, x_* is an equilibrium point. From Problem 4.1.62 we conclude that x_* is a stable equilibrium for Φ. The definition of a stable equilibrium implies that $\delta > 0$ is such that each element $\{x_n\}$ converges to x_* provided that $|x_* - x_1| < \delta$.

23. From Problem 4.7.21 we conclude that the Newton-Raphson iterations converge to x_*, if x_* is stable equilibrium. The definition of stable equilibrium implies that $\{x_n\}$ converges to x_* if x_1 is chosen sufficiently close to x_*, so that $|x_* - x_1| < \delta$, where $\delta > 0$.

Calculator/Computer Exercises

25. $x_1 = 1$, $N = 4$, $x \approx 1.08891683$

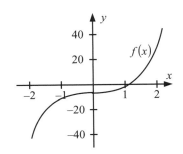

27. $x_1 = 2$, $N = 5$, $x = 1.8$

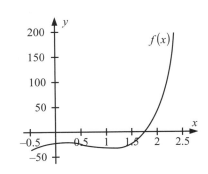

29. $f'(x) = 1 + 4.6\cos\dfrac{4.6x}{x} - \sin\dfrac{4.6x}{x^2}$

$x_1 = 1.1$

$x_2 = 1.053186231$

$x_3 = 1.007844617$

$x_4 = 0.9965113072$

$x_5 = 0.9956047642$

$x_6 = 0.9955987723$

The largest real root of $f(x)$ is $x \approx 0.9955988$.

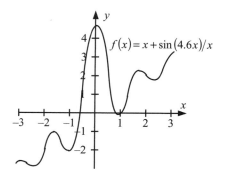

31. $f(x) = \sqrt{x^3 + 1} - \sqrt[3]{x^2 + 1}$

$f'(x) = \dfrac{3x^2}{2\sqrt{x^3 + 1}} + \dfrac{2x}{3(x^2 + 1)^{2/3}}$

$x_1 = 1.0$

$x_2 = 0.759176$

$x_3 = 0.657953$

$x_4 = 0.631090$

$x_5 = 0.629030$

$x_6 = 0.629018$

$x_7 = 0.629018$

The nonzero approximation is $x \approx 0.629$.

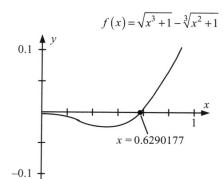

33. $p'(x) = 4x^3 - 6x^2 + 6x - 4$; $c = 1.0$

$x_1 = 1.5$

$x_2 = 1.28570$

$x_3 = 1.15696$

$x_4 = 1.08253$

$x_5 = 1.04240$

$x_6 = 1.02150$

$x_7 = 1.01083$

$x_8 = 1.00543$

$x_9 = 1.00272$

$x_{10} = 1.00136$

$x_{11} = 1.00068$

$x_{12} = 1.00034$

$N = 12$

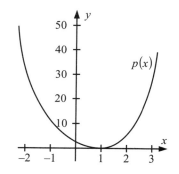

35. $p'(x) = 16x^3 - 12x^2 - 70x + 36$; $c = 0.5$

$x_1 = 1.0$

$x_2 = 0.73333$

$x_3 = 0.61456$

$x_4 = 0.55666$

$x_5 = 0.52823$

$x_6 = 0.51409$

$x_7 = 0.50704$

$x_8 = 0.50352$

$x_9 = 0.50176$

$x_{10} = 0.50088$

$x_{11} = 0.50044$

$N = 11$

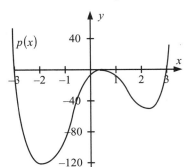

37.   For Problem 4.7.33 $x_4 = 1.00000$; $M = 4$.

For Problem 4.7.34 $x_4 = -2.00000$; $M = 4$.

For Problem 4.7.35 $x_3 = 0.49994$; $M = 3$.

For Problem 4.7.36 $x_4 = 1.99999$ $M = 4$.

For all these problems, $M \le 4$, as compared to $N \ge 10$ when the unmodified Newton-Raphson Method is used.

39.   We apply the Newton-Raphson Method to $f(\varphi) = 1.856\left(1 + \dfrac{1}{16}\varphi^2 + \dfrac{11}{3072}\varphi^4\right) - 2$. We have

$$\Phi(\varphi) = \varphi - \frac{f(\varphi)}{f'(\varphi)} = \varphi - \frac{1.856\left(1 + \frac{1}{16}\varphi^2 + \frac{11}{3072}\varphi^4\right) - 2}{1.856\left(\frac{1}{8}\varphi + \frac{11}{768}\varphi^3\right)}.$$ Choosing $x_1 = 1.0$, we calculate

$x_2 = \Phi(1.0) = 1.083$, $x_3 = \Phi(1.083) = 1.079$, and $x_4 = \Phi(1.079) = 1.079$.

41. $V - P\left((1+x)^{-n} + \dfrac{r}{2} \cdot \dfrac{1 - (1+x)^{-n}}{\sqrt{1+x} - 1}\right) = 0$

     a.   7.75%

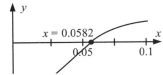

     b.   5.82%

43.   Price $8,681, coupon rate = 5% has an effective yield of 6.25%, whereas price $10,674, coupon rate = 7% has an effective yield of 6.50%.

45.   Price $10,679, coupon rate $= 8\%$ has an effective yield of 7.00%, whereas price $11,052, coupon rate $= 9\%$ has an effective yield of 7.38%.

47.   We calculate: $x_2 = x_1 - \dfrac{f(x_1)}{f'(x_1)}$, $x_3 = x_2 - \dfrac{f(x_2)}{f'(x_2)} = \left(x_1 - \dfrac{f(x_1)}{f'(x_1)} \right) - \dfrac{f(x_1) - f(x_1)/f'(x_1)}{f'(x_1 - f(x_1)/f'(x_1))}$.

The phenomenon of cycling is observed when $x_1 = x_3$, or $x_3 - x_1 = 0$. Calculations in MAPLE show that $x_1 = 1.8030059$ leads to cycling.

49.   The function $\dfrac{Y}{x}$ has a maximum when $\left(\dfrac{Y}{x} \right)' = 0$. This maximum is found by using the Newton-Raphson Method for $f(x) = \dfrac{Y}{x}$. Using MAPLE to calculate the maximum of $\dfrac{Y}{x}$, we find that the absolute maximum value is 1.0288 and that it occurs at $x = 0.51282$.

Section 4.8   Antidifferentiation and Applications

Problems for Practice

1. $\int \left(x^2 - 5x \right) dx = \int x^2 dx - 5 \int x \, dx = \left(\dfrac{1}{3} \right) x^3 - \left(\dfrac{5}{2} \right) x^2 + C$

3. $\int e^x dx = e^x + C$

5.   We calculate the integral as $\int \sqrt{x+2} \, dx = \dfrac{(x+2)^{3/2}}{3/2} = \left(\dfrac{2}{3} \right)(x+2)^{3/2} + C$, because

$$\tfrac{d}{dx}(x+2)^{3/2} = \left(\dfrac{3}{2} \right) \sqrt{x+2} .$$

7. $\int \dfrac{x^2 + x^{-3}}{x^4} dx = \int x^{-2} dx + \int x^{-7} dx = -x^{-1} - \left(\dfrac{1}{6} \right) x^{-6} + C$

9.   We calculate the integral as $\int (x+1)^2 \, dx = \left(\dfrac{1}{3} \right)(x+1)^3 + C$, because $\tfrac{d}{dx}(x+1)^3 = 3(x+1)^2$.

11.   We calculate the integral as $\int \exp(e \cdot x) \, dx = \left(\dfrac{1}{e} \right) \exp(e \cdot x) + C$, because

$$\tfrac{d}{dx} \exp(e \cdot x) = e \cdot \exp(e \cdot x).$$

13. $\int \left(x^{-7/3} - 4x^{-2/3} \right) dx = \left(-\dfrac{3}{4} \right) x^{-4/3} - 12x^{1/3} + C$

15.   We calculate the integral as

$$\int \left(3\cos(4x) + 2x \right) dx = 3\int \cos(4x)\, dx + 2\int x\, dx = \left(\frac{3}{4} \right)\sin(4x) + x^2 + C,$$

because $\frac{d}{dx}\sin(4x) = 4\cos(4x)$.

17.   We calculate the integral as $\int \sec^2(8x)\, dx = \left(\frac{1}{8} \right)\tan(8x) + C$, because $\frac{d}{dx}\tan(8x) = 8\sec^2(8x)$.

19.   We calculate the integral as $\int (3x-2)^3\, dx = \left(\frac{1}{12} \right)(3x-2)^4 + C$, because

$$\frac{d}{dx}(3x-2)^4 = 12(3x-2)^3.$$

21.   **a.**   For some constant C, we have $F(x) = 2x^3 + C$. Given that $3 = F(1) = 2 + C$, we see that $C = 1$. Therefore, $F(x) = 2x^3 + 1$ and $F(0) = 1$.

      b.   For some constant C, we have $F(x) = \sin(x) + C$. Given that

$$-1 = F\left(\frac{\pi}{2} \right) = \sin\left(\frac{\pi}{2} \right) + C,$$ we see that $C = -2$. Therefore, $F(x) = \sin(x) - 2$ and

$$F\left(\frac{\pi}{6} \right) = \sin\left(\frac{\pi}{6} \right) - 2 = -\frac{3}{2}.$$

      c.   For some constant C, we have $F(x) = x^2 + 3x + C$. Given that $2 = F(1) = 1^2 + 3(1) + C$, we see that $C = -2$. Therefore, $F(x) = x^2 + 3x - 2$ and $F(-1) = (-1)^2 + 3(-1) - 2 = -4$.

      d.   For some constant C, we have $F(x) = 3e^{2x} + C$. Given that $-1 = F(0) = 3e^0 + C$, we see that $C = -4$. Therefore, $F(x) = 3e^{2x} - 4$ and $F\left(\frac{1}{2} \right) = 3e - 4$.

23.   Given that $a(t) = 4$ we calculate $v(t) = 4t + C$ and $p(t) = 2t^2 + Ct + D$. From the initial conditions $v(0) = 0$, $p(0) = 0$ we conclude $C = D = 0$. Hence $p(t) = 20$ for $t = \sqrt{10}$ s. At this instant, the velocity is $v\left(\sqrt{10} \right) = 4\sqrt{10}$ m/s.

25.   As $v(t) = 100 - 32t$ and $h(0) = 0$, we deduce $h(t) = 100t - 16t^2$. The baseball changes its direction of motion at the moment t_0 when $v(t_0) = 0$, that is, at $t_0 = \dfrac{625}{200}$ s. At this moment, the trajectory is at its highest point, i.e., $h_{\max} = h(t_0) = \dfrac{31250}{200} \approx 156.25$ ft.

From $h(t) = 25$ on the way down we obtain $t_1 = \sqrt{\dfrac{2(156.25 - 25)}{32}} = 2.86$ s. At this moment $v(t_1) = 32t_1 \approx 91.65$ ft/s.

27. We calculate $v(t) = 50 + at$ mi/h , $p(t) = \left(\dfrac{5}{22}\right) - 50t - \dfrac{at^2}{2}$ mi and for some t_0 the values

are: $v(t_0) = 0$, $p(t_0) = 0$. Therefore, $t_0 = \dfrac{1}{110}$ h and $a = -5500$ mi/h^2 . In different units:

$a = -\dfrac{121}{54}$ ft/s^2 .

29. The height is $h(t) = 16t^2$. The time it takes for the ball to fall from a building 375 ft tall is:

$t = \sqrt{\dfrac{375}{16}} = \dfrac{5\sqrt{15}}{4} \approx 4.84$ s .

Further Theory and Practice

31. $\displaystyle\int \sin(x)\cos(x)\,dx = \left(\dfrac{1}{2}\right)\int \sin(2x)\,dx = -\left(\dfrac{1}{4}\right)\cos(2x) + C$

33. $\displaystyle\int \tan^2(x)\,dx = \int\left(\sec^2(x) - 1\right)dx = \tan(x) - x + C$

35. $\displaystyle\int \cos^2(x)\,dx = \int\left(\dfrac{1 + \cos(2x)}{2}\right)dx = \left(\dfrac{1}{2}\right)\left(x + \left(\dfrac{1}{2}\right)\sin(2x)\right) + C$

37. No, because $\dfrac{d}{dx}G(f(x)) = g(f(x)) \cdot \dfrac{d}{dx}f(x) = (g \circ f)(x) \cdot \dfrac{d}{dx}f(x)$.

39. **a.** The result of integration is $\displaystyle\int x\left(x^2 + 1\right)^{100}dx = \dfrac{\left(x^2 + 1\right)^{101}}{202} + C$, because

$\dfrac{d}{dx}\left(x^2 + 1\right)^{101} = 202x\left(x^2 + 1\right)^{100}$.

b. The result of integration is $\displaystyle\int \cos(x)\sin(x)\,dx = \dfrac{\sin^2(x)}{2} + C$, because

$\dfrac{d}{dx}\sin^2(x) = 2\sin(x)\cos(x)$. Note: $-\dfrac{1}{2}\cos^2(x) + C$ and $-\dfrac{1}{4}\cos(2x) + C$ are also

correct answers.

c. The result of integration is $\displaystyle\int \sin^2(x)\cos(x)\,dx = \dfrac{\sin^3(x)}{3} + C$, because

$\dfrac{d}{dx}\sin^3(x) = 3\sin^2(x)\cos(x)$.

d. The result of integration is $\displaystyle\int x\exp\left(x^2\right)dx = \dfrac{\exp\left(x^2\right)}{2} + C$, because

$\dfrac{d}{dx}\exp\left(x^2\right) = 2x\exp\left(x^2\right)$.

e. The result of integration is $\displaystyle\int \cos(\sin(x))\cos(x)\,dx = \sin(\sin(x)) + C$, because

$\dfrac{d}{dx}\sin(\sin(x)) = \cos(\sin(x))\cos(x)$.

41.   We calculate the derivative to be $\frac{d}{dx}\left(1+e^{kt}\right)^{n+1} = (n+1)\,ke^{kt}\left(1+e^{kt}\right)^{n}$, hence $A = \dfrac{1}{k(n+1)}$ for $n \neq -1$ and $k \neq 0$.

43.   As $\mathrm{erf}'(x) = \dfrac{2\exp\left(-x^2\right)}{\sqrt{\pi}}$ is positive, the error function is everywhere increasing.

We calculate $\mathrm{erf}''(x) = -4x\dfrac{\exp\left(-x^2\right)}{\sqrt{\pi}}$, which is positive for $x < 0$ and negative for $x > 0$.

Therefore, the error function is concave up on $(-\infty,\, 0)$ and concave down on $(0,\, \infty)$, with $(0,\, 0)$ being an inflection point.

45.   Horizontal component of velocity is constant and equals $190 \cdot \cos(60°) = 95$ ft/s. Vertical component of this velocity at the start equals $190 \cdot \sin(60°) = 95\sqrt{3}$ ft/s. For a vertical velocity and height dependence on time we calculate $v(t) = 95\sqrt{3} - 32t$, $h(t) = 95\sqrt{3}t - 16t^2$.

Hence $h_{\max} = h\left(\dfrac{95\sqrt{3}}{32}\right) = \dfrac{27075}{64} \approx 423$ ft is the highest point on the trajectory.

Horizontal velocity remains at 95 ft/s.

47.   If t is the time in seconds of the falling stone, the depth of the well is $h(t) = 16t^2$. The speed of sound is 968 ft/s, and it takes the sound wave $\dfrac{h(t)}{968}$ min to reach the woman's ears. Hence, we find $t + \dfrac{16t^2}{968} = 6$ and $t = 5.5$ s. The depth of the well is 484 ft.

49.   The initial velocity is $\dfrac{3}{2}$ mi/min. Let t be the time of acceleration. Car A passes at

$$p_A(t) = \left(\frac{3}{2}\right)t + 3t^2 \text{ mi};$$

car B passes at $p_B(t) = \left(\dfrac{3}{2}\right)t + \dfrac{9t^2}{2}$ mi. Car B laps car A three times after the start of acceleration when $p_B(t) = p_A(t) + 6$, i.e., at $t = 2$ min.

51. During the first 8 s the rocket will climb 800 ft. After that, for time t it will climb an additional $h(t) = 100t - 16t^2$ ft, $h_{max} = h\left(\dfrac{100}{32}\right) = 168.75$ ft. Therefore the maximum height is 968.75 ft.

Climb time is $8 + \dfrac{100}{32} \approx 11.125$ s, descent time τ can be found from the condition

$$h_{max} = 16\tau^2,$$

i.e. $\tau \approx 7.78$ s. The total flight time is 18.905 s and the final velocity is 248.96 ft/s.

53. We calculate $m(t) = \left(\dfrac{A}{k}\right) e^{-kt} + C$. We substitute $t = 0$ to obtain $m(0) = \dfrac{A}{k} + C$. If

$$m(0) = \dfrac{A}{k},$$

then $C = 0$ and $m(t) = m(0) e^{-kt}$. We find that $m(t) = \dfrac{m(0)}{2}$ when $e^{-kt} = \dfrac{1}{2}$, or $t = \dfrac{\ln(2)}{k}$.

55. Denote the height at time t by $y(t)$. Then $y(t) = -kg\left(t + ke^{-t/k}\right) + C$. Substituting $t = 0$ we obtain $H = -k^2 g + C$, or $C = k^2 g + H$. Therefore, $y(t) = -kg\left(t + ke^{-t/k}\right) + k^2 g + H$, or $y(t) = H - kgt + k^2 g\left(1 - e^{-t/k}\right)$.

Calculator/Computer Exercises

57. We calculate

$$m(t) = -0.1213 \int e^{-0.0001213t} \, dt = -0.1213 \frac{e^{-0.0001213t}}{-0.0001213} + C = 1000 e^{-0.0001213t} + C.$$

As $m(0) = 1000 e^{-(0.0001213)(0)} + C = 1000 + C$, we use the given initial value, $m(0) = 1000$, to conclude that $C = 0$. Thus, $m(t) = 1000 e^{-0.0001213 \cdot t}$. The required time, τ, therefore, satisfies $1000 e^{-0.0001213 \cdot \tau} = 800$ or, equivalently, $-0.0001213 \cdot \tau = \ln\left(\dfrac{800}{1000}\right)$. The value is $\tau = 1839.6$ years.

59. In Problem 4.8.54 we calculated $T(t) = 40 + 10e^{-0.02t}$ and $\lim_{t \to \infty} T(t) = 40$. First, we solve for $46 = T(t) = 40 + 10e^{-0.02t}$ to find that 25.54 seconds must elapse for the object to cool 4°C. Then, we solve for $44 = T(t) = 40 + 10e^{-0.02t}$ to find that the temperature is 44°C when $t = 45.81$ s. This means that after the initial drop, $45.81 - 25.54 = 20.27$ seconds elapse before the object cools a further two degrees. Finally, we solve for $43 = T(t) = 40 + 10e^{-0.02t}$ to find $t = 50.20$ seconds elapse from the initial time until the object reaches 43°C. Therefore, for the object to cool from 44°C to 43°C, an elapsed time of $60.20 - 45.81 = 14.39$ s is required.

61. The velocity is $v(t) = 8.855\left(1 + e^{4.427t}\right)^{-1} - 4.4275$; $v_\infty = -4.4275$; $v(t) = 0.999v_\infty$ when $t = 1.7168$.

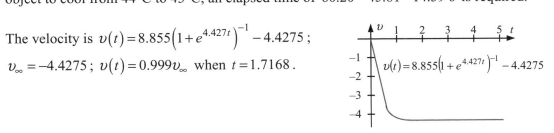

63. By definition let $\text{Si}'(x) = \dfrac{\sin(x)}{x}$. Thus, for $x \geq 0$, the signs of $\text{Si}'(x)$ and $\sin(x)$ are the same. We deduce that $\text{Si}(x)$ increases on the intervals on which $\sin(x)$ is positive and decreases on intervals on which $\sin(x)$ is negative. To be specific, $\text{Si}(x)$ increases on $[0, \pi]$, decreases on $[\pi, 2\pi]$, increases on $[2\pi, 3\pi]$, and decreases on $[3\pi, 4\pi]$. We conclude that $\text{Si}(x)$ has a local maximum at $\pi = x$, a local minimum at $x = 2\pi$, a local maximum at $x = 3\pi$, and a local minimum at $x = 4\pi$. The second derivative,

$$\text{Si}''(x) = \frac{x\cos(x) - \sin(x)}{x^2},$$

has the same sign as $x\cos(x) - \sin(x)$. We can reduce our analysis to the positive interval $[0, 4\pi]$. That is, because $\dfrac{\sin(x)}{x}$ is an even function:

$$\frac{\sin(-x)}{-x} = \frac{-\sin(x)}{-x} = \frac{\sin(x)}{x}.$$

Therefore the sine integral is an odd function.

Section 4.9   Enrichment: Applications to Economics

Problems for Practice

1. As $C'(x) = 2$, the marginal cost at $x = 1000$ is $C'(1000) = 2$.

3. As $C'(x) = 20\left(1 - \dfrac{1}{2\sqrt{x}}\right)$, the marginal cost at $x = 100$ is $C'(100) = 19$.

5.   **a.**   Fixed cost: $C'(99) = 0.01$, $C_0 = C(0) = 2000$

      b.   Cost of producing the 100th unit: $C(100) - C(99) = 3$

      c.   Marginal cost: $C'(x) = 3$, $C'(100) = 3$

      d.   Average cost: $\overline{C}(100) = \dfrac{C(100)}{100} = 23$

7.   **a.**   Fixed cost: $C_0 = C(0) = 1200$

      b.   Cost of producing the 100th unit:

$$C(100) - C(99) = 12000\left(\frac{101}{20} - \frac{100}{3\sqrt{10} + 10} \right) \approx 449.25$$

      c.   Marginal cost: $C'(x) = 6000 \cdot \dfrac{x + 20\sqrt{x} - 1}{\sqrt{x}\left(\sqrt{x} + 10\right)^2}$, $C'(100) = 448.5$

      d.   Average cost: $\overline{C}(100) = \dfrac{C(100)}{100} = 606$

9.   The profit function is $P(x) = -8x^2 + 1600x - 1000$. Its derivative is $P'(x) = -16x + 1600$ and $x = 100$ is a critical point. At this point, $P''(x) = -16 < 0$, hence $x = 100$ leads to the maximum profit.

11.   In this case $x = D(p) = 51 - \dfrac{p}{8}$ and the profit function is

$$P(p) = p\left(51 - \frac{p}{8}\right) - 1200 - 8\left(51 - \frac{p}{8}\right),$$

or $P(p) = -\dfrac{p^2}{8} + 52p - 1608$. Then $P'(p) = 52 - \dfrac{p}{4}$, $p = 208$ is a critical point, $P''(p) = -\dfrac{1}{4} < 0$, hence P has a maximum at $p = 208$, hence $x = 25$ leads to the maximum profit.

13.   Here $c = 3$ and $\mu N = 200$. Therefore, $x_* = \sqrt{\dfrac{1.20 \cdot 12000}{2 \cdot 3}} \approx 48.99$. Since

$$TC(48) = 3(48) + 200 + 1.20 \cdot \frac{12000}{2(48)} = 494.0$$

is greater than $TC(49) = 3(49) + 200 + 1.20 \cdot \dfrac{12000}{2(49)} = 496.94$, we conclude that $TC(x)$ is minimized for $x = 49$.

15. Here $c = 8$ and $\mu N = 9000$. Therefore, $x_* = \sqrt{\dfrac{2 \cdot 64000}{2 \cdot 8}} = 40\sqrt{5} = 89.4\ldots$. We test $x = 89$

and $x = 90$: $TC(89) = 8(89) + 9000 + \dfrac{2 \cdot 64000}{2 \cdot 89} = 10431.404\ldots$ and

$$TC(90) = 8(90) + 9000 + \dfrac{2 \cdot 64000}{2 \cdot 90} = 10431.111\ldots .$$

Since $TC(89) < TC(90)$, $x = 89$ is the economic lot size.

17. Let $x_\alpha = \sqrt{\dfrac{6 \cdot 12000}{200 \cdot 90}} = 2$. Hence the optimal number of withdrawals that Alfred should make is twice per year.

19. Let $\sqrt{\dfrac{10 \cdot 10000}{200 \cdot 45}} = \dfrac{10}{3} = 3.3\ldots$. First, we calculate the transaction cost for 3 withdrawals:

$3 \cdot 45 + \dfrac{10 \cdot 10000}{200 \cdot 3} = 301.67$. We then compare this with the transaction costs for 4 withdrawals:

$4 \cdot 45 + \dfrac{10 \cdot 10000}{200 \cdot 4} = 305.0$. Hence, the optimal number of withdrawals that Orville should make is 3 times per year.

21. At production level x, the revenue is Axe^{-kx}. The derivative of revenue is $Ae^{-kx}(1 - xk)$,

which is positive for $x < \dfrac{1}{k}$ and negative for $x > \dfrac{1}{k}$. Maximum revenue occurs at $x = \dfrac{1}{k}$, or, if

x is regarded as an integer-valued variable, at either $\left[\dfrac{1}{k}\right]$ or $\left[\dfrac{1}{k}\right] + 1$.

23. As $\overline{C}'(x) = \dfrac{x^2 - 32000000}{2000x^2} < 0$ for $0 < x < \sqrt{32000000} \approx 5656.9$ and $\overline{C}'(x) > 0$ for $x \geq 5657$,

as $\overline{C}(5656) = 5.65685431\ldots$, and because $\overline{C}(5657) = 5.65685425\ldots$, we conclude that $\overline{C}(x)$ is minimized at $x = 5657$.

25. As $\overline{C}'(x) = 0.00025 \dfrac{\left(x^3 - 64 \cdot 10^6\right)}{x^2} < 0$ for $0 < x < \sqrt[3]{64 \cdot 10^6} = 400$ and $\overline{C}'(x) > 0$ for $x > 400$,

we see that $\overline{C}(x)$ is minimized at $x = 400$.

27. We calculate: $E(p) = -\dfrac{p \mathrm{D}'(p)}{\mathrm{D}(p)} = \dfrac{\sqrt{p}}{8 - 2\sqrt{p}}$.

Demand is inelastic for $0 < p < \dfrac{64}{9}$.

Demand is elastic for $\dfrac{64}{9} < p < 16$.

29.   We solve $E(p) = 1$. Thus revenue is maximized when $p = 16$.

31.   We solve $E(p) = 1$. Thus revenue is maximized when $p = 40$.

33.   The percent change in p is $+20$. Therefore the percent change in x is approximately

$$-\left(\frac{1}{2}\right) \cdot 20 = -10,$$

or 10% decrease.

35.   The percent change in p is -10. Therefore the percent change x is approximately

$$-(1)(-10) = +10,$$

or 10% increase.

37.   **a.**   In point-slope form: $x = \left(\dfrac{2000 - 1700}{3.75 - 4.50}\right)(p - 3.75) + 2000$, or $x = 3500 - 400p$.

b.   As a function of price, the revenue is $(3500 - 400p)\,p$. Therefore, the change in revenue is $(3500 - 400(4.50))(4.50) - (3500 - 400(3.75))(3.75)$, or 150.

c.   We calculate: $E(p) = \dfrac{4p}{35 - 4p}$. Therefore, the elasticity of demand at \$3.75 per

hamburger is: $E(3.75) = \dfrac{4(3.75)}{35 - 4(3.75)} = 0.75$. The elasticity of demand at \$4.50 per

hamburger is $E(4.50) = \dfrac{4(4.50)}{35 - 4(4.50)} = 1.06$.

d.   Given that $E(4.50) > 1$, raising the price beyond 4.50 will decrease the revenue because, when $E > 1$, revenue is a decreasing function.

e.   At $p = 4.50$, a 10% increase in price results in a change in demand of about $-1.06(10)\%$. Therefore, demand would fall to approximately $1700(1 - 0.106) \approx 1520$.

Further Theory and Practice

39.   **Length:** $\left|\overline{OA}\right| = C_0 -$ fixed cost.

Slope $\left(\overline{OB}\right) = \overline{C} -$ average cost.

Since the marginal cost is the first derivative of cost, the rate of change of the marginal cost is the second derivative of the cost, which must be 0 at P since P is an inflection point.

41. Because of fixed costs, the profit when $x = 0$ is negative: $P(0) = -C(0) < 0$. By assumption, there is an N such that $P(x) < 0$ for $x \geq N$. Since P is continuous on $[0, N]$, there is a value $x_0 \in [0, N]$ at which $P(x)$ has a maximum value. By Fermat's Theorem, $P'(x_0) = 0$. That is, the marginal profit is 0 at x_0. According to the Maximum Profit Principle, marginal cost and marginal revenue are equal at x_0.

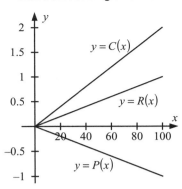

43. Because the cost is an increasing function of the production level, we have $C'(x) \geq 0$. Thus, ξ is a point at which C' has a minimum. If C' is reasonably behaved, then $C'(x)$ decreases as $x \to \xi$ (from either side). In other words, C' decreases just to the left of ξ and increases just to the right of ξ. This means that $(\xi, C(\xi))$ is a point of inflection for the graph of C.

45. Notice that, because $\lim\limits_{x \to 0^+} C(x) = C(0) > 0$, we have $\lim\limits_{x \to 0^+} \overline{C}(x) = \infty$. Also observe that, for every fixed $N > 0$ and positive integer n, the average cost function $\overline{C}$ is continuous on the closed interval $I_n = \left[\dfrac{1}{n}, N\right]$. By the Extreme Value Theorem, C has a minimum value m_n at a point x_n in I_n. If the set $\{m_n\}_{n=1}^{\infty}$ has a minimum m, then m is the minimum value of C on $(0, N]$. Otherwise, for every x_{n_k}, we would be able to find a $x_{n_{k+1}}$ with $x_{n_{k+1}} < x_{n_k}$ and $m_{n_{k+1}} < m_{n_k}$. But this is impossible since $\lim\limits_{k \to \infty} x_{n_k} = 0$ and

$$\lim_{k \to \infty} m_{n_k} = \lim_{k \to \infty} \overline{C}(x_{n_k}) = \lim_{k \to \infty} \frac{C(x_{n_k})}{x_{n_k}} = \infty.$$

In all these exercises it is implicitly assumed that the constants are positive. Without such an assumption, the function C does not resemble a realistic cost function.

47. The derivatives are $C'(x) = 2bx$, $C''(x) = 2b$ and there are no inflection points. On the other hand, $\overline{C}(x) = \dfrac{a}{x} + bx$, $\overline{C}'(x) = -\dfrac{a}{x^2} + b$ and we solve $\overline{C}'(x) = 0$ for $x = \sqrt{\dfrac{a}{b}}$, $\dfrac{a}{b} > 0$. As $\overline{C}''(x) = \dfrac{2a}{x^3} > 0$ for $x > 0$, the local minimum of $\overline{C}$ is at $x = \sqrt{\dfrac{a}{b}}$.

49. Given that $\overline{C}(x) = \dfrac{10}{x} - \dfrac{1}{\sqrt{x}} + 2$, we derive

$$\overline{C}'(x) = -10x^{-2} + \left(\frac{1}{2}\right)x^{-3/2} = \left(\frac{1}{2}\right)x^{-2}\left(-20 + x^{1/2}\right).$$

This is 0 when $x = 400$.

51.   Let $\bar{C}(x) = \dfrac{1000}{x} + x^{1/2}$ and $\bar{C}'(x) = \dfrac{x^{3/2} - 2000}{2x^2}$, which is 0 when $x = 2000^{2/3} \approx 158.7$. Given

that $\bar{C}(158) = \dfrac{1000}{158} + 158^{1/2} = 18.8989\ldots$ and $\bar{C}(159) = \dfrac{1000}{159} + 159^{1/2} = 18.8988\ldots$ we

conclude that $x = 159$ minimizes at $\bar{C}(x)$.

53.   Given that $\bar{C}(x) = \dfrac{a}{x} + b + cx$, it follows that $\bar{C}'(x) = -\dfrac{a}{x^2} + c$. This is 0 when $x = \sqrt{\dfrac{a}{c}}$ and

$\dfrac{a}{c} > 0$, and does not depend on b.

55.   From Problem 4.9.54 we calculate $p_1 = \dfrac{a + b\mu}{2b}$, $p_2 = \dfrac{a + b(\mu + \delta)}{2b}$ and $\Delta p = p_2 - p_1 = \dfrac{\delta}{2}$.

The manufacturer needs to increase his price by $\dfrac{\delta}{2}$.

57. $C_1(x) = c_0 + m(x) -$ initial cost function , $C_2(x) = c_0 + m(x) + \mu x -$ modified cost function .

$\bar{C}_1(x) = \dfrac{c_0 + m(x)}{x}$, $\bar{C}_1'(x) = \dfrac{xm'(x) - c_0 - m(x)}{x^2}$, $\bar{C}_2(x) = \dfrac{c_0 + m(x) + \mu x}{x}$,

$$\bar{C}_2'(x) = \dfrac{xm'(x) - c_0 - m(x)}{x^2} = \bar{C}_1'(x).$$

Therefore, $\bar{C}_1(x)$ has the same point of extrema as $\bar{C}_2(x)$, i.e. $x = \xi$ minimizes $\bar{C}_2(x)$.

59.   Let $TC(x) = cx + \mu N + \dfrac{sN}{2x}$ and $TC'(x) = c - \dfrac{sN}{2x^2}$, which is 0 when $x_* = \sqrt{\dfrac{sN}{2c}}$. Given that

$TC''(x) = \dfrac{sN}{x^3} > 0$, total costs have a minimum at x_*. If this value is a positive integer, then

$\dfrac{N}{x_*}$ is the economic order quantity. Otherwise, if $TC\left(\lfloor x^* \rfloor\right) < TC\left(\lfloor x^* \rfloor + 1\right)$, then $\dfrac{N}{\lfloor x^* \rfloor}$ is the

economic order quantity and if $TC\left(\lfloor x^* \rfloor\right) > TC\left(\lfloor x^* \rfloor + 1\right)$, then $\dfrac{N}{\lfloor x^* \rfloor + 1}$ is the economic

order quantity.

61.   Given that $\tau = \dfrac{1}{a} > 0$ and $\dfrac{rN}{200} = a$, the minimum occurs at $t_* = \sqrt{\dfrac{rN}{200\tau}} = \sqrt{a^2} = a$.

Direct calculations lead to: if $f(L) = \dfrac{L}{a} + \dfrac{a}{L}$, $a > 0$, then $f'(L) = \dfrac{1}{a} - \dfrac{a}{L^2} = \dfrac{L^2 - a^2}{L^2}$. This is 0

when $L = a$; $f''(L) = \dfrac{2a}{L^3} > 0$ and local minimum is at $L = a$.

63. Since $R(p) = pD(p)$, we have $R(0) = 0$. As p begins to increase through positive values, the revenue typically begins to increase through positive values. However, the factor $D(p)$ of $R(p)$ typically decreases. Usually there is a value p_0 such that for $p > p_0$, the decrease of $D(p)$ overcomes the increase of p and the product $R(p) = pD(p)$ decreases. In other words, the graph of a typical revenue function rises on an interval $(0, p_0)$, is maximized at p_0, and falls on the interval (p_0, ∞). See Figure 9 of the text (p. 334). Thus, $R' > 0$ on $(0, p_0)$ and $R' < 0$ on (p_0, ∞). Now

$$R'(p) = \frac{d}{dp}\left(pD(p)\right) = D(p) + pD'(p) = D(p) - D(p)E(p) = D(p)\left(1 - E(p)\right),$$

from which we infer $E(p) < 1$ for p in $(0, p_0)$ and $E(p) > 1$ for p in (p_0, ∞).

65. From the equation $R(p) = pD(p)$, we obtain $R''(p) = 2D'(p) + pD''(p)$. But, by differentiating each side of the equation $pD'(p) = -E(p)D(p)$ with respect to p, we obtain $D'(p) + pD''(p) = -E'(p)D(p) - E(p)D'(p)$, or

$$pD''(p) = -D'(p) - E'(p)D(p) - E(p)D'(p).$$

By substituting this expression for $pD''(p)$ into the formula for $R''(p)$, we see that

$$R''(p) = 2D'(p) - D'(p) - E'(p)D(p) - E(p)D'(p) = \left(1 - E(p)\right)D'(p) - E'(p)D(p).$$

If $E(p) < 1$, if D is decreasing, and if E is increasing, then

$$\operatorname{signum}\left(\left(1 - E(p)\right)D'(p)\right) = (+1)(-1) = -1$$

and $\operatorname{signum}\left(-E'(p)D(p)\right) = -(+1)(+1) = -1$. Under these hypotheses, it follows that the sign of $R''(p)$ is negative.

67. Given that $R(p) = pD(p)$, we calculate $R'(p) = D(p) + pD'(p)$ and

$$1 - \frac{R'(p)}{D(p)} = \frac{D(p) - R'(p)}{D(p)} = -p\frac{D'(p)}{D(p)}.$$

Therefore, definitions $E(p) = 1 - \dfrac{R'(p)}{D(p)}$ and $E(p) = -p\dfrac{D'(p)}{D(p)}$ are equivalent.

69.   **a.**   Let $E_D(p) = -p\dfrac{D'(p)}{D(p)}$, $E_{\mathbf{D}}(p) = -p\dfrac{\mathbf{D}'(p)}{\mathbf{D}(p)}$. Using $\mathbf{D}(p) = A \cdot D(p)$, we calculate

$$E_{\mathbf{D}}(p) = -p\frac{\mathbf{D}'(p)}{\mathbf{D}(p)} = -\frac{pAD'(p)}{AD(p)} = -p\frac{D'(p)}{D(p)} = E_D(p).$$

b.   As $E_{\mathbf{D}}(p) = E_D(p)$, we calculate $\dfrac{\mathbf{D}'(p)}{\mathbf{D}(p)} = \dfrac{D'(p)}{D(p)}$ and $\displaystyle\int\left(\dfrac{\mathbf{D}'(p)}{\mathbf{D}(p)}\right)dp = \int\left(\dfrac{D'(p)}{D(p)}\right)dp$.

From Problem 4.9.68 we deduce $\ln\left(\mathbf{D}(p)\right) = \ln\left(D(p)\right) + C$, or $\mathbf{D}(p) = A \cdot D(p)$.

Calculator/Computer Exercises

71.   The range of the allowed values of x is $[0, 124721]$. For this range, there are two points where
$C(x) = R(x)$: $x_1 = 918.36$, $x_2 = 96211.97$. At $c = 38958.35... \approx 38958$ we have a local maximum.

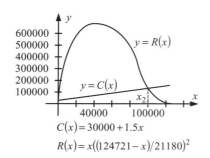

$C(x) = 30000 + 1.5x$

$R(x) = x\left((124721 - x)/21180\right)^2$

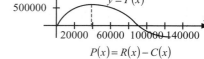

$P(x) = R(x) - C(x)$

73.   We calculate:

$$R(x) = 10xe^{-x/40000}$$

$$P(x) = 10xe^{-x/40000} - 30000 - 2x$$

$$P_{\max} = P(25039.33)$$

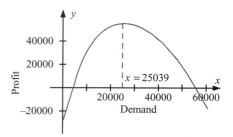

As $P(25039) > P(25040)$, the maximum profit corresponds to the demand of 25039 units.

75.   The demand x that minimizes the average cost of production is 24495; it is the demand x at which $\overline{C}(x)$ is minimized.

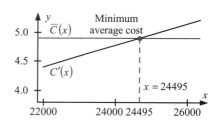

77.   We calculate

$$E(p) = \frac{10590\sqrt{p}}{124721 - 21180\sqrt{p}}.$$

As $E(p) < 1$, the demand is inelastic for

$$0 < p < 15.41.$$

$E(p) = 10590\sqrt{p}\,/\left(124721 - 21180\sqrt{p}\right)$

As $E(p) > 1$, the demand is elastic for $15.41 < p < 34.68$; maximum revenue is at $p_0 = 15.41$.

The Integral

Introduction to Integration—The Area Problem

Problems for Practice

1. $3+6+9+12+15+18=63$

3. $\dfrac{4}{5}+\dfrac{5}{6}+\dfrac{6}{7}=\dfrac{523}{210}$

5. $(-4)+9+40=45$

7. $\sin\left(\dfrac{\pi}{2}\right)+2\sin(\pi)+3\sin\left(\dfrac{3\pi}{2}\right)+4\sin(2\pi)+5\sin\left(\dfrac{5\pi}{2}\right)=3$

9. $\displaystyle\sum_{j=2}^{6} j$ **11.** $\displaystyle\sum_{j=1}^{6}(5+4j)$ **13.** $\displaystyle\sum_{j=4}^{8}\left(\dfrac{1}{j}\right)$

15. $x_0=3$, $x_1=4.5$, $x_2=6$, $\Delta x=1.5$; $f(x_1)=11.25$, $f(x_2)=24$,
$A_1+A_2=f(x_1)\Delta x+f(x_2)\Delta x=52.875$

17. $x_0=\dfrac{\pi}{2}$, $x_1=\pi$, $x_2=\dfrac{3\pi}{2}$, $x_3=2\pi$, $\Delta x=\dfrac{\pi}{2}$; $f(x_1)=f(x_2)=f(x_3)=2$,
$A_1+A_2+A_3=f(x_1)\Delta x+f(x_2)\Delta x+f(x_3)\Delta x=3\pi$

19. $x_0=-5$, $x_1=-4.25$, $x_2=-3.5$, $x_3=-2.75$, $x_4=-2$, $\Delta x=0.75$; $f(x_1)=\dfrac{17}{13}$, $f(x_2)=\dfrac{7}{5}$,
$f(x_3)=\dfrac{11}{7}$, $f(x_4)=2$,

$$A_1+A_2+A_3+A_4=f(x_1)\Delta x+f(x_2)\Delta x+f(x_3)\Delta x+f(x_4)\Delta x=\dfrac{2857}{455}$$

21. $x_0=-\pi$, $x_1=-\dfrac{\pi}{2}$, $x_2=0$, $x_3=\dfrac{\pi}{2}$, $x_4=\pi$, $\Delta x=\dfrac{\pi}{2}$; $f(x_1)=f(x_3)=\dfrac{\pi}{2}$,
$f(x_2)=f(x_4)=0$, $A_1+A_2+A_3+A_4=f(x_1)\Delta x+f(x_2)\Delta x+f(x_3)\Delta x+f(x_4)\Delta x=\dfrac{\pi^2}{2}$

23. $\displaystyle\sum_{j=1}^{16} j^2 = \frac{16 \cdot 17 \cdot 33}{6} = 1496$

25. $\displaystyle\sum_{j=3}^{40}\left(4j^2 + j\right) = 4\sum_{j=3}^{40} j^2 + \sum_{j=3}^{40} j = 4\left(\sum_{j=1}^{40} j^2 - \sum_{j=1}^{2} j^2\right) + \left(\sum_{j=1}^{40} j - \sum_{j=1}^{2} j\right)$

$$= 4\left(\frac{40 \cdot 41 \cdot 81}{6} - 5\right) + \left(\frac{40 \cdot 41}{2} - 3\right) = 4 \cdot 22135 + 817 = 89357$$

Further Theory and Practice

27. With m being a midpoint, we find $m = \dfrac{a+b}{2} = 1$, $\Delta x = 2$, $f(a) = 0$, $f(m) = 2$, $f(b) = 28$.

Right endpoint approximation: $A_1 + A_2 = f(m)\Delta x + f(b)\Delta x = 60$.

Trapezoidal approximation:

$$A_1 + A_2 = \left(\frac{f(a) + f(m)}{2}\right) \cdot \Delta x + \left(\frac{f(m) + f(b)}{2}\right) \cdot \Delta x = \left(f(a) + 2f(m) + f(b)\right)\frac{\Delta x}{2} = 32.$$

29. With m being a midpoint, we find $m = 0.5$, $\Delta x = 0.5$, $f(a) = 0$, $f(m) = 1.5$, $f(b) = 1$.

Right endpoint approximation: $A_1 + A_2 = f(m)\Delta x + f(b)\Delta x = \dfrac{5}{4}$.

Trapezoidal approximation: $A_1 + A_2 = \left(f(a) + 2f(m) + f(b)\right)\dfrac{\Delta x}{2} = 1$.

31. The uniform partition of the interval $[0, 2]$ is $\{x_j\}$, $0 \le j \le N$, where $x_j = j \cdot \Delta x$ and $\Delta x = \dfrac{2}{N}$. We provide the following approximation for the area:

$$\sum_{j=1}^{N} f(x_j) \cdot \Delta x = \sum_{j=1}^{N}\left(4 - x_j^2\right) \cdot \Delta x = \sum_{j=1}^{N}\left(4 - \frac{4j^2}{N^2}\right)\left(\frac{2}{N}\right) = \left(\frac{8}{N}\right)\sum_{j=1}^{N} 1 - \left(\frac{8}{N^3}\right)\sum_{j=1}^{N} j^2$$

$$= \left(\frac{8}{N}\right)N - \left(\frac{8}{N^3}\right)\left(\frac{N(N+1)2N+1}{6}\right) = 8 - \left(\frac{4}{3}\right)\left(1 + \frac{1}{N}\right)\left(2 + \frac{1}{N}\right)$$

The area of the region is calculated as a limit at $N \to \infty$:

$$\lim_{N \to \infty}\sum_{j=1}^{N} f(x_j) \cdot \Delta x = 8 - \lim_{N \to \infty}\left(\frac{4}{3}\right)\left(1 + \frac{1}{N}\right)\left(2 + \frac{1}{N}\right) = 8 - \frac{8}{3} = \frac{16}{3}.$$

33.   We compare logarithms of the left- and right-hand sides and find $\ln\left(e^p\right)=p$,

$$\ln\left(e\cdot e^2\cdot e^3\cdot\ldots\cdot e^{100}\right)=1+2+3+\ldots+100, \text{ therefore } p=\sum_{j=1}^{100}j=\frac{100\cdot101}{2}=5050.$$

As $e^{\ln(q)}=q$, $e^{\sum_{n=1}^{100}\ln(n)}=1\cdot2\cdot\ldots\cdot100=100!$, we find $q=100!$.

35. $S=\sum_{j=1}^{N}(2j)^2=4\sum_{j=1}^{N}j^2=4\left(\frac{N\cdot(N+1)\cdot(2N+1)}{6}\right)=2N(N+1)\frac{2N+1}{3}$. Then

$$\sum_{j=1}^{N}(2j-1)^2=\sum_{j=1}^{2N}j^2-\sum_{j=1}^{N}(2j)^2=\frac{2N\cdot(2N+1)\cdot(4N+1)}{6}-2N(N+1)\frac{2N+1}{3}$$

$$=2N(2N+1)\frac{4N+1-2N-2}{6}=N(2N+1)\frac{2N-1}{3}.$$

37. $S=\sum_{j=M}^{N}\left(a_j-a_{j-1}\right)=\sum_{j=M}^{N}a_j-\sum_{j=M}^{N}a_{j-1}=\sum_{j=M}^{N}a_j-\sum_{j=M-1}^{N-1}a_j=a_N+\sum_{j=M}^{N-1}a_j-\sum_{j=M}^{N-1}a_j-a_{M-1}$

$=a_N-a_{M-1}$

The term "collapsing or telescoping sum" stems from the fact that in the process of summation all intermediate terms of the sequence cancel out; the sum depends only on the first and last terms of the sequence and is independent of all other terms.

39.
$$S=\quad 1\ +\quad 2+\ \ldots\ +\quad N$$
$$S=\quad N\ +N-1+\ \ldots\ +\quad 1$$

$$2S=N+1+N+1+\ \ldots\ +N+1$$

From the last equation we find $2S=N(N+1)$ so that $S=N\dfrac{N+1}{2}$.

41. The sum $\sum_{j=1}^{N} \left(3j^2 - 3j + 1\right)$ can be written as $\sum_{j=1}^{N}\left(3j^2 - 3j + 1\right) = 3S - 3\sum_{j=1}^{N} j + \sum_{j=1}^{N} 1$, where

$S = \sum_{j=1}^{N} j^2$.

Due to the fact that $\sum_{j=1}^{N} \left(3j^2 - 3j + 1\right) = \sum_{j=1}^{N} \left(j^3 - (j-1)^3\right)$ and the sum on the right-hand side is

the collapsing sum, we obtain $\sum_{j=1}^{N} \left(3j^2 - 3j + 1\right) = N^3 - 0 = N^3$. Therefore, we have

$3S - 3\sum_{j=1}^{N} j + \sum_{j=1}^{N} 1 = N^3$.

It follows from $\sum_{j=1}^{N} j = \dfrac{N(N+1)}{2}$ and from $\sum_{j=1}^{N} 1 = N$ that

$S = \dfrac{1}{3}\left[3\dfrac{N(N+1)}{2} - N + N^3\right] = \dfrac{N(N+1)(2N+1)}{6}$.

43. Let $x_j = j \cdot \Delta x$, $0 \le j \le N$, $\Delta x = \dfrac{b}{N}$, and let $S(0, b)$ be the area under the curve $f(x) = x^3$
from $x = 0$ to $x = b$. Then

$$\sum_{j=1}^{N} f\left(x_j\right) \cdot \Delta x = \sum_{j=1}^{N} j^3 \left(\Delta x\right)^4 = \left(\dfrac{b}{N}\right)^4 \left(N^2 \dfrac{(N+1)^2}{4}\right) = \left(\dfrac{b^4}{4}\right)\left(\dfrac{N+1}{N}\right)^2$$

and $S(0, b) = \lim_{N \to \infty} \sum_{j=1}^{N} f\left(x_j\right) \cdot \Delta x = \left(\dfrac{b^4}{4}\right) \lim_{N \to \infty}\left(1 + \dfrac{1}{N}\right)^2 = \dfrac{b^4}{4}$.

For $S(a, b)$ we have: $S(a, b) = S(0, b) - S(0, a) = \dfrac{b^4}{4} - \dfrac{a^4}{4} = \dfrac{b^4 - a^4}{4}$.

45.   Let $x_j = j \cdot \Delta x$, $0 \le j \le N$, $\Delta x = \dfrac{b}{N}$. Then the following sum is an approximation for the area.

$$\sum_{j=1}^{N} f(x_j) \cdot \Delta x = \sum_{j=1}^{N} x_j^2 \cdot \Delta x = \sum_{j=1}^{N} j^2 (\Delta x)^3 = \left(\frac{b}{N}\right)^3 \left(\frac{N(N+1)(2N+1)}{6}\right)$$

$$= \left(\frac{b^3}{6}\right)\left(1+\frac{1}{N}\right)\left(2+\frac{1}{N}\right).$$

The area found as a limit of the above approximation is:

$$A(b) = \lim_{N \to \infty}\left(\left(\frac{b^3}{6}\right)\left(1+\frac{1}{N}\right)\left(2+\frac{1}{N}\right)\right) = \left(\frac{b^3}{6}\right)\lim_{N \to \infty}\left(\left(1+\frac{1}{N}\right)\left(2+\frac{1}{N}\right)\right) = \frac{b^3}{3}.$$

Then $A'(b) = b^2 = f(b)$.

47.   **a.**   We see that $\lim\limits_{N \to \infty} \dfrac{S_N(k)}{N^{k+1}} = \lim\limits_{N \to \infty}\left(\dfrac{1}{k+1} + \dfrac{P_k(N)}{N^{k+1}}\right) = \dfrac{1}{k+1}$, because the degree of the

polynomial $P_k(N)$ in the numerator is smaller than the degree of the polynomial

N^{k+1} in the denominator.

b.   Let $x_j = j \cdot \Delta x$, $0 \le j \le N$, $\Delta x = \dfrac{b}{N}$. Between $x = 0$ and $x = b$ the function is

$f(x) = x^k$. Within this interval, the area $S(0, b)$ under the curve $f(x)$ can be
approximated as

$$\sum_{j=1}^{N} f(x_j) \cdot \Delta x = \sum_{j=1}^{N} j^k (\Delta x)^{k+1} = \left(\frac{b}{N}\right)^{k+1}\left(\frac{N^k}{k+1} + P_k(N)\right) = \frac{b^{k+1}}{k+1} + b^{k+1}\frac{P_k(N)}{N^{k+1}},$$

and the area found as a limit of the above approximation is:

$$S(0, b) = \lim_{N \to \infty}\sum_{j=1}^{N} f(x_j) \cdot \Delta x = \frac{b^{k+1}}{k+1}.$$

c. $S(a, b) = S(0, b) - S(0, a) = \dfrac{b^{k+1} - a^{k+1}}{k+1}$

49.   See the figures. The requested area is $ab - c$.

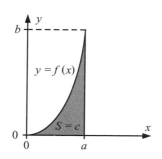
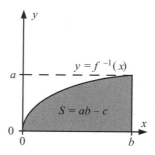

51.   For the function $f(x) = \cos(x)$ we select the partition $x_j = j \cdot \Delta x$, $0 \le j \le N$, $\Delta x = \dfrac{b}{N} = t$,
Then, using the last formula from Problem 5.1.50, we find

$$\sum_{j=1}^{N} f(x_j) \cdot \Delta x = \left(\frac{b}{N}\right) \sum_{j=1}^{N} \cos(jt) = \left(\frac{b}{N}\right) \sin\left(\frac{Nt}{2}\right) \frac{\cos((N+1)t/2)}{\sin(t/2)}$$

$$= \left(\frac{b}{N}\right) \sin\left(\frac{b}{2}\right) \frac{\cos(b/2 + b/(2N))}{\sin(b/(2N))}.$$

The area is

$$A(b) = \lim_{N \to \infty} \left[\left(\frac{b}{N}\right) \sin\left(\frac{b}{2}\right) \frac{\cos(b/2 + b/(2N))}{\sin(b/(2N))}\right]$$

$$= 2\sin\left(\frac{b}{2}\right) \lim_{N \to \infty} \frac{b/(2N)}{\sin(b/(2N))} \cdot \lim_{N \to \infty} \cos\left(\frac{b}{2} + \frac{b}{(2N)}\right)$$

$$= 2\sin\left(\frac{b}{2}\right) \cos\left(\frac{b}{2}\right) = \sin(b).$$

Calculator/Computer Exercises

53. $N = 25$: $0.165...$; $N = 50$: $0.166...$; $N = 75$: $0.166...$

55. $N = 25$: $1.997...$; $N = 50$: $1.999...$; $N = 75$: $1.999...$

57. $a = -1.8414$, $b = 1.1462$; $N = 25$: $1.945...$; $N = 50$: $1.948...$; $N = 75$: $1.948...$

59. $a = 0.1010$, $b = 1.6796$; $N = 25$: $0.306...$; $N = 50$: $0.307...$; $N = 75$: $0.307...$

Section 5.2   The Riemann Integral

Problems for Practice

1. **a.** $R(f,S) = (f(2)+f(6)+f(10))\cdot\Delta x = (2+6+10)\cdot 4 = 18\cdot 4 = 72$

b. $R = \int_{0}^{12} x\,dx = \left.\frac{x^2}{2}\right|_{0}^{12} = 72-0 = 72$

3. **a.** $R(f,S) = (f(-1)+f(1)+f(3)+f(5))\cdot\Delta x = (1+1+9+25)\cdot 2 = 36\cdot 2 = 72$

b. $R = \int_{-2}^{6} x^2\,dx = \left.\frac{x^3}{3}\right|_{-2}^{6} = 72-\left(-\frac{8}{3}\right) = \frac{224}{3}$

5. **a.** $R(f,S) = (f(1)+f(3)+f(5))\cdot\Delta x = (5+39+145)\cdot 2 = 189\cdot 2 = 378$

b. $R = \int_{1}^{7}(x^3+4x)\,dx = \left.\left(\frac{x^4}{4}+2x^2\right)\right|_{1}^{7} = \frac{2401}{4}+98-\left(\frac{1}{4}+2\right)=696$

7. **a.** $R(f,S) = (f(-7)+f(-4)+f(-1)+f(2))\cdot\Delta x = (7+4+1+2)\cdot 3 = 14\cdot 3 = 42$

b. $R = \int_{-7}^{5}|x|\,dx = \int_{-7}^{0}(-x)\,dx + \int_{0}^{5}x\,dx = \left.-\frac{x^2}{2}\right|_{-7}^{0} + \left.\frac{x^2}{2}\right|_{0}^{5} = \frac{49}{2}+\frac{25}{2} = 37$

9. **a.** $R(f,S) = (f(-2)+f(0)+f(2))\cdot\Delta x = (\exp(-2)+\exp(0)+\exp(2))\cdot 2$
$= 2\exp(-2)+2+2\exp(2)$

b. $R = \int_{-4}^{2}\exp(x)\,dx = \left.\exp(x)\right|_{-4}^{2} = \exp(2)-\exp(-4)$

11. **a.** $R(f,S) = \left(f\left(-\frac{\pi}{3}\right)+f\left(\frac{\pi}{3}\right)+f(\pi)+f(5\pi/3)\right)\cdot\Delta x = \left(\frac{1}{2}+\frac{1}{2}+(-1)+\frac{1}{2}\right)\cdot\frac{2\pi}{3}$
$= \frac{1}{2}\cdot\frac{2\pi}{3} = \frac{\pi}{3}$

b. $R = \int_{-\pi}^{5\pi/3}\cos(x)\,dx = \left.\sin(x)\right|_{-\pi}^{5\pi/3} = -\frac{\sqrt{3}}{2}-0 = -\frac{\sqrt{3}}{2}$

13. $\int_{1}^{3}\sqrt{2}\,dx = \sqrt{2}\cdot(3-1) = \sqrt{2}\cdot 2 = 2\sqrt{2}$

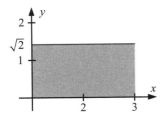

15. $\int\limits_{-1}^{3}|x|dx = \int\limits_{-1}^{0}|x|dx + \int\limits_{0}^{3}|x|dx$

$$= \frac{1}{2}\cdot 1\cdot 1 + \frac{1}{2}\cdot 3\cdot 3 = \frac{1}{2} + \frac{9}{2} = 5$$

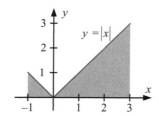

17. $\int\limits_{-2}^{1}(x+1)dx = \int\limits_{-2}^{-1}(x+1)dx + \int\limits_{-1}^{1}(x+1)dx$

$$= -\frac{1}{2}\cdot 1\cdot 1 + \frac{1}{2}\cdot 2\cdot 2$$

$$= -\frac{1}{2} + 2 = \frac{3}{2}$$

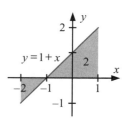

19. $\int\limits_{-2}^{3}|x-1|dx = \int\limits_{-2}^{1}|x-1|dx + \int\limits_{1}^{3}|x-1|dx = \frac{1}{2}\cdot 3\cdot 3 + \frac{1}{2}\cdot 2\cdot 2$

$$= \frac{9}{2} + 2 = \frac{13}{3}$$

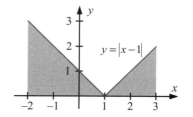

21. $\int\limits_{1}^{2}(6x^2 - 2x)dx = (2x^3 - x^2)\Big|_{1}^{2} = 16 - 4 - (2-1) = 11$

23. $\int\limits_{1}^{4}\sqrt{x}dx = \frac{2x^{3/2}}{3}\Big|_{1}^{4} = \frac{16}{3} - \frac{2}{3} = \frac{14}{3}$

25. $\int\limits_{0}^{\pi/4}\sec^2(x)dx = \tan(x)\Big|_{0}^{\pi/4} = 1 - 0 = 1$

27. $\int\limits_{0}^{\pi/3}\sec(x)\tan(x)dx = \sec x\Big|_{0}^{\pi/3} = 2 - 1 = 1$

29. $\int\limits_{8}^{27}x^{1/3}dx = \frac{3x^{4/3}}{4}\Big|_{8}^{27} = \frac{243}{4} - 12 = \frac{195}{4}$

31. $\int\limits_{1}^{e}\frac{1}{x}dx = \ln(x)\Big|_{1}^{e} = 1 - 0 = 1$

Further Theory and Practice

33. **a.** $R(f,U) = f(1)\cdot 1 = e$; $R(f,L) = f(0)\cdot 1 = 1$

   b. $R(f,S) = f\left(\frac{1}{2}\right)\cdot 1 = e^{1/2} = \sqrt{e}$

   c. $1 \le \sqrt{e} \le e$

35. **a.** $R(f,U) = (f(1) + f(2)) \cdot \Delta x = \left(\dfrac{1}{2} + \dfrac{1}{3}\right) \cdot 1 = \dfrac{5}{6}$;

$R(f,L) = (f(2) + f(3)) \cdot \Delta x = \left(\dfrac{1}{3} + \dfrac{1}{4}\right) \cdot 1 = \dfrac{7}{12}$

b. $R(f,S) = \left(f\left(\dfrac{3}{2}\right) + f\left(\dfrac{5}{2}\right)\right) \cdot \Delta x = \left(\dfrac{2}{5} + \dfrac{2}{7}\right) \cdot 1 = \dfrac{24}{35}$

c. $\dfrac{7}{12} \le \dfrac{24}{35} \le \dfrac{5}{6}$

37. **a.** $R(f,U) = (f(1) + f(2) + f(3)) \cdot \Delta x = \left(\sqrt{2} + 3 + 2\sqrt{7}\right) \cdot 1 = \sqrt{2} + 3 + 2\sqrt{7}$;

$R(f,L) = (f(0) + f(1) + f(2)) \cdot \Delta x = \left(1 + \sqrt{2} + 3\right) \cdot 1 = 4 + \sqrt{2}$

b. $R(f,S) = \left(f\left(\dfrac{1}{2}\right) + f\left(\dfrac{3}{2}\right) + f\left(\dfrac{5}{2}\right)\right) \cdot \Delta x = \left(\dfrac{3\sqrt{2}}{4} + \dfrac{\sqrt{70}}{4} + \dfrac{\sqrt{266}}{4}\right) \cdot 1$

$= \dfrac{3\sqrt{2} + \sqrt{70} + \sqrt{266}}{4}$

c. $4 + \sqrt{2} \le \dfrac{3\sqrt{2} + \sqrt{70} + \sqrt{266}}{4} \le \sqrt{2} + 3 + 2\sqrt{7}$

39. $R(f,U) = \left(f\left(\dfrac{\pi}{3}\right) + f\left(\dfrac{\pi}{2}\right) + f\left(\dfrac{2\pi}{3}\right)\right) \cdot \Delta x = \left(\dfrac{\sqrt{3}}{2} + 1 + \dfrac{\sqrt{3}}{2}\right) \cdot \dfrac{\pi}{3} = \left(\sqrt{3} + 1\right)\dfrac{\pi}{3}$;

$R(f,L) = \left(f(0) + f\left(\dfrac{\pi}{3}\right) + f(\pi)\right) \cdot \Delta x = \left(0 + \dfrac{\sqrt{3}}{2} + 0\right) \cdot \dfrac{\pi}{3} = \sqrt{3}\dfrac{\pi}{6}$; $c_1 = 1$; $b - a = \pi$;

$R(f,U) - R(f,L) = \left(\sqrt{3} + 2\right)\dfrac{\pi}{6} \le 1 \cdot \dfrac{\pi^2}{3} = \dfrac{\pi^2}{3}$

41. $R(f,U) = (f(-2) + f(-1) + f(1)) \cdot \Delta x = \left(e^2 + e + e\right) \cdot 1 = e^2 + 2e$;

$R(f,L) = (f(-1) + f(0) + f(0)) \cdot \Delta x = (e + 1 + 1) \cdot 1 = e + 2$; $c_1 = e^2$; $b - a = 3$;

$R(f,U) - R(f,L) = e^2 + e - 2 \le e^2 \cdot \dfrac{3^2}{3} = 3e^2$

43. $R(f,U) = (f(1) + f(1) + f(3)) \cdot \Delta x = (6 + 6 + 10) \cdot 1 = 22$;

$R(f,L) = (f(0) + f(2) + f(2)) \cdot \Delta x = (1 + 5 + 5) \cdot 1 = 11$

$\displaystyle\int_0^3 f(x)\,dx = \left(\dfrac{x^4}{2} - 3x^3 + 6x^2 + x\right)\Bigg|_0^3 = \dfrac{33}{2} \in (11,\ 22)$

45.   The name of the variable does not matter, therefore $\int_a^b f(t)\,dt = \int_a^b f(u)\,du = F(b) - F(a)$, where F is an antiderivative for f. On the other hand,

$$\int_a^b f(u)\,dt = (b-a)\cdot f(u)$$

$$\int_a^b f(t)\,du = (b-a)\cdot f(t)$$

and they are unequal.

47.   For any continuous function the following is true:

$$\int_{-b}^b f(x)\,dx = \int_{-b}^0 f(x)\,dx + \int_0^b f(x)\,dx = -\int_b^0 f(-x)\,dx + \int_0^b f(x)\,dx = \int_0^b f(-x)\,dx + \int_0^b f(x)\,dx.$$

If the function is odd, then $\int_0^b f(-x)\,dx = -\int_0^b f(x)\,dx$ and $\int_{-b}^b f(x)\,dx = 0$.

If the function is even, then $\int_0^b f(-x)\,dx = \int_0^b f(x)\,dx$ and $\int_{-b}^b f(x)\,dx = 2\int_0^b f(x)\,dx$.

49.   We compute $f(s_i)\Delta x = s_i^2 \cdot 1 = s_i^2$, $F(x) = \dfrac{x^3}{3}$.

$$F(x_1) - F(x_0) = \frac{1}{3} - 0 = \frac{1}{3}$$

$$F(x_2) - F(x_1) = \frac{8}{3} - \frac{1}{3} = \frac{7}{3}$$

Therefore, $s_1 = \dfrac{\sqrt{3}}{3}$ and $s_2 = \dfrac{\sqrt{7}}{\sqrt{3}} = \dfrac{\sqrt{21}}{3}$.

$$R(f,\{s_1,s_2\}) = \left(f\!\left(\frac{\sqrt{3}}{3}\right) + f\!\left(\frac{\sqrt{21}}{3}\right) \right)\cdot \Delta x = \left(\frac{1}{3} + \frac{7}{3}\right)\cdot 1 = \frac{8}{3} = \int_0^2 x^2\,dx$$

51.   Let $y_E = b\sqrt{1 - \dfrac{x^2}{a^2}} = b\sqrt{\dfrac{a^2 - x^2}{a^2}} = \dfrac{b}{a}\sqrt{a^2 - x^2} = \dfrac{b}{a}y_C$. Comparing the Riemann sums for the ellipse and for the circle and using $y_E = \dfrac{b}{a}y_C$ we find that the area enclosed by the ellipse is equal to $\dfrac{b}{a}\cdot \pi a^2 = \pi ab$.

53. The exact value is $\int_0^1 x^2 dx = \frac{1}{3}$. The Riemann sum is

$$\frac{1}{N}\sum_{i=1}^{N} f\left(\frac{j+r-1}{N}\right) = \frac{1}{N}\sum_{i=1}^{N}\left(\frac{j+r-1}{N}\right)^2 = \frac{1}{N^3}\sum_{i=1}^{N}(j+r-1)^2$$

$$= \frac{1}{N^3}\sum_{i=1}^{N}j^2 + \frac{1}{N^3}\sum_{i=1}^{N}2j(r-1) + \frac{1}{N^3}\sum_{i=1}^{N}(r-1)^2$$

$$= \frac{(N+1)(2N+1)}{6N^2} + \frac{(N+1)(r-1)}{N^2} + \frac{(r-1)^2}{N^2}$$

$$= \frac{1}{3} - \frac{1}{2N} + \frac{1}{6N^2} + \frac{(N-1)r}{N^2} + \frac{r^2}{N^2}.$$

Thus, from the equation $-\frac{1}{2N} + \frac{1}{6N^2} + \frac{(N-1)r}{N^2} + \frac{r^2}{N^2} = 0$ we have $r = \frac{1-N+\sqrt{N^2+1/3}}{2}$.

55. By the Mean Value Theorem, there exists $c \in (\ell_j, u_j)$ such that

$$f(u_j) - f(\ell_j) = f'(c)(u_j - \ell_j).$$

Since $|u_j - \ell_j| \le \Delta x$ and $|f'(c)| \le C_1$, $f(u_j) - f(\ell_j) \le C_1 \Delta x$. Then

$$R(f, U_N) - R(f, L_N) = \frac{b-a}{N}\sum_{j=1}^{N}\big(f(u_j) - f(\ell_j)\big) \le \frac{b-a}{N}\cdot\sum_{j=1}^{N}C_1\frac{b-a}{N} = \frac{C_1(b-a)^2}{N^2}\cdot N$$

$$= \frac{C_1(b-a)^2}{N}$$

Calculator/Computer Exercises

57. 0.375

59. 3.14195

61. The upper Riemann sum is 0.55563.
The lower Riemann sum is 0.55469.

63. The upper Riemann sum is 0.40977.
The lower Riemann sum is 0.40549.

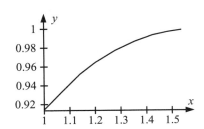

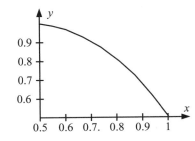

Section 5.3   Rules for Integration

Problems for Practice

1. $\displaystyle\int_1^7 f(x)\,dx = \int_1^3 f(x)\,dx + \int_3^7 f(x)\,dx = -8 + 12 = 4$

3. $\displaystyle\int_{-7}^3 \left(4f(x) - 9g(x)\right)dx = 4\int_{-7}^3 f(x)\,dx - 9\int_{-7}^3 g(x)\,dx = 4\cdot(-7) - 9\cdot(-4) = 8$

5. $\displaystyle\int_9^{-2} f(x)\,dx = -\int_{-2}^9 f(x)\,dx = -\left(\int_{-2}^7 f(x)\,dx + \int_7^9 f(x)\,dx\right) = -\left(-\int_7^{-2} f(x)\,dx + \int_7^9 f(x)\,dx\right)$
$\qquad\qquad = -(-6-4) = 10$

7. $\displaystyle\int_4^9 \left(6f(x) - 7g(x)\right)dx = 6\int_4^9 f(x)\,dx - 7\int_4^9 g(x)\,dx = -6\int_9^4 f(x)\,dx + 7\int_9^4 f(x)\,dx = -6\cdot5 + 7\cdot15 = 75$

9. $\displaystyle\int_5^{-3} \left(6f(x) + 1\right)dx = 6\int_5^{-3} f(x)\,dx + \int_5^{-3} dx = 6\cdot\left(-\frac{28}{3}\right) + (-8) = -64$

11. $\displaystyle\int_4^1 3f(x)\,dx = -9\int_1^4 \frac{f(x)}{3}\,dx = -9\cdot2 = -18$

13. $\displaystyle\int_2^1 \left(f(x) - 3g(x) + 5\right)dx = \int_2^1 f(x)\,dx - 3\int_2^1 g(x)\,dx + 5\int_2^1 dx = 5\int_2^1 dx = 5\cdot(-1) = -5$

15.   Average value:

$$f_{ave} = \frac{1}{5-1}\int_1^5 (1+x)\,dx = \frac{1}{4}\left(x + \frac{x^2}{2}\right)\Bigg|_{x=1}^{x=5} = \frac{1}{4}\cdot16 = 4.$$

Geometric significance: c is the point such that
$f(c) = f_{ave}$.

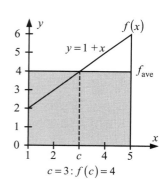

$c = 3: f(c) = 4$

17.   Average value:

$$f_{\text{ave}} = \frac{1}{4-1}\int_1^4 \sqrt{x}\,dx = \frac{1}{3}\left(\left.\frac{x^{3/2}}{3/2}\right|_{x=1}^{x=4}\right) = \frac{1}{3}\cdot\frac{14}{3} = \frac{14}{9}.$$

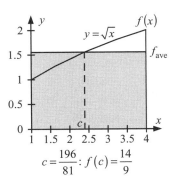

Geometric significance: c is the point such that
$f(c) = f_{\text{ave}}$.

$$c = \frac{196}{81}: f(c) = \frac{14}{9}$$

19.   Average value:

$$f_{\text{ave}} = \frac{1}{3-0}\int_0^3\left(12x^2+5\right)dx = \frac{1}{3}\left(\left.\left(4x^3+5x\right)\right|_{x=0}^{x=3}\right)$$

$$= \frac{1}{3}\cdot 123 = 41$$

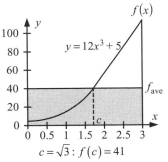

Geometric significance: c is the point such that
$f(c) = f_{\text{ave}}$.

$$c = \sqrt{3}: f(c) = 41$$

Further Theory and Practice

21.   Let $I = \int_a^b f(x)\,dx$ and $J = \int_a^b g(x)\,dx$. Then $\int_a^b\left(f(x)-3g(x)\right)dx = I - 3J$ and

$\int_a^b\left(-6g(x)+9f(x)\right)dx = -6J+9I$. The solution to the system: $\begin{cases} I - 3J = 3 \\ -6J + 9I = 6 \end{cases}$ is $I = 0$,
$J = -1$.

23.   Let $I = \int_a^b f(x)\,dx$ and $J = \int_a^b g(x)\,dx$. Then $\int_a^b\left(f(x)+2g(x)\right)dx = I + 2J$,

$\int_a^b\left(g(x)-6f(x)\right)dx = J - 6I$. The solution to the system: $\begin{cases} I + 2J = -9 \\ J - 6I = 4 \end{cases}$ is $I = -\frac{17}{13}$, $J = -\frac{50}{13}$.

25. $\int_1^3\left(\frac{3x-5}{x}\right)dx = 3\int_1^3 dx - 5\int_1^3\left(\frac{1}{x}\right)dx = \left.\left(3x - 5\ln(x)\right)\right|_{x=1}^{x=3} = 6 - 5\ln(3)$

27. $\int_{-1}^0\left(6x^2+\exp(x)\right)dx = 6\int_{-1}^0 x^2 dx + \int_{-1}^0\exp(x)\,dx = \left.\left(2x^3+\exp(x)\right)\right|_{-1}^0 = 3 - \frac{1}{e}$

29. $\int_0^1 3\exp(-x)\left(2-\exp(x)\right)dx = 6\int_0^1\exp(-x)\,dx - 3\int_0^1 dx = \left.\left(-6\exp(-x)-3x\right)\right|_{x=0}^{x=1} = 3 - \frac{6}{e}$

31.   It is true when $\int_a^b f(x)\,dx = 0$ or $\int_a^b 1\cdot dx = 1$, that is $b = a+1$.

33.   Given that $\sqrt{1+8x^5}$ increases on $[0,1]$, we find $1=\sqrt{1+8\cdot 0^5}\le\sqrt{1+8x^5}\le\sqrt{1+8\cdot 1^5}=3$, and

$m=1$, $M=3$, $b-a=1$. Therefore, $1\le\int_0^1\sqrt{1+8x^5}\,dx\le 3$ and $A=1$, $B=3$.

35.   Given that $\min\limits_{x\in[1,4]}\left(x^2-4x+5\right)^{1/3}=\left(2^2-4\cdot 2+5\right)^{1/3}=1$ and

$$\max\limits_{x\in[1,4]}\left(x^2-4x+5\right)^{1/3}=\left(4^2-4\cdot 4+5\right)^{1/3}=5^{1/3},$$

we find $m=1$, $M=5^{1/3}$, $b-a=3$ and $3\le\int_1^4\left(x^2-4x+5\right)^{1/3}dx\le 3\cdot 5^{1/3}$. Therefore $A=3$,

$B=3\cdot 5^{1/3}$.

37.   The area under $y=\cos^2(x)$ (light gray zone) is equal
to the area under $y=\sin^2(x)$ (dark gray zone), their

sum being π. Therefore, each area is $\dfrac{\pi}{2}$.

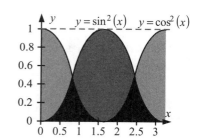

The same is clear from the integration of the
trigonometric identity:

$$\int_0^\pi\cos^2(x)\,dx+\int_0^\pi\sin^2(x)\,dx=\int_0^\pi dx=\pi,\text{ so that}$$

$$\int_0^\pi\cos^2(x)\,dx=\int_0^\pi\sin^2(x)\,dx=\frac{\pi}{2}.$$

39.   We compute as follows: $F(x+h)-F(x)=\int_x^{x+h}\pi\,dt=\pi\big|_x^{x+h}=\pi h$.

Then $\dfrac{F(x+h)-F(x)}{h}=\pi=f(x)$ for all x and $F'(x)=f(x)$.

41.   We compute as follows: $F(x+h)-F(x)=\int_x^{x+h}t^2\,dt=\dfrac{t^3}{3}\Big|_x^{x+h}=\dfrac{(x+h)^3-x^3}{3}=x^2h+xh^2+\dfrac{h^3}{3}$.

Then $\dfrac{F(x+h)-F(x)}{h}=x^2+xh+\dfrac{h^2}{3}\to x^2$ as $h\to 0$, and $F'(x)=f(x)$.

43.   The definition of the integral does not imply that $\int_a^b f(u)g(u)\,du\ne\int_a^b f(u)\,du\cdot\int_a^b g(u)\,du$.

The correct evaluation is $\int_0^1\sqrt{u}\,(u-1)\,du=\int_0^1\left(u^{3/2}-u^{1/2}\right)du=\left(\dfrac{u^{5/2}}{5/2}-\dfrac{u^{3/2}}{3/2}\right)\Big|_0^1=-\dfrac{4}{15}$.

Calculator/Computer Exercises

45. $\min_{x\in[0,2]}\left(\dfrac{1+x}{1+x^4}\right)=\dfrac{3}{17}$, $\max_{x\in[0,2]}\left(\dfrac{1+x}{1+x^4}\right)\approx1.42032$, $b-a=2$

The lower estimate is $\ell=\dfrac{6}{17}$. The upper estimate is $u=2.84064$.

47. $\min_{x\in[0,\,\pi/2]}\left(3\sin\left(x^2\right)+\cos\left(x\right)\right)=1$, $\max_{x\in[0,\,\pi/2]}\left(3\sin\left(x^2\right)+\cos\left(x\right)\right)\approx3.33673$, $b-a=\dfrac{\pi}{2}$

The lower estimate is $\ell=\dfrac{\pi}{2}$. The upper estimate is $u=5.24132$.

49. Average value:

$$f_{ave}=\frac{2}{\pi}\int_0^{\pi/2}\sin\left(x\right)dx=\frac{2}{\pi}\left(-\cos\left(x\right)\Big|_0^{\pi/2}\right)=\frac{2}{\pi}\,.$$

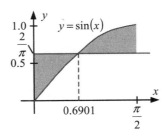

Geometric significance is $c=0.69011$, $f\left(c\right)=\dfrac{2}{\pi}$.

51. Average value:

$$f_{ave}=\frac{1}{4}\int_0^4\left(\frac{x\sqrt{9+x^2}}{8}\right)dx=\frac{1}{4\cdot8\cdot3}\left(\left(9+x^2\right)^{3/2}\Big|_0^4\right)$$

$$=\frac{49}{48}$$

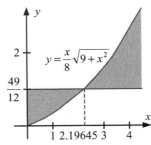

Geometric significance is $c=2.19645$, $f\left(c\right)=\dfrac{49}{48}$.

Section 5.4   The Fundamental Theorem of Calculus

Problems for Practice

1. $\displaystyle\int_0^2\left(x^4+2\right)dx=\left(\frac{x^5}{5}+2x\right)\Bigg|_0^2=\frac{32}{5}+4-0=\frac{52}{5}$

3. $\displaystyle\int_{\pi/2}^{\pi}3\cos\left(x\right)dx=3\sin\left(x\right)\Big|_{\pi/2}^{\pi}=0-3=-3$

5. $\displaystyle\int_{-2}^{3}\left(t^2-1\right)^3dt=\left(\frac{t^7}{7}-\frac{3t^5}{5}+t^3-t\right)\Bigg|_{-2}^{3}=\frac{6672}{35}-\left(-\frac{178}{35}\right)=\frac{1370}{7}$

7. $\displaystyle\int_{-2}^{9}\exp(-x)\,dx=-\exp(-x)\Big|_{-2}^{9}=e^2-e^{-9}$

9. $\displaystyle\int_{-\pi/3}^{\pi/2}4\sin(2x)\,dx=-2\cos(2x)\Big|_{-\pi/3}^{\pi/2}=2-$

11. $\displaystyle S=\int_{3}^{7}(3x^2+2)\,dx=\left(x^3+2x\right)\Big|_{3}^{7}$

$\qquad =357-33=324$

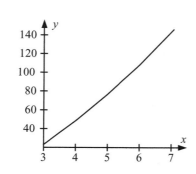

13. $\displaystyle S=\int_{1}^{2}(x^6-x^2)\,dx=\left(\frac{x^7}{7}-\frac{x^3}{3}\right)\Big|_{1}^{2}$

$\qquad =\left(\frac{128}{7}-\frac{8}{3}\right)-\left(\frac{1}{7}-\frac{1}{3}\right)=\frac{332}{21}$

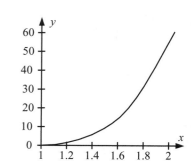

15. $\displaystyle S=\int_{-27}^{-8}(-s^{-1/3})\,ds=\left(-\frac{3s^{2/3}}{2}\right)\Big|_{-27}^{-8}$

$\qquad =-6-\left(-\frac{27}{2}\right)=\frac{15}{2}$

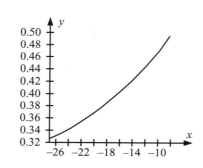

17. $\displaystyle S=\int_{-1}^{1}\exp(x)\,dx=\exp(x)\Big|_{-1}^{1}=e-\frac{1}{e}$

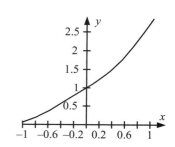

19. $\displaystyle F(x)=\int_{-1}^{x}(3t^2+1)\,dt=\left(t^3+t\right)\Big|_{-1}^{x}=\left(x^3+x\right)-(-1-1)=x^3+x+2$

21. $\displaystyle F(x)=\int_{8}^{x}t^{1/3}\,dt=\left(\frac{3t^{4/3}}{4}\right)\Big|_{8}^{x}=\frac{3x^{4/3}}{4}-12$

23. $\displaystyle F(x)=\int_{0}^{x}\frac{e^t+e^{-t}}{2}\,dt=\frac{e^t-e^{-t}}{2}\Big|_{0}^{x}=\frac{e^x-e^{-x}}{2}$

25. $\displaystyle F(x)=\int_{0}^{x}\sec(t)\tan(t)\,dt=\sec(t)\Big|_{0}^{x}=\sec(x)-1$

27. $\displaystyle F'(x)=\frac{d}{dx}\int_{0}^{x}\tan(t^2)\,dt=\tan(x^2)$

29. $\displaystyle F'(x)=\frac{d}{dx}\int_{\pi/4}^{3x}\cot(3t)\,dt=3\cot(3x)$

31. $\displaystyle F'(x)=\frac{d}{dx}\int_{-1}^{2x}\sqrt{1+t^2}\,dt=2\sqrt{1+4x^2}$

33. $F'(x) = x(x-1)$ and $F''(x) = 2x - 1$

Increasing on the intervals: $(-\infty,\ 0) \cup (1,\ \infty)$

Decreasing on the interval: $(0,\ 1)$

Concave up on the interval: $\left(\dfrac{1}{2},\ \infty\right)$

Concave down on the interval: $\left(-\infty,\ \dfrac{1}{2}\right)$

35. $F'(x) = x\ln(x)$ and $F''(x) = \ln(x) + 1$

Increasing on the interval: $(1,\ \infty)$

Decreasing on the interval: $(0,\ 1)$

Concave up on the interval: $\left(\dfrac{1}{e},\ \infty\right)$

Concave down on the interval: $\left(0,\ \dfrac{1}{e}\right)$

Further Theory and Practice

37. $F'(x) = \dfrac{d}{dx}\displaystyle\int_{2x}^{3x} \left(t^3 + t\right)^{1/4} dt = \left((3x)^3 + 3x\right)^{1/4} \dfrac{d}{dx}(3x) - \left((2x)^3 + 2x\right)^{1/4} \dfrac{d}{dx}(2x)$

$= 3\left(27x^3 + 3x\right)^{1/4} - 2\left(8x^3 + 2x\right)^{1/4}$

39. $F'(x) = \dfrac{d}{dx}\displaystyle\int_{x^{-2}}^{x^{-1}} \dfrac{1}{t}\, dt = \dfrac{1}{x^{-1}} \cdot \dfrac{d}{dx}\left(x^{-1}\right) - \dfrac{1}{x^{-2}} \cdot \dfrac{d}{dx}\left(x^{-2}\right) = x\left(-\dfrac{1}{x^2}\right) - x^2\left(-\dfrac{2}{x^3}\right) = -\dfrac{1}{x} + \dfrac{2}{x} = \dfrac{1}{x}$

41.   We first compute the derivative:

$$F'(x) = \dfrac{d}{dx}\displaystyle\int_{\ln(1/x)}^{\ln(x)} \sin\left(t^3\right) dt = \sin\left(\ln^3(x)\right) \cdot \dfrac{1}{x} - \sin\left(\ln^3\left(\dfrac{1}{x}\right)\right)\left(-\dfrac{1}{x}\right)$$

$$= \dfrac{1}{x}\left(\sin\left(\ln^3(x)\right) + \sin\left(-\ln^3(x)\right)\right) = 0$$

As the derivative is identically 0, $F(x) = $ const. Because $\ln\left(\dfrac{1}{x}\right) = -\ln(x)$, and the integrand is an odd function, $F(x) = 0$.

43. $\displaystyle\int_a^b v(t)\, dt = \int_a^b p'(t)\, dt = p(t)\Big|_a^b = p(b) - p(a)$

45.   Point $(e, 1)$ is on the graph because

$$F(e) = \int_1^e \frac{1}{t}\, dt = \ln(t)\big|_1^e = 1 - 0 = 1 .$$

$$y = 1 + \frac{1}{e}(x - e) = \frac{x}{e}$$

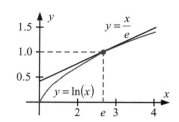

47.   Point $(\pi/6,\ 1/2)$ is on the graph becau

$$F\left(\frac{\pi}{6}\right) = \int_0^{\pi/6} \cos(t)\, dt = \sin(t)\big|_0^{\pi/6} = \frac{1}{2} - 0 =$$

$$y = \frac{1}{2} + \frac{\sqrt{3}}{2}\left(x - \frac{\pi}{6}\right)$$

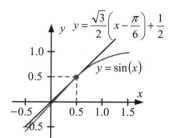

49. $f(t) = \begin{cases} 0 & \text{if} \quad -1 < t \le 0 \\ 2 & \text{if} \quad\ \ 0 < t \le 1 \\ 0 & \text{if} \quad -1 \le t \le 0 \\ -1 & \text{if} \quad\ \ 2 < t \le 3 \end{cases}$

51.   The result is incorrect because $f(t) = t^{-4}$ is not continuous at $x = 0 \in (-2,\ 2)$ and the integral is not defined.

53. $\dfrac{d}{dx} \displaystyle\int_a^{g(h(x))} f(t)\, dt = f\big(g(h(x))\big)\, g'(h(x))\, h'(x)$

55.   Average rate of change of the function f over the interval $[a,\ b]$ is: $\dfrac{f(b) - f(a)}{b - a}$.

The average value of f' is $\dfrac{\displaystyle\int_a^b f'(x)\, dx}{b - a}$.

These two averages have the same value.

57.   We denote $h(x) = g(x) - f(x)$. According to the conditions of the problem, $h(a) \ge 0$, and $h'(x) \ge 0$ for all $x \in [a,\ b]$. Therefore $h(x) = h(a) + \displaystyle\int_a^x h'(t)\, dt \ge 0$ or $f(x) \le g(x)$ for all $x \in [a,\ b]$.

59. $F'(x) = 1 - 2xF(x)$

61. Differentiation of $C(x)$ gives us $C'(x) = \dfrac{d}{dx} \displaystyle\int_0^x \dfrac{t^4}{\sqrt{1+t^2}}\, dt = \dfrac{x^4}{\sqrt{1+x^2}} > 0$, so the function is

increasing on the interval $[0,\ \infty)$.

Differentiating $C'(x)$ gives us $C''(x) = \left(\dfrac{x^4}{\sqrt{1+x^2}}\right)' = \dfrac{x^3\left(4+3x^2\right)}{\sqrt{\left(1+x^2\right)^3}} > 0$ when $x > 0$, so the

graph is concave up on the interval $[0,\ \infty)$.

63. We write $F(x) = \displaystyle\int_b^x f(t)\, dt + \int_a^b f(t)\, dt = G(x) + \int_a^b f(t)\, dt$, where b is an arbitrary number. The

last term is a constant whose derivative is 0. Therefore $F'(x) = G'(x)$. However, $G(x)$ and

hence $G'(x)$ do not depend on a. Then $F'(x)$ does not depend on a either.

65. $A_f(x) = 5 \displaystyle\int_{x-1/10}^{x+1/10} |t|\, dt = \left.\dfrac{5t|t|}{2}\right|_{x-1/10}^{x+1/10}$

$= \begin{cases} -x & \text{if} & x \le -1/10 \\ 5x^2 + 1/20 & \text{if} & -1/10 < x \le 1/10 \\ x & \text{if} & 1/10 < x \end{cases}$

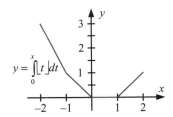

$y = A_f(x),\ f(x) = |x|$

67. F is continuous at all points.

f is continuous except at $-1,\ 0,\ 1$.

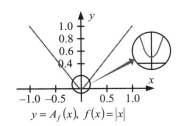

$y = \displaystyle\int_0^x \lfloor t \rfloor\, dt$

Calculator/Computer Exercises

69. $F'(x) = x^3 - 3x^2 + 3x + 4$; $F''(x) = 3x^2 - 6x + 3$

Increasing on the interval: $(-0.71,\ \infty)$

Decreasing on the interval: $(-\infty,\ -0.71)$

Concave up on the interval: $(-\infty,\ \infty)$

Concave down on the interval: The function is not

concave down on any interval.

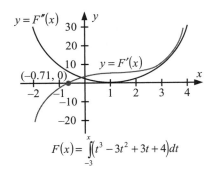

$F(x) = \displaystyle\int_{-3}^x \left(t^3 - 3t^2 + 3t + 4\right) dt$

71. $F'(x) = (x^3 - x)\exp(x)$; $F''(x) = (x^3 + 3x^2 - x - 1)\exp(x)$

Increasing on the intervals: $(-1,\ 0) \cup (1,\ \infty)$

Decreasing on the intervals: $(-\infty,\ -1) \cup (0,\ 1)$

Concave up on the intervals: $(-3.2143,\ -0.46081) \cup (0.67513,\ \infty)$

Concave down on the intervals: $(-\infty,\ -3.2143) \cup (-0.46081,\ 0.67513)$

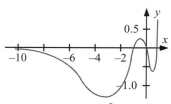

$y = F'(x)$ where $F(x) = \int_0^x (t^3 - t)\exp(t)\,dt$

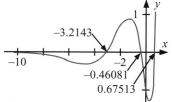

$y = F''(x)$ where $F(x) = \int_0^x (t^3 - t)\exp(t)\,dt$

73. $R(f, S_2) = \left(f(-0.01) + f(0.01)\right)\dfrac{h}{2} = (0.9999 + 0.9999)0.02 = 0.039996$

The approximation is $F'(0) \approx \dfrac{1}{h} \cdot R(f, S_2) = 0.9999$.

The exact value is $F'(0) = \exp(0) = 1$.

75. $R(f, S_2) = \left(f(0.99975) + f(1.00025)\right)\dfrac{h}{2} = (0.706974 + 0.707239)0.0005 = 0.0007071$

The approximation is $F'(1) \approx \dfrac{1}{h} \cdot R(f, S_2) = 0.70710677$.

The exact value is $F'(1) = 1/\sqrt{2} \approx 0.70710678$.

77. The approximation is $F'(\sqrt{3}) = 2$.

The exact value is

$$F'(\sqrt{3}) = f(\sqrt{3}) = \sqrt{1 + 3} = 2.$$

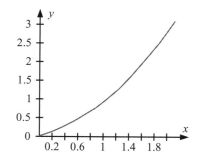

79. $365 - N(37) \approx 6$ days

This number is the number of days when the temperature reached $37°C$ or higher.

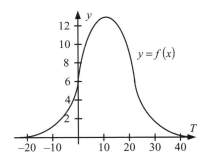

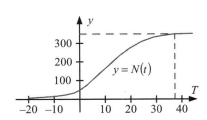

81.   We compute the first and second derivatives of the sine-integral

$$f'(x) = \left(\int_0^x \frac{\sin t}{t} dt \right)' = \frac{\sin x}{x}$$

$$f''(x) = -\frac{\sin x}{x^2} + \frac{\cos x}{x}.$$

Solving for $f''(x) < 0$ where the function is concave down we obtain an interval $(0, a)$ where $a = 4.49341$. Solving for $f''(x) > 0$ where the function is concave up we obtain an interval (a, b) where $b = 7.72525$.

Section 5.5   Integration by Substitution

Problems for Practice

1.   We change variables $u = 3x$, which gives $du = 3dx$. The integral now has the form
$\int \sin(u) \left(\frac{1}{3} \right) du = -\left(\frac{1}{3} \right) \cos(u) + C$. Therefore, $\int \sin(3x) dx = -\left(\frac{1}{3} \right) \cos(3x) + C$.

3.   We change variables $u = x^8 + 1$, which gives $du = 8x^7 dx$. Therefore,

$$\int (x^8 + 1)^{-5} x^7 dx = \int u^{-5} \left(\frac{1}{8} \right) du = \left(\frac{1}{8} \right) \left(\frac{u^{-4}}{-4} \right) + C = -\left(\frac{1}{32} \right) u^{-4} + C = -\left(\frac{1}{32} \right) (x^8 + 1)^{-4} + C.$$

5.   We change variables $u = x^3 - 5$, which gives $du = 3x^2 dx$. Therefore,

$$\int (x^3 - 5)^{3/2} 3x^2 dx = \int u^{3/2} du = \left(\frac{2}{5} \right) u^{5/2} + C = \left(\frac{2}{5} \right) (x^3 - 5)^{5/2} + C.$$

7. We change variables $u = \cos(s)$, which gives $du = -\sin(s)\,ds$. Therefore,

$$\int \sin(s)\cos^4(s)\,ds = -\int u^4\,du = -\left(\frac{1}{5}\right)u^5 + C = -\left(\frac{1}{5}\right)\cos^5(s) + C.$$

9. We change variables $u = \dfrac{\pi}{t}$, which gives $du = -\left(\dfrac{\pi}{t^2}\right)dt$. Therefore,

$$\int 8\left(\frac{\sin(\pi/t)}{t^2}\right)dt = \left(\frac{-8}{\pi}\right)\int \sin(u)\,du = \left(\frac{8}{\pi}\right)\cos(u) + C = \left(\frac{8}{\pi}\right)\cos\left(\frac{\pi}{t}\right) + C.$$

11. We change variables $u = x^2 + 1$, which gives $du = 2x\,dx$. Therefore,

$$\int \left(x^2 + 1\right)^7 2x\,dx = \int u^7\,du = \left(\frac{1}{8}\right)u^8 + C = \frac{1}{8}\left(x^2 + 1\right)^8 + C.$$

13. We change variables $u = 1 + x^2$, which gives $du = 2x\,dx$. Therefore,

$$\int x\sqrt{1 + x^2}\,dx = \left(\frac{1}{2}\right)\int \sqrt{u}\,du = \left(\frac{1}{3}\right)u^{3/2} + C = \left(\frac{1}{3}\right)\sqrt{1 + x^2} + C = \left(\frac{1}{3}\right)\left(1 + x^2\right)^{3/2} + C.$$

15. We change variables $u = \cos(x)$, which gives $du = -\sin(x)\,dx$. Therefore,

$$\int \left(\frac{\sin(x)}{\cos^2(x)}\right)dx = -\int u^{-2}\,du = \frac{1}{u} + C = \frac{1}{\cos(x)} + C = \sec(x) + C.$$

17. We change variables $u = \sin(2x)$, which gives $du = 2\cos(2x)\,dx$. Therefore,

$$\int \sin^5(2x)\cos(2x)\,dx = \left(\frac{1}{2}\right)\int u^5\,du = \left(\frac{1}{12}\right)u^6 + C = \left(\frac{1}{12}\right)\sin^6(2x) + C.$$

19. We change variables $u = 1 + x^2$, which gives $du = 2x\,dx$. Therefore,

$$\int \left(\frac{5x}{\sqrt{1 + x^2}}\right)dx = \left(\frac{5}{2}\right)\int u^{-1/2}\,du = 5u^{1/2} + C = 5\sqrt{1 + x^2} + C.$$

21. We change variables $u = \sqrt{x}$, which gives $du = \left(\dfrac{1}{2}\right)x^{-1/2}\,dx$. Therefore,

$$\int \left(\frac{\sin\sqrt{x}}{\sqrt{x}}\right)dx = 2\int \sin(u)\,du = -2\cos(u) + C = -2\cos\sqrt{x} + C.$$

23.   We change variables $u = t^2 - t$, which gives $du = (2t-1)dt$. For $t = 1$, $u = 0$ and for $t = 2$,

$u = 2$. Therefore, $\int\limits_1^2 (t^2 - t)^5 (2t - 1)dt = \int\limits_0^2 u^5 du = \left(\dfrac{1}{6}\right)u^6\Big|_0^2 = \dfrac{32}{3}$.

25.   We change variables $u = x^2 + 5$, which gives $du = 2xdx$. For $x = 0$, $u = 5$, and for $x = 2$,

$u = 9$. Therefore, $\int\limits_0^2 x\sqrt{x^2 + 5}dx = \left(\dfrac{1}{2}\right)\int\limits_5^9 \sqrt{u}\,du = \left(\dfrac{1}{3}\right)u^{3/2}\Big|_5^9 = 9 - \dfrac{5\sqrt{5}}{3}$.

27.   We change variables $u = 4x^2 - \pi$, which gives $du = 8xdx$. For $x = \sqrt{\pi}$, $u = 3\pi$, and for

$x = \dfrac{\sqrt{\pi}}{2}$, $u = 0$. Therefore, $\int\limits_{\sqrt{\pi}}^{\sqrt{\pi}/2} 2x\cos(4x^2 - \pi)dx = \left(\dfrac{1}{4}\right)\int\limits_{3\pi}^0 \cos(u)\,du = \left(\dfrac{1}{4}\right)\sin(u)\Big|_{3\pi}^0 = 0$.

29.   We change variables $u = x^3 - 1$, which gives $du = 3x^2 dx$. For $x = -1$, $u = -2$, and for $x = 0$,

$u = -1$. Therefore, $\int\limits_{-1}^0 \left(\dfrac{x^2}{(x^3 - 1)^5}\right)dx = \left(\dfrac{1}{3}\right)\int\limits_{-2}^{-1} u^{-5} du = -\left(\dfrac{1}{12}\right)u^{-4}\Big|_{-2}^{-1} = -\dfrac{5}{64}$.

31.   We change variables $u = \sqrt{x}$, which gives $du = \left(\dfrac{1}{2}\right)x^{-1/2}dx$. For $x = \dfrac{\pi^2}{4}$, $u = \dfrac{\pi}{2}$, and for

$x = \pi^2$, $u = \pi$. Therefore, $\int\limits_{\pi^2/4}^{\pi^2} \left(\dfrac{\cos(\sqrt{x})}{\sqrt{x}}\right)dx = 2\int\limits_{\pi/2}^{\pi} \cos(u)\,du = 2\sin(u)\Big|_{\pi/2}^{\pi} = -2$.

33.   We change variables $u = \sec(t)$, which gives $du = \sec(t)\tan(t)dt$. For $t = 0$, $u = 1$, and for

$t = \dfrac{\pi}{4}$, $u = \sqrt{2}$. Therefore,

$\int\limits_0^{\pi/4} \sec^3(t)\tan(t)dt = \int\limits_0^{\pi/4} \sec^2(t)\left[\sec(t)\tan(t)\right]dt = \int\limits_1^{\sqrt{2}} u^2 du = \left(\dfrac{1}{3}\right)u^3\Big|_1^{\sqrt{2}} = \left(\dfrac{2}{3}\right)\sqrt{2} - \dfrac{1}{3}$.

35.   We change variables $u = 1 + \tan(u)$, which gives $du = \sec^2(x)dx$. For $x = 0$, $u = 1$, and for

$x = \dfrac{\pi}{6}$, $u = 1 + \dfrac{1}{\sqrt{3}}$. Therefore, $\int\limits_0^{\pi/6} \left(\dfrac{\sec^2(x)}{(1 + \tan(x))^2}\right)dx = \int\limits_1^{1+1/\sqrt{3}} u^{-2} du = -u^{-1}\Big|_1^{1+1/\sqrt{3}} = \dfrac{1}{1 + \sqrt{3}}$.

Further Theory and Practice

37.   We change variables $x = a\sin(\theta)$, then $\sqrt{a^2 - x^2} = a\cos(\theta)$ and $dx = a\cos(\theta)d\theta$. For $x = -a$, $\theta = -\dfrac{\pi}{2}$, and for $x = a$, $\theta = \dfrac{\pi}{2}$. This yields

$$\int\limits_{-a}^{a} \sqrt{a^2 - x^2}\, dx = a^2 \int\limits_{-\pi/2}^{\pi/2} \cos^2(\theta)\, d\theta = \left(\frac{a^2}{2}\right) \int\limits_{-\pi/2}^{\pi/2} (1 + \cos(2\theta))\, d\theta$$

$$= \left(\frac{a^2}{2}\right)\left(\theta + \frac{\sin(2\theta)}{2}\right)\Bigg|_{-\pi/2}^{\pi/2} = \frac{\pi a^2}{2}$$

This is the familiar formula for the area of a semicircle.

39.   We change variables $u = 2x + 3$, then $x = \dfrac{u - 3}{2}$ and $dx = \dfrac{du}{2}$. This yields

$$\int x(2x+3)^{1/2}\, dx = \left(\frac{1}{4}\right)\int (u-3)u^{1/2}\, du = \left(\frac{1}{4}\right)\int \left(u^{3/2} - 3u^{1/2}\right) du = \left(\frac{1}{4}\right)\left(\left(\frac{2}{5}\right)u^{5/2} - 2u^{3/2}\right) + C$$

$$= \left(\frac{1}{10}\right)(2x+3)^{5/2} - \left(\frac{1}{2}\right)(2x+3)^{3/2} + C.$$

41.   We change variables $u = x + 3$, then $x = u - 3$, $dx = du$. This yields

$$\int \left(\frac{x}{\sqrt{x+3}}\right) dx = \int \left(\frac{u-3}{\sqrt{u}}\right) du = \int \sqrt{u} - \frac{3}{\sqrt{u}}\, du = \left(\frac{2}{3}\right)u^{3/2} - 6u^{1/2} + C$$

$$= \left(\frac{2}{3}\right)(x+3)^{3/2} - 6(x+3)^{1/2} + C.$$

43.   We change variables $u = 6 + s^2$, then $s^2 = u - 6$ and $du = 2s\,ds$. This yields

$$\int s^5 \left(6+s^2\right)^{-1/2} ds = \left(\frac{1}{2}\right)\int (u-6)^2 u^{-1/2}\, du = \left(\frac{1}{2}\right)\int \left(u^{3/2} - 12u^{1/2} + 36u^{-1/2}\right) du$$

$$= \left(\frac{1}{2}\right)\left(\left(\frac{2}{5}\right)u^{5/2} - 8u^{3/2} + 72u^{1/2}\right) + C$$

$$= \left(6+s^2\right)^{1/2}\left(\frac{\left(6+s^2\right)^2}{5} - 4\left(6+s^2\right) + 36\right) + C$$

$$= \left(6+s^2\right)^{1/2}\left(\frac{s^4 - 8s^2 + 96}{5}\right) + C.$$

45. We change variables $u = s - 5$, then $s = u + 5$ and $du = ds$. This yields

$$\int (s+2)\sqrt{s-5}\,ds = \int (u+7)\sqrt{u}\,du = \int \left(u^{3/2} + 7u^{1/2}\right)du = \left(\frac{2}{5}\right)u^{5/2} + \left(\frac{14}{3}\right)u^{3/2} + C$$

$$= \left(\frac{2}{5}\right)(s-5)^{5/2} + \left(\frac{14}{3}\right)(s-5)^{3/2} + C.$$

47. We change variables $u = x^4 + x^2 + 1$, then $du = \left(4x^3 + 2x\right)dx$. This yields

$$\int \left(\frac{2x^3 + x}{\sqrt{x^4 + x^2 + 1}}\right)dx = \left(\frac{1}{2}\right)\int \left(\frac{1}{\sqrt{u}}\right)du = u^{1/2} + C = \sqrt{x^4 + x^2 + 1} + C.$$

49. We change variables $u = \cos(x)$, then $du = -\sin(x)\,dx$. For $x = 0$, $u = 1$, and for $x = \dfrac{\pi}{3}$,

$u = \dfrac{1}{2}$. This yields $\displaystyle\int_0^{\pi/3} \left(\frac{\sin(x)}{1 - \sin^2(x)}\right)dx = \int_0^{\pi/3} \left(\frac{\sin(x)}{\cos^2(x)}\right)dx = -\int_1^{1/2} u^{-2}\,du = -u^{-1}\Big|_1^{1/2} = 1.$

51. We change variables $u = 1 - x$, then $x = 1 - u$ and $du = dx$. For $x = 0$, $u = 1$, and for $x = 1$, $u = 0$. This yields

$$\int_0^1 x\sqrt[3]{1-x}\,dx = -\int_1^0 (1-u)\sqrt[3]{u}\,du = \int_0^1 (1-u)\sqrt[3]{u}\,du = \int_0^1 \left(u^{1/3} - u^{4/3}\right)du$$

$$= \left(\left(\frac{3}{4}\right)u^{4/3} - \left(\frac{3}{7}\right)u^{7/3}\right)\Bigg|_0^1 = \frac{9}{28}.$$

53. We change variables $u = x^2 + 5$, then $du = 2x\,dx$. For $x = -3$, $u = 14$, and for $x = 2$, $u = 9$, and for $x = 0$, $u = 5$. This yields:

$$\int_{-3}^2 x\sqrt{x^2 + 5}\,dx = \left(\frac{1}{2}\right)\int_{14}^9 \sqrt{u}\,du = \left(\frac{1}{3}\right)u^{3/2}\Big|_{14}^9 = 9 - \frac{14\sqrt{14}}{3}$$

$$\int_{-3}^0 x\sqrt{x^2 + 5}\,dx = \left(\frac{1}{2}\right)\int_{14}^5 \sqrt{u}\,du = \left(\frac{1}{3}\right)u^{3/2}\Big|_{14}^5 = -\frac{14\sqrt{14}}{3} + \frac{5\sqrt{5}}{3}$$

$$\int_0^2 x\sqrt{x^2 + 5}\,dx = \left(\frac{1}{2}\right)\int_5^9 \sqrt{u}\,du = \left(\frac{1}{3}\right)u^{3/2}\Big|_5^9 = 9 - \frac{5\sqrt{5}}{3}$$

We compare the result of adding the last two lines with the first line:

$$9 - \frac{14\sqrt{14}}{3} = \left(-\frac{14\sqrt{14}}{3} + \frac{5\sqrt{5}}{3}\right) + \left(9 - \frac{5\sqrt{5}}{3}\right),$$

which is true by inspection.

55.   We use the substitution $u = -x$ and the property $f(-u) = -f(u)$ of odd functions.

$$\int_{-a}^{a} f(x)\,dx = \int_{-a}^{0} f(x)\,dx + \int_{0}^{a} f(x)\,dx = -\int_{a}^{0} f(-u)\,du + \int_{0}^{a} f(x)\,dx = \int_{0}^{a} f(-u)\,du + \int_{0}^{a} f(x)\,dx$$

$$= -\int_{0}^{a} f(u)\,du + \int_{0}^{a} f(x)\,dx = 0$$

57.   We change variables $g(x) = x \cdot f(\cos(x))$, then

$$g(-x) = -x \cdot f(\cos(-x)) = -x \cdot f(\cos(x)) = -g(x).$$

This function is odd. From Problem 5.5.55: $\int_{-\pi/2}^{\pi/2} x \cdot f(\cos(x))\,dx = 0$.

59.   We change variables $u = a + b - x$, then $du = -dx$. For $x = a$, $u = b$, and for $x = b$, $u = a$.

This yields $\int_{a}^{b} f(x)\,dx = -\int_{b}^{a} f(a+b-u)\,du = \int_{a}^{b} f(a+b-u)\,du = \int_{a}^{b} f(a+b-x)\,dx$.

61.   We change variables $u = \tan(x)$, then $du = \sec^2(x)\,dx$. Therefore,

$$\int \sec^2(x)\tan(x)\,dx = \int u\,du = \frac{u^2}{2} + C_1 = \frac{\tan^2(x)}{2} + C_1.$$

We change variables $v = \sec(x)$, then $dv = \sec(x)\tan(x)\,dx$ and

$$\int \sec^2(x)\tan(x)\,dx = \int v\,dv = \frac{v^2}{2} + C_2 = \frac{\sec^2(x)}{2} + C_2.$$

We observe that $\dfrac{\tan^2(x)}{2} + C_1 = \dfrac{1 - \cos^2(x)}{2\cos^2(x)} + C_1 = \dfrac{\sec^2(x)}{2} + C_1 - \dfrac{1}{2}$. The same results are

obtained for $C_2 = C_1 - \dfrac{1}{2}$.

63. Consider the substitution $\dfrac{2t^3 + 3t^2 + 2t - 1}{(t^2 + 1)(t+1)^2} = \dfrac{At}{t^2 + 1} + \dfrac{B}{(t+1)^2}$. Then reducing to a common

denominator leads to $2t^3 + 3t^2 + 2t - 1 = At^3 + 2At^2 + At + Bt^2 + B$, or

$$2t^3 + 3t^2 + 2t - 1 = At^3 + (2A + B)t^2 + At + B.$$

Equating the coefficients for the same powers of variable t: $\begin{cases} 2 = A \\ 3 = 2A + B \\ 2 = A \\ -1 = B \end{cases}$ we obtain $A = 2$,

$B = -1$. Then $\displaystyle\int \dfrac{2t^3 + 3t^2 + 2t - 1}{(t^2 + 1)(t+1)^2}\, dt = 2\int \dfrac{t}{t^2 + 1}\, dt - \int \dfrac{1}{(t+1)^2}\, dt = \ln(t^2 + 1) + \dfrac{1}{t+1} + C$.

Note that we have used the substitution $u = t^2 + 1$ to evaluate $\displaystyle\int\left(\dfrac{t}{t^2 + 1}\right) dt$.

Calculator/Computer Exercises

65. We change variables $u = x^4 + x^2$, then $du = (4x^3 + 2x)\, dx$. For $x = 0$, $u = 0$, and for $x = b$,

$u = \pi$. This yields $b^4 + b^2 = \pi$ and $b = \left(\dfrac{1}{2}\right)\sqrt{-2 + 2\sqrt{1 + 4\pi}}$.

67. We change variables $u = \dfrac{1}{x} - \ln(x)$, then $du = \left(-\dfrac{1}{x^2} - \dfrac{1}{x}\right) dx = -\left(\dfrac{x+1}{x^2}\right) dx$. For $x = 1$, $u = 1$,

and for $x = b$, $u = \dfrac{1}{4}$. This yields $\dfrac{1}{b} - \ln(b) = \dfrac{1}{4}$ and $b \approx 1.510129$.

Section 5.6   More on the Calculation of Area

Problems for Practice

1.   We calculate the area between the curve and the x-axis on the given interval as follows:

$f(x) \geq 0$ for $x \in \left[\dfrac{\pi}{4}, \dfrac{\pi}{2} \right]$ and $f(x) \leq 0$ for $x \in \left[\dfrac{\pi}{2}, \dfrac{2\pi}{3} \right]$,

$$A = \int_{\pi/4}^{\pi/2} \cos(x)\,dx - \int_{\pi/2}^{2\pi/3} \cos(x)\,dx = \sin(x)\big|_{\pi/4}^{\pi/2} - \sin(x)\big|_{\pi/2}^{2\pi/3} = 1 - \frac{\sqrt{2}}{2} - \left(\frac{\sqrt{3}}{2} - 1 \right)$$

$$= 2 - \frac{\sqrt{2} + \sqrt{3}}{2}$$

3.   We calculate the area between the curve and the x-axis on the given interval as follows:
$h(x) \geq 0$ for $x \in [-5, -2]$ or $x \in [2, 7]$ and $h(x) \leq 0$ for $x \in [-2, 2]$,

$$A = \int_{-5}^{-2} \left(2x^2 - 8 \right) dx - \int_{-2}^{2} \left(2x^2 - 8 \right) dx + \int_{2}^{7} \left(2x^2 - 8 \right) dx$$

$$= \left(\frac{2x^3}{3} - 8x \right)\Bigg|_{-5}^{-2} - \left(\frac{2x^3}{3} - 8x \right)\Bigg|_{-2}^{2} + \left(\frac{2x^3}{3} - 8x \right)\Bigg|_{2}^{7}$$

$$= \left(\frac{32}{3} - \left(-\frac{130}{3} \right) \right) - \left(-\frac{32}{3} - \frac{32}{3} \right) + \left(\frac{518}{3} - \left(-\frac{32}{3} \right) \right) = \frac{776}{3}$$

5.   We calculate the area between the curve and the x-axis on the given interval as follows:

$g(x) \geq 0$ for $x \in \left[-\dfrac{\pi}{4}, \dfrac{3\pi}{4} \right]$ and $g(x) \leq 0$ for $x \in \left[\dfrac{3\pi}{4}, \dfrac{3\pi}{2} \right]$ or $x \in \left[-\pi, -\dfrac{\pi}{4} \right]$,

$$A = -\int_{-\pi}^{-\pi/4} \left(\sin(x) + \cos(x) \right) dx + \int_{-\pi/4}^{3\pi/4} \left(\sin(x) + \cos(x) \right) dx - \int_{3\pi/4}^{3\pi/2} \left(\sin(x) + \cos(x) \right) dx$$

$$= -\left(\sin(x) - \cos(x) \right)\big|_{-\pi}^{-\pi/4} + \left(\sin(x) - \cos(x) \right)\big|_{-\pi/4}^{3\pi/4} - \left(\sin(x) - \cos(x) \right)\big|_{3\pi/4}^{3\pi/2}$$

$$-\left(-\sqrt{2} - 1 \right) + \left(\sqrt{2} - \left(-\sqrt{2} \right) \right) - \left(-1 - \sqrt{2} \right) = 4\sqrt{2} + 2$$

7. We calculate the area between the curve and the x-axis on the given interval as follows:

$f(x) \geq 0$ for $x \in [0, 1]$ and $f(x) \leq 0$ for $x \in \left[-\dfrac{1}{2}, 0\right]$,

$$A = -\int_{-1/2}^{0} x\left(1 - x^2\right)^2 dx + \int_{0}^{1} x\left(1 - x^2\right)^2 dx = \left(-\dfrac{\left(1 - x^2\right)^3}{6}\Bigg|_{-1/2}^{0}\right) + \left(-\dfrac{\left(1 - x^2\right)^3}{6}\Bigg|_{0}^{1}\right)$$

$$= \left(\dfrac{1}{6} - \dfrac{(3/4)^3}{6}\right) + \left(-0 + \dfrac{1}{6}\right) = \dfrac{101}{384}.$$

9. We calculate the area between the curve and the x-axis on the given interval as follows:

$h(x) \geq 0$ for $x \in \left[0, \dfrac{1}{2}\right]$ and $h(x) \leq 0$ for $x \in \left[\dfrac{1}{2}, \dfrac{3}{2}\right]$,

$$A = \int_{0}^{1/2} \left(\sin(\pi x)\right)^2 \cos(\pi x)\, dx - \int_{1/2}^{3/2} \left(\sin(\pi x)\right)^2 \cos(\pi x)\, dx = \left(\dfrac{\sin(\pi x)^3}{3\pi}\Bigg|_{0}^{1/2}\right) - \left(\dfrac{\sin(\pi x)^3}{3\pi}\Bigg|_{1/2}^{3/2}\right)$$

$$= \left(\dfrac{1}{3\pi} - 0\right) - \left(-\dfrac{1}{3\pi} - \dfrac{1}{3\pi}\right) = \dfrac{1}{\pi}.$$

11. We calculate the area between the curve and the x-axis on the given interval as follows:

$g(x) \geq 0$ for $x \in \left[-\dfrac{\pi}{3}, -\dfrac{\pi}{6}\right]$ or $x \in \left[0, \dfrac{\pi}{6}\right]$ and $g(x) \leq 0$ for $x \in \left[-\dfrac{\pi}{2}, -\dfrac{\pi}{3}\right]$ or

$x \in \left[-\dfrac{\pi}{6}, 0\right]$ or $x \in \left[\dfrac{\pi}{6}, \dfrac{\pi}{3}\right]$,

$$A = -\int_{-\pi/2}^{-\pi/3} \sin(3x)\cos(3x)\, dx + \int_{-\pi/3}^{-\pi/6} \sin(3x)\cos(3x)\, dx - \int_{-\pi/6}^{0} \sin(3x)\cos(3x)\, dx$$

$$+ \int_{0}^{\pi/6} \sin(3x)\cos(3x)\, dx - \int_{\pi/6}^{\pi/3} \sin(3x)\cos(3x)\, dx = -\left(-\dfrac{\cos(6x)}{12}\Bigg|_{-\pi/2}^{-\pi/3}\right) + \left(-\dfrac{\cos(6x)}{12}\Bigg|_{-\pi/3}^{-\pi/6}\right)$$

$$-\left(-\dfrac{\cos(6x)}{12}\Bigg|_{-\pi/6}^{0}\right) + \left(-\dfrac{\cos(6x)}{12}\Bigg|_{0}^{\pi/6}\right) - \left(-\dfrac{\cos(6x)}{12}\Bigg|_{\pi/6}^{\pi/3}\right)$$

$$= -\left(-\dfrac{1}{12} - \dfrac{1}{12}\right) + \left(\dfrac{1}{12} - \left(-\dfrac{1}{12}\right)\right) - \left(-\dfrac{1}{12} - \dfrac{1}{12}\right) + \left(\dfrac{1}{12} - \left(-\dfrac{1}{12}\right)\right) - \left(-\dfrac{1}{12} - \dfrac{1}{12}\right) = \dfrac{5}{6}.$$

13.   We calculate the area between the curve and the x-axis on the given interval as follows:

$f(x) \geq 0$ for $x \in \left[\dfrac{3}{2},\ 4\right]$ and $f(x) \leq 0$ for $x \in \left[-3,\ \dfrac{3}{2}\right]$,

$$A = -\int_{-3}^{1}\left(-x^2\right)dx - \int_{1}^{3/2}(2x-3)\,dx + \int_{3/2}^{4}(2x-3)\,dx = \left.\frac{x^3}{3}\right|_{-3}^{1} - \left.\left(x^2 - 3x\right)\right|_{1}^{3/2} + \left.\left(x^2 - 3x\right)\right|_{3/2}^{4}$$

$$= \left(\frac{1}{3} - (-9)\right) - \left(-\frac{9}{4} - (-2)\right) + \left(4 - \left(-\frac{9}{4}\right)\right) = \frac{95}{6}.$$

15.   We calculate the area between the curve and the x-axis on the given interval as follows:

$f(x) \geq 0$ for $x \in \left[0,\ \dfrac{\pi}{2}\right]$ and $f(x) \leq 0$ for $x \in \left[\dfrac{\pi}{2},\ \pi\right]$,

$$A = \int_{0}^{\pi/4}\sin(x)\,dx + \int_{\pi/4}^{\pi/2}\cos(x)\,dx - \int_{\pi/2}^{\pi}\cos(x)\,dx = \left.-\cos(x)\right|_{0}^{\pi/4} + \left.\sin(x)\right|_{\pi/4}^{\pi/2} - \left.\sin(x)\right|_{\pi/2}^{\pi}$$

$$= \left(-\frac{\sqrt{2}}{2} - (-1)\right) + \left(1 - \frac{\sqrt{2}}{2}\right) - (0-1) = 3 - \sqrt{2}.$$

17. $A = \displaystyle\int_{-7}^{4}(4-x)\,dx + \int_{4}^{6}(x-4)\,dx = \left.\left(4x - \frac{x^2}{2}\right)\right|_{-7}^{4} + \left.\left(\frac{x^2}{2} - 4x\right)\right|_{4}^{6} = \left(8 - \left(-\frac{105}{2}\right)\right) + \left(-6 - (-8)\right)$

$= \dfrac{125}{2}$

19.   To obtain the points of intersection for the two curves we solve $x^2 + x + 1 = 2x^2 + 3x - 7$, and find that the roots are $x = -4$ and $x = 2$.

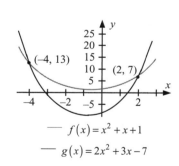

We calculate the area of the region bounded by these curves as follows: On the interval $[-4,\ 2]$,

$f(x) \geq g(x)$,

$$A = \int_{-4}^{2}\big(f(x) - g(x)\big)\,dx = \int_{-4}^{2}\left(-x^2 - 2x + 8\right)dx = \left.\left(-\frac{x^3}{3} - x^2 + 8x\right)\right|_{x=-4}^{x=2}$$

$$= \frac{28}{3} - \left(-\frac{80}{3}\right) = 36$$

21.   To obtain the points of intersection for the two curves we solve $x^2 + 5 = 2x^2 + 1$, and find that the roots are $x = -2$ and $x = 2$.

We calculate the area of the region bounded by these curves as follows: On the interval $\left[-2,\ 2\right]$, $f(x) \ge g(x)$,

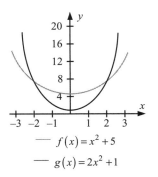

— $f(x) = x^2 + 5$

— $g(x) = 2x^2 + 1$

$$A = \int_{-2}^{2} \left(f(x) - g(x)\right) dx = \int_{-2}^{2} \left(-x^2 + 4\right) dx = \left(-\frac{x^3}{3} + 4x\right)\Bigg|_{x=-2}^{x=2} = \frac{16}{3} - \left(-\frac{16}{3}\right) = \frac{32}{3}$$

23.   To obtain the points of intersection for the two curves we solve $x^3 - 3x^2 - x + 4 = -3x + 4$, and find that the roots are $x = 0$, $x = 1$ and $x = 2$.

We calculate the area of the region bounded by these curves as follows: On the interval $\left[0,\ 1\right]$, $f(x) \ge g(x)$ and $f(x) \le g(x)$ on the interval $\left[1,\ 2\right]$,

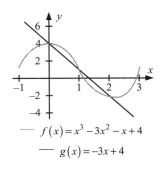

— $f(x) = x^3 - 3x^2 - x + 4$

— $g(x) = -3x + 4$

$$A = \int_{0}^{1}\left(f(x) - g(x)\right) dx + \int_{1}^{2}\left(g(x) - f(x)\right) dx = \int_{0}^{1}\left(x^3 - 3x^2 + 2x\right) dx + \int_{1}^{2}\left(-x^3 + 3x^2 - 2x\right) dx$$

$$= \left(\frac{x^4}{4} - x^3 + x^2\right)\Bigg|_{x=0}^{x=1} + \left(-\frac{x^4}{4} + x^3 - x^2\right)\Bigg|_{x=1}^{x=2} = \left(\frac{1}{4} - 0\right) + \left(0 - \left(-\frac{1}{4}\right)\right) = \frac{1}{2}$$

25.   To obtain the points of intersection for the two curves we solve $-x^4 + x^2 + 16 = 2x^4 - 2x^2 - 20$, and find that the roots are $x = -2$ and $x = 2$.

We calculate the area of the region bounded by these curves as follows: On the interval $\left[-2,\ 2\right]$, $f(x) \ge g(x)$,

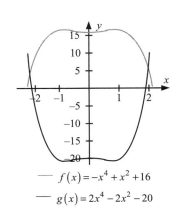

— $f(x) = -x^4 + x^2 + 16$

— $g(x) = 2x^4 - 2x^2 - 20$

$$A = \int_{-2}^{2}\left(f(x) - g(x)\right) dx = \int_{-2}^{2}\left(-3x^4 + 3x^2 + 36\right) dx$$

$$= \left(-\frac{3x^5}{5} + x^3 + 36x\right)\Bigg|_{x=-2}^{x=2} = \frac{304}{5} - \left(-\frac{304}{5}\right) = \frac{608}{5}$$

27.   To obtain the points of intersection we solve $-\sqrt{3}\cos(x)=\sin(x)$. The roots on the interval are

$$x=-\pi \quad \text{and} \quad x=\frac{2\pi}{3}.$$

We compute the area between the two curves as follows: On the interval $\left[-\pi, \dfrac{2\pi}{3}\right]$,

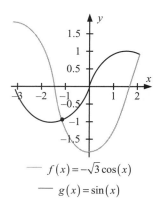

$f(x)=-\sqrt{3}\cos(x)$

$g(x)=\sin(x)$

$$A = \int_{-\pi}^{-\pi/3} \left(f(x)-g(x)\right)dx + \int_{-\pi/3}^{2\pi/3} \left(g(x)-f(x)\right)dx$$

$$= \int_{-\pi}^{-\pi/3} \left(-\sqrt{3}\cos x - \sin x\right)dx + \int_{-\pi/3}^{2\pi/3} \left(\sin(x)+\sqrt{3}\cos(x)\right)dx$$

$$= \left(-\sqrt{3}\sin(x)+\cos(x)\right)\Big|_{-\pi}^{-\pi/3} + \left(-\cos(x)+\sqrt{3}\sin x\right)\Big|_{-\pi/3}^{2\pi/3}$$

$$= \left(\left[-\sqrt{3}\left(-\frac{\sqrt{3}}{2}\right)+\frac{1}{2}\right]-[0-1]\right)+\left(\left[\frac{1}{2}+\sqrt{3}\frac{\sqrt{3}}{2}\right]-\left[-\frac{1}{2}+\sqrt{3}\left(-\frac{\sqrt{3}}{2}\right)\right]\right)=3+4=7$$

29.   To obtain the points of intersection we solve $\sin(2x)=\cos(x)$. The roots on the interval are

$$x=0 \quad \text{and} \quad x=\frac{\pi}{2}.$$

We compute the area between the two curves as follows: On the interval $\left[0, \dfrac{\pi}{2}\right]$,

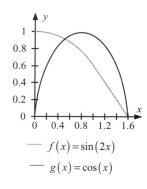

$f(x)=\sin(2x)$

$g(x)=\cos(x)$

$$A = \int_{0}^{\pi/6} \left(g(x)-f(x)\right)dx + \int_{\pi/6}^{\pi/2} \left(f(x)-g(x)\right)dx$$

$$= \int_{0}^{\pi/6} \left(\cos(x)-\sin(2x)\right)dx + \int_{\pi/6}^{\pi/2} \left(\sin(2x)-\cos(x)\right)dx$$

$$= \left(\sin(x)+\frac{\cos(2x)}{2}\right)\Big|_{0}^{\pi/6} + \left(-\frac{\cos(2x)}{2}-\sin(x)\right)\Big|_{\pi/6}^{\pi/2}$$

$$= \left[\left(\frac{1}{2}+\frac{1/2}{2}\right)-\left(0+\frac{1}{2}\right)\right]+\left[\left(-\frac{(-1)}{2}-1\right)-\left(-\frac{1/2}{2}-\frac{1}{2}\right)\right]=\left[\frac{1}{2}\right]+\left[-\frac{1}{2}+\frac{1}{4}+\frac{1}{2}\right]=\frac{3}{4}.$$

Further Theory and Practice

31. To obtain the points of intersection we solve

$\dfrac{x}{1+x^2} = \dfrac{x}{2}$. The roots on the interval are $x = 0$ and $x = 1$.

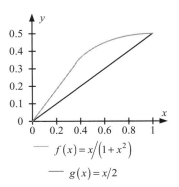

$f(x) = x/(1+x^2)$

$g(x) = x/2$

We compute the area of the region between the two curves as follows: $f(x) \geq g(x)$ on the interval $[0, 1]$,

$$A = \int_0^1 (f(x) - g(x))\,dx = \int_0^1 \left(\frac{x}{1+x^2} - \frac{x}{2} \right) dx = \left(\frac{\ln(x^2+1)}{2} - \frac{x^2}{4} \right)\Bigg|_{x=0}^{x=1} = \left(\frac{\ln(2)}{2} - \frac{1}{4} \right) - (0 - 0)$$

$$= \frac{\ln(2)}{2} - \frac{1}{4}$$

33. To obtain the points of intersection we solve $2\sin(x) = \sin(2x)$. On the interval the roots are $x = 0$ and $x = \pi$.

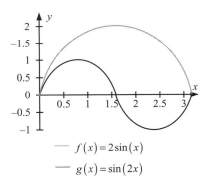

$f(x) = 2\sin(x)$

$g(x) = \sin(2x)$

We compute the area between the two curves as follows: $f(x) \geq g(x)$ on the interval $[0, \pi]$,

$$A = \int_0^\pi (f(x) - g(x))\,dx = \int_0^\pi (2\sin(x) - \sin(2x))\,dx$$

$$= \left(-2\cos(x) + \frac{\cos(2x)}{2} \right)\Bigg|_{x=0}^{x=\pi} = \frac{5}{2} - \left(-\frac{3}{2} \right) = 4$$

35. To obtain the points of intersection we solve $y^2 + 6 = -y^2 + 14$. The roots are $y = 2$ and $y = -2$.

We compute the area between the pair of curves as follows:

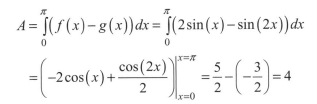

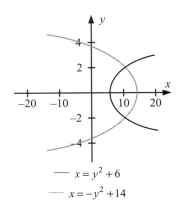

$x = y^2 + 6$

$x = -y^2 + 14$

$$A = \int_{-2}^{2} \left((-y^2 + 14) - (y^2 + 6) \right) dy = \int_{-2}^{2} (-2y^2 + 8)\,dy$$

$$= \left(-\frac{2y^3}{3} + 8y \right)\Bigg|_{y=-2}^{y=2} = \frac{32}{3} - \left(-\frac{32}{3} \right) = \frac{64}{3}$$

37.   To obtain the points of intersection we solve $y^2 = y^3$.
The roots are $y = 0$ and $y = 1$.

We compute the area between the two curves as
follows:

$$A = \int_0^1 \left(y^2 - y^3 \right) dy = \left(\frac{y^3}{3} - \frac{y^4}{4} \right)\Bigg|_{y=0}^{y=1} = \frac{1}{3} - \frac{1}{4} = \frac{1}{12}.$$

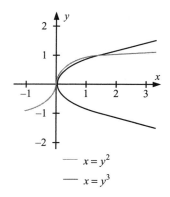

$x = y^2$

$x = y^3$

39. $\int_0^{\pi} \sqrt{2}\, dy$

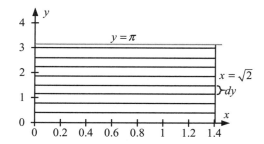

41. $\int_0^1 \left(e - \exp(x) \right) dx = e - \int_0^1 \exp(x)\, dx$

$$= e - \left[1 + \int_1^e \left(1 - \ln(y) \right) dy \right]$$

$$= \int_1^e \ln(y)\, dy$$

$y = e$

$y = \exp(x)$

$\big\}\, dy$

$x = 1$

43. $\int_0^4 \left(\sqrt{x} - \frac{x}{2} \right) dx = \int_0^2 \left(4 - y^2 \right) dy - \int_0^2 \left(4 - 2y \right) dy$

$$= \int_0^2 \left(2y - y^2 \right) dy$$

— $y = \sqrt{x}$ or $x = y^2$

— $y = x/2$ or $x = 2y$

45.
$$\int_{-3}^{3}\left(3-|x|\right)dx = \int_{0}^{3}-y+3\,dy + \int_{0}^{3}0-(y-3)\,dy$$
$$= \int_{0}^{3}-2y+6\,dy$$

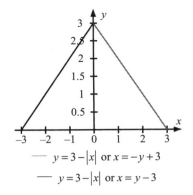

——— $y = 3-|x|$ or $x = -y+3$

——— $y = 3-|x|$ or $x = y-3$

47.
$$A = \int_{1}^{2}(x-1)\,dx + \int_{2}^{3}(3-x)\,dx$$

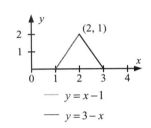

——— $y = x-1$

——— $y = 3-x$

49. To find the point of intersection we solve $2-x^2 = |x|$. The root is $x=1$ for $x>0$ and $x=-1$ for $x<0$; hence the area is

$$A = \int_{-1}^{0}\left((2-x^2)-(-x)\right)dx + \int_{0}^{1}\left((2-x^2)-x\right)dx$$
$$= \int_{-1}^{0}\left(2-x^2+x\right)dx + \int_{0}^{1}\left(2-x^2-x\right)dx.$$

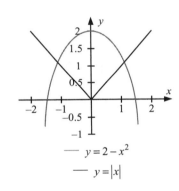

——— $y = 2-x^2$

——— $y = |x|$

51.
$$A = \int_{0}^{1}\left((3-y)-(y+1)\right)dy$$

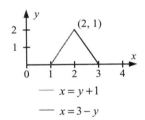

——— $x = y+1$

——— $x = 3-y$

53.
$$\int_{0}^{1}\left(y-(-y)\right)dy + \int_{1}^{2}\left(\sqrt{2-y}-(-\sqrt{2-y})\right)dy$$

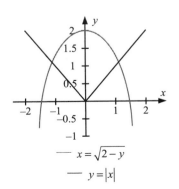

——— $x = \sqrt{2-y}$

——— $y = |x|$

55.   First, we find two intersection points. From the first
equation we have $x = 4y + 1$, and after substituting in
the second equation we obtain the following equation
for the intersection points: $25y^2 + 5y - 1 = 0$.

Therefore, the intersection points are $y_{1,2} = \dfrac{-1 \mp \sqrt{5}}{10}$.

Next, we plot the curves.

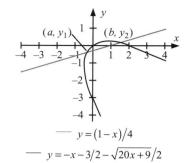

$$y = (1 - x)/4$$
$$y = -x - 3/2 - \sqrt{20x + 9}/2$$

The x-coordinates of the intersection points are $a = \dfrac{3 - 2\sqrt{5}}{5}$ and $b = \dfrac{3 + 2\sqrt{5}}{5}$.

Clearly, it is a lot easier to find the area between the two curves by integrating over
x-coordinate:

$$A = \int_{(3-2\sqrt{5})/5}^{(3+2\sqrt{5})/5} \left(-x - \frac{3}{2} + \frac{\sqrt{20x+9}}{2} - \frac{1-x}{4} \right) dx = \left(\frac{-5x^2}{8} - \frac{5x}{4} + \frac{(20x+9)^{3/2}}{60} \right) \Bigg|_{(3-2\sqrt{5})/5}^{(3+2\sqrt{5})/5} = \frac{\sqrt{5}}{6}$$

The second curve when expressed as a function $x = g(y)$, is not single-valued within
the range of y between two intersection points. To find the area using integration over
y-coordinate, we have to calculate the coordinates of the maximum of the second curve and
compute two integrals, each over the branches of the curve where $g(y)$ is single-valued.

To calculate the coordinate of the maximum, we differentiate the equation of the curve with
respect to x and obtain: $2x + 2y + 2x\dfrac{dy}{dx} + 2y\dfrac{dy}{dx} - 2 + 3\dfrac{dy}{dx} = 0$, hence $\dfrac{dy}{dx} = \dfrac{2x + 2y - 2}{2x + 2y + 3}$.

Setting $\dfrac{dy}{dx} = 0$, we find $x = 1 - y$. Substituting this expression in the equation for the curve

we find for the maximum $x_0 = \dfrac{4}{5}, y_0 = \dfrac{1}{5}$.

Next, we solve the equation for the curve to find: $x = g(y) = 1 - y \pm \sqrt{-5y + 1}$.

From the graph above, the area between the curves is calculated as a sum of the following
integrals:

$$A = \int_{y_1}^{y_0} \left((4y+1) - \left(1 - y - \sqrt{-5y+1}\right) \right) dy - \int_{y_2}^{y_0} \left((4y+1) - \left(1 - y + \sqrt{-5y+1}\right) \right) dy$$

$$= \int_{(-1-\sqrt{5})/10}^{1/5} \left((4y+1) - \left(1 - y - \sqrt{-5y+1}\right) \right) dy - \int_{(-1+\sqrt{5})/10}^{1/5} \left((4y+1) - \left(1 - y + \sqrt{-5y+1}\right) \right) dy$$

$$= \left(\frac{5}{2}y^2 - \frac{2}{15} \cdot (1 - 5y)^{3/2} \right) \Bigg|_{(-1-\sqrt{5})/10}^{1/5} - \left(\frac{5}{2}y^2 + \frac{2}{15} \cdot (1 - 5y)^{3/2} \right) \Bigg|_{(-1+\sqrt{5})/10}^{1/5} = \frac{\sqrt{5}}{6}$$

Calculator/Computer Exercises

For Problems 56–63 extra graphs are provided to demonstrate the intersection points graphically and with sufficient accuracy.

57.

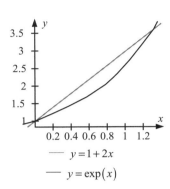

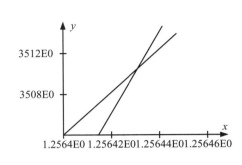

The abscissas are $a = 0$ and $b = 1.25642$.

The bounded area is $\displaystyle\int_0^{1.25642} (1 + 2x - \exp(x))\,dx = \left(x + x^2 - \exp(x)\right)\Big|_{x=0}^{x=1.25642} = 0.32219$.

59.

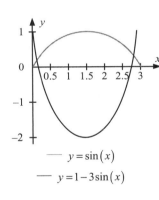

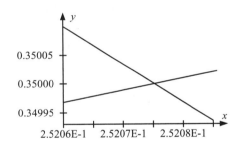

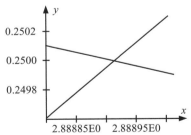

The abscissas are $a = 0.25268$ and $b = 2.88891$.

The bounded area is $\displaystyle\int_{0.25268}^{2.8889} \left(\sin(x) - (1 - 3\sin(x))\right)dx = \left(-4\cos(x) - x\right)\Big|_{x=0.25268}^{x=2.8889} = 5.1097$.

61.

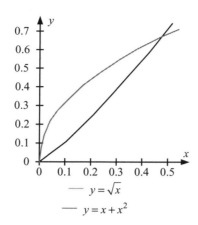

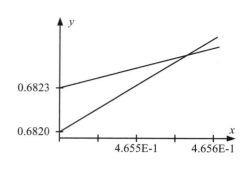

The abscissas are $a = 0$ and $b = 0.46557$.

The bounded area is $\displaystyle\int_0^{0.46557}\left(\sqrt{x}-\left(x+x^2\right)\right)dx = \left(\frac{2x^{3/2}}{3}-\frac{x^3}{3}-\frac{x^2}{2}\right)\Bigg|_{x=0}^{x=0.46557} = 0.069765$.

63.   From the plot, the smallest positive root is $b \approx 0.93257$

The area between the curves is $A = \displaystyle\int_0^b\left(\cos\left(x^2\right)-\sin^2\left(x\right)\right)dx \approx \sum_{j=1}^{10}\left(\left(\cos\left(x_j^2\right)-\sin^2\left(x_j\right)\right)\ x\right)$,

where $x = \dfrac{b}{10}$ and $x_j = \ x\cdot j$. The area is approximately 0.589.

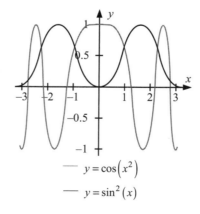

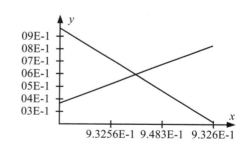

Problems for Practice

1.   The partition is $\Delta x = \dfrac{3}{2}$, $x_0 = 0$, $x_1 = \dfrac{3}{2}$, $x_2 = 3$, $\tilde{x}_1 = \dfrac{3}{4}$, $\tilde{x}_2 = \dfrac{9}{4}$.

Midpoint Rule: $M_2 = \Delta x \cdot \left(f(\tilde{x}_1) + f(\tilde{x}_2) \right) = \left(\dfrac{3}{2} \right) \left(\dfrac{9}{16} + \dfrac{81}{16} \right) = \dfrac{135}{16}$

Trapezoidal Rule: $T_2 = \left(\dfrac{\Delta x}{2} \right) \left(f(x_0) + 2f(x_1) + f(x_2) \right) = \left(\dfrac{3}{4} \right) \left(0 + 2 \cdot \dfrac{9}{4} + 9 \right) = \dfrac{81}{8}$

Simpson's Rule: $S_2 = \left(\dfrac{\Delta x}{3} \right) \left(f(x_0) + 4f(x_1) + f(x_2) \right) = \left(\dfrac{1}{2} \right) \left(0 + 4 \cdot \dfrac{9}{4} + 9 \right) = 9$

3.   The partition is $\Delta x = \dfrac{\pi}{4}$, $x_0 = \pi$, $x_1 = \dfrac{5\pi}{4}$, $x_2 = \dfrac{3\pi}{2}$, $x_3 = \dfrac{7\pi}{4}$, $x_4 = 2\pi$, $\tilde{x}_1 = \dfrac{9\pi}{8}$, $\tilde{x}_2 = \dfrac{11\pi}{8}$,

$\tilde{x}_3 = \dfrac{13\pi}{8}$, $\tilde{x}_4 = \dfrac{15\pi}{8}$.

Midpoint Rule: $M_4 = \Delta x \cdot \left(f(\tilde{x}_1) + f(\tilde{x}_2) + f(\tilde{x}_3) + f(\tilde{x}_4) \right)$

$\qquad = \left(\dfrac{\pi}{4} \right) \left(\sin\left(\dfrac{9\pi}{8} \right) + \sin\left(\dfrac{11\pi}{8} \right) + \sin\left(\dfrac{13\pi}{8} \right) + \sin\left(\dfrac{15\pi}{8} \right) \right)$

$\qquad = -\pi \dfrac{\sqrt{2-\sqrt{2}} + \sqrt{2+\sqrt{2}}}{4} \approx -2.0523$

Trapezoidal Rule: $T_4 = \left(\dfrac{\Delta x}{2} \right) \left(f(x_0) + 2f(x_1) + 2f(x_2) + 2f(x_3) + f(x_4) \right)$

$\qquad = \left(\dfrac{\pi}{8} \right) \left(\sin(\pi) + 2\sin\left(\dfrac{5\pi}{4} \right) + 2\sin\left(\dfrac{3\pi}{2} \right) + 2\sin\left(\dfrac{7\pi}{4} \right) + \sin(2\pi) \right)$

$\qquad = -\pi \dfrac{1+\sqrt{2}}{4} \approx -1.8961$

Simpson's Rule: $S_4 = \left(\dfrac{\Delta x}{3} \right) \left(f(x_0) + 4f(x_1) + 2f(x_2) + 4f(x_3) + f(x_4) \right)$

$\qquad = \left(\dfrac{\pi}{12} \right) \left(\sin(\pi) + 4\sin\left(\dfrac{5\pi}{4} \right) + 2\sin\left(\dfrac{3\pi}{2} \right) + 4\sin\left(\dfrac{7\pi}{4} \right) + \sin(2\pi) \right)$

$\qquad = -\pi \dfrac{1+2\sqrt{2}}{6} \approx -2.0046$

5.   The partition is $\Delta x = \dfrac{5}{4}$, $x_0 = 3$, $x_1 = \dfrac{17}{4}$, $x_2 = \dfrac{11}{2}$, $x_3 = \dfrac{27}{4}$, $x_4 = 8$, $\tilde{x}_1 = \dfrac{29}{8}$, $\tilde{x}_2 = \dfrac{39}{8}$,

$\tilde{x}_3 = \dfrac{49}{8}$, $\tilde{x}_4 = \dfrac{59}{8}$.

Midpoint Rule: $M_4 = \Delta x \cdot \left(f(\tilde{x}_1) + f(\tilde{x}_2) + f(\tilde{x}_3) + f(\tilde{x}_4) \right)$

$$= \left(\frac{5}{4}\right)\left(\sqrt{\frac{37}{8}} + \sqrt{\frac{47}{8}} + \sqrt{\frac{57}{8}} + \sqrt{\frac{67}{8}} \right) \approx 12.6721$$

Trapezoidal Rule: $T_4 = \left(\dfrac{\Delta x}{2}\right)\left(f(x_0) + 2f(x_1) + 2f(x_2) + 2f(x_3) + f(x_4) \right)$

$$= \left(\frac{5}{8}\right)\left(2 + \sqrt{21} + \sqrt{26} + \sqrt{31} + 3\right) \approx 12.6558$$

Simpson's Rule: $S_4 = \left(\dfrac{\Delta x}{3}\right)\left(f(x_0) + 4f(x_1) + 2f(x_2) + 4f(x_3) + f(x_4) \right)$

$$= \left(\frac{5}{12}\right)\left(2 + 2\sqrt{21} + \sqrt{26} + 2\sqrt{31} + 3\right) \approx 12.6665$$

7.   The exact integral is $\displaystyle\int_0^3 x^2\,dx = \left(\frac{1}{3}\right)x^3\Big|_0^3 = 9$, $f''(x) = 2$, $f^{(4)}(x) = 0$, $b - a = 3$.

Midpoint Rule: error $\dfrac{9}{16}$, error bound $\dfrac{2}{24}\cdot\dfrac{3^3}{2^2} = \dfrac{9}{16}$

Trapezoidal Rule: error $\dfrac{9}{8}$, error bound $\dfrac{2}{12}\cdot\dfrac{3^3}{2^2} = \dfrac{9}{8}$

Simpson's Rule: error 0, error bound 0

9.   The exact integral is $\displaystyle\int_\pi^{2\pi} \sin(x)\,dx = -\cos(x)\Big|_\pi^{2\pi} = -2$, $f''(x) = -\sin(x)$, $|f''(x)| \le 1$,

$f^{(4)}(x) = \sin(x)$, $\left|f^{(4)}(x)\right| \le 1$, $b - a = \pi$.

Midpoint Rule: error 0.0523 , error bound $\dfrac{1}{24}\cdot\dfrac{\pi^3}{4^2} = \dfrac{\pi^3}{384}$

Trapezoidal Rule: error 0.1039, error bound $\dfrac{1}{12}\cdot\dfrac{\pi^3}{4^2} = \dfrac{\pi^3}{192}$

Simpson's Rule: error 0.0046, error bound $\dfrac{1}{180}\cdot\dfrac{\pi^5}{4^4} = \dfrac{\pi^5}{2880}$

11. The exact integral is $\int_3^8 \sqrt{1+x}\,dx = \left(\frac{2}{3}\right)(1+x)^{3/2}\Big|_3^8 = \frac{38}{3}$, $f''(x) = -\left(\frac{1}{4}\right)(1+x)^{-3/2}$,

$\left|f''(x)\right| \le \frac{1}{32}$, $f^{(4)}(x) = -\left(\frac{15}{16}\right)(1+x)^{-7/2}$, $\left|f^{(4)}(x)\right| \le \frac{15}{2048}$, $b-a = 5$.

Midpoint Rule: error 0.0054, error bound $\dfrac{1}{32 \cdot 24} \cdot \dfrac{5^3}{4^2} = \dfrac{125}{12288}$

Trapezoidal Rule: error 0.0109, error bound $\dfrac{1}{32 \cdot 24} \cdot \dfrac{5^3}{4^2} = \dfrac{125}{6144}$

Simpson's Rule: error 0.0002, error bound $\dfrac{15}{2048 \cdot 180} \cdot \dfrac{5^5}{4^4} = \dfrac{46875}{94371840}$

13. The Lorenz function is $L(0) = 0$, $L(20) = 5$, $L(40) = 20$, $L(60) = 30$, $L(80) = 55$, $L(100) = 100$. The partition is $N = 5$, $\Delta x = 20$. Using the Trapezoidal Rule we obtain:

$$\int_0^{100} L(x)\,dx \approx \frac{20}{2}(0 + 2 \cdot 5 + 2 \cdot 20 + 2 \cdot 30 + 2 \cdot 55 + 100) = 3200 .$$

Therefore the coefficient of inequality for Country A is approximately $1 - \dfrac{3200}{5000} = 0.36$.

15. The partition is $N = 4$, $\Delta x = 25$. Using the Simpson's Rule we obtain

$$\int_0^{100} L(x)\,dx \approx \frac{25}{3}(0 + 4 \cdot 15 + 2 \cdot 25 + 4 \cdot 40 + 100) = \frac{9250}{3} \approx 3083.33 .$$

Therefore the coefficient of inequality for Country B is approximately $1 - \dfrac{3083.33}{5000} \approx 0.383$

Further Theory and Practice

17. To evaluate the accuracy of the Trapezoidal and Simpson's rules we compute

$f''(x) = \left(\frac{1}{4}\right)e^{-x/2}$, $\left|f''(x)\right| \le \frac{1}{4\sqrt{e}}$, $f^{(4)}(x) = \left(\frac{1}{16}\right)e^{-x/2}$, $\left|f^{(4)}(x)\right| \le \frac{1}{16\sqrt{e}}$, $b-a = 8$.

For the accuracy of the Trapezoidal Rule we obtain $\dfrac{1}{4\sqrt{e} \cdot 12} \cdot \dfrac{8^3}{N^2} \le 10^{-4}$, hence $N \ge 255$.

For the accuracy of Simpson's Rule we obtain: $\dfrac{1}{16\sqrt{e} \cdot 180} \cdot \dfrac{8^5}{N^4} \le 10^{-4}$, and because N must be even, then $N \ge 18$.

19.   According to Simpson's rule

$$\int_0^8 c(t)\,dt \approx \frac{1}{3}\left(0 + 4 \cdot 1.92 + 2 \cdot 5.74 + 4 \cdot 9.00 + 2 \cdot 10.24 + 4 \cdot 9.02 + 2 \cdot 5.78 + 4 \cdot 2.00 + 0\right)$$

$$\approx 43.76$$

The cardiac output taken after the injection of dye is approximately $\dfrac{5}{43.76} \approx 0.1143$ L/s or 6.86 L/min .

21.   If p is a polynomial of degree three or less, then $p^{(4)}(x) \equiv 0$, $C = 0$, and the error bound is 0. Therefore, Simpson's Rule for the integral approximation is exact.

23.   The tangent line for $y = \sqrt{1 - x^2}$ is vertical at $x = 1$, so y' is unbounded. The same is true for all other derivatives. In addition, the error of Simpson's Rule is unbounded. However, the approximating function is a parabola, and all of its derivatives are bounded. Therefore, parabolas are not very good for approximating functions with vertical tangents. The error bound is infinite, meaning that there is no bound. This reflects the fact that the approximation is not useful in this case.

25.   Denote $g(x) = f'''(x)$, then by the Mean Value Theorem there exists $c \in (a, b)$ such that $g(b) - g(a) = g'(c)(b - a)$. Hence

$$\left| g(b) - g(a) \right| \le (b - a) \max_{x \in [a,b]} g'(x) = (b - a) \max_{x \in [a,b]} f^{(4)}(x) = (b - a)C.$$

Using Chevilliet's form of the error of Simpson's rule we obtain

$$\frac{(b - a)^4}{180 N^4}\left(f'''(\xi) - f'''(\eta)\right) \le \frac{(b - a)^5}{180 N^4}C,$$

which is Stirling's form of the error.

27.   We evaluate the integral of the Lorentz function using trapezoidal approximation:

$$\int_0^{100} L(x)\,dx \approx \left(L(0) + L(16)\right)\left(\frac{16}{2}\right) + \left(L(16) + L(28)\right)\left(\frac{12}{2}\right) + \left(L(28) + L(51)\right)\left(\frac{23}{2}\right)$$

$$+ \left(L(51) + L(75)\right)\left(\frac{24}{2}\right) + \left(L(75) + L(88)\right)\left(\frac{13}{2}\right) + \left(L(88) + L(97)\right)\left(\frac{9}{2}\right)$$

$$+ \left(L(97) + L(100)\right)\left(\frac{3}{2}\right) = 2752$$

The coefficient of inequality is approximately is $1 - \dfrac{2752}{5000} = 0.4496$.

29.   Let S_S and S_N be the areas of southern and northern parts respectively.

$$S_S = 0.5(450 + 465)(1225 - 780) = 203587.5 \text{ km}^2.$$

We estimate S_N and $S = S_S + S_N$ using the following different numerical procedures. Following are estimates of the area of the northern portion of the province using:

Simpson's Rule:
$$S_N \approx 26(420 + 4 \cdot 435 + 2 \cdot 510 + 4 \cdot 530 + 2 \cdot 755 + 4 \cdot 730 + 2 \cdot 675 + 4 \cdot 630 + 2 \cdot 585$$
$$+ 4 \cdot 515 + 450) = 449280 \text{ km}^2$$

The total area is $S \approx 652867.5 \text{ km}^2$ with 0.45% error.

Trapezoidal Rule:
$$S_N \approx 39(420 + 2 \cdot 435 + 2 \cdot 510 + 2 \cdot 530 + 2 \cdot 755 + 2 \cdot 730 + 2 \cdot 675 + 2 \cdot 630 + 2 \cdot 585 + 2 \cdot 515 + 450)$$
$$= 452400 \text{ km}^2$$

The total area is $S \approx 655987.5 \text{ km}^2$ with 0.93% error.

Midpoint Rule: $S_N \approx 156(435 + 530 + 730 + 630 + 515) = 443040 \text{ km}^2$

The total area is $S \approx 646627.5 \text{ km}^2$ with 0.51% error.

31.   Using equation $S_N = \dfrac{1}{3}T_{N/2} + \dfrac{2}{3}M_{N/2}$, we find that $3S_N = T_{N/2} + 2M_{N/2}$, hence

$$S_N = T_{N/2} + 2M_{N/2} - 2S_N \text{ and } S_N - T_{N/2} = 2(M_{N/2} - S_N).$$

33.   Statement 1: We compute $M_1 = 4\sqrt{2} \approx 5.66$, $T_1 = 4$, $S_2 = 2\dfrac{4\sqrt{2} + 2}{3} \approx 5.1$ and $T_1 < S_2 < M_1$.

Statement 2: Using these values we obtain $\left|S_2 - T_1\right| = 8\dfrac{\sqrt{2} - 1}{3}$, $\left|S_2 - M_1\right| = 4\dfrac{\sqrt{2} - 1}{3}$, so

$\left|S_2 - T_1\right| = 2\left|S_2 - M_1\right|$.

35.   Statement 1: We compute $M_2 = \pi\sqrt{2}/2 \approx 2.22$, $T_2 = \pi/2 \approx 1.57$, $S_4 = \pi(2\sqrt{2} + 1)/6 \approx 2.005$ and $T_2 < S_4 < M_2$.

Statement 2: Using these values we obtain $\left|S_4 - T_2\right| = \pi\dfrac{\sqrt{2} - 1}{3}$, $\left|S_4 - M_2\right| = \pi\dfrac{\sqrt{2} - 1}{6}$, so

$\left|S_4 - T_2\right| = 2\left|S_4 - M_2\right|$.

37.   We denote $f_p(x) = \dfrac{x^p}{p^2}$, $0 \le x \le 1$, $p > 0$. Then $\displaystyle\int_0^1 f_p(x)\,dx = \dfrac{x^{p+1}}{p^2(p+1)}\bigg|_0^1 = \dfrac{1}{p^2(p+1)}$ and

$\displaystyle\lim_{p\to\infty}\int_0^1 f_p(x)\,dx = 0$.

After computing the derivatives $f''(x) = (p-1)\dfrac{x^{p-2}}{p}$, $f^{(4)}(x) = (p-1)(p-2)(p-3)\dfrac{x^{p-4}}{p}$, we

obtain for $p > 4$ and $0 \le x \le 1$ $\left|f''(x)\right| \le \dfrac{p-1}{p} = M_p$ and $\left|f^{(4)}(x)\right| \le (p-1)(p-2)\dfrac{p-3}{p} = N_p$

$\displaystyle\lim_{p\to\infty} M_p = 1$, $\displaystyle\lim_{p\to\infty} N_p = \infty$.

The error bounds are as follows:

Midpoint Rule: $\dfrac{M_p}{24N^2} \to \dfrac{1}{24N^2}$ as $p \to \infty$

Trapezoidal Rule: $\dfrac{M_p}{12N^2} \to \dfrac{1}{12N^2}$ as $p \to \infty$

Simpson's Rule: $\dfrac{N_p}{180N^4} \to \infty$ as $p \to \infty$

The error bound for Simpson's Rule is very conservative; it grows unlimited at large p.

Calculator/Computer Exercises

39.   The upper bound on the second derivative is
$\left|f''(x)\right| < 1.49$ (see figure). We set $C = 1.49$, and to
guarantee three-decimal places accuracy the condition

on N is $1.49 \cdot \dfrac{3.75^3}{24N^2} < 5 \cdot 10^{-4}$, or $N \ge 115$.

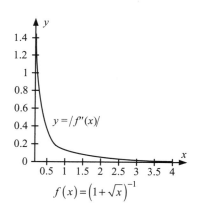

$f(x) = \left(1 + \sqrt{x}\right)^{-1}$

41. The upper bound on the second derivative is $\left|f^{(4)}(x)\right|<0.282$ (see the figure). We set $C=0.3$, then N should be even and to guarantee three-decimal places accuracy the condition on N is

$$0.282\cdot\frac{8^5}{180\cdot N^4}<5\cdot10^{-4},\text{ or }N\geq18.$$

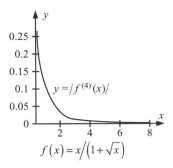

$$f(x)=x/(1+\sqrt{x})$$

43. We denote $f(x)=\sqrt{1-x^2}-\dfrac{\sqrt{3}}{4}$, and find an upper bound for its fourth derivative $\left|f^{(4)}(x)\right|<16.5$ (see figure). For $C=16.5$ we compute

$$16.5\cdot\frac{0.5^5}{180\cdot N^4}<5\cdot\frac{10^{-6}}{12},$$

or $N\geq10$. Applying Simpson's Rule to this formula, we solve to find $\int\limits_0^{1/2}f(x)\,dx\approx0.261799$. Then π can be approximated as $\pi\approx0.261799\cdot12=3.14159$.

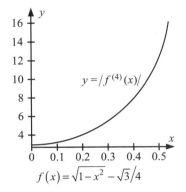

$$f(x)=\sqrt{1-x^2}-\sqrt{3}/4$$

45. Exact value of the integral is $\int\limits_1^e x^{-1}dx=\ln(x)\big|_1^e=1$.

Simpson's Rule approximation of order N is $\int\limits_1^e x^{-1}dx\approx1.00081$.

The absolute error is $\epsilon=0.000181$.

From the equation $\epsilon=(e-1)^5\dfrac{\left|f^{(4)}(c)\right|}{180N^4}$ we obtain $c=1.57303$.

47. Exact value of the integral is $\int\limits_8^{15}\left(\dfrac{1}{\sqrt{1+x}}\right)dx=2\sqrt{1+x}\,\Big|_8^{15}=2$.

Simpson's Rule approximation of order N is $\int\limits_8^{15}\left(\dfrac{1}{\sqrt{1+x}}\right)dx\approx2.000036$.

Absolute error is $\epsilon=0.000036$.

From the equation $\epsilon=7^5\dfrac{\left|f^{(4)}(c)\right|}{180N^4}$ we obtain $c=10.79562$.

49. For $\mu = 80$ (mean deviation) and $\sigma = 20$ (standard deviation) we determine the probability as

$$P(60,\ 100) = \frac{1}{20\sqrt{2\pi}} \int_{60}^{100} \exp\left(-\frac{1}{2}\left(\frac{x-80}{20}\right)^2\right) dx.$$ Using Simpson's Rule, we approximate this

probability as $P(60,\ 100) \approx 0.6828$.

51. **a.**   Using Simpson's Rule to approximate gives $F(46) \approx 364.998$ (using $N = 50$).

The exact value is 365.

b.   Let $F(36) \approx 357$. There are about 8 days a year when heat alerts are issued.

c.   Let $\displaystyle\int_{-25}^{36} 12.66\exp\left(-\frac{(x-16)^2}{265.8}\right) dx \approx 351$. There will be about 14 heat alert days per

year, which is equivalent to a 75% increase.

Differential Equations and Transcendental Functions

Section 6.1 **First Order Differential Equations**

Problems for Practice

1. The left-hand side of the equation is $\frac{dy}{dx} = Ce^{x^2/2}x$; the right-hand side of the equation is

$xy = Cxe^{x^2/2}$. They are equal.

3. The left-hand side of the equation is $\frac{dy}{dx} = \frac{1}{3} - 3Ce^{-3x}$; the right-hand side of the equation is

$x - 3y = x - x + \frac{1}{3} - 3e^{-3x} = \frac{1}{3} - 3e^{-3x}$. They are equal.

5. The left-hand side of the equation is $\frac{dy}{dx} = Ce^x - 1$; the right-hand side of the equation is

$x + y = x + Ce^x - x - 1 = Ce^x - 1$. They are equal.

7. The left-hand side of the equation is $\frac{dy}{dx} = Ce^x - 2x - 2$; the right-hand side of the equation is

$y + x^2 = Ce^x - x^2 - 2x - 2 + x^2 = Ce^x - 2x - 2$. They are equal.

9. As $F(x,y) = 2x \cdot 1$, this is a separable differential equation. Then $\frac{dy}{dx} = 2x$, $dy = 2xdx$ and

$\int dy = \int 2xdx$ or $y = x^2 + C$. As $y(1) = 1 + C = 3$, we have $C = 2$ and $y = x^2 + 2$.

11. We rewrite the equation in the form $\frac{dy}{dx} = \cos(x)$, to obtain $dy = \cos(x)dx$, $\int dy = \int \cos(x)dx$

and $y = \sin(x) + C$. As $y(0) = C = 2$, we have $y(x) = \sin(x) + 2$.

13. We rewrite the equation in the form $\frac{dy}{dx} = \frac{x}{y}$, to obtain $ydy = xdx$, $\int ydy = \int xdx$ and

$\frac{y^2}{2} = \frac{x^2}{2} + C$. Therefore, $C = \frac{1}{2}$ and $y^2 = x^2 + 1$. As $y(0) > 0$, we have $y(x) = \sqrt{x^2 + 1}$.

15.   We rewrite the equation in the form $\frac{dy}{dx} = y^2 \sin(x)$, to obtain $\frac{dy}{y^2} = \sin(x)\,dx$,

$\int y^{-2} dy = \int \sin(x)\,dx$ and $-\frac{1}{y} = -\cos(x) + C$. From the initial condition we have

$-\frac{1}{2} = -1 + C$, $C = \frac{1}{2}$ and $y = \frac{2}{2\cos(x) - 1}$.

17.   We rewrite the equation in the form $y\frac{dy}{dx} = \frac{x}{\sqrt{1+y^2}}$, to obtain $y\sqrt{1+y^2}\,dy = xdx$,

$\int y\sqrt{1+y^2}\,dy = \int xdx$ and $\frac{1}{3}\left(1+y^2\right)^{3/2} = \frac{x^2}{2} + C$. From the initial condition we see that

$C = \frac{2^{3/2}}{3}$ and $2\left(1+y^2\right)^{3/2} = 3x^2 + 4\sqrt{2}$.

19.   We rewrite the equation in the form $\frac{dy}{dx} = y + \frac{1}{y}$, to obtain $\frac{ydy}{y^2 + 1} = dx$, $\int\left(\frac{y}{y^2+1}\right)dy = \int dx$

and $\left(\frac{1}{2}\right)\ln\left(y^2 + 1\right) = x + C$. From the initial condition we see that $C = \frac{\ln(10)}{2}$ and

$y^2 + 1 = 10e^{2x}$. As $y(0) > 0$, we solve to find $y(x) = \sqrt{10e^{2x} - 1}$.

21.   We rewrite the equation in the form $x^2\frac{dy}{dx} = y\sin\left(\frac{1}{x}\right)$, to obtain $\frac{dy}{y} = \left(\frac{\sin(1/x)}{x^2}\right)dx$,

$\int y^{-1} dy = \int \sin\left(\frac{1}{x}\right)x^{-2}dx$, $\ln(y) = \cos\left(\frac{1}{x}\right) + C$. The initial condition leads to $\ln(3) = C$ and

$y(x) = 3\exp\left(\cos\left(\frac{1}{x}\right)\right)$.

23.   We rewrite the equation in the form $\frac{dy}{dx} - xy^2 = xy(4x - y)$, to obtain $\frac{dy}{y} = 4x^2 dx$,

$\int y^{-1} dy = 4\int x^2 dx$ and $\ln(y) = \frac{4x^3}{3} + C$. From the initial condition: $C = \ln(2)$ and

$y(x) = 2\exp\left(\frac{4x^3}{3}\right)$.

Further Theory and Practice

25.   From the Chain Rule we find $y(x) = g(f(x)) + C$. For $x = 0$, $g(f(0)) = g(3) = 2$. Moreover, because $y(0) = 6$, we see that $C = 4$ and $y(x) = g(f(x)) + 4$.

27. Let $\frac{d}{dx}P(x) = p(x)$. Then $\frac{d}{dx}\left(e^{P(x)}y(x)\right) = e^{P(x)}p(x)y(x) + y'(x)e^{P(x)}$. This differential

equation is equivalent to the equation $e^{-P(x)}\frac{d}{dx}\left(e^{P(x)}y(x)\right) = q(x)$, or

$$\frac{d}{dx}\left(e^{P(x)}y(x)\right) = e^{P(x)}q(x).$$

Hence $e^{P(x)}y(x) = \int e^{P(x)}q(x)dx + C$ or $y(x) = e^{-P(x)}\left(\int e^{P(x)}q(x)dx + C\right)$.

For the particular equation $\frac{dy}{dx} = e^{-x} - y$ we see that $p(x) = 1$ and $q(x) = e^{-x}$. Then

$P(x) = x$, $\int e^{P(x)}q(x)dx = \int e^{x}e^{-x}dx = x$ and $y(x) = e^{-x}(x + C)$.

29. If $w(x) \cdot x = y(x)$ and $\frac{dy}{dx} = \phi(y/x)$, we have $\frac{dw}{dx}x + w = \phi(w)$. We can rewrite this in the

form: $\frac{dw}{dx} = (\phi(w) - w)x^{-1}$. This is a separable differential equation. For the given equation

$\frac{dy}{dx} = \frac{xy}{x^2 + y^2}$, we have $\phi(w) = \frac{w}{1 + w^2}$ and $-\left(1 + w^2\right)\frac{dw}{w^3} = \frac{dx}{x}$. Then $\frac{1}{2w^2} - \ln(w) + C = \ln(x)$,

or $\frac{1}{2w^2} + C = \ln(wx)$. Therefore, $2\ln(y) = \left(\frac{x}{y}\right)^2 + C$.

31. Let $PdP = CT^{7.5}dT$, $\frac{P^2}{2} = \frac{CT^{8.5}}{8.5} + C_1$. From the initial condition $C_1 = 0$ and

$$P(T) = 2\sqrt{\frac{C}{17}}T^{17/4}.$$

33. Assuming that $\rho(r) = \rho$, $m(r) = 4\pi r^3 \frac{\rho}{3}$ and $\frac{dP}{dr} = -4\pi G\rho^2\frac{r}{3}$. Then $dP = -\left(4\pi G\frac{\rho^2}{3}\right)rdr$

and $P = -2\pi G\rho^2\frac{r^2}{3} + C$. Given that $P(R) = 0$, we see that $C = 2\pi G\rho^2\frac{R^2}{3}$ and

$$P(r) = 2G\pi\rho^2\frac{R^2 - r^2}{3}.$$

Then $P(0) = 2\pi G\rho^2\frac{R^2}{3}$, or, taking into account that $M = 4\pi R^3\frac{\rho}{3}$,

$$P(0) = \left(\frac{\pi}{6}\right)^{1/3}GM^{2/3}\rho^{4/3}.$$

35. The velocity $v(t)$ satisfies equation $v'(t) = -12\sqrt{v(t)}$, or $\int v^{-1/2}dv = -12\int dt$. Solving this equation, we obtain $2\sqrt{v} = C - 12t$, or $v(t) = (C - 6t)^2$. As $v(0) = 1250$, we get

$$v(t) = \left(\sqrt{1250} - 6t\right)^2.$$

The bullet's velocity is 0 (at rest) when $\sqrt{1250} - 6t = 0$, or $t = 25\dfrac{\sqrt{2}}{6}$ s.

37. Let $y = [C_{12}H_{22}O_{11}]$. Then $\frac{dy}{dt} = -5.7\times 10^{-5}y$ and $y = Ce^{-5.7\times 10^{-5}t}$. We need to solve the equation $Ce^{-5.7\times 10^{-5}t} = \dfrac{C}{3}$. The value of t for which the sucrose concentration is equal to one-third its initial value is $t = \dfrac{\ln(3)}{5.7\times 10^{-5}} \approx 19273.9$.

39. Let $y = [N_2O_5]$. Then $\frac{dy}{dt} = -ky$ and $y = Ce^{-kt}$. The system:

$$\begin{cases} y(0) = 2.32 \\ y(3000) = 0.37 \end{cases}$$

leads to $C = 2.32$ and

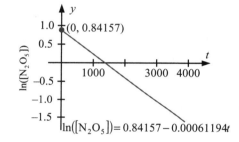

$$k = -\dfrac{\ln(0.37/2.32)}{3000} = 6.1194\times 10^{-4}.$$

Therefore, $y = 2.32\exp(-0.00061194t)$; coefficient k is the negative of the slope of the line $\ln([N_2O_5]) = 0.84157 - 0.00061194t$.

41. From $\frac{df}{dx} = -xf(x)$ we see that $\int \left(\dfrac{1}{f}\right)df = -\int x\,dx$, or $\ln(f(x)) = -\dfrac{x^2}{2} + C_1$ and $f(x) = Ce^{-x^2/2}$, $C = e^{C_1}$. As $f(0) = \dfrac{1}{\sqrt{2\pi}}$, $C = \dfrac{1}{\sqrt{2\pi}}$ and $f(x) = \left(\dfrac{1}{\sqrt{2\pi}}\right)\exp\left(-\dfrac{x^2}{2}\right)$.

43.   It follows from $\frac{dy}{dt} = \left(\frac{\lambda}{p}\right)t^{p-1}(A-y)$ that $\frac{dy}{A-y} = \left(\frac{\lambda}{p}\right)t^{p-1}dt$ and $-\ln(A-y) = \frac{\lambda t^{p}}{p^{2}} + C_{1}$.

We may rewrite this as follows: $y(t) = A + C \cdot \exp\left(-\frac{\lambda t^{p}}{p^{2}}\right)$, $C = -e^{-C_{1}}$.

The limiting size of the species is $\lim\limits_{t \to \infty} y(t) = A$, hence it is called "the size of the adult."

45.   Gompertz's assumption leads to the equation $-\dfrac{L'(x)}{L(x)} = Bg^{x}$, or $-\frac{d}{dx}\left(\ln\left(L(x)\right)\right) = Bg^{x}$. Then

$-\ln\left(L(x)\right) = \dfrac{Bg^{x}}{\ln(g)} + C_{1}$, or $L(x) = \dfrac{C}{\exp\left(Bg^{x}/\ln(g)\right)}$.

47.   Starting with $\frac{dT}{dt} = -\dfrac{kT}{\sqrt{t}}$, $\dfrac{dT}{T} = -\dfrac{kdt}{\sqrt{t}}$, we obtain $\ln(T) = -2k\sqrt{t} + \tilde{C}$. From the initial

condition we solve to find $\ln(\tau) = \tilde{C}$ and $T(t) = \tau \cdot \exp\left(-2k\sqrt{t}\right)$.

49.   From $\frac{da}{dx} = -\dfrac{\gamma a}{x + x_{0}}$, $\dfrac{da}{a} = -\dfrac{\gamma dx}{x + x_{0}}$, we obtain

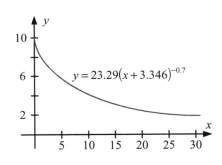

$$\ln(a) = -\gamma\ln(x + x_{0}) + C_{1}$$

or

$$a(x) = C(x + x_{0})^{-\gamma}.$$

As $a(30) = 2$, for the given values we have the system:

$$\begin{cases} Cx_{0}^{-0.7} = 10 \\ C(30 + x_{0})^{-0.7} = 2. \end{cases}$$

Then $C = 23.29$ and $x_{0} = 3.346$. We obtain the following function for the horn:

$$a(x) = 23.29(x + 3.346)^{-0.7}.$$

51.   The differential equation, $y'(x) + 2xy = 1$ is linear. Its solution is $y(x) = e^{-x^2}\left(\int e^{x^2} dx + C\right)$, or

$$y(x) = \exp\left(-x^2\right)\left(\int_0^x \exp\left(t^2\right) dt + C\right). \text{ As } y(0) = 0, \text{ we find that } C = 0 \text{ and}$$

$$y(x) = \exp\left(-x^2\right)\int_0^x \exp\left(t^2\right) dt, \quad y(0) = 0$$

(verifying Dawson's integral).

Calculator/Computer Exercises

53.   From $\frac{dy}{dx} = \frac{1+x^2}{1+y^4}$ we obtain $\left(1+y^4\right) dy = \left(1+x^2\right) dx$ and $\frac{y^{(5)}(x)}{5} + y(x) = x + \frac{x^3}{3} + C$. As

$y(0) = 0$, we see that $C = 0$. The resulting equation is $\frac{y^{(5)}(x)}{5} + y(x) = x + \frac{x^3}{3}$.

The value at $y(2)$ is $y(2) = 1.713385$.

55.   From $y^2 \frac{dy}{dx} = \frac{1+y}{1+x}$ we obtain $\left(\frac{y^2}{1+y}\right) dy = \frac{dx}{1+x}$. To calculate the integral over y, we use the

decomposition $\frac{y^2}{1+y} = y - 1 + \frac{1}{1+y}$. Then $\int \frac{y^2}{1+y} dy = \int\left(y - 1 + \frac{1}{1+y}\right) dy = \frac{y^2}{2} - y + \ln(1+y)$,

$\int\left(\frac{1}{1+x}\right) dx = \ln(1+x) + C$, $\frac{y^{(2)}(x)}{2} - y(x) + \ln(1+y(x)) = \ln(x+1) + C$. Since $y(0) = 0$,

we see that $C = 0$, and the resulting equation is $\frac{y^{(2)}(x)}{2} - y(x) + \ln(1+y(x)) = \ln(x+1)$.

The value at $y(2)$ is $y(2) = 2.0$.

57.   Let $\int \dfrac{1}{5000-y}\,dy = 0.09 \int \sqrt{t}\,dt$. Therefore, $-\ln\left(5000-y\right) = 0.06t^{3/2} + C_1$, or

$$y(t) = 5000 - C\exp\left(-0.06t^{3/2}\right)$$

where $C = \exp\left(-C_1\right)$ is a positive constant that can be determined from the condition

$$y(0) = 175 .$$

We see that $C = 5000 - 175 = 4825$ and $y(t) = 5000 - 4825\exp\left(-0.06t^{3/2}\right)$. The mature weight of the goose is $\lim_{t\to\infty} y(t) = 5000$. We solve the equation

$$5000 - 4825\exp\left(-0.06t^{3/2}\right) = \left(\dfrac{1}{2}\right)5000$$

to find that the goose is half its mature weight after 4.9337 weeks.

Section 6.2   A Calculus Approach to the Logarithm

Problems for Practice

1. $\left(\ln\left(3x\right)\right)' = \dfrac{1}{3x}\cdot 3 = \dfrac{1}{x}$

3. $\left(\ln\left(5x+2\right)\right)' = \dfrac{1}{5x+2}\cdot 5 = \dfrac{5}{5x+2}$

5. $\left(\ln\left(\sin\left(x\right)\right)\right)' = \dfrac{1}{\sin\left(x\right)}\cdot\cos\left(x\right) = \cot\left(x\right)$

7. $\left(\ln\left(\dfrac{1}{x}\right)\right)' = x\cdot\left(-\dfrac{1}{x^2}\right) = -\dfrac{1}{x}$

9. $\left(\tan\left(2x\right)\cdot\ln\left(x\right)\right)' = \sec^2\left(2x\right)\cdot 2\cdot\ln\left(x\right) + \tan\left(2x\right)\cdot\dfrac{1}{x} = 2\sec^2\left(2x\right)\ln\left(x\right) + \dfrac{\tan\left(2x\right)}{x}$

11. $\left(\ln^3\left(x\right)\right)' = 3\ln^2\left(x\right)\cdot\dfrac{1}{x} = \dfrac{3\ln^2\left(x\right)}{x}$

13. $u = x+1$, $du = dx$

$$\int \frac{1}{x+1}dx = \int \frac{1}{u}du = \ln|u| + C = \ln|x+1| + C$$

15. $u = x^2 +4$, $du = 2xdx$

$$\int \frac{x}{x^2+4}dx = \frac{1}{2}\int \frac{1}{u}du = \frac{1}{2}\ln|u| + C = \frac{1}{2}\ln\left|x^2 +4\right| + C$$

17. $u = \ln(x)$, $du = \dfrac{dx}{x}$

$$\int \frac{1}{x\ln(x)}dx = \int \frac{1}{u}du = \ln|u| + C = \ln\left|\ln(x)\right| + C$$

19. $u = \ln(x)$, $du = \dfrac{dx}{x}$

$$\int \frac{\sqrt{\ln(x)}}{x}dx = \int \sqrt{u}\,du = \frac{2}{3}\sqrt{u^3} + C = \frac{2}{3}\sqrt{\ln^3(x)} + C$$

21. $u = \tan(x)$, $du = \sec^2(x)dx$

$$\int \frac{\sec^2(x)}{\tan(x)}dx = \int \frac{1}{u}du = \ln|u| + C = \ln\left|\tan(x)\right| + C$$

23. $A = 3$, $B = -2$, $C = -2$, $D = -3$, $E = \left(\dfrac{1}{2}\right)\ln(2)$

25. $\ln(\sec(x)) + \ln(\cos(x)) = \ln(\sec(x) \cdot \cos(x)) = \ln(1) = 0$

Further Theory and Practice

27.   The formula for the mortality rate associated with several cancers may be written in the form: $\ln(R(t)) = \ln(B) + k \cdot \ln(t)$.

The plot may be described as a straight line with slope k.

29. $\displaystyle\int_0^1 \frac{2x+3}{x+1}dx = \int_0^1 \frac{2x+2+1}{x+1}dx = \int_0^1 \left(2 + \frac{1}{x+1}\right)dx = \left(2x + \ln(x+1)\right)\Big|_0^1 = 2 + \ln(2)$

31. $\int \dfrac{1}{\sqrt{x}\left(2+\sqrt{x}\right)}dx = 2\int \dfrac{1}{2+\sqrt{x}}d\left(2+\sqrt{x}\right) = 2\ln\left|2+\sqrt{x}\right| + C$

33. $\displaystyle\int_{1}^{e}\left(\ln(x)+\ln\left(\dfrac{1}{x}\right)\right)dx = \int_{1}^{e}\ln\left(x\cdot\dfrac{1}{x}\right)dx = \int_{1}^{e}\ln(1)\,dx = \int_{1}^{e}0\,dx = 0$

35. Domain: $(0,\ \infty)$

Increasing on the interval: $(1,\ \infty)$

Decreasing on the interval: $(0,\ 1)$

Concave up on the interval: $(0,\ \infty)$

Concave down on the interval: Nowhere

Inflection points: None

Vertical asymptote: $x=0$

Horizontal asymptote: None

Skew asymptote: None

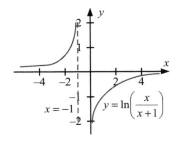

37. Domain: $(-\infty,\ -1)\cup(0,\ \infty)$

Increasing on the intervals: $(-\infty,\ -1)\cup(0,\ \infty)$

Decreasing on the interval: Nowhere

Concave up on the interval: $(-\infty,\ -1)$

Concave down on the interval: $(0,\ \infty)$

Inflection points: None

Vertical asymptotes: $x=-1$, $x=0$

Horizontal asymptote: None

Skew asymptote: None

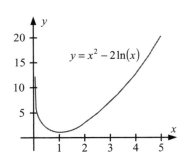

39. The derivative is $\dfrac{dy}{dx} = \dfrac{x-xy-y^2}{x\left((x+y)\ln(x)-1\right)}$. When $x=1$ and $y=0$, $\dfrac{dy}{dx} = -1$.

41. $g(-x) = \ln\left(\dfrac{A-f(-x)}{A+f(-x)}\right) = \ln\left(\dfrac{A+f(x)}{A-f(x)}\right) = -\ln\left(\dfrac{A-f(x)}{A+f(x)}\right) = -g(x)$

43. $g'(x) = \dfrac{f'(x)}{f(x)}$

$g'(2.90) \approx \dfrac{1}{f(2.90)}\cdot\dfrac{f(2.95)-f(2.85)}{0.1} = \dfrac{1}{0.1}\cdot\dfrac{0.5-(-0.1)}{0.1} = \dfrac{0.6}{0.01} = 60$

45. The derivative is $f'(x) = 1 - \dfrac{1}{x} = \dfrac{x-1}{x}$. We observe that $f'(x) = 0$ when $x = 1$, $f'(x) < 0$ when $x < 1$ and $f'(x) > 0$ when $x > 1$. Therefore, $x = 1$ is a minimum. The minimum value is $m = f(1) = 1$.

We deduce $x - \ln(x) \ge 1$ or $\ln(x) \le x - 1$. So $\ln(x+1) \le x$ when $x > -1$ and $\ln(x) \le x - 1 < x$ when $x > 0$. Since $\ln(x)$ is only defined for $x > 0$, it follows that there is no x for which $\ln(x) = x$.

47.
$$y' = -A\sin(\ln(x)) \cdot \frac{1}{x} + B\cos(\ln(x)) \cdot \frac{1}{x} = \frac{1}{x}\left(B\cos(\ln(x)) - A\sin(\ln(x))\right)$$
$$y'' = -\frac{1}{x^2}\left(B\cos(\ln(x)) - A\sin(\ln(x))\right) + \frac{1}{x}\left(-B\sin(\ln(x)) \cdot \frac{1}{x} - A\cos(\ln(x))\frac{1}{x}\right)$$
$$= \frac{1}{x^2}\left((A-B)\sin(\ln(x)) - (A+B)\cos(\ln(x))\right)$$
$$x^2 y'' + xy' + y = (A-B)\sin(\ln(x)) - (A+B)\cos(\ln(x)) + B\cos(\ln(x)) - A\sin(\ln(x))$$
$$+ A\cos(\ln(x)) + B\sin(\ln(x)) = 0$$

49. We start by performing the integration
$$\int \ln(x)\,dx = \ln(x) \cdot x - \int x \cdot d(\ln(x)) = x\ln(x) - \int x \cdot \frac{1}{x}\,dx = x\ln(x) - \int dx = x\ln(x) - x + C.$$

By comparison to the original formula we see that $A = 1$ and $B = -1$.

51. From the initial inequality, we find $0 \le \left(\sqrt{t} - 1\right)^2 = t - 2\sqrt{t} + 1$, therefore $2\sqrt{t} \le t + 1$. From the above calculation, we deduce that $\dfrac{1}{t} \le \dfrac{1}{2}\dfrac{t+1}{t^{3/2}} = \dfrac{1}{2\sqrt{t}} + \dfrac{1}{2t\sqrt{t}}$.

This verifies the inequality $\dfrac{d}{dt}(\ln(t)) \le \dfrac{d}{dt}\left(\sqrt{t} - \dfrac{1}{\sqrt{t}}\right)$.

Using the last inequality rewritten in the form of integrals, we obtain
$$\int_1^x \frac{1}{t}\,dt = \ln(x) \le \int_1^x \left(\frac{1}{2\sqrt{t}} + \frac{1}{2t\sqrt{t}}\right)dt = \sqrt{x} - \frac{1}{\sqrt{x}}.$$

53.

a. $f'(xy) \cdot y = f'(x) + 0 = f'(x)$

b. $f'(1) \cdot \dfrac{1}{t} = f'(t)$

c. $f(t) = f'(1) \cdot \ln(t) + C$, $f(1) = f'(1) \cdot \ln(1) + C = 0 + C = C$

d. We can now deduce that $f(1) = f(1 \cdot 1) = f(1) + f(1)$ hence $f(1) = 0$. Therefore $C = 0$ and $f(x) = f'(1) \cdot \ln(x)$.

55.

a. Let $t = \dfrac{y}{x}$. Substituting this into the logarithm and square root formulas, we deduce:

$$\ln(t) = \ln\left(\frac{y}{x}\right) = \ln(y) - \ln(x)$$

$$\sqrt{t} - \frac{1}{\sqrt{t}} = \sqrt{\frac{y}{x}} - \sqrt{\frac{x}{y}} = \frac{y-x}{\sqrt{xy}}$$

$$\ln(y) - \ln(x) \le \frac{y-x}{\sqrt{xy}}$$

$$\sqrt{xy} \le \frac{y-x}{\ln(y) - \ln(x)}$$

b. Using the inequality from Exercise 52, we deduce:

$$2\frac{y}{x} \le \left(\frac{y}{x} + 1\right)\ln\left(\frac{y}{x}\right) + 2$$

$$2y \le (y+x)(\ln(y) - \ln(x)) + 2x$$

$$\frac{y-x}{\ln(y) - \ln(x)} \le \frac{y+x}{2}$$

Calculator/Computer Exercises

57. Use Simpson's Rule from Chapter 5 $\ln(2) = \displaystyle\int_1^2 \frac{1}{x}\,dx = \int_1^2 f(x)\,dx$; $f(x) = \dfrac{1}{x}$.

$x_0 = 1.0; x_1 = 1.16667; x_2 = 1.33333; x_3 = 1.5; x_4 = 1.66667; x_5 = 1.83333; x_6 = 2.0$; $f(x_0) = 1.0$;

$f(x_1) = 0.85714$; $f(x_2) = 0.75$; $f(x_3) = 0.66667$; $f(x_4) = 0.58823$; $f(x_5) = 0.6$;

$f(x_6) = 0.5$; $\Delta x = 0.16667$;

$$\int_1^2 \frac{1}{x}\,dx \approx \frac{\Delta x}{3}\left(f(x_0) + 4f(x_1) + 2f(x_2) + 4f(x_3) + 2f(x_4) + 4f(x_5) + f(x_6)\right) = 0.69317$$

59.

x	$x \ln(x)$
0.1	-0.230259
0.01	-0.046051
0.001	-0.006907
0.0001	-0.000921
0.00001	-0.000115
0.000001	-0.000013

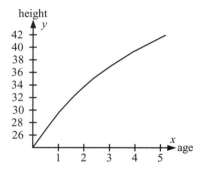

The limit appears to be 0.

61.   As we see on the graph, $H(3.75) = 40$, the typical child reaches 40 inches at age 3.75 yrs.

63. $f(x) = \dfrac{1}{\ln(x)}$

$x_0 = 50000$; $x_1 = 52500$; $x_2 = 55000$; $x_3 = 57500$; $x_4 = 60000$; $f(x_0) = 0.09242$; $f(x_1) = 0.09200$; $f(x_2) = 0.09161$; $f(x_3) = 0.09124$; $f(x_4) = 0.09089$; $\Delta x = 2500$;

$$\int_{50000}^{60000} f(x)\,dx \approx \frac{\Delta x}{3}\left(f(x_0) + 4f(x_1) + 2f(x_2) + 4f(x_3) + f(x_4)\right) = 916.3$$

Using Simpson's Rule, there are 916 primes that lie in the interval [50000, 60000].

Section 6.3   The Exponential Function

Problems for Practice

1. $\frac{d}{dx}\exp(-x) = \frac{d}{dx}\exp(u) = \exp(u)\frac{du}{dx} = \exp(-x)\frac{d}{dx}(-x) = -\exp(-x)$

3. $\frac{d}{dx}\left(\sin(x)\exp(x)\right) = \exp(x)\frac{d}{dx}\sin(x) + \sin(x)\frac{d}{dx}\exp(x) = \exp(x)\cos(x) + \sin(x)\exp(x)$
$$= \exp(x)\left(\cos(x) + \sin(x)\right)$$

5. $\frac{d}{dx}\exp\left(\sin(x)\right) = \frac{d}{dx}\exp(u) = \exp(u)\frac{du}{dx} = \exp\left(\sin(x)\right)\frac{d}{dx}\sin(x) = \cos(x)\exp\left(\sin(x)\right)$

7. $\frac{d}{dx}\ln\left(1+\exp(x)\right)=\frac{d}{dx}\ln\left(u\right)=\left(\frac{1}{u}\right)\frac{du}{dx}=\left(\frac{1}{1+\exp(x)}\right)\frac{d}{dx}\left(1+\exp(x)\right)=\frac{\exp(x)}{1+\exp(x)}$

9.   As of now:

$$\frac{d}{dx}\exp\left(\exp(x)\right)=\frac{d}{dx}\exp(u)=\exp(u)\frac{du}{dx}=\exp\left(\exp(x)\right)\frac{d}{dx}\exp(x)=\exp\left(\exp(x)\right)\cdot\exp(x).$$

11. $\frac{d}{dx}\exp\left(1+\ln(x)\right)=\frac{d}{dx}\left(e\cdot\exp\left(\ln(x)\right)\right)=e\frac{dx}{dx}=e$

13.   Let $u=-x$. Then $du=-dx$ and $\int\exp(-x)\,dx=-\int\exp(u)\,du=-\exp(u)+C=-\exp(-x)+C$.

15.   Let $u=\pi x$; $v=-\pi x$. Then $dx=\dfrac{du}{\pi}=-\dfrac{dv}{\pi}$ and

$$\int\left(\exp(\pi x)-\exp(-\pi x)\right)dx=\int\exp(\pi x)\,dx-\int\exp(-\pi x)\,dx$$
$$=\left(\frac{1}{\pi}\right)\int\exp(u)\,du+\left(\frac{1}{\pi}\right)\int\exp(v)\,dv$$
$$=\frac{\exp(\pi x)+\exp(-\pi x)}{\pi}+C.$$

17.   Let $u=1+\exp(x)$. Then $du=\exp(x)\,dx$ and

$$\int\left(\frac{\exp(x)}{1+\exp(x)}\right)dx=\int\left(\frac{1}{u}\right)du=\ln(u)+C=\ln\left(1+\exp(x)\right)+C.$$

19.   Let $u=1+\exp(x)$. Then $du=\exp(x)\,dx$ and

$$\int\exp(x)\cdot\sqrt{1+\exp(x)}\,dx=\int\sqrt{u}\,du=\left(\frac{2}{3}\right)u^{3/2}+C=\left(\frac{2}{3}\right)\left(1+\exp(x)\right)^{3/2}+C.$$

21.   Let $u=\exp(x)$. Then $du=\exp(x)\,dx$ and

$$\int\exp(x)\cdot\sin\left(\exp(x)\right)dx=\int\sin(u)\,du=-\cos(u)+C=-\cos\left(\exp(x)\right)+C.$$

23. Let $u = 3\exp(x) + 1$. Then $du = 3\exp(x)\,dx$ and

$$\int (3 + \exp(-x))^{-1}\,dx = \int \left(\frac{\exp(x)}{3\exp(x) + 1} \right) dx = \left(\frac{1}{3} \right) \int \left(\frac{1}{u} \right) du = \frac{\ln(u)}{3} + C = \frac{\ln(3\exp(x) + 1)}{3} + C$$

$$= \frac{\ln(3 + \exp(-x)) + x}{3} + C.$$

25. $\exp\left(\ln\left(a^3\right) - \ln\left(b^2\right) \right) = \exp\left(\ln\left(\frac{a^3}{b^2} \right) \right) = \frac{a^3}{b^2}$ for $a > 0$ and $b \neq 0$

27. $\exp\left(\ln(ac) - 4\ln\left(b^2 c^{-5}\right) \right) = \exp(\ln(ac)) \cdot \left(\exp\left(\ln\left(b^2 c^{-5}\right) \right) \right)^{-4} = ac\left(b^2 c^{-5}\right)^{-4} = ab^{-8}c^{21} = \frac{ac^{21}}{b^8}$

for $ac > 0$ and $b^2 c^{-5} > 0$

29. $\ln\left(\frac{\exp(a)\exp(2b)}{\exp\left(c^{-2}\right)} \right) = \ln(\exp(a)) + \ln(\exp(2b)) - \ln\left(\exp\left(c^{-2}\right)\right) = a + 2b - c^{-2} = \frac{ac^2 + 2bc^2 - 1}{c^2}$

31. $\ln\left(e^3 \cdot 5\right) = \ln\left(e^3\right) + \ln(5) = 3\ln(e) + \ln(5) = 3 + \ln(5)$

33. $\ln\left(\frac{e}{5} \right) = \ln(e) - \ln(5) = 1 - \ln(5)$

Further Theory and Practice

35. Calculating the derivative of $g(x)$, we obtain

$$g'(x) = \frac{f'(x)\exp(x) - f(x)\exp(x)}{\exp(2x)} = \frac{f'(x) - f(x)}{\exp(x)}.$$

If $f'(x) = f(x)$, we conclude that $g'(x) = 0$ and, from Theorem 4, Section 4.1, $g(x) = C$.

As $f(x) = C\exp(x)$, we can write the constant C as $f(0)$.

We, therefore, deduce that $f(x) = f(0)\exp(x)$.

37.   **a.**   See figure.
      b.   See figure.
      c.   See figure.
      d.   See figure.
      e.   See figure.

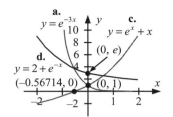

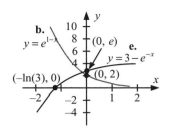

39.

$$\frac{d}{dx}\ln\left(\exp(x)\right)=\frac{d}{dx}x$$

$$\left(\frac{1}{\exp(x)}\right)\frac{d}{dx}\exp(x)=1$$

$$\frac{d}{dx}\exp(x)=\exp(x)$$

41.   The limit as n approaches infinity may be calculated as follows:

$$\lim_{n\to\infty}\left(\frac{n}{n+1}\right)^n=\lim_{n\to\infty}\frac{1}{\left((n+1)/n\right)^n}=\frac{1}{\lim_{n\to\infty}\left((n+1)/n\right)^n}=\frac{1}{e}.$$

43.   We evaluate the limit as follows: $\lim_{n\to\infty}\left(1+\frac{1}{n^2}\right)^{n^2}=\lim_{m\to\infty}\left(1+\frac{1}{m}\right)^m=e$, where we take $m=n^2$.

45.   As $x\to\infty$, we find $\exp(-\gamma x)\to0$, $\exp(-\beta\exp(-\gamma x))\to1$ and $G(x)\to\alpha$.

47.   Using implicit differentiation, let $y'\exp(2y-1)+2yy'\exp(2y-1)=2\exp(3x)+6x\exp(3x)$.

Then $y'=2\exp(3x)\dfrac{1+3x}{\exp(2y-1)(1+2y)}$. Substituting $x=1$ and $y=2$ into this equation

gives us $y'=\dfrac{8}{5}$.

49.   If $f(x)=\dfrac{x}{\ln(x)}$, we see that $f'(x)=\dfrac{\ln(x)-1}{\ln^2(x)}$ and $f''(x)=\dfrac{2-\ln(x)}{x\ln^3(x)}$.

The inflection point is at $x=e^2$. The graph of f is concave up for $1<x<e^2$ and concave down for $x>e^2$.

The only critical point is at $x=e$. Given that $f''(e)>0$, the local minimum is at $x=e$. The minimum value is $f_{\min}=f(e)=e$.

51.   If $f(x) = \ln(x)$, we calculate $f'(x) = \dfrac{1}{x}$. The equation of the tangent line is

$$y - \ln(c) = \frac{x-c}{c}$$

or $y = \dfrac{x}{c} + \ln(c) - 1$. The tangent point is $(c, \ln(c))$. We then solve to find $\ln(c) - 1 = 0$, $c = e$

and $m = \dfrac{1}{e}$.

53.   From Problem 6.3.52, let $u = \sqrt{x}$. Then $\exp(u) > u^2$, or $\exp(\sqrt{x}) > x$ and $\sqrt{x} > \ln(x)$ for $x > 0$.

55.   Using the results obtained from Problem 6.3.54, we apply the substitution $u = x^p$ to derive

$$\lim_{x\to\infty} \frac{\ln(x)}{x^p} = \frac{1}{p}\lim_{x\to\infty}\frac{\ln(x^p)}{x^p} = \frac{1}{p}\lim_{u\to\infty}\frac{\ln(u)}{u} = \frac{1}{p}\cdot 0 = 0.$$

57.   Using the results obtained from Problem 6.3.56, we apply the substitution $u = \exp(x)$ to derive

$$\lim_{x\to\infty}\frac{x^r}{\exp(x)} = \lim_{u\to\infty}\frac{\left(\ln(u)\right)^r}{u} = 0.$$

59.   Let $u = \dfrac{1}{x}$. As $x \to \infty$ we have $u \to 0^+$. Using the results obtained from Problem 6.3.55 we

find $\lim_{u\to 0^+} u^p \ln(u) = \lim_{x\to\infty}\left(x^{-p}\right)\ln\left(\dfrac{1}{x}\right) = -\lim_{x\to\infty}\left(\dfrac{\ln(x)}{x^p}\right) = 0$ for any positive constant p.

61.   The condition $T \to 0$ is equivalent to $k \to 0$. Differentiating, we deduce that

$$\frac{d}{dT}\exp\left(-\frac{E_A}{T}\right) = \exp\left(-\frac{E_A}{T}\right)\left(\frac{E_A}{T^2}\right).$$

Using the results obtained in Exercise 57 we see that $x = \dfrac{E_A}{T}$ and this gives us:

$$\lim_{T\to 0^+}\frac{d}{dt}\exp\left(-\frac{E_A}{T}\right) = \left(\frac{1}{E_A}\right)\lim_{T\to 0^+}\frac{(E_A/T)^2}{\exp(E_A/T)} = 0.$$

63.   Using L'Hôpital's Rule we derive the following:

$$\lim_{v\to 0^+}\frac{2h}{c^2}\frac{v^3}{e^{hv/(kT)}-1}=\frac{2h}{c^2}\lim_{v\to 0^+}\frac{3v^2}{(h/(kT))\exp(hv/(kT))}=\frac{6kT}{c^2}\lim_{v\to 0^+}\frac{v^2}{\exp(hv/(kT))}=0\,.$$

65.   Using Rolle's Theorem: $g'(x)=\exp(-x)(f'(x)-f(x))$ and $g(a)=g(b)=0$. We therefore see that there exists a point $c\in(a,\,b)$ such that $g'(c)=0$, that is $f'(c)=f(c)$.

Calculator/Computer Exercises

67.   The original equation $2\exp(-2x)=\exp(4x)$ can be written in the form: $\exp(6x)-2=0$. If we denote $f(x)=\exp(6x)-2$, then its derivative is $f'(x)=6\exp(6x)$, and $x_1=1$. The

following recursive relation holds: $x_{j+1}=x_j-\dfrac{\exp(6x_j)-2}{6\exp(x_j)}$, so that $x_2=0.834159...$,

$x_3=0.669727...$, $x_4=0.509055...$, $x_5=0.358106...$, $x_6=0.230320...$, $x_7=0.147352...$,
$x_8=0.118379...$, $x_9=0.115548...$, $x_{10}=0.115524...$, $x_{11}=0.115524....$ Hence $x=0.1155$.

69.   Denote: $T=\tau\times 10^9$ to calculate: $0.00725\dfrac{\exp(\tau/1.015)-1}{\exp(\tau/6.45)-1}=0.641$, or

$$\exp\left(\frac{\tau}{1.015}\right)-1-88.4138\left(\exp\left(\frac{\tau}{6.45}\right)-1\right)=0\,.$$

If $f(x)=\exp\left(\dfrac{x}{1.015}\right)-1-88.4138\left(\exp\left(\dfrac{x}{6.45}\right)-1\right)$ and $x_1=4$, then Newton-Raphson tells us that the following recursive relation holds:

$$x_{j+1}=x_j-\frac{\exp(x_j/1.015)-1-88.4138(\exp(x_j/6.45)-1)}{\exp(x_j/1.015)/1.015-88.4138\exp(x_j/6.45)/6.45}$$

so that: $x_2=5.01126...$, $x_3=4.69025...$, $x_4=4.60582...$, $x_5=4.600732...$, $x_6=4.600714...$,
$x_7=4.600714...$, $x_8=4.600714....$ Hence $T=4.6007\times 10^9$ yr.

71.   Let

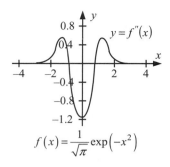

$$P(r) = \left(\frac{1}{\sqrt{\pi}}\right) \int_{-r}^{r} \exp\left(-t^2\right) dt .$$

To estimate this integral we use the Midpoint Rule. Since $C_2 < 1.2$ (see figure), we can find N from the condition

$$\frac{C_2 \cdot 10^3}{24 N^2} < \varepsilon ,$$

i.e. $N \geq 317$.

The probability that the arrow will land in the circle of radius 1 is $P(1) = 0.8427$.

The probability that the arrow will land in the circle of radius 2 is $P(2) = 0.9953$.

The probability that the arrow will land in the circle of radius 5 is $P(5) = 1.0000$.

73.   Let $f(x) = \exp\left(-x^2\right) \int_{0}^{x} \exp\left(t^2\right) dt$, then

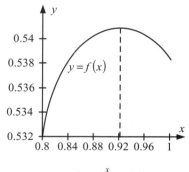

$$f'(x) + 2xf(x) - 1 = 0$$

and $f'(c) = 0$ for $c = 0.92413887$. As $f''(c) < 0$, c is a maximum of f.

 Section 6.4   **Logarithms and Powers with Arbitrary Bases**

Problems for Practice

1. $\left(3^x\right)' = 3^x \ln(3)$

3. $\left(x^{-\sqrt{7}}\right)' = -\sqrt{7} x^{-\sqrt{7}-1}$

5. $\left(\left(1+\sqrt{e}\right)^x\right)' = \left(1+\sqrt{e}\right)^x \ln\left(1+\sqrt{e}\right)$

7. $\left(4^{\left(x^2\right)}\right)' = 4^{\left(x^2\right)} \cdot \ln(4) \cdot 2x = 2\ln(4)x4^{\left(x^2\right)}$

9. $\left(\dfrac{\ln(x)}{3^x+1}\right)' = \dfrac{3^x+1-x3^x\ln(x)\ln(3)}{x\left(3^x+1\right)^2}$

11. $\left(\dfrac{3^x+4^x}{5^x}\right)' = \left(\left(\dfrac{3}{5}\right)^x + \left(\dfrac{4}{5}\right)^x\right)' = \left(\dfrac{3}{5}\right)^x \ln\left(\dfrac{3}{5}\right) + \left(\dfrac{4}{5}\right)^x \ln\left(\dfrac{4}{5}\right)$

13. $\displaystyle\int x^{-\sqrt{3}}\,dx = \dfrac{x^{-\sqrt{3}+1}}{-\sqrt{3}+1} + C$

15. $\displaystyle\int 5^x\,dx = \dfrac{5^x}{\ln(5)} + C$

17. $\displaystyle\int \sec^2(x)6^{\tan(x)}\,dx = \int 6^{\tan(x)}\,d\left(\tan(x)\right) = \dfrac{6^{\tan(x)}}{\ln(6)} + C$

19. $\displaystyle\int 10^x\left(10^x+7\right)^{5/2}\,dx = \dfrac{1}{\ln(10)}\int\left(10^x+7\right)^{5/2}d\left(10^x+7\right) = \dfrac{2\left(10^x+7\right)^{7/2}}{7\ln(10)} + C$

21. $\displaystyle\int x\cdot 3^{\left(x^2\right)}\,dx = \dfrac{1}{2}\int 3^{\left(x^2\right)}d\left(x^2\right) = \dfrac{1}{2}\cdot 3^{\left(x^2\right)}\cdot\dfrac{1}{\ln(3)} = \dfrac{3^{\left(x^2\right)}}{2\ln(3)} + C$

23. $\displaystyle\int \dfrac{5^x}{1+5^x}\,dx = \dfrac{1}{\ln(5)}\int\dfrac{1}{1+5^x}d\left(1+5^x\right) = \dfrac{\ln\left(1+5^x\right)}{\ln(5)} + C$

25. $\left(\log_2(5x)\right)' = \dfrac{5}{5x\cdot\ln(2)} = \dfrac{1}{x\ln(2)}$

27. $\left(\dfrac{1}{5+\log_{10}(x)}\right)' = -\dfrac{1}{\left(5+\log_{10}(x)\right)^2}\cdot\dfrac{1}{x\ln(10)} = -\dfrac{1}{x\left(5+\log_{10}(x)\right)^2\ln(10)}$

29. $\left(\dfrac{\log_3(x)}{\log_5(x)+\log_9(x)}\right)' = \dfrac{\frac{\log_5(x)+\log_9(x)}{x\ln(3)} - \log_3(x)\left(\frac{1}{x\ln(5)}+\frac{1}{x\ln(9)}\right)}{\left(\log_5(x)+\log_9(x)\right)^2} = \dfrac{\frac{1}{x\ln(3)}\left(\frac{\ln(x)}{\ln(5)}+\frac{\ln(x)}{\ln(9)}\right) - \frac{\ln(x)}{\ln(3)}\left(\frac{1}{x\ln(5)}+\frac{1}{x\ln(9)}\right)}{\left(\log_5(x)+\log_9(x)\right)^2}$

$= \dfrac{\frac{\ln(x)}{x\ln(3)\ln(5)} + \frac{\ln(x)}{x\ln(5)\ln(9)} - \frac{\ln(x)}{x\ln(3)\ln(5)} - \frac{\ln(x)}{x\ln(3)\ln(9)}}{\left(\log_5(x)+\log_9(x)\right)^2} = 0$

31. $4^x \cdot 5^{2x} = 8 \cdot 6^{-x}$, $\ln\left(4^x\right)+\ln\left(5^{2x}\right)=\ln(8)+\ln\left(6^{-x}\right)$, $x\ln(4)+2x\ln(5)=\ln(8)-x\ln(6)$,

$x\left(\ln(4)+2\ln(5)+\ln(6)\right)=\ln(8)$,

$x = \dfrac{\ln(8)}{\ln(4)+2\ln(5)+\ln(6)} = \dfrac{\ln(8)}{\ln(4)+\ln(25)+\ln(6)} = \dfrac{\ln(8)}{\ln(4\cdot25\cdot6)} = \dfrac{\ln(8)}{\ln(600)}$

33. $\log_4\left(16^x\right)-\log_3(27)+4^{\log_4(5)} = \log_4\left(4^{2x}\right)-3+5 = 2x+2$

35. $\log_8\left(64\cdot4^{2x}\cdot2^{-6}\right)=\log_8\left(2^6\cdot4^{2x}\cdot2^{-6}\right)=\log_8\left(4^{2x}\right)=\log_8\left(\left(8^{2/3}\right)^{2x}\right)=\log_8\left(8^{(2/3\cdot2x)}\right)$

$= \dfrac{2}{3}\cdot2x = \dfrac{4x}{3}$

Further Theory and Practice

37.   **a.** $\dfrac{P'(t)}{P(t)} = \dfrac{87\cdot2^{t/1.048}\cdot\ln(2)}{87\cdot2^{t/1.048}\cdot1.048} = \dfrac{\ln(2)}{1.048} = const$

b.   We compute the derivative: $P'(t) = \dfrac{87\cdot2^{t/1.048}\cdot\ln(2)}{1.048}$ number of diagnosed cases per half-year.

c.   We evaluate: $P'(2) = \dfrac{87\cdot2^{2/1.048}\cdot\ln(2)}{1.048} \approx 216.007$.

d.   We evaluate the function $P(24)=87\cdot2^{24/1.048}\approx681296000$ number of cases diagnosed by the end of 1993 which, of course does not make any sense.

39. $\left(x^{3x}\right)' = \left(e^{3x\ln(x)}\right)' = e^{3x\ln(x)}\left(3\ln(x)+3x\cdot\dfrac{1}{x}\right) = x^{3x}\left(3\ln(x)+3\right)$

41. Let $\ln\left(f(x)\right) = \left(\dfrac{x}{2}\right)\ln(x)$. Calculating the derivative of this expression, we conclude:

$\dfrac{f'(x)}{f(x)} = \left(\dfrac{1}{2}\right)\left(\ln(x)+1\right)$ and $f'(x) = \left(\dfrac{\sqrt{x}^{\,x}}{2}\right)\left(\ln(x)+1\right)$.

43. Let $\ln\left(f(x)\right)=3^x\ln\left(\cos(x)\right)$. Calculating the derivative of this expression, we conclude:

$$\frac{f'(x)}{f(x)}=\ln(3)3^x\ln\left(\cos(x)\right)-3^x\tan(x) \text{ and } f'(x)=3^x\cos^{3x}(x)\left(\ln(3)\ln\left(\cos(x)\right)-\tan(x)\right).$$

45. $\displaystyle\int\frac{\ln(x)}{\ln\left(x^5\right)}dx=\int\frac{\ln(x)}{5\ln(x)}dx=\int\frac{1}{5}dx=\frac{x}{5}+C$

47. $\displaystyle\int\frac{4^{7x}}{5^{2x}}dx=\int\left(\frac{4^7}{5^2}\right)^x dx=\frac{\left(4^7/5^2\right)^x}{7\ln(4)-2\ln(5)}+C$

49. $\displaystyle\int\frac{\log_2(5x)\log_7(3x)}{x}dx=\int\frac{\ln(5x)\ln(3x)}{x\ln(2)\ln(7)}dx=\int\frac{\left(\ln(5)+\ln(x)\right)\left(\ln(3)+\ln(x)\right)}{x\ln(2)\ln(7)}dx$

$$=\int\frac{\ln(5)\ln(3)+\left(\ln(5)+\ln(3)\right)\ln(x)+\ln^2(x)}{x\ln(2)\ln(7)}dx$$

$$=\int\frac{\ln(5)\ln(3)+\left(\ln(5)+\ln(3)\right)\ln(x)+\ln^2(x)}{\ln(2)\ln(7)}d\left(\ln(x)\right)$$

$$=\frac{\ln(5)\ln(3)}{\ln(2)\ln(7)}\ln(x)+\frac{\ln(5)+\ln(3)}{2\ln(2)\ln(7)}\ln^2(x)+\frac{1}{3\ln(2)\ln(7)}\ln^3(x)+C$$

51. $\displaystyle\int\frac{1}{x}\cdot\log_{10}(x)dx=\int\frac{\ln(x)}{x\ln(10)}dx=\int\frac{\ln(x)}{\ln(10)}d\left(\ln(x)\right)=\frac{\ln^2(x)}{2\ln(10)}+C$

53.

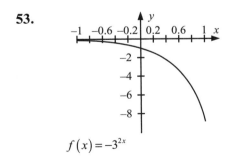

$f(x)=-3^{2x}$

55.

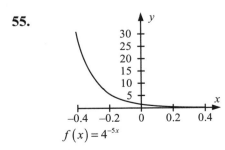

$f(x)=4^{-5x}$

57.

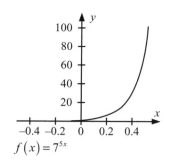

$f(x) = 7^{5x}$

59.   Let $\log_{10}(M) = 16 + (3/2)M_W$ and therefore $M = 10^{16+(3/2)M_W} = 10^{16}10^{(3/2)M_W}$ dyn-cm.

61.   Given that $M = 10^{16}10^{(3/2)M_W}$, we calculate the derivative $M'(M_W) = 10^{16} \cdot 10^{(3/2)M_W}\ln(10)$. Using the Method of Increments, we calculate:

$$M(M_W + \Delta M_W) \approx M(M_W) + 10^{16} \cdot 10^{(3/2)M_W}\ln(10)\Delta M_W.$$

In particular,

$$M(8.4) = M(8.3 + 0.1) \approx 2.82 \times 10^{28} + 10^{16} \cdot 10^{3 \cdot 8.3/2}\ln(10) \cdot 0.1 \approx 3.469 \times 10^{28} \text{ dyn-cm.}$$

63.   Calculating the derivative of the suggested function we derive:

$$f'(x) = \left(\frac{e^x}{x^e}\right)' = \frac{e^x x^e - e^x e x^{e-1}}{x^{2e}} = \frac{e^x(x-e)}{x^{e+1}}$$

Then $f(x)$ is increasing when $x > e$. As $f(e) = 1$, $f(\pi) = \frac{e^\pi}{\pi^e} > 1$ because $\pi > e$, so $e^\pi > \pi^e$.

65.   We calculate $f'(x) = x^{1/x^2}\dfrac{1 - 2\ln(x)}{x^3}$ from which we conclude that $f(x)$ has a local maximum at $x = e^{1/2}$. The maximum value is $e^{\frac{1}{2e}}$.

67. $\displaystyle\lim_{x \to 0}\frac{a^x - 1}{x} = \lim_{x \to 0}\frac{e^{x\ln(a)} - 1}{x} = \lim_{x \to 0}\left(\ln(a) \cdot \frac{e^{x\ln(a)} - 1}{x\ln(a)}\right) = \ln(a) \cdot \lim_{x \to 0}\left(\frac{e^{x\ln(a)} - 1}{x\ln(a)}\right)$
    $= \ln(a) \cdot 1 = \ln(a)$

69. $x^y = \left(t^{1/(t-1)}\right)^{t^{t/(t-1)}} = t^{\frac{1}{t-1}t^{t/(t-1)}} = t^{\frac{t}{t-1}t^{t/(t-1)-1}} = t^{\frac{t}{t-1}t^{1/(t-1)}} = \left(t^{t/(t-1)}\right)^{t^{1/(t-1)}} = y^x$

71.   Maximum value: $t_M = \dfrac{\ln(a) - \ln(b)}{a - b}$

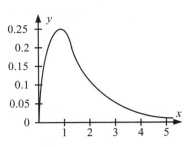

Inflection point: $t_I = \dfrac{\ln(a^2) - \ln(b^2)}{a - b}$

Limit as t approaches infinity: $\lim_{t \to \infty} c(t) = 0$

73.   From the initial condition, the derivative at $t = 0$ is $h(t) = 0$. At any subsequent time, $h(t) > 0$. The fraction

$$\left(\frac{M}{h(t)} \right)^q$$

is large at small t because M is a positive constant and $h(t)$ is close to 0. Therefore, the function $h'(t)$ starts out positive leading to the growth of $h(t)$. The second derivative is:

$$h''(t) = C \cdot h'(t) \left(\left(\frac{M}{h(t)} \right)^q (1 - q) - 1 \right),$$

and it also starts out positive because $q < 1$.

Therefore, the function $h(t)$ is concave down at the beginning. However, if it were always increasing and concave down, at some instant it would become equal to M. At this point

$$h'(t) = 0.$$

However, there is a contradiction here: if $h'(t)$ changes sign and becomes negative, $h(t)$ has to start decreasing from its value M. On the other hand, if

$$h(t) < M,$$

$h'(t) > 0$. If alternatively $h'(t)$ does not change sign and remains positive, $h(t)$ has to continue growing so that $h'(t)$ has to be negative. We conclude that $h(t)$ can never reach M, which is only possible if it becomes concave up, that is, if there is an inflection point. This immediately leads to the graph of $h(t)$ having an S-shape.

75.   Approximate tree growth in the juvenile stage:

$$h(t) = M\left(1 - e^{-rt}\right)^p \approx M\left(1 - (1 - rt)\right)^p = M(rt)^p.$$

77. Using the Method of Separation of Variables, we find the following explicit formula:

$$P'(t) = \alpha \cdot e^{-\beta t} \cdot P(t)$$

$$\frac{P'(t)}{P(t)} = \alpha \cdot e^{-\beta t}$$

$$\ln(P(t)) = -\frac{\alpha}{\beta} \cdot e^{-\beta t} + \ln(C)$$

$$P(t) = Ce^{-\frac{\alpha}{\beta} \cdot e^{-\beta t}}.$$

From this explicit formula we obtain the *carrying capacity*, such that

$$\lim_{t \to \infty} P(t) = \lim_{t \to \infty} Ce^{-\frac{\alpha}{\beta} \cdot e^{-\beta t}} = C = P_\infty$$

$$\frac{\alpha}{\beta} e^{-\beta t} = \ln(C) - \ln(P(t)) = \ln(P_\infty) - \ln(P(t)) = \ln\left(\frac{P_\infty}{P(t)}\right).$$

We now write the Gompertz growth equation in the form

$$P'(t) = \frac{1}{\beta} \cdot \ln\left(\frac{P_\infty}{P(t)}\right) \cdot P(t) = k \cdot P(t) \cdot \ln\left(\frac{P_\infty}{P(t)}\right).$$

Calculator/Computer Exercises

79. The function $k(x) = \left(1 + \dfrac{x}{100}\right)^{100}$ is much closer to

$f(x) = e^x$ than $g(x) = \left(1 + \dfrac{x}{10}\right)^{10}$. It is understandable

because $\lim_{n \to \infty} \left(1 + \dfrac{x}{n}\right)^n = e^x$.

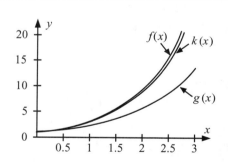

81. **a.** $x_{n+1} = x_n - \dfrac{4^{x_n} - 18}{4^{x_n} \ln(4)}$, $x_0 = 2$, $x_1 = 2.09017$, $x_2 = 2.08498$, $x_3 = 2.08496$,

$x_4 = 2.08496$

b. $\log_4(18) = \dfrac{\ln(18)}{\ln(4)} = \dfrac{2.89037}{1.38629} = 2.08496$

83. The equation of the normal line to the curve at $x = x_0$ is $y - \left(\dfrac{1}{3}\right)^{x_0} = \dfrac{3^{x_0}}{\ln(3)}(x - x_0)$.

Substituting $x = 0$ and $y = 0$, we get $\dfrac{1}{3^{x_0}} = \dfrac{3^{x_0} x_0}{\ln(3)}$, from which $x_0 = 0.4285$. So the point on

the curve at which the normal line passes through the origin is $\left(0.4285, \left(\dfrac{1}{3}\right)^{0.4285}\right)$.

85. Local maximum: $\left(e, e^{1/e}\right)$

Local minimum: None

Inflection points: $\left(0.5819327, f(0.5819327)\right)$ and

$\left(4.36777, f(4.36777)\right)$

Increasing on the interval: $(0, e)$

Decreasing on the interval: (e, ∞)

Concave up on the intervals: $(0, 0.58193) \cup (4.36777, \infty)$

Concave down on the intervals: $(0.58193, 4.36777)$

Horizontal asymptote: $y = 1$

Vertical asymptote: $x = 0$

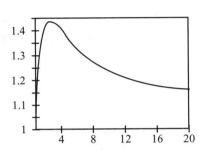

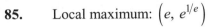**Section 6.5**   **Applications of the Exponential Function**

Problems for Practice

1. Let $P(t) = 5000e^{kt}$ and $P(3) = 8000$. We solve $8000 = 5000e^{3k}$ to find $k = \left(\dfrac{1}{3}\right)\ln\left(\dfrac{8}{5}\right)$ and

$P(8) = 5000e^{(8/3)\ln(8/5)} \approx 17510$ bacteria.

3. Let $P(t) = Ae^{kt}$. Given that $P(7) = 6500$ and $P(9) = 8000$, we obtain $\dfrac{8000}{6500} = \dfrac{e^{9k}}{e^{7k}}$. First, we

find $k = \ln\left(\dfrac{4}{\sqrt{13}}\right)$. Using this value of k, we calculate $8000 = Ae^{9\ln\left(4/\sqrt{13}\right)}$, or

$$A = \dfrac{3570125\sqrt{13}}{4096}.$$

Finally, we solve to find $P(8) = \left(\dfrac{3570125\sqrt{13}}{4096}\right)\exp\left(8\ln\left(\dfrac{4}{\sqrt{13}}\right)\right) \approx 7211$ bacteria.

5. Let $5 = Ae^{-1945\lambda}$ and $4 = Ae^{-1986\lambda}$. So, $\dfrac{5}{4} = e^{(1986-1945)\lambda}$, or $\lambda = \left(\dfrac{1}{41}\right)\ln\left(\dfrac{5}{4}\right)$. It follows that $A \approx 1.9783 \times 10^5$ and $R(2000) \approx 1.9783 \times 10^5 e^{-(2000/41)\ln(5/4)} \approx 3.7066$ grams of radio isotopes present in the year 2000.

7. Let $12 = Ae^{-2018\lambda}$ and $10 = Ae^{-2030\lambda}$. So, $\dfrac{12}{10} = e^{(2030-2018)\lambda}$, or $\lambda = \left(\dfrac{1}{12}\right)\ln\left(\dfrac{6}{5}\right)$. It follows that $A \approx 2.48215 \times 10^{14}$ and $R(2000) \approx 2.48215 \times 10^{14} e^{-(2000/12)\ln(6/5)} \approx 15.7744$ grams of radio isotopes present in the year 2000.

9. Let $12 = Ae^{-1945\lambda}$ and $8 = Ae^{-1985\lambda}$. So $\dfrac{12}{8} \approx e^{(1985-1945)\lambda}$, or $\lambda = \left(\dfrac{1}{40}\right)\ln\left(\dfrac{3}{2}\right)$. The half-life of the radioactive substance is $\tau = \dfrac{40\ln(2)}{\ln(3/2)} = 68.34825$ years.

11. Let $14 = Ae^{-1880\lambda}$ and $10 = Ae^{-1980\lambda}$. Therefore $\dfrac{14}{10} = e^{(1980-1880)\lambda}$, or $\lambda = \left(\dfrac{1}{100}\right)\ln\left(\dfrac{7}{5}\right)$. The half-life of the radioactive substance is $\tau = \dfrac{100\ln(2)}{\ln(7/5)} = 205.973$ years.

13. Let $400 = (T_0 - T_\infty)$ and $200 = (T_0 - T_\infty)e^{-5K}$. We solve it to find $K = \left(\dfrac{1}{5}\right)\ln(2)$. Therefore, $T(t) - T_\infty = 100$ when $100 = 400e^{-(t/5)\ln(2)}$, or $t = 10$. The temperature difference reaches $100°F$ at 12:40 PM.

15. Given that $R(v) = -Kv$ is a force that carries units of mlt^{-2} we see that the dimensions of the drag coefficient of K are $mlt^{-2}l^{-1}t$, or m/t.

17. Let $-0.99\dfrac{mg}{K} = -\left(\dfrac{mg}{K}\right)\left(1 - e^{-2\cdot10/m}\right)$, or $0.99 = 1 - e^{-2\cdot10/m}$. We solve to find the mass of the object to be $m = 4.343$ kg.

19. Given that $P'(t) = \left(\dfrac{\ln(2)}{50}\right)P(t)$, we find the instantaneous rate of change of the $E\text{-}coli$ population to be $P'(t) = \left(\dfrac{\ln(2)}{50}\right)100000 \approx 1386.3$ cells/minute when $P(t) = 100000$.

Further Theory and Practice

21. Given that $T'(t) = K \cdot (T_\infty - T(t))$, the closer $T(t)$ comes to T_∞, the slower its rate of change will be.

23. Let $R(t) = R(0) e^{-(t/1620)\ln(2)}$ be the amount of ^{226}Ra. We solve $\dfrac{R(0)}{10} = R(0) e^{-(t/1620)\ln(2)}$ to find that the radiation level will remain unsafe for 5381.5 years.

25. **a.** If $k > K$, then $p(0)\exp(kt) > P(0)\exp(Kt)$ for sufficiently large values of t and subpopulation becomes larger than population; that is impossible.

  b. Let $p'(t) = k \cdot p(t)$ and $P'(t) = K \cdot P(t)$. Therefore, the differential equation that $\dfrac{p}{P}$ satisfies is

$$\left(\frac{p}{P}\right)'(t) = \frac{P(t)p'(t) - p(t)P'(t)}{P(t)^2} = \frac{P(t)k \cdot p(t) - p(t)K \cdot P(t)}{P(t)^2} = (k - K)\left(\frac{p}{P}\right)(t).$$

To deduce the same equation in a more simple way, we calculate:

$$\left(\frac{p}{P}\right)(t) = \left(\frac{p}{P}\right)(0)\exp\big((k - K)t\big),$$

and so $\left(\dfrac{p}{P}\right)'(t) = (k - K)\left(\dfrac{p}{P}\right)(t)$.

27. **a.** To authenticate the Shroud the percentage of m should have been

$$100\exp\left(-(1988 - 33)\frac{\ln(2)}{5700}\right),$$

or approximately 79%.

  b. The percentage measured was $100\exp\left(-(1988 - 1325)\dfrac{\ln(2)}{5700}\right)$, or 92%.

29. The percentage measured for the pigments was approximately

$$100m\frac{\exp(-32410\ln(2)/5700)}{m},$$

or 1.94%.

31. If t is the age of the fossil, then $0.026 = \exp\left(\dfrac{-t\ln(2)}{5700}\right)$, or $t \approx 30012$. The European Neanderthals and Cro-Magnons coexisted for at least 10000 years.

33. The maximum percentage of pigments to be found for one of Vermeer's paintings is approximately 96%.

In the year 2004 the answer would be

$$100\exp\left(-(2004-1675)\frac{\ln(2)}{5700}\right) = 100\exp\left(-\frac{329}{5700}\ln(2)\right) = 50 \cdot 2^{\frac{5371}{5700}} \approx 96.08,$$

whereas, for example, in the year 2010 it would be

$$100\exp\left(-(2010-1675)\frac{\ln(2)}{5700}\right) = 100\exp\left(-\frac{67}{1140}\ln(2)\right) = 50 \cdot 2^{\frac{1073}{1140}} \approx 96.01.$$

The minimum percentage of pigments to be found for one of Meegeren's paintings is approximately 99%.

If Meegeren forged the Vermeer in 1930, then in 2004 the percentage of m would be $100\exp\left(-(2004-1930)\dfrac{\ln(2)}{5700}\right)$, or 99.1%.

For the year 2010 the percentage of m would be $100\exp\left(-(2010-1930)\dfrac{\ln(2)}{5700}\right)$, or 99.0%.

35. Let $A(t) = A(0)\exp\left(-\dfrac{t}{K}\right)$. If $K = \dfrac{1}{2}$ and $A(0) = 0.0012$, then $A(t) = 0.0006$ when

$$0.0006 = 0.0012\exp(-2t),$$

or it would take $t \approx 0.35$ hr for his blood alcohol content to reduce from 0.12% to 0.06%.

37. According to Newton's law, the temperature $T(t)$ of the iron bar is $T(t) = 70 + 80e^{-Kt}$ for some K. Given $T(4) = 125$, we solve to find $\left(\dfrac{1}{4}\right)\ln\left(\dfrac{16}{11}\right)$. Therefore, the temperature of the bar after ten minutes is $T(10) = 70 + 80e^{-(10/4)\ln(16/11)} \approx 101.35°$.

Now, let $T(t) = 90 + 70e^{-Kt}$ be the temperature of the rock. We solve the equation

$$T(20) = 150$$

to obtain $K = \left(\dfrac{1}{20}\right)\ln\left(\dfrac{7}{6}\right)$. We substitute this value into the equation $T(t) = 100$ and solve

$t = \dfrac{20\ln(7)}{\ln(7/6)} \approx 252.5$. In other words, it takes an additional 232.5 minutes for the rock to cool to 100°.

39. The differential equation is satisfied as follows:

$$H'(r) = \lambda^2 e^{-\lambda r} = \lambda\big(\lambda - H(r)\big) = \lambda^2 - \lambda H(r) = k \cdot H(r) + b$$

for $k = -\lambda$ and $b = \lambda^2$.

41. The number $N(t)$ of weapons satisfies $N'(t) = \mu - \delta N(t)$. From Theorem 2 with $b = \mu$ and $k = \delta$, the number of weapons at time t is $N(t) = \left(N_0 - \dfrac{\mu}{\delta}\right)\exp(-\delta t) + \dfrac{\mu}{\delta}$.

When t is large, the number of weapons on hand will be

$$N(t) \approx \lim_{t \to \infty}\left(\left(N_0 - \frac{\mu}{\delta}\right)\exp(-\delta t) + \frac{\mu}{\delta}\right) = \frac{\mu}{\delta}.$$

43. Using Theorem 2, we set $b = K \cdot c_o$ and $k = -K$. We find the concentration of a solute inside the cell to be $c(t) = \big(c(0) - c_o\big)\exp(-Kt) + c_o$.

45. Setting $b = A$ and $k = -A \cdot K$ in Theorem 2, we obtain

$$N(t) = \left(N(0) - \frac{1}{K}\right)\exp(-A \cdot K \cdot t) + \frac{1}{K}.$$

47. **a.** At time t hours, the tank holds $L + (r_{in} - r_{out})t$ liters. In a small time interval Δt the amount of solution that is drained is $r_{out}\Delta t$, and the amount Δm of salt in this solution is $\left(\dfrac{r_{out}\Delta t}{L + (r_{in} - r_{out})t}\right)m(t)$. Therefore, $\dfrac{\Delta m}{\Delta t} = -\left(\dfrac{r_{out}}{L + (r_{in} - r_{out})t}\right)m(t)$. The differential equation is obtained by making Δt arbitrarily close to zero.

b. If $r_{in} = r_{out}$, then $m'(t) = -\left(\dfrac{r_{out}}{L}\right)m(t)$ and $m(t) = A\exp\left(\dfrac{-r_{out}t}{L}\right)$.

c. Let $\int \dfrac{1}{m}dm = -\int \dfrac{r_{out}}{L + (r_{in} - r_{out})t}dt + C'$. Therefore,

$$\ln\big(m(t)\big) = C' - \dfrac{r_{out}}{(r_{in} - r_{out})}\ln\big(L + (r_{in} - r_{out})t\big).$$

We exponentiate, letting $C = \exp(C')$ to obtain

$$m(t) = C\big(L + (r_{in} - r_{out})t\big)^{-r_{out}/(r_{in} - r_{out})}.$$

Finally, $m(0) = A$ so $C = A \cdot L^{r_{out}/(r_{in} - r_{out})}$ and $m(t) = A \cdot \left(\dfrac{L}{L + (r_{in} - r_{out})t}\right)^{r_{out}/(r_{in} - r_{out})}$.

It is instructive to observe that the answer to part **b** results from letting $r_{out} \to r_{in}$ in this last formula.

49. For the equation $m\dfrac{d\sigma}{dt} = -mg - k\sigma$, $\sigma(\tau) = v_\tau$, using Theorem 2 we derive

$$\sigma(t) = \left(v_\tau + \dfrac{mg}{k}\right)e^{-g(t-\tau)/m} - \dfrac{mg}{k}$$

for $t \geq \tau$.

For the equation $\dfrac{d\eta}{dt} = \sigma(t)$, $\eta(\tau) = y_\tau$, the solution is

$$\eta(t) = y(\tau) + \int_\tau^t \sigma(u)\,du = y(\tau) + \int_\tau^t \left(\left(v_\tau + \dfrac{mg}{k}\right)e^{-g(u-\tau)/m} - \dfrac{mg}{k}\right)du$$

$$= y(\tau) - mg\dfrac{t-\tau}{k} + m\dfrac{mg + kv_\tau}{k^2}\left(1 - e^{-k(t-\tau)/m}\right)$$

$$= H - mg\dfrac{\tau}{K} + \left(m^2\dfrac{g}{K^2}\right)\left(1 - e^{-K\tau/m}\right) - mg\dfrac{t-\tau}{k} + m\dfrac{mg + kv_\tau}{k^2}\left(1 - e^{-k(t-\tau)/m}\right)$$

51. Given that $y(t) = \int_0^t v(u)\,du$, we calculate

$$y(t) = \int_0^t \left(\frac{mg}{K}\right)\left(\left(1 + \frac{Kv_0}{mg}\right)e^{-Ku/m} - 1\right)du = \frac{mg}{K^2}\left(\left(m + \frac{Kv_0}{g}\right)\left(1 - e^{-Kt/m}\right) - Kt\right)$$

for $0 \le t \le T_u + T_d$. Given that $e^{KT_u/m} = e^{\ln(1 + Kv_0/(mg))} = 1 + \frac{Kv_0}{mg}$, we derive

$$y(t) = \left(\frac{mg}{K^2}\right)\left(me^{KT_u/m}\left(1 - e^{-Kt/m}\right) - Kt\right).$$

At the highest point,

$$H = y(T_u) = \left(\frac{mg}{K^2}\right)\left(\left(m + \frac{Kv_0}{g}\right)\left(1 - e^{-KT_u/m}\right) - KT_u\right) = \left(\frac{mg}{K^2}\right)\left(\frac{Kv_0}{g} - m\ln\left(1 + \frac{Kv_0}{mg}\right)\right).$$

53. Given that $y(T_u + T_d) = 0$, we calculate $\left(\frac{mg}{K^2}\right)\left(me^{KT_u/m}\left(1 - e^{-K(T_u + T_d)/m}\right) - K(T_u + T_d)\right) = 0$

and $e^{KT_u/m} - e^{-KT_d/m} = \frac{KT_u}{m} + \frac{KT_d}{m}$. Using the results of Exercise 78 from Section 6.4, we

conclude that $\frac{KT_d}{m} > \frac{KT_u}{m}$, or $T_d > T_u$.

55. Given that $\tau > 0$, we calculate $\lim_{t \to \infty} \exp\left(-\frac{t}{\tau}\right) = 0$. Therefore, $v_\infty = \lim_{t \to \infty} v(t) = P\tau$.

57. From the given formula we obtain $N(t_0 + \Delta t) = N(t_0) \cdot 10^{\Delta t/(3.3\tau)}$. With $t_0 = 0$ and $\Delta t = t$,

the formula $N(t) = N(0) \cdot 10^{t/(3.3\tau)} = N(0) \cdot \exp\left(\frac{t\ln(10)}{3.3\tau}\right)$ results. In the *Britannica* formula,

the growth constant is $\frac{\ln(10)}{3.3\tau}$.

The exact growth constant k is related to the doubling time τ by $k = \frac{\ln(2)}{\tau} = \frac{\ln(10)}{(\ln(10)/\ln(2))\tau}$.

Britannica has rounded $\frac{\ln(10)}{\ln(2)} = 3.321928\ldots$ to 3.3.

Calculator/Computer Exercises

59.   Substituting the given data into the formula $v(t) = -\left(\dfrac{mg}{K}\right)\left(1 - \exp\left(-\dfrac{Kt}{m}\right)\right)$, we obtain

$-9.4 = -\left(\dfrac{9.81}{K}\right)(1 - \exp(-K))$. We solve this equation to obtain $K = 0.86$. The terminal

velocity is $-\dfrac{9.81}{0.86} = -11.4$ m/s.

61.   **a.**   From the formula of Exercise 48 and the given data, we obtain

$$0 = y(26) = 3170 - (800)(9.81)\dfrac{26}{K} + (800)^2 (9.81)\dfrac{1 - \exp(-K(26)/800)}{K^2}.$$

We solve to find $K = 4.1977$ kg/s.

b.   The terminal velocity was $(800)\dfrac{9.81}{4.1977} = 1870$ m/s.

c.   We substitute $m = 800$, $g = 9.81$,   **d.**
$K = 4.1977$, and $t = 26$ into the
formula

$$v(t) = -\left(\dfrac{mg}{K}\right)\left(1 - \exp\left(-\dfrac{Kt}{m}\right)\right).$$

The resulting value is 238 m/s.

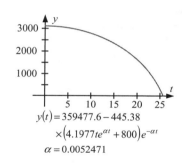

$y(t) = 359477.6 - 445.38$
$\times \left(4.1977te^{\alpha t} + 800\right)e^{-\alpha t}$
$\alpha = 0.0052471$

e.   Time of fall: 25.422 s; velocity at impact: 249.4 m/s

63.   We have $m = 0.1$ kg, $v_0 = 20$ m/s, $K = 0.03$ kg/s. From the Exercise 51:

$$H = \left(\dfrac{mg}{K^2}\right)\left(\dfrac{Kv_0}{g} - m\ln\left(1 + \dfrac{Kv_0}{mg}\right)\right) = 14.6 \text{ m},$$

from the Exercise 50: $T_u = \left(\dfrac{m}{K}\right)\ln\left(1 + \dfrac{Kv_0}{mg}\right) = 1.59$ s, from the Exercise 53:

$$\left(\dfrac{mg}{K^2}\right)\left(me^{KT_u/m}\left(1 - e^{-K(T_u + T_d)/m}\right) - K(T_u + T_d)\right) = 0$$

and $T_u + T_d = 3.48$, $T_d = 1.89$ s, from the Exercise 52: $v_I = \left(\dfrac{mg}{K}\right)\left(e^{-KT_d/m} - 1\right) = -14.16$ m/s.

65.   a.   Given that $x(t) = v_\infty \cdot \left(\tau - r\left(1 - \exp\left(-\dfrac{t}{\tau}\right)\right)\right)$, we calculate, for Johnson,

$$100 = 11.8 \cdot \left(9.83 - \tau\left(1 - \exp\left(-\dfrac{9.83}{\tau}\right)\right)\right).$$

We solve to find $\tau = 1.356$ and $P = \dfrac{v_\infty}{\tau} = \dfrac{11.8}{1.356} \approx 8.702$. For Lewis we estimate

$$100 = 11.8 \cdot \left(9.93 - \tau\left(1 - \exp\left(-\dfrac{9.93}{\tau}\right)\right)\right),$$

which leads to $\tau = 1.457$ and $P = \dfrac{v_\infty}{\tau} = \dfrac{11.8}{1.457} \approx 8.099$. Johnson's victory was attributable to both greater propulsive force and lesser resistance to motion.

b.

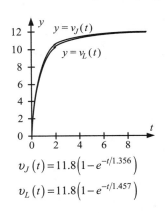

$$v_J(t) = 11.8\left(1 - e^{-t/1.356}\right)$$
$$v_L(t) = 11.8\left(1 - e^{-t/1.457}\right)$$

c.   For Lewis,

$$x(9.83) = 11.8\left(9.83 - 1.457\left(1 - \exp\left(\dfrac{-9.83}{1.457}\right)\right)\right)$$
$$= 98.82$$

Johnson's winning margin was $100 - 98.82 = 1.18$ m.

Section 6.6   Inverse Trigonometric Functions

Problems for Practice

1.   Given that $\sin\left(\dfrac{\pi}{2}\right) = 1$ and $\dfrac{\pi}{2}$ is inside $\left[-\dfrac{\pi}{2}, \dfrac{\pi}{2}\right]$, $\arcsin(1) = \dfrac{\pi}{2}$.

3. Given that $\sin\left(-\dfrac{\pi}{2}\right) = -1$ and $-\dfrac{\pi}{2}$ is inside $\left[-\dfrac{\pi}{2}, \dfrac{\pi}{2}\right]$, $\arcsin(-1) = -\dfrac{\pi}{2}$.

5. Given that $\cos\left(\dfrac{\pi}{3}\right) = \dfrac{1}{2}$ and $\dfrac{\pi}{3}$ is inside $[0, \pi]$, $\arccos\left(\dfrac{1}{2}\right) = \dfrac{\pi}{3}$.

7. Given that $\cos\left(\dfrac{\pi}{4}\right) = \dfrac{\sqrt{2}}{2}$ and $\dfrac{\pi}{4}$ is inside $[0, \pi]$, $\arccos\left(\dfrac{\sqrt{2}}{2}\right) = \dfrac{\pi}{4}$.

9. Given that $\tan\left(-\dfrac{\pi}{3}\right) = -\sqrt{3}$ and $-\dfrac{\pi}{3}$ is inside $\left[-\dfrac{\pi}{2}, \dfrac{\pi}{2}\right]$, $\arctan\left(-\sqrt{3}\right) = -\dfrac{\pi}{3}$.

11. Given that $\sec\left(\dfrac{\pi}{4}\right) = \sqrt{2}$ and $\dfrac{\pi}{4}$ is inside $[0, \pi]$, $\operatorname{arcsec}\left(\sqrt{2}\right) = \dfrac{\pi}{4}$.

13. Given that $\csc\left(-\dfrac{\pi}{4}\right) = -\sqrt{2}$ and $-\dfrac{\pi}{4}$ is inside $\left[-\dfrac{\pi}{2}, \dfrac{\pi}{2}\right]$, $\operatorname{arcsec}\left(-\sqrt{2}\right) = -\dfrac{\pi}{4}$.

15. $\dfrac{d}{dx}\arcsin(3x) = \left(\dfrac{1}{\sqrt{1-(3x)^2}}\right) \cdot \dfrac{d}{dx}(3x) = \dfrac{3}{\sqrt{1-9x^2}}$

17. $\dfrac{d}{dx}\arcsin(\ln(x)) = \left(\dfrac{1}{\sqrt{1-(\ln(x))^2}}\right) \cdot \dfrac{d}{dx}(\ln(x)) = \dfrac{1}{x\sqrt{1-\ln^2(x)}}$

19. $\dfrac{d}{dt}\arcsin(\sin^2(t)) = \left(\dfrac{1}{\sqrt{1-(\sin^2(t))^2}}\right) \cdot \dfrac{d}{dt}(\sin^2(t)) = 2\sin(t)\dfrac{\cos(t)}{\sqrt{1-\sin^4(t)}} = \dfrac{\sin(2t)}{\sqrt{1-\sin^4(t)}}$

21. $\dfrac{d}{dx}\arccos(x^4) = \left(-\dfrac{1}{\sqrt{1-(x^4)^2}}\right) \cdot \dfrac{d}{dx}(x^4) = -\dfrac{4x^3}{\sqrt{1-x^8}}$

23. $\dfrac{d}{dx}(\arcsin(x) \cdot \arccos(x)) = \arccos(x) \cdot \left(\dfrac{1}{\sqrt{1-(x)^2}}\right) + \arcsin(x) \cdot \left(-\dfrac{1}{\sqrt{1-(x)^2}}\right)$

$\qquad = \dfrac{\arccos(x) - \arcsin(x)}{\sqrt{1-x^2}}$

25. $\frac{d}{dx}\arctan\left(e^x\right)=\left(\frac{1}{1+\left(e^x\right)^2}\right)\cdot\frac{d}{dx}\left(e^x\right)=\frac{e^x}{1+e^{2x}}$

27. $\frac{d}{dx}\left(\tan\left(x\right)\cdot\arctan\left(x\right)\right)=\arctan\left(x\right)\frac{d}{dx}\tan\left(x\right)+\tan\left(x\right)\frac{d}{dx}\arctan\left(x\right)=\sec^2\left(x\right)\arctan\left(x\right)+\frac{\tan\left(x\right)}{1+x^2}$

29. $\frac{d}{dx}\cos\left(\arctan\left(x\right)\right)=-\sin\left(\arctan\left(x\right)\right)\cdot\frac{d}{dx}\arctan\left(x\right)=-\frac{x/\sqrt{1+x^2}}{1+x^2}=-\frac{x}{\left(1+x^2\right)^{3/2}}$

31. $\frac{d}{dt}\operatorname{arc\,sec}\left(2+\sin\left(t\right)\right)=\dfrac{1}{\left(2+\sin\left(t\right)\right)\sqrt{\left(2+\sin\left(t\right)\right)^2-1}}\cdot\frac{d}{dt}\left(2+\sin\left(t\right)\right)$

$\qquad\qquad = \dfrac{\cos\left(t\right)}{\left(2+\sin\left(t\right)\right)\sqrt{3+4\sin\left(t\right)+\sin^2\left(t\right)}}$

33. $\frac{d}{dx}\left(\operatorname{arc\,csc}\left(x^2\right)\right)=\frac{d}{dx}\operatorname{arc\,csc}\left(x^2\right)\cdot\frac{d}{dx}\left(x^2\right)=-\dfrac{2}{x\sqrt{x^4-1}}$

35. $\frac{d}{dt}\operatorname{arc\,cot}\left(\frac{1}{t^2}\right)=\left(-\dfrac{1}{1+\left(1/t^2\right)^2}\right)\cdot\frac{d}{dt}\left(\frac{1}{t^2}\right)=\dfrac{2}{t^3\left(1+1/t^4\right)}=\dfrac{2t}{1+t^4}$

37. $\displaystyle\int_0^{1/2}\frac{1}{\sqrt{1-x^2}}\,dx=\arcsin\left(x\right)\Big|_0^{1/2}=\arcsin\left(\frac{1}{2}\right)-\arcsin\left(0\right)=\frac{\pi}{6}-0=\frac{\pi}{6}$

39. $\displaystyle\int_0^{\sqrt{3}}\frac{1}{\sqrt{4-x^2}}\,dx=\int_0^{\sqrt{3}}\frac{1/2}{\sqrt{1-\left(x/2\right)^2}}\,dx=\arcsin\left(\frac{x}{2}\right)\Big|_0^{\sqrt{3}}=\arcsin\left(\frac{\sqrt{3}}{2}\right)-\arcsin\left(0\right)=\frac{\pi}{3}-0=\frac{\pi}{3}$

41. $\displaystyle\int_0^3\frac{1}{9+x^2}\,dx=\int_0^3\frac{1/9}{1+\left(x/3\right)^2}\,dx=\left(\frac{1}{3}\right)\arctan\left(\frac{x}{3}\right)\Big|_0^3=\left(\frac{1}{3}\right)\left(\arctan\left(1\right)-\arctan\left(0\right)\right)=\frac{\pi/4-0}{3}=\frac{\pi}{12}$

43. We use the substitution $u=x^2$, $du=2x\,dx$ to calculate the integral:

$$\left(\frac{1}{2}\right)\int_0^1\left(\frac{1}{1+u^2}\right)du=\left(\frac{1}{2}\right)\arctan\left(x\right)\Big|_0^1=\frac{\arctan\left(1\right)-\arctan\left(0\right)}{2}=\frac{\pi/4-0}{2}=\frac{\pi}{8}.$$

45.   We use the substitution $u = \ln(x)$, $du = \dfrac{dx}{x}$ to calculate the integral:

$$\int \left(\frac{1}{\sqrt{1-u^2}} \right) du = \arcsin(u) + C. \text{ Therefore, } \int \left(\frac{1}{x\sqrt{1-\ln^2(x)}} \right) dx = \arcsin(\ln(x)) + C.$$

47.   We use the substitution $u = \arctan(x)$, $du = \dfrac{dx}{1+x^2}$ to calculate the integral:

$$\int_{\arctan(0)}^{\arctan(1)} u\, du = \left(\frac{u^2}{2} \right)\Big|_0^{\pi/4} = \frac{\pi^2}{32}.$$

49.   We use the substitution $u = \tan(x)$, $du = \sec^2(x)\, dx$ to calculate the integral:

$$\int \left(\frac{1}{\sqrt{1-u^2}} \right) du = \arcsin(u) + C. \text{ Therefore, } \int \left(\frac{\sec^2(x)}{\sqrt{1-\tan^2(x)}} \right) dx = \arcsin(\tan(x)) + C.$$

51.   We use the substitution $u = e^x$, $du = e^x dx$ to calculate the integral:

$$\int \left(\frac{1}{\sqrt{1-u^2}} \right) du = \arcsin(u) + C. \text{ Therefore, } \int \left(\frac{e^x}{\sqrt{1-e^{2x}}} \right) dx = \arcsin(e^x) + C.$$

Further Theory and Practice

53.   Using the method from Section 3.6:

$$\frac{d}{dx}\operatorname{arc\,cot}(x) = \frac{1}{\frac{d}{ds}\cot(s)\big|_{s=\operatorname{arc\,cot}(x)}} = -\frac{1}{\csc^2(s)}\Big|_{s=\operatorname{arc\,cot}(x)} = -\frac{1}{\csc^2}\big(\operatorname{arc\,cot}(x)\big).$$

Given that $\csc^2(t) = 1 + \cot^2(t)$, we derive:

$$\frac{d}{dx}\operatorname{arc\,cot}(x) = -\frac{1}{\csc^2}\big(\operatorname{arc\,cot}(x)\big) = -\frac{1}{1+\cot^2\big(\operatorname{arc\,cot}(x)\big)} = -\frac{1}{1+x^2},$$

which is formula (6.61).

55.   Using the method from Section 3.6:

$$\frac{d}{dx}\operatorname{arc\,csc}(x) = \frac{1}{\frac{d}{ds}\csc(s)\big|_{s=\operatorname{arc\,csc}(x)}} = -\frac{1}{\big(\csc(s)\cot(s)\big)\big|_{s=\operatorname{arc\,csc}(x)}}$$

$$= -\frac{1}{\csc\big(\operatorname{arc\,csc}(x)\big)\cot\big(\operatorname{arc\,csc}(x)\big)} = -\frac{1}{x\cdot\cot\big(\operatorname{arc\,csc}(x)\big)}$$

Given that $\cot(t) = \sqrt{\csc^2(t)-1}$, we derive:

$$\frac{d}{dx}\operatorname{arc\,csc}(x) = -\frac{1}{x\cdot\cot\big(\operatorname{arc\,csc}(x)\big)} = -\frac{1}{x\cdot\sqrt{\csc^2\big(\operatorname{arc\,csc}(x)\big)-1}} = -\frac{1}{x\sqrt{x^2-1}},$$

which is formula (6.63).

57.   Let θ be a viewing angle and $y \in [-75,\ 225]$ be the distance. Then $\theta = \pi - \arctan\left(\dfrac{y}{150}\right)$ and

$\dfrac{d\theta}{dt} = -150\dfrac{dy/dt}{150^2+y^2}$. This expression is maximized when y^2 is minimized. Therefore, $\dfrac{d\theta}{dt}$

is greatest when $y=0$. The height at which the elevator appears to be moving most quickly is 75 feet.

59.   Let $\arctan(x) = s$ and $\operatorname{arc\,cot}\left(\dfrac{1}{x}\right) = t$. Then $x = \tan(s)$, $\dfrac{1}{x} = \cot(t)$ and $\tan(s) = \dfrac{1}{\cot(t)}$. This

is true if $t = s$, or $\arctan(x) = \operatorname{arc\,cot}\left(\dfrac{1}{x}\right)$. The domain is $x \neq 0$.

61.   By completing the square we obtain:

$$\int\left(\frac{1}{\sqrt{x-x^2}}\right)dx = \int\left(\frac{1}{\sqrt{1/4-\left(1/4-x+x^2\right)}}\right)dx = \int\left(\frac{1}{\sqrt{1/4-(1/4)(2x-1)^2}}\right)dx$$

$$= \int\left(\frac{2}{\sqrt{1-(2x-1)^2}}\right)dx$$

We use the substitution $u = 2x-1$, $du = 2dx$. Then the integral takes the form:

$\int\left(\dfrac{1}{\sqrt{1-u^2}}\right)du = \arcsin(u)+C$. Therefore, $\int\left(\dfrac{1}{\sqrt{x-x^2}}\right)dx = \arcsin(2x-1)+C$.

63.   By completing the square we obtain:

$$\int\left(\frac{1}{2+2x+x^2}\right)dx = \int\left(\frac{1}{1+\left(1+2x+x^2\right)}\right)dx = \int\left(\frac{1}{1+\left(x+1\right)^2}\right)dx.$$

We use the substitution $u = x+1$, $du = dx$. The integral takes the form:

$$\int\left(\frac{1}{1+u^2}\right)du = \arctan(u)+C. \text{ Therefore, } \int\left(\frac{1}{2+2x+x^2}\right)dx = \arctan(x+1)+C.$$

65.   At $x=c$ we have $\frac{d}{dx}\arcsin(x)\Big|_{x=c} = \frac{1}{\sqrt{1-x^2}}\Big|_{x=c} = \frac{1}{\sqrt{1-c^2}}$, the linearization is then

$$y = \frac{x-c}{\sqrt{1-c^2}} + \arcsin(c).$$

Linearization for $c = \frac{1}{2}$: $\arcsin\left(\frac{1}{2}\right)+\frac{\left(x+\frac{1}{2}\right)}{\sqrt{\frac{3}{4}}} = \frac{\pi}{6}+\left(\frac{\sqrt{3}}{3}\right)(2x-1)$

Linearization for $c = -\frac{1}{\sqrt{2}}$: $\arcsin\left(\frac{-1}{\sqrt{2}}\right)+\frac{\left(x+\frac{1}{\sqrt{2}}\right)}{\sqrt{1-\frac{1}{2}}} = -\frac{\pi}{4}+\frac{\sqrt{2}}{2}\left(x+\frac{1}{\sqrt{2}}\right) = -\frac{\pi}{4}+\frac{\left(x\sqrt{2}+1\right)}{2}$

67. $\frac{d}{dx}\arcsin(x)\Big|_{x=1/2} = \frac{1}{\sqrt{1-x^2}}\Big|_{x=1/2} = \frac{2\sqrt{3}}{3}$

$$\frac{d^2}{dx^2}\arcsin(x)\Big|_{x=1/2} = \frac{d}{dx}\left(\frac{1}{\sqrt{1-x^2}}\right)\Big|_{x=1/2} = \left(x\left(1-x^2\right)^{-3/2}\right)\Big|_{x=1/2} = \frac{4\sqrt{3}}{9}$$

$$T_2(x) = \frac{\pi}{6}+\left(\frac{2\sqrt{3}}{3}\right)\left(x-\frac{1}{2}\right)+\left(\frac{2\sqrt{3}}{9}\right)\left(x-\frac{1}{2}\right)^2$$

69. $\frac{d}{dx}\operatorname{arc\,sec}(x)\Big|_{x=2} = \frac{1}{x\sqrt{x^2-1}}\Big|_{x=2} = \frac{\sqrt{3}}{6}$

$$\frac{d^2}{dx^2}\operatorname{arc\,sec}(x)\Big|_{x=2} = \frac{d}{dx}\left(\frac{1}{x\sqrt{x^2-1}}\right)\Big|_{x=2} = \left(-\frac{2-1/x^2}{\left(x^2-1\right)^{3/2}}\right)\Big|_{x=2} = -\frac{7\sqrt{3}}{36}$$

$$T_2(x) = \frac{\pi}{3}+\left(\frac{\sqrt{3}}{6}\right)(x-2)-\left(\frac{7\sqrt{3}}{72}\right)(x-2)^2$$

71.   Let $\int \dfrac{x^0}{a^2+x^2}dx = \left(\dfrac{1}{a}\right)\arctan\left(\dfrac{x}{a}\right)+C$ and $\int \dfrac{x^1}{a^2+x^2}dx = \left(\dfrac{1}{2}\right)\ln\left(a^2+x^2\right)+C$. For $k \geq 2$ we

divide. Thus, $\dfrac{x^k}{a^2+x^2} = x^{k-2} - a^2\dfrac{x^{k-2}}{a^2+x^2}$ and

$$\int \dfrac{x^k}{a^2+x^2}dx = \int \left(x^{k-2} - \dfrac{a^2 x^{k-2}}{a^2+x^2}\right)dx = \int x^{k-2}dx - \int \left(\dfrac{a^2 x^{k-2}}{a^2+x^2}\right)dx + C.$$

We have a recurrent formula for $k \geq 2$: $\int \dfrac{x^k}{a^2+x^2}dx = \dfrac{x^{k-1}}{k-1} - \int \left(\dfrac{a^2 x^{k-2}}{a^2+x^2}\right)dx + C$. For example,

$$\int \dfrac{x^2}{a^2+x^2}dx = x - a\arctan\left(\dfrac{x}{a}\right)+C \text{ and } \int \dfrac{x^3}{a^2+x^2}dx = \dfrac{x^2}{2} - \left(\dfrac{a^2}{2}\right)\ln\left(a^2+x^2\right)+C.$$

73.   For $x > 0$ we calculate: $\dfrac{d}{dx}\left(2\arctan\left(\sqrt{x}\right)\right) = \dfrac{1}{\sqrt{x}(x+1)} = \dfrac{d}{dx}\arcsin\left(\dfrac{x-1}{x+1}\right)$. According to

Theorem 5 from Section 4.1 there is a constant C such that $2\arctan\left(\sqrt{x}\right) = \arcsin\left(\dfrac{x-1}{x+1}\right)+C$

for $x > 0$. Setting $x = 1$ we see that $C = \dfrac{\pi}{2}$. Clearly the identity also holds at the endpoint

$x = 0$.

75.   First limit: $\displaystyle\lim_{x\to 0}\left(\dfrac{x-\arcsin(x)}{x^3}\right) = \lim_{x\to 0}\dfrac{\frac{d}{dx}\left(x-\arcsin(x)\right)}{\frac{d}{dx}x^3} = \dfrac{1}{3}\lim_{x\to 0}\dfrac{\sqrt{1-x^2}-1}{x^2\sqrt{1-x^2}}$ which equals

$$\dfrac{1}{3}\lim_{x\to 0}\dfrac{\frac{d}{dx}\left(\sqrt{1-x^2}-1\right)}{\frac{d}{dx}\left(x^2\sqrt{1-x^2}\right)} = \dfrac{1}{3}\lim_{x\to 0}\dfrac{-1}{2-3x^2} = -\dfrac{1}{6}.$$

Second limit:

$$\lim_{x\to 0}\left(\dfrac{x\sqrt{1-x^2}-\arcsin(x)}{x^3}\right) = \lim_{x\to 0}\dfrac{\frac{d}{dx}\left(x\sqrt{1-x^2}-\arcsin(x)\right)}{\frac{d}{dx}x^3} = \lim_{x\to 0}\dfrac{-2x^2/\sqrt{1-x^2}}{3x^2} = -\dfrac{2}{3}.$$

Calculator/Computer Exercises

77.   Given that $c = 0.8$ and $h = 0.0002$, we approximate as follows:

$$\dfrac{f(c+h/2)-f(c-h/2)}{h} = 2.082316894.$$

We calculate exactly: $f'(c) = \dfrac{2c}{\sqrt{1-c^4}} = 2.082316825$.

79. Given that $c = 0.65$ and $h = 0.0002$, we approximate as follows:

$$\frac{f(c+h/2) - f(c-h/2)}{h} = 0.3758628342.$$

We calculate exactly: $f'(c) = \dfrac{1}{2\sqrt{c}(1+c)} = 0.3758628321$.

81. $f(x) = \arcsin\left(\dfrac{x}{1+x^2}\right)$

$f'(x) = \dfrac{1-x^2}{(1+x^2)\sqrt{1+x^2+x^4}}$; $f'(x) = 0$ at $x = \pm 1$

$$f''(x) = \frac{-2x(1+x^2)\sqrt{1+x^2+x^4}}{(1+x^2)^2 + (1+x^2+x^4)} - \frac{(1-x^2)(2x\sqrt{1+x^2+x^4})}{(1+x^2)^2(1+x^2+x^4)}$$

$$- \frac{(1-x^4)\left(\frac{1}{2}(1+x^2x^4)^{-1/2}\right)(2x+4x^3)}{(1+x^2)^2(1+x^2+x^4)}$$

$$= \frac{-4x\sqrt{1+x^2+x^4} - \frac{1}{2}(1-x^4)(2x+4x^3)(1+x^2+x^4)}{(1+x^2)^2(1+x^2+x^4)}$$

$$= -\frac{8x(1+x^2+x^4) + (1-x^4)(2x+4x^3)}{2(1+x^2)^2(1+x^2+x^4)^{3/2}}$$

$$= \frac{4x^7 - 3x^5 - 12x^3 - 10x}{2(1+x^2)^2(1+x^2+x^4)^{3/2}}$$

$f''(x) = 0$ at $x = \pm 1.690141555,\ 0$

Domain: $(-\infty,\ \infty)$

Local maximum: $\left(1,\ \dfrac{\pi}{6}\right)$

Local minimum: $\left(-1,\ -\dfrac{\pi}{6}\right)$

Increasing on the interval: $(-1,\ 1)$

Decreasing on the intervals: $(-\infty,\ -1) \cup (1,\ \infty)$

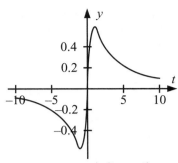

$f(x) = \arcsin\left(x/(1+x^2)\right)$

Inflection points: $(-1.690141555,\ -0.4536496788)$, $(0,\ 0)$, and

$(1.690141555,\ 0.4536496788)$

Concave up on the intervals: $(-1.690141555,\ 0) \cup (1.690141555,\ \infty)$

Concave down on the intervals: $(-\infty,\ -1.690141555) \cup (0,\ 1.690141555)$

Horizontal asymptote: $y = 0$

Vertical asymptote: None

Symmetry: about the origin

83.   The point of intersection is the solution of equation

$\dfrac{b}{\pi} = \dfrac{b^2}{b^2 + 1}$, i.e. $b = \dfrac{\pi - \sqrt{\pi^2 - 4}}{2} = 0.3594330037$.

We calculate the area of the region between these two graphs and over the interval as follows:

$$\text{Area} = \int_0^b \left(\frac{x}{\pi} - \frac{x^2}{x^2 + 1}\right) dx$$

$$= \int_0^b \left(\frac{x}{\pi} - 1 + \frac{1}{x^2 + 1}\right) dx$$

$$= \left(\frac{x^2}{2\pi} - x + \arctan(x)\right)\Bigg|_0^b$$

$$= 0.006182100655$$

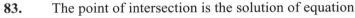

$\quad\text{---}\ \ y = x/\pi$

$\quad\text{---}\ \ y = x^2/(x^2 + 1)$

85.   The points of intersection in the second (and then the first) quadrant are, respectively:

$$a = -1.302775638$$

and

$$b = 2.302775638 \,.$$

We calculate the area of the region enclosed by the two curves as follows:

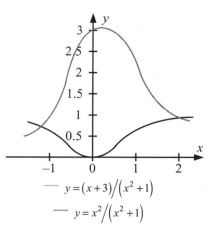

$$y = (x+3)/(x^2+1)$$

$$y = x^2/(x^2+1)$$

$$\text{Area} = \int_a^b \left(\frac{x+3}{x^2+1} - \frac{x^2}{x^2+1} \right) dx$$

$$= \int_a^b \left(\frac{x}{x^2+1} - 1 + \frac{4}{x^2+1} \right) dx$$

$$= \left(\frac{\ln\left(x^2+1\right)}{2} - x + 4\arctan(x) \right) \Bigg|_a^b$$

$$= 5.127796048$$

Section 6.7   Enrichment: The Hyperbolic Functions

Problems for Practice

1. $f'(x) = \cosh(3x) \cdot (3x)' = 3\cosh(3x)$

3. $f'(x) = \left(\dfrac{1}{\sinh(x)} \right) \cdot (\sinh(x))' = \dfrac{\cosh(x)}{\sinh(x)} = \coth(x)$

5. $f'(x) = -\mathrm{sech}\left(\sqrt{x}\right)\tanh\left(\sqrt{x}\right)\left(\sqrt{x}\right)' = -\mathrm{sech}\left(\sqrt{x}\right)\dfrac{\tanh\left(\sqrt{x}\right)}{2\sqrt{x}}$

7. $f'(x) = \sinh\left(x^2 - 5x\right) \cdot \left(x^2 - 5x\right)' = (2x - 5)\sinh\left(x^2 - 5x\right)$

9. $f'(x) = \left(\dfrac{1}{\tanh(x)} \right) \cdot (\tanh(x))' = \dfrac{\mathrm{sech}^2(x)}{\tanh(x)} = \dfrac{1}{\sinh(x)\cosh(x)} = \coth(x) - \tanh(x)$

11. $\int \cosh(3x)\,dx = \left(\dfrac{1}{3}\right)\int \cosh(u)\,du = \left(\dfrac{1}{3}\right)\sinh(u) + C = \left(\dfrac{1}{3}\right)\sinh(3x) + C$, where $u = 3x$.

13. $\int\left(\dfrac{\cosh(2x)}{\sqrt{\sinh(2x)}}\right)dx = \left(\dfrac{1}{2}\right)\int\left(\dfrac{1}{\sqrt{u}}\right)du = \sqrt{u} + C = \sqrt{\sinh(2x)} + C$, where $u = \sinh(2x)$.

15. $\int x^2 \operatorname{sech}\left(x^3\right)\tanh\left(x^3\right)dx = \left(\dfrac{1}{3}\right)\int \operatorname{sech}(u)\tanh(u)\,du = -\left(\dfrac{1}{3}\right)\operatorname{sech}(u) + C$

$$= -\left(\dfrac{1}{3}\right)\operatorname{sech}\left(x^3\right) + C$$

where $u = x^3$.

17. $\int\tanh^3(x)\operatorname{sech}^2(x)\,dx = \int u^3\,du = \left(\dfrac{1}{4}\right)u^4 + C = \left(\dfrac{1}{4}\right)\tanh^4(x) + C$, where $u = \tanh(x)$.

19. $\int\cosh(2x+5)\,dx = \left(\dfrac{1}{2}\right)\int\cosh(u)\,du = \left(\dfrac{1}{2}\right)\sinh(u) + C = \left(\dfrac{1}{2}\right)\sinh(2x+5) + C$, where $u = 2x + 5$.

21. $f'(x) = \dfrac{d}{dx}\ln\left(x + \sqrt{x^2 - 1}\right) = \dfrac{1}{x + \sqrt{x^2 - 1}}\cdot\left(1 + \dfrac{x}{\sqrt{x^2 - 1}}\right) = \dfrac{1}{\sqrt{x^2 - 1}}$

23. $h'(x) = \dfrac{d}{dx}\ln\left(\dfrac{1}{x} + \dfrac{\sqrt{1 + x^2}}{|x|}\right)$.

For $x > 0$:

$$h'(x) = \dfrac{d}{dx}\ln\left(\dfrac{1}{x} + \dfrac{\sqrt{1 + x^2}}{x}\right) = \dfrac{x}{1 + \sqrt{1 + x^2}}\cdot\left(-\dfrac{1}{x^2} - \dfrac{1}{x^2\sqrt{1 + x^2}}\right) = -\dfrac{1}{x\sqrt{1 + x^2}}.$$

For $x < 0$: $h'(x) = \dfrac{d}{dx}\ln\left(\dfrac{1}{x} - \dfrac{\sqrt{1 + x^2}}{x}\right) = \dfrac{x}{1 - \sqrt{1 + x^2}}\cdot\left(-\dfrac{1}{x^2} + \dfrac{1}{x^2\sqrt{1 + x^2}}\right) = \dfrac{1}{x\sqrt{1 + x^2}}$. In any

case $h'(x) = -\dfrac{1}{|x|\sqrt{1 + x^2}}$.

25. $g'(x) = \tanh^{-1}(2x)\dfrac{d}{dx}\operatorname{sech}^{-1}(x) + \operatorname{sech}^{-1}(x)\dfrac{d}{dx}\tanh^{-1}(2x)$

$$= \tanh^{-1}(2x)\left(-\dfrac{1}{|x|\sqrt{1 - x^2}}\right) + \operatorname{sech}^{-1}(x)\left(\dfrac{2}{1 - 4x^2}\right)$$

Further Theory and Practice

27. $\dfrac{d}{dx}\tanh(x) = \dfrac{d}{dx}\dfrac{\sinh(x)}{\cosh(x)} = \dfrac{\cosh(x)\frac{d}{dx}\sinh(x) - \sinh(x)\frac{d}{dx}\cosh(x)}{\cosh^2(x)} = \dfrac{\cosh^2(x) - \sinh^2(x)}{\cosh^2(x)}$

$$= \dfrac{1}{\cosh^2(x)} = \operatorname{sech}^2(x)$$

$\dfrac{d}{dx}\coth(x) = \dfrac{d}{dx}\dfrac{\cosh(x)}{\sinh(x)} = \dfrac{\sinh(x)\frac{d}{dx}\cosh(x) - \cosh(x)\frac{d}{dx}\sinh(x)}{\sinh^2(x)} = \dfrac{\sinh^2(x) - \cosh^2(x)}{\sinh^2(x)}$

$$= -\dfrac{1}{\sinh^2(x)} = -\operatorname{csch}^2(x)$$

$\dfrac{d}{dx}\operatorname{sech}(x) = \dfrac{d}{dx}\dfrac{1}{\cosh(x)} = -\dfrac{1}{\cosh^2(x)}\dfrac{d}{dx}\cosh(x) = -\dfrac{\sinh(x)}{\cosh^2(x)} = -\dfrac{1}{\cosh(x)}\cdot\dfrac{\sinh(x)}{\cosh(x)}$

$$= -\operatorname{sech}(x)\cdot\tanh(x)$$

$\dfrac{d}{dx}\operatorname{csch}(x) = \dfrac{d}{dx}\dfrac{1}{\sinh(x)} = -\dfrac{1}{\sinh^2(x)}\dfrac{d}{dx}\sinh(x) = -\dfrac{\cosh(x)}{\sinh^2(x)} = -\dfrac{1}{\sinh(x)}\cdot\dfrac{\cosh(x)}{\sinh(x)}$

$$= -\operatorname{csch}(x)\cdot\coth(x)$$

29. Let $f(x) = \coth(x)$; then $f(x) = \dfrac{e^x + e^{-x}}{e^x - e^{-x}}$ and the

domain of f is $\{x \in \mathbb{R} : x \neq 0\}$.

This function is odd.

As $f'(x) = -\operatorname{csch}^2(x) = -\dfrac{1}{\sinh^2(x)} < 0$, this function

decreases on the intervals $(-\infty, 0)$ and $(0, \infty)$.

There are no extrema.

As $f''(x) = 2\operatorname{csch}^2(x)\coth(x)$, the second derivative $f''(x) > 0$ and the function is concave

up on the interval $(0, \infty)$, $f''(x) < 0$ and concave down on the interval $(-\infty, 0)$.

There are no points of inflection.

By analogy with the Problem 6.7.28, the horizontal asymptotes are $\lim\limits_{x \to \infty} f(x) = 1$ and

$\lim\limits_{x \to -\infty} f(x) = -1$.

There is a vertical asymptote at $x = 0$.

31. Let $f(x) = \operatorname{csch}(x)$, or $f(x) = \dfrac{2}{e^x - e^{-x}}$.

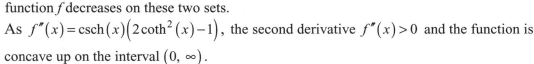

The domain of this function is $\{x \in \mathbb{R} : x \neq 0\}$.

The function is odd.

As $f'(x) = -\operatorname{csch}(x)\coth(x)$, there are no extrema.

As $f'(x) < 0$ on the intervals $(-\infty, 0)$ and $(0, \infty)$, the function f decreases on these two sets.

As $f''(x) = \operatorname{csch}(x)\left(2\coth^2(x) - 1\right)$, the second derivative $f''(x) > 0$ and the function is concave up on the interval $(0, \infty)$.

The second derivative $f''(x) < 0$ and concave down on the interval $(-\infty, 0)$.

There are no points of inflection.

Horizontal asymptotes: $y = 0$

Vertical asymptote: $x = 0$

33. $\sinh(x)\cosh(y) + \cosh(x)\sinh(y)$

$= \left(\dfrac{1}{2}\right)\left(e^x - e^{-x}\right)\left(\dfrac{1}{2}\right)\left(e^y + e^{-y}\right) + \left(\dfrac{1}{2}\right)\left(e^x + e^{-x}\right)\left(\dfrac{1}{2}\right)\left(e^y - e^{-y}\right)$

$= \left(\dfrac{1}{4}\right)\left(e^{x+y} - e^{-x+y} + e^{x-y} - e^{-x-y} + e^{x+y} + e^{-x+y} - e^{x-y} - e^{-x-y}\right) = \left(\dfrac{1}{4}\right)\left(2e^{x+y} - 2e^{-x-y}\right)$

$= \sinh(x + y)$

35. $2\sinh^2(x) + 1 = 2\left(\dfrac{1}{4}\right)\left(e^x - e^{-x}\right)^2 + 1 = \left(\dfrac{1}{2}\right)\left(e^{2x} - 2 + e^{-2x} + 2\right) = \left(\dfrac{1}{2}\right)\left(e^{2x} + e^{-2x}\right) = \cosh(2x)$

37. Let $y(x) = \sinh(\omega x)$. Then $y'(x) = \omega\cosh(\omega x)$, $y''(x) = \omega^2 \sinh(\omega x)$ and

$$y'' - \omega^2 y = \omega^2 \sinh(\omega x) - \omega^2 \sinh(\omega x) = 0.$$

For $y(x) = \cosh(\omega x)$ the second derivative $y''(x) = \omega^2 \cosh(\omega x)$ and $y'' - \omega^2 y = 0$ for all x.

39. $\left(\cosh(x) + \sinh(x)\right)^n = \left(\left(\dfrac{1}{2}\right)\left(e^x + e^{-x}\right) + \left(\dfrac{1}{2}\right)\left(e^x - e^{-x}\right)\right)^n = e^{nx}$

$= \left(\dfrac{1}{2}\right)\left(e^{nx} + e^{-nx}\right) + \left(\dfrac{1}{2}\right)\left(e^{nx} - e^{-nx}\right) = \cosh(nx) + \sinh(nx)$

41. $\dfrac{d}{dx}\left(\dfrac{1}{a}\tanh^{-1}\left(\dfrac{x}{a}\right)+C\right)=\dfrac{d}{dx}\left(\dfrac{1}{2a}\ln\left(\dfrac{1+x/a}{1-x/a}\right)+C\right)=\dfrac{d}{dx}\left(\dfrac{1}{2a}\ln\left(\dfrac{x+a}{a-x}\right)\right)=\dfrac{1}{2a}\cdot\dfrac{a-x}{x+a}\cdot\dfrac{2a}{(a-x)^2}$

$$=\dfrac{1}{a^2-x^2}$$

43.   As $\dfrac{d}{dx}\tanh(x)\Big|_{x=0}=1$, for small x we derive $\tanh(x)\approx x$. Then $v\approx\sqrt{\dfrac{g\lambda}{2\pi}\cdot\dfrac{2\pi d}{\lambda}}=\sqrt{gd}$ when

d is small compared to λ. From $\lim\limits_{x\to\infty}\tanh(x)=1$, for large d we obtain $v\approx\sqrt{\dfrac{g\lambda}{2\pi}}$.

45.   Using the explicit form of $T(x)$ we write: $y(x)=a\left(a\ln\left(\dfrac{1+\sqrt{1-(x/a)^2}}{x/a}\right)-\sqrt{a^2-x^2}\right)$. Its

derivative is:

$$\dfrac{dy}{dx}=a\left(\dfrac{\left(-\dfrac{1}{a\sqrt{1-(x/a)^2}}-\dfrac{\left(1+\sqrt{1-(x/a)^2}\right)a}{x^2}\right)x}{1+\sqrt{1-(x/a)^2}}+\dfrac{x}{\sqrt{a^2-x^2}}\right).$$

For $a>0$, we calculate $a\sqrt{1-(x/a)^2}=\sqrt{a^2-x^2}$ so that $\dfrac{dy}{dx}=-\dfrac{\sqrt{a^2-x^2}}{x}$. For $a<0$, the

calculations lead to $a\sqrt{1-(x/a)^2}=-\sqrt{a^2-x^2}$ and the equation is not correct:

$$\dfrac{dy}{dx}=\dfrac{a\left(a^2+x^2\right)}{x\sqrt{a^2-x^2}}.$$

47.   As shown in Problem 6.1.44, the boat moves along the solution of the differential equation

$\dfrac{dy}{dx}=-\dfrac{\sqrt{a^2-x^2}}{x}$, i.e. along the tractrix.

49.   The shape of the cable is a curve: $y=h+a\cosh\left(\dfrac{x}{a}\right)$. Given that $w=3$, $s=4$, $a=4$, we

calculate $h=h_0-\dfrac{T_1}{w}=1-4=-3$ and the equation is: $y=4\cosh\left(\dfrac{x}{4}\right)-3$.

51. Assuming the quadratic drag law with proportionality we write $v(t) = \sqrt{\dfrac{mg}{K}}\tanh\left(-t\sqrt{\dfrac{Kg}{m}}\right)$.

Then, using the Fundamental Theorem of Calculus we write

$$y(t) = y(0) + \int_0^t v(u)\,du = H + \sqrt{\frac{mg}{K}}\int_0^t \tanh\left(-u\sqrt{\frac{Kg}{m}}\right)du$$

$$= H + \sqrt{\frac{mg}{K}}\left(-\sqrt{\frac{m}{Kg}}\right)\ln\left(\cosh\left(-u\sqrt{\frac{Kg}{m}}\right)\right)\Bigg|_0^t = H - \left(\frac{m}{K}\right)\ln\left(\cosh\left(t\sqrt{\frac{Kg}{m}}\right)\right)$$

53. Given $\tau = \sqrt{\dfrac{m}{Kg}}\ln\left(e^{HK/m} + \sqrt{e^{2HK/m} - 1}\right)$ (see Problem 6.7.52), we write

$$v_1(H) = \sqrt{\frac{mg}{K}}\tanh\left(-\ln\left(e^{HK/m} + \sqrt{e^{2HK/m} - 1}\right)\right).$$

This leads to the speed of impact: $v_1(H) = -\sqrt{\dfrac{mg}{K}}\left(1 - \dfrac{1}{e^{2HK/m} + e^{HK/m}\sqrt{e^{2HK/m} - 1}}\right).$

55. If we set $x = \sinh(u)$ and $y = \sinh(v)$, then

$$\sinh^{-1}\left(x\sqrt{1+y^2} + y\sqrt{1+x^2}\right) = \sinh^{-1}\left(\sinh(u)\cosh(v) + \sinh(v)\cosh(u)\right)$$

$$= \sinh^{-1}\left(\sinh(u+v)\right)$$

This shows that $\sinh^{-1}(x) + \sinh^{-1}(y) = \sinh^{-1}\left(x\sqrt{1+y^2} + y\sqrt{1+x^2}\right)$. The first identity is an immediate consequence of this one.

57. $\dfrac{d}{du}gd(u) = \left(\dfrac{1}{1+\sinh^2(u)}\right)\cdot\cosh(u) = \dfrac{\cosh(u)}{\cosh^2(u)} = \mathrm{sech}(u)$

59.   Let $u = x\tan(\alpha)$, then the derivative is $\frac{dy}{du} = \cos(y)$, so that $\frac{dy}{\cos(y)} = du$ and $y(0) = 0$. As

$$\int \frac{dy}{\cos(y)} = \ln\left(\left|\sec(y) + \tan(y)\right|\right) + C = gd^{-1}(y) + C, \text{ we calculate } C = 0 \text{ and } u = gd^{-1}(y),$$

that is $y = gd\left(x\tan(\alpha)\right)$.

Calculator/Computer Exercises

61.   We calculate the derivatives:

$$f'(x) = -\cosh\left(\frac{x-1}{x^2+1}\right)\frac{x^2 - 2x - 1}{\left(x^2+1\right)^2}$$

$$f''(x) = -\sinh\left(\frac{x-1}{x^2+1}\right)\frac{x^2 - 2x - 1}{\left(x^2+1\right)^2}$$

$$+ 2\cosh\left(\frac{x-1}{x^2+1}\right)\frac{x^3 - 3x^2 - 3x + 1}{\left(x^2+1\right)^3}$$

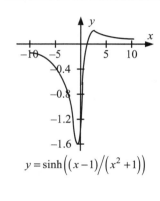

$$y = \sinh\left((x-1)\big/\left(x^2+1\right)\right)$$

Local maximum: $\left(1+\sqrt{2},\ 0.2086\right)$

Local minimum: $\left(1-\sqrt{2},\ -1.5224\right)$

Points of inflection: $\left(-0.8972,\ -1.2556\right)$, $\left(0.1374,\ -0.9514\right)$, $\left(-0.8814,\ 0.1843\right)$

Increasing on the interval: $\left(1-\sqrt{2},\ 1+\sqrt{2}\right)$

Decreasing on the intervals: $\left(-\infty,\ 1-\sqrt{2}\right)$ and $\left(1+\sqrt{2},\ \infty\right)$

Concave up on the intervals: $\left(-0.8972,\ 0.1374\right)$ and $\left(3.7211,\ \infty\right)$

Concave down on the intervals: $\left(-\infty,\ -0.8972\right)$ and $\left(0.1374,\ 3.7211\right)$

Horizontal asymptote: $y = 0$

Vertical asymptote: None

63.   We calculate the derivatives:

$$f'(x) = 1 + \operatorname{sech}(x)\tanh(x) - \tanh^2(x)$$

$$f''(x) = -2\tanh(x)\left(1 - \tanh^2(x)\right)$$

$$-\operatorname{sech}(x)\tanh^2(x) + \operatorname{sech}(x)\left(1 - \tanh^2(x)\right)$$

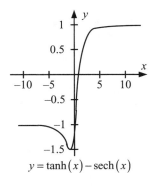

$$y = \tanh(x) - \operatorname{sech}(x)$$

Increasing on the interval: $\left(\ln\left(\sqrt{2}-1\right), \infty\right)$

Decreasing on the interval: $\left(-\infty, \ln\left(\sqrt{2}-1\right)\right)$

Local minimum: $\left(\ln\left(\sqrt{2}-1\right), -\sqrt{2}\right)$

Points of inflection: $(-1.6149, -1.307)$, $(0.4032, -0.5412)$

Concave up on the interval: $(-1.6149, 0.4032)$

Concave down on the intervals: $(-\infty, -1.6149)$ and $(0.4032, \infty)$

Horizontal asymptotes: $y = -1$, $y = 1$

Vertical asymptote: None

65.   The seven points of intersection in the first quadrant are: (see figure)

$$P = \left(\ln\left(1+\sqrt{2}\right), \sqrt{2}\right)$$

$$Q = \left(\ln(\sigma), \frac{1+\rho}{\sigma}\right)$$

$$R = \left(\ln\left(\rho + \sqrt{\rho}\right), \frac{1+\rho}{\sigma}\right)$$

$$S = \left(\ln\left(1+\sqrt{2}\right), 1\right)$$

$$T = \left(\ln(\sigma), \frac{\rho}{\sigma}\right)$$

$$U = \left(\ln\left(\rho + \sqrt{\rho}\right), \frac{\rho}{\sigma}\right)$$

$$V = \left(\ln\left(1+\sqrt{2}\right), \frac{1}{\sqrt{2}}\right)$$

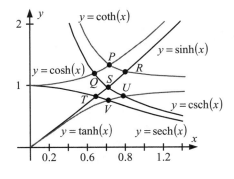

where $\rho = \dfrac{1+\sqrt{5}}{2}$ and $\sigma = \sqrt{1+2\rho}$.

67.   **a.**   See figure.
          b.   See figure.
          c.   See figure.

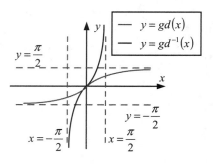

Techniques of Integration

Integration by Parts

Problems for Practice

1. Let $u = x$, $dv = e^x dx$, $du = dx$, $v = e^x$. Then

$$\int xe^x dx = xe^x - \int e^x dx = xe^x - e^x + C.$$

3. Let $u = x$, $dv = \sin(x) dx$, $du = dx$, $v = -\cos(x)$. Then

$$\int x\sin(x) dx = -x\cos(x) + \int \cos(x) dx = -x\cos(x) + \sin(x) + C.$$

5. Let $u = x$, $dv = \cos(2x) dx$, $du = dx$, $v = \dfrac{\sin(2x)}{2}$. Then

$$\int x\cos(2x) dx = x\frac{\sin(2x)}{2} - \left(\frac{1}{2}\right) \int \sin(2x) dx = x\frac{\sin(2x)}{2} + \frac{\cos(2x)}{4} + C.$$

7. Let $u = x$, $dv = e^x dx$, $du = dx$, $v = e^x$. Then $\displaystyle\int_0^1 xe^x dx = xe^x \Big|_0^1 - \int_0^1 e^x dx = xe^x \Big|_0^1 - e^x \Big|_0^1 = 1$.

9. Let $u = x$, $dv = \cos(2x) dx$, $du = dx$, $v = \dfrac{\sin(2x)}{2}$. Then

$$\int_0^{\pi/4} x\cos(2x) dx = x\frac{\sin(2x)}{2}\Bigg|_0^{\pi/4} - \left(\frac{1}{2}\right) \int_0^{\pi/4} \sin(2x) dx = x\frac{\sin(2x)}{2}\Bigg|_0^{\pi/4} + \frac{\cos(2x)}{4}\Bigg|_0^{\pi/4} = \frac{\pi}{8} - \frac{1}{4}.$$

11.   Let $u = \ln(3x)$, $dv = xdx$, $du = \dfrac{dx}{x}$, $v = \dfrac{x^2}{2}$. Then

$$\int\limits_1^{e/3} x\ln(3x)\,dx = x^2\frac{\ln(3x)}{2}\bigg|_1^{e/3} - \left(\frac{1}{2}\right)\int\limits_1^{e/3} xdx = x^2\frac{\ln(3x)}{2}\bigg|_1^{e/3} - \frac{x^2}{4}\bigg|_1^{e/3} = \frac{e^2}{36} - \frac{\ln(3)}{2} + \frac{1}{4}.$$

13.   Let $u_1 = x^2$, $dv_1 = e^x dx$, $du_1 = 2xdx$, $v_1 = e^x$, $u_2 = x$, $dv_2 = e^x dx$, $du_2 = dx$, $v_2 = e^x$. Applying Integration by Parts formula twice, we obtain

$$\int x^2 e^x dx = x^2 e^x - \int 2xe^x dx = x^2 e^x - 2\left(xe^x - \int e^x dx\right) = x^2 e^x - 2xe^x + 2e^x + C$$

$$= \left(x^2 - 2x + 2\right)e^x + C$$

15.   Let $u_1 = x^2$, $dv_1 = \sin(3x)\,dx$, $du_1 = 2xdx$, $v_1 = -\left(\dfrac{1}{3}\right)\cos(3x)$, $u_2 = x$, $dv_2 = \cos(3x)\,dx$,

$du_2 = dx$, $v_2 = \left(\dfrac{1}{3}\right)\sin(3x)$. Applying Integration by Parts formula twice, we obtain

$$\int x^2 \sin(3x)\,dx = -\left(\frac{1}{3}\right)x^2\cos(3x) + \left(\frac{2}{3}\right)\int x\cos(3x)\,dx$$

$$= -\left(\frac{1}{3}\right)x^2\cos(3x) + \left(\frac{2}{3}\right)\left(\left(\frac{1}{3}\right)x\sin(3x) - \left(\frac{1}{3}\right)\int\sin(3x)\,dx\right)$$

$$= -\left(\frac{1}{3}\right)x^2\cos(3x) + \left(\frac{2}{9}\right)x\sin(3x) + \left(\frac{1}{9}\right)\cos(3x) + C$$

17.   Let $u_1 = x^3$, $dv_1 = e^{-2x}dx$, $du_1 = 3x^2 dx$, $v_1 = -\frac{1}{2}e^{-2x}$, $u_2 = x^2$, $dv_2 = e^{-2x}\,dx$,

$du_2 = 2xdx$, $v_2 = -\frac{1}{2}e^{-2x}$, $u_3 = x$, $dv_3 = e^{-2x}\,dx$, $du_3 = dx$, $v_3 = -\frac{1}{2}e^{-2x}\,dx$. Applying

Integration by Parts formula consecutively three times, we obtain

$$\int x^3 e^{-2x}dx = -\left(\frac{1}{2}\right)x^3 e^{-2x} + \left(\frac{3}{2}\right)\int x^2 e^{-2x}dx = -\left(\frac{1}{2}\right)x^3 e^{-2x} + \left(\frac{3}{2}\right)\left(-\frac{x^2}{2}e^{-2x} + \int xe^{-2x}dx\right)$$

$$= -\left(\frac{1}{2}\right)x^3 e^{-2x} - \frac{3}{4}x^2 e^{-2x} + \frac{3}{2}\left(-\frac{x}{2}e^{-2x} + \frac{1}{2}\int e^{-2x}dx\right)$$

$$= -\left(\frac{1}{2}\right)x^3 e^{-2x} - \left(\frac{3}{4}\right)x^2 e^{-2x} - \left(\frac{3}{4}\right)xe^{-2x} - \left(\frac{3}{8}\right)e^{-2x} + C$$

$$= -\left(4x^3 + 6x^2 + 6x + 3\right)\frac{e^{-2x}}{8} + C$$

19. Using $\ln\left(\sqrt{x}\right)=\left(\dfrac{1}{2}\right)\ln\left(x\right)$, and $u=\ln\left(x\right)$, $dv=dx$, $du=\dfrac{dx}{x}$, $v=x$, we integrate by parts:

$$\int\ln\left(\sqrt{x}\right)dx=\left(\frac{1}{2}\right)\int\ln\left(x\right)dx=\left(\frac{1}{2}\right)\left(x\ln\left(x\right)-\int dx\right)=\left(\frac{1}{2}\right)\left(x\ln\left(x\right)-x\right)+C=x\frac{\ln\left(x\right)-1}{2}+C.$$

21. Using $\ln\left(\dfrac{1}{x}\right)=-\ln\left(x\right)$, for $u=\ln\left(x\right)$, $dv=x^3dx$, $du=\dfrac{dx}{x}$ and $v=\dfrac{x^4}{4}$ we integrate by parts:

$$\int x^3\ln\left(\frac{1}{x}\right)dx=-\int x^3\ln\left(x\right)dx=-\left(\frac{1}{4}\right)\left(x^4\ln\left(x\right)-\int x^3dx\right)=-\left(\frac{1}{4}\right)x^4\ln\left(x\right)+\frac{x^4}{16}+C$$

$$=\left(\frac{1}{4}\right)x^4\ln\left(\frac{1}{x}\right)+\frac{x^4}{16}+C$$

23. Let $u=\arctan\left(x\right)$, $dv=dx$, $du=\dfrac{dx}{x^2+1}$, $v=x$. Then

$$\int_0^1\arctan\left(x\right)dx=\left(x\arctan\left(x\right)\right)\Big|_0^1-\int_0^1\frac{x}{x^2+1}dx=\left(x\arctan\left(x\right)\right)\Big|_0^1-\left(\frac{1}{2}\right)\ln\left(x^2+1\right)\Big|_0^1=\frac{\pi}{4}-\frac{\ln\left(2\right)}{2}.$$

25. Let $u=\text{arcsec}\left(x\right)$, $dv=xdx$, $du=\dfrac{dx}{x\sqrt{x^2-1}}$, $v=\dfrac{x^2}{2}$. Then

$$\int_1^{\sqrt{2}}x\,\text{arcsec}\left(x\right)dx=\left(\frac{1}{2}\right)\left(x^2\,\text{arcsec}\left(x\right)\Big|_1^{\sqrt{2}}-\int_1^{\sqrt{2}}\frac{x}{\sqrt{x^2-1}}dx\right)=\left(\frac{1}{2}\right)\left(x^2\,\text{arcsec}\left(x\right)\Big|_1^{\sqrt{2}}-\sqrt{x^2-1}\Big|_1^{\sqrt{2}}\right)$$

$$=\frac{\pi}{4}-\frac{\sqrt{3}}{2}$$

27. There are two intersection points: $\left(0,\,0\right)$ and $\left(1,\,\dfrac{\pi}{2}\right)$.

The area, A, bounded from above by $y=\dfrac{\pi x}{2}$ and from below by $y=\arcsin\left(x\right)$ is

$$A=\int_0^1\left(\frac{\pi x}{2}-\arcsin\left(x\right)\right)dx$$

$$=\left(\frac{\pi}{2}\right)\left(\frac{x^2}{2}\right)\Big|_0^1-\int_0^1\arcsin\left(x\right)dx=1-\frac{\pi}{4}$$

— $y=\pi x/2$

— $y=\arcsin\left(x\right)$

(We used the result of Problem 7.1.24.)

Further Theory and Practice

29. Using $u = x$, $dv = 2^x dx$, we obtain $du = dx$, $v = \dfrac{2^x}{\ln(2)}$. We then evaluate the integral as

follows: $\displaystyle\int x 2^x dx = \frac{x 2^x}{\ln(2)} - \left(\frac{1}{\ln(2)}\right)\int 2^x dx = \frac{x 2^x}{\ln(2)} - \frac{2^x}{\ln^2(2)} + C = -2^x \frac{1/\ln(2) - x}{\ln(2)} + C$.

31. Using $u = \ln(x)$, $dv = \dfrac{dx}{\sqrt{x}}$, we obtain $du = \dfrac{dx}{x}$, $v = 2\sqrt{x}$. We then evaluate the integral as

follows: $\displaystyle\int\left(\frac{\ln(x)}{\sqrt{x}}\right) dx = 2\sqrt{x}\ln(x) - 2\int\left(\frac{1}{\sqrt{x}}\right) dx = 2\sqrt{x}\ln(x) - 4\sqrt{x} + C = 2\sqrt{x}\left(\ln(x) - 2\right) + C$.

33. Using $u = x$, $dv = \sec(x)\tan(x) dx$, we obtain $du = dx$, $v = \sec(x)$. We then evaluate the integral as follows:

$$\int x \sec(x)\tan(x) dx = x\sec(x) - \int\sec(x) dx = x\sec(x) - \ln\left(\left|\sec(x) + \tan(x)\right|\right) + C.$$

35. Using $u = \ln(1 + x^2)$, $dv = dx$, we obtain $du = \left(\dfrac{2x}{1 + x^2}\right) dx$, $v = x$. We then evaluate the integral as follows:

$$\int \ln(1 + x^2) dx = x\ln(1 + x^2) - 2\int\frac{x^2}{1 + x^2} dx = x\ln(1 + x^2) - 2\left(\int dx - \int\frac{dx}{1 + x^2}\right)$$

$$= x\ln(1 + x^2) - 2x + 2\arctan(x) + C$$

37. We integrate by parts

$$I = \int(x - 2)^2 \ln(x) dx = \int x^2 \ln(x) dx - 4\int x\ln(x) dx + 4\int\ln(x) dx = I_1 - 4I_2 + 4I_3.$$

To evaluate I_3 we take $u = \ln(x)$, $dv = dx$, $du = \dfrac{dx}{x}$, $v = x$:

$$I_3 = \int\ln(x) dx = x\ln(x) - \int dx = x\ln(x) - x + C_1.$$

To compute the integral I_2 we use $u = \ln(x)$, $dv = x dx$, $du = \dfrac{dx}{x}$, $v = \dfrac{x^2}{2}$:

$$I_2 = \int x\ln(x) dx = \left(\frac{1}{2}\right)\left(x^2 \ln(x) - \int x dx\right) = \left(\frac{1}{2}\right)\left(x^2 \ln(x) - \frac{x^2}{2}\right) + C_2.$$

To compute the integral I_1 we use $u = \ln(x)$, $dv = x^2 dx$, $du = \dfrac{dx}{x}$, $v = \dfrac{x^3}{3}$:

$$I_1 = \int x^2 \ln(x)\, dx = \left(\frac{1}{3}\right)\left(x^3 \ln(x) - \int x^2 dx\right) = \left(\frac{1}{3}\right)\left(x^3 \ln(x) - \frac{x^3}{3}\right) + C_3.$$

Then

$$I = \left(\frac{1}{3}\right)\left(x^3 \ln(x) - \frac{x^3}{3}\right) - 2\left(x^2 \ln(x) - \frac{x^2}{2}\right) + 4\left(x\ln(x) - x\right) + C$$

$$= \left(\frac{1}{3}\right)x^3 \ln(x) - \left(\frac{1}{9}\right)x^3 - 2x^2 \ln(x) + x^2 + 4x\ln(x) - 4x + C$$

39.   Using $u = x$, $dv = \dfrac{dx}{\sqrt{x+3}}$, we obtain $du = dx$, $v = 2\sqrt{x+3}$. We then evaluate the integral as follows:

$$\int_{-1}^{1}\left(\frac{x}{(x+3)^{1/2}}\right)dx = 2x\sqrt{x+3}\Big|_{-1}^{1} - 2\int_{-1}^{1}\sqrt{x+3}\,dx = 2x\sqrt{x+3}\Big|_{-1}^{1} - \left(\frac{4}{3}\right)(x+3)^{3/2}\Big|_{-1}^{1} = \frac{2\left(7\sqrt{2}-10\right)}{3}.$$

41.   Using $t = \sqrt{x}$, we obtain $x = t^2$, $dx = 2t\,dt$. Taking into account the result of Problem 7.1.3, we integrate by parts:

$$\int \sin\left(\sqrt{x}\right)dx = 2\int t\sin(t)\,dt = 2\left(\sin(t) - t\cos(t)\right) + C = 2\left(\sin\left(\sqrt{x}\right) - \sqrt{x}\cos\left(\sqrt{x}\right)\right) + C.$$

43.   Using $t = x^2$, we obtain $dt = 2x\,dx$. Taking into account the result from Problem 7.1.13, we integrate by parts:

$$\int x^5 e^{x^2}\,dx = \left(\frac{1}{2}\right)\int t^2 e^t\,dt = \left(\frac{1}{2}\right)\left(t^2 - 2t + 2\right)e^t + C = \left(\frac{1}{2}\right)\left(x^4 - 2x^2 + 2\right)e^{x^2} + C$$

$$= \left(\frac{x^4}{2} - x^2 + 1\right)e^{x^2} + C = \left(\frac{1}{2}\right)x^4 e^{x^2} - x^2 e^{x^2} + e^{x^2} + C$$

45.   Using $t = x^3$, we obtain $dt = 3x^2 dx$. Taking into account the result from Problem 7.1.3, we integrate by parts:

$$\int x^5 \sin\left(x^3\right)dx = \left(\frac{1}{3}\right)\int t\sin(t)\,dt = \left(\frac{1}{3}\right)\left(\sin(t) - t\cos(t)\right) + C = \left(\frac{1}{3}\right)\left(\sin\left(x^3\right) - x^3 \cos\left(x^3\right)\right) + C.$$

47. We integrate directly: $\int 2^x e^x dx = \int (2e)^x \, dx = \dfrac{(2e)^x}{\ln(2e)} + C.$

49. We denote $I = \int \cos(x)\sin(3x)\,dx$. Using $u_1 = \sin(3x)$, $dv_1 = \cos(x)\,dx$, we obtain $du_1 = 3\cos(3x)\,dx$, $v_1 = \sin(x)$, and $I = \sin(3x)\sin(x) - 3\int \sin(x)\cos(3x)\,dx$. For the new integral we use $u_2 = \cos(3x)$, $dv_2 = \sin(x)\,dx$, and compute $du_2 = -3\sin(3x)\,dx$, $v_2 = -\cos(x)$. Then

$$I = \sin(3x)\sin(x) - 3\int \sin(x)\cos(3x)\,dx$$
$$= \sin(3x)\sin(x) - 3\left(-\cos(x)\cos(3x) - 3\int \cos(x)\sin(3x)\,dx\right)$$
$$= \sin(3x)\sin(x) + 3\cos(x)\cos(3x) + 9I.$$

As a result,

$$I = -\left(\frac{3}{8}\right)\cos(x)\cos(3x) - \left(\frac{1}{8}\right)\sin(x)\sin(3x) + C.$$

51. First we apply Problem 1.6.40 to obtain:
$\sin(ax)\cos(bx) = \left(\dfrac{1}{2}\right)\big(\sin((a-b)x) + \sin((a+b)x)\big).$ We then derive

$$\int \sin(ax)\cos(bx)\,dx = \left(\frac{1}{2}\right)\int \sin((a-b)x)\,dx + \left(\frac{1}{2}\right)\int \sin((a+b)x)\,dx$$
$$= -\frac{\cos((a-b)x)}{2(a-b)} - \frac{\cos((a+b)x)}{2(a+b)} + C$$

53. First we apply Problem 1.6.39 to obtain:
$\cos(ax)\cos(bx) = \left(\dfrac{1}{2}\right)\big(\cos((a-b)x) + \cos((a+b)x)\big).$ We then derive

$$\int \cos(ax)\cos(bx)\,dx = \left(\frac{1}{2}\right)\int \cos((a-b)x)\,dx + \left(\frac{1}{2}\right)\int \cos((a+b)x)\,dx$$
$$= \frac{\sin((a-b)x)}{2(a-b)} + \frac{\sin((a+b)x)}{2(a+b)} + C$$

55. We denote $I_k = \int_0^1 x^k f^{(k)}(x)\,dx$. Using $u = x^k$, $dv = f^{(k)}(x)\,dx$, we obtain $du = kx^{k-1}\,dx$,

$v = f^{(k-1)}(x)$, and $I_k = x^k f^{(k-1)}(x)\big|_0^1 - k\int_0^1 x^{k-1} f^{(k-1)}(x)\,dx = f^{(k-1)}(1) - kI_{k-1}$. Applying

$I_{k-1} = f^{(k-2)}(1) - (k-1)I_{k-2}$ as a recursive relation, we obtain

$$I_k = f^{(k-1)}(1) - kf^{(k-2)}(1) + k(k-1)f^{(k-3)}(1) - \cdots + (-1)^{k-1} k! f^{(0)}(1)$$
$$+ (-1)^k k! \int_0^1 f(x)\,dx.$$

57. Using $u = e^{ax}$, $dv = \sin(bx)\,dx$, we obtain $du = ae^{ax}\,dx$, $v = -\left(\frac{1}{b}\right)\cos(bx)$. Applying the

Integration by Parts Formula and the result of Problem 7.1.56 (or using the Integration by Parts Formula once again), we find that

$$I = \int e^{ax}\sin(bx)\,dx = -\left(\frac{1}{b}\right)e^{ax}\cos(bx) + \left(\frac{a}{b}\right)\int e^{ax}\cos(bx)\,dx$$

$$= -\left(\frac{1}{b}\right)e^{ax}\cos(bx) + \left(\frac{a}{b}\right)\frac{e^{ax}(b\sin(bx) + a\cos(bx))}{a^2 + b^2} = e^{ax}\frac{a\sin(bx) - b\cos(bx)}{a^2 + b^2} + C$$

59. a. Using $u = \text{FresnelS}(x)$, $dv = dx$, we obtain $du = \sin\left(\frac{\pi x^2}{2}\right)dx$, $v = x$. Then, with the

substitution $t = \frac{\pi}{2}x^2$, $dt = \pi x\,dx$,

$$\int \text{FresnelS}(x)\,dx = x\,\text{FresnelS}(x) - \int x\sin\left(\frac{\pi x^2}{2}\right)dx = x\,\text{FresnelS}(x) - \left(\frac{1}{\pi}\right)\int \sin(t)\,dt$$

$$= x\,\text{FresnelS}(x) + \left(\frac{1}{\pi}\right)\cos(t) + C$$

$$= x\,\text{FresnelS}(x) + \left(\frac{1}{\pi}\right)\cos\left(\frac{\pi x^2}{2}\right) + C$$

b. Using $u = \text{FresnelC}(x)$, $dv = dx$, we obtain $du = \cos\left(\dfrac{\pi x^2}{2}\right)dx$, $v = x$. Then, with the substitution $t = \dfrac{\pi}{2}x^2$, $dt = \pi x\,dx$,

$$\int \text{FresnelC}(x)\,dx = x\,\text{FresnelC}(x) - \int x\cos\left(\frac{\pi x^2}{2}\right)dx = x\,\text{FresnelC}(x) - \left(\frac{1}{\pi}\right)\int \cos(t)\,dt$$

$$= x\,\text{FresnelC}(x) - \left(\frac{1}{\pi}\right)\sin(t) + C$$

$$= x\,\text{FresnelC}(x) - \left(\frac{1}{\pi}\right)\sin\left(\frac{\pi x^2}{2}\right) + C$$

c. Using $u = \text{FresnelS}(x)$, $dv = x\,dx$, we obtain $du = \sin\left(\dfrac{\pi x^2}{2}\right)dx$, $v = \dfrac{x^2}{2}$. Then

$$\int x\,\text{FresnelS}(x)\,dx = \left(\frac{1}{2}\right)\left(x^2\,\text{FresnelS}(x) - \int x^2 \sin\left(\frac{\pi x^2}{2}\right)dx\right). \text{ To evaluate the last}$$

integral, we apply the Integration by Parts Formula with $u = x$, $dv = x\sin\left(\dfrac{\pi x^2}{2}\right)dx$,

$du = dx$ and $v = -\left(\dfrac{1}{\pi}\right)\cos\left(\dfrac{\pi x^2}{2}\right)$, and obtain

$$\int x^2 \sin\left(\frac{\pi x^2}{2}\right)dx = -\left(\frac{1}{\pi}\right)\left(x\cos\left(\frac{\pi x^2}{2}\right) - \int \cos\left(\frac{\pi x^2}{2}\right)dx\right)$$

$$= -\left(\frac{1}{\pi}\right)\left(x\cos\left(\frac{\pi x^2}{2}\right) - \text{FresnelC}(x)\right)$$

Finally, we combine our results to obtain

$$\int x\,\text{FresnelS}(x) = x^2\,\frac{\text{FresnelS}(x)}{2} + \frac{x\cos \pi x^2/2}{2\pi} - \frac{\text{FresnelC}(x)}{2\pi} + C.$$

Following a similar procedure to that above we obtain

$$\int x\,\text{FresnelC}(x)\,dx = x^2\,\frac{\text{FresnelC}(x)}{2} - \frac{x\sin \pi x^2/2}{2\pi} - \frac{\text{FresnelS}(x)}{2\pi} + C.$$

Calculator/Computer Problems

61.   Using $\int_0^{\rho} 4r^2 e^{-2r}\, dr = -\left(2r^2 + 2r + 1\right)e^{-2r}\Big|_0^{\rho} = 1 - \left(2\rho^2 + 2\rho + 1\right)e^{-2\rho}$, we solve

$$1 - \left(2\rho^2 + 2\rho + 1\right)e^{-2\rho} = \frac{1}{2}$$

to find $\rho \approx 1.337$. Therefore the electron is within a distance $1.337 \cdot \alpha_0$ of a nucleus with the probability $\dfrac{1}{2}$.

63.   We integrate by parts: $\lim\limits_{x \to 0^+} x \ln(x) = 0$, then

$$\lim\limits_{x \to 0^+} x^2 \ln(x) = \lim\limits_{x \to 0^+} x \cdot \lim\limits_{x \to 0^+} x \ln(x) = 0.$$

The area of the region is

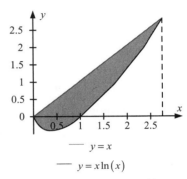
— $y = x$
— $y = x\ln(x)$

$$A = \int_0^{e} \left(x - x\ln(x)\right) dx = \lim\limits_{a \to 0^+} \int_a^{e} \left(x - x\ln(x)\right) dx = \lim\limits_{a \to 0^+} \left(\left(\frac{3}{4}\right)x^2 - \left(\frac{1}{2}\right)x^2 \ln(x)\right)\Bigg|_a^{e}$$

$$= \lim\limits_{a \to 0^+} \left(\left(\frac{3}{4}\right)e^2 - \left(\frac{1}{2}\right)e^2 - \left(\frac{3}{4}\right)a^2 + \left(\frac{1}{4}\right)a^2 \ln(a)\right) = \left(\frac{1}{4}\right)e^2$$

65.   We solve $\arctan(x) - 1 = \dfrac{\ln(x)}{x}$ and find that it has two roots: $a \approx 0.7671$ and $b \approx 4.2597$ (see figure). The area of the region is:

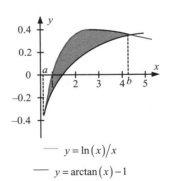
······ $y = \ln(x)/x$
— $y = \arctan(x) - 1$

$$A = \int_a^{b} \left(\frac{\ln(x)}{x} - \left(\arctan(x) - 1\right)\right) dx$$

$$= \left(\left(\frac{1}{2}\right)\ln^2(x) + \left(\frac{1}{2}\right)\ln\left(1 + x^2\right) - x\arctan(x) + x\right)\Bigg|_a^{b}$$

$$= 0.5452$$

67.   For $f(x) = x^2$ we can use substitution $t = nx$ and the result of Problem 7.1.14 to obtain

$$a_0 = \left(\frac{1}{\pi}\right) \int_{-\pi}^{\pi} x^2 dx = \frac{2\pi^2}{3}.$$

For $n \geq 1$ we integrate by parts:

$$a_n = \left(\frac{1}{\pi}\right) \int_{-\pi}^{\pi} x^2 \cos(nx) dx$$

$$= \left(\frac{1}{\pi n^3}\right) \left(n^2 x^2 \sin(nx) - 2\sin(nx) + 2nx\cos(nx)\right)\Big|_{-\pi}^{\pi}$$

$$= \left(\frac{4}{n^2}\right) \cos(n\pi) = (-1)^n \frac{4}{n^2}$$

It follows that $a_1 = -4$, $a_2 = 1$, $a_3 = -\frac{4}{9}$, $a_4 = \frac{1}{4}$, $a_5 = -\frac{4}{25}$ and we integrate by parts: Fourier cosine-approximation polynomials for $N = 3$ and $N = 5$ in the form

$$F_3 = \frac{\pi^2}{3} - 4\cos(x) + \cos(2x) - \left(\frac{4}{9}\right)\cos(3x)$$

and $F_5 = \frac{\pi^2}{3} - 4\cos(x) + \cos(2x) - \left(\frac{4}{9}\right)\cos(3x) + \left(\frac{1}{4}\right)\cos(4x) - \left(\frac{4}{25}\right)\cos(5x)$.

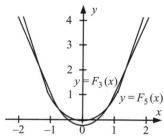

$$F_3 = \pi^2/3 - 4\cos(x) + \cos(2x) - (4/9)\cos(3x)$$
$$F_5 = \pi^2/3 - 4\cos(x) + \cos(2x) - (4/9)\cos(3x) + (1/4)\cos(4x) - (4/25)\cos(5x)$$

Section 7.2   Partial Fractions—Linear Factors

Problems for Practice

1.   The degree of the numerator is less than that of the denominator. The partial fraction decomposition of the rational function is $\dfrac{x+2}{x(x+1)} = \dfrac{A}{x} + \dfrac{B}{x+1}$.

3. The degree of the numerator is less than that of the denominator, $x^2 - 4 = (x-2)(x+2)$. The partial fraction decomposition of the rational function is $\dfrac{3x+1}{x^2-4} = \dfrac{A}{x-2} + \dfrac{B}{x+2}$.

5. The degree of the numerator is greater than that of the denominator. The partial fraction decomposition of the rational function is $\dfrac{x^3}{(2x-5)(x+4)} = Ax + B + \dfrac{C}{2x-5} + \dfrac{D}{x+4}$.

7. The degree of the numerator is less than that of the denominator. The partial fraction decomposition of the rational function is $\dfrac{x^3+7}{x^2(x+1)(x-2)} = \dfrac{A}{x} + \dfrac{B}{x^2} + \dfrac{C}{x+1} + \dfrac{D}{x-2}$.

9. The degree of the numerator is greater than that of the denominator. The partial fraction decomposition of the rational function is

$$\frac{x^7 + x^6 - 17}{(x-5)^2(x+3)^3} = Ax^2 + Bx + C + \frac{D}{x-5} + \frac{E}{(x-5)^2} + \frac{F}{x+3} + \frac{G}{(x+3)^2} + \frac{H}{(x+3)^3}.$$

11. The degree of the numerator is less than that of the denominator. The partial fraction decomposition is $\dfrac{1}{x(x+1)} = \dfrac{A}{x} + \dfrac{B}{x+1}$. We put the terms on the right over the common denominator $\dfrac{1}{x(x+1)} = \dfrac{A(x+1)+Bx}{x(x+1)}$. Then we have the identity $1 = A(x+1) + Bx$ or $1 = (A+B)x + A$.

Comparing coefficients with like powers of x we obtain a system $\begin{cases} 0 = A+B \\ 1 = A \end{cases}$. Its solution is $A = 1$, $B = -1$. Therefore $\dfrac{1}{x(x+1)} = \dfrac{1}{x} - \dfrac{1}{x+1}$.

13. The degree of the numerator is less than that of the denominator. The partial fraction decomposition is $\dfrac{x+6}{x^2-4} = \dfrac{A}{x-2} + \dfrac{B}{x+2}$. We put the terms on the right over the common denominator $\dfrac{x+6}{(x-2)(x+2)} = \dfrac{A(x+2)+B(x-2)}{(x-2)(x+2)}$. Then we have the identity $x+6 = A(x+2) + B(x-2)$ or $x+6 = (A+B)x + (2A-2B)$.

Comparing coefficients with like powers of x we obtain a system $\begin{cases} 1 = A+B \\ 6 = 2A-2B \end{cases}$. Its solution is $A = 2$, $B = -1$. Therefore $\dfrac{x+6}{x^2-4} = \dfrac{2}{x-2} - \dfrac{1}{x+2}$.

15.   The degree of the numerator is less than that of the denominator. The partial fraction

decomposition is $\dfrac{26x}{(2x-5)(x+4)} = \dfrac{A}{2x-5} + \dfrac{B}{x+4}$. We put the terms on the right over the

common denominator $\dfrac{26x}{(2x-5)(x+4)} = \dfrac{A(x+4)+B(2x-5)}{(2x-5)(x+4)}$. Then we have the identity

$26x = A(x+4)+B(2x-5)$ or $26x = (A+2B)x+(4A-5B)$.

Comparing coefficients with like powers of x we obtain a system $\begin{cases} 26 = A+2B \\ 0 = 4A-5B \end{cases}$. Its solution is

$A=10$, $B=8$. Therefore $\dfrac{26x}{(2x-5)(x+4)} = \dfrac{10}{2x-5} + \dfrac{8}{x+4}$.

17.   The degree of the numerator is less than that of the denominator. The partial fraction

decomposition is $\dfrac{2}{x^2(x-1)(x+2)} = \dfrac{A}{x} + \dfrac{B}{x^2} + \dfrac{C}{x-1} + \dfrac{D}{x+2}$. We put the terms on the right

over the common denominator

$$\frac{2}{x^2(x-1)(x+2)} = \frac{Ax(x-1)(x+2)+B(x-1)(x+2)+Cx^2(x+2)+Dx^2(x-1)}{x^2(x-1)(x+2)}.$$

Then we have the identity $2 = Ax(x-1)(x+2)+B(x-1)(x+2)+Cx^2(x+2)+Dx^2(x-1)$ or

$2 = (A+C+D)x^3 + (A+B+2C-D)x^2 + (-2A+B)x - 2B$.

Comparing coefficients with like powers of x we obtain a system $\begin{cases} 0 = A+C+D \\ 0 = -A+B-2C+D \\ 0 = -2A-B \\ 2 = -2B \end{cases}$. Its

solution is $A = -\dfrac{1}{2}$, $B=-1$, $C = \dfrac{2}{3}$, $D = -\dfrac{1}{6}$. Therefore

$$\frac{2}{x^2(x+1)(x-2)} = -\frac{1}{2}\frac{1}{x} - \frac{1}{x^2} + \frac{2}{3}\frac{1}{x+1} - \frac{1}{6}\frac{1}{x-2}.$$

19.   The degree of the numerator is less than that of the denominator. The partial fraction

decomposition is $\dfrac{2x^3 - 19x^2 + 8x + 313}{(x-5)^2(x+3)^2} = \dfrac{A}{x-5} + \dfrac{B}{(x-5)^2} + \dfrac{C}{x+3} + \dfrac{D}{(x+3)^2}$. We put the terms

on the right over the common denominator

$$\frac{2x^3 - 19x^2 + 8x + 313}{(x-5)^2(x+3)^2} = \frac{A(x-5)(x+3)^2 + B(x+3)^2 + C(x-5)^2(x+3) + D(x-5)^2}{(x-5)^2(x+3)^2}.$$

Then we have the identity

$$2x^3 - 19x^2 + 8x + 313 = A(x-5)(x+3)^2 + B(x+3)^2 + C(x-5)^2(x+3) + D(x-5)^2$$

or

$$2x^3 - 19x^2 + 8x + 313 = (A+C)x^3 + (A+B-7C+D)x^2 + (-21A+6B-5C-10D)x$$
$$+ (-45A+9B+75C+25D)$$

Comparing coefficients with like powers of x we obtain a system

$$\begin{cases} 2 = A+C \\ -19 = A+B-7C+D \\ 8 = -21A+6B-5C-10D \\ 313 = -45A+9B+75C+25D \end{cases}.$$

Its solution is $A = -1$, $B = 2$, $C = 3$, $D = 1$. Therefore

$$\frac{2x^3 - 19x^2 + 8x + 313}{(x-5)^2(x+3)^2} = -\frac{1}{x-5} + \frac{2}{(x-5)^2} + \frac{3}{x+3} + \frac{1}{(x+3)^2}.$$

21.   The partial fraction decomposition of the integrand is $\dfrac{3x+1}{x^2-1} = \dfrac{A}{x-1} + \dfrac{B}{x+1}$. Putting the right

side over the common denominator leads to $3x+1 = A(x+1) + B(x-1)$. Substituting $x = 1$

and $x = -1$ we obtain $A = 2$, $B = 1$. Therefore

$$\int \frac{3x+1}{x^2-1}dx = \int \frac{2}{x-1}dx + \int \frac{1}{x+1}dx = 2\ln\left(|x-1|\right) + \ln\left(|x+1|\right) + C.$$

23. The partial fraction decomposition of the integrand is $\dfrac{9x+18}{(x-3)(x+6)} = \dfrac{A}{x-3} + \dfrac{B}{x+6}$. Putting the right side over the common denominator leads to $9x+18 = A(x+6) + B(x-3)$. Substituting $x=3$ and $x=-6$ we obtain $A=5$, $B=4$. Therefore

$$\int \frac{9x+18}{(x-3)(x+6)}\,dx = \int \frac{5}{x-3}\,dx + \int \frac{4}{x+6}\,dx = 5\ln\big(|x-3|\big) + 4\ln\big(|x+6|\big) + C.$$

25. After dividing we obtain $\dfrac{x^3-4x^2+6}{x^2-3x-4} = x-1+\dfrac{2+x}{x^2-3x-4}$. The partial fraction decomposition

is $\dfrac{2+x}{x^2-3x-4} = \dfrac{A}{x-4} + \dfrac{B}{x+1}$. Putting the right side over the common denominator leads to

$2+x = A(x+1) + B(x-4)$. Substituting $x=4$ and $x=-1$ we obtain $A=\dfrac{6}{5}$, $B=-\dfrac{1}{5}$.

Therefore

$$\int \frac{x^3-4x^2+6}{x^2-3x-4}\,dx = \int x\,dx + \int(-1)\,dx + \int\left(\frac{6}{5}\right)\frac{1}{x-4}\,dx + \int\left(-\frac{1}{5}\right)\frac{1}{x+1}\,dx$$

$$= \frac{x^2}{2} - x + \frac{6}{5}\ln\big(|x-4|\big) - \frac{1}{5}\ln\big(|x+1|\big) + C$$

27. The partial fraction decomposition of the integrand is

$$\frac{9x}{(x-2)^2(x+1)} = \frac{A}{(x-2)^2} + \frac{B}{x-2} + \frac{C}{x+1}.$$

Putting the right side over the common denominator leads to

$$9x = A(x+1) + B(x-2)(x+1) + C(x-2)^2.$$

Substituting $x=2$ and $x=-1$ we obtain $A=6$, $C=-1$. Comparing the coefficients at x^2 we get $B+C=0$ or $B=-C=1$. Therefore

$$\int \frac{9x}{(x-2)^2(x+1)}\,dx = \int \frac{6}{(x-2)^2}\,dx + \int \frac{1}{x-2}\,dx + \int \frac{(-1)}{x+1}\,dx$$

$$= -\frac{6}{x-2} + \ln\big(|x-2|\big) - \ln\big(|x+1|\big) + C$$

29. The partial fraction decomposition of the integrand is $\dfrac{x^2+1}{x^3+x^2}=\dfrac{A}{x^2}+\dfrac{B}{x}+\dfrac{C}{x+1}$. Putting the

right side over the common denominator leads to $x^2+1=A(x+1)+Bx(x+1)+Cx^2$.

Substituting $x=0$ and $x=-1$ we obtain $A=1$, $C=2$. Comparing the coefficients at x^2 we

get $B+C=1$ or $B=1-C=-1$. Therefore

$$\int\frac{x^2+1}{x^3+x^2}\,dx=\int\frac{1}{x^2}\,dx+\int\frac{(-1)}{x}\,dx+\int\frac{2}{x+1}\,dx=-\frac{1}{x}-\ln\left(|x|\right)+2\ln\left(|x+1|\right)+C\,.$$

31. After dividing we obtain $\dfrac{x^2+3x+6}{(x-2)(x+5)}=1+\dfrac{16}{(x-2)(x+5)}$. The partial fraction

decomposition is $\dfrac{16}{(x-2)(x+5)}=\dfrac{A}{x-2}+\dfrac{B}{x+5}$. Putting the right side over the common

denominator leads to $16=A(x+5)+B(x-2)$. Substituting $x=2$ and $x=-5$ we obtain

$A=\dfrac{16}{7}$, $B=-\dfrac{16}{7}$. Therefore

$$\int_3^4\frac{x^2+3x+6}{(x-2)(x+5)}\,dx=\int_3^4 1\,dx+\int_3^4\left(\frac{16}{7}\right)\frac{1}{x-2}\,dx+\int_3^4\left(-\frac{16}{7}\right)\frac{1}{x+5}\,dx$$

$$=\left(x+\frac{16}{7}\ln\left(|x-2|\right)-\frac{16}{7}\ln\left(|x+5|\right)\right)\bigg|_3^4$$

$$=4-3+\frac{16}{7}\ln(2)-\frac{16}{7}\ln(9)+\frac{16}{7}\ln(8)=1+\frac{16}{7}\ln\left(\frac{16}{9}\right)$$

33. The partial fraction decomposition of the integrand is $\dfrac{x+2}{(2x-1)(3x+4)}=\dfrac{A}{2x-1}+\dfrac{B}{3x+4}$.

Putting the right side over the common denominator leads to $x+2=A(3x+4)+B(2x-1)$.

Substituting $x=\dfrac{1}{2}$ and $x=-\dfrac{4}{3}$ we obtain $A=\dfrac{5}{11}$, $B=-\dfrac{2}{11}$. Therefore

$$\int_1^4\frac{x+2}{(2x-1)(3x+4)}\,dx=\int_1^4\left(\frac{5}{11}\right)\frac{1}{2x-1}\,dx+\int_1^4\left(-\frac{2}{11}\right)\frac{1}{3x+4}\,dx$$

$$=\left(\frac{5}{22}\ln\left(|2x-1|\right)-\frac{2}{33}\ln\left(|3x+4|\right)\right)\bigg|_1^4$$

$$=\frac{5}{22}\ln(7)-\frac{2}{33}\ln(16)+\frac{2}{33}\ln(7)=\frac{19}{66}\ln(7)-\frac{2}{33}\ln(16)$$

Further Theory and Practice

35.   Using $x = e^t$, $dx = e^t dt$ we obtain $\int \dfrac{e^t}{e^{2t}-1}dx = \int \dfrac{dx}{x^2-1}$. The partial fraction decomposition of

the integrand is $\dfrac{1}{(x-1)(x+1)} = \dfrac{A}{x-1} + \dfrac{B}{x+1}$. Putting the right side over the common

denominator leads to $1 = A(x+1) + B(x-1)$. Substituting $x = 1$ and $x = -1$ we obtain $A = \dfrac{1}{2}$,

$B = -\dfrac{1}{2}$. Therefore

$$\int \frac{dx}{x^2-1} = \int \frac{1}{2}\frac{1}{x-1}dt - \int \frac{1}{2}\frac{1}{x+1}dt = \frac{1}{2}\ln\left(\left|x-1\right|\right) - \frac{1}{2}\ln\left(\left|x+1\right|\right) + C$$

$$= \frac{1}{2}\ln\left(\left|e^t - 1\right|\right) - \frac{1}{2}\ln\left(\left|e^t + 1\right|\right) + C$$

37.   Using $\sin(x) = t$, $dt = \cos(x)dx$ we obtain $\int \dfrac{16\cos(x)}{(3+\sin(x))^2(1-\sin(x))}dx = \int \dfrac{16dt}{(3+t)^2(1-t)}$.

The partial fraction decomposition of the integrand is $\dfrac{1}{(3+t)^2(1-t)} = \dfrac{A}{(3+t)^2} + \dfrac{B}{3+t} + \dfrac{C}{1-t}$.

Putting the right side over the common denominator leads to

$$16 = A(1-t) + B(3+t)(1-t) + C(3+t)^2.$$

Substituting $t = 1$ and $t = -3$ we obtain $A = 4$, $C = 1$. Comparing the coefficients at x^2 we
get $-B + C = 0$ or $B = C = 1$. Therefore

$$\int \frac{16dt}{(3+t)^2(1-t)} = \int \frac{4}{(3+t)^2}dt + \int \frac{1}{3+t}dt + \int \frac{1}{1-t}dt = -\frac{4}{3+t} + \ln\left(\left|3+t\right|\right) - \ln\left(\left|t-1\right|\right) + C$$

$$= -\frac{4}{3+\sin(x)} + \ln\left(\left|3+\sin(x)\right|\right) - \ln\left(\left|\sin(x)-1\right|\right) + C$$

39.   Using $u = x^{1/3}$, $du = \left(\dfrac{1}{3}\right)x^{-2/3}dx$ we obtain $\int \dfrac{1}{x^{4/3}-x^{2/3}}dx = 3\int \dfrac{du}{u^2-1}$. The partial fraction

decomposition of the integrand is $\dfrac{1}{u^2-1} = \dfrac{A}{u-1} + \dfrac{B}{u+1}$. Putting the right side over the

common denominator leads to $1 = A(u+1) + B(u-1)$. Substituting $u = 1$ and $u = 0$ we obtain
$A = \tfrac{1}{2}$, $B = -\tfrac{1}{2}$. Therefore

$$3\int \frac{du}{u^2-1} = 3\int \left(\frac{1}{2}\right)\frac{1}{u-1}dt + 3\int \left(-\frac{1}{2}\right)\frac{1}{u+1}dt = \frac{3}{2}\ln\left(\left|u-1\right|\right) - \frac{3}{2}\ln\left(\left|u+1\right|\right) + C$$

$$= \frac{3}{2}\ln\left(\left|x^{1/3} - 1\right|\right) - \frac{3}{2}\ln\left(\left|x^{1/3} + 1\right|\right) + C$$

41. **a.**   The correct form of the decomposition is $\dfrac{5x^2-4x+2}{x^2(x-1)^2}=\dfrac{A}{x^2}+\dfrac{B}{(x-1)^2}+\dfrac{C}{x}+\dfrac{D}{x-1}$.

b.   The coefficients C and D are equal to 0 in this case.

c.   The partial fractions decomposition is $\dfrac{5x^2-4x+1}{x^2(x-1)^2}=\dfrac{A}{x^2}+\dfrac{B}{(x-1)^2}$. Putting the terms

over the common denominator leads to the identity $5x^2-4x+1=A(x-1)^2+Bx^2$. Comparing the coefficients with like powers of x we obtain the system

$$\begin{cases}5=A+B\\-4=-2A\\1=A\end{cases}$$

which does not have a solution. The correct partial decomposition is

$$\frac{5x^2-4x+1}{x^2(x-1)^2}=\frac{1}{x^2}+\frac{2}{(x-1)^2}-\frac{2}{x}+\frac{2}{x-1}.$$

43. **a.**   False; $P'(t)<0$

b.   False; $P(t)<15000$ for all t

c.   True; $\lim_{t\to\infty}P(t)=15000$

d.   True; $P'(0)=10^{-5}\left(7500^2-\left(P(0)-7500\right)^2\right)$

45. $P(t_I)=\dfrac{P_0P_\infty}{P_0+(P_\infty-P_0)\exp(-kP_\infty t_I)}=\dfrac{1}{2}P_\infty,\ \ P_0+(P_\infty-P_0)\exp(-kP_\infty t_I)=2P_0,$

$\exp(-kP_\infty t_I)=\dfrac{P_0}{P_\infty-P_0},\ \ t_I=-\dfrac{1}{kP_\infty}\ln\left(\dfrac{P_0}{P_\infty-P_0}\right)=\dfrac{1}{kP_\infty}\ln\left(\dfrac{P_\infty-P_0}{P_0}\right)$

47.   From Exercise 7.2.46 we obtain $P(t)=\dfrac{P_\infty}{1+\exp(-kP_\infty(t-t_I))}$. On the other hand, from the

data shown in the problem statement, i.e., $P(0)=3.6$, $t_I=9$ and $\frac{1}{2}P_\infty=25$, it follows then

that $k=\dfrac{\ln(50-3.6)-\ln(3.6)}{9\cdot50}\approx5.6808\times10^{-3}$.

a.   Logistic differential equation:

$$P'(9)=kP(9)(P_\infty-P_0)=5.6808\times10^{-3}\cdot25\cdot(50-3.6)\approx6.5897.$$

b.   Central difference quotient approximation: $P'(9)\approx\dfrac{P(10)-P(8)}{10-8}=\dfrac{29.8-21.2}{2}=4.3$.

Calculator/Computer Exercises

49. The equation $\dfrac{2x^2+5}{(x+2)(x+3)(x+5)}=\dfrac{x^3+2}{(x+2)(x+3)(x+5)}$ is satisfied when $x=2.3593$.

Therefore

$$\frac{2x^2+5}{(x+2)(x+3)(x+5)}-\frac{x^3+2}{(x+2)(x+3)(x+5)}=\frac{-x^3+2x^2+3}{(x+2)(x+3)(x+5)}=-1+\frac{-2x^3-8x^2-31x-27}{(x+2)(x+3)(x+5)}$$

$$=-1+\frac{A}{x+2}+\frac{B}{x+3}+\frac{C}{x+5}$$

Putting the right side over the common denominator leads to

$$-2x^3-8x^2-31x-27=A(x+3)(x+5)+B(x+2)(x+5)+C(x+2)(x+3).$$

Substituting $x=-2$, $x=-3$ and $x=-5$ we obtain $A=\dfrac{19}{3}$, $B=-24$, $C=\dfrac{89}{3}$. Therefore, the area is equal to

$$\int_0^{2.3593}\left(-1+\frac{19}{3}\frac{1}{x+2}-\frac{24}{x+3}+\frac{89}{3}\frac{1}{x+5}\right)dx=\left(-x+\frac{19}{3}\ln(x+2)-24\ln(x+3)+\frac{89}{3}\ln(x+5)\right)\Big|_0^{2.3593}$$

$$=25.887-\left(\frac{19}{3}\right)\ln(2)+24\ln(3)-\left(\frac{89}{3}\right)\ln(5)=0.117066$$

51. The points of intersection are $x=1.39915$ and $x=2.17953$.

Using partial decomposition we obtain

$$\frac{x^3}{x^2+3x+2}=-x+3+\frac{7x+6}{x^2+3x+2}$$

$$=x-3+\frac{A}{x+1}+\frac{B}{x+2}$$

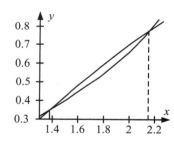

Putting the terms over the common denominator leads to $7x+6=A(x+2)+B(x+1)$.

Substituting $x=-1$ and $x=-2$ we obtain $A=-1$, $B=8$. Therefore, the area enclosed by the two curves is equal to

$$\int_{1.39915}^{2.17953}\left(\ln(x)-\frac{x^3}{x^2+3x+2}\right)dx=\int_{1.39915}^{2.17953}\ln(x)\,dx-\int_{1.39915}^{2.17953}x\,dx-\int_{1.39915}^{2.17953}(-3)\,dx$$

$$-\int_{1.39915}^{2.17953}\frac{(-1)}{x+1}\,dx-\int_{1.39915}^{2.17953}\frac{8}{x+2}\,dx$$

$$=\left(x\ln(x)-x-\frac{x^2}{2}+3x+\ln(x+1)-8\ln(x+2)\right)\Big|_{1.39915}^{2.17953}$$

$$=0.02079$$

53.

a.
$$-\frac{1}{2}\cdot\frac{1}{x+1}+\frac{1}{x+2}-\frac{3}{2}\cdot\frac{1}{x+3}+\frac{2}{x+3}$$

b.
$$3+2x-\frac{3}{(x-2)^2}+\frac{7}{x-2}+\frac{9}{x+2}+\frac{3}{(x+3)^2}-\frac{2}{x+3}$$

c.
$$-\frac{12}{(x-1)^2}+\frac{5}{x-1}+\frac{2}{(x+1)^2}-\frac{3}{(x+2)^3}$$

d.
$$-\frac{13}{(x-1)^3}+\frac{11}{x-1}-\frac{7}{(x+1)^4}+\frac{2}{(x+1)^2}$$

Section 7.3   Powers and Products of Trigonometric Functions

Problems for Practice

1. Using the relations for trigonometric functions of a double argument we evaluate the integral as follows:

$$\int\cos^2(4\theta)\,d\theta=\left(\frac{1}{2}\right)\int(1+\cos(8\theta))\,d\theta=\left(\frac{1}{2}\right)\left(\theta+\left(\frac{1}{8}\right)\sin(8\theta)\right)+C$$

$$=\left(\frac{1}{2}\right)\left(\theta+\left(\frac{1}{4}\right)\sin(4\theta)\cos(4\theta)\right)+C=\frac{\theta}{2}+\sin(4\theta)\frac{\cos(4\theta)}{8}+C$$

3. Given that $\cos^2(t)+\sin^2(t)=1$, we evaluate the integral as follows:

$$\int\left(\cos^2(t)+\sin^2(t)\right)dt=\int 1\cdot dt=t+C.$$

5. Given that $\cos^2(2x)+\sin^2(2x)=1$, we evaluate the integral as follows:

$$\int\sin^3(2x)\,dx=\int\left(1-\cos^2(2x)\right)\sin(2x)\,dx=-\left(\frac{1}{2}\right)\int\left(1-u^2\right)du=-\left(\frac{1}{2}\right)\left(u-\frac{u^3}{3}\right)+C$$

$$=-\left(\frac{1}{2}\right)\left(\cos(2x)-\frac{\cos^3(2x)}{3}\right)+C=-\left(\frac{1}{6}\right)\cos(2x)\left(3-\cos^2(2x)\right)+C$$

$$=-\left(\frac{1}{6}\right)\cos(2x)\left(2+\sin^2(2x)\right)+C=-\left(\frac{1}{3}\right)\cos(2x)\left(1+\frac{\sin^2(2x)}{2}\right)+C$$

where $u=\cos(2x)$.

7.   Given that $\cos^2(x+2)+\sin^2(x+2)=1$, we evaluate the integral as follows:

$$\int \sin^3(x+2)\,dx = \int\left(1-\cos^2(x+2)\right)\sin(x+2)\,dx = -\int\left(1-u^2\right)du = -u+\frac{u^3}{3}+C$$

$$= -\cos(x+2)+\frac{\cos^3(x+2)}{3}+C = -\left(\frac{1}{3}\right)\cos(x+2)\left(2+\sin^2(x+2)\right)+C$$

where $u=\cos(x+2)$.

9.   Using $u=\cos(t)$, $du=-\sin(t)\,dt$, we evaluate the integral as follows:

$$\int_{0}^{\pi/2}\sin(t)\cos^4(t)\,dt = -\int_{1}^{0}u^4\,du = -\frac{u^5}{5}\bigg|_{1}^{0} = \frac{1}{5}.$$

11.   Using $u=\cos(\theta)$, $du=-\sin(\theta)\,d\theta$, and $\cos^2(\theta)+\sin^2(\theta)=1$,   we evaluate the integral as follows:

$$\int_{0}^{\pi}\cos^4(\theta)\sin^3(\theta)\,d\theta = -\int_{1}^{-1}u^4\left(1-u^2\right)du = \int_{-1}^{1}\left(u^4-u^6\right)du = \left(\frac{u^5}{5}-\frac{u^7}{7}\right)\bigg|_{-1}^{1} = \frac{4}{35}.$$

13.   We apply formulas (7.23) and (7.20) from the textbook to evaluate the integral as follows:

$$\int\cos^4(x)\,dx = \left(\frac{1}{4}\right)\sin(x)\cos^3(x)+\left(\frac{3}{4}\right)\int\cos^2(x)\,dx$$

$$= \left(\frac{1}{4}\right)\sin(x)\cos^3(x)+\left(\frac{3}{8}\right)(x+\sin(x)\cos(x))+C$$

$$= \frac{3x}{8}+3\cos(x)\frac{\sin(x)}{8}+\sin(x)\frac{\cos^3(x)}{4}+C$$

15.   We apply formulas (7.22) and (7.19) from the textbook to obtain

$$\int_{\pi}^{2\pi}\sin^4(x)\,dx = \left(\frac{1}{4}\right)\sin^3(x)\cos(x)\bigg|_{\pi}^{2\pi}+\left(\frac{3}{4}\right)\int_{\pi}^{2\pi}\sin^2(x)\,dx$$

$$= \left(\frac{1}{4}\right)\sin^3(x)\cos(x)\bigg|_{\pi}^{2\pi}+\left(\frac{3}{8}\right)(x-\sin(x)\cos(x))\bigg|_{\pi}^{2\pi} = \frac{3\pi}{8}$$

17. $\int_0^{\pi/4} \cos^2(t)\sin^2(t)\,dt = \left(\frac{1}{4}\right)\int_0^{\pi/4}(1+\cos(2t))(1-\cos(2t))\,dt = \left(\frac{1}{4}\right)\int_0^{\pi/4}(1-\cos^2(2t))\,dt$

$$= \left(\frac{1}{4}\right)\int_0^{\pi/4}\left(1-\left(\frac{1}{2}\right)(1+\cos(4t))\right)dt = \left(\frac{1}{8}\right)\int_0^{\pi/4}(1-\cos(4t))\,dt$$

$$= \left(\frac{1}{8}\right)\left(t-\left(\frac{1}{4}\right)\sin(4t)\right)\Big|_0^{\pi/4} = \frac{\pi}{32}$$

19. $\int_{\pi/6}^{\pi/3} \cos^2(x)\sin^4(x)\,dx = \left(\frac{1}{8}\right)\int_{\pi/6}^{\pi/3}(1+\cos(2x))(1-\cos(2x))^2\,dx$

$$= \left(\frac{1}{8}\right)\int_{\pi/6}^{\pi/3}\left(1-\cos(2x)-\cos^2(2x)+\cos^3(2x)\right)dx$$

$$= \left(\frac{1}{8}\right)\int_{\pi/6}^{\pi/3}\left(1-\cos(2x)-\left(\frac{1}{2}\right)(1+\cos(4x))\right)dx$$

$$+ \left(\frac{1}{8}\right)\int_{\pi/6}^{\pi/3}\left(1-\sin^2(2x)\right)\cos(2x)\,dx$$

$$= \left(\frac{1}{8}\right)\left(\frac{x}{2}-\frac{\sin(2x)}{2}-\frac{\sin(4x)}{8}\right)\Big|_{\pi/6}^{\pi/3} + \left(\frac{1}{16}\right)\left(\sin(2x)-\frac{\sin^3(2x)}{3}\right)\Big|_{\pi/6}^{\pi/3}$$

$$= \frac{\pi}{96}+\frac{\sqrt{3}}{64}$$

21. Using $u = \cos(t),\ du = -\sin(t)\,dt,$ we obtain

$$\int \tan(t)\sec^3(t)\,dt = \int \frac{\sin(t)}{\cos^4(t)}\,dt = -\int \frac{du}{u^4} = \frac{1}{3u^3}+C = \frac{1}{3\cos^3(t)}+C = \frac{\sec^3(t)}{3}+C\,.$$

23. Using $u = \cos(t),\ du = -\sin(t)\,dt,$ and $\cos^2(t)+\sin^2(t)=1$, we obtain

$$\int \tan^3(t)\,dt = \int \frac{\sin^3(t)}{\cos^3(t)}\,dt = \int \frac{\left(1-\cos^2(t)\right)\sin(t)}{\cos^3(t)}\,dt = -\int \frac{1-u^2}{u^3}\,du = -\int\left(\frac{1}{u^3}-\frac{1}{u}\right)du$$

$$= \left(\frac{1}{2u^2}+\ln(|u|)\right)+C = \frac{1}{2\cos^2(t)}+\ln(|\cos(t)|)+C = \frac{\sec^2(t)}{2}+\ln(|\cos(t)|)+C\,.$$

25.   Using $\cos^2(t) + \sin^2(t) = 1$, we obtain

$$\int \tan(t)\sin(t)\,dt = \int \frac{\sin^2(t)}{\cos(t)}\,dt = \int \frac{1-\cos^2(t)}{\cos(t)}\,dt = \int \sec(t)\,dt - \int \cos(t)\,dt$$
$$= \ln\left(\left|\sec(t) + \tan(t)\right|\right) - \sin(t) + C$$

27.   Using $\cos^2(\theta) + \sin^2(\theta) = 1$, we obtain

$$\int \frac{1}{1+\cos(\theta)}\,d\theta = \int \frac{1-\cos(\theta)}{1-\cos^2(\theta)}\,d\theta = \int \frac{1-\cos(\theta)}{\sin^2(\theta)}\,d\theta = \int \frac{1}{\sin^2(\theta)}\,d\theta - \int \frac{\cos(\theta)}{\sin^2(\theta)}\,d\theta$$
$$= -\cot(\theta) + \csc(\theta) + C$$

29.   Using $\cos^2(\theta) + \sin^2(\theta) = 1$, we obtain

$$\int \frac{1}{\sec(\theta)-1}\,d\theta = \int \frac{\sec(\theta)+1}{\sec^2(\theta)-1}\,d\theta = \int \frac{\sec(\theta)+1}{\tan^2(\theta)}\,d\theta = \int \frac{\cos(\theta)}{\sin^2(\theta)}\,d\theta + \int \frac{\cos^2(\theta)}{\sin^2(\theta)}\,d\theta$$
$$= \int \frac{\cos(\theta)}{\sin^2(\theta)}\,d\theta + \int \frac{1-\sin^2(\theta)}{\sin^2(\theta)}\,d\theta = -\csc(\theta) - \cot(\theta) - \theta + C$$

Further Theory and Practice

31. $\displaystyle\int \cos^6(t)\,dt = \int \left(\cos^2(t)\right)^3\,dt = \left(\frac{1}{8}\right)\int \left(1+\cos(2t)\right)^3\,dt$

$$= \left(\frac{1}{8}\right)\int \left(1 + 3\cos(2t) + 3\cos^2(2t) + \cos^3(2t)\right)\,dt$$

$$= \left(\frac{1}{8}\right)\left(t + 3\frac{\sin(2t)}{2} + \left(\frac{3}{2}\right)\int \left(1+\cos(4t)\right)\,dt + \int \left(1-\sin^2(2t)\right)\cos(2t)\,dt\right)$$

$$= \left(\frac{1}{8}\right)\left(t + 3\frac{\sin(2t)}{2} + \frac{3t}{2} + 3\frac{\sin(4t)}{8} + \frac{\sin(2t)}{2} - \frac{\sin^3(2t)}{6}\right) + C$$

$$= \frac{5t}{16} + \frac{\sin(2t)}{4} + 3\frac{\sin(4t)}{64} - \frac{\sin^3(2t)}{48} + C$$

$$= \frac{5t}{16} + 11\frac{\sin(2t)}{48} + 3\frac{\sin(4t)}{64} + \sin(2t)\frac{\cos^2(2t)}{48} + C$$

33. $\int \cos^2(t)\sin^4(t)\,dt = \left(\dfrac{1}{8}\right)\int (1+\cos(2t))(1-\cos(2t))^2\,dt$

$$= \left(\dfrac{1}{8}\right)\int \left(1-\cos(2t)-\cos^2(2t)+\cos^3(2t)\right)dt$$

$$= \left(\dfrac{1}{8}\right)\int \left(1-\cos(2t)-\left(\dfrac{1}{2}\right)(1+\cos(4t))\right)dt + \left(\dfrac{1}{8}\right)\int \left(1-\sin^2(2t)\right)\cos(2t)\,dt$$

$$= \left(\dfrac{1}{8}\right)\left(\dfrac{t}{2}-\dfrac{\sin(2t)}{2}-\dfrac{\sin(4t)}{8}\right) + \left(\dfrac{1}{16}\right)\left(\sin(2t)-\dfrac{\sin^3(2t)}{3}\right)+C$$

$$= \dfrac{1}{16}-\dfrac{\sin(4t)}{64}-\dfrac{\sin^3(2t)}{48}+C$$

35.   **a.** $\int \cos^2(x)\sin^2(x)\,dx = \left(\dfrac{1}{4}\right)\int \sin^2(2x)\,dx = \left(\dfrac{1}{8}\right)\int (1-\cos(4x))\,dx$

$$= \left(\dfrac{1}{8}\right)\left(x-\dfrac{\sin(4x)}{4}\right)+C$$

b. $\int \cos^4(x)\sin^4(x)\,dx = \left(\dfrac{1}{16}\right)\int \sin^4(2x)\,dx = \left(\dfrac{1}{64}\right)\int (1-\cos(4x))^2\,dx$

$$= \left(\dfrac{1}{64}\right)\int \left(1-2\cos(4x)+\cos^2(4x)\right)dx$$

$$= \left(\dfrac{1}{64}\right)\left(x-\left(\dfrac{1}{2}\right)\sin(4x)+\left(\dfrac{1}{2}\right)\int (1+\cos(8x))\,dx\right)$$

$$= \left(\dfrac{1}{64}\right)\left(x-\dfrac{\sin(4x)}{2}+\dfrac{x}{2}+\dfrac{\sin(8x)}{16}\right)+C$$

$$= \dfrac{3x}{128}-\dfrac{\sin(4x)}{128}+\dfrac{\sin(8x)}{1024}+C$$

37.   We denote $f(x)=\cos^{2m}(x)\sin^{2n}(x)$, then $f(-x)=f(x)$ and

$$\int_{-\pi}^{\pi} f(x)\,dx = \int_{-\pi}^{0} f(x)\,dx + \int_{0}^{\pi} f(x)\,dx = -\int_{\pi}^{0} f(u)\,du + \int_{0}^{\pi} f(x)\,dx = \int_{0}^{\pi} f(u)\,du + \int_{0}^{\pi} f(x)\,dx$$

$$= 2\int_{0}^{\pi} f(x)\,dx$$

where we changed variables by $u=-x$. As $f(x)>0$ for $x\in\left(0,\dfrac{\pi}{2}\right)\cup\left(\dfrac{\pi}{2},\pi\right)$, we find that

$\int_{0}^{\pi} f(x)\,dx > 0$ and hence $\int_{-\pi}^{\pi} f(x)\,dx > 0$.

39. We compute $\int \sin(x)\cos(x)\,dx = \int u\,du = \left(\frac{1}{2}\right)u^2 + C_1 = \left(\frac{1}{2}\right)\sin^2(x) + C_1$, where we changed variables $u = \sin(x)$. Next, we integrate

$$\int \sin(x)\cos(x)\,dx = -\int v\,dv = -\left(\frac{1}{2}\right)v^2 + C_2 = -\left(\frac{1}{2}\right)\cos^2(x) + C_2,$$

where we changed variables $v = \cos(x)$. Also, we note that

$$\int \sin(x)\cos(x)\,dx = \left(\frac{1}{2}\right)\int \sin(2x)\,dx = -\left(\frac{1}{4}\right)\cos(2x) + C_3,$$

hence all calculations are correct. At the same time,

$$\left(\frac{1}{2}\right)\sin^2(x) + C_1 = \left(\frac{1}{2}\right)\left(1 - \cos^2(x)\right) + C_1 = -\left(\frac{1}{2}\right)\cos^2(x) + C_1 + \frac{1}{2},$$

and we see that the second result coincides with the first one for $C_2 = C_1 + \frac{1}{2}$. As

$$\left(\frac{1}{2}\right)\sin^2(x) + C_1 = \left(\frac{1}{4}\right)\left(1 - \cos(2x)\right) + C_1 = -\left(\frac{1}{4}\right)\cos(2x) + C_1 + \frac{1}{4},$$ the third result is the

same as the first one for $C_3 = C_1 + \frac{1}{4}$. All constants C_i are arbitrary, so these three results are correct.

41. We use the Integration by Parts Formula to evaluate

$$\int x\cos(2x)\,dx = \left(\frac{1}{2}\right)\left(x\sin(2x) - \int \sin(2x)\,dx\right) = \left(\frac{1}{2}\right)\left(x\sin(2x) + \left(\frac{1}{2}\right)\cos(2x)\right) + C.$$

Applying this result, we compute the integral

$$\int_0^{\pi/2} x\cos^2(x)\,dx = \left(\frac{1}{2}\right)\int_0^{\pi/2} x\left(1 + \cos(2x)\right)dx = \left(\frac{1}{2}\right)\left(\frac{x^2}{2} + \frac{\cos(2x)}{4} + \frac{x\sin(2x)}{2}\right)\Bigg|_0^{\pi/2}$$

$$= \frac{\pi^2}{16} - \frac{1}{4}$$

43. $\displaystyle\int_0^{\pi/4} \sin^2(x)\tan(x)\,dx = \int_0^{\pi/4} \left(1-\cos^2(x)\right)\tan(x)\,dx = \int_0^{\pi/4} \tan(x)\,dx - \int_0^{\pi/4} \sin(x)\cos(x)\,dx$

$$= -\ln\left(\left|\cos(x)\right|\right)\Big|_0^{\pi/4} - \left(\frac{1}{2}\right)\sin^2(x)\Big|_0^{\pi/4} = -\ln\left(\frac{\sqrt{2}}{2}\right) - \frac{1}{4} = \ln\left(\sqrt{2}\right) - \frac{1}{4}$$

45. We denote $u = \ln^n(x)$, $dv = dx$, $du = \left(\dfrac{n\ln^{n-1}(x)}{x}\right)dx$, $v = x$. Then, using the Integration by

Parts Formula, we find that $\displaystyle\int \ln^n(x)\,dx = x\ln^n(x) - n\int \ln^{n-1}(x)\,dx$.

47. $\displaystyle\int \tan^2(x)\,dx = \int \left(\dfrac{1-\cos^2(x)}{\cos^2(x)}\right)dx = \int \sec^2(x)\,dx - \int dx = \tan(x) - x + C$

49. We denote $I_n = \displaystyle\int \sec^n(x)\,dx$, and use $u = \sec^{n-2}(x)$, $dv = \sec^2(x)\,dx$ to find

$$du = (n-2)\sec^{n-2}(x)\tan(x)\,dx,$$

$v = \tan(x)$. Then applying the Integration by Parts Formula we find

$$I_n = \int \sec^n(x)\,dx = \sec^{n-2}(x)\tan(x) - (n-2)\int \sec^{n-2}(x)\tan^2(x)\,dx$$
$$= \sec^{n-2}(x)\tan(x) - (n-2)\int \left(\sec^n(x) - \sec^{n-2}(x)\right)dx$$
$$= \sec^{n-2}(x)\tan(x) - (n-2)I_n + (n-2)\int \sec^{n-2}(x)\,dx$$

and

$$I_n = \left(\frac{1}{n-1}\right)\sec^{n-2}(x)\tan(x) + \left(\frac{n-2}{n-1}\right)\int \sec^{n-2}(x)\,dx$$
$$= \left(\frac{1}{n-1}\right)\sec^{n-2}(x)\tan(x) + \left(\frac{n-2}{n-1}\right)I_{n-2}$$

51.   We denote $I_{n,m} = \int \sin^n(x)\cos^m(x)dx$, and use $u = \sin^{n-1}(x)$, $dv = \cos^m(x)\sin(x)dx$, to

find $du = (n-1)\sin^{n-2}(x)\cos(x)dx$, $v = -\left(\dfrac{1}{m+1}\right)\cos^{m+1}(x)$. Then, applying the Integration

by Parts Formula, we find that

$$I_{n,m} = -\left(\frac{1}{m+1}\right)\left(\sin^{n-1}(x)\cos^{m+1}(x) - (n-1)\int\sin^{n-2}(x)\cos^{m+2}(x)dx\right)$$

$$= -\left(\frac{1}{m+1}\right)\left(\sin^{n-1}(x)\cos^{m+1}(x) - (n-1)\int\sin^{n-2}(x)\cos^m(x)\left(1-\sin^2(x)\right)dx\right)$$

$$= -\left(\frac{1}{m+1}\right)$$

$$\times\left(\sin^{n-1}(x)\cos^{m+1}(x) - (n-1)\int\sin^{n-2}(x)\cos^m(x)dx + (n-1)\int\sin^n(x)\cos^m(x)dx\right)$$

$$= -\left(\frac{1}{m+1}\right)\left(\sin^{n-1}(x)\cos^{m+1}(x) - (n-1)\int\sin^{n-2}(x)\cos^m(x)dx + (n-1)I_{n,m}\right)$$

or

$$I_{n,m} = -\left(\frac{1}{m+n}\right)\sin^{n-1}(x)\cos^{m+1}(x) + \left(\frac{n-1}{m+n}\right)\int\sin^{n-2}(x)\cos^m(x)dx$$

$$= -\left(\frac{1}{m+n}\right)\sin^{n-1}(x)\cos^{m+1}(x) + \left(\frac{n-1}{m+n}\right)I_{n-2,m}$$

53.   We denote $I_n = \int\dfrac{1}{\left(x^2+a^2\right)^n}dx$, and use $u = \dfrac{1}{\left(x^2+a^2\right)^n}$, $dv = dx$ to find

$$du = \frac{-2nx}{\left(x^2+a^2\right)^{n+1}}dx,$$

$v = x$. Then applying the Integration by Parts Formula we have

$$I_n = \frac{x}{\left(x^2+a^2\right)^n} + 2n\int\frac{x^2}{\left(x^2+a^2\right)^{n+1}}dx = \frac{x}{\left(x^2+a^2\right)^n} + 2n\int\left(\frac{1}{\left(x^2+a^2\right)^n} - \frac{a^2}{\left(x^2+a^2\right)^{n+1}}\right)dx$$

$$= \frac{x}{\left(x^2+a^2\right)^n} + 2nI_n - 2a^2nI_{n+1}$$

Therefore $I_{n+1} = \dfrac{1}{2a^2 n} \cdot \dfrac{x}{\left(x^2 + a^2\right)^n} + \dfrac{2n-1}{2a^2 n} \cdot I_n$, or

$$\int \frac{1}{\left(x^2 + a^2\right)^{n+1}} dx = \frac{1}{2a^2 n} \cdot \frac{x}{\left(x^2 + a^2\right)^n} + \frac{2n-1}{2a^2 n} \int \frac{1}{\left(x^2 + a^2\right)^n} dx \, .$$

55.   We denote $\dfrac{A}{\sqrt{A^2 + B^2}} = \cos(\phi)$, $\dfrac{B}{\sqrt{A^2 + B^2}} = \sin(\phi)$, then

$$\int \frac{1}{A\cos(x) + B\sin(x)} dx = \int \frac{1}{\sqrt{A^2 + B^2} \, \cos(x+\phi)} dx$$

$$= \frac{1}{\sqrt{A^2 + B^2}} \ln\left(\sec(x+\phi) + \tan(x+\phi)\right) + C .$$

Then

$$\int \frac{1}{\left(A\cos(x) + B\sin(x)\right)^2} dx = \frac{1}{A^2 + B^2} \int \frac{1}{\cos^2(x+\phi)} dx = \frac{1}{A^2 + B^2} \tan(x+\phi) + C .$$

Calculator/Computer Problems

57.   We solve $\sin^2(x)\cos^3(x) = \cos^2(x)\sin^5(x)$ and find $a = 0$, $b = 0.9720$. To find the area between the two curves we compute

$$A = \int_a^b \left(\sin^2(x)\cos^3(x) - \cos^2(x)\sin^5(x)\right) dx$$

$$= 0.07428$$

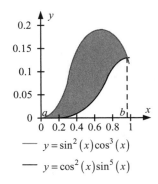

— $y = \sin^2(x)\cos^3(x)$

— $y = \cos^2(x)\sin^5(x)$

59. As $y' = 4\sin^3(x)\cos(x)$, the equation of the tangent line at point $\left(x_0, \sin^4(x_0)\right)$ is

$$y - \sin^4(x_0) = 4\sin^3(x_0)\cos(x_0)(x - x_0).$$

This line passes through the origin when

$$\sin^4(x_0) = 4\sin^3(x_0)\cos(x_0)x_0,$$

i.e., when $x_0 = 1.393249$. The tangent line is $y = 0.673668x$ and the area is

$$A = \int_0^{x_0} \left(0.673668x - \sin^4(x)\right)dx = 0.238669.$$

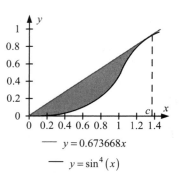

$\quad\quad$ $y = 0.673668x$

$\quad\quad$ $y = \sin^4(x)$

Section 7.4 $\quad$ Integrals Involving Quadratic Expressions

Problems for Practice

1. Using substitution $x = \sin(\theta)$, $dx = \cos(\theta)d\theta$, we find

$$\int\left(\sin(\theta)\cdot\frac{\cos(\theta)}{1-\sin^2(\theta)}\right)d\theta = \int\left(\frac{\sin(\theta)}{\cos(\theta)}\right)d\theta = -\ln\left(|\cos(\theta)|\right) + C.$$

For the right triangle with the hypotenuse 1 we have $\cos(\theta) = \sqrt{1-x^2}$, and

$$\int\left(\frac{x}{1-x^2}\right)dx = -\ln\left(\sqrt{1-x^2}\right) + C = -\left(\frac{1}{2}\right)\ln\left(|1-x^2|\right) + C.$$

3. Using substitution $x = 2\sin(\theta)$, $dx = 2\cos(\theta)d\theta$, we find

$$\int\left(4\cdot\frac{2\cos(\theta)}{\sqrt{4-4\sin^2(\theta)}}\right)d\theta = 4\int\left(\frac{\cos(\theta)}{|\cos(\theta)|}\right)d\theta.$$

The argument θ ranges from $-\frac{\pi}{2}$ to $\frac{\pi}{2}$, so that $\cos(\theta) > 0$. Therefore, we find

$$4\int\left(\frac{\cos(\theta)}{|\cos(\theta)|}\right)d\theta = 4\theta + C = 4\arcsin\left(\frac{x}{2}\right) + C.$$

5. Using substitution $x = \tan(\theta)$, $dx = \sec^2(\theta)d\theta$, we find

$$\int \left(\tan(\theta) \cdot \frac{\sec^2(\theta)}{\tan^2(\theta)+1} \right) d\theta = \int \tan(\theta)d\theta = -\ln\left(\left|\cos(\theta)\right|\right) + C.$$

Recall that $\cos(\theta) = \sqrt{\dfrac{1}{\tan^2(\theta)+1}}$, or $\cos(\theta) = \sqrt{\dfrac{1}{x^2+1}}$, thus

$$\int \left(\frac{x}{x^2+1} \right) d\theta = -\ln\left(\sqrt{\frac{1}{x^2+1}} \right) + C = \left(\frac{1}{2}\right)\ln\left(x^2+1\right) + C.$$

7. Using substitution $x = \left(\dfrac{1}{3}\right)\tan(\theta)$, $dx = \left(\dfrac{1}{3}\right)\sec^2(\theta)d\theta$, we obtain

$$\int \left(2 \cdot \frac{(1/3)\sec^2(\theta)}{\tan^2(\theta)+1} \right) d\theta = \frac{2}{3}\int d\theta = \left(\frac{2}{3}\right)\theta + C.$$

Therefore, we find $\int \left(\dfrac{2}{9x^2+1} \right) dx = \left(\dfrac{2}{3}\right)\arctan(3x) + C$.

9. Using substitution $x = \sec(\theta)$, $dx = \sec(\theta)\tan(\theta)d\theta$, we find

$$\int_{\mathrm{arc\,sec}(2)}^{\mathrm{arc\,sec}(3)} \left(\frac{\sec(\theta)\tan(\theta)}{\sec(\theta)\sqrt{\sec^2(\theta)-1}} \right) d\theta = \int_{\mathrm{arc\,sec}(2)}^{\mathrm{arc\,sec}(3)} \left(\frac{\tan(\theta)}{\left|\tan(\theta)\right|} \right) d\theta = \theta\Big|_{\mathrm{arc\,sec}(2)}^{\mathrm{arc\,sec}(3)} = \mathrm{arc\,sec}(3) - \frac{\pi}{3}.$$

11. Using substitution $x = 2\sec(\theta)$, $dx = 2\sec(\theta)\tan(\theta)d\theta$, we find

$$\int_{\mathrm{arc\,sec}\left(\frac{\sqrt{5}}{2}\right)}^{\mathrm{arc\,sec}\left(\sqrt{2}\right)} \left(\frac{2\sec(\theta)\tan(\theta)}{2\sec(\theta)\sqrt{4\sec^2(\theta)-4}} \right) d\theta = \int_{\mathrm{arc\,sec}\left(\frac{\sqrt{5}}{2}\right)}^{\mathrm{arc\,sec}\left(\sqrt{2}\right)} \left(\frac{\tan(\theta)}{2\left|\tan(\theta)\right|} \right) d\theta = \int_{\mathrm{arc\,sec}\left(\frac{\sqrt{5}}{2}\right)}^{\frac{\pi}{4}} \frac{1}{2} d\theta = \frac{\theta}{2}\Big|_{\mathrm{arc\,sec}\left(\frac{\sqrt{5}}{2}\right)}^{\frac{\pi}{4}}$$

$$= \left(\frac{1}{2}\right)\left(\frac{\pi}{4} - \mathrm{arc\,sec}\left(\frac{\sqrt{5}}{2}\right)\right) = \frac{\pi}{8} - \frac{1}{2}\mathrm{arc\,sec}\left(\frac{\sqrt{5}}{2}\right).$$

13.   Using substitution $x = \sin(\theta)$, $dx = \cos(\theta)d\theta$, we obtain

$$\int\left(\frac{(2\sin(\theta)+3)\cos(\theta)}{\sqrt{1-\sin^2(\theta)}}\right)d\theta = \int\left(2\sin(\theta)\frac{\cos(\theta)}{|\cos(\theta)|}+3\frac{\cos(\theta)}{|\cos(\theta)|}\right)d\theta.$$

The argument θ ranges from $-\dfrac{\pi}{2}$ to $\dfrac{\pi}{2}$ so that $\cos(\theta)>0$. Therefore we find

$$\int\left(2\sin(\theta)\frac{\cos(\theta)}{|\cos(\theta)|}+\left(\frac{3\cos(\theta)}{|\cos(\theta)|}\right)\right)d\theta = \int 2\sin(\theta)d\theta + \int 3d\theta = -2\cos(\theta)+3\theta+C.$$

As $\cos\theta = \sqrt{1-\sin^2\theta} = \sqrt{1-x^2}$, we obtain $\int\left(\dfrac{2x+3}{\sqrt{1-x^2}}\right)dx = -2\sqrt{1-x^2}+3\arcsin(x).$

15.   We compute $\int\left(\dfrac{x^2+2}{x^2+1}\right)dx = \int\left(1+\dfrac{1}{x^2+1}\right)dx = \int dx + \int\left(\dfrac{1}{x^2+1}\right)dx$. Using substitution

$x = \tan(\theta)$, $dx = \sec^2(\theta)d\theta$ we find $\int\left(\dfrac{1}{\tan^2(\theta)+1}\right)\sec^2(\theta)d\theta = \int d\theta = \theta+C$. Therefore,

we obtain $\int\left(\dfrac{x^2+2}{x^2+1}\right)dx = x + \arctan(x)+C.$

17.   Using substitution $x = \sin(\theta)$, $dx = \cos(\theta)d\theta$, we find

$$\int\left(\sin^2(\theta)\frac{\cos(\theta)}{\sqrt{1-\sin^2(\theta)}}\right)d\theta = \int\left(\sin^2(\theta)\frac{\cos(\theta)}{|\cos(\theta)|}\right)d\theta.$$

The argument θ ranges from $-\dfrac{\pi}{2}$ to $\dfrac{\pi}{2}$ so that $\cos(\theta)>0$. Therefore, we obtain

$$\int\left(\frac{\sin^2(\theta)\cos(\theta)}{|\cos(\theta)|}\right)d\theta = \int\sin^2(\theta)d\theta = \left(\frac{1}{2}\right)\theta-\left(\frac{1}{4}\right)\sin(2\theta)+C,$$

$$\int\left(\frac{x^2}{\sqrt{1-x^2}}\right)dx = \frac{\arcsin(x)-x\sqrt{1-x^2}}{2}+C.$$

19.   Using substitution $x = \sin(\theta)$, $dx = \cos(\theta)\,d\theta$, we find

$$\int\left(\frac{\sin^2(\theta)\cos(\theta)}{\left(1-\sin^2(\theta)\right)^{3/2}}\right)d\theta = \int\left(\frac{\sin^2(\theta)\cos(\theta)}{\left|\cos(\theta)\right|^3}\right)d\theta\,.$$

The argument θ ranges from $-\dfrac{\pi}{2}$ to $\dfrac{\pi}{2}$ so that $\cos(\theta) > 0$. Therefore, we obtain

$$\int\left(\sin^2(\theta)\frac{\cos(\theta)}{\left|\cos(\theta)\right|^3}\right)d\theta = \int\tan^2(\theta)\,d\theta = \tan(\theta)-\theta+C\,,$$

$$\int\left(\frac{x^2}{\left(1-x^2\right)^{3/2}}\right)dx = \frac{x}{\sqrt{1-x^2}} - \arcsin(x)+C\,.$$

21.   Using substitution $x = \tan(\theta)$, $dx = \sec^2(\theta)\,d\theta$, we find

$$\int\left(\frac{\sec^2(\theta)}{\left(1+\tan^2(\theta)\right)^{3/2}}\right)d\theta = \int\left(\frac{1}{\sec(\theta)}\right)d\theta = \sin(\theta)+C\,.$$

Therefore, we obtain $\displaystyle\int\left(\frac{1}{\left(1+x^2\right)^{3/2}}\right)dx = \frac{x}{\sqrt{x^2+1}} + C\,.$

23.   Using substitution $x = \tan(\theta)$, $dx = \sec^2(\theta)\,d\theta$, we find

$$\int\left(\tan^2(\theta)\frac{\sec^2(\theta)}{\sqrt{1+\tan^2(\theta)}}\right)d\theta = \int\tan^2(\theta)\sec(\theta)\,d\theta = \int\left(\sec^2(\theta)-1\right)\sec(\theta)\,d\theta$$

$$= \int\left(\sec^3(\theta)-\sec(\theta)\right)d\theta = \left(\frac{1}{2}\right)\sec(\theta)\tan(\theta)$$

$$+\left(\frac{1}{2}\right)\ln\left(\left|\sec(\theta)+\tan(\theta)\right|\right)-\ln\left(\left|\sec(\theta)+\tan(\theta)\right|\right)$$

$$= \frac{\sec(\theta)\tan(\theta)-\ln\left(\left|\sec(\theta)+\tan(\theta)\right|\right)}{2}+C$$

Therefore we obtain $\displaystyle\int\left(\frac{x^2}{\sqrt{1+x^2}}\right)dx = \frac{x\sqrt{x^2+1}-\ln\left(x+\sqrt{x^2+1}\right)}{2}+C\,.$

25.   Using substitution $x = \sec(\theta)$, $dx = \sec(\theta)\tan(\theta)d\theta$, we find

$$\int \left(\sec(\theta) \frac{\sec(\theta)\tan(\theta)}{\sqrt{\sec^2(\theta) - 1}} \right) d\theta = \int \sec^2(\theta)d\theta = \tan(\theta) + C.$$

Therefore, we obtain $\int \left(\dfrac{x^2}{\sqrt{x^2 - 1}} \right) dx = \sqrt{x^2 - 1} + C$.

27.   Using substitution $x = \sec(\theta)$, $dx = \sec(\theta)\tan(\theta)d\theta$, we find

$$\int \left(\sec(\theta) \frac{\tan(\theta)}{\left(\sec^2(\theta) - 1\right)^{3/2}} \right) d\theta = \int \left(\frac{\sec(\theta)}{\tan^2(\theta)} \right) d\theta = \int \left(\frac{\cos(\theta)}{\sin^2(\theta)} \right) d\theta = -\sin^{-1}(\theta) = -\csc(\theta).$$

Therefore, we obtain $\int \left(\dfrac{1}{\left(\sqrt{x^2 - 1}\right)^{3/2}} \right) dx = -\dfrac{x}{\sqrt{x^2 - 1}}$.

29.   Using substitution $x = \sqrt{\dfrac{8}{7}}\tan\theta$, $dx = \sqrt{\dfrac{8}{7}}\sec^2 d\theta$, we find

$$\int \left(\frac{1}{7 \cdot (8/7)\tan^2(\theta) + 8} \right) \cdot \sqrt{\frac{8}{7}}\sec^2(\theta)d\theta = \sqrt{\frac{1}{56}} \int \left(\frac{\sec^2\theta}{\sec^2\theta} \right) d\theta = \frac{\theta}{\sqrt{56}} + C.$$

Therefore, we obtain $\displaystyle\int_0^2 \left(\frac{1}{7x^2 + 8} \right) dx = \left(\frac{\sqrt{14}}{28} \right)\arctan\left(\frac{x\sqrt{14}}{4} \right)\Bigg|_0^2 = \left(\frac{\sqrt{14}}{28} \right)\arctan\left(\frac{\sqrt{14}}{2} \right)$.

31.   Using substitution $x = \left(\dfrac{4}{3} \right)\tan(\theta)$, $dx = \left(\dfrac{4}{3} \right)\sec^2(\theta)d\theta$, we find

$$\int \left(\left(\frac{1}{3} \right) \frac{\sec^2(\theta)}{\sqrt{1 + \tan^2(\theta)}} \right) d\theta = \left(\frac{1}{3} \right) \int \sec(\theta)d\theta = \left(\frac{1}{3} \right)\ln\left(\left|\sec(\theta) + \tan(\theta)\right|\right) + C.$$

Therefore, we obtain

$$\int_0^1 \left(\frac{1}{\sqrt{16 + 9x^2}} \right) dx = \left(\frac{1}{3} \right)\ln\left(\left(\frac{3}{4} \right)x + \sqrt{\frac{9x^2}{16} + 1} \right)\Bigg|_0^1 = \left(\frac{1}{3} \right)\ln\left(\frac{3}{4} + \frac{5}{4} \right) = \frac{\ln(2)}{3}.$$

33.   Using substitution $x = \left(\dfrac{3}{2}\right)\sin(\theta)$, $dx = \left(\dfrac{3}{2}\right)\cos(\theta)d\theta$, we find

$$\left(\frac{9}{8}\right)\int\left(\sin^2(\theta)\frac{\cos(\theta)}{\sqrt{1-\sin^2(\theta)}}\right)d\theta = \left(\frac{9}{8}\right)\int\sin^2(\theta)d\theta = \left(\frac{9}{16}\right)(\theta - \sin(\theta)\cos(\theta)).$$

Therefore we obtain

$$\int_0^1\left(\frac{x^2}{\sqrt{9-4x^2}}\right)dx = \left(\frac{9}{16}\right)\left(\arcsin\left(\frac{2x}{3}\right) - \left(\frac{2x}{3}\right)\sqrt{1-\frac{4x^2}{9}}\right)\Bigg|_0^1$$

$$= \left(\frac{9}{16}\right)\left(\arcsin\left(\frac{2}{3}\right) - \frac{2\sqrt{5}}{9}\right) = \left(\frac{9}{16}\right)\arcsin\left(\frac{2}{3}\right) - \frac{\sqrt{5}}{8}$$

35.   Use the substitution $u = \arcsin(x)$, $du = \left(\dfrac{1}{\sqrt{1-x^2}}\right)dx$ and $v = \dfrac{x^2}{2}$, $dv = xdx$. Then

$$\int_0^{1/\sqrt{2}} x\cdot\arcsin(x)dx = \left(\frac{x^2}{2}\right)\arcsin(x)\Big|_0^{1/\sqrt{2}} - \left(\frac{1}{2}\right)\int_0^{1/\sqrt{2}}\left(\frac{x^2}{\sqrt{1-x^2}}\right)dx$$

$$= \left(\frac{1}{4}\right)\arcsin\left(\frac{1}{\sqrt{2}}\right) - \left(\frac{1}{2}\right)\int_0^{1/\sqrt{2}}\left(\frac{x^2}{\sqrt{1-x^2}}\right)dx.$$

Using the solution of Problem 7.4.17 we find that

$$\int_0^{1/\sqrt{2}} x\cdot\arcsin(x)dx = \frac{\pi}{16} - \left(\frac{1}{4}\right)\left(\arcsin(x) - x\sqrt{1-x^2}\right)\Big|_0^{1/\sqrt{2}} = \frac{\pi}{16} - \frac{\pi}{16} + \frac{1/\sqrt{2}\cdot 1/\sqrt{2}}{4} = \frac{1}{8}.$$

37.   Use the substitution $u = \operatorname{arc\,sec}(x)$, $du = \left(\dfrac{1}{x\sqrt{x^2-1}}\right)dx$ and $v = \dfrac{x^3}{3}$, $dv = x^2dx$. Thus

$$\int_1^{\sqrt{2}} x^2\operatorname{arc\,sec}(x)dx = \left(\frac{x^3}{3}\right)\operatorname{arcsec}(x)\Big|_1^{\sqrt{2}} - \left(\frac{1}{3}\right)\int_1^{\sqrt{2}}\left(\frac{x^3}{x\sqrt{x^2-1}}\right)dx$$

$$= \left(\frac{\sqrt{2}\pi}{6}\right) - \left(\frac{1}{3}\right)\int_1^{\sqrt{2}}\left(\frac{x^2}{\sqrt{x^2-1}}\right)dx.$$

Using substitution $x = \sec(\theta)$, $dx = \sec(\theta)\tan(\theta)d\theta$, we obtain

$$\int_{\mathrm{arcsec}(1)}^{\mathrm{arcsec}(\sqrt{2})} \left(\sin^2(\theta)\sec(\theta)\frac{\tan(\theta)}{\sqrt{\sec^2(\theta)-1}} \right) d\theta = \int_{\mathrm{arcsec}(1)}^{\mathrm{arcsec}(\sqrt{2})} \sec^3(\theta)d\theta = \left(\frac{1}{2}\right)\sec(\theta)\tan(\theta)$$

$$+ \left(\frac{1}{2}\right)\ln\left(\left|\sec(x)+\tan(x)\right|\right)\Big|_{\mathrm{arcsec}(1)}^{\mathrm{arcsec}(\sqrt{2})}$$

$$= \left(\frac{\sqrt{2}}{2}\right) + \left(\frac{1}{2}\right)\ln\left(1+\sqrt{2}\right).$$

Therefore the integral is $\displaystyle\int_1^{\sqrt{2}} x^2 \, \mathrm{arcsec}(x)dx = \frac{\sqrt{2}\pi - \sqrt{2} - \ln\left(1+\sqrt{2}\right)}{6}$.

39. We compute $A = \displaystyle\int_1^2 (2-x)\sqrt{x^2-1}dx = 2\int_1^2 \sqrt{x^2-1} \, dx - \int_1^2 x\sqrt{x^2-1}dx$. From Problem 7.4.28 we have for the first integral

$$2\int_0^1 \sqrt{x^2-1} \, dx = 2 \cdot \left(\frac{1}{2}\right)\left(x\sqrt{x^2-1} - \ln\left(x+\sqrt{x^2-1}\right)\right)\Big|_1^2 = 2\sqrt{3} - \ln\left(2+\sqrt{3}\right).$$

Using substitution $x = \sec(\theta)$, $dx = \sec(\theta)\tan(\theta)d\theta$ to calculate the second integral we obtain

$$\int \sec(\theta)\sqrt{\sec^2(\theta)-1} \, \sec(\theta)\tan(\theta)d\theta = \int \sec^2(\theta)\tan^2(\theta)d\theta = \frac{\tan^3(\theta)}{3}.$$

Therefore, we find the area under the graph is

$$A = 2\sqrt{3} - \ln\left(2+\sqrt{3}\right) - \left(\frac{1}{3}\right)\left(x^2-1\right)^{3/2}\Big|_1^2 = 2\sqrt{3} - \ln\left(2+\sqrt{3}\right) - \sqrt{3} = \sqrt{3} - \ln\left(2+\sqrt{3}\right).$$

Further Theory and Practice

41. Using substitution: $x = \tan(\theta)$, we find $dx = \sec^2(\theta)d\theta$, and

$$\int\left(\frac{\sec^2(\theta)}{\left(\tan^2(\theta)+1\right)^2}\right)d\theta = \int\left(\frac{\sec^2(\theta)}{\sec^4(\theta)}\right)d\theta = \int\cos^2(\theta)d\theta = \left(\frac{1}{2}\right)\theta + \left(\frac{1}{4}\right)\sin(2\theta) + C.$$

Therefore we evaluate as follows: $\displaystyle\int\frac{1}{\left(x^2+1\right)^2}dx = \frac{\arctan(x)}{2} + \frac{x}{2\left(x^2+1\right)}$.

43.   Using substitution: $x = \tan(\theta)$, we find $dx = \sec^2(\theta)d\theta$, and

$$\int\left(\sqrt{1+\tan^2(\theta)}\,\frac{\sec^2(\theta)}{\tan(\theta)}\right)d\theta = \int\left(\frac{\sec^3(\theta)}{\tan(\theta)}\right)d\theta = \int\left(\frac{1}{\sin(\theta)\cos^2(\theta)}\right)d\theta$$

$$= \int\frac{\sin^2(\theta)+\cos^2(\theta)}{\sin^2(\theta)\cos(\theta)}d\theta$$

$$= \int\left(\frac{\sin^2(\theta)}{\cos^2(\theta)\sin(\theta)}\right)d\theta + \int\left(\frac{\cos^2(\theta)}{\cos^2(\theta)\sin(\theta)}\right)d\theta$$

$$= \int\left(\frac{\sin(\theta)}{\cos^2(\theta)}\right)d\theta + \int\csc(\theta)d\theta$$

For the first integral we use substitution: $u = \dfrac{1}{\cos(\theta)}$, hence $du = \left(\dfrac{\sin(\theta)}{\cos^2(\theta)}\right)d\theta$. For the

second integral we use formula (6.19) to find $\int\csc(\theta)d\theta = \sec(\theta) - \ln\left(\left|\csc(\theta)+\cot(\theta)\right|\right)$.

Therefore we obtain

$$\int\left(\frac{\sqrt{x^2+1}}{x}\right)dx = \sqrt{x^2+1} - \ln\left(\left|\frac{\sqrt{x^2+1}}{x}+\frac{1}{x}\right|\right) + C = \sqrt{x^2+1} - \ln\left(\frac{1+\sqrt{x^2+1}}{|x|}\right) + C.$$

45.   Using substitution: $x = \tan(\theta)$, we find $dx = \sec^2(\theta)d\theta$, and

$$\int\left(\sqrt{1+\tan^2(\theta)}\,\frac{\sec^2(\theta)}{\tan^2(\theta)}\right)d\theta = \int\left(\frac{\sec^3(\theta)}{\tan^2(\theta)}\right)d\theta$$

$$= \int\left(\frac{1}{\sin^2(\theta)}\cos(\theta)\right)d\theta$$

$$= \int\left(\frac{\sin^2(\theta)+\cos^2(\theta)}{\sin^2(\theta)\cos(\theta)}\right)d\theta$$

$$= \int\left(\frac{\cos^2(\theta)}{\sin^2(\theta)\cos(\theta)}\right)d\theta + \int\left(\frac{\sin^2(\theta)}{\sin^2(\theta)\cos(\theta)}\right)d\theta$$

$$= \int\left(\frac{\cos(\theta)}{\sin^2(\theta)}\right)d\theta + \int\sec(\theta)d\theta$$

For the first integral we use the substitution: $u = \dfrac{1}{\sin(\theta)}$, hence $du = -\left(\dfrac{\cos(\theta)}{\sin^2(\theta)}\right)d\theta$. For the

second integral we use formula (6.19) to obtain $\int\sec(\theta)d\theta = -\csc(\theta) + \ln\left(\left|\sec(\theta)+\tan(\theta)\right|\right)$.

Therefore, we find $\int\left(\dfrac{\sqrt{x^2+1}}{x^2}\right)dx = \ln\left(x+\sqrt{x^2+1}\right) - \dfrac{\sqrt{x^2+1}}{x} + C$.

47.   First, we complete the square for the polynomial $2x - x^2 = 1 - (x-1)^2$. To evaluate the integral we make a direct substitution: $u = x-1$, $du = dx$. The integral then becomes $\int\left(\dfrac{1}{\sqrt{1-u^2}}\right)du$.

Using substitution $u = \sin(\theta)$, $du = \cos(\theta)d\theta$ we find

$$\int\left(\frac{1}{\sqrt{1-u^2}}\right)du = \int\left(\frac{\cos(\theta)}{\sqrt{1-\sin^2(\theta)}}\right)d\theta = \int d\theta = \theta + C.$$

Therefore, we obtain $\int\left(\dfrac{1}{\sqrt{2x-x^2}}\right)dx = \arcsin(u) + C = \arcsin(x-1) + C$.

49.   First we complete the square for the polynomial $x^2 + 2x + 2 = (x+1)^2 + 1$. To evaluate this integral we make the direct substitution: $u = x+1$, hence $du = dx$, and find

$$\int\left(\frac{u-1}{\sqrt{u^2+1}}\right)du = \int\left(\frac{u}{\sqrt{u^2+1}}\right)du - \int\left(\frac{1}{\sqrt{u^2+1}}\right)du.$$

Using substitution $u = \tan(\theta)$, hence $du = \sec^2(\theta)d\theta$, we compute

$$\int\left(\tan(\theta)\frac{\sec^2(\theta)}{\sqrt{\tan^2(\theta)+1}}\right)d\theta - \int\left(\frac{\sec^2(\theta)}{\sec(\theta)}\right)d\theta = \sec(\theta) - \ln\left(\left|\sec(\theta)+\tan(\theta)\right|\right) + C.$$

Therefore we finally obtain

$$\int\left(\frac{x}{\sqrt{x^2+2x+2}}\right)dx = \sqrt{1+u^2} - \ln\left(u+\sqrt{1+u^2}\right) + C$$

$$= \sqrt{x^2+2x+2} - \ln\left(\sqrt{x^2+2x+2}+x+1\right) + C$$

51.   First we complete the square for the polynomial $x^2 - 2x + 2 = (x-1)^2 + 1$. To evaluate this integral we make the direct substitution: $u = x - 1$, hence $du = dx$, and compute

$$\int \left(\frac{1}{\left(u^2 + 1\right)^2} \right) du .$$ Use the substitution $u = \tan(\theta)$, $du = \sec^2(\theta) d\theta$. Therefore we find that

the integral becomes $\int \left(\frac{1}{\left(u^2 + 1\right)^2} \right) du = \int \left(\frac{\sec^2(\theta)}{\sec^4(\theta)} \right) d\theta = \int \cos^2(\theta) d\theta = \frac{\theta}{2} - \frac{\sin(2\theta)}{4} + C$. Thus

we find

$$\int \left(\frac{1}{\left(x^2 - 2x + 2\right)^2} \right) dx = \frac{\arctan(u) - u/u^2 + 1}{2} + C = \frac{\arctan(x-1) - (x-1)/\left(x^2 - 2x + 2\right)}{2} + C .$$

53.   The first step is to complete the square: $x^2 - 14x + 58 = (x-7)^2 + 9$. To evaluate this integral we make a direct substitution: $u = x - 7$, hence $du = dx$, and find

$$\int \left(\frac{u+7}{u^2 + 9} \right) du = \int \left(\frac{u}{u^2 + 9} \right) du + 7 \int \left(\frac{1}{u^2 + 9} \right) du .$$ Use the substitution $u = 3\tan(\theta)$,

$du = 3\sec^2(\theta) d\theta$. We find that the integral becomes

$$3 \int \left(\frac{\tan(\theta) \cdot \sec^2(\theta)}{\tan^2(\theta) + 1} \right) d\theta + \left(\frac{7}{3} \right) \int \left(\frac{\sec^2(\theta)}{\tan^2(\theta) + 1} \right) d\theta = \int \tan(\theta) d\theta + \left(\frac{7}{3} \right) \int d\theta$$

$$= -\ln\left(\left|\cos(\theta)\right|\right) + \left(\frac{7}{3} \right) \theta + C$$

Therefore we conclude

$$\int \left(\frac{x}{\sqrt{x^2 - 14x + 58}} \right) dx = -\ln\left(\sqrt{\frac{1}{(u/3)^2 + 1}} \right) + \left(\frac{7}{3} \right) \arctan\left(\frac{u}{3} \right) + C$$

$$= \left(\frac{1}{2} \right) \ln\left(\left|u^2 + 9\right|\right) + \left(\frac{7}{3} \right) \arctan\left(\frac{u}{3} \right) + C$$

$$= \left(\frac{1}{2} \right) \ln\left(\left|x^2 - 7x + 58\right|\right) + \left(\frac{7}{3} \right) \arctan\left(\frac{x-7}{3} \right) + C .$$

55. First we complete the square for the polynomial

$$\frac{x^2 - 6x + 8}{x^2 + 4} = 1 - \frac{6x}{x^2 + 4} + \frac{4}{x^2 + 4}.$$

Using the substitution $x = 2\tan(\theta)$, $dx = 2\sec^2(\theta)d\theta$, we obtain

$$\int dx + 6\int\left(\tan(\theta)\frac{\sec^2(\theta)}{\tan^2(\theta) + 1}\right)d\theta + 2\int\left(\frac{\sec^2(\theta)}{\tan^2(\theta) + 1}\right)d\theta = x - 6\int\tan(\theta)d\theta + 2\theta$$

$$= x + 6\ln\left(\left|\cos(\theta)\right|\right) + 2\theta + C.$$

Therefore we conclude

$$\int\left(\frac{x^2 - 6x + 8}{x^2 + 4}\right)dx = x + 6\ln\left(\frac{1}{\sqrt{4 + x^2}}\right) + 2\arctan\left(\frac{x}{2}\right) + C$$

$$= x - 3\ln\left(x^2 + 4\right) + 2\arctan\left(\frac{x}{2}\right) + C.$$

57. First, we complete the square for the polynomial:

$$2x^2 + 6x + 5 = 2\left(x^2 + 3x + \frac{5}{2}\right) = 2\left(\left(x + \frac{3}{2}\right)^2 + \frac{1}{4}\right).$$

To evaluate the integral, we make a direct substitution: $u = x + \frac{3}{2}$, $du = dx$, and find

$$\left(\frac{3}{2}\right)\int\left(\frac{u - 3/2}{u^2 + 1/4}\right)du = 6\int\left(\frac{u}{4u^2 + 1}\right)du - 9\int\left(\frac{1}{4u^2 + 1}\right)du.$$

Using the substitution $u = \left(\frac{1}{2}\right)\tan(\theta)$, $du = \left(\frac{1}{2}\right)\sec^2(\theta)d\theta$, we find

$$\int\left(\frac{u}{4u^2 + 1}\right)du = \left(\frac{1}{4}\right)\int\left(\tan(\theta)\frac{\sec^2(\theta)}{\tan^2(\theta) + 1}\right)d\theta = \left(\frac{1}{4}\right)\int\tan(\theta)d\theta = -\left(\frac{1}{4}\right)\ln\left(\left|\cos(\theta)\right|\right) + C_1$$

$$\int\left(\frac{1}{4u^2 + 1}\right)du = \left(\frac{1}{2}\right)\int\left(\frac{\sec^2(\theta)}{\tan^2(\theta) + 1}\right) = \left(\frac{1}{2}\right)\theta + C_2$$

Therefore we conclude

$$\int\left(\frac{3x}{2x^2 + 6x + 5}\right)du = \left(\frac{3}{4}\right)\ln\left(\left|4u^2 + 1\right|\right) - \left(\frac{9}{2}\right)\arctan(2u) + C$$

$$= \left(\frac{3}{4}\right)\ln\left(\left|2x^2 + 6x + 5\right|\right) - \left(\frac{9}{2}\right)\arctan(2x + 3) + C$$

59.   First, we complete the square for the polynomial $x^2 + 2x + 2 = (x+1)^2 + 1$. To evaluate this integral we make the direct substitution: $u = x+1$, $du = dx$, and find

$$\int \left(\frac{u-1}{\left(u^2+1\right)^3} \right) du = \int \left(\frac{u}{\left(u^2+1\right)^3} \right) du - \int \left(\frac{1}{\left(u^2+1\right)^3} \right) du \,.$$

Using the substitution: $u = \tan(\theta)$, $du = \sec^2(\theta)d\theta$ we compute

$$\int \left(\frac{u}{\left(u^2+1\right)^3} \right) du = \int \left(\tan(\theta) \frac{\sec^2(\theta)}{\left(\tan^2(\theta)+1\right)^3} \right) d\theta = \int \left(\tan(\theta) \frac{\sec^2(\theta)}{\sec^6(\theta)} \right) d\theta = \int \left(\frac{\tan(\theta)}{\sec^4(\theta)} \right) d\theta$$

$$= \int \sin(\theta)\cos^3(\theta)d\theta = -\frac{\cos^4(\theta)}{4} + C_1$$

$$\int \left(\frac{1}{\left(u^2+1\right)^3} \right) du = \int \left(\frac{\sec^2(\theta)}{\left(\tan^2(\theta)+1\right)^3} \right) d\theta = \int \left(\frac{\sec^2(\theta)}{\sec^6(\theta)} \right) d\theta = \int \cos^4(\theta)d\theta$$

$$= \left(\frac{1}{4} \right) \sin(\theta)\cos^3(\theta) + \left(\frac{3}{4} \right) \left(\left(\frac{\theta}{2} \right) + \sin(\theta)\frac{\cos(\theta)}{2} \right) + C_2$$

Therefore we conclude that

$$\int \left(\frac{x}{x^2+2x+2} \right) dx = -\frac{\cos^4(\theta)}{4} - \left(\frac{1}{4} \right) \sin(\theta)\cos^3(\theta) - \left(\frac{3}{8} \right) \theta - \left(\frac{3}{8} \right) \sin(\theta)\cos(\theta)$$

$$= -\left(\frac{1}{4} \right) \frac{x+2}{\left(x^2+2x+2\right)^2} - \left(\frac{3}{8} \right) \frac{x+1}{x^2+2x+2} - \left(\frac{3}{8} \right) \arctan(x+1) + C.$$

61.   First we complete the square for the polynomial $10x - x^2 = 25 - (x-5)^2$. To evaluate the integral we make the direct substitution: $u = x-5$, $du = dx$, and the integral becomes

$$\int \left(\frac{u+6}{\sqrt{25-u^2}} \right) du = \int \left(\frac{u}{\sqrt{25-u^2}} \right) du + 6\int \left(\frac{1}{\sqrt{25-u^2}} \right) du \,.$$ Changing variables with $u = 5\sin(\theta)$,

$du = 5\cos(\theta)d\theta$, we compute

$$\int \left(\frac{u}{\sqrt{25-u^2}} \right) du = \int \left(5\sin(\theta) \frac{\cos(\theta)}{\sqrt{1-\sin^2(\theta)}} \right) d\theta = 5\int \sin(\theta)d\theta = -5\cos(\theta) + C_1,$$

$$\int \left(\frac{1}{\sqrt{25-u^2}} \right) du = \int \left(\frac{\cos(\theta)}{\sqrt{1-\sin^2(\theta)}} \right) d\theta = \theta + C_2.$$

Therefore we conclude that

$$\int \left(\frac{x+1}{\sqrt{10x-x^2}} \right) dx = -5\sqrt{1-\frac{u^2}{25}} + 6\arcsin\left(\frac{u}{5}\right) + C = -\sqrt{25-(x-5)^2} + 6\arcsin\left(\frac{x-5}{5}\right) + C$$

$$= -\sqrt{10x-x^2} + 6\arcsin\left(\frac{x}{5}-1\right) + C.$$

63. Let $c = \int_0^1 x(1-x)^{3/2}\, dx$. Changing variables: $x = \sin^2(\theta)$, hence $dx = 2\sin(\theta)\cos(\theta)d\theta$, we

find $c = \int_0^{\pi/2} \sin^2(\theta)\left(1-\sin^2(\theta)\right)^{3/2} 2\sin(\theta)\cos(\theta)d\theta = \int_0^{\pi/2} 2\sin^3(\theta)\cos^4(\theta)d\theta$. Using the

substitution $u = \cos(\theta)$, $du = -\sin(\theta)d\theta$, we conclude that

$$c = -\int_1^0 2u^4\left(1-u^2\right)du = 2\left(\frac{u^7}{7} - \frac{u^5}{5}\right)\Bigg|_{u=1}^{u=0} = -2\left(\frac{1}{7} - \frac{1}{5}\right) = \frac{4}{35}.$$

65. First, divide the integral:

$$\int\left(\frac{1+x^2}{4+x^2}\right)dx = \int\left(1 - \frac{3}{4+x^2}\right)dx = \int dx - 3\int\left(\frac{1}{4+x^2}\right)dx.$$

Changing variables: $x = 2\tan(\theta)$, hence $dx = 2\sec^2(\theta)d\theta$. For the second integral we find that

$$\int\left(\frac{2\sec^2(\theta)}{4(1+\tan^2(\theta))}\right)d\theta = \left(\frac{1}{2}\right)\int\left(\frac{\sec^2(\theta)}{\sec^2(\theta)}\right)d\theta = \left(\frac{1}{2}\right)\theta + C_1.$$

Therefore we conclude that $\int\left(\frac{1+x^2}{4+x^2}\right)dx = x - \left(\frac{3}{2}\right)\arctan\left(\frac{x}{2}\right) + C$.

67. First, divide the integral:

$$\int\left(\frac{1+x^2}{5+4x+x^2}\right)dx = \int\left(1 - 4\frac{x+1}{5+4x+x^2}\right)dx = \int dx - 4\int\frac{x+1}{(x+2)^2+1}dx.$$

To evaluate the integral we make the direct substitution $u = x+2$, $du = dx$ in the second

integral and find that $\int\frac{u-1}{u^2+1}du = \int\frac{u}{u^2+1}du - \int\frac{1}{u^2+1}du$. Both these integrals are easy.

We conclude that

$$\int\left(\frac{1+x^2}{5+4x+x^2}\right)dx = x - 2\ln\left(5+4x+x^2\right) + 4\arctan(x+2) + C.$$

69. To evaluate this integral we make the substitution $x = \tan^4(\theta)$, hence

$$dx = 4\tan^3(\theta)\sec^2(\theta)\,d\theta,$$

and find

$$\int\left(4\tan^3(\theta)\frac{\sec^2(\theta)}{1+\tan^2(\theta)}\right)d\theta = 4\int\tan^3(\theta)\,d\theta = 4\left(\frac{\tan^2(\theta)}{2} - \int\tan(\theta)\,d\theta\right)$$

$$= 2\tan^2(\theta) + 4\ln\left(\left|\cos(\theta)\right|\right) = 2\tan^2(\theta) + 2\ln\left(\cos^2(\theta)\right) + C$$

Therefore, we conclude: $\displaystyle\int\left(\frac{1}{1+\sqrt{x}}\right)dx = 2\sqrt{x} - 2\ln\left(1+\sqrt{x}\right) + C$.

71. To evaluate the integral we make the substitution $x = \tan^4(\theta)$, hence
$dx = 4\tan^3(\theta)\sec^2(\theta)\,d\theta$. We calculate:

$$\int\left(4\tan^3(\theta)\sec^2(\theta)\frac{1-\tan^2(\theta)}{1+\tan^2(\theta)}\right)d\theta = 4\int\left(\tan^3(\theta) - \tan^5(\theta)\right)d\theta$$

$$= 4\int\tan^3(\theta)\,d\theta - 4\left(\frac{\tan^4(\theta)}{4} - \int\tan^3(\theta)\,d\theta\right)$$

$$= -\tan^4(\theta) + 8\int\tan^3(\theta)\,d\theta$$

$$= -\tan^4(\theta) + 8\left(\frac{\tan^2(\theta)}{2} - \int\tan(\theta)\,d\theta\right)$$

$$= -\tan^4(\theta) + 4\tan^2(\theta) + 8\ln\left(\left|\cos(\theta)\right|\right) + C$$

$$= -\tan^4(\theta) + 4\tan^2(\theta) + 4\ln\left(\cos^2(\theta)\right) + C$$

Therefore, we conclude: $\displaystyle\int\left(\frac{1-\sqrt{x}}{1+\sqrt{x}}\right)dx = -x + 4\sqrt{x} - 4\ln\left(1+\sqrt{x}\right) + C$.

73. To evaluate this integral we make the indirect substitution $x = \tan^2(\theta)$, hence
$dx = 2\tan(\theta)\sec^2(\theta)\,d\theta$, and compute

$$\int\left(\sqrt{1+\tan^2(\theta)}\cdot 2\tan(\theta)\frac{\sec^2(\theta)}{\tan^2(\theta)}\right)d\theta = 2\int\left(\frac{\sec^3(\theta)}{\tan(\theta)}\right)d\theta.$$

Using the calculation of this integral in Problem 7.4.43, we find that

$$2\int\left(\frac{\sec^3(\theta)}{\tan(\theta)}\right)d\theta = 2\Big(\sec(\theta) - \ln\big(\big|\csc(\theta) + \cot(\theta)\big|\big)\Big).$$

Therefore we obtain

$$\int\left(\frac{\sqrt{1+x}}{x}\right)dx = 2\sqrt{1+x} - 2\ln\left(\left|\frac{\sqrt{x+1}}{\sqrt{x}} + \frac{1}{\sqrt{x}}\right|\right) + C = 2\sqrt{1+x} - \ln\left(\frac{x+2+2\sqrt{1+x}}{|x|}\right) + C.$$

75. Using substitution $x = a\sin(\theta)$, $dx = a\cos(\theta)d\theta$, we obtain

$$\int\left(\frac{a\cos(\theta)}{a^2 - a^2\sin^2(\theta)}\right)d\theta = \left(\frac{1}{a}\right)\int\sec(\theta)d\theta = \left(\frac{1}{a}\right)\ln\big(\big|\sec(\theta) + \tan(\theta)\big|\big) + C.$$

Therefore we find that

$$\int\left(\frac{1}{a^2 - x^2}\right)dx = \left(\frac{1}{a}\right)\ln\left(\frac{a}{\sqrt{a^2 - x^2}} + \frac{x}{\sqrt{a^2 - x^2}}\right) = \left(\frac{1}{a}\right)\ln\left(\frac{x+a}{\sqrt{a^2 - x^2}}\right).$$

Using the substitution $x = a\sec(\theta)$, hence $dx = a\tan(\theta)\sec(\theta)d\theta$, we compute

$$\int\left(a\sec(\theta)\frac{\tan(\theta)}{a^2 - a^2\sec^2(\theta)}\right)d\theta = \left(\frac{1}{a}\right)\int\frac{\sec(\theta)}{\tan(\theta)}d\theta = \left(\frac{1}{a}\right)\int\csc(\theta)d\theta$$

$$= \left(\frac{1}{a}\right)\ln\big(\big|\csc(\theta) + \cot(\theta)\big|\big) + C$$

Therefore we calculate

$$\int\left(\frac{1}{a^2 - x^2}\right)dx = \left(\frac{1}{a}\right)\ln\left(\frac{x}{\sqrt{x^2 - a^2}} + \frac{a}{\sqrt{x^2 - a^2}}\right) = \left(\frac{1}{a}\right)\ln\left(\frac{x+a}{\sqrt{x^2 - a^2}}\right).$$

To find $\displaystyle\int_0^{1/2}\left(\frac{1}{1-x^2}\right)dx$ we use the first formula because the limits of integration are less than 1. Thus, $\displaystyle\int_0^{1/2}\left(\frac{1}{1-x^2}\right)dx = \ln\left(\frac{x+1}{\sqrt{1-x^2}}\right)\Bigg|_0^{1/2} = \ln\left(\frac{3/2}{\sqrt{1-1/4}}\right) - 0 = \ln\left(\sqrt{3}\right) = \left(\frac{1}{2}\right)\ln(3).$

To calculate $\int_{\sqrt{2}}^{2}\left(\dfrac{1}{1-x^2}\right)dx$ we use the second formula because the limits of integration are greater than 1. We now find

$$\int_{\sqrt{2}}^{2}\left(\frac{1}{1-x^2}\right)dx = \ln\left(\frac{x+1}{\sqrt{x^2-1}}\right)\Bigg|_{\sqrt{2}}^{2} = \ln\left(\frac{3}{\sqrt{4-1}}\right) - \ln\left(\frac{\sqrt{2}}{\sqrt{2-1}}\right) = \ln\left(\sqrt{3}\right) - \ln\left(\sqrt{2}+1\right).$$

Calculator/Computer Exercises

77.

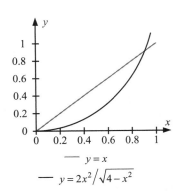
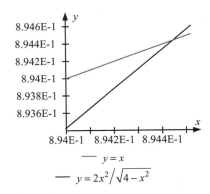

The points of intersection are $x=0$ and $x=0.89443$. Therefore, we compute

$$A = \int_{0}^{0.89443}\left(x - \frac{2x^2}{\sqrt{4-x^2}}\right)dx = \int_{0}^{0.89443} x\,dx - 2\int_{0}^{0.89443}\left(\frac{x^2}{\sqrt{4-x^2}}\right)dx$$

$$= \left(\frac{x^2}{2}\right)\Bigg|_{0}^{0.89443} - 2\int_{0}^{0.89443}\left(\frac{x^2}{\sqrt{4-x^2}}\right)dx$$

Using substitution: $x = 2\sin(\theta)$, hence $dx = 2\cos(\theta)d\theta$, we find

$$\int_{0}^{\arcsin(0.89443/2)}\left(4\sin^2(\theta)\frac{\cos(\theta)}{\sqrt{1-\sin^2(\theta)}}\right)d\theta = \int_{0}^{\arcsin(0.89443/2)} 4\sin^2(\theta)d\theta$$

$$= 4\left(\frac{\theta}{2} - \frac{\sin(2\theta)}{4}\right)\Bigg|_{0}^{0.46365}$$

Therefore the area under the two curves is $A = \left(\frac{x^2}{2}\right)\Bigg|_{0}^{0.89443} - \left(4\theta - 2\sin(2\theta)\right)\Big|_{0}^{0.46365} \approx 0.1454$.

79.

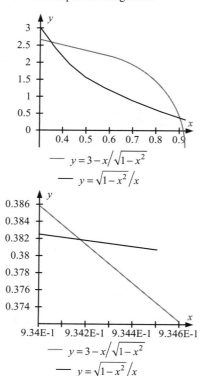

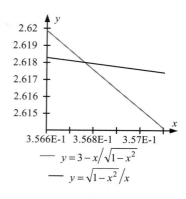

The points of intersection are $x = 0.93417$ and 0.35682. Therefore we find that

$$A = \int_{0.35682}^{0.93417}\left(3 - \frac{x}{\sqrt{1-x^2}} - \frac{\sqrt{1-x^2}}{x}\right)dx = 3\int_{0.35682}^{0.93417}dx - \int_{0.35682}^{0.93417}\left(\frac{x}{\sqrt{1-x^2}}\right)dx - \int_{0.35682}^{0.93417}\left(\frac{\sqrt{1-x^2}}{x}\right)dx$$

$$= 3x\Big|_{0.35682}^{0.93417} - \int_{0.35682}^{0.93417}\left(\frac{x}{\sqrt{1-x^2}}\right)dx - \int_{0.35682}^{0.93417}\left(\frac{\sqrt{1-x^2}}{x}\right)dx$$

Using the substitution $x = \sin(\theta)$, $dx = \cos(\theta)d\theta$, we obtain for the first and second integrals respectively:

$$\int_{\arcsin(0.35682)}^{\arcsin(0.93417)}\left(\sin(\theta)\frac{\cos(\theta)}{\sqrt{1-\sin^2(\theta)}}\right)d\theta = \int_{\arcsin(0.35682)}^{\arcsin(0.93417)}\sin(\theta)d\theta = -\cos(\theta)\Big|_{0.36486}^{1.20593} \approx 0.57735,$$

and

$$\int_{\arcsin(0.35682)}^{\arcsin(0.93417)}\left(\sqrt{1-\sin^2(\theta)}\frac{\cos(\theta)}{\sin(\theta)}\right)d\theta = \int_{\arcsin(0.35682)}^{\arcsin(0.93417)}\frac{\cos^2(\theta)}{\sin(\theta)}d\theta$$

$$= \left(\cos(\theta) + \ln\left(\frac{1-\cos(\theta)}{\sin(\theta)}\right)\right)\Bigg|_{0.36486}^{1.20593} \approx 0.73961.$$

Therefore the area enclosed by the two graphs is $A \approx 0.41509$.

81.

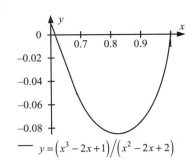

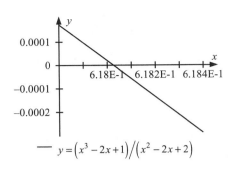

The points of intersection are $a = 0.61803$ and $a = 1$. Therefore, we find that

$$A = \int_{0.61803}^{1} \left(-\frac{x^3 - 2x + 1}{x^2 - 2x + 2} \right) dx = -\int_{0.61803}^{1} \left((x+2) - \frac{3}{x^2 - 2x + 2} \right) dx$$

$$= -\left(\frac{x^2}{2} + 2x \right) \Bigg|_{0.61803}^{1} - \int_{0.61803}^{1} \left(\frac{3}{x^2 - 2x + 2} \right) dx = -1.07296 - \int_{0.61803}^{1} \left(\frac{3}{x^2 - 2x + 2} \right) dx$$

To evaluate this integral we make a substitution: $u = x - 1$, hence $du = dx$. The integral

becomes: $\int_{0.38197}^{0} \left(\frac{3}{u^2 + 1} \right) du = 3 \arctan(u) \Big|_{0.38197}^{0} \approx 1.09460$. Therefore, the area enclosed by

this curve is $A \approx 0.02164$.

Section 7.5   **Partial Fractions—Irreducible Quadratic Factors**

Problems for Practice

1.   The degree of the numerator is lower than that of the denominator. Moreover, $x^2 + 1$ and
$x^2 + 4$ cannot be factored into linear factors with real coefficients. Thus, the partial fraction
decomposition is $\dfrac{2x^3 + x + 1}{\left(x^2 + 1 \right)\left(x^2 + 4 \right)} = \dfrac{Ax + B}{x^2 + 1} + \dfrac{Cx + D}{x^2 + 4}$.

3.   The degree of the numerator is lower than that of the denominator. Moreover, $x^2 + 1$ and
$x^2 + x + 1$ cannot be factored into linear factors with real coefficients. Thus, the partial
fraction decomposition is $\dfrac{2x^3 + x + 1}{\left(x^2 + 1 \right)\left(x^2 + x + 1 \right)^2} = \dfrac{Ax + B}{x^2 + 1} + \dfrac{Cx + D}{x^2 + x + 1} + \dfrac{Ex + F}{\left(x^2 + x + 1 \right)^2}$.

5.   The degree of the numerator is lower than that of the denominator. Moreover, $x^2 + x + 3$ cannot be factored into linear factors with real coefficients. Thus, the partial fraction decomposition is $\dfrac{2x+1}{\left(x^2+x+3\right)\left(x-4\right)} = \dfrac{A}{x-4} + \dfrac{Bx+C}{x^2+x+3}$.

7.   The degree of the numerator is lower than that of the denominator. Moreover, $x^2 + 4$ cannot be factored into linear factors with real coefficients. Thus, the partial fraction decomposition is $\dfrac{2x^6}{\left(x^2+4\right)^3\left(x-2\right)} = \dfrac{A}{(x-2)} + \dfrac{Bx+C}{x^2+4} + \dfrac{Dx+E}{\left(x^2+4\right)^2} + \dfrac{Fx+G}{\left(x^2+4\right)^3}$.

9.   The degree of the numerator is lower than that of the denominator. Moreover,

$$x^2 - 1 = (x-1)(x+1)$$

and $x^2 + 1$ cannot be factored into linear factors with real coefficients. Thus, the partial fraction decomposition is

$$\frac{3x^5+x+1}{x^2\left(x^2-1\right)\left(x^2+1\right)} = \frac{3x^5+x+1}{x^2\left(x-1\right)\left(x+1\right)\left(x^2+1\right)} = \frac{A}{x} + \frac{B}{x^2} + \frac{C}{x-1} + \frac{D}{x+1} + \frac{Ex+F}{x^2+1}.$$

11.   The degree of the numerator is lower than that of the denominator. Moreover, $x^2 + 1$ cannot be factored into linear factors with real coefficients. Thus, the partial fraction decomposition is $\dfrac{3x^2-5x+4}{\left(x-1\right)\left(x^2+1\right)} = \dfrac{A}{x-1} + \dfrac{Bx+C}{x^2+1}$. We put the terms on the right over a common denominator:

$\dfrac{3x^2-5x+4}{\left(x-1\right)\left(x^2+1\right)} = \dfrac{A\left(x^2+1\right)+\left(Bx+C\right)\left(x-1\right)}{\left(x-1\right)\left(x^2+1\right)}$. Then we have the identity

$$3x^2 - 5x + 4 = A\left(x^2+1\right) + \left(Bx+C\right)\left(x-1\right)$$

or $3x^2 - 5x + 4 = (A+B)x^2 + (C-B)x + A - C$. Comparing coefficients of the terms with like powers of x, we obtain the following system of equations for the coefficients

$$\begin{cases} 3 = A+B \\ -5 = C-B \\ 4 = A-C \end{cases}.$$

Its solution is $A = 1$, $B = 2$, $C = -3$. Therefore

$$\frac{3x^2-5x+4}{\left(x-1\right)\left(x^2+1\right)} = \frac{1}{x-1} + \frac{2x-3}{x^2+1}.$$

13.   The degree of the numerator is lower than that of the denominator. Moreover, $x^2 + 2$ and $x^2 + 1$ cannot be factored into linear factors with real coefficients. Thus, the partial fraction decomposition is

$$\frac{7x^3 + 9x - 3x^2 - 6}{\left(x^2 + 2\right)\left(x^2 + 1\right)} = \frac{Ax + B}{x^2 + 2} + \frac{Cx + D}{x^2 + 1}.$$

We put the terms on the right over a common denominator:

$$\frac{7x^3 + 9x - 3x^2 - 6}{\left(x^2 + 2\right)\left(x^2 + 1\right)} = \frac{(Ax + B)\left(x^2 + 1\right) + (Cx + D)\left(x^2 + 2\right)}{\left(x^2 + 2\right)\left(x^2 + 1\right)}$$

and obtain the identity $7x^3 + 9x - 3x^2 - 6 = (Ax + B)\left(x^2 + 1\right) + (Cx + D)\left(x^2 + 2\right)$ or

$$7x^3 - 3x^2 + 9x - 6 = (A + C)x^3 + (B + D)x^2 + (A + 2C)x + B + 2D.$$

Comparing coefficients of the terms with like powers of x, we obtain the following system of equations for the coefficients

$$\begin{cases} 7 = A + C \\ -3 = B + D \\ 9 = A + 2C \\ -6 = B + 2D \end{cases}.$$

Its solution is $A = 5$, $B = 0$, $C = 2$, $D = -3$. Therefore,

$$\frac{7x^3 + 9x - 3x^2 - 6}{\left(x^2 + 2\right)\left(x^2 + 1\right)} = \frac{5x}{x^2 + 2} + \frac{2x - 3}{x^2 + 1}.$$

15.   The degree of the numerator is lower than that of the denominator. Moreover, $x^2 + 1$ cannot be factored into linear factors with real coefficients. Thus, the partial fraction decomposition is

$$\frac{x^3 - x}{\left(x^2 + 1\right)^2} = \frac{Ax + B}{x^2 + 1} + \frac{Cx + D}{\left(x^2 + 1\right)^2}.$$

We put the terms on the right over a common denominator:

$$\frac{x^3 - x}{\left(x^2 + 1\right)^2} = \frac{\left(Ax + B\right)\left(x^2 + 1\right) + Cx + D}{\left(x^2 + 1\right)^2}$$

and obtain the identity

$$x^3 - x = \left(Ax + B\right)\left(x^2 + 1\right) + Cx + D$$

or $x^3 - x = Ax^3 + Bx^2 + \left(A + C\right)x + B + D$. Comparing coefficients of the terms with like powers of x, we obtain the following system of equations for the coefficients

$$\begin{cases} 1 = A \\ 0 = B \\ -1 = A + C \\ 0 = B + D \end{cases}.$$

Its solution is $A = 1$, $B = 0$, $C = -2$, $D = 0$. Therefore,

$$\frac{x^3 - x}{\left(x^2 + 1\right)^2} = \frac{x}{x^2 + 1} - \frac{2x}{\left(x^2 + 1\right)^2}.$$

17.   The degree of the numerator is equal to the degree of the denominator. After the division we obtain

$$\frac{x^3 + 12x^2 - 9x + 48}{\left(x - 3\right)\left(x^2 + 4\right)} = 1 + \frac{15x^2 - 13x + 60}{\left(x - 3\right)\left(x^2 + 4\right)}.$$

Given that $x^2 + 4$ cannot be factored into linear factors with real coefficients, the partial fraction decomposition for the second fraction is

$$\frac{15x^2 - 13x + 60}{\left(x - 3\right)\left(x^2 + 4\right)} = \frac{A}{x - 3} + \frac{Bx + C}{x^2 + 4}.$$

To find coefficients of this decomposition we put the terms over the common denominators and require that numerators be equal:

$$15x^2 - 13x + 60 = A\left(x^2 + 4\right) + \left(Bx + C\right)\left(x - 3\right),$$

or $15x^2 - 13x + 60 = \left(A + B\right)x^2 + \left(C - 3B\right)x + 4A - 3C$. Comparing coefficients of the terms with like powers of x, we obtain the following system of equations for the coefficients

$$\begin{cases} 15 = A + B \\ -13 = C - 3B \\ 60 = 4A - 3C \end{cases}$$

Then $A = 12$, $B = 3$, $C = -4$ and

$$\frac{x^3 + 12x^2 - 9x + 48}{(x-3)(x^2+4)} = 1 + \frac{12}{x-3} + \frac{3x-4}{x^2+4}.$$

19. The degree of the numerator is lower than that of the denominator. Moreover, $x^2 + 4$ and $x^2 + 3$ cannot be factored into linear factors with real coefficients. Thus, the partial fraction decomposition is

$$\frac{3x^3 - 5x^2 + 10x - 19}{(x^2+4)(x^2+3)} = \frac{Ax+B}{x^2+4} + \frac{Cx+D}{x^2+3}.$$

To find coefficients of this decomposition we put the terms over the common denominators and require that numerators be equal:

$$3x^3 - 5x^2 + 10x - 19 = (Ax+B)(x^2+3) + (Cx+D)(x^2+4)$$

or $3x^3 - 5x^2 + 10x - 19 = (A+C)x^3 + (B+D)x^2 + (3A+4C)x + 3B + 4D$. Comparing coefficients of the terms with like powers of x, we obtain the following system of equations for the coefficients

$$\begin{cases} 3 = A + C \\ -5 = B + D \\ 10 = 3A + 4C \\ -19 = 3B + 4D \end{cases}.$$

Therefore $A = 2$, $B = -1$, $C = 1$, $D = -4$ and

$$\frac{3x^3 - 5x^2 + 10x - 19}{(x^2+4)(x^2+3)} = \frac{2x-1}{x^2+4} + \frac{x-4}{x^2+3}.$$

21.   From Problem 7.5.11 we have

$$\int \frac{3x^2 - 5x + 4}{(x-1)(x^2+1)}dx = \int\left(\frac{1}{x-1} + \frac{2x-3}{x^2+1}\right)dx = \int\frac{1}{x-1}dx + \int\frac{2x}{x^2+1}dx - 3\int\frac{1}{x^2+1}dx$$

$$= \ln\left(|x-1|\right) + \ln\left(\left|x^2+1\right|\right) - 3\arctan(x) + C$$

$$= \ln\left(|x-1|\right) + \ln\left(x^2+1\right) - 3\arctan(x) + C$$

We substituted $u = x^2 + 1$ and used the fact that u is positive to evaluate the second integral.

23.   From Problem 7.5.13 we have

$$\int\frac{7x^3 + 9x - 3x^2 - 6}{(x^2+2)(x^2+1)}dx = \int\left(\frac{5x}{x^2+2} + \frac{2x-3}{x^2+1}\right)dx = 5\int\frac{x}{x^2+2}dx + 2\int\frac{x}{x^2+1}dx - 3\int\frac{1}{x^2+1}dx$$

$$= \left(\frac{5}{2}\right)\ln\left(\left|x^2+2\right|\right) + \ln\left(\left|x^2+1\right|\right) - 3\arctan(x)$$

$$= \left(\frac{5}{2}\right)\ln\left(x^2+2\right) + \ln\left(x^2+1\right) - 3\arctan(x) + C$$

We made substitutions $u = x^2 + 2$ and $v = x^2 + 1$ in order to evaluate the first and second integrals respectively. Given that u and v are positive, $\ln\left(|u|\right) = \ln(u)$ and $\ln\left(|v|\right) = \ln(v)$.

25.   From Problem 7.5.15 we have

$$\int\frac{x^3 - x}{\left(x^2+1\right)^2}dx = \int\left(\frac{x}{x^2+1} - \frac{2x}{\left(x^2+1\right)^2}\right)dx = \frac{1}{2}\int\frac{2x}{x^2+1}dx - \int\frac{2x}{\left(x^2+1\right)^2}dx.$$

Let $u = x^2 + 1$, $u > 0$, $du = 2xdx$. Then

$$\int\frac{2x}{x^2+1}dx = \int\frac{1}{u}du = \ln\left(|u|\right) + C_1 = \ln(u) + C_1 = \ln\left(x^2+1\right) + C_1$$

and $\int\frac{2x}{\left(x^2+1\right)^2}dx = \int\frac{1}{u^2}du = -\frac{1}{u} + C_2 = -\frac{1}{x^2+1} + C_2$. Finally we obtain

$$\int\frac{x^3 - x}{\left(x^2+1\right)^2}dx = \frac{1}{2}\ln\left(x^2+1\right) + \frac{1}{x^2+1} + C.$$

27.   From Problem 7.5.17 we have

$$\int \frac{x^3 + 12x^2 - 9x + 48}{(x-3)(x^2+4)} dx = \int \left(1 + \frac{12}{x-3} + \frac{3x-4}{x^2+4}\right) dx$$

$$= \int dx + 12\int \frac{1}{x-3} dx + \frac{3}{2}\int \frac{2x}{x^2+4} dx - 4\int \frac{1}{x^2+4} dx$$

$$= x + 12\ln\left(|x-3|\right) + \frac{3}{2}\ln\left(\left|x^2+4\right|\right) - 2\arctan\left(\frac{x}{2}\right) + C$$

$$= x + 12\ln\left(|x-3|\right) + \frac{3}{2}\ln\left(x^2+4\right) - 2\arctan\left(\frac{x}{2}\right) + C$$

To evaluate the third integral we used the substitution $u = x^2 + 4$.

29.   From Problem 7.5.19 we have

$$\int \frac{3x^3 - 5x^2 + 10x - 19}{(x^2+4)(x^2+3)} dx = \int \left(\frac{2x-1}{x^2+4} + \frac{x-4}{x^2+3}\right) dx$$

$$= \int \frac{2x}{x^2+4} dx - \int \frac{1}{x^2+4} dx + \frac{1}{2}\int \frac{2x}{x^2+3} dx - 4\int \frac{1}{x^2+3} dx$$

$$= \ln\left(\left|x^2+4\right|\right) - \frac{1}{2}\arctan\left(\frac{x}{2}\right) + \frac{1}{2}\ln\left(\left|x^2+3\right|\right) - \frac{4}{\sqrt{3}}\arctan\left(\frac{x}{\sqrt{3}}\right) + C$$

$$= \ln\left(x^2+4\right) - \frac{1}{2}\arctan\left(\frac{x}{2}\right) + \frac{1}{2}\ln\left(x^2+3\right) - \frac{4}{\sqrt{3}}\arctan\left(\frac{x}{\sqrt{3}}\right) + C$$

We use substitutions $u = x^2 + 4$ and $v = x^2 + 3$ in order to evaluate the first and third integrals respectively. Given that u and v are positive, $\ln\left(|u|\right) = \ln(u)$ and $\ln\left(|v|\right) = \ln(v)$.

Further Theory and Practice

31.   Given that partial fraction decomposition is unique, we obtain $A_i = 0$ for $1 \le i \le 4$, $B_j = 0$ for $1 \le j \le 3$, $B_4 = 1$.

33.   Decomposition of the integrand into elementary partial fractions using the expansion $x^3 + 1 = (x+1)(x^2 - x + 1)$ leads to the equation

$$\frac{5x^2 - 2x + 2}{x^3 + 1} = \frac{A}{x+1} + \frac{Bx + C}{x^2 - x + 1} .$$

Therefore $5x^2 - 2x + 2 = A(x^2 - x + 1) + (Bx + C)(x+1)$. This may also be written as

$$5x^2 - 2x + 2 = (A + B)x^2 + (C + B - A)x + A + C .$$

Comparing coefficients of the terms with like powers of x, we obtain the following system of equations for the coefficients

$$\begin{cases} A + B = 5 \\ C + B - A = -2 \\ C + A = 2 \end{cases} .$$

Its solution is $A = 3$, $B = 2$, $C = -1$. The integral is now readily calculated by changing variables to $u = x + 1$ and $v = x^2 - x + 1$:

$$\begin{aligned} \int \frac{5x^2 - 2x + 2}{x^3 + 1} dx &= 3 \int \frac{dx}{x+1} + \int \frac{2x - 1}{x^2 - x + 1} dx \\ &= 3 \int \frac{du}{u} + \int \frac{dv}{v} \\ &= 3 \ln \left(|x + 1| \right) + \ln \left(|x^2 - x + 1| \right) + C \\ &= 3 \ln \left(|x + 1| \right) + \ln \left(x^2 - x + 1 \right) + C \end{aligned}$$

35.   The integrand is a repeated quadratic building block, so we use the substitution $x = \tan(\theta)$, $dx = \sec^2(\theta)\,d\theta$. Then

$$\int \frac{48}{\left(x^2+1\right)^4}\,dx = 48\int \frac{\sec^2(\theta)}{\left(\tan^2(\theta)+1\right)^4}\,d\theta = 48\int \frac{\sec^2(\theta)}{\sec^8(\theta)}\,d\theta = 48\int \cos^6(\theta)\,d\theta$$

$$= 48\left(\frac{5}{16}\theta + \frac{1}{6}\sin(\theta)\cos^5(\theta) + \frac{5}{24}\sin(\theta)\cos^3(\theta) + \frac{5}{16}\sin(\theta)\cos(\theta)\right) + C$$

where we used Example 4 of Section 7.3 to evaluate $\int \cos^6(\theta)\,d\theta$. Given that

$$\sin(\theta) = \frac{x}{\sqrt{x^2+1}}, \quad \cos(\theta) = \frac{1}{\sqrt{x^2+1}}, \text{ we find}$$

$$\int \frac{48}{\left(x^2+1\right)^4}\,dx = 48\left(\frac{5}{16}\theta + \frac{1}{6}\sin(\theta)\cos^5(\theta) + \frac{5}{24}\sin(\theta)\cos^3(\theta) + \frac{5}{16}\sin(\theta)\cos(\theta)\right) + C$$

$$= 15\arctan(x) + 8\frac{x}{\left(x^2+1\right)^3} + 10\frac{x}{\left(x^2+1\right)^2} + 15\frac{x}{x^2+1} + C$$

Calculator/Computer Problems

37.   We solve $x^4 + 2 = \dfrac{5x^3 + 3x + 2}{x^3 + x^2 + x + 1}$ and find $a = 0$, $b = 0.7585$. Then the area is

$$A = \int_a^b \left(\frac{5x^3 + 3x + 2}{x^3 + x^2 + x + 1} - x^4 - 2\right)dx$$

$$= \int_a^b \frac{5x^3 + 3x + 2}{x^3 + x^2 + x + 1}\,dx - \left(\frac{1}{5}x^5 - 2x\right)\Big|_a^b$$

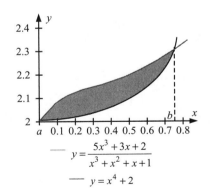

$$y = \frac{5x^3 + 3x + 2}{x^3 + x^2 + x + 1}$$

$$y = x^4 + 2$$

We have to evaluate $\int\limits_a^b \dfrac{5x^3+3x+2}{x^3+x^2+x+1}\,dx$. We see that the degree of the numerator is equal to the degree of the denominator, and after division we obtain

$$\frac{5x^3+3x+2}{x^3+x^2+x+1} = 5 - \frac{5x^2+2x+3}{x^3+x^2+x+1}.$$

Given that $x^3+x^2+x+1=(x+1)\left(x^2+1\right)$, the partial fraction decomposition takes the form

$$\frac{5x^2+2x+3}{x^3+x^2+x+1} = \frac{B}{x+1} + \frac{Cx+D}{x^2+1}.$$

The identity $5x^2+2x+3 = B\left(x^2+1\right)+(Cx+D)(x+1)$ leads to the system

$$\begin{cases} 5 = B+C \\ 2 = C+D \\ 3 = B+D \end{cases}.$$

Its solution is $B=3$, $C=2$, $D=0$. Then

$$\int\frac{5x^2+2x+3}{x^3+x^2+x+1}dx = \int\left(\frac{3}{x+1}+\frac{2x}{x^2+1}\right)dx = 3\ln\left(|x+1|\right)+\ln\left(x^2+1\right)+\tilde{C}$$

and

$$\int\limits_a^b\frac{5x^2+2x+3}{x^3+x^2+x+1}dx = \left(3\ln\left(|x+1|\right)+\ln\left(x^2+1\right)\right)\Bigg|_a^b.$$

Finally we conclude that the area between the two curves is

$$A = \int\limits_a^b\left(\frac{5x^3+3x+2}{x^3+x^2+x+1}-x^4-2\right)dx = \left(5x-3\ln\left(|x+1|\right)-\ln\left(x^2+1\right)\right)\Bigg|_a^b - \left(\frac{1}{5}x^5-2x\right)\Bigg|_a^b$$

$$= 0.07744$$

39. Given that $x^5 - 7x^4 + 4x^3 - 23x^2 + 20 = \left(x^2 + 4\right)\left(x^3 - 7x^2 + 5\right)$, we solve $x^3 - 7x^2 + 5 = 0$ and find the following three roots: $x_1 = -0.80061$, $x_2 = 0.90579$, and $x_3 = 6.89482$. The partial fraction decomposition then takes the form:

$$\frac{2x+4}{x^5 - 7x^4 + 4x^3 - 23x^2 + 20} = \frac{A}{x - x_1} + \frac{B}{x - x_2} + \frac{C}{x - x_3} + \frac{Dx + E}{x^2 + 4}.$$

The identity

$$2x + 4 = A\left(x - x_2\right)\left(x - x_3\right)\left(x^2 + 4\right) + B\left(x - x_1\right)\left(x - x_3\right)\left(x^2 + 4\right)$$

$$+ C\left(x - x_1\right)\left(x - x_2\right)\left(x^2 + 4\right) + \left(Dx + E\right)\left(x^3 - 7x^2 + 5\right)$$

leads to the system

$$\begin{cases} 0 = A + B + C + D \\ 0 = -7D + E - A\left(x_2 + x_3\right) - B\left(x_1 + x_3\right) - C\left(x_1 + x_2\right) \\ 0 = 4A - 7E + 4C + 4B + Ax_2x_3 + Bx_1x_3 + Cx_1x_2 \\ 2 = 4A\left(x_2x_3 - x_2 - x_3\right) + 4B\left(x_1x_3 - x_1 - x_3\right) + 4C\left(x_1x_2 - x_1 - x_2\right) + 5D \\ 4 = 4Ax_2x_3 + 4Bx_1x_3 + 4Cx_1x_2 + 5E \end{cases}.$$

This system can be solved completely using computer algebra. Some of its coefficients, however, are easier to find by the other means. We substitute $x = x_1$, then

$$2x_1 + 4 = A\left(x_1 - x_2\right)\left(x_1 - x_2\right)\left(x_1^2 + 4\right).$$

For $x_1 = -0.80061$ this leads to $A = 0.03936$. After substitution $x_2 = 0.90579$ and $x_3 = 6.89482$ we obtain $B = -0.11797$ and $C = 0.00749$. Then $D = 0.08673$ and $E = 0.07112$. We, therefore, have partial fraction decomposition:

$$\frac{2x+4}{x^5 - 7x^4 + 4x^3 - 23x^2 + 20} = \frac{0.03936}{x + 0.80061} - \frac{0.11767}{x - 0.90579} + \frac{0.00749}{x - 6.89482} + \frac{0.08673x + 0.07112}{x^2 + 4}$$

Finally, we can evaluate the integral as follows:

$$\int_1^2 \frac{2x+4}{x^5 - 7x^4 + 4x^3 - 23x^2 + 20}\, dx = \int_1^2 \frac{0.03936}{x + 0.80061}\, dx - \int_1^2 \frac{0.11767}{x - 0.90579}\, dx + \int_1^2 \frac{0.00749}{x - 6.89482}\, dx$$

$$+ \int_1^2 \frac{0.08673x + 0.07112}{x^2 + 4}\, dx = 0.03936 \ln\left(\left|x + 0.80061\right|\right)\Big|_1^2 - 0.11767 \ln\left(\left|x - 0.90579\right|\right)\Big|_1^2$$

$$+ 0.00749 \ln\left(\left|x - 6.89482\right|\right)\Big|_1^2 + 0.04337 \ln\left(x^2 + 4\right)\Big|_1^2$$

$$+ 0.03556 \arctan\left(\frac{x}{2}\right)\Big|_1^2 = -0.24263$$

41.   **a.** $$\frac{2x^5 + 13x^4 - 28x^3 + 30x^2 - 37x - 8}{x^6 - x^5 + 2x^4 - x^3 + 2x^2 - x + 1} = -\frac{3x - 14}{x^2 + x + 1} + \frac{5x - 7}{x^2 - x + 1} + \frac{4x - 15}{\left(x^2 - x + 1\right)^2}$$

b. $$\frac{17x^5 + 39x^4 + 140x^3 + 140x^2 + 199x + 33}{x^6 + 3x^5 + 12x^4 + 19x^3 + 36x^2 + 27x + 27} = \frac{17x + 5}{x^2 + x + 3} + \frac{11x - 8}{\left(x^2 + x + 3\right)^2} + \frac{12 - 9x}{\left(x^2 + x + 3\right)^3}$$

Applications of the Integral

Section 8.1 Volumes

Problems for Practice

1. $V = \lim\limits_{N\to\infty} \sum\limits_{j=1}^{N} \pi y_j^2 \Delta x = \lim\limits_{N\to\infty} \pi \sum\limits_{j=1}^{N} \left(3x_j^{1/3}\right)^2 \Delta x = 9\pi \int\limits_{1}^{8} x^{2/3} dx = 9\pi \left(\frac{3}{5}\right) x^{5/3} \Big|_{1}^{8} = 837\frac{\pi}{5}$

3. $V = \lim\limits_{N\to\infty} \sum\limits_{j=1}^{N} \pi y_j^2 \Delta x = \lim\limits_{N\to\infty} \pi \sum\limits_{j=1}^{N} \left(\sqrt{\sin(x_j)}\right)^2 \Delta x = \pi \int\limits_{0}^{\pi} \sin(x) dx = -\pi \cos(x)\Big|_{0}^{\pi} = 2\pi$

5. $V = \lim\limits_{N\to\infty} \sum\limits_{j=1}^{N} \pi x_j^2 \Delta y = \lim\limits_{N\to\infty} \pi \sum\limits_{j=1}^{N} y_j^6 \Delta y = \pi \int\limits_{0}^{1} y^6 dy = \pi \frac{y^7}{7}\Big|_{0}^{1} = \frac{\pi}{7}$

7. Given that $y(1)=0$, $y(e)=1$, and $x=\exp(y)$, the volume obtained by rotating the planar region is

$$V = \lim\limits_{N\to\infty} \sum\limits_{j=1}^{N} \pi x_j^2 \Delta y = \lim\limits_{N\to\infty} \pi \sum\limits_{j=1}^{N} \left(\exp(y_j)\right)^2 \Delta y = \pi \int\limits_{0}^{1} \exp(2y) dy = \pi \frac{\exp(2y)}{2}\Big|_{0}^{1} = \pi \frac{e^2-1}{2}.$$

9. Given that $x^2 \le x$ for $0 \le x \le 1$, we find that $U(x)=x$, $L(x)=x^2$. The volume obtained by

rotating the planar region is $V = \pi \int\limits_{0}^{1} \left(x^2 - \left(x^2\right)^2\right) dx = \pi \left(\frac{x^3}{3} - \frac{x^5}{5}\right)\Big|_{0}^{1} = \frac{2\pi}{15}.$

11.   Let $U(x)=4-x^2$, $L(x)=x+2$. They intersect at $x=-2$ and $x=1$. The volume obtained by rotating the planar region is then

$$V = \int_{-2}^{1} \pi\left(\left(4-x^2\right)^2 - (x+2)^2\right)dx = \pi\left(16x + \frac{x^5}{5} - \frac{8x^3}{3} - \frac{(x+2)^3}{3}\right)\Bigg|_{-2}^{1} = \frac{108\pi}{5}.$$

13.   Given that $y^2 \le 2y$ for $0 \le y \le 2$, we find that $U(y)=2y$, $L(y)=y^2$. The volume obtained

by rotating the planar region is $V = \pi\int_{0}^{2}\left((2y)^2 - \left(y^2\right)^2\right)dy = \frac{64\pi}{15}$.

15.   Two curves $x=y^2+2$ and $x=4-y$ intersect at $y=-2$ and $y=1$. Given that $U(y)=4-y$, $L(y)=y^2+2$, the volume obtained by rotating the planar region is

$$V = \int_{-2}^{1} \pi\left((4-y)^2 - \left(y^2+2\right)^2\right)dy = \frac{162\pi}{5}.$$

17. $V = 2\pi\int_{0}^{1} x\left(x^2\right)dx = \frac{\pi}{2}$

19. $V = 2\pi\int_{2}^{4} x\left(x^2+1\right)dx = 132\pi$

21. $V = 2\pi\int_{0}^{4} y\left(\sqrt{y}\right)dy = \frac{128\pi}{5}$

23. $V = 2\pi\int_{0}^{2} y\left(2+y+y^2\right)dy = \frac{64\pi}{3}$

25.   Two curves $y=x$ and $y=x^2$ intersect at $x=0$ and $x=1$. For $0 \le x \le 1$ the following inequality is valid: $x^2 \le x$. The volume obtained by rotating the planar region is

$$V = 2\pi\int_{0}^{1} x\left(x-x^2\right)dx = \frac{\pi}{6}.$$

27. $V = 2\pi \int\limits_{2}^{4} x\left(\left(x^2+1\right)-x\right) dx = \dfrac{284\pi}{3}$

29. For $x=4$, we find $y=2$, $g(y)=y^2$. The volume obtained by rotating the planar region is

$V = 2\pi \int\limits_{0}^{2} y\left(4-y^2\right) dy = 8\pi$.

31. The two curves $y=2\sqrt{x}$ and $y=3x-1$ intersect at $y=2$. For $1 \le y \le 2$ we find

$$g_1(y) = \dfrac{y^2}{4} \le g_2(y) = \dfrac{y+1}{3} .$$

The volume obtained by rotating the planar region is $V = 2\pi \int\limits_{1}^{2} y\left(\dfrac{y+1}{3} - \dfrac{y^2}{4}\right) dy = \dfrac{49\pi}{72}$.

Further Theory and Practice

33. $\int\limits_{0}^{\pi} \pi \sin^2(x)\, dx = \dfrac{\pi}{2}\int\limits_{0}^{\pi}(1-\cos(2x))\, dx = \dfrac{\pi}{2}\left(\dfrac{x-\sin(2x)}{2}\right)\Bigg|_{0}^{\pi} = \dfrac{\pi^2}{2}$

35. The x-intercept of the line $y = \sqrt{x}\exp(x)$ is $x=0$. The volume obtained by rotating the

planar region is $V = \int\limits_{0}^{1} \pi\left(\sqrt{x}\cdot\exp(x)\right)^2 dx = \pi\int\limits_{0}^{1} x\exp(2x)\, dx = \pi e^{2x}\dfrac{(2x-1)}{4}\Bigg|_{0}^{1} = \dfrac{\pi\left(e^2+1\right)}{4}$.

37. The volume is $V = \int\limits_{1}^{e} \pi \ln^2(y)\, dy$. By making the substitution $y=e^w$, $dy=e^w dw$, we see that

$V = \pi \int\limits_{0}^{1} w^2 e^w dw$. We calculate this integral by integrating by parts twice, setting $dv = e^w dw$ in

each application:

$$V = \pi \int\limits_{0}^{1} w^2 e^x dw = \pi\left(e - \int\limits_{0}^{1} 2we^w dw\right) = \pi\left(e - \left(2e - \int\limits_{0}^{1} 2e^w dw\right)\right) = \pi(e-2) .$$

39. Let $g(y) = \sqrt{\dfrac{y}{y+1}}$. Given that $g(y) = 0$ for $y = 0$, the volume obtained by rotating the planar region is

$$V = \int_0^{1/3} \pi \left(\sqrt{\frac{y}{y+1}} \right)^2 dy = \pi \int_0^{1/3} \left(1 - \frac{1}{y+1} \right) dy = \pi \left(y - \ln(y+1) \right) \Big|_0^{1/3} = \pi \left(\frac{1}{3} - \ln\left(\frac{4}{3} \right) \right).$$

41. The two curves meet at $x = 2 - \sqrt{5}$ and $x = 2 + \sqrt{5}$. In the region between these points, the following inequality is valid: $-x^2 + 4x - 3 \geq x^2 - 4x - 5$. We then calculate the integral using cylindrical shells. The volume obtained by rotating the planar region is

$$V = 2\pi \int_{2-\sqrt{5}}^{2+\sqrt{5}} (x+3)\left(\left(-x^2 + 4x - 3 \right) - \left(x^2 - 4x - 5 \right) \right) dx = 400 \frac{\pi\sqrt{5}}{3}.$$

43. The two curves intersect at $x = -2$ and $x = 3$. In the region between these points, the following inequality is valid: $-x^2 + x + 7 \geq x^2 - x - 5$. The volume obtained by rotating the planar region is $V = 2\pi \int_{-2}^{3} (6-x)\left(\left(-x^2 + x + 7 \right) - \left(x^2 - x - 5 \right) \right) dx = \dfrac{1375\pi}{3}$.

45. We have two points of intersection at $y = 0$ and $y = 1$. In the region between these points, $0 \leq y \leq 1$, the following inequality is valid: $\sqrt{y} \geq y^4$. The volume obtained by rotating the planar region is $V = 2\pi \int_0^1 (1+y)\left(\sqrt{y} - y^4 \right) dy = \dfrac{7\pi}{5}$.

47. We have two curves that meet at the points $x = -2$ and $x = 3$. In the region between these points, $-2 \leq x \leq 3$, the following inequality is valid: $-x^2 + 6 \geq -x$. Using the Method of Cylindrical Shells we obtain the following volume of the solid when the region is rotated about the line: $V = 2\pi \int_{-2}^{3} (x+4)\left(-x^2 + 6 - (-x) \right) dx = \dfrac{375\pi}{2}$.

49. For $\dfrac{\pi}{2} \leq x \leq \dfrac{3\pi}{2}$, we find $0 \geq \cos(x)$. Using the Method of Cylindrical Shells we obtain the following volume of the solid when the region is rotated about the line:

$$V = 2\pi \int_{\pi/2}^{3\pi/2} (0 - \cos(x))(3\pi - x) \, dx = 8\pi^2.$$

51. Given that $x = 0$ for $y = 0$, using the Method of Cylindrical Shells, we obtain the following volume of the solid when the region is rotated about the line: $V = 2\pi \int\limits_0^{\pi/2} (4 - y)\sin(y)\,dy = 6\pi$.

53. Given that $x^3 + x = 0$ for $x = 0$, we use the Method of Cylindrical Shells to obtain the following volume of the solid when the region is rotated about the line:

$$V = 2\pi \int\limits_0^1 x\left(x^3 + x\right)dx = \frac{16\pi}{15}.$$

55. The volume is $V = \int\limits_{-r}^{r} 2\pi(R - x) \cdot 2\sqrt{r^2 - x^2}\,dx$. Making the substitution $x = r\sin(\theta)$, $dx = r\cos(\theta)\,d\theta$, we have

$$V = 4\pi r^2 \int\limits_{-\pi/2}^{\pi/2} \left(R - r\sin(\theta)\right)\cos^2(\theta)\,d\theta$$

$$= 4\pi r^2 R \int\limits_{-\pi/2}^{\pi/2} \cos^2(\theta)\,d\theta - 4\pi^3 \int\limits_{-\pi/2}^{\pi/2} \sin(\theta)\cos^2(\theta)\,d\theta.$$

The second of these integrals evaluates to 0 and the first can be evaluated using formula (7.21). There results $V = 2\pi^2 R r^2 = (2\pi R)\left(\pi r^2\right)$.

57. By Exercise 8.1.55, the volume of the torus is $(2\pi \cdot 7)\left(\pi 2^2\right) = 56\pi^2$. Cut-out from that solid is a volume equal to $\int\limits_{-1}^{1} 2\pi(2)(7 - y)\,dy = 56\pi$. Thus, we calculate the volume to be

$$56\pi^2 - 56\pi = 56\pi(\pi - 1).$$

59. The volume is $\int\limits_{-5}^{5} \left(2\sqrt{25 - x^2}\right)^2 dx = \frac{2000}{3}$.

61.   Let a be the *unspecified* radius of the sphere. Let $2h$ denote the length of the hole that is drilled. The sphere with the hole may be realized as the solid that results when the region that lies to the right of the vertical line $x = \sqrt{a^2 - h^2}$ and to the left of $x = \sqrt{a^2 - y^2}$ is revolved about the y-axis. Using the Method of Cylindrical Shells we calculate the volume:

$$V = \int_{\sqrt{a^2-h^2}}^{a} 2\pi x \cdot 2\sqrt{a^2 - x^2}\, dx = -\frac{4}{3}\pi\left(a^2 - x^2\right)^{3/2}\Big|_{x=\sqrt{a^2-h^2}}^{x=a} = \frac{4}{3}\pi h^3 \, .$$

Notice that the volume does not depend on the unspecified radius a! For $2h = 10$ in , the volume is $\frac{4}{3}\pi 5^3$ in^3 , or $\dfrac{500\pi}{3}$ cubic inches.

Calculator/Computer Exercises

63.   The solutions of $x\exp(x) = \sqrt{x}$ are $x = 0$ and $x = 0.42630275$. Using cylindrical shells, the volume is

$$2\pi \int_{0}^{0.42630275} x\left(\sqrt{x} - x\exp(x)\right) dx \approx 0.0741 \, .$$

Instructors may point out that the integral can also be evaluated using integration by parts. The antiderivative may then be evaluated at the approximate value 0.42630275. Or, the integral may be numerically approximated straight-away.

65.   The mass is

$$3.743 \int_{0}^{2.5} \pi\left(\left(2.530\sqrt{x} + 0.300\right)^2 - \left(2.530\sqrt{x}\right)^2\right) dx$$

$$= 49.685 \text{ grams}$$

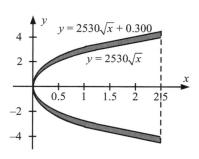

Section 8.2   Arc Length and Surface Area

Problems for Practice

1. $\displaystyle\int_{3}^{8} \sqrt{1 + 6^2}\, dx = 5\sqrt{37} = \sqrt{(48-18)^2 + (8-3)^2}$

3. $\int_{-1}^{2} \sqrt{1+\left(\frac{d}{dx}(3-x)\right)^2}\, dx = \sqrt{2}x\Big|_{-1}^{2}\ 3\sqrt{2} = \sqrt{(1-4)^2+(2-(-1))^2}$

5. $\int_{1}^{4} \sqrt{1+\left(\frac{d}{dx}\left(2+x^{3/2}\right)\right)^2}\, dx = \left(\frac{1}{2}\right)\int_{1}^{4}\sqrt{4+9x}\, dx = \dfrac{(4+9x)^{3/2}}{27}\Bigg|_{1}^{4} = \dfrac{80\sqrt{10}-13\sqrt{13}}{27}$

7. $\int_{0}^{4} \sqrt{1+\left(\frac{d}{dx}\left(3+(2x+1)^{3/2}\right)\right)^2}\, dx = \sqrt{2}\int_{0}^{4}\sqrt{5+9x}\, dx = \dfrac{2\sqrt{2}(5+9x)^{3/2}}{27}\Bigg|_{0}^{4} = \sqrt{2}\,\dfrac{82\sqrt{41}-10\sqrt{5}}{27}$

9. $2\pi\int_{-2}^{3}\left(2x^3\right)\left(1+\left(\frac{d}{dx}\left(2x^3\right)\right)^2\right)^{1/2} dx = 2\pi\int_{-2}^{3} 2x^3\sqrt{1+36x^4}\, dx = \dfrac{\pi\left(1+36x^4\right)^{3/2}}{54}\Bigg|_{-2}^{3}$

$$= 2\pi\,\dfrac{2917\sqrt{2917}-577\sqrt{577}}{108}$$

11. $2\pi\int_{1}^{3}\left(\dfrac{x^3}{3}+\dfrac{1}{4x}\right)\left(1+\left(\frac{d}{dx}\left(\dfrac{x^3}{3}+\dfrac{1}{4x}\right)\right)^2\right)^{1/2} dx = 2\pi\int_{1}^{3}\left(\dfrac{4x^4+3}{12x}\right)\left(\dfrac{\left(1-2x+2x^2\right)^2\left(1+2x+2x^2\right)^2}{16x^4}\right)^{1/2} dx$

$$= \dfrac{\pi}{24}\int_{1}^{3}x^{-3}\left(4x^4+3\right)\left(1-2x+2x^2\right)\left(1+2x+2x^2\right)dx$$

$$= \dfrac{\pi}{24}\int_{1}^{3}\left(16x+16x^5+3x^{-3}\right)dx$$

$$= \dfrac{\pi}{24}\left(8x^2+\dfrac{8x^6}{3}-\dfrac{3x^{-2}}{2}\right)\Bigg|_{1}^{3} = \dfrac{1505\pi}{18}$$

13. $2\pi\int_{1}^{4}(3x-1)\left(1+\left(\frac{d}{dx}(3x-1)\right)^2\right)^{1/2} dx = 2\pi\sqrt{10}\int_{1}^{4}(3x-1)\, dx = \pi\sqrt{10}\left(3x^2-2x\right)\Big|_{1}^{4} = 39\pi\sqrt{10}$

15. $2\pi\int_{-1}^{-1/2}\left(x^4+\dfrac{x^{-2}}{32}\right)\left(1+\left(\dfrac{d}{dx}\left(x^4+\dfrac{x^{-2}}{32}\right)\right)^2\right)^{1/2} dx = 2\pi\int_{-1}^{-1/2}\left(x^4+\dfrac{x^{-2}}{32}\right)\left(4x^3+\dfrac{1}{16x^3}\right)dx$

$$= 2\pi\int_{-1}^{-1/2}\left(4x^7+\dfrac{3}{16}x+\dfrac{x^{-5}}{512}\right)dx$$

$$= 2\pi\left(\dfrac{x^8}{2}+\dfrac{3}{32}x^2-\dfrac{1}{2048}x^{-4}\right)\Bigg|_{-1}^{1/2} = \dfrac{1179}{1024}\pi$$

17.   The inverse function of f is $g(y) = y^3$. The area of the surface is

$$2\pi \int_2^3 (g(y)) \left(1 + \left(\tfrac{d}{dy}(g(y)) \right)^2 \right)^{1/2} dy = 2\pi \int_2^3 y^3 \left(1 + \left(\tfrac{d}{dy} y^3 \right)^2 \right)^{1/2} dy$$

$$= 2\pi \int_2^3 y^3 \left(1 + 9y^4 \right)^{1/2} dy = \frac{1}{27} \pi \left(1 + 9y^4 \right)^{3/2} \Big|_2^3$$

$$= 2\pi \left(\tfrac{365}{27} \sqrt{730} - \tfrac{145}{54} \sqrt{145} \right).$$

19.   The same surface is generated by rotating $f(x) = x^2$ for $2 \le x \le 6$. The inverse function of f is $g(y) = \sqrt{y}$. The surface area is

$$2\pi \int_4^{36} \sqrt{y} \sqrt{1 + \left(\tfrac{d}{dy} \sqrt{y} \right)^2} \, dy = 2\pi \int_4^{36} \sqrt{y + \frac{1}{4}} \, dy = \frac{1}{6} \pi (4y+1)^{3/2} \Big|_4^{36}$$

$$= \pi \frac{145\sqrt{145} - 17\sqrt{17}}{6}.$$

Further Theory and Practice

21.   The arc length of the graph is $\displaystyle \int_0^1 \sqrt{1 + \left(\tfrac{d}{dx} \frac{(x^2+2)^{3/2}}{3} \right)^2} \, dx = \int_0^1 (1 + x^2) \, dx = \frac{4}{3}.$

23.   The arc length of the graph is $\displaystyle \int_0^1 \sqrt{1 + f'(x)^2} \, dx = \left(\frac{1}{2} \right) \int_0^1 \left(e^x + e^{-x} \right) dx = \frac{e - e^{-1}}{2}.$

25.   As $\displaystyle \sqrt{1 + f'(x)^2} = \frac{1 + 256x^6}{32|x|^3}$, the arc length of the graph is

$$\int_{-2}^{-1} \frac{1 + 256x^6}{32|x|^3} \, dx = \frac{1}{32} \int_{-2}^{-1} \left(|x|^{-3} + 256|x|^3 \right) dx = \frac{1}{32} \int_{-2}^{-1} \left((-x)^3 + 256(-x)^3 \right) dx$$

$$= \left(\frac{x^{-2}}{64} - 2x^4 \right) \Big|_{-2}^{-1} = \frac{7683}{256}.$$

27. We solve $y = \dfrac{x^5}{30} + \dfrac{1}{2x^3}$ and simplify $\sqrt{1 + \left(\dfrac{dy}{dx}\right)^2}$ to $\dfrac{9 + x^8}{6x^4}$. The arc length of the graph is

$\displaystyle\int_{0.1}^{1} \dfrac{9 + x^8}{6x^4}\, dx \approx 499.53$. (If the abscissa of P were exactly 1/10, then the exact length would be

499533333/1000000 .)

29. We have $y = b\sqrt{1 - \dfrac{x^2}{a^2}}$ on the upper half of the ellipse. We calculate

$$\sqrt{1 + \left(\dfrac{dy}{dx}\right)^2} = \sqrt{1 + \dfrac{b^2}{a^2 - x^2}\dfrac{x^2}{a^2}} = \sqrt{\dfrac{a^4 - \left(a^2 - b^2\right)x^2}{\left(a^2 - x^2\right)a^2}} = \sqrt{\dfrac{a^2 - \varepsilon^2 x^2}{a^2 - x^2}}$$

and the given formula follows. We now make the change of variable $x = a\sin(t)$,

$dx = a\cos(t)$ to obtain $\displaystyle\int_{0}^{\xi} \sqrt{\dfrac{a^2 - \varepsilon^2 x^2}{a^2 - x^2}}\, dx = a \int_{0}^{\arcsin(\xi/a)} \sqrt{1 - \varepsilon^2 \sin^2(t)}\, dt$. Instructors may point

out that the first integral requires special attention at $\xi = a$ and that techniques for handling such integrals will be developed in Section 8.6. Given that $\varepsilon < 1$, the second integral does not present any difficulties at $\xi = a$ and may be used as the basis of a formula for the arc length of the entire ellipse.

31. We have $\displaystyle\int_{0}^{1} \sqrt{\left(\tfrac{d}{dt}\left(\exp(t)\cos(t)\right)\right)^2 + \left(\tfrac{d}{dt}\left(\exp(t)\sin(t)\right)\right)^2}\, dt = \int_{0}^{1} \sqrt{2}e^t\, dt = \sqrt{2}\left(e - 1\right)$.

33. We have $\displaystyle\int_{0}^{1} \sqrt{\left(\tfrac{d}{dt}\left(2\arctan(t)\right)\right)^2 + \left(\tfrac{d}{dt}\left(\ln\left(1 + t^2\right)\right)\right)^2}\, dt = 2\int_{0}^{1}\left(1 + t^2\right)^{-1/2}\, dt = 2\ln\left(1 + \sqrt{2}\right)$.

35. The arc length of the sector formed by cutting and flattening the cone is $\dfrac{2\pi\ell}{2\pi}\theta = \ell\theta$. On the

other hand, it is $2\pi R$. So $\ell\theta = 2\pi R$ or $\theta = \dfrac{2\pi R}{\ell}$. The surface area of the cone is equal to the

area of the sector, which is $\dfrac{\pi\ell^2}{2\pi}\theta = \dfrac{\ell^2}{2}\cdot\dfrac{2\pi R}{\ell} = \pi R\ell$. We have for the frustum

$\theta = \dfrac{2\pi r}{\ell - s} = \dfrac{2\pi R}{\ell}$ or $\ell = \dfrac{Rs}{R - r}$. The surface area of the frustum is

$$\pi R\ell - \pi r\left(\ell - s\right) = \pi R\dfrac{Rs}{R - r} - \pi r\left(\dfrac{Rs}{R - r} - s\right) = \dfrac{\pi R^2 s}{R - r} - \dfrac{\pi r^2 s}{R - r} = \dfrac{\pi\left(R^2 - r^2\right)s}{R - r} = 2\pi\left(\dfrac{R + r}{2}\right)s\,.$$

37.   The area is $2\pi \int\limits_{\pi/4}^{3\pi/4} \sin(x)\sqrt{1+\cos^2(x)}\,dx = 2\pi \int\limits_{-1/\sqrt{2}}^{1/\sqrt{2}} \sqrt{1+u^2}\,du$, after making the change of

variable $u = \cos(x)$, $du = -\sin(x)\,dx$. The calculation is concluded by making the second

change of variable, $u = \tan(\theta)$, $du = \sec^2(\theta)\,d\theta$, and then using the integration formula that

is stated in Exercise 26 of Section 7.1. The result is

$$2\pi \int\limits_{-1/\sqrt{2}}^{1/\sqrt{2}} \sqrt{1+u^2}\,du = \pi\left(\sqrt{3} - \ln\left(-1+\sqrt{3}\right) + \ln\left(1+\sqrt{3}\right)\right).$$

39.   The surface area is $2\pi \int\limits_{0}^{2} e^x \sqrt{1+\left(e^x\right)^2}\,dx = 2\pi \int\limits_{1}^{e^2} \sqrt{1+u^2}\,du$, after making the change of variable

$u = e^x$, $du = e^x\,dx$. The calculation is concluded by making the second change of variable,

$u = \tan(\theta)$, $du = \sec^2(\theta)\,d\theta$, and then using the integration formula that is stated in

Exercise 26 of Section 7.1. The result is

$$2\pi \int\limits_{1}^{e^2} \sqrt{1+u^2}\,du = \pi\left(e^2\sqrt{1+e^4} + \ln\left(e^2 + \sqrt{1+e^4}\right) - \sqrt{2} - \ln\left(1+\sqrt{2}\right)\right).$$

Calculator/Computer Exercises

41.   Let L be the arc length of $y = 4\sqrt{1-x^2/4}$, $0 \le x \le 2$. That is, L is the length of that part of the
ellipse that lies in the first quadrant. We may quadruple an approximation of L to obtain an
estimate of the total elliptic arc length. To obtain two decimal places, our approximation of L
should not have an error exceeding $\epsilon = 5/4 \times 10^{-3} = 0.00125$, since we will be quadrupling this
estimate. We calculate the integrand $\sqrt{1+(dy/dx)^2} = \sqrt{4+3x^2}/\sqrt{4-x^2}$. We are not able to
apply Simpson's Rule to this integrand on the interval $[0, 2]$. That is because the integrand is
not defined at the right endpoint $x = 2$. However, we can apply Simpson's Rule very
efficiently by breaking the elliptic arc at the point $\left(1, 2\sqrt{3}\right)$. Apply Simpson's Rule to

estimate the arc length of $y = 4\sqrt{1-x^2/4}$, $0 \le x \le 1$. The Simpson's Rule error estimate shows
that $N = 4$ subdivisions suffice to obtain an accuracy within $\epsilon/2$. The estimate for this part of
the arc is 1.170721. Now apply Simpson's Rule to calculate the length of that part of the arc of
$x = 2\sqrt{1-y^2/16}$ that lies over $\left[0, 2\sqrt{3}\right]$. We find that 24 subdivisions suffice. Our estimate
for this arc is 3.673596. Our estimate for the entire elliptic arc length is therefore

$$4(1.170721 + 3.673596) = 19.377268,$$

or 19.38 to two decimal places. Another approach is to parameterize the curve by $x = 2\cos(t)$, $y = 4\sin(t)$ $(0 \le t \le 2\pi)$. The formula given in the instructions to Exercises 30–33 gives us

$$4\int_0^{\pi/2} \sqrt{\left(\tfrac{d}{dt}(2\cos(t))\right)^2 + \left(\tfrac{d}{dt}(4\sin(t))\right)^2}\, dt = 4\int_0^{\pi/2} \sqrt{4\sin^2(t) + 16\cos^2(t)}\, dt$$

$$= 4\int_0^{\pi/2} \sqrt{4 + 12\cos^2(t)}\, dt$$

for the arc length of the ellipse. No special measures are needed when we apply Simpson's Rule to approximate this integrand.

43.   As $T'(x) = 12x^2 - 3$, the length is $\displaystyle\int_{-1}^{1} \sqrt{1 + \left(12x^2 - 3\right)^2}\, dx = \int_{-1}^{1} \sqrt{144x^4 - 72x^2 + 10}\, dx \approx 6.5186$.

45.   Let $a = 299.2239$, $k = 0.0100333$, and

$$f(x) = 693.8597 - 34.38365\left(\exp(kx) + \exp(-kx)\right).$$

Set $p(x) = Ax^2 + Bx + C$ with the three unknown coefficients A, B, and C to be determined by the three equations $p(-a) = 0$, $p(a) = 0$, and $p(0) = f(0)$. We solve that $p(x) = 625.092 - 0.00698155x^2$. The length of the arch is

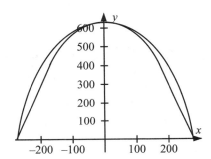

$$\int_{-a}^{a} \sqrt{1 + \left(f'(x)\right)^2}\, dx = \int_{-a}^{a} \sqrt{1 + (34.38365k)^2 \left(e^{kx} - e^{-kx}\right)^2}\, dx \approx 1480.28 \text{ feet}.$$

The length of the parabola is $\displaystyle\int_{-a}^{a} \sqrt{1 + \left(p'(x)\right)^2}\, dx = \int_{-a}^{a} \sqrt{1 + 0.000194968x^2}\, dx \approx 1438.54 \text{ feet}.$

Section 8.3   The Average Value of a Function

Problems for Practice

1. $\displaystyle f_{\text{ave}} = \left(\frac{\pi}{2}\right)^{-1} \int_0^{\pi/2} \cos(x)\, dx = \frac{2}{\pi}$

3. $f_{ave} = (4-1)^{-1} \int\limits_{1}^{4} \frac{1}{x} dx = \left(\frac{1}{3}\right) \ln(4)$

5. $f_{ave} = \pi^{-1} \int\limits_{0}^{\pi} \sin(x) dx = \frac{2}{\pi}$

7. $f_{ave} = (5-2)^{-1} \int\limits_{2}^{5} (x-1)^{1/2} dx = \frac{14}{9}$

9. $P(1 \le X \le 2) = \int\limits_{1}^{2} \frac{3x^2}{8} dx = \frac{7}{8}$

11. $P\left(\frac{\pi}{4} \le X \le \frac{\pi}{2}\right) = \int\limits_{\pi/4}^{\pi/2} \left(\frac{1}{2}\right) \sin(x) dx = \frac{\sqrt{2}}{4}$

13. $P\left(0 \le X \le \frac{1}{2}\right) = \int\limits_{0}^{1/2} \frac{e^{1-x}}{e-1} dx = -\frac{e}{(e-1)} e^{-x}\Big|_{0}^{1/2} = \frac{e-e^{1/2}}{e-1} = \frac{\sqrt{e}}{\sqrt{e}+1}$

15. $\mu = \int\limits_{0}^{1} x \cdot 3x^2 dx = \frac{3}{4}$

17. $\mu = \int\limits_{1/4}^{4} \frac{x}{3\sqrt{x}} dx = \frac{7}{4}$

19. $\mu = \int\limits_{-1}^{2} x \cdot \frac{x^2}{3dx} = \frac{5}{4}$

21.   The average temperature is $T_{ave} = \left(\frac{1}{3}\right) \int\limits_{0}^{3} \left(99.6 - t + 0.8t^2\right) dt = 100.5$ degrees Fahrenheit.

Further Theory and Practice

23. $\left(\frac{1}{\pi}\right) \int\limits_{0}^{\pi} x \sin(x^2) dx = -\frac{1}{2\pi} \cos(x^2)\Big|_{0}^{\pi} = \left(\frac{1}{2\pi}\right)\left(1 - \cos(\pi^2)\right)$

25. $\left(\frac{4}{\pi}\right) \int\limits_{0}^{\pi/4} \tan(x) dx = \frac{4}{\pi} \ln(\sec(x))\Big|_{0}^{\pi/4} = \left(\frac{2}{\pi}\right) \ln(2)$

27. $\left(\dfrac{1}{\pi}\right)\displaystyle\int_0^{\pi} x\sin(x)\,dx = \dfrac{1}{\pi}\left(\sin(x)-x\cos(x)\right)\Big|_0^{\pi} = 1$

29.   We solve $\left(\dfrac{3}{\pi}\right)\displaystyle\int_0^{\pi/3}(x-c)\sin(x)\,dx = -\dfrac{1}{2}$. The equation simplifies to

$$\left(\dfrac{3}{2\pi}\right)\left(\sqrt{3}-c-\dfrac{\pi}{3}\right) = -\dfrac{1}{2},$$

or $c = \sqrt{3}$.

31.   The averages for the three consecutive 24 hour periods are

$$\left(\dfrac{1}{24}\right)\displaystyle\int_0^{24}\left(40-\dfrac{(t-45)^2}{200}\right)dt = \dfrac{6863}{200},$$

$$\left(\dfrac{1}{24}\right)\displaystyle\int_{24}^{48}\left(40-\dfrac{(t-45)^2}{200}\right)dt = \dfrac{7871}{200},\ \text{and}\ \left(\dfrac{1}{24}\right)\displaystyle\int_{48}^{72}\left(40-\dfrac{(t-45)^2}{200}\right)dt = \dfrac{7727}{200}.$$

The number of degree days is $3\cdot 68-\left(\dfrac{6863}{200}+\dfrac{7871}{200}+\dfrac{7727}{200}\right)=\dfrac{18339}{200}.$

The fuel cost is $\$0.30\left(\dfrac{18339}{200}\right)=\$27.51$.

33.   Suppose that $I=[\alpha,\beta]$. Choose $\alpha_n=\alpha$ for $\xi=\alpha$, $\alpha_n=\beta-\frac{1}{n}$ for $\xi=\beta$, and $\alpha_n=\dfrac{\xi-1}{2n}$

for $\xi\in(\alpha,\beta)$. It follows that in each case $\alpha_n\to\xi$ and $\alpha_n+\dfrac{1}{n}\to\xi$ as $n\to\infty$. The

existence of $c_n\in\left(\alpha_n,\alpha_n+\dfrac{1}{n}\right)$ such that $f(c_n)=n\displaystyle\int_{\alpha_n}^{\alpha_n+1/n}f(x)\,dx$ follows from the Mean

Value Theorem for Integrals. Given that c_n lies in an interval the endpoints of which both

tend to ξ we deduce that $c_n\to\xi$. As $f(c_n)=nP\left(\alpha_n\le X\le\alpha_n+\dfrac{1}{n}\right)\ge 0$, and because

$f(c_n)\to f(\xi)$, we deduce that $f(\xi)\ge 0$.

In Exercises 34–39 the normalization requirement $\displaystyle\int_I f(x)\,dx=1$ is used to find the constant

$c=\displaystyle\int_I g(x)\,dx$.

35. $c = \int_0^1 x^3 (1-x)^2 \, dx = \int_0^1 \left(x^5 - 2x^4 + x^3 \right) dx = \left(\frac{1}{6} x^6 - \frac{2}{5} x^5 + \frac{1}{4} x^4 \right)\Big|_0^1 = \frac{1}{60}$

37. Making the substitution $x = 3\tan(\theta)$, $dx = 3\sec^2(\theta)\,d\theta$, we obtain

$$c = \int_0^1 \left(9 + x^2 \right)^{-1/2} dx = \int_0^{\arctan(4/3)} \sec(\theta)\,d\theta = \ln\left(\sec(\theta) + \tan(\theta) \right)\Big|_0^{\arctan(4/3)} = \ln(3).$$

39. Making the substitution $u = 1-x$, $du = -dx$, we obtain

$$c = \int_0^1 x(1-x)^{-1/2} \, dx = -\int_0^1 \left(u^{1/2} - u^{-1/2} \right) du = \left(2u^{1/2} - \frac{2}{3} u^{3/2} \right)\Big|_0^1 = \frac{4}{3}.$$

41. We solve $\exp(b) = \left(\frac{1}{c} \right) \int_0^c \exp(x)\,dx$, or $b = \ln\left(\frac{e^c - 1}{c} \right)$.

43. Integrating by parts with $u = \ln(x)$ and $dv = x\,dx$, we obtain

$$E(X) = \int_1^c x\ln(x)\,dx = \frac{1}{2} x^2 \ln(x)\Big|_1^e - \frac{1}{2} \int_1^e x\,dx = \left(\frac{1}{4} \right)\left(e^2 + 1 \right)$$

45. Integrating by parts with $u = x$ and $dv = \left(\exp\left(1 - \frac{x}{e} \right) - 1 \right) dx$, we obtain

$$E(X) = \left(\frac{1}{e^2 - 2e} \right) \int_0^e x\left(\left(\exp\left(1 - \frac{x}{e} \right) \right) - 1 \right) dx$$

$$= \left(\frac{1}{e^2 - 2e} \right)\left(-x\left(\exp\left(2 - \frac{x}{e} \right) + x \right) + \frac{1}{2} x^2 - \exp\left(3 - \frac{x}{e} \right) \right)\Big|_0^e$$

$$= e\frac{(2e - 5)}{2(e - 2)}.$$

47. The solution of $\int_0^m \cos(x)\,dx = \frac{1}{2}$ is $m = \frac{\pi}{6}$.

49. The solution of $\int_0^m \frac{e^{1-x}}{e-1}\,dx = \frac{1}{2}$, or $\frac{e - e^{1-m}}{e-1} = \frac{1}{2}$, is $m = 1 - \ln\left(\frac{e+1}{2} \right)$.

51.   We have $g(0)=1+0-2^0=1$ and $g(1)=1+1-2^1=0$. Also, $g'(u)=1-2^u\ln(2)$, so g

decreases on $\left(0, \dfrac{-\ln(\ln(2))}{\ln(2)}\right)$ and increases on $\left(\dfrac{-\ln(\ln(2))}{\ln(2)}, 1\right)$. It follows that

$$g\left(\frac{-\ln(\ln(2))}{\ln(2)}\right)>0$$

is the minimum value of g on the interval $(0, 1)$. Therefore, $g(u)>0$ for $u\in(0, 1)$. In

particular, for any $p>0$, we have $g\left(\dfrac{1}{p+1}\right)>0$. Thus, $\dfrac{p+2}{p+1}>2^{1/(p+1)}$, and

$$\frac{p+1}{p+2}<\left(\frac{1}{2}\right)^{1/(p+1)}.$$

If X is a random variable with p.d.f. $f(x)=(p+1)x^p$ for $0\le x\le 1$, we have

$$\overline{X}=\int_0^1 x\cdot(p+1)x^p\,dx=\frac{p+1}{p+2}.$$

The median m of X is the solution of $\displaystyle\int_0^m(p+1)x^p\,dx=\frac{1}{2}$, namely $m=\left(\dfrac{1}{2}\right)^{1/(p+1)}$. Thus,

$\overline{X}<m$. In particular, the mean and the median of X are unequal.

Calculator/Computer Exercises

53.   Let $F(x)=\left(\dfrac{1}{2h}\right)\displaystyle\int_{x-h}^{x+h}|t|\,dt$. If $x-h>0$, we have

$$F(x)=\left(\frac{1}{2h}\right)\int_{x-h}^{x+h}|t|\,dt=\left(\frac{1}{2h}\right)\int_{x-h}^{x+h}t\,dt=\frac{2xh}{2h}=x\,;\ \text{if}$$

$x+h<0$, we have

$$F(x)=\left(\frac{1}{2h}\right)\int_{x-h}^{x+h}|t|\,dt=\left(\frac{1}{2h}\right)\int_{x-h}^{x+h}(-t)\,dt$$

$$=-\frac{2xh}{2h}=-x\,;$$

if $x-h\le 0\le x+h$, we have

$$F(x)=\left(\frac{1}{2h}\right)\int_{x-h}^{x+h}|t|\,dt=\left(\frac{1}{2h}\right)\left(\int_{x-h}^{0}(-t)\,dt+\int_{0}^{x+h}t\,dt\right)=\frac{x^2+h^2}{2h}$$

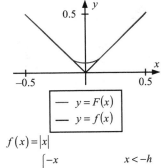

$f(x)=|x|$

$$F(x)=\begin{cases}-x & x<-h\\ \left(x^2+h^2\right)\big/(2h) & -h\le x\le h\\ x & x>h\end{cases}$$

55. **a.** See graph.

b. $f_{ave} = \int_0^1 \exp(-x^2) dx = \sqrt{\pi} \dfrac{\text{erf}(1)}{2} \approx 0.7468$

c. $c = \sqrt{\ln\left(\dfrac{2}{\sqrt{\pi}\,\text{erf}(1)}\right)} \approx 0.5403$

d. See graph.

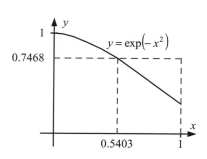

57. Let $\mu = 2$ and $\sigma = 0.5$, then

$$P(-0.5 \le X \le 4.5) = \left(\dfrac{1}{0.5\sqrt{2\pi}}\right) \int_{-0.5}^{4.5} \exp\left(-\dfrac{(x-2)^2}{0.5}\right) dx \approx \dfrac{1.253315419\sqrt{2}}{\sqrt{\pi}}$$

$$\approx 0.9999994265 > 0.999999$$

Generally, we can use the substitution $t = \dfrac{x-\mu}{\sigma}$ to obtain

$$P(\mu - 5\sigma \le X \le \mu + 5\sigma) = \left(\dfrac{1}{\sigma\sqrt{2\pi}}\right) \int_{\mu-5\sigma}^{\mu+5\sigma} \exp\left(-\dfrac{(x-\mu)^2}{2\sigma^2}\right) dx = \left(\dfrac{1}{\sqrt{2\pi}}\right) \int_{-5}^{5} \exp\left(-\dfrac{t^2}{2}\right) dt$$

$$= \text{erf}\left(\dfrac{5}{\sqrt{2}}\right) \approx 0.9999994265 > 0.999999$$

Section 8.4   Center of Mass

Problems for Practice

1. $M_{x=3} = \int_0^6 (x-3)\left(2 - \dfrac{x}{3}\right) dx = \int_0^6 \left(-\dfrac{1}{3}x^2 + 3x - 6\right) dx = -6$

3. $M_{y=0} = \int_0^2 (y-0)(6-3y) dy = 4$, or, alternatively, $M_{y=0} = \left(\dfrac{1}{2}\right)\int_0^6 \left(2 - \dfrac{x}{3}\right)^2 dx = 4$.

5. $M_{x=3} = \int_0^2 (x-3)\left(4x - x^3\right) dx = \int_0^2 \left(-x^4 + 3x^3 + 4x^2 - 12x\right) dx = -\dfrac{116}{15}$

7. $M_{x=3} = \int_1^2 (x+3)\left(\dfrac{1}{x}\right) dx = \left(x + 3\ln(x)\right)\Big|_1^2 = 1 + 3\ln(2)$

In the solutions to Exercises 9–20, we assume without loss of generality that the uniform density is equal to 1.

9.   The total mass of the region is 6. Therefore, we find for the center of mass of the region

$$\overline{x} = \left(\frac{1}{6}\right)\int_0^6 x\left(2 - \frac{x}{3}\right)dx = 2 \text{ and } \overline{y} = \left(\frac{1}{12}\right)\int_0^6\left(2 - \frac{x}{3}\right)^2 dx = \frac{2}{3}.$$

11.   The total mass of the region is $\int_{-a}^a \sqrt{a^2 - x^2}\,dx = \pi\frac{a^2}{2}$. Therefore, we find for the center of

mass of the region $\overline{x} = \frac{2}{\pi a^2}\int_{-a}^a x\sqrt{a^2 - x^2}\,dx = 0$ and $\overline{y} = \frac{1}{\pi a^2}\int_{-a}^a\left(\sqrt{a^2 - x^2}\right)^2 dx = \frac{4a}{3\pi}.$

13.   The total mass of the region is $\int_{-4}^2\left(8 - 2x - x^2\right)dx = 36$. Therefore, we find for the center of

mass of the region $\overline{x} = \left(\frac{1}{36}\right)\int_{-4}^2 x\left(8 - 2x - x^2\right)dx = \left(-\frac{1}{144}x^4 - \frac{1}{54}x^3 + \frac{1}{9}x^2\right)\Big|_{-4}^2 = -1$ and

$$\overline{y} = \left(\frac{1}{72}\right)\int_{-4}^2\left(8 - 2x - x^2\right)^2 dx = \frac{1}{73}\int_{-4}^2\left(x^4 + 4x^3 - 12x^2 - 32x + 64\right)dx = \frac{18}{5}.$$

15.   The total mass of the region is $\int_1^2\left(\frac{1}{x}\right)dx = \ln(2)$. Therefore, we find for the center of mass of

the region $\overline{x} = \frac{1}{\ln(2)}\int_1^2 x\left(\frac{1}{x}\right)dx = \frac{1}{\ln(2)}$ and $\overline{y} = \frac{1}{2\ln(2)}\int_1^2\left(\frac{1}{x}\right)^2 dx = \frac{1}{4\ln(2)}.$

17.   The total mass of the region is $\int_1^4\left(x^2 - 4x + 5\right)dx = 6$. Therefore, we find for the center of mass

of the region $\overline{x} = \left(\frac{1}{6}\right)\int_1^4 x\left(x^2 - 4x + 5\right)dx = \frac{23}{8}$ and

$$\overline{y} = \left(\frac{1}{12}\right)\int_1^4\left(x^2 - 4x + 5\right)^2 dx = \frac{1}{12}\int_1^4\left(x^4 - 8x^3 + 26x^2 - 40x + 25\right)dx = \frac{13}{10}.$$

19.   The total mass of the region is $\frac{243\pi}{8}$. Therefore, we find for the center of mass of the region

$$\overline{x} = \left(\frac{8}{243\pi}\right)\int_{-3}^3 x\left(9 - x^2\right)^{3/2}dx = 0 \text{ and}$$

$$\overline{y} = \left(\frac{4}{243\pi}\right)\int_{-3}^3\left(\left(9 - x^2\right)^{3/2}\right)^2 dx = \frac{4}{243\pi}\int_{-3}^3\left(729 - 243x^2 + 27x^4 - x^6\right)dx = \frac{1152}{35\pi}.$$

Further Theory and Practice

21. The total mass of the region is $\displaystyle\int_0^{\pi/2} \cos(x)\,dx = 1$. Therefore, by integrating by parts with $u = x$

and $dv = \cos(x)\,dx$, we have $\displaystyle \overline{x} = \left(\frac{1}{1}\right)\int_0^{\pi/2} x\cos(x)\,dx = \left(x\sin(x) + \cos(x)\right)\Big|_0^{\pi/2} = \frac{\pi}{2} - 1$. Also,

$\displaystyle \overline{y} = \left(\frac{1}{2}\right)\int_0^{\pi/2}\left(\cos(x)\right)^2 dx = \left(\frac{1}{4}\right)\int_0^{\pi/2}\left(1 + \cos(2x)\right)dx = \frac{\pi}{8}$. The center of mass is $\displaystyle\left(\frac{\pi}{2} - 1,\ \frac{\pi}{8}\right)$.

23. The total mass of the region is $\displaystyle\int_0^1 x\,dx + \int_1^{\sqrt{2}}\left(2 - x^2\right)dx = \frac{4\sqrt{2}}{3} - \frac{7}{6}$. Therefore, we find for the

center of mass of the region $\displaystyle \overline{x} = \left(\frac{4\sqrt{2}}{3} - \frac{7}{6}\right)^{-1}\left(\int_0^1 x^2\,dx + \int_1^{\sqrt{2}}\left(2x - x^3\right)dx\right) = \frac{7}{16\sqrt{2} - 14}$ and

$\displaystyle \overline{y} = \left(\frac{8\sqrt{2}}{3} - \frac{7}{3}\right)^{-1}\left(\int_0^1 x^2\,dx + \int_1^{\sqrt{2}}\left(2 - x^2\right)^2 dx\right) = 2\,\frac{16\sqrt{2} - 19}{5\left(8\sqrt{2} - 7\right)}$.

25. The total mass of the region is

$$\int_0^1\left(\frac{1}{\sqrt{1 + x^2}}\right)dx = \int_0^1 \frac{1}{\sqrt{1 + x^2}} \cdot \frac{1 + \frac{x}{\sqrt{1+x^2}}}{1 + \frac{x}{\sqrt{1+x^2}}}\,dx = \int_0^1 \frac{1 + \frac{x}{\sqrt{1+x^2}}}{x + \sqrt{1 + x^2}}\,dx.$$

Substituting $u = x + \sqrt{1 + x^2}$; $du = \left(1 + \dfrac{x}{\sqrt{1+x^2}}\right)^0 dx$, we get $\displaystyle\int_1^{1+\sqrt{2}} \frac{du}{u} = \ln\left(1 + \sqrt{2}\right)$. Therefore,

we find for the center of mass of the region $\displaystyle \overline{x} = \left(\frac{1}{\ln\left(1 + \sqrt{2}\right)}\right)\int_0^1\left(\frac{x}{\sqrt{1 + x^2}}\right)dx$. Substituting

$u = 1 + x^2$, $du = 2x\,dx$, we have

$$\left(\frac{1}{\ln\left(1 + \sqrt{2}\right)}\right)\int_0^1 \frac{x}{\sqrt{1 + x^2}}\,dx = \left(\frac{1}{\ln\left(1 + \sqrt{x}\right)}\right)\int_1^2 \frac{1}{2}u^{-1/2}\,du = \left(\frac{1}{\ln\left(t + \sqrt{2}\right)}\right)u^{1/2}\Bigg|_1^2 = \frac{\sqrt{2} - 1}{\ln\left(1 + \sqrt{2}\right)}.$$

and $\displaystyle \overline{y} = \left(\frac{1}{2\ln\left(1 + \sqrt{2}\right)}\right)\int_0^1\left(\frac{1}{\sqrt{1 + x^2}}\right)^2 dx$. Substituting $x = \tan u$, $1 + x^2 = \sec^2 u$,

$dx = \sec^2 u\,du$, we have

$$\left(\frac{1}{2\ln\left(1 + \sqrt{2}\right)}\right)\int_0^1 \frac{dx}{1 + x^2} = \left(\frac{1}{2\ln\left(1 + \sqrt{2}\right)}\right)\int_0^{\pi/4} \frac{\sec^2 u}{\sec^2 u}\,du = \left(\frac{1}{2\ln\left(1 + \sqrt{2}\right)}\right)\cdot\frac{\pi}{4} = \frac{\pi}{8\ln\left(1 + \sqrt{2}\right)}.$$

27. The total mass of the region is $\int_1^e \ln(x)\,dx = 1$. Therefore, we find for the center of mass of the

region $\overline{x} = \left(\dfrac{1}{1}\right)\int_1^e x\ln(x)\,dx = \dfrac{e^2+1}{4}$ and $\overline{y} = \left(\dfrac{1}{2}\right)\int_1^e (\ln(x))^2\,dx = \dfrac{e}{2}-1$, using the integration in

Problem 8.1.37.

29. The total mass of the region is $\int_0^2 x^2\,dx + \int_2^3 4(x-3)^2\,dx = 4$. Therefore, we find for the center of

mass of the region $\overline{x} = \left(\dfrac{1}{4}\right)\left(\int_0^2 x^3\,dx + \int_2^3 4x(x-3)^2\,dx\right) = \dfrac{7}{4}$ and

$$\overline{y} = \left(\dfrac{1}{8}\right)\left(\int_0^2 x^4\,dx + \int_2^3 16(x-3)^4\,dx\right) = \dfrac{6}{5}.$$

31. We calculate for the center of mass of the region bounded by the graph of f, by the lines $x=a$, $x=b$ and by the x-axis as follows:

$$\int_a^b xf(x)\,dx = \overline{x}\int_a^b (f(x)-g(x))\,dx + \int_a^b xg(x)\,dx = M\cdot\overline{x} + \int_a^b xg(x)\,dx$$

$$\overline{x} = \dfrac{1}{M}\left(\int_a^b xf(x)\,dx - \int_a^b xg(x)\,dx\right) = \dfrac{1}{M}\int_a^b x(f(x)-g(x))\,dx$$

$$\dfrac{1}{2}\int_a^b f(x)^2\,dx = \overline{y}\int_a^b (f(x)-g(x))\,dx + \dfrac{1}{2}\int_a^b g(x)^2\,dx = M\cdot\overline{y} + \dfrac{1}{2}\int_a^b g(x)^2\,dx$$

$$\overline{y} = \dfrac{1}{2M}\left(\int_a^b f(x)^2\,dx - \int_a^b g(x)^2\,dx\right) = \dfrac{1}{2M}\int_a^b \left(f(x)^2 - g(x)^2\right)\,dx$$

33. The total mass of the region is $\int_1^2 \left((2+x)-x^2\right)\,dx = \dfrac{7}{6}$. Therefore, we find for the center of

mass of the region $\overline{x} = \left(\dfrac{6}{7}\right)\int_1^2 x\left((2+x)-x^2\right)\,dx = \dfrac{19}{14}$ and

$$\overline{y} = \left(\dfrac{3}{7}\right)\int_1^2 \left((2+x)^2 - \left(x^2\right)^2\right)\,dx = \dfrac{92}{35}.$$

35.   The total mass of the region is $\int_0^1 \left(\left(5-x^2\right) - (x-1)^2 \right) dx = \dfrac{13}{3}$. Therefore, we find for the center

of mass of the region $\bar{x} = \left(\dfrac{3}{13}\right)\int_0^1 \left(\left(5-x^2\right) - (x-1)^2 \right) dx = \dfrac{1}{2}$ and

$$\bar{y} = \left(\frac{3}{26}\right)\int_0^1 \left(\left(5-x^2\right)^2 - (x-1)^4 \right) dx = \frac{5}{2}.$$

37.   The mean is $\mu_X = \int_0^1 3x^3 dx = \dfrac{3}{4}$, and the variance is $\text{Var}(X) = \int_0^1 \left(x - \dfrac{3}{4} \right)^2 3x^2 dx = \dfrac{3}{80}$.

39.   If $q > 0$ then $\int_0^1 x^q (1-x) dx = \dfrac{1}{(q+1)(q+2)}$. It follows that $c = (p+1)(p+2)$ and

$$\mu_X = \frac{(p+1)(p+2)}{(p+2)(p+3)} = \frac{p+1}{p+3}.$$

Therefore, $\text{Var}(X) = (p+1)(p+2)\int_0^1 \left(x - \dfrac{(p+1)}{(p+3)} \right)^2 x^p (1-x) dx = \dfrac{2(p+1)}{(p+3)^2 (p+4)}$.

Calculator/Computer Exercises

41.   The total mass of the region is $M = \int_0^{\sqrt{\pi}} \sin\left(\pi - x^2\right) dx \approx 0.89483$. Therefore, we find for the

center of mass of the region $\bar{x} \approx \dfrac{1}{0.89483} \int_0^{\sqrt{\pi}} x \sin\left(\pi - x^2\right) dx \approx \dfrac{1}{0.89483} \approx 1.1175$ and

$$\bar{y} \approx \frac{1}{0.89483} \cdot \frac{1}{2} \int_0^{\sqrt{\pi}} \sin^2\left(\pi - x^2\right) dx \approx \frac{0.33494}{0.89483} \approx 0.37431.$$

43.   Let $c = 1.306559$ be the solution of $x = 5 - \exp(x)$. The total mass of the region is

$$\int_0^{1.306559} x\,dx + \int_{1.306559}^{\ln(5)} (5 - \exp(x))\,dx \approx 5\ln(5) - 6.985804 \approx 1.0613856.$$

Therefore, we find for the center of mass of the region

$$\bar{x} = \left(\frac{1}{1.0613856}\right)\left(\int_0^{1.306559} x^2\,dx + \int_{1.306559}^{\ln(5)} x(5 - \exp(x))\,dx\right) = 0.9765796$$

and $\bar{y} = \left(\frac{1}{(2)1.0613856}\right)\left(\int_0^{1.306559} x^2\,dx + \int_{1.306559}^{\ln(5)} (5 - \exp(x))^2\,dx\right) = 0.4376893.$

45.   We calculate for each of the specified values as follows: $g(0.2) = 0.70524$, $g(1) = 0.14105$, and $g(2.0) = 0.070524$. The points $\left(\ln(\sigma),\ \ln(g(\sigma))\right)$ for these three values of σ appear in the accompanying figure. They are very nearly collinear. In general, we have

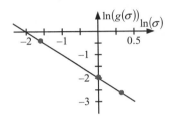

$$g(\sigma) = \frac{\displaystyle\int_{-10}^{10} f(x)^2\,dx}{2\displaystyle\int_{-10}^{10} f(x)\,dx} = \frac{1}{4\sqrt{\pi}\sigma}\frac{\operatorname{erf}(10/\sigma)}{\operatorname{erf}(5\sqrt{2}/\sigma)}.$$

Applying the logarithm we obtain $\ln(g(\sigma)) = -\ln(4\sqrt{\pi}) - \ln(\sigma) + \dfrac{\ln(\operatorname{erf}(10/\sigma))}{\operatorname{erf}(5\sqrt{2}/\sigma)}$. The last

summand is negligible for values of $\sigma \le 3$ (as can be seen by plotting the expression). Therefore, $\ln(g(\sigma)) \approx -\ln(4\sqrt{\pi}) - \ln(\sigma) \approx -1.958659 - \ln(\sigma)$, as can be seen from the figure.

Section 8.5   Work

Problems for Practice

1.   At time $t \in [0,\ 60]$ the weight of the gravel is $500 - t$ and the gravel is $y = \dfrac{4t}{3}$ feet above the ground. Thus, the weight is $500 - \dfrac{3y}{4}$ as a function of height. The work done is

$$\int_0^{80} \left(500 - \frac{3y}{4}\right)dy = 37600 \text{ ft-lb.}$$

3.   When the rocket is $y \in [0, 20]$ miles high, its weight is $7000 - 30y$ pounds. The work done is therefore

$$\int_0^{20} (7000 - 30y) \, dy = 134000 \text{ ft-lb.}$$

5.   At time $t \in [0, 10]$ (in minutes), the weight of the load is $80000(2.2046) - (50)(62.428)t$ pounds. The load is $y = 5t$ feet above the ground. Thus, the weight is

$$80000(2.2046) - (50)(62.428)\frac{y}{5}$$

as a function of y. The work done is

$$\int_0^{50} \left(80000 \, (2.2046) - (50)(62.428)\frac{y}{5} \right) dy,$$

or 8.03805×10^6 ft-lb.

7.   The bottom 70 feet of the cable weighs $(70)(20)$ pounds. It and the 300 pound load are lifted 30 feet, amounting to $((70)(20) + 300)(30) = 51000$ ft-lb. The work done lifting the upper 30 feet of cable is $\int_0^{30} y \cdot 20 \, dy = 9000$ ft-lb. Altogether, the work done is

$$51000 + 9000 = 60000 \text{ ft-lb.}$$

9.   In foot-pounds the work is $\int_0^{7/12} 96x \, dx$, or $\dfrac{49}{3}$.

11.   The spring constant is 840 pounds per foot. The work done is $\int_{1/3}^{2/3} 840x \, dx$, or 140 ft-lb.

13.   The percentage of the total amount of work is $100 \cdot \dfrac{\int_0^{L/2} kx \, dx}{\int_0^L kx \, dx} = 25\%$.

15. The amount of work performed by the pump in emptying the pool is calculated as follows:
$\int_{0}^{10} 62.428\left(15^2\right) y\,dy$, or 7.02315×10^5 ft-lb.

17. The amount of work done is $\int_{2}^{6} 62.428(16 \times 30) y\,dy$, or 4.7945×10^5 ft-lb.

19. Let the positive y-axis be oriented downward along the central axis of the cone, with $y = 100$ at the vertex. A thin disk y feet from the top of the reservoir has radius $100 - y$ and weighs $62.428 \cdot \pi \cdot (100 - y)^2 \cdot \Delta y$ where Δy feet is the thickness. To calculate the work, we sum these contributions from $y = 0$ to $y = 50$ and let $\Delta y \to 0$. The result is

$$\int_{0}^{50} 62.428 \cdot \pi \cdot (100 - y)^2\,dy = 5.7203 \times 10^7 \text{ ft-lb.}$$

Further Theory and Practice

21. At time $t \in [0, 3]$ the weight of the sack is $50 - 2t$ and the mason is $y = 20t$ feet above the ground. Thus, the weight is $50 - y/10$ as a function of height. The work done for the first 60 feet of the ascent is $(180)(60) + \int_{0}^{60} (50 - y/10)\,dy = 13620$ ft-lb. At time $t \in [3, 6]$ the weight of the sack is $44 - 3(t - 3)$ and the mason is $y = 20t$ feet above the ground. Thus, the weight is $44 - 3(y/20 - 3)$ as a function of height. The work done for the second 60 feet of the ascent is $(180)(60) + \int_{60}^{120} (44 - 3(y/20 - 3))\,dy = 13170$ ft-lb. The total work is $13620 + 13170 = 26790$ ft-lb.

23. We have $pv^{1.4} = \left(50 \cdot 12^2\right)\left(\dfrac{20}{12^3}\right)^{1.4} = 14$, or $p(v) = 14v^{-1.4} = 14\left(\left(\dfrac{2}{12^2}\right)x\right)^{-1.4}$, with x measured in feet. By Exercise 22 the work is $W = \left(\dfrac{2}{12^2}\right)\int_{2/12}^{10/12} 14\left(\left(\dfrac{2}{12^2}\right)x\right)^{-1.4}\,dx = 188.2$ ft-lb.

25. Let $\eta = -y$. Then the η-axis is directed downward and the curve that generates the tank has equation $\eta = 18 - 2x^2$ in the $x\eta$-plane. A thin slice of the tank at height η has approximate volume $\pi(9 - \eta/2)^2 \Delta y$, approximate weight $62.428 \cdot \pi(9 - \eta/2)^2 \Delta y$, and the work done pumping this slice to the top of the tank is $62.428 \cdot \pi(9 - \eta/2)^2 \eta \Delta\eta$. Let $\Delta\eta \to 0^+$. We calculate that the work done is $\int_{0}^{15} 62.428 \cdot \pi(9 - \eta/2)^2 \eta\,d\eta = 435490.3$ ft-lb.

27. The work done is $\int_{14}^{18} 62.428 \cdot \pi (9 - \eta/2)^2 \, \eta \, d\eta = 15689.87$ ft-lb.

29. According to Newton's Law, $F(x) = m\dfrac{dv}{dt} = m\dfrac{dv}{dx}\dfrac{dx}{dt} = m \cdot v \cdot \dfrac{dv}{dx}$. Therefore

$$W(b) = \int_0^b F(x)\, dx = m\int_0^b v \cdot \frac{dv}{dx}\, dx = \frac{1}{2} mv(b)^2 .$$

31. Given that $F(x) = kx$ and $W(x) = \int_0^x k\xi\, d\xi = \dfrac{kx^2}{2} = \dfrac{F(x)^2}{2k}$, the parametric curve is the right

half of the parabola $W = \dfrac{F^2}{2k}$.

Calculator/Computer Exercises

33. Let $\eta = 10\left(\exp\left(-\dfrac{x^2}{2}\right) - \dfrac{1}{e^2}\right)$. Then $x^2 = -2\ln\left(\dfrac{\eta}{10} + \dfrac{1}{e^2}\right)$ and the work done is

$$\int_0^{6-10/e^2} 62.428 \cdot \pi \left(-2\ln\left(\frac{\eta}{10} + \frac{1}{e^2}\right)\right)\eta\, d\eta = 3582 \text{ ft-lb.}$$

35. The work done is $\displaystyle\int_0^{10-10/e^2} 62.428 \cdot \pi \left(-2\ln\left(\dfrac{\eta}{10} + \dfrac{1}{e^2}\right)\right)\eta\, d\eta = 5754.9$ ft-lb.

Section 8.6   Improper Integrals—Unbounded Integrands

Problems for Practice

1. The integral is improper with infinite integrand at $x = 5$. We compute:

$$\lim_{\varepsilon \to 0^+} \int_1^{5-\varepsilon} (x-5)^{-4/3}\, dx = \lim_{\varepsilon \to 0^+} \left(\frac{-(x-5)^{-1/3}}{1/3}\right)\Bigg|_1^{5-\varepsilon} = 3\lim_{\varepsilon \to 0^+}\left(\varepsilon^{-1/3} - 4^{-1/3}\right) = \infty .$$

Integral diverges.

3. The integral is improper with infinite integrand at $x = 4$. We compute:

$$\lim_{\varepsilon \to 0^+} \int_2^{4-\varepsilon} (4-x)^{-0.9}\, dx = \lim_{\varepsilon \to 0^+} \left(\frac{-(4-x)^{1/10}}{0.1}\right)\Bigg|_2^{4-\varepsilon} = \lim_{\varepsilon \to 0^+}\left(10\left(2^{1/10} - \varepsilon^{1/10}\right)\right) = 10 \cdot 2^{1/10} .$$

5.   The integral is improper with infinite integrand at $x = \dfrac{\pi}{2}$. We compute:

$$\lim_{\varepsilon \to 0^+} \int_0^{\pi/2-\varepsilon} \tan(x)\,dx = \lim_{\varepsilon \to 0^+} \left(-\ln\left(\left|\cos(x)\right|\right)\right)\Big|_0^{\pi/2-\varepsilon} = \lim_{\varepsilon \to 0^+} \left(\left(\ln(1) - \ln\left(\cos\left(\frac{\pi}{2}-\varepsilon\right)\right)\right)\right) = \infty .$$

Integral diverges.

7.   The integral is improper with infinite integrand at $x = 1$. We compute:

$$\lim_{\varepsilon \to 0^+} \int_0^{1-\varepsilon} \frac{x}{\left(1-x^2\right)^{1/4}}\,dx = \lim_{\varepsilon \to 0^+} \left(-\left(\frac{1}{2}\right)\frac{\left(1-x^2\right)^{3/4}}{3/4}\right)\Bigg|_0^{1-\varepsilon} = \left(\frac{2}{3}\right)\lim_{\varepsilon \to 0^+} \left(1-\left(1-(1-\varepsilon)^2\right)^{3/4}\right)$$

$$= \left(\frac{2}{3}\right)\lim_{\varepsilon \to 0^-} \left(1-\left(2\varepsilon-\varepsilon^2\right)^{3/4}\right) = \frac{2}{3}$$

9.   The integral is improper with infinite integrand at $x = -3$. We compute:

$$\lim_{\varepsilon \to 0^+} \int_{-3+\varepsilon}^{2} (x+3)^{-1.1}\,dx = \lim_{\varepsilon \to 0^+} \left(\frac{-(x+3)^{-0.1}}{0.1}\right)\Bigg|_{-3+\varepsilon}^{2} = 10\lim_{\varepsilon \to 0^+} \left(\varepsilon^{-0.1} - 5^{-0.1}\right) = \infty .$$

Integral diverges.

11.   The integral is improper with infinite integrand at $x = 0$. We compute:

$$\lim_{\varepsilon \to 0^+} \int_\varepsilon^8 x^{-1/3}\,dx = \lim_{\varepsilon \to 0^+} \left(\frac{x^{2/3}}{2/3}\right)\Bigg|_\varepsilon^8 = \left(\frac{3}{2}\right)\lim_{\varepsilon \to 0^+} \left(8^{2/3} - \varepsilon^{2/3}\right) = 6 .$$

13.   The integral is improper with infinite integrand at $x = 0$. We compute:

$$\lim_{\varepsilon \to 0^+} \int_\varepsilon^1 \frac{\ln(x)}{x}\,dx = \lim_{\varepsilon \to 0^+} \left(\frac{\ln^2(x)}{2}\right)\Bigg|_\varepsilon^1 = \left(\frac{1}{2}\right)\lim_{\varepsilon \to 0^+} \left(\ln^2(1) - \ln^2(\varepsilon)\right) = -\infty .$$

Integral diverges.

15.   The integral is improper with infinite integrand at $x = 0$. We compute:

$$\lim_{\varepsilon \to 0^+} \int_\varepsilon^3 x^{-1/2}(x+1)\,dx = \lim_{\varepsilon \to 0^+} \left(2x^{1/2}(x+1)\Big|_\varepsilon^3 - \int_\varepsilon^3 2x^{1/2}\,dx \right)$$

$$= \lim_{\varepsilon \to 0^+} \left(8\sqrt{3} - 2(\varepsilon+1)\varepsilon^{1/2} \right) - 2\lim_{\varepsilon \to 0^+} \left(\frac{x^{3/2}}{3/2} \right)\Big|_\varepsilon^3$$

$$= 8\sqrt{3} - \left(\frac{4}{3} \right)\lim_{\varepsilon \to 0^+} \left(3\sqrt{3} - \varepsilon^{3/2} \right) = 8\sqrt{3} - 4\sqrt{3} = 4\sqrt{3}$$

17.   The integrand is singular at $x = 1$. We break the integral into two parts and compute:

$$\lim_{\varepsilon \to 0^+} \int_0^{1-\varepsilon} \frac{1}{x-1}\,dx + \lim_{\varepsilon \to 0^+} \int_{1+\varepsilon}^2 \frac{1}{x-1}\,dx = \lim_{\varepsilon \to 0^+} \left(\ln\left(|x-1|\right)\right)\Big|_0^{1-\varepsilon} + \lim_{\varepsilon \to 0^+} \left(\ln\left(|x-1|\right)\right)\Big|_{1+\varepsilon}^2$$

$$= \lim_{\varepsilon \to 0^+} \left(\left(\ln(\varepsilon) - \ln(1) \right) \right) + \lim_{\varepsilon \to 0^+} \left(\left(\ln(2) - \ln(\varepsilon) \right) \right)$$

Both limits diverge, therefore the integral diverges.

19.   The integrand is singular at $x = -3$. We break the integral into two parts and compute:

$$\lim_{\varepsilon \to 0^+} \int_{-5}^{-3-\varepsilon} \frac{1}{(x+3)^{2/5}}\,dx + \lim_{\varepsilon \to 0^+} \int_{-3+\varepsilon}^{-1} \frac{1}{(x+3)^{2/5}}\,dx = \lim_{\varepsilon \to 0^+} \left(\frac{(x+3)^{3/5}}{3/5} \right)\Big|_{-5}^{-3-\varepsilon} + \lim_{\varepsilon \to 0^+} \left(\frac{(x+3)^{3/5}}{3/5} \right)\Big|_{-3+\varepsilon}^{-1}$$

$$= \left(\frac{5}{3} \right)\lim_{\varepsilon \to 0^+} \left(\varepsilon^{3/5} - (-2)^{3/5} \right)$$

$$+ \left(\frac{5}{3} \right)\lim_{\varepsilon \to 0^+} \left(2^{3/5} + \varepsilon^{3/5} \right)$$

$$= \left(\frac{5}{3} \right)\left(2^{3/5} + 2^{3/5} \right) = 10 \cdot \frac{2^{3/5}}{3}$$

21.   The integrand is singular at $x = -1$. We break the integral into two parts and compute:

$$\lim_{\varepsilon \to 0^+} \int_{-2}^{-1-\varepsilon} (x+1)^{-2/3}\,dx + \lim_{\varepsilon \to 0^+} \int_{-1+\varepsilon}^4 (x+1)^{-2/3}\,dx = \lim_{\varepsilon \to 0^+} \left(\frac{(x+1)^{1/3}}{1/3} \right)\Big|_{-2}^{-1-\varepsilon} + \lim_{\varepsilon \to 0^+} \left(\frac{(x+1)^{1/3}}{1/3} \right)\Big|_{-1+\varepsilon}^4$$

$$= 3\lim_{\varepsilon \to 0^+} \left(\varepsilon^{1/3} - (-1)^{1/3} \right) + 3\lim_{\varepsilon \to 0^+} \left(5^{1/3} - \varepsilon^{1/3} \right)$$

$$= 3\left(1 + 5^{1/3} \right) = 3 + 3\sqrt[3]{5}$$

23. The integrand is singular at $x = \sqrt{2}$. We break the integral into two parts and compute:

$$\lim_{\varepsilon \to 0^+} \int_0^{\sqrt{2}-\varepsilon} \frac{x}{x^2-2}dx + \lim_{\varepsilon \to 0^+} \int_{\sqrt{2}+\varepsilon}^3 \frac{x}{x^2-2}dx = \lim_{\varepsilon \to 0^+} \left(\frac{\ln\left(\left|x^2-2\right|\right)}{2} \right)\Bigg|_0^{\sqrt{2}-\varepsilon} + \lim_{\varepsilon \to 0^+} \left(\frac{\ln\left(\left|x^2-2\right|\right)}{2} \right)\Bigg|_{\sqrt{2}+\varepsilon}^3$$

$$= \lim_{\varepsilon \to 0^+} \left(\frac{\ln\left(\left|\varepsilon^2-2\varepsilon\sqrt{2}\right|\right)}{2} - \frac{\ln(2)}{2} \right)$$

$$+ \lim_{\varepsilon \to 0^+} \left(\frac{\ln(7)}{2} - \frac{\ln\left(\left|\varepsilon^2+2\varepsilon\sqrt{2}\right|\right)}{2} \right)$$

Both limits diverge, therefore the integral diverges.

25. We obtain the solid by rotating the region under the graph of the function around the x-axis. The volume of the solid is $V = \int_0^2 \pi f^2(x)\,dx = \pi \int_0^2 x^{-2/3}dx$. Given that this integral is improper with infinite integrand at 0, we compute the volume as a limit:

$$V = \pi \lim_{\varepsilon \to 0^+} \int_\varepsilon^2 x^{-2/3}dx = \pi \lim_{\varepsilon \to 0^+} \left(\frac{x^{1/3}}{1/3} \right)\Bigg|_\varepsilon^2 = 3\pi \lim_{\varepsilon \to 0^+} \left(2^{1/3} - \varepsilon^{1/3} \right) = 3 \cdot 2^{1/3} \cdot \pi.$$

27. We obtain the solid by rotating the region under the graph of the function around the x-axis. The volume of the solid is $V = \int_0^1 \pi f^2(x)\,dx = \pi \int_0^1 \left(1-x^2\right)^{-1/2} dx$. Given that this integral is improper with infinite integrand at 1, we compute the volume as a limit:

$$V = \pi \lim_{\varepsilon \to 0^+} \int_0^{1-\varepsilon} \left(1-x^2\right)^{-1/2} dx = \pi \lim_{\varepsilon \to 0^+} \left(\arcsin(x) \right)\Big|_0^{1-\varepsilon} = \pi \lim_{\varepsilon \to 0^+} \left(\arcsin(1-\varepsilon) - \arcsin(0) \right) = \frac{\pi^2}{2}.$$

Further Theory and Practice

29. The integral is improper with infinite integrand at $x = -5$. We compute:

$$\lim_{\varepsilon \to 0^+} \int_{-5+\varepsilon}^2 \ln(x+5)\,dx = \lim_{\varepsilon \to 0^+} \left((x+5)\ln(x+5) - (x+5) \right)\Big|_{-5+\varepsilon}^2$$

$$= \lim_{\varepsilon \to 0^+} \left(7\ln(7) - 7 - \varepsilon\ln(\varepsilon) + \varepsilon \right) = 7\left(\ln(7) - 1\right)$$

31.   The integral is improper with infinite integrand at $x = -2$. We compute:

$$\lim_{\varepsilon \to 0^+} \int_{-2+\varepsilon}^{-3/2} \frac{1}{(x+2)\ln(x+2)}\,dx = \lim_{\varepsilon \to 0^+} \ln\left(\left|\ln\left(|x+2|\right)\right|\right)\Big|_{-2+\varepsilon}^{-3/2}$$

$$= \lim_{\varepsilon \to 0^+}\left(\ln\left(\left|\ln\left(\frac{1}{2}\right)\right|\right) - \ln\left(\left|\ln(\varepsilon)\right|\right)\right) = -\infty$$

The integral diverges.

33.   The integral is improper with infinite integrand at $x = 0$. We compute:

$$\lim_{\varepsilon \to 0^+} \int_{\varepsilon}^{7} x^{-1/2}\ln(x)\,dx = 2\lim_{\varepsilon \to 0^+}\left(\left(x^{1/2}\ln(x)\right)\Big|_{\varepsilon}^{7} - \int_{\varepsilon}^{7}\left(\frac{x^{1/2}}{x}\right)dx\right) = 2\lim_{\varepsilon \to 0^+}\left(x^{1/2}\ln(x) - 2x^{1/2}\right)\Big|_{\varepsilon}^{7}$$

$$= \lim_{\varepsilon \to 0^+}\left(2\sqrt{7}\left(\ln(7) - 2\right) - 2\sqrt{\varepsilon}\left(\ln(\varepsilon) - 2\right)\right) = 2\sqrt{7}\left(\ln(7) - 2\right)$$

35.   The integral is singular at $x = 0$. We break the integral into two parts and compute:

$$\int_{-1}^{3} \ln\left(|x|\right)dx = \lim_{\varepsilon \to 0^+} \int_{-1}^{-\varepsilon} \ln(-x)\,dx + \lim_{\varepsilon \to 0^+} \int_{\varepsilon}^{3} \ln(x)\,dx$$

$$= \lim_{\varepsilon \to 0^-}\left(-x\ln(-x) + x\right)\Big|_{-1}^{-\varepsilon} + \lim_{\varepsilon \to 0^+}\left(x\ln(x) - x\right)\Big|_{\varepsilon}^{3}$$

$$= \lim_{\varepsilon \to 0^+}\left(\varepsilon\ln(\varepsilon) + \varepsilon + \ln(1) - 1\right) + \lim_{\varepsilon \to 0^+}\left(3\ln(3) - 3 - \varepsilon\ln(\varepsilon) + \varepsilon\right) = 3\ln(3) - 4$$

37.   If $p \neq 1$ we have:

$$\int_{0}^{b}\left(\frac{1}{x^p}\right)dx = \lim_{\varepsilon \to 0^+}\int_{\varepsilon}^{b}\left(\frac{1}{x^p}\right)dx = \lim_{\varepsilon \to 0^+}\left(\frac{x^{-p+1}}{-p+1}\right)\Big|_{\varepsilon}^{b} = \frac{1}{-p+1}\lim_{\varepsilon \to 0^+}\left(b^{-p+1} - \varepsilon^{-p+1}\right)$$

$$= \frac{b^{-p+1}}{-p+1} - \frac{1}{-p+1}\lim_{\varepsilon \to 0^+}\varepsilon^{-p+1}$$

The limit $\lim_{\varepsilon \to 0^+}\varepsilon^{-p+1}$ is equal to 0 if $(-p+1) > 0$, and diverges if $(-p+1) < 0$. Therefore, the integral is convergent if $p < 1$, and the integral diverges if $p > 1$. If $p = 1$, we have:

$$\int_{0}^{b}\left(\frac{1}{x}\right)dx = \lim_{\varepsilon \to 0^+}\int_{\varepsilon}^{b}\left(\frac{1}{x}\right)dx = \lim_{\varepsilon \to 0^+}\ln(x)\Big|_{\varepsilon}^{b} = \lim_{\varepsilon \to 0^+}\left(\ln(b) - \ln(\varepsilon)\right) = \infty.$$

Therefore, the integral converges only if $p < 1$.

39.   If $p \neq 1$, we have:

$$\int_1^e \left(\frac{1}{x \ln^p(x)} \right) dx = \lim_{\varepsilon \to 0^+} \int_{1+\varepsilon}^e \left(\frac{\ln^{-p}(x)}{x} \right) dx = \lim_{\varepsilon \to 0^+} \left(\frac{\ln^{-p+1}(x)}{-p+1} \right) \Big|_{1+\varepsilon}^e$$

$$= \frac{1}{-p+1} \lim_{\varepsilon \to 0^+} \left(1 - \ln^{-p+1}(\varepsilon) \right) = \frac{1}{-p+1} - \frac{1}{-p+1} \lim_{\varepsilon \to 0^+} \ln^{-p+1}(\varepsilon)$$

The limit $\lim_{\varepsilon \to 0^+} \ln^{-p+1}(\varepsilon)$ is equal to 0 if $(-p+1) < 0$, and diverges if $(-p+1) > 0$. Therefore, the integral converges if $p > 1$ and the integral diverges if $p < 1$. If $p = 1$, we have:

$$\int_1^e \left(\frac{1}{\ln(x) x} \right) dx = \lim_{\varepsilon \to 0^+} \int_{1+\varepsilon}^e \left(\frac{1}{\ln(x) x} \right) dx = \lim_{\varepsilon \to 0^+} \ln(\ln(x)) \Big|_{1+\varepsilon}^e = \lim_{\varepsilon \to 0^+} \left(-\ln(\ln(\varepsilon)) \right) = -\infty.$$

Therefore, the integral converges only if $p < 1$.

41.   Let $f(x) = \dfrac{1}{2(2-x)^{4/3}}$ and $g(x) = \dfrac{1}{x(2-x)^{4/3}}$. Both functions $f(x)$ and $g(x)$ are continuous on $[1, 2)$, unbounded at $x = 2$, and the inequality $f(x) \leq g(x)$ holds for $1 < x < 2$. From Problem 8.6.37 we conclude that the integral $\int_1^2 f(x) dx$ diverges, and therefore the integral $\int_1^2 g(x) dx$ diverges as well.

43.   Let $f(x)$ be the integrand of the given integral and $g(x) = x^{-1/2} + (4-x)^{-1/2}$. Both functions $f(x)$ and $g(x)$ are continuous on $(0, 4)$, unbounded at $x = 0$ and $x = 4$, and the inequality $f(x) \leq g(x)$ holds for $0 < x < 4$. From Problems 8.6.37 and 8.6.38 we conclude that $\int_0^4 g(x) dx$ is convergent; therefore, the integral $\int_0^4 g(x) dx$ is convergent as well.

45.   Let $g(x)$ be the integrand of the given integral and $f(x) = \dfrac{1}{2x} + \dfrac{1}{100(3-x)}$. Both functions $f(x)$ and $g(x)$ are continuous on $(0, 3)$, unbounded at $x = 0$ and $x = 3$, and the inequality $f(x) \leq g(x)$ holds for $0 < x < 3$. From Problems 8.6.37 and 8.6.38 we conclude that the integral $\int_0^3 f(x) dx$ diverges; hence, the integral $\int_0^3 g(x) dx$ diverges as well.

47.   Let $f(x)$ be the integrand of the given integral and $g(x) = e^x + x^{-1/2}$. Both functions $f(x)$ and $g(x)$ are continuous on $(0, 1]$, unbounded at $x = 0$, $x = 1$, and the inequality $f(x) \leq g(x)$ holds for $0 < x < 1$. Then

$$\int_0^1 g(x)dx = \int_0^1 \left(e^x + x^{-1/2}\right)dx = e^x\Big|_0^1 + \int_0^1 x^{-1/2}dx = (e - 1) + \int_0^1 x^{-1/2}dx.$$

From Problem 8.6.38 we conclude that the integral $\int_0^1 g(x)dx$ is convergent; hence, the integral

$\int_0^1 f(x)dx$ is convergent as well.

Calculator/Computer Exercises

49.   If we define $g(x) = x^{-1/2}$, then $f(x) \leq g(x)$ for $0 < x \leq 1$. We compute $\int_0^1 f(x)dx \approx 1.9$;

$$\int_0^1 g(x)dx = \lim_{\varepsilon \to 0^+} \int_\varepsilon^1 x^{-1/2}dx = 2 \lim_{\varepsilon \to 0^+} x^{1/2}\Big|_\varepsilon^1 = 2 \lim_{\varepsilon \to 0^+} (1 - \varepsilon) = 2.$$

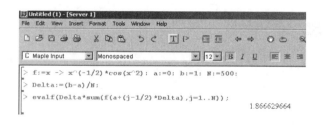

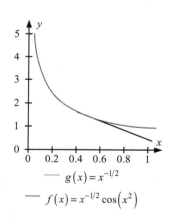

$$—\ g(x) = x^{-1/2}$$

$$—\ f(x) = x^{-1/2} \cos\left(x^2\right)$$

51.   If we define $g(x) = e^2 x^{-1/3}$, then $f(x) \le g(x)$ for $0 < x \le 1$. We compute $\int_0^1 f(x)dx \approx 3.5$;

$$\int_0^1 g(x)dx = e^2 \lim_{\varepsilon \to 0^+} \int_\varepsilon^1 x^{-1/3} dx = \left(\frac{3e^2}{2}\right) \lim_{\varepsilon \to 0^+} x^{2/3}\Big|_\varepsilon^1 = \left(\frac{3e^2}{2}\right) \lim_{\varepsilon \to 0^+} (1 - \varepsilon) = \frac{3e^2}{2} \approx 11.08358415.$$

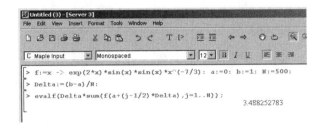

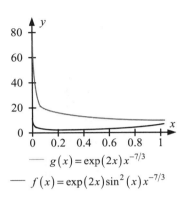

$g(x) = \exp(2x) x^{-7/3}$

$f(x) = \exp(2x) \sin^2(x) x^{-7/3}$

Section 8.7   Improper Integrals—Unbounded Intervals

Problems for Practice

1.   The improper integral converges. We evaluate as follows:

$$\int_3^\infty x^{-3/2} dx = \lim_{N \to \infty} \int_3^N x^{-3/2} dx = \lim_{N \to \infty}\left(-2x^{-1/2}\Big|_3^N\right) = \lim_{N \to \infty}\left(-2N^{-1/2} + \frac{2}{\sqrt{3}}\right) = \frac{2}{\sqrt{3}}.$$

3.   The improper integral converges. We evaluate as follows:

$$\int_{-1}^\infty \left(\frac{1}{(3+x)^{3/2}}\right) dx = \lim_{N \to \infty} \int_{-1}^N \left(\frac{1}{(3+x)^{3/2}}\right) dx = \lim_{N \to \infty}\left(-2(3+x)^{-1/2}\Big|_{-1}^N\right)$$

$$= \lim_{N \to \infty}\left(-2(3+N)^{-1/2} + \frac{2}{\sqrt{2}}\right) = \sqrt{2}.$$

5.   The improper integral converges. We evaluate as follows:

$$\int_0^\infty \left(\frac{1}{1+x^2}\right) dx = \lim_{N \to \infty} \int_0^N \left(\frac{1}{1+x^2}\right) dx = \lim_{N \to \infty}\left(\arctan(x)\Big|_0^N\right)$$

$$= \lim_{N \to \infty}\left(\arctan(N) - \arctan(0)\right) = \frac{\pi}{2}.$$

7.   The improper integral converges. We evaluate as follows:

$$\int_0^\infty \left(\frac{x}{\left(1+x^2\right)^2} \right) dx = \lim_{N\to\infty} \int_0^N \left(\frac{x}{\left(1+x^2\right)^2} \right) dx = \lim_{N\to\infty} \left(\frac{-1}{2\left(1+x^2\right)} \Bigg|_0^N \right)$$

$$= \lim_{N\to\infty} \left(\frac{-1}{2\left(1+N^2\right)} + \frac{1}{2} \right) = \frac{1}{2}.$$

9.   The improper integral diverges. We reach this conclusion as follows:

$$\int_0^\infty \left(\frac{e^x}{e^x+1} \right) dx = \lim_{N\to\infty} \int_0^N \left(\frac{e^x}{e^x+1} \right) dx = \lim_{N\to\infty} \left(\ln\left(e^x+1\right) \Big|_0^N \right)$$

$$= \lim_{N\to\infty} \left(\ln\left(e^N+1\right) - \ln(2) \right) = \infty.$$

11.   The improper integral converges. We evaluate as follows:

$$\int_1^\infty x e^{-3x^2} \, dx = \lim_{N\to\infty} \int_1^N x e^{-3x^2} \, dx = \lim_{N\to\infty} \left(\frac{-e^{-3x^2}}{6} \Bigg|_1^N \right) = \lim_{N\to\infty} \left(\frac{-e^{-3x^2}}{6} + \frac{e^{-3}}{6} \right) = \frac{e^{-3}}{6}.$$

13.   The improper integral diverges. We reach this conclusion as follows:

$$\int_e^\infty \left(\frac{1}{x\ln(x)} \right) dx = \lim_{N\to\infty} \int_e^N \left(\frac{1}{x\ln(x)} \right) dx = \lim_{N\to\infty} \left(\ln\big|\ln(x)\big| \Big\|_e^N \right) = \lim_{N\to\infty} \left(\ln\left(\ln(N)\right) \right) = \infty.$$

15.   The improper integral converges. We evaluate as follows:

$$\int_1^\infty \left(\frac{\arctan(x)}{1+x^2} \right) dx = \lim_{N\to\infty} \int_1^N \left(\frac{\arctan(x)}{1+x^2} \right) dx$$

$$= \lim_{N\to\infty} \left(\frac{\arctan^2(x)}{2} \Bigg|_1^N \right)$$

$$= \lim_{N\to\infty} \left(\frac{\arctan^2(N)}{2} - \frac{\arctan^2(1)}{2} \right) = \frac{3\pi^2}{32}.$$

17.   The improper integral diverges. We reach this conclusion as follows:

$$\int_{-\infty}^{-2} x^{-1/3}\,dx = \lim_{M\to-\infty}\int_{M}^{-2} x^{-1/3}\,dx = \lim_{M\to-\infty}\left(\frac{3x^{2/3}}{2}\Big|_{M}^{-2}\right) = \lim_{M\to-\infty}\left(3\cdot 2^{-1/3} - \frac{3M^{2/3}}{2}\right) = -\infty\,.$$

19.   The improper integral diverges. We reach this conclusion as follows:

$$\int_{-\infty}^{-100}\left(\frac{1}{\sqrt{|x|}}\right)dx = \int_{100}^{\infty}\left(\frac{1}{\sqrt{x}}\right)dx = \lim_{N\to\infty}\int_{100}^{N}\left(\frac{1}{\sqrt{x}}\right)dx = \lim_{N\to\infty}\left(2\sqrt{x}\,\Big|_{100}^{N}\right)$$
$$= \lim_{N\to\infty}\left(2\sqrt{N}-20\right) = \infty\,.$$

21.   The improper integral converges. We evaluate as follows:

$$\int_{-\infty}^{4} e^{x/3}\,dx = \lim_{M\to-\infty}\int_{M}^{4} e^{x/3}\,dx = \lim_{M\to-\infty}\left(3e^{x/3}\Big|_{M}^{4}\right) = \lim_{M\to-\infty}\left(3e^{4/3}-3e^{M/3}\right) = 3e^{4/3}\,.$$

23.   The improper integral converges. We evaluate as follows:

$$\int_{-\infty}^{-5}\left(\frac{e^{x}}{e^{2x}+1}\right)dx = \lim_{M\to-\infty}\int_{M}^{-5}\left(\frac{e^{x}}{e^{2x}+1}\right)dx = \lim_{M\to-\infty}\left(\arctan\left(e^{x}\right)\Big|_{M}^{-5}\right)$$
$$= \lim_{M\to-\infty}\left(\arctan\left(e^{-5}\right)-\arctan\left(e^{M}\right)\right) = \arctan\left(e^{-5}\right).$$

25.   The improper integral diverges. We reach this conclusion as follows:

$$\int_{-\infty}^{\infty} e^{-x}\,dx = \lim_{N\to\infty}\lim_{M\to-\infty}\int_{M}^{N} e^{-x}\,dx = \lim_{N\to\infty}\lim_{M\to-\infty}\left(-e^{-x}\Big|_{M}^{N}\right) = \lim_{N\to\infty}\lim_{M\to-\infty}\left(-e^{-N}+e^{-M}\right)$$
$$= -\lim_{N\to\infty}e^{-N} + \lim_{M\to-\infty}e^{-M} = 0+\infty = \infty.$$

27.   The improper integral converges. We evaluate as follows:

$$\int_{-\infty}^{\infty}\left(\frac{1}{1+x^{2}}\right)dx = \lim_{N\to\infty}\lim_{M\to-\infty}\int_{M}^{N}\left(\frac{1}{1+x^{2}}\right)dx = \lim_{N\to\infty}\lim_{M\to-\infty}\left(\arctan(x)\Big|_{M}^{N}\right)$$
$$= \lim_{N\to\infty}\lim_{M\to-\infty}\left(\arctan(N)-\arctan(M)\right)$$
$$= \lim_{N\to\infty}\arctan(N) - \lim_{M\to-\infty}\arctan(M)$$
$$= \frac{\pi}{2}-\left(-\frac{\pi}{2}\right) = \pi.$$

29. The improper integral converges. We evaluate as follows:

$$\int_{-\infty}^{\infty}\left(\frac{x}{\left(1+x^2\right)^2}\right)dx = \lim_{N\to\infty}\lim_{M\to-\infty}\int_M^N\left(\frac{x}{\left(1+x^2\right)^2}\right)dx = \lim_{N\to\infty}\lim_{M\to-\infty}\left(\frac{-1}{2\left(1+x^2\right)}\Bigg|_M^N\right)$$

$$= \lim_{N\to\infty}\lim_{M\to-\infty}\left(\frac{-1}{2\left(1+N^2\right)}+\frac{1}{2\left(1+M^2\right)}\right)$$

$$= -\lim_{N\to\infty}\left(\frac{1}{2\left(1+N^2\right)}\right)+\lim_{M\to-\infty}\left(\frac{1}{2\left(1+M^2\right)}\right) = 0.$$

Further Theory and Practice

31. By making the substitution $x = e^y$ and then integrating by parts with $u = y$ and $dv = e^{-y}dy$, we have $\int ye^{-y}dy =$ is true

$$\int_1^N x^{-2}\ln(x)\,dx = \int_0^{\ln(N)} ye^{-y}\,dy = -e^{-y}(y+1)\Big|_0^{\ln(N)} = 1 - \frac{1}{N}\left(1+\ln(N)\right)$$

Therefore, using L'Hôpital's Rule, we see that the given improper integral converges and

$$\int_1^\infty x^{-2}\ln(x)\,dx = \lim_{N\to\infty}\left(1-\frac{1}{N}\left(1+\ln(N)\right)\right) = 1 - \lim_{N\to\infty}\left(\frac{\ln(N)}{N}\right) = 1 - \lim_{N\to\infty}\left(\frac{\frac{d}{dN}\ln(N)}{\frac{d}{dN}N}\right) = 1.$$

33. The improper integral converges. Noting that $\int x^2 e^{x+1}dx = e\int x^2 e^x\,dx = e^{x+1}\left(x^2-2x+2\right)$ by Exercise 13 from Section 7.1, we have:

$$\int_{-\infty}^0 x^2 e^{x+1}dx = \lim_{M\to-\infty}\int_M^0 x^2 e^{x+1}dx = \lim_{M\to-\infty}\left(e^{x+1}\left(x^2-2x+2\right)\Big|_M^0\right)$$

$$= \lim_{M\to-\infty}\left(2e-e^{M+1}\left(M^2-2M+2\right)\right) = 2e.$$

35. The improper integral converges. We evaluate as follows:

$$\int_0^\infty\left(\frac{e^{-1/x}}{x^2}\right)dx = \lim_{N\to\infty}\lim_{\varepsilon\to0^+}\int_\varepsilon^N\left(\frac{e^{-1/x}}{x^2}\right)dx = \lim_{N\to\infty}\lim_{\varepsilon\to0^+}\left(e^{-1/x}\Big|_\varepsilon^N\right) = \lim_{N\to\infty}\lim_{\varepsilon\to0^+}\left(e^{-1/N}-e^{-1/\varepsilon}\right)$$

$$= \lim_{N\to\infty}e^{-1/N} - \lim_{\varepsilon\to0^+}e^{-1/\varepsilon} = 1-0 = 1.$$

37. The improper integral diverges. We evaluate as follows:

$$\int_1^\infty \left(\frac{1}{x\ln^2(x)}\right)dx = \lim_{N\to\infty}\lim_{\varepsilon\to 0^+}\int_{1+\varepsilon}^N \left(\frac{1}{x\ln^2(x)}\right)dx = \lim_{N\to\infty}\lim_{\varepsilon\to 0^+}\left(\frac{-1}{\ln(x)}\Big|_{1+\varepsilon}^N\right)$$

$$= \lim_{N\to\infty}\lim_{\varepsilon\to 0^+}\left(\frac{-1}{\ln(N)}+\frac{1}{\ln(1+\varepsilon)}\right) = \lim_{N\to\infty}\left(\frac{-1}{\ln(N)}\right)+\lim_{\varepsilon\to 0^+}\left(\frac{1}{\ln(1+\varepsilon)}\right)$$

$$= 0+\infty = \infty$$

39. $V = \int_1^\infty \pi f^2(x)\,dx = \pi\int_1^\infty\left(\frac{1}{x^3}\right)dx = \lim_{N\to\infty}\pi\int_1^N\left(\frac{1}{x^3}\right)dx = \pi\lim_{N\to\infty}\left(-\frac{1}{2x^2}\Big|_1^N\right) = \left(\frac{\pi}{2}\right)\lim_{N\to\infty}\left(-\frac{1}{N^2}+1\right)$

$$= \frac{\pi}{2}$$

41. $V = \int_0^\infty \pi f^2(x)\,dx = \pi\int_0^\infty\left(\frac{1}{(1+x)^2}\right)dx = \pi\lim_{N\to\infty}\int_0^N\left(\frac{1}{(1+x)^2}\right)dx = \pi\lim_{N\to\infty}\left(\frac{-1}{1+x}\Big|_0^N\right)$

$$= \pi\lim_{N\to\infty}\left(\frac{-1}{1+N}+1\right) = \pi$$

43. We have for $p\neq 1$

$$\int_0^\infty\left(\frac{1}{x^p}\right)dx = \lim_{N\to\infty}\lim_{\varepsilon\to 0^+}\int_\varepsilon^N\left(\frac{1}{x^p}\right)dx = \lim_{N\to\infty}\lim_{\varepsilon\to 0^+}\left(\frac{x^{1-p}}{1-p}\Big|_\varepsilon^N\right) = \lim_{N\to\infty}\lim_{\varepsilon\to 0^+}\left(\frac{N^{1-p}}{1-p}-\frac{\varepsilon^{1-p}}{1-p}\right)$$

$$= \lim_{N\to\infty}\left(\frac{N^{1-p}}{1-p}\right)-\lim_{\varepsilon\to 0^+}\left(\frac{\varepsilon^{1-p}}{1-p}\right)$$

For $p<1$ the first limit is ∞; the second limit is 0. For $p>1$ the first limit is 0; the second is ∞. In both cases the integral diverges. For $p=1$ we have

$$\int_0^\infty\left(\frac{1}{x}\right)dx = \lim_{N\to\infty}\lim_{\varepsilon\to 0^+}\int_\varepsilon^N\left(\frac{1}{x}\right)dx = \lim_{N\to\infty}\lim_{\varepsilon\to 0^+}\left(\ln(|x|)\Big|_\varepsilon^N\right) = \lim_{N\to\infty}\lim_{\varepsilon\to 0^+}\left(\ln(N)-\ln(\varepsilon)\right)$$

$$= \lim_{N\to\infty}\ln(N)-\lim_{\varepsilon\to 0^+}\ln(\varepsilon) = \infty$$

and the integral also diverges.

45.   As $0 < \dfrac{x}{1+x^3} < \dfrac{1}{x^2}$ for $1 \le x < \infty$, we have

$$\int_1^\infty \left(\frac{x}{1+x^3}\right)dx < \int_1^\infty \left(\frac{1}{x^2}\right)dx = \lim_{N\to\infty}\int_1^N \left(\frac{1}{x^2}\right)dx = \lim_{N\to\infty}\left(-\frac{1}{x}\Big|_1^N\right) = \lim_{N\to\infty}\left(-\frac{1}{N}+1\right) = 1$$

and the original integral converges.

47.   As $0 < \dfrac{1}{\sqrt{1+x^{5/2}}} < \dfrac{1}{x^{5/4}}$ for $1 \le x < \infty$, we have

$$\int_1^\infty \left(\frac{1}{\sqrt{1+x^{5/2}}}\right)dx < \int_1^\infty \left(\frac{1}{x^{5/4}}\right)dx = \lim_{N\to\infty}\int_1^N \left(\frac{1}{x^{5/4}}\right)dx = \lim_{N\to\infty}\left(-4x^{-1/4}\Big|_1^N\right) = \lim_{N\to\infty}\left(-4N^{-1/4}+4\right) = 4$$

and the original integral converges.

49.   As $0 < \dfrac{e^{-x}}{\sqrt{x}} \le \dfrac{e^{-\sqrt{x}}}{\sqrt{x}}$ for $1 \le x < \infty$, we have $\displaystyle\int_1^\infty \left(\frac{e^{-x}}{\sqrt{x}}\right)dx \le \int_1^\infty \left(\frac{e^{-\sqrt{x}}}{\sqrt{x}}\right)dx < \int_0^\infty \left(\frac{e^{-\sqrt{x}}}{\sqrt{x}}\right)dx = 2$ (see

Exercise 8.7.36) and the original integral converges.

51.   As $0 < \dfrac{1}{\sqrt{2x}} \le \dfrac{1}{\sqrt{1+x^2}}$ for $1 \le x < \infty$, we have

$$\int_1^\infty \left(\frac{1}{\sqrt{1+x^2}}\right)dx \ge \int_1^\infty \left(\frac{1}{\sqrt{2x}}\right)dx = \lim_{N\to\infty}\int_1^N \left(\frac{1}{\sqrt{2x}}\right)dx = \lim_{N\to\infty}\left(\frac{\ln(|x|)}{\sqrt{2}}\Big|_1^N\right)$$

$$= \lim_{N\to\infty}\left(\frac{\ln(N)}{\sqrt{2}}-0\right) = \infty$$

and the original integral diverges.

53.   As $0 < \dfrac{x}{x^{3/2}} \le \dfrac{\sin^2(x)+x}{x^{3/2}}$ for $1 \le x < \infty$, we have

$$\int_1^\infty \left(\frac{\sin^2(x)+x}{x^{3/2}}\right)dx > \int_1^\infty x^{-1/2}dx = \lim_{N\to\infty}\int_1^N x^{-1/2}dx = \lim_{N\to\infty}\left(2\sqrt{x}\Big|_1^N\right) = \lim_{N\to\infty}\left(2\sqrt{N}-2\right) = \infty$$

and the original integral diverges.

55.   As $0 < \dfrac{1}{x} < \dfrac{1}{x - \exp(-x)} = \dfrac{\exp(x)}{x\exp(x) - 1}$ for $1 \le x < \infty$, we have

$$\int_1^\infty \left(\frac{\exp(x)}{x\exp(x) - 1} \right) dx > \int_1^\infty \left(\frac{1}{x} \right) dx = \lim_{N \to \infty} \int_1^N \left(\frac{1}{x} \right) dx = \lim_{N \to \infty} \left(\ln(|x|) \Big|_1^N \right) = \lim_{N \to \infty} \left(\ln(N) - 0 \right) = \infty$$

and the original integral diverges.

57.   The present value is $\displaystyle\sum_{j=1}^\infty 10000 e^{-0.05j}$, or \$195,041.66.

The present value of the continuous income stream is $\displaystyle\int_0^\infty 10000 e^{-0.05t} \, dt$, or \$200000.

59.   The present value of one share's worth 3 years hence is $25 + \displaystyle\int_0^3 (2 + 0.35t)\exp(-0.08t)\,dt$, or \$31.68.

If the company could sustain its expected growth rate in perpetuity, the present value of one share would be $25 + \displaystyle\int_0^\infty (2 + 0.35t)\exp(-0.08t)\,dt$, or \$104.69.

61.   Let $f(x) = \left(\dfrac{1}{\sqrt{2\pi}\sigma} \right) \exp\left(-\dfrac{(x-\mu)^2}{2\sigma^2} \right)$, then $f'(x) = \left(\dfrac{\mu - x}{\sqrt{2\pi}\sigma^3} \right) \exp\left(-\dfrac{(x-\mu)^2}{2\sigma^2} \right)$ and

$$f''(x) = \left(\frac{(x-\mu)^2 - \sigma^2}{\sqrt{2\pi}\sigma^5} \right) \exp\left(-\frac{(x-\mu)^2}{2\sigma^2} \right).$$

The critical point is $x = \mu$ and the points of inflection are $x = \mu \pm \sigma$. The graph is concave up on the interval at $(-\infty, \mu - \sigma)$ and $(\mu + \sigma, \infty)$; concave down on the interval at $(\mu - \sigma, \mu + \sigma)$; maximum value: $\dfrac{1}{\sqrt{2\pi}\sigma}$, occurring at $x = \mu$.

63.   Let $u = \dfrac{x - \mu}{\sigma}$, then

$$P(\mu - \lambda\sigma < X < \mu + \lambda\sigma) = \int_{\mu - \lambda\sigma}^{\mu + \lambda\sigma} \left(\frac{1}{\sqrt{2\pi}\sigma} \right) \exp\left(-\frac{(x-\mu)^2}{2\sigma^2} \right) dx = \left(\frac{1}{\sqrt{2\pi}} \right) \int_{-\lambda}^{\lambda} \exp\left(-\frac{u^2}{2} \right) du$$

$$= P(-\lambda < X_0 < \lambda)$$

65. Given that $f(t) = \exp(-t^2)$ is an even function, from Exercise 8.7.64 it follows that

$$\int_0^\infty f(t)\,dt = \left(\frac{1}{2}\right)\int_{-\infty}^\infty f(t)\,dt = \frac{\sqrt{\pi}}{2} \text{ and } \lim_{x\to\infty}\text{erf}(x) = \lim_{x\to\infty}\left(\frac{2}{\sqrt{\pi}}\right)\int_0^x f(t)\,dt = \left(\frac{2}{\sqrt{\pi}}\right)\int_0^\infty f(t)(dt) = 1.$$

By the substitution $u = \dfrac{x-\mu}{\sigma}$ we have $P(X < \mu + \lambda\sigma) = P(X_0 < \lambda) = \dfrac{1+\text{erf}\left(\lambda/\sqrt{2}\right)}{2}$.

67. **a.** Let $u = x^s$, $dv = e^{-x}dx$, $du = sx^{s-1}dx$, $v = -e^{-x}$, then applying the Integration by Parts Formula and L'Hôpital's Rule we obtain

$$\Gamma(s+1) = \int_0^\infty x^s e^{-x}dx = \lim_{N\to\infty}\int_0^N x^s e^{-x}dx = \lim_{N\to\infty}\left(-sx^{s-1}e^{-x}\Big|_0^N + \int_0^N sx^{s-1}e^{-x}dx\right)$$

$$= -s\lim_{N\to\infty}\left(N^{s-1}e^{-N}\right) + s\lim_{N\to\infty}\int_0^N x^{s-1}e^{-x}dx = 0 + s\int_0^\infty x^{s-1}e^{-x}dx = s\Gamma(s)$$

b. $\Gamma(1) = \int_0^\infty e^{-x}dx = 1$, $\Gamma(2) = 1\cdot\Gamma(1) = 1$, $\Gamma(3) = 2\cdot\Gamma(2) = 2$, $\Gamma(4) = 3\cdot\Gamma(3) = 6$

c. We'll use the Method of Induction. For small values of n the statement is true. Let $\Gamma(n) = (n-1)!$, then $\Gamma(n+1) = n\cdot\Gamma(n) = n\cdot(n-1)! = n!$, and the statement is true for all positive integer n.

d. We have $\int_0^\infty cx^{s-1}e^{-\lambda x}dx = \int_0^\infty c\left(\dfrac{u}{\lambda}\right)^{s-1}e^{-u}\left(\dfrac{du}{\lambda}\right) = \left(\dfrac{c}{\lambda^s}\right)\Gamma(s) = 1$, where $u = \lambda x$;

therefore, $c = \dfrac{\lambda^s}{\Gamma(s)}$.

Calculator/Computer Exercises

69. Let $\xi = \dfrac{1}{x}$, then $\int_1^\infty\left(\dfrac{1}{\sqrt{1+x^3}}\right)dx = \int_0^1\left(\dfrac{1}{\sqrt{\xi^4 + \xi}}\right)d\xi \approx 1.89$.

71. Let $\xi = \dfrac{1}{x}$, then $\int_1^\infty\left(\dfrac{1}{x\sqrt{1+x}}\right)dx = \int_0^1\left(\dfrac{1}{\sqrt{\xi + \xi^2}}\right)d\xi \approx 1.76$.

Infinite Series

Section 9.1   Series

Problems for Practice

1. The first five partial sums are 1, $\dfrac{3}{2}$, $\dfrac{11}{6}$, $\dfrac{25}{12}$, and $\dfrac{137}{60}$.

3. The first five partial sums are $\dfrac{5}{4}$, $\dfrac{37}{16}$, $\dfrac{213}{64}$, $\dfrac{1109}{256}$, and $\dfrac{5461}{1024}$.

5. The first five partial sums are $\dfrac{2}{3}$, $\dfrac{10}{9}$, $\dfrac{38}{27}$, $\dfrac{130}{81}$, and $\dfrac{422}{243}$.

7. The first five partial sums are 1, $\dfrac{3}{4}$, $\dfrac{31}{36}$, $\dfrac{115}{144}$, and $\dfrac{3019}{3600}$.

9. To calculate the partial sum we use equation (9.1): $S_N = \dfrac{1}{7} - \left(\dfrac{8}{7}\right)8^{-(N+1)}$. Then, we find the

sum of the series as follows: $\displaystyle\sum_{n=1}^{\infty} 8^{-n} = \lim_{N\to\infty}\sum_{n=1}^{N} 8^{-n} = \lim_{N\to\infty} S_N = \dfrac{1}{7}$.

11. To calculate the partial sum we use equation (9.1): $S_N = \dfrac{4}{5} - \left(\dfrac{9}{5}\right)\left(\dfrac{4}{9}\right)^{N+1}$. Then, we find the

sum of the series as follows: $\displaystyle\sum_{n=1}^{\infty}\left(\dfrac{2}{3}\right)^{2n} = \sum_{n=1}^{\infty}\left(\dfrac{4}{9}\right)^{n} = \lim_{N\to\infty}\sum_{n=1}^{N}\left(\dfrac{4}{9}\right)^{n} = \lim_{N\to\infty} S_N = \dfrac{4}{5}$.

13.   To calculate the partial sum we use equation (9.1): $S_N = \dfrac{1}{7^{1/3}-1} + \dfrac{7^{1/3}\left(7^{-1/3}\right)^{N+1}}{1-7^{1/3}}$. Then, we find the sum of the series as follows:

$$\sum_{n=1}^{\infty} 7^{-n/3} = \sum_{n=1}^{\infty} \left(7^{-1/3}\right)^n = \lim_{N \to \infty} \sum_{n=1}^{N} \left(7^{-1/3}\right)^n = \lim_{N \to \infty} S_N = \frac{1}{7^{1/3}-1} .$$

15.   To calculate the partial sum we use equation (9.1): $S_N = -\left(\dfrac{6}{5}\right)6^{-N-1} - \left(\dfrac{7}{6}\right)7^{-N-1} + \dfrac{11}{30}$. Then, we find the sum of the series as follows:

$$\sum_{n=1}^{\infty} \left(2^{-n} \cdot 3^{-n} + 7^{-n}\right) = \sum_{n-1}^{\infty} \left(6^{-n} + 7^{-n}\right) = \lim_{N \to \infty} \sum_{n-1}^{\infty} \left(6^{-n} + 7^{-n}\right) = \lim_{N \to \infty} S_N = \frac{11}{30} .$$

17. $8\sum_{n=1}^{\infty} \left(\dfrac{1}{10}\right)^n = \dfrac{8}{9}$

19. $17\sum_{n=1}^{\infty} \left(\dfrac{1}{1000}\right)^n = \dfrac{17}{999}$

Further Theory and Practice

21.   Let $a_n = \dfrac{n}{n+1} \ge \dfrac{1}{2}$ for $n \ge 1$, then we have $S_N \ge \left(\dfrac{1}{2}\right)N$ and this series diverges.

23.   Let $a_n = \dfrac{n}{\sqrt{n^2+1}} \ge \dfrac{1}{\sqrt{2}}$ for $n \ge 1$, then we have $S_N \ge \left(\dfrac{1}{\sqrt{2}}\right) \cdot N$ and this series diverges.

25.   The Nth partial sum is $S_N = 1 - \dfrac{1}{N+1}$. The series sum is $\sum_{n=1}^{\infty}\left(\dfrac{1}{n} - \dfrac{1}{n+1}\right) = \lim_{N \to \infty} S_N = 1$.

27.   The Nth partial sum is $S_N = \dfrac{-2N}{\left(N+1\right)^3}$. The series sum is

$$\sum_{n=1}^{\infty}\left(\frac{2n-2}{n^3} - \frac{2n}{\left(n+1\right)^3}\right) = \lim_{N \to \infty} S_N = 0 .$$

29.   The Nth partial sum is $S_N = 1 - \dfrac{1}{\sqrt{N+1}}$. The series sum is $\displaystyle\sum_{n=1}^{\infty}\left(\dfrac{1}{\sqrt{n}} - \dfrac{1}{\sqrt{n+1}}\right) = \lim_{N\to\infty} S_N = 1$.

31.   Let $\dfrac{1}{n(n+2)} = \dfrac{1}{2}\left(\dfrac{1}{n} - \dfrac{1}{n+2}\right) = \dfrac{1}{2}\left(\left(\dfrac{1}{n} - \dfrac{1}{n+1}\right) + \left(\dfrac{1}{n+1} - \dfrac{1}{n+2}\right)\right)$. Then

$$S_N = \sum_{n=1}^{N}\frac{1}{n(n+2)} = \left(\frac{1}{2}\right)\left(\sum_{n=1}^{N}\left(\frac{1}{n} - \frac{1}{n+1}\right) + \sum_{n=1}^{N}\left(\frac{1}{n+1} - \frac{1}{n+2}\right)\right)$$

$$= \left(\frac{1}{2}\right)\left(1 - \frac{1}{N+1} + \frac{1}{2} - \frac{1}{N+2}\right) = \frac{3}{4} - \left(\frac{1}{2}\right)\left(\frac{1}{N+1} + \frac{1}{N+2}\right) = \frac{3}{4} - \frac{2N+3}{2(N+1)(N+2)}$$

and $\displaystyle\sum_{n=1}^{\infty}\frac{1}{n(n+2)} = \lim_{N\to\infty} S_N = \frac{3}{4}$.

33.   Let $\dfrac{1}{(2n+1)(2n+3)} = \left(\dfrac{1}{2}\right)\left(\dfrac{1}{2n+1} - \dfrac{1}{2n+3}\right) = \left(\dfrac{1}{2}\right)\left(\dfrac{1}{2n+1} - \dfrac{1}{2(n+1)+1}\right)$. Then

$$S_N = \sum_{n=1}^{N}\frac{1}{(2n+1)(2n+3)} = \left(\frac{1}{2}\right)\sum_{n=1}^{N}\left(\frac{1}{2n+1} - \frac{1}{2(n+1)+1}\right) = \left(\frac{1}{2}\right)\left(\frac{1}{3} - \frac{1}{2N+3}\right) = \frac{1}{6} - \frac{1}{2(2N+3)}$$

and $\displaystyle\sum_{n=1}^{\infty}\frac{1}{(2n+1)(2n+3)} = \lim_{N\to\infty} S_N = \frac{1}{6}$.

35.   As $\ln\left(\dfrac{n}{n+1}\right) = \ln(n) - \ln(n+1)$, we have

$$S_N = \sum_{n=1}^{N}\ln\left(\frac{n}{n+1}\right) = \sum_{n=1}^{N}\left(\ln(n) - \ln(n+1)\right) = \ln(1) - \ln(N+1) = -\ln(N+1),$$

$\displaystyle\lim_{N\to\infty} S_N = -\infty$ and this series diverges.

37.   Let $b_n = \dfrac{r^n}{(r^n-1)(1-r)}$, then $a_n = \dfrac{r^n}{(r^n-1)(r^{n+1}-1)} = b_{n+1} - b_n$ and

$$S_N = \sum_{n=1}^{N} a_n = b_{N+1} - b_1 = \frac{r^{N+1}}{(r^{N+1}-1)(1-r)} + \frac{r}{(1-r)^2} = r\frac{r^N-1}{(r^{N+1}-1)(r-1)^2}.$$

Then the sum of the series $\displaystyle\sum_{n=1}^{\infty} \frac{r^n}{(r^n-1)(r^{n+1}-1)}$ is $\displaystyle\lim_{N\to\infty} S_N = \frac{1}{(r-1)^2}$.

39. $\dfrac{1}{2^2} + \dfrac{1}{4^2} + \dfrac{1}{6^2} + \dfrac{1}{8^2} + ... = \displaystyle\sum_{n=1}^{\infty} \frac{1}{(2n)^2} = \left(\frac{1}{4}\right)\sum_{n=1}^{\infty}\frac{1}{n^2} = \left(\frac{1}{4}\right)\left(\frac{\pi^2}{6}\right) = \frac{\pi^2}{24};$

$1 + \dfrac{1}{3^2} + \dfrac{1}{5^2} + \dfrac{1}{7^2} + ... = \displaystyle\sum_{n=1}^{\infty}\frac{1}{n^2} - \sum_{k=1}^{\infty}\frac{1}{(2k)^2} = \frac{\pi^2}{6} - \frac{\pi^2}{24} = \frac{\pi^2}{8}$

41.   As $\ln\left(\dfrac{1}{n}\right) = -\ln(n) < -\ln(2)$ for $n > 2$, we have

$$S_N = \sum_{n=1}^{N} \ln\left(\frac{1}{n}\right) = -\sum_{n=1}^{N} \ln(n) < -\ln(2)(N-1)$$

for $N > 2$ and $\displaystyle\lim_{N\to\infty} S_N = -\infty$, so the series $\displaystyle\sum_{n=1}^{\infty} \ln\left(\frac{1}{n}\right)$ diverges.

43.   We calculate as follows:

$$16000\left(0.9 + 0.9^2 + 0.9^3 + ...\right) = 16000\sum_{n=1}^{\infty} 0.9^n = 16000\left(\sum_{n=0}^{\infty} 0.9^n - 1\right) = 16000\left(\frac{1}{1-0.9} - 1\right)$$
$$= 144000.$$

Therefore, the spending that results is \$144000.

45.   The heart patient's body after N days maintains $0.5 \cdot 0.1^{N-1}$ mg of the first dose of medication, $0.5 \cdot 0.1^{N-2}$ mg of the second dose of medication and so on. The total maintained amount is $S_N = 0.5 \cdot 0.1 + 0.5 \cdot 0.1^2 + ... + 0.5 \cdot 0.1^{N-1} = 5\displaystyle\sum_{n=2}^{N} 0.1^n$.

The limit of the amount of medicine maintained in the patient's body is $S = \displaystyle\lim_{N\to\infty} S_N = \frac{1}{18}$ mg.

47. The area between $y = x^{n-1}$ and $y = x^n$ is

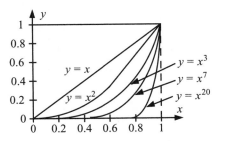

$$A_{n-1} = \int_0^1 \left(x^{n-1} - x^n \right) dx = \left(\frac{x^n}{n} - \frac{x^{n+1}}{n+1} \right) \Bigg|_0^1 = \frac{1}{n(n+1)}.$$

Then, $\displaystyle\sum_{n=1}^{\infty} \frac{1}{n(n+1)} = \sum_{n=1}^{\infty} A_{n-1}$. The right side is the sum

of the areas of all regions between all the curves of

form $y = x^n$ for $n \geq 0$, i.e., the area of the square.

Therefore, $\displaystyle\sum_{n=1}^{\infty} \frac{1}{n(n+1)} = 1$.

49. Let m_n be the abscissa of the center of mass of the nth domino, $m_n = m_{n-1} + \dfrac{1}{2(N-n+1)}$ or

$m_n = \dfrac{1}{2} + \displaystyle\sum_{k=2}^{N} \dfrac{1}{2(N-k+1)}$. The center of mass of the tower of dominos has the abscissa

$$M_N = \left(\frac{1}{N} \right) \sum_{n=1}^{N} m_n = \frac{1}{2} + \left(\frac{1}{2N} \right) \sum_{n=2}^{N} \sum_{k=2}^{n} \frac{1}{N-k+1}$$

$$= \frac{1}{2} + \left(\frac{1}{2N} \right) \left(\frac{1}{N-1} + \left(\frac{1}{N-1} + \frac{1}{N-2} \right) + \left(\frac{1}{N-1} + \frac{1}{N-2} + \frac{1}{N-3} \right) + ... + 1 \right)$$

$$= \frac{1}{2} + \left(\frac{1}{2N} \right) \left(\frac{N-1}{N-1} + \frac{N-2}{N-2} + \frac{N-3}{N-3} + .. + 1 \right) = \frac{1}{2} + \frac{N-1}{2N} = 1 - \frac{1}{2N} < 1$$

Therefore, the center of mass of this tower lies over the bottom domino and the stack is stable.

The total length of this stack is $1 + \displaystyle\sum_{n=2}^{N} \frac{1}{2(N-n+1)}$. For sufficiently large N, it can be as long

as we want due to the divergence of the harmonic series.

For the 10 ft length it will take about $1.2924243 \times 10^{103}$ dominoes; or about 6.5×10^{84} light-years high.

Calculator/Computer Exercises

51.   The series converges. We compute the partial sum as follows: $S_{100} = 0.4586751453...$, hence

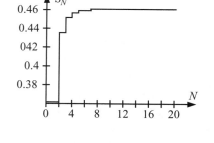

$$\sum_{n=1}^{\infty} \frac{e^{-n}}{n} \approx 0.458675.$$

Since $e^{-22} < 5 \times 10^{-10}$ we conclude that

$$\sum_{n=1}^{22} \frac{e^{-n}}{n} = 0.4586751453...$$

gives $\displaystyle\sum_{n=1}^{22} \frac{e^{-n}}{n}$ correct to 9 decimal places.

53.   The series converges. We compute the partial sum as follows: $S_{100} = 1.478071271...$, hence

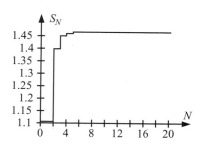

$$\sum_{n=1}^{\infty} \left(\frac{1.1}{n}\right)^{n} \approx 1.478071.$$

Notice that for $N \geq 10$,

$$\sum_{n=N+1}^{\infty} \left(\frac{1.1}{n}\right)^{n} < \sum_{n=N+1}^{\infty} \left(\frac{1.1}{n}\right)^{n} = \frac{1}{9}\left(\frac{1}{10}\right)^{N}.$$

Since $\dfrac{1}{9}\left(\dfrac{1}{10}\right)^{10} < 5 \cdot 10^{-10}$, we conclude that $\displaystyle\sum_{n=1}^{10} \left(\frac{1.1}{n}\right)^{n} = 1.458071271...$ gives $\displaystyle\sum_{n=1}^{\infty} \left(\frac{1.1}{n}\right)^{n}$ correct to 9 decimal places.

55.   From (9.1) and (9.2) we have $S_N = \dfrac{r^{N+1}-1}{r-1}$ and $S = \dfrac{1}{1-r}$. Then

$$S - S_N = \frac{r^{N+1}}{1-r} = \frac{10^{50}}{\left(1+10^{-50}\right)^{N}}.$$

For $N = 1000000$, $S - S_{1000000} = \dfrac{10^{50}}{\left(1+10^{-50}\right)^{1000000}}$. The inequality $S - S_N < 0.1$ leads to

$$N > 1.17432 \times 10^{52}.$$

Section 9.2   Determining Convergence

Problems for Practice

---

1.   The limit of the summand is $\lim_{n \to \infty}\left(ne^{-n}\right) = 0$. No conclusion can be drawn.

3.   The limit of the summand is $\lim_{n \to \infty}\left(\dfrac{n^2}{n^2+1}\right) = 1 \neq 0$. The series diverges.

5.   The limit of the summand is $\lim_{n \to \infty}\left(\dfrac{3^n}{4^n+3}\right) = 0$. No conclusion can be drawn.

7.   The limit of the summand is $\lim_{n \to \infty}\left(\dfrac{1}{1+1/n}\right) = 1 \neq 0$. The series diverges.

9.   The limit of the summand is $\lim_{n \to \infty}\left(\dfrac{3^n+5^n}{8^n}\right) = 0$. No conclusion can be drawn.

11.   The limit of the summand is $\lim_{n \to \infty}\left(\dfrac{2^n+1}{2^n+n^2}\right) = 1 \neq 0$. The series diverges.

13.   The limit of the summand is $\lim_{n \to \infty}\left(\dfrac{2n}{3n^2+1}\right) = 0$. No conclusion can be drawn.

15.   The limit of the summand is $\lim_{n \to \infty}\left(n^{1/n}\right) = \lim_{n \to \infty}\left(e^{\ln(n)/n}\right) = 1 \neq 0$. The series diverges.

17.   The limit of the summand is $\lim_{n \to \infty}\left(\dfrac{\pi}{2} - \arctan(n)\right) = 0$. No conclusion can be drawn.

19.   The limit of the summand is $\lim_{n \to \infty}\left(\sin\left(\dfrac{1}{n}\right)\right) = 0$. No conclusion can be drawn.

21. $\displaystyle\sum_{n=0}^{\infty}(0.1)^n = \dfrac{1}{1-0.1} = \dfrac{1}{0.9} = \dfrac{10}{9}$

23. $\displaystyle\sum_{n=3}^{\infty}\frac{(0.1)^n}{(0.2)^{n+2}}=\frac{1}{(0.2)^2}\sum_{n=3}^{\infty}(0.5)^n=\frac{1}{0.04}\cdot\frac{(0.5)^3}{1-0.5}=\frac{0.125}{0.02}=\frac{25}{4}$

25. $\displaystyle\sum_{n=-1}^{\infty}\left(\frac{2}{3}\right)^{2n+1}=\frac{2}{3}\sum_{n=-1}^{\infty}\left(\frac{4}{9}\right)^n=\frac{2}{3}\cdot\frac{9}{4}\cdot\frac{1}{1-4/9}=\frac{27}{10}$

27. A series with the summands tending to 0 *may* converge. For example, the series $\displaystyle\sum_{n=0}^{\infty}\frac{1}{2^n}$ converges, and the series $\displaystyle\sum_{n=1}^{\infty}\frac{1}{n}$ diverges.

29. If a series diverges, then the Divergence Test *may* succeed in proving the divergence. For example, $\displaystyle\lim_{n\to\infty}\left(\frac{n^2}{n^2+1}\right)=1\neq 0$ and the series $\displaystyle\sum_{n=1}^{\infty}\frac{n^2}{n^2+1}$ diverges. However, the series $\displaystyle\sum_{n=1}^{\infty}\frac{1}{n}$ diverges, but $\displaystyle\lim_{n\to\infty}\frac{1}{n}=0$.

Further Theory and Practice

31. The limit of the summand is $\displaystyle\lim_{n\to\infty}\left(\frac{1}{n}\right)^{1/n}=\lim_{n\to\infty}e^{\ln(1/n)/n}=\lim_{n\to\infty}e^{-\ln(n)/n}=1\neq 0$. The series diverges.

33. The limit of the summand is $\displaystyle\lim_{n\to\infty}\left(1+\frac{1}{n}\right)^n=e\neq 0$. The series diverges.

35. The limit of the summand is $\displaystyle\lim_{n\to\infty}\left(\frac{\ln(n^3)}{n}\right)=\lim_{n\to\infty}\left(\frac{3\ln(n)}{n}\right)=0$. No conclusion can be drawn.

37. The limit of the summand is $\displaystyle\lim_{n\to\infty}\left(\left(1+\frac{1}{n}\right)^n-n\right)=-\infty\neq 0$. The series diverges.

39. The limit of the summand is $\displaystyle\lim_{n\to\infty}\left(n\sin\left(\frac{1}{n}\right)\right)=\lim_{n\to\infty}\left(n\cdot\frac{1}{n}\right)=1\neq 0$. The series diverges.

41. The limit of the summand is $\lim\limits_{n\to\infty}\left(\dfrac{1}{\arctan(n)}\right)=\dfrac{2}{\pi}\neq 0$. The series diverges.

43. The limit of the summand is $\lim\limits_{n\to\infty}\left(\arctan(n)-\dfrac{\pi}{2}\right)=0$. No conclusion can be drawn.

45. Given that $\left(\dfrac{1}{2}\right)^{n^2}<\left(\dfrac{1}{2}\right)^{n}$ for any $n>1$, we conclude that $\sum\limits_{n=1}^{N}\left(\dfrac{1}{2}\right)^{n^2}<\sum\limits_{n=1}^{N}\left(\dfrac{1}{2}\right)^{n}$ for $N>1$. The

series $\sum\limits_{n=1}^{\infty}\left(\dfrac{1}{2}\right)^{n}$ converges; its partial sums are bounded above. The partial sums of the series

$\sum\limits_{n=1}^{\infty}\left(\dfrac{1}{2}\right)^{n^2}$ are also bounded above and the series converges.

47. Let $\sum\limits_{n=1}^{\infty}\exp\left(-2n\ln(n)+7\right)=e^{7}\sum\limits_{n=1}^{\infty}\exp\left(-2n\ln(n)\right)$. Given that $2n\ln(n)>n$ for $n>1$, we

conclude that $\exp\left(-2n\ln(n)\right)<\exp(-n)$ and $\sum\limits_{n=1}^{N}\exp\left(-2n\ln(n)\right)<\sum\limits_{n=1}^{N}\exp(-n)$ for $N>1$. The

series $\sum\limits_{n=1}^{\infty}\exp(-n)$ converges, so its partial sums are bounded above. Hence, the partial sums

of the series $\sum\limits_{n=1}^{\infty}\exp\left(-2n\ln(n)\right)$ are bounded above, and the series $e^{7}\sum\limits_{n=1}^{\infty}\exp\left(-2n\ln(n)\right)$

converges.

49. Let $\dfrac{\sqrt{n}}{4^{n}}=\dfrac{1}{2^{n}}\cdot\dfrac{\sqrt{n}}{2^{n}}$. Given that $\sqrt{n}<2^{n}$ for all $n>1$, $\dfrac{\sqrt{n}}{2^{n}}<1$ and $\dfrac{\sqrt{n}}{4^{n}}<\dfrac{1}{2^{n}}$. Therefore, we

conclude that $\sum\limits_{n=1}^{N}\dfrac{\sqrt{n}}{4^{n}}<\sum\limits_{n=1}^{N}\dfrac{1}{2^{n}}$ for $N>1$. The partial sums of the series $\sum\limits_{n=1}^{\infty}\dfrac{1}{2^{n}}$ are bounded

above by 1, so the partial sums of the series $\sum\limits_{n=1}^{\infty}\dfrac{\sqrt{n}}{4^{n}}$ are bounded above by 1 and the series is

convergent.

51. Let $\sum\limits_{n=M}^{N}r^{n}=\sum\limits_{n=M}^{N}r^{M+(n-M)}=r^{M}\sum\limits_{n=M}^{N}r^{n-M}=r^{M}\sum\limits_{n=0}^{N-M}r^{n}$. Given that $\sum\limits_{n=0}^{N-M}r^{n}=\dfrac{r^{N-M+1}-1}{r-1}$ by (9.1),

we conclude that $\sum\limits_{n=M}^{N}r^{n}=r^{M}\dfrac{\left(r^{N-M+1}-1\right)}{r-1}$. Taking the limit $N\to\infty$, we get $\sum\limits_{n=M}^{\infty}r^{n}=\dfrac{r^{M}}{1-r}$.

Calculator/Computer Exercises

53.   We plot the summand $y = \sin\left(\dfrac{1}{n}\right)\csc\left(\dfrac{2}{n}\right)$ and find that it does not tend to 0. The series diverges.

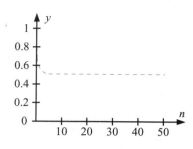

55.   We plot the summand $y = \cos\left(\dfrac{1}{n}\right) - \sec\left(\dfrac{1}{n}\right)$ and find that it tends to 0. No conclusion can be drawn.

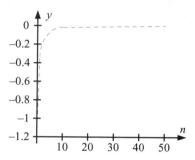

57.   We plot the summand $y = \sqrt{n^2 + 3n} - n$ and find that it does not tend to 0. The series diverges.

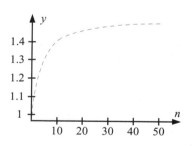

Section 9.3   Series with Nonnegative Terms—The Integral Test

Problems for Practice

1.   Let $f(x) = e^{-x}$. This function is positive, continuous, and decreasing on the interval $[1, \infty)$.

Given that $\displaystyle\int_1^\infty e^{-x}\,dx = \lim_{N \to \infty} \int_1^N e^{-x}\,dx = \lim_{N \to \infty}\left(e^{-1} - e^{-N}\right) = e^{-1}$, the convergence of the improper

integral $\displaystyle\int_1^\infty e^{-x}\,dx$ implies the convergence of the series $\displaystyle\sum_{n=1}^\infty e^{-n}$.

3. Let $f(x) = \dfrac{1}{x^2 + 4}$. This function is positive, continuous, and decreasing on the interval $[1, \infty)$. Given that

$$\int_1^\infty \left(\frac{1}{x^2 + 4} \right) dx = \lim_{N \to \infty} \int_1^N \left(\frac{1}{x^2 + 4} \right) dx = \lim_{N \to \infty} \left(\frac{1}{2} \right) \left(\arctan\left(\frac{N}{2} \right) - \arctan\left(\frac{1}{2} \right) \right)$$

$$= \left(\frac{1}{2} \right) \left(\frac{\pi}{2} - \arctan\left(\frac{1}{2} \right) \right),$$

the convergence of the integral $\displaystyle\int_1^\infty \left(\frac{1}{x^2 + 4} \right) dx$ implies the convergence of the series

$$\sum_{n=1}^\infty \frac{1}{n^2 + 4}.$$

5. Let $f(x) = \dfrac{1}{x + 3}$. This function is positive, continuous, and decreasing on the interval $[1, \infty)$.

Given that $\displaystyle\int_1^\infty \left(\frac{1}{x + 3} \right) dx = \lim_{N \to \infty} \int_1^N \left(\frac{1}{x + 3} \right) dx = \lim_{N \to \infty} \left(\ln(N + 3) - \ln(4) \right) = \infty$, the divergence of

the integral $\displaystyle\int_1^\infty \left(\frac{1}{x + 3} \right) dx$ implies the divergence of the series $\displaystyle\sum_{n=1}^\infty \frac{1}{n + 3}$.

7. Let $f(x) = \dfrac{2x^2}{x^3 + 4}$. This function is positive, continuous, and decreasing on the interval

$[2, \infty)$. Given that $\displaystyle\int_1^\infty \left(\frac{2x^2}{x^3 + 4} \right) dx = \lim_{N \to \infty} \int_1^N \left(\frac{2x^2}{x^3 + 4} \right) dx = \lim_{N \to \infty} \left(\frac{2}{3} \right) \left(\ln(N^3 + 4) - \ln(5) \right) = \infty$, the

divergence of the integral $\displaystyle\int_1^\infty \left(\frac{2x^2}{x^3 + 4} \right) dx$ implies the divergence of the series $\displaystyle\sum_{n=1}^\infty \frac{2n^2}{n^3 + 4}$.

9. Let $f(x) = \dfrac{\ln(x)}{x}$. This function is positive, continuous, and decreasing on the interval

$[e, \infty)$. Given that $\displaystyle\int_1^\infty \left(\frac{\ln(x)}{x} \right) dx = \lim_{N \to \infty} \int_1^N \left(\frac{\ln(x)}{x} \right) dx = \lim_{N \to \infty} \left(\frac{1}{2} \right) \ln^2(N) = \infty$, the divergence of

the integral $\displaystyle\int_1^\infty \left(\frac{\ln(x)}{x} \right) dx$ implies the divergence of the series $\displaystyle\sum_{n=1}^\infty \frac{\ln(n)}{n}$.

11. Let $f(x) = \dfrac{e^x}{\left(1+e^x\right)^2}$. This function is positive, continuous, and decreasing on the interval $[8, \infty)$. Given that

$$\int_8^\infty \left(\frac{e^x}{\left(1+e^x\right)^2}\right) dx = \lim_{N \to \infty} \int_8^N \left(\frac{e^x}{\left(1+e^x\right)^2}\right) dx = \lim_{N \to \infty} \left(\frac{1}{1+e^8} - \frac{1}{1+e^N}\right) = \frac{1}{1+e^8},$$

the convergence of the integral $\displaystyle\int_8^\infty \left(\frac{e^x}{\left(1+e^x\right)^2}\right) dx$ implies the convergence of the series

$$\sum_{n=8}^\infty \frac{e^n}{\left(1+e^n\right)^2}.$$

13. Let $f(x) = \dfrac{3}{x^2 + x}$. This function is positive, continuous, and decreasing on the interval $[2, \infty)$. Given that

$$\int_2^\infty \left(\frac{3}{x^2+x}\right) dx = \lim_{N \to \infty} \int_2^N \left(\frac{3}{x^2+x}\right) dx = \lim_{N \to \infty} 3\int_2^N \left(\frac{1}{x} - \frac{1}{x+1}\right) dx$$

$$= \lim_{N \to \infty} 3\left(\ln(N) - \ln(N+1) + \ln(3) - \ln(2)\right) = 3\ln\left(\frac{3}{2}\right),$$

the convergence of the integral $\displaystyle\int_2^\infty \left(\frac{3}{x^2+x}\right) dx$ implies the convergence of the series

$$\sum_{n=2}^\infty \frac{3}{n^2 + n}.$$

15. Let $f(x) = \dfrac{1}{(x+3)^{5/4}}$. This function is positive, continuous, and decreasing on the interval $[1, \infty)$. Given that

$$\int_1^\infty \left(\frac{1}{(x+3)^{5/4}}\right) dx = \lim_{N \to \infty} \int_1^N \left(\frac{1}{(x+3)^{5/4}}\right) dx = \lim_{N \to \infty} 4\left(\frac{1}{4^{1/4}} - \frac{1}{(N+3)^{1/4}}\right) = 4^{3/4} = 2\sqrt{2},$$

the convergence of the integral $\displaystyle\int_1^\infty \left(\frac{1}{(x+3)^{5/4}}\right) dx$ implies the convergence of the series

$$\sum_{n=1}^\infty \frac{1}{(n+3)^{5/4}}.$$

17. Let $f(x) = xe^{-2x}$. This function is positive, continuous, and decreasing on the interval $[1, \infty)$.

Given that $\int_1^\infty xe^{-2x}\,dx = \lim_{N\to\infty} \int_1^N xe^{-2x}\,dx = \lim_{N\to\infty}\left(\frac{1}{4}\right)\left(3e^{-2} - (1+2N)e^{-2N}\right) = \frac{3}{4e^2}$, the

convergence of the integral $\int_1^\infty xe^{-2x}\,dx$ implies the convergence of the series $\sum_{n=1}^\infty ne^{-2n}$.

19. Let $f(x) = \dfrac{1}{x\sqrt{x^2 - 1}}$. This function is positive, continuous, and decreasing on the interval

$[2, \infty)$. Given that

$$\int_2^\infty \left(\frac{1}{x\sqrt{x^2-1}}\right)dx = \lim_{N\to\infty}\int_2^N\left(\frac{1}{x\sqrt{x^2-1}}\right)dx = \lim_{N\to\infty}\left(\operatorname{arc\,sec}(2) - \operatorname{arc\,sec}(N)\right) = \frac{\pi}{6},$$

the convergence of the integral $\int_2^\infty\left(\dfrac{1}{x\sqrt{x^2-1}}\right)dx$ implies the convergence of the series

$\displaystyle\sum_{n=2}^\infty \frac{1}{n\sqrt{n^2-1}}$.

21. The series diverges: $p = \dfrac{1}{2} < 1$.

23. The series converges : $p = \sqrt{2} > 1$.

25. We represent the original series as a sum of two p-series $\displaystyle\sum_{n=1}^\infty \frac{\sqrt{n}+5}{n^2} = \sum_{n=1}^\infty \frac{1}{n^{3/2}} + 5\sum_{n=1}^\infty \frac{1}{n^2}$. For

the first series: $p_1 = \dfrac{3}{2} > 1$ and for the second series: $p_2 = 2 > 1$. Both series converge;
therefore, the original series converges as well.

27. We represent the original series as a sum of two p-series $\displaystyle\sum_{n=1}^\infty \frac{n^2+11}{n^4} = \sum_{n=1}^\infty \frac{1}{n^2} + 11\sum_{n=1}^\infty \frac{1}{n^4}$. For

the first series: $p_1 = 2 > 1$ and for the second series : $p_2 = 4 > 1$. Both series converge;
therefore, the original series converges.

Further Theory and Practice

29. We have $\displaystyle\int_1^\infty f(x)\,dx = \lim_{N\to\infty}\int_1^N f(x)\,dx = \lim_{N\to\infty}\left(\frac{N}{2}-\left(\frac{1}{2\pi}\right)\cos(\pi N)\sin(\pi N)-\frac{1}{N}+\frac{1}{2}\right)=\infty$ and

$\displaystyle\int_1^\infty f(x)\,dx$ diverges. For $f(n)=\dfrac{1}{n^2}$, the series $\displaystyle\sum_{n=1}^\infty f(n)$ converges as the p-series with

$p=2>1$. The Integral Test is not applicable because $f(x)$ is not a decreasing function.

31. For large values of n we have $\left(\dfrac{n}{n^2+1}\right)^a \approx \dfrac{1}{n^a}$. Therefore, the series $\displaystyle\sum_{n=1}^\infty \left(\dfrac{n}{n^2+1}\right)^a$ converges

if and only if the p-series for $p=a$ converges as well. This happens if and only if $a>1$.

33. **a.** Let $f(x)=\dfrac{x^4}{2^x}$, $f'(x)=x^3\dfrac{4-x\ln(2)}{2^x}$ and $f_{\max}=f\left(\dfrac{4}{\ln(2)}\right)=\left(\dfrac{4}{e\ln(2)}\right)^4$ on $[1,\infty)$.

The function is positive, continuous, and decreasing on the interval $\left[\dfrac{4}{\ln(2)},\infty\right)$.

Therefore,

$$\int_{4/\ln(2)}^\infty f(x)\,dx = \lim_{N\to\infty}\int_{4/\ln(2)}^N f(x)\,dx = \lim_{N\to\infty}\left(\frac{1}{\ln^5(2)}\right.$$
$$\left.\times\left(\frac{842}{e^4}-\frac{24+24N\ln(2)+12N^2\ln^2(2)+4N^3\ln^3(2)+N^4\ln(2)}{2^N}\right)\right)$$
$$=\frac{824}{e^4\ln^5(2)}$$

The convergence of $\displaystyle\int_{4/\ln(2)}^\infty f(x)\,dx$ implies the convergence of $\displaystyle\int_1^\infty f(x)\,dx$ and of

$\displaystyle\sum_{n=1}^\infty\frac{n^4}{2^n}$.

b.   Let $f(x) = \dfrac{\ln(x)}{x^2}$, $f'(x) = \dfrac{1 - 2\ln(x)}{x^3}$ and $f_{max} = f(\sqrt{e}) = \dfrac{1}{2e}$ on $[1, \infty)$. The

function is positive, continuous, and decreasing on the interval $\left[\dfrac{1}{\sqrt{e}}, \infty\right)$. Therefore,

$$\int_{\sqrt{e}}^{\infty} f(x)\,dx = \lim_{N \to \infty} \int_{\sqrt{e}}^{N} f(x)\,dx = \lim_{N \to \infty}\left(\dfrac{3}{2\sqrt{e}} - \dfrac{1 + \ln(N)}{N}\right) = \dfrac{3}{2\sqrt{e}}$$

The convergence of $\displaystyle\int_{\sqrt{e}}^{\infty} f(x)\,dx$ implies the convergence of $\displaystyle\int_{1}^{\infty} f(x)\,dx$ and of the

series $\displaystyle\sum_{n=1}^{\infty} \dfrac{\ln(n)}{n^2}$.

c.   Let $f(x) = \dfrac{1}{1 + \sqrt{x}}$. The function is positive, continuous, and decreasing on the

interval $[1, \infty)$. Therefore,

$$\int_{1}^{\infty} f(x)\,dx = \lim_{N \to \infty} \int_{1}^{N} f(x)\,dx = \lim_{N \to \infty}\left(2 - 2\ln(2) + 2\sqrt{N} + 2\ln\left(\sqrt{N} + 1\right)\right) = \infty.$$

The divergence of the integral implies the divergence of the series $\displaystyle\sum_{n=1}^{\infty} \dfrac{1}{1 + \sqrt{n}}$.

d.   Let $f(x) = \ln\left(\dfrac{x+3}{x}\right)$. The function is positive, continuous, and decreasing on the

interval $[1, \infty)$. Therefore,

$$\int_{1}^{\infty} f(x)\,dx = \lim_{N \to \infty} \int_{1}^{N} f(x)\,dx$$

$$= \lim_{N \to \infty}\left((N+3)\ln\left(\dfrac{N+3}{N}\right) - 3\ln\left(\dfrac{3}{N}\right) + 3\ln(3) - 4\ln(4)\right) = \infty$$

The divergence of the integral implies the divergence of the series $\displaystyle\sum_{n=1}^{\infty} \ln\left(\dfrac{n+3}{n}\right)$.

e. Let $f(x) = \ln\left(\dfrac{x^2+1}{x^2}\right)$. The function is positive, continuous, and decreasing on the interval $[1, \infty)$. Therefore,

$$\int_1^\infty f(x)\,dx = \lim_{N\to\infty} \int_1^N f(x)\,dx$$

$$= \lim_{N\to\infty}\left(2\arctan(N) + N\ln\left(\frac{N^2+1}{N^2}\right) - 2\arctan(1) - \ln(2)\right) = \frac{\pi}{2} - \ln(2)$$

The convergence of the integral implies the convergence of the series $\displaystyle\sum_{n=1}^\infty \ln\left(\frac{n^2+1}{n^2}\right)$.

f. Let $f(x) = x^3 e^{-2x}$, $f'(x) = x^2(3-2x)e^{-2x}$, $f_{max} = f\left(\dfrac{3}{2}\right) = \dfrac{27}{8e^3}$ on $[1, \infty)$. The function is positive, continuous, and decreasing on the interval $\left[\dfrac{3}{2}, \infty\right)$. Therefore,

$$\int_{3/2}^\infty f(x)\,dx = \lim_{N\to\infty} \int_{3/2}^N f(x)\,dx = \lim_{N\to\infty}\left(\frac{1}{8}\right)\left(39e^{-3} - e^{-2N}\left(3 + 6N + 6N^2 + 4N^3\right)\right)$$

$$= \frac{39}{8e^3}$$

The convergence of $\displaystyle\int_{3/2}^\infty f(x)\,dx$ implies the convergence of $\displaystyle\int_1^\infty f(x)\,dx$ and $\displaystyle\sum_{n=1}^\infty n^3 e^{-2n}$.

35. Let $a_n \geq 0$ and $\displaystyle\sum_{n=1}^\infty a_n$ converges, and set $b_n = a_n + \dfrac{1}{n^2}$; then $\displaystyle\sum_{n=1}^\infty b_n = \sum_{n=1}^\infty a_n + \sum_{n=1}^\infty \frac{1}{n^2}$ converges.

Let $c_n > 0$ and $\displaystyle\sum_{n=1}^\infty c_n$ diverges, and set $d_n = \dfrac{c_n}{2}$; then $\displaystyle\sum_{n=1}^\infty d_n$ diverges.

37. Because $\displaystyle\int_n^{2n} e^{-x}\,dx = e^{-n} - e^{-2n}$, $\displaystyle\sum_{n=2}^\infty \int_n^{2n} e^{-x}\,dx = \sum_{n=2}^\infty \left(e^{-n} - e^{-2n}\right)$. Let $f(x) = e^{-x} - e^{-2x}$. The function is positive, continuous, and decreasing on the interval $[2, \infty)$. For

$$\int_2^\infty f(x)\,dx = \lim_{N\to\infty} \int_2^N f(x)\,dx = \lim_{N\to\infty}\left(\frac{e^{-2N}}{2} - e^{-N} - \frac{1}{2e^4} + \frac{1}{e^2}\right) = \frac{1}{e^2} - \frac{1}{2e^4},$$

the convergence of the integral implies the convergence of the series $\displaystyle\sum_{n=2}^\infty \int_n^{2n} e^{-x}\,dx$.

39.   Let $f(x) = \dfrac{1}{x \ln(x) \ln^2(\ln(x))}$. The function is positive, continuous, and decreasing on the interval $[20, \infty)$. For

$$\int_{20}^{\infty} f(x)\,dx = \lim_{N \to \infty} \int_{20}^{N} f(x)\,dx = \lim_{N \to \infty}\left(\frac{1}{\ln(\ln(20))} - \frac{1}{\ln(\ln(N))} \right) = \frac{1}{\ln(\ln(20))},$$

the convergence of the integral implies the convergence of the series $\displaystyle\sum_{n=1}^{\infty} \frac{1}{n \ln(n) \ln^2(\ln(n))}$.

41. $0 < a_n = S_{CEF} < S_{BEFC} = S_{AEFD} - S_{ABCD} = \dfrac{1}{n} - \dfrac{1}{n+1}$

Note that S_G is the area of the region G (see figure).

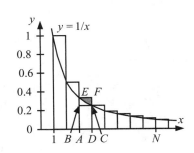

43.   If $f(x) = x + \ln(1-x)$, we have $f'(x) = 1 - \dfrac{1}{1-x} = \dfrac{x}{1-x} < 0$ for $0 < x < 1$ and f decreases on $(0, 1)$. Then $f(x) < f(0) = 0$ for $0 < x < 1$. We calculate

$$A_{N+1} - A_N = \frac{1}{N+1} - \ln(N+1) + \ln(N) = \frac{1}{N+1} + \ln\left(\frac{N}{N+1} \right)$$

$$= \frac{1}{N+1} + \ln\left(1 - \frac{1}{N+1} \right) = f\left(\frac{1}{N+1} \right) < 0$$

Therefore $\{A_N\}$ is positive, decreasing (hence it is bounded), and convergent.

Enrichment Exercises

45.   We solve $10^{-M} < 0.005$ to find $M = 3$ and

$$A_3 = \frac{1}{10} + \frac{1}{100} + \int_{3}^{\infty} 10^{-x}\,dx = \frac{11}{100} + \frac{1}{1000 \ln(10)} \approx 0.1104.$$

47. We solve $\dfrac{M}{\left(M^2+1\right)^3} < 0.005$ to find $M=3$ and

$$A_3 = \frac{1}{8} + \frac{2}{125} + \int_3^\infty \left(\frac{x}{\left(x^2+1\right)^3}\right) dx = \frac{141}{1000} + \frac{1}{400} = \frac{287}{2000} = 0.1435.$$

49. Given that $f(x)$ is decreasing, we have $\displaystyle\int_n^{n+1} f(x)\,dx \le f(n) \le \int_{n-1}^n f(x)\,dx$ and

$$\sum_{n=L}^M \int_n^{n+1} f(x)\,dx \le \sum_{n=L}^M f(n) \le \sum_{n=L}^M \int_{n-1}^n f(x)\,dx.$$ Taking into account that $\displaystyle\int_M^{M+1} f(x)\,dx < f(M)$

and $\displaystyle\int_{L-1}^L f(x)\,dx > f(L)$, we derive $\displaystyle\sum_{n=L}^M \int_n^{n+1} f(x)\,dx = \int_L^{M+1} f(x)\,dx < \int_L^M f(x)\,dx + f(M)$,

$$\sum_{n=L}^M \int_{n-1}^n f(x)\,dx = \int_{L-1}^M f(x)\,dx < \int_L^M f(x)\,dx + f(L) \text{ and}$$

$$f(M) + \int_L^M f(x)\,dx \le \sum_{n=L}^M f(n) \le f(L) + \int_L^M f(x)\,dx.$$

51. We solve $\dfrac{1}{M^2} < 0.0001$ to find $M=101$ and $n^{-2}(n+1)^{-1} < 0.0001$ to find $M=22$,

$$\int_{22}^\infty x^{-2}(x+1)^{-1}\,dx = \lim_{N\to\infty} \int_{22}^N x^{-2}(x+1)^{-1}\,dx$$

$$= \lim_{N\to\infty}\left(\ln(N+1) - \ln(N) - \frac{1}{N} - \ln(23) + \ln(22) + \frac{1}{22}\right) = \ln\left(\frac{22}{23}\right) + \frac{1}{22}.$$

Calculator/Computer Exercises

53. $\displaystyle\sum_{n=1}^\infty f(n) = 6.79\times10^{-2}$

$$\int_1^\infty \frac{\exp(x)}{\left(1+\exp(x)\right)^3}\,dx = \frac{1}{2(1+e)^2} \approx 3.62\times10^{-2}.$$

As $f(1) = \dfrac{e}{(1+e)^3} \approx 5.29\times10^{-2}$, the inequality (9.4) states

$3.62\times10^{-2} < 6.79\times10^{-2} < 5.29\times10^{-2} + 3.62\times10^{-2}$, which is true.

55. $\sum\limits_{n=1}^{\infty} f(n) = 0.13$

$\int\limits_{1}^{\infty} \dfrac{n^2}{\left(1+n^3\right)^3}\, dn = \dfrac{1}{24} \approx 0.04$

As $f(1) = \dfrac{1}{8} = 0.125$, the inequality (9.4) states $0.04 < 0.13 < 0.125 + 0.04$, which is true.

Enrichment Exercises

57.   We solve $\dfrac{M}{2^{M^3}} < 0.0005$ to find $M = 4$, $A_4 = \dfrac{1}{2} + \dfrac{1}{8} + \dfrac{3}{512} + \int\limits_{4}^{\infty}\left(\dfrac{x}{2^{x^2}}\right) dx = 0.63087\ldots$.

59.   We solve $\left(1 + M + M^3\right)^{-1} < 0.0005$ to find $M = 13$,

$$A_{13} = \sum\limits_{n=1}^{12}\left(1 + n + n^3\right)^{-1} + \int\limits_{13}^{\infty}\left(1 + x + x^3\right)^{-1} dx = 0.49449\ldots.$$

61.   From Exercise 9.3.49 we have

$$\sum\limits_{n=20}^{N} f(n) \le f(20) + \int\limits_{20}^{N} f(x)\, dx = \dfrac{1}{20\ln(20)\ln(\ln(20))} + \ln\big(\ln(\ln(N))\big) - \ln\big(\ln(\ln(20))\big)$$

$$= \dfrac{1}{20\ln(20)\ln(\ln(20))} + \ln\big(100\ln(10) + \ln(\ln(10))\big) - \ln\big(\ln(\ln(20))\big)$$

$$= 5.365279\ldots < 6.$$

On the other hand

$$\sum\limits_{n=20}^{N} f(n) \ge f(N) + \int\limits_{20}^{N} f(x)\, dx = \dfrac{1}{N\ln(N)\ln(\ln(N))} + \ln\big(\ln(\ln(N))\big) - \ln\big(\ln(\ln(20))\big) > 6$$

or $N > \exp\big(\exp\big(\exp(6.1)\big)\big)$.

Section 9.4   Series with Nonnegative Terms—The Comparison Test

Problems for Practice

1.   We observe that $0 \le \dfrac{n}{n^3 + 1} < \dfrac{n}{n^3} = \dfrac{1}{n^2}$ for every n. We note that $\displaystyle\sum_{n=1}^{\infty} \dfrac{1}{n^2}$ is a convergent

      p-series, because $p = 2 > 1$. From the Comparison Test for Convergence, we conclude that the series converges.

3.   We observe that $0 \le \dfrac{2 + \sin(n)}{n^4} < \dfrac{3}{n^4}$ for every n. We note that $\displaystyle\sum_{n=1}^{\infty} \dfrac{3}{n^4}$ is a convergent

      p-series, because $p = 4 > 1$. From the Comparison Test for Convergence, we conclude that the series converges.

5.   We observe that $0 \le \dfrac{n-2}{2n^{5/2} + 3} < \dfrac{n}{2n^{5/2}} = \dfrac{1}{2n^{3/2}}$ for every n. We note that $\displaystyle\sum_{n=1}^{\infty} \dfrac{1}{2n^{3/2}}$ is a

      convergent p-series, because $p = \dfrac{3}{2} > 1$. From the Comparison Test for Convergence, we conclude that the series converges.

7.   We observe that $0 \le \dfrac{2^n}{n \cdot 3^n} = \dfrac{(2/3)^n}{n} < \left(\dfrac{2}{3}\right)^n$ for every n. We note that $\displaystyle\sum_{n=1}^{\infty} \left(\dfrac{2}{3}\right)^n$ is a convergent

      geometric series, because $r = \dfrac{2}{3} < 1$. From the Comparison Test for Convergence, we conclude that the series converges.

9.   We observe that $n! \ge 2^{n-1}$ for every n and conclude that $0 \le \dfrac{1}{n!} < 2\left(\dfrac{1}{2}\right)^n$. We note that

      $\displaystyle\sum_{n=1}^{\infty} 2\left(\dfrac{1}{2}\right)^n$ is a convergent geometric series, because $r = \dfrac{1}{2} < 1$. From the Comparison Test for

      Convergence, we conclude that the series converges.

11.   We observe that $2n\sqrt{n} - 1 \ge n\sqrt{n} = n^{3/2}$ and conclude that $0 \le \dfrac{1}{2n\sqrt{n} - 1} < \dfrac{1}{n^{3/2}}$ for every n.

      We note that $\displaystyle\sum_{n=1}^{\infty} \dfrac{1}{n^{3/2}}$ is a convergent p-series, because $p = \dfrac{3}{2} > 1$. From the Comparison Test

      for Convergence, we conclude that the series converges.

13.   We observe that $0 \le \dfrac{n^3}{n^{4.01}+1} < \dfrac{n^3}{n^{4.01}} = \dfrac{1}{n^{1.01}}$ for every n. We note that $\displaystyle\sum_{n=1}^{\infty}\dfrac{1}{n^{1.01}}$ is a convergent

p-series, because $p = 1.01 > 1$. From the Comparison Test for Convergence, we conclude that the series converges.

15.   We observe that $0 \le \dfrac{\sqrt{n^2+1}}{\left(n\ln(n)\right)^2} \le \dfrac{\sqrt{n^2}+\sqrt{n^2}}{\left(n\ln(n)\right)^2} = \dfrac{2}{n\ln^2(n)}$ for $n \ge 2$. For $n > 2$ we have

$$\dfrac{2}{n\ln^2(n)} < \dfrac{2}{n\ln(n)}.$$

It was shown in Example 4, Section 9.3 that $\displaystyle\sum_{n=2}^{\infty}\dfrac{2}{n\ln(n)}$ is a convergent series. From the

Comparison Test for Convergence, we conclude that the series converges.

17.   We observe that $\ln(n) < n$ for $n \ge 2$. Therefore, $\dfrac{1}{\ln(n)} > \dfrac{1}{n}$. We note that $\displaystyle\sum_{n=2}^{\infty}\dfrac{1}{n}$ is a divergent

harmonic series. From the Comparison Test for Divergence, we conclude that the series diverges.

19.   We observe that $n^2 + 10 < 16n^2$ for every n. Therefore, $\dfrac{1}{\sqrt{n^2+1}} > \dfrac{1}{4n}$. We note that $\displaystyle\sum_{n=1}^{\infty}\dfrac{1}{4n}$ is

a divergent harmonic series. From the Comparison Test for Divergence, we conclude that the series diverges.

21.   We observe that $2n^{3/2} + 3 < 2n^{3/2} + 3n^{3/2}$ for every n. Therefore for large n the following is

true: $\dfrac{n+2}{2n^{3/2}+3} > \dfrac{n}{5n^{3/2}} = \dfrac{1}{5n^{1/2}}$. We note that $\displaystyle\sum_{n=1}^{\infty}\dfrac{1}{5n^{1/2}}$ is a divergent p-series, because

$p = \dfrac{1}{2} < 1$. From the Comparison Test for Divergence, we conclude that the series diverges.

23.   We first rewrite $\dfrac{3^n+n}{\sqrt{n}3^n+1} = \dfrac{1+n/3^n}{\sqrt{n}+n^1/3^n}$. We then observe that $1 + \dfrac{n}{3^n} > 1$ and $\sqrt{n} + \dfrac{1}{3^n} < 2\sqrt{n}$

for every n. Therefore, $\dfrac{3^n+n}{\sqrt{n}3^n+1} > \dfrac{1}{2\sqrt{n}}$. We note that $\displaystyle\sum_{n=1}^{\infty}\dfrac{1}{2n^{1/2}}$ is a divergent p-series because

$p = \dfrac{1}{2} < 1$. From the Comparison Test for Divergence, we conclude that the series diverges.

25.   The n-th summand a_n approaches $\left(\dfrac{2}{3}\right)^n$ for large n. We apply the Limit Comparison Test to

the series $\sum\limits_{n=1}^{\infty} b_n = \sum\limits_{n=1}^{\infty} \left(\dfrac{2}{3}\right)^n$ and compute as follows:

$$\lim_{n\to\infty}\left(\frac{a_n}{b_n}\right) = \lim_{n\to\infty} \frac{\left(2^n + 11\right)\big/\left(3^n - 1\right)}{\left(2/3\right)^n} = \lim_{n\to\infty}\left(\frac{1 + 11/2^n}{1 - 1/3^n}\right) = 1 \, .$$

The limit is finite and positive and $\sum\limits_{n=1}^{\infty} b_n$ converges as a geometric series with $r < 1$; the

series $\sum\limits_{n=1}^{\infty} a_n$ converges.

27.   The n-th summand a_n approaches $\left(\dfrac{2}{3}\right)^n$ for large n. We apply the Limit Comparison Test to

the series $\sum\limits_{n=1}^{\infty} b_n = \sum\limits_{n=1}^{\infty} \left(\dfrac{2}{3}\right)^n$ and compute as follows:

$$\lim_{n\to\infty}\left(\frac{a_n}{b_n}\right) = \lim_{n\to\infty} \frac{\left(2^n + n^3\right)\big/\left(3^n + n^2\right)}{\left(2/3\right)^n} = \lim_{n\to\infty}\left(\frac{1 + n^3/2^n}{1 + n^2/3^n}\right) = 1 \, .$$

The limit is finite and positive and $\sum\limits_{n=1}^{\infty} b_n$ converges as a geometric sequence with $r < 1$; the

series $\sum\limits_{n=1}^{\infty} a_n$ converges.

29.   The n-th summand a_n approaches $\dfrac{1}{n^2}$ for large n. We apply the Limit Comparison Test to the

series $\sum\limits_{n=1}^{\infty} b_n = \sum\limits_{n=1}^{\infty} \dfrac{1}{n^2}$ and compute using L'Hôpital's rule as follows:

$$\lim_{n\to\infty}\left(\frac{a_n}{b_n}\right) = \lim_{n\to\infty} \frac{\left(n + \ln\left(n\right)\right)\big/n^3}{1/n^2} = \lim_{n\to\infty}\left(1 + \frac{\ln\left(n\right)}{n}\right) = 1 \, .$$

The limit is finite and positive and $\sum\limits_{n=1}^{\infty} b_n$ converges as a p-series with $p > 1$; the series $\sum\limits_{n=1}^{\infty} a_n$

converges.

31. The n-th summand a_n approaches $\dfrac{1}{n^2}$ for large n. We apply the Limit Comparison Test to the

series $\displaystyle\sum_{n=1}^{\infty} b_n = \sum_{n=1}^{\infty} \dfrac{1}{n^2}$ and compute as follows:

$$\lim_{n\to\infty}\left(\frac{a_n}{b_n}\right) = \lim_{n\to\infty}\frac{n^2/\left(n^2+1\right)^2}{1/n^2} = \lim_{n\to\infty}\left(\frac{1}{1+2/n^2+1/n^4}\right) = 1 \,.$$

The limit is finite and positive and $\displaystyle\sum_{n=1}^{\infty} b_n$ converges as a p-series with $p>1$; the series $\displaystyle\sum_{n=1}^{\infty} a_n$

converges.

Further Theory and Practice

33. We rewrite the terms of the series as $\dfrac{\ln\left(n^3\right)}{n^3} = \dfrac{3\ln(n)}{n^3} = \dfrac{3\ln(n)}{n^{1/2}}\cdot\dfrac{1}{n^{5/2}}$. Using formula (9.8),

we deduce that there is some integer M such that $\dfrac{\ln(x)}{x^{1/2}} \le 1$ for $x \ge M$. Thus, for $n \ge M$ we

have $\dfrac{\ln\left(n^3\right)}{n^3} < \dfrac{3}{n^{5/2}}$. This p-series is convergent, because $p = \dfrac{5}{2} > 1$. From the Comparison

Test for Convergence, we conclude that the series converges.

35. We observe that $2 + \sin(n) \ge 1$ for every n. Therefore, $\dfrac{2+\sin(n)}{\sqrt{n}} > \dfrac{1}{\sqrt{n}}$. We note that $\displaystyle\sum_{n=1}^{\infty}\dfrac{1}{\sqrt{n}}$

is a divergent p-series, because $p = \dfrac{1}{2} < 1$. From the Comparison Test for Divergence, we

conclude that the series diverges.

37. We apply the Limit Comparison Test to series $\displaystyle\sum_{n=1}^{\infty} b_n = \sum_{n=1}^{\infty}\dfrac{1}{\sqrt{n}}$ and compute as follows:

$$\lim_{n\to\infty}\left(\frac{a_n}{b_n}\right) = \lim_{n\to\infty}\frac{(n+2)/\left(2n^{3/2}+3\right)}{1/n^{1/2}} = \lim_{n\to\infty}\left(\frac{1+2/n}{2+3/n^{3/2}}\right) = \frac{1}{2}\,.$$

Given that the limit is finite and positive and $\displaystyle\sum_{n=1}^{\infty} b_n$ diverges as a p-series with $p \le 1$, we see

that the series diverges.

39. We apply the Limit Comparison Test to the series $\sum_{n=1}^{\infty} b_n = \sum_{n=1}^{\infty} \frac{1}{n}$ and compute as follows:

$$\lim_{n \to \infty} \left(\frac{a_n}{b_n} \right) = \lim_{n \to \infty} \left(\frac{\sin(1/n)}{1/n} \right) = 1.$$

The limit is finite and positive and $\sum_{n=1}^{\infty} b_n$ diverges as a harmonic series; the series $\sum_{n=1}^{\infty} a_n$ diverges.

41. We apply the Limit Comparison Test to the series $\sum_{n=1}^{\infty} b_n = \sum_{n=1}^{\infty} \frac{1}{n}$ and compute as follows:

$$\lim_{n \to \infty} \left(\frac{a_n}{b_n} \right) = \lim_{n \to \infty} \left(\frac{(1+1/n)/n}{1/n} \right) = \lim_{n \to \infty} \left(1 + \frac{1}{n} \right) = 1.$$

The limit is finite and positive and $\sum_{n=1}^{\infty} b_n$ diverges as a harmonic series; the series $\sum_{n=1}^{\infty} a_n$ diverges.

43. We apply the Limit Comparison Test to the series $\sum_{n=1}^{\infty} b_n = \sum_{n=1}^{\infty} \frac{1}{n^2}$ and compute as follows:

$$\lim_{n \to \infty} \left(\frac{a_n}{b_n} \right) = \lim_{n \to \infty} \left(\frac{\tan(1/n)/(1+n)}{1/n^2} \right) = \lim_{n \to \infty} \left(\left(\frac{\tan(1/n)}{n} \right) \cdot \left(\frac{n}{1+n} \right) \right) = 1.$$

The limit is finite and positive and $\sum_{n=1}^{\infty} b_n$ converges as a p-series with $p > 1$; the series $\sum_{n=1}^{\infty} a_n$ converges.

45. We apply the Limit Comparison Test to the series $\sum_{n=1}^{\infty} b_n = \sum_{n=1}^{\infty} \frac{1}{n}$ and compute as follows:

$$\lim_{n \to \infty} \left(\frac{a_n}{b_n} \right) = \lim_{n \to \infty} \left(\frac{\arctan(n)/n}{1/n} \right) = \lim_{n \to \infty} \arctan(n) = \frac{\pi}{2}.$$

The limit is finite and positive and $\sum_{n=1}^{\infty} b_n$ diverges as a harmonic series; the original series diverges.

47. Let $\lim\limits_{n\to\infty}\left(\dfrac{a_n}{b_n}\right)=0$, then for every $\epsilon>0$ and sufficiently large n, the inequality $\dfrac{a_n}{b_n}<\epsilon$ holds,

so that $0<a_n<\epsilon b_n$. The series $\sum\limits_{n=1}^{\infty}b_n$ converges. From the Comparison Test, we obtain the

convergence of the series $\sum\limits_{n=1}^{\infty}a_n$.

49. To show the convergence of $\sum\limits_{n=1}^{\infty}a_n^2$ compute as follows: $\lim\limits_{n\to\infty}\left(\dfrac{a_n^2}{a_n}\right)=\lim\limits_{n\to\infty}a_n=0$, since if

$\lim\limits_{n\to\infty}a_n\neq0$ then $\sum\limits_{n=1}^{\infty}a_n$ would diverge by the Divergence Test. We note that $\sum\limits_{n=1}^{\infty}a_n$ converges

and according to Exercise 9.4.47, $\sum\limits_{n=1}^{\infty}a_n^2$ also converges.

51. If $p=0$, the series becomes $\sum\limits_{n=1}^{\infty}r^n$ and converges if $0<r<1$. To determine the convergence

of $\sum\limits_{n=1}^{\infty}n^p r^n$ if $p\neq0$, we apply the Limit Comparison Test to the series $\sum\limits_{n=1}^{\infty}\left(r^n\right)^2$ and compute

as follows:

$$\lim_{n\to\infty}\left(\frac{n^p r^n}{\left(r^n\right)^2}\right)=\lim_{n\to\infty}\left(\frac{n^p}{r^n}\right).$$

Let $p<0$. Then $\lim\limits_{n\to\infty}\left(\dfrac{n^p}{r^n}\right)=0$. If $p>0$, calculate this limit using L'Hôpital's Rule to obtain:

$$\lim_{x\to\infty}\left(\frac{x^p}{r^x}\right)=\lim_{x\to\infty}\left(\frac{\frac{d}{dx}x^p}{\frac{d}{dx}r^x}\right)=\frac{px^{p-1}}{r^x\ln(r)}.$$

We apply this Rule M times, where M is such that $p-M\leq0$. Then

$$\lim_{x\to\infty}\left(\frac{x^p}{r^x}\right)=\lim_{x\to\infty}\left(\frac{\frac{d}{dx}x^p}{\frac{d}{dx}r^x}\right)=p(p-1)\ldots(p-m)\frac{x^{p-m}}{r^x\ln^m(r)}.$$

If $p-M=0$, we have: $\lim\limits_{n\to\infty}\left(\dfrac{n^p}{r^n}\right)=0$. If $p-M<0$, we have case (a) again and

$$\lim_{n\to\infty}\left(\frac{n^p}{r^n}\right)=0.$$

If $0 < r < 1$, then $\sum_{n=1}^{\infty} (r^n)$ converges. So, by Exercise 9.4.49, $\sum_{n=1}^{\infty} (r^n)^2$ also converges, and according to Exercise 9.4.47 $\sum_{n=1}^{\infty} n^p r^n$ converges as well.

53.   Let

$$\frac{n!}{n^n} = \frac{1}{n} \cdot \frac{2}{n} \cdot \frac{3}{n} \cdot \ldots \cdot \frac{n-1}{n} \cdot \frac{n}{n} < \frac{1}{n} \cdot \frac{2}{n} = \frac{2}{n^2}$$

for any $n > 3$. The series $\sum_{n=1}^{\infty} \frac{2}{n^2}$ converges as a p-series with $p = 2 > 1$; therefore, the series $\sum_{n=1}^{\infty} \frac{n!}{n^n}$ also converges.

Calculator/Computer Exercises

55.   The inequality $n^{100} \cdot 0.99^n > 0.995^n$ is equivalent to $100 \ln(n) + n \ln(0.99) > n \ln(0.995)$, or

$$100 \ln(n) + n \ln\left(\frac{0.99}{0.995}\right) > 0 .$$

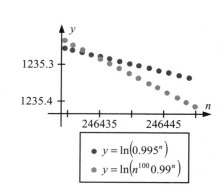

$\bullet \ y = \ln(0.995^n)$
$\bullet \ y = \ln(n^{100} 0.99^n)$

Using Maple we find $n > 246435$, i.e. $M = 246435$.

Therefore $\sum_{n=M}^{\infty} n^{100} \cdot 0.99^n < \sum_{n=M}^{\infty} 0.995^n$ and $\sum_{n=M}^{\infty} 0.995^n$

is a convergent geometric series, because $r = 0.995 < 1$. From the Comparison Test for

Convergence, we conclude that the series $\sum_{n=M}^{\infty} n^{100} \cdot 0.99^n$ converges and

$$\sum_{n=M}^{\infty} 0.995^n = \frac{1}{1-0.995} - \frac{(0.995)^M}{0.995-1} = 200(0.995)^M .$$

Problems for Practice

1. The hypotheses of the Alternating Series Test are satisfied: $\lim\limits_{n\to\infty}\left(\dfrac{1}{n^3+1}\right)=0$;

$$\frac{1}{n^3+1}>\frac{1}{\left(n+1\right)^3+1}$$

for $n\geq 1$.

3. The hypotheses of the Alternating Series Test are satisfied: $\lim\limits_{n\to\infty}\left(\dfrac{1}{\sqrt{n^2+1}}\right)=0$;

$$\frac{1}{\sqrt{n^2}}>\frac{1}{\sqrt{\left(n+1\right)^2+1}}$$

for $n\geq 1$.

5. The hypotheses of the Alternating Series Test are satisfied: $\lim\limits_{n\to\infty}\left(\dfrac{2}{3}\right)^n=0$; $\left(\dfrac{2}{3}\right)^n>\left(\dfrac{2}{3}\right)^{n+1}$ for $n\geq 1$.

7. The hypotheses of the Alternating Series Test are satisfied:

$$\sum_{n=1}^{\infty}\left(-\frac{4}{5}\right)^n\cdot\frac{1}{n+2}=\sum_{n=1}^{\infty}(-1)^n\cdot\left(\frac{4}{5}\right)^n\cdot\frac{1}{n+2};$$

$\lim\limits_{n\to\infty}\left(\left(\dfrac{4}{5}\right)^n\cdot\dfrac{1}{n+2}\right)=0$; $\left(\dfrac{4}{5}\right)^n\cdot\dfrac{1}{n+2}>\left(\dfrac{4}{5}\right)^{n+1}\cdot\dfrac{1}{n+3}$ for $n\geq 1$.

9. The hypotheses of the Alternating Series Test are satisfied: $\lim\limits_{n\to\infty}\left(\dfrac{n^2}{e^n}\right)=0$;

$$\frac{d}{dx}\left(\frac{x^2}{e^x}\right)=\frac{e^x\left(2x-x^2\right)}{e^{2x}}<0$$

when $x>2$, so $\dfrac{n^2}{e^n}>\dfrac{\left(n+1\right)^2}{e^{n+1}}$ for $n\geq 2$.

11.   The hypotheses of the Alternating Series Test are satisfied:

$$\sum_{n=1}^{\infty} \cos(\pi n) \cdot \sin\left(\frac{\pi}{n}\right) = \sum_{n=1}^{\infty} (-1)^n \cdot \sin\left(\frac{\pi}{n}\right);$$

$$\lim_{n\to\infty}\left(\sin\left(\frac{\pi}{n}\right)\right) = 0 \; ; \; \sin\left(\frac{\pi}{n}\right) > \sin\left(\frac{\pi}{n+1}\right) \text{ for } n \geq 2 \, .$$

13.   Given that $\dfrac{1}{n^3+1} < \dfrac{1}{n^3}$ for $n \geq 1$ and the series $\displaystyle\sum_{n=1}^{\infty} \dfrac{1}{n^3}$ converges, the series $\displaystyle\sum_{n=1}^{\infty} \dfrac{1}{n^3+1}$ also

converges. The series $\displaystyle\sum_{n=1}^{\infty} \dfrac{(-1)^n}{n^3+1}$ converges absolutely.

15.   The series $\displaystyle\sum_{n=1}^{\infty} \dfrac{(-1)^n n}{n^2+1}$ converges per Exercise 9.5.8. Given that $\displaystyle\lim_{n\to\infty}\left(\dfrac{n/(n^2+1)}{1/n}\right) = 1$ and the

series $\displaystyle\sum_{n=1}^{\infty} \dfrac{1}{n}$ is the divergent harmonic series, the series $\displaystyle\sum_{n=1}^{\infty} \dfrac{n}{n^2+1}$ also diverges. Therefore the

series $\displaystyle\sum_{n=1}^{\infty} \dfrac{(-1)^n n}{n^2+1}$ converges conditionally.

17.   We have $\displaystyle\lim_{n\to\infty}\left(\dfrac{1}{\ln(n)}\right) = 0$ and $\dfrac{1}{\ln(n)} > \dfrac{1}{\ln(n+1)}$ for $n \geq 2$. Therefore the series $\displaystyle\sum_{n=2}^{\infty} \dfrac{(-1)^n}{\ln(n)}$

converges. Given that $\dfrac{1}{\ln(n)} > \dfrac{1}{n}$ for $n \geq 2$ and the series $\displaystyle\sum_{n=2}^{\infty} \dfrac{1}{n}$ diverges, the series $\displaystyle\sum_{n=2}^{\infty} \dfrac{1}{\ln(n)}$

also diverges. The series $\displaystyle\sum_{n=2}^{\infty} \dfrac{(-1)^n}{\ln(n)}$ converges conditionally.

19.   Given that $\displaystyle\lim_{n\to\infty}\left(1-\dfrac{1}{n}\right)^n = \lim_{n\to\infty}\left(\left(1-\dfrac{1}{n}\right)^{-n}\right)^{-1} = e^{-1} \neq 0$, so $(-1)^n\left(1-\dfrac{1}{n}\right)^n$ does not tend to 0, the

series $\displaystyle\sum_{n=1}^{\infty} (-1)^n\left(1-\dfrac{1}{n}\right)^n$ diverges.

21. Let $\lim_{n\to\infty}\left(\dfrac{n+\ln(n)}{n^{3/2}}\right)=0$. We compute $\dfrac{d}{dx}\left(\dfrac{x+\ln(x)}{x^{3/2}}\right)=\dfrac{2-x-3\ln(x)}{x^{5/2}}<0$ when $x>2$, so

$\dfrac{n+\ln(n)}{n^{3/2}}>\dfrac{n+1+\ln(n+1)}{(n+1)^{3/2}}$ for $n\geq 2$. Therefore, the series $\sum_{n=1}^{\infty}(-1)^{n+1}\dfrac{n+\ln(n)}{n^{3/2}}$ converges.

Given that $\lim_{n\to\infty}\left(\dfrac{(n+\ln(n))/n^{3/2}}{1/\sqrt{n}}\right)=1$ and the series $\sum_{n=1}^{\infty}\dfrac{1}{\sqrt{n}}$ diverges, the series $\sum_{n=1}^{\infty}\dfrac{n+\ln(n)}{n^{3/2}}$

also diverges. The series $\sum_{n=1}^{\infty}(-1)^{n+1}\dfrac{n+\ln(n)}{n^{3/2}}$ converges conditionally.

23. Given that $\left|\dfrac{\sin(n)}{e^n}\right|\leq\dfrac{1}{e^n}$ for $n>1$ and the series $\sum_{n=1}^{\infty}\dfrac{1}{e^n}$ converges, the series $\sum_{n=1}^{\infty}\dfrac{\sin(n)}{e^n}$

converges absolutely.

25. As $|a_{101}|=\dfrac{1}{101}<0.01$, $M=100$.

27. As $|a_{201}|=\dfrac{4}{401}<0.01$, $M=200$.

29. As $a_6=\dfrac{1}{126}<0.01$, $M=5$.

Further Theory and Practice

31. As $\dfrac{d}{dx}\left(\dfrac{1}{x^2-20x+101}\right)=\dfrac{20-2x}{\left(x^2-20x+101\right)^2}<0$ for $x>10$, we note that

$a_n=\dfrac{1}{n^2-20n+101}$ decreases for $n\geq 10$. The Alternating Series Test may be applied to the

tail $\sum_{n=M}^{\infty}(-1)^n a_n$ with $M=10$.

33. As $\dfrac{d}{dn}\left(\dfrac{n^5}{2^n}\right)=\left(\dfrac{n^4}{2^n}\right)(5-n\ln(2))$, we see that $a_n=\dfrac{n^5}{2^n}$ decreases for $n\geq\left\lfloor\dfrac{5}{\ln(2)}\right\rfloor+1=8$.

L'Hôpital's Rule shows that $a_n\to 0$. The Alternating Series Test may be applied to the tail

$\sum_{n=M}^{\infty}(-1)^n a_n$ with $M=8$.

35. Since $\dfrac{1}{7!} < 5 \times 10^{-4}$, the partial sum $\displaystyle\sum_{n=1}^{6} \dfrac{(-1)^n}{n!}$ approximates $\displaystyle\sum_{n=1}^{\infty} \dfrac{(-1)^n}{n!}$ to three decimal places.

Thus, $\displaystyle\sum_{n=1}^{\infty} \dfrac{(-1)^n}{n!} \approx -0.632$.

37. $\displaystyle\sum_{n=1}^{\infty} (-1)^n \dfrac{n!}{(2n)!} \approx \sum_{n=1}^{4} (-1)^n \dfrac{n!}{(2n)!} = -0.424$

Calculator/Computer Exercises

39.

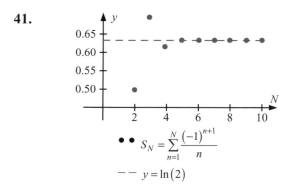

$$\bullet\bullet \quad S_N = \sum_{n=1}^{N} \frac{(-1)^{n+1}}{n}$$

$$-- \quad y = \ln(2)$$

41.

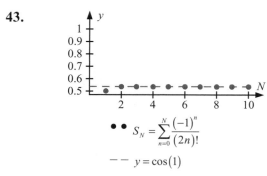

$$\bullet\bullet \quad S_N = \sum_{n=1}^{N} \frac{(-1)^{n+1}}{n}$$

$$-- \quad y = \ln(2)$$

43.

$$\bullet\bullet \quad S_N = \sum_{n=0}^{N} \frac{(-1)^n}{(2n)!}$$

$$-- \quad y = \cos(1)$$

45.   As $\dfrac{d}{dx}\left(\dfrac{9x^2+13}{x^3+55x+60}\right)=\dfrac{-9x^4+456x^2+1080x-715}{\left(x^3+55x+60\right)^2}<0$ for $x>8.024$, we see that

$$a_n=\frac{9n^2+13}{n^3+55n+60}$$

decreases for $n\geq 9$. The Alternating Series Test may be applied to the tail $\displaystyle\sum_{n=M}^{\infty}(-1)^n\,a_n$ with $M=9$.

47.   As $\dfrac{d}{dx}\left(\dfrac{100x^{9/4}+x}{150+x^{5/2}}\right)=\dfrac{-50x^{15/4}-3x^{5/2}+67500x^{5/4}+300}{2\left(150+x^{5/2}\right)^2}<0$ for $x>17.859$, we see that

$$a_n=\frac{100n^{9/4}+n}{150+n^{5/2}}$$

decreases for $n\geq 18$. The Alternating Series Test may be applied to the tail $\displaystyle\sum_{n=M}^{\infty}(-1)^n\,a_n$ with $M=18$.

Section 9.6   The Ratio and Root Tests

Problems for Practice

1.   We apply the Ratio Test to find:

$$L=\lim_{n\to\infty}\left|\frac{a_{n+1}}{a_n}\right|=\lim_{n\to\infty}\left(\frac{(n+1)/e^{n+1}}{n/e^n}\right)=\lim_{n\to\infty}\left(\frac{1+1/n}{e}\right)=\frac{1}{e}<1.$$

The series converges.

3.   We apply the Ratio Test to find:

$$L=\lim_{n\to\infty}\left|\frac{a_{n+1}}{a_n}\right|=\lim_{n\to\infty}\left(\frac{2^{n+1}/(n+1)^3}{2^n/n^3}\right)=\lim_{n\to\infty}\left(\frac{2}{(1+1/n)^3}\right)=2>1.$$

The series diverges.

5. We apply the Ratio Test to find:

$$L = \lim_{n \to \infty} \left| \frac{a_{n+1}}{a_n} \right| = \lim_{n \to \infty} \left(\frac{(n+1)^{100}/(n+1)!}{n^{100}/n!} \right) = \lim_{n \to \infty} \left(\frac{(1+1/n)^{100}}{n+1} \right) = 0 < 1.$$

The series converges.

7. We apply the Ratio Test to find:

$$L = \lim_{n \to \infty} \left| \frac{a_{n+1}}{a_n} \right| = \lim_{n \to \infty} \left(\frac{(n+1)!/((n+1)3^{n+1})}{n!/(n3^n)} \right) = \lim_{n \to \infty} \left(\frac{n}{3} \right) = \infty > 1.$$

The series diverges.

9. We apply the Ratio Test to find:

$$L = \lim_{n \to \infty} \left| \frac{a_{n+1}}{a_n} \right| = \lim_{n \to \infty} \left(\frac{2^{n+1}\sqrt{n+1}/3^{n+1}}{2^n\sqrt{n}/3^n} \right) = \lim_{n \to \infty} \frac{2\sqrt{1+1/n}}{3} = \frac{2}{3} < 1.$$

The series converges.

11. We first apply the Ratio Test to find:

$$L = \lim_{n \to \infty} \left| \frac{a_{n+1}}{a_n} \right| = \lim_{n \to \infty} \left(\frac{1/\ln\left((n+1)^2\right)}{1/\ln\left(n^2\right)} \right) = \lim_{n \to \infty} \left(\frac{\ln\left(n^2\right)}{\ln\left((n+1)^2\right)} \right) = 1.$$

The Ratio Test tells us nothing about the convergence of the series. Next, we apply the Comparison Test for Divergence and find: $|a_n| = \dfrac{1}{\ln\left(n^2\right)} = \dfrac{1}{2\ln(n)} > \dfrac{1}{2n}$ for $n > 2$. Given that $\displaystyle\sum_{n=1}^{\infty} \frac{1}{2n}$ diverges, we see that $\displaystyle\sum_{n=2}^{\infty} |a_n|$ also diverges, and that the series does not converge absolutely. Finally, we apply the Alternating Series Test to obtain: $a_2 \geq a_3 \geq a_4 \ldots$ and $\displaystyle\lim_{n \to \infty} \left(\frac{1}{\ln\left(n^2\right)} \right) = 0$. From this test, we conclude the series converges conditionally.

13.   We rewrite $a_n = \dfrac{(-3)^n}{n^3 + 3^n} = \dfrac{(-1)^n}{n^3/3^n + 1}$. First, we apply the Ratio Test to find:

$$L = \lim_{n\to\infty} \left| \frac{a_{n+1}}{a_n} \right| = \lim_{n\to\infty} \frac{1/\left((n+1)^3/3^{n+1} + 1\right)}{1/\left(n^3/3^n + 1\right)} = \lim_{n\to\infty} \left(\frac{n^3/3^n + 1}{(n+1)^3/3^{n+1} + 1} \right) = 1.$$

The Ratio Test tells us nothing about the convergence of the series. Next, we apply the Divergence Test to obtain: $\lim\limits_{n\to\infty} |a_n| = \lim\limits_{n\to\infty} \left(\dfrac{1}{n^3/3^n + 1} \right) = 1 \neq 0$. The series diverges.

15.   We first apply the Ratio Test to find:

$$L = \lim_{n\to\infty} \left| \frac{a_{n+1}}{a_n} \right| = \lim_{n\to\infty} \frac{\ln(n+1)/(n+1)^2}{\ln(n)/n^2} = \lim_{n\to\infty} \left(\frac{\ln(n+1)/\ln(n)}{(1+1/n)^2} \right) \lim_{n\to\infty} \frac{1+1/(n+1)}{1+1/(n+1)^3} = 1.$$

by L'Hôpital's Rule. The Ratio Test tells us nothing about the convergence of the series. From Example 8 of Section 9.4, we see that the series $\sum\limits_{n=3}^{\infty} \left(\dfrac{\ln(n)}{n^2} \right)$ converges. We conclude that $\sum\limits_{n=2}^{\infty} |a_n|$ converges and the original series converges absolutely.

17.   We apply the Root Test to find:

$$L = \lim_{n\to\infty} |a_n|^{1/n} = \lim_{n\to\infty} \left(\frac{10^n}{n^{10}} \right)^{1/n} = \lim_{n\to\infty} \left(\frac{10}{n^{10/n}} \right) = 10 \lim_{n\to\infty} \exp\left(\frac{-10\ln(n)}{n} \right) = 10e^0 = 10 > 1.$$

The series diverges.

19.   We apply the Root Test to find: $L = \lim\limits_{n\to\infty} |a_n|^{1/n} = \lim\limits_{n\to\infty} \left(\left(\dfrac{37}{n} \right)^n \right)^{1/n} = \lim\limits_{n\to\infty} \left(\dfrac{37}{n} \right) = 0 < 1$. The series (absolutely) converges.

21.   We apply the Root Test to find:

$$L = \lim_{n\to\infty} |a_n|^{1/n} = \lim_{n\to\infty} \left(\left(\frac{n^2 + 7n + 13}{2n^2 + 1} \right)^n \right)^{1/n} = \lim_{n\to\infty} \left(\frac{n^2 + 7n + 13}{2n^2 + 1} \right) = \lim_{n\to\infty} \left(\frac{1 + 7/n + 13/n^2}{2 + 1/n^2} \right)$$

$$= \frac{1}{2} < 1.$$

The series (absolutely) converges.

Further Theory and Practice

23.   We apply the Ratio Test to find:

$$L = \lim_{n\to\infty} \left| \frac{a_{n+1}}{a_n} \right| = \lim_{n\to\infty} \left(\frac{(2(n+1))!/(3(n+1))!}{(2n)!/(3n)!} \right) = \lim_{n\to\infty} \left(\frac{(2n+1)(2n+2)}{(3n+1)(3n+2)(3n+3)} \right) = 0 < 1.$$

The series converges.

25.   We apply the Ratio Test to find:

$$L = \lim_{n\to\infty} \left| \frac{a_{n+1}}{a_n} \right| = \lim_{n\to\infty} \left(\frac{(2(n+1))!/((n+1)! \cdot 2^{n+1})}{(2n)!/(n! \cdot 2^n)} \right) = \lim_{n\to\infty} (2n+1) \frac{2n+2}{2(n+1)} = \lim_{n\to\infty} (2n+1)$$

$$= \infty > 1.$$

The series diverges.

27.   We apply the Ratio Test to find:

$$L = \lim_{n\to\infty} \left| \frac{a_{n+1}}{a_n} \right| = \lim_{n\to\infty} \left(\frac{((n+1) + 3^{n+1})/((n+1)^3 + 2^{n+1})}{(n + 3^n)/(n^3 + 2^n)} \right)$$

$$= \lim_{n\to\infty} \left(\frac{3^{n+1}((n+1)/3^{n+1} + 1)/(2^{n+1}((n+1)^3/2^{n+1} + 1))}{3^n(n/3^n + 1)/(2^n(n^3/2^n + 1))} \right) = \frac{3}{2} > 1.$$

The series diverges.

29.   We apply the Root Test to find:

$$L = \lim_{n \to \infty} |a_n|^{1/n} = \lim_{n \to \infty} \left(\left(\frac{4}{n^{1/n} + 2} \right)^n \right)^{1/n} = \lim_{n \to \infty} \left(\frac{4}{n^{1/n} + 2} \right) = \frac{4}{3} > 1 \, .$$

The series diverges.

31.   We apply the Root Test to find:

$$L = \lim_{n \to \infty} |a_n|^{1/n} = \lim_{n \to \infty} \left(\frac{2^n}{1 + \ln^n (n)} \right)^{1/n} = \lim_{n \to \infty} \left(\frac{2}{\left(1 + \ln^n (n) \right)^{1/n}} \right) = 0 < 1 \, .$$

The series converges.

33.   We first apply the Ratio Test to find:

$$L = \lim_{n \to \infty} \left| \frac{a_{n+1}}{a_n} \right| = \lim_{n \to \infty} \left(\frac{1/\sqrt{(n+1) + 10}}{1/\sqrt{n+10}} \right) = \lim_{n \to \infty} \sqrt{\frac{n+10}{n+11}} = 1 \, .$$

The Ratio Test tells us nothing about the convergence of the series. Next, we apply the Limit Comparison Test and find: $\lim_{n \to \infty} \left(\frac{1/\sqrt{n+10}}{1/\sqrt{n}} \right) = \lim_{n \to \infty} \sqrt{\frac{n}{n+10}} = 1$. Given that the limit is finite and positive and $\sum_{n=1}^{\infty} \frac{1}{\sqrt{n}}$ diverges, we see that $\sum_{n=2}^{\infty} |a_n|$ diverges as well, and the series does not converge absolutely. Finally, we apply the Alternating Series Test to obtain: $a_1 \geq a_2 \geq a_3 \ldots$ and $\lim_{n \to \infty} \left(\frac{1}{\sqrt{n+10}} \right) = 0$. From this test, we conclude that the series converges. Since it does not converge absolutely, it must converge conditionally.

35.   We apply the Ratio Test and find: $L = \lim_{n \to \infty} \left| \frac{a_{n+1}}{a_n} \right| = \lim_{n \to \infty} \frac{(n+1)!/3^{n+1}}{n!/3^n} = \lim_{n \to \infty} \left(\frac{n+1}{3} \right) = \infty > 1 \, .$

From this test, we conclude that the series diverges.

37. We first apply the Ratio Test to find:

$$L = \lim_{n\to\infty}\left|\frac{a_{n+1}}{a_n}\right| = \lim_{n\to\infty}\left(\frac{(n+1)/\sqrt{(n+1)^2-11}}{n/\sqrt{n^2-11}}\right) = \lim_{n\to\infty}\left(\left(1+\frac{1}{n}\right)\left(\frac{\sqrt{n^2-11}}{\sqrt{(n+1)^2-11}}\right)\right) = 1.$$

The Ratio Test tells us nothing about the convergence of the series. Next, we apply the Comparison Test for Divergence and find: $\dfrac{n}{n^2-11} > \dfrac{n}{n^2} = \dfrac{1}{n}$. Given that the series $\displaystyle\sum_{n=1}^{\infty}\frac{1}{n}$ diverges, then $\displaystyle\sum_{n=2}^{\infty}|a_n|$ diverges as well, and the series does not converge absolutely. Finally, we apply the Alternating Series Test to obtain: $a_3 \ge a_4 \ge a_5 \ldots$ and $\displaystyle\lim_{n\to\infty}\left(\frac{n}{n^2+10}\right) = 0$. From this test, we conclude the series converges. Since it does not converge absolutely it must converge conditionally.

39. We first apply the Ratio Test to find:

$$L = \lim_{n\to\infty}\left|\frac{a_{n+1}}{a_n}\right| = \lim_{n\to\infty}\left(\frac{(n+1)^{1/(n+1)}/\ln(n+1)}{n^{1/n}/\ln(n)}\right) = \frac{n^{1/n}/\ln(n)}{(n+1)^{1/(n+1)}/\ln(n+1)} = 1$$

The Ratio Test tells us nothing about the convergence of the series. Next, we apply the Limit Comparison Test for Divergence and find: $\displaystyle\lim_{n\to\infty}\left(\frac{n^{1/n}/\ln(n)}{1/\ln(n)}\right) = \lim_{n\to\infty} n^{1/n} = 1$. Given that the limit is finite and positive and the series $\displaystyle\sum_{n=1}^{\infty}\frac{1}{\ln(n)}$ diverges, we see that $\displaystyle\sum_{n=2}^{\infty}|a_n|$ diverges as well, so the original series does not converge absolutely. Finally, we apply the Alternating Series Test to obtain: $a_2 \ge a_3 \ge a_4 \ldots$ and $\displaystyle\lim_{n\to\infty}\left(\frac{n^{1/n}}{\ln(n)}\right) = 0$. From this test we conclude that the series converges. Since it does not converge absolutely, it must converge conditionally.

41. We apply the Test for Divergence to find: $\displaystyle\lim_{n\to\infty}\left(\frac{n^{1/n}}{1+1/n}\right) = \frac{\displaystyle\lim_{n\to\infty} n^{1/n}}{\displaystyle\lim_{n\to\infty}(1+1/n)} = 1 \ne 0$. The series diverges.

43.   We first apply the Ratio Test to find:

$$L = \lim_{n \to \infty} \left| \frac{a_{n+1}}{a_n} \right| = \lim_{n \to \infty} \frac{e^{1/(n+1)} / (n+1)^e}{e^{1/n} / n^e} = \lim_{n \to \infty} \left(e^{1/(n+1) - 1/n} \left(\frac{1}{1 + 1/n} \right) \right) = 1.$$

The Ratio Test tells us nothing about the convergence of the series. Next, we apply the Limit Comparison Test and find: $\lim_{n \to \infty} \dfrac{e^{1/n} / n^e}{1/n^e} = \lim_{n \to \infty} e^{1/n} = 1$. Given that the limit is finite and positive and $\sum_{n=1}^{\infty} \dfrac{1}{n^e}$ converges, we see that $\sum_{n=2}^{\infty} |a_n|$ converges as well, and the series converges absolutely.

45.   We first apply the Ratio Test to find:

$$L = \lim_{n \to \infty} \left| \frac{a_{n+1}}{a_n} \right| = \lim_{n \to \infty} \left(\ln \frac{1 + 1/(n+1)}{\ln(1 + 1/n)} \right) = 1.$$

The Ratio Test tells us nothing about the convergence of the series. Next, we apply the Comparison Test for Divergence and find: $\ln\left(1 + \dfrac{1}{n}\right) > \dfrac{1}{2n}$ for $n > 2$. Given that $\sum_{n=1}^{\infty} \dfrac{1}{2n}$ diverges, we see that $\sum_{n=2}^{\infty} |a_n|$ diverges as well, and the series does not converge absolutely.

Finally, we apply the Alternating Series Test to obtain: $a_1 \geq a_2 \geq a_3 \ldots$ and

$$\lim_{n \to \infty} \left(\ln\left(1 + \frac{1}{n} \right) \right) = 0.$$

From this test we conclude that the series converges. Since it does not converge absolutely, it must converge conditionally.

47.   We apply the Root Test to find:

$$L = \lim_{n \to \infty} |a_n|^{1/n} = \lim_{n \to \infty} \left(\left(1 + \frac{1}{n} \right)^n \right)^{1/n} = \lim_{n \to \infty} \left(1 + \frac{1}{n} \right) = 1.$$

The Root Test tells us nothing about the convergence of the series. Next, we apply the Test for Divergence and obtain: $\lim_{n \to \infty} \left(1 + \dfrac{1}{n} \right)^n = e \neq 0$. From this test, we conclude that the series diverges.

49.   We Apply the Ratio Test to the series $\sum\limits_{n=1}^{\infty} n^p a_n$, to obtain

$$\lim_{n\to\infty}\left(\frac{(n+1)^p a_{n+1}}{n^p a_n}\right) = \lim_{n\to\infty}\left(\frac{(n+1)^p}{n^p}\right)\cdot\lim_{n\to\infty}\left(\frac{a_{n+1}}{a_n}\right) = \lim_{n\to\infty}\left(\frac{a_{n+1}}{a_n}\right),$$

so it gives the same result as the Ratio Test applied to the series $\sum\limits_{n=1}^{\infty} a_n$. Therefore, if the terms

of the series can be written in the form $n^p a_n$, one only needs to perform a Ratio Test for the

series $\sum\limits_{n=1}^{\infty} a_n$, which saves a lot of work.

For any p-series we find: $\lim\limits_{n\to\infty}\left(\frac{(n+1)^p}{n^p}\right) = \lim\limits_{n\to\infty}\left(\left(1+\frac{1}{n}\right)^p\right) = 1$, and the Ratio Test tells us

nothing about the convergence of the series.

51. $\lim\limits_{n\to\infty}\left(\frac{\ln^p(n+1)a_{n+1}}{\ln^p(n)a_n}\right) = \lim\limits_{n\to\infty}\left(\frac{\ln^p(n+1)}{\ln^p(n)}\right)\cdot\lim\limits_{n\to\infty}\left(\frac{a_{n+1}}{a_n}\right) = \lim\limits_{n\to\infty}\left(\frac{a_{n+1}}{a_n}\right)$

Calculator/Computer Exercises

53.

$$\bullet\bullet \quad y=\frac{a_{n+1}}{a_n}$$

$$\bullet\bullet \quad y=a_n^{1/2}$$

55.

$$\bullet\bullet \quad y=\frac{a_{n+1}}{a_n}$$

$$\bullet\bullet \quad y=a_n^{1/2}$$

Taylor Series

Section 10.1   Introduction to Power Series

Problems for Practice

1. Given that $|a_n| = 2^n$, we compute $l = \lim_{n \to \infty} \left(2^n \right)^{1/n} = \lim_{n \to \infty} 2 = 2$, so that the radius of

convergence is $R = \dfrac{1}{l} = \dfrac{1}{2}$. Hence the series converges absolutely for $|x| < \dfrac{1}{2}$ and diverges for

$|x| > \dfrac{1}{2}$. When $x = \dfrac{1}{2}$, the series becomes $\displaystyle\sum_{n=0}^{\infty} 1$, which diverges. When $x = -\dfrac{1}{2}$, the series

becomes $\displaystyle\sum_{n=0}^{\infty} (-1)^n$, which does not have a limit as well. Therefore we determine that the

interval of convergence is $\left(-\dfrac{1}{2}, \dfrac{1}{2} \right)$.

3. Given that $|a_n| = n^3$, we compute $l = \lim_{n \to \infty} n^{3/n} = \left(\lim_{n \to \infty} n^{1/n} \right)^3 = 1$, so that the radius of

convergence is $R = \dfrac{1}{l} = 1$. Hence the series converges absolutely for $|x| < 1$ and diverges for

$|x| > 1$. When $x = 1$, the series becomes $\displaystyle\sum_{n=0}^{\infty} n^3$, which diverges. When $x = -1$, the series

becomes $\displaystyle\sum_{n=0}^{\infty} n^3 (-1)^n$, which diverges as well. Therefore we determine that the interval of

convergence is $(-1, 1)$.

5. Given that $|a_n| = \dfrac{1}{3^n}$, we compute $l = \lim_{n \to \infty} \left(\dfrac{1}{3^n} \right)^{1/n} = \dfrac{1}{3}$, so that the radius of convergence is

$R = 3$. Hence the series converges absolutely for $|x| < 3$ and diverges for $|x| > 3$. When $x = 3$,

the series becomes $\displaystyle\sum_{n=0}^{\infty} (-1)^n$, which diverges. When $x = -3$, the series becomes $\displaystyle\sum_{n=0}^{\infty} 1$, which

diverges as well. Therefore we determine that the interval of convergence is $(-3, 3)$.

7.   Given that $|a_n| = n^n$, we compute $l = \lim_{n\to\infty} (n^n)^{1/n} = \infty$, so that the radius of convergence is $R = 0$. The interval of convergence is the one-point set $\{0\}$.

9.   Given that $|a_n| = \left(1 + \dfrac{1}{n}\right)^{n^2}$, we compute $l = \lim_{n\to\infty} \left(\left(1 + \dfrac{1}{n}\right)^{n^2}\right)^{1/n} = \lim_{n\to\infty} \left(1 + \dfrac{1}{n}\right)^n = e$, so that the radius of convergence is $R = 1/e$. Hence the series converges absolutely for $|x| < 1/e$ and diverges for $|x| > 1/e$. When $x = 1/e$, the series becomes $\displaystyle\sum_{n=1}^{\infty} \left(1 + \dfrac{1}{n}\right)^{n^2} \left(\dfrac{1}{e}\right)^n$, which diverges because

$$\lim_{n\to\infty} \left(1 + \frac{1}{n}\right)^{n^2} \left(\frac{1}{e}\right)^n = 1 \neq 0.$$

When $x = -1/e$, the series becomes $\displaystyle\sum_{n=1}^{\infty} (-1)^n \left(1 + \dfrac{1}{n}\right)^{n^2} \left(\dfrac{1}{e}\right)^n$, which diverges for the same reason. Therefore we determine that the interval of convergence is $(-1/e, \, 1/e)$.

11.   Given that $|a_n| = e^n$, we compute $l = \lim_{n\to\infty} (e^n)^{1/n} = e$, so that the radius of convergence is $R = \dfrac{1}{e}$. Hence the series converges absolutely for $|x| < \dfrac{1}{e}$ and diverges for $|x| > \dfrac{1}{e}$. When $x = \dfrac{1}{e}$, the series becomes $\displaystyle\sum_{n=0}^{\infty} 1$, which diverges. When $x = -\dfrac{1}{e}$, the series becomes $\displaystyle\sum_{n=0}^{\infty} (-1)^n$, which diverges as well. Therefore we determine that the interval of convergence is

$$\left(-\frac{1}{e}, \, \frac{1}{e}\right).$$

13.   Given that $|a_n| = \dfrac{3n+5}{n-2}$, we compute

$$l = \lim_{n\to\infty} \left(\frac{3n+5}{n-2}\right)^{1/n} = 1,$$

so that the radius of convergence is $R = 1$. Hence the series converges absolutely for $|x| < 1$ and diverges for $|x| > 1$. When $x = 1$, the series becomes $\displaystyle\sum_{n=3}^{\infty} \dfrac{3n+5}{n-2}$, which diverges because $\lim_{n\to\infty} \left(\dfrac{3n+5}{n-2}\right) = 3 \neq 0$. When $x = -1$, the series becomes $\displaystyle\sum_{n=3}^{\infty} (-1)^n \dfrac{3n+5}{n-2}$, which diverges for the same reason. Therefore we determine that the interval of convergence is $(-1, \, 1)$.

15. Given that $\left|a_n\right| = \dfrac{n^3}{3^n}$, we compute

$$l = \lim_{n\to\infty}\left(\frac{n^3}{3^n}\right)^{1/n} = \frac{1}{3},$$

so that the radius of convergence is $R = 3$. Hence the series converges absolutely for $|x| < 3$ and diverges for $|x| > 3$. When $x = 3$, the series becomes $\displaystyle\sum_{n=1}^{\infty} n^3$, which diverges. When $x = -3$, the series becomes $\displaystyle\sum_{n=1}^{\infty}(-1)^n\, n^3$, which diverges as well. Therefore we determine that the interval of convergence is $(-3,\, 3)$.

17. Given that $\left|a_n\right| = \dfrac{n}{\sqrt{n+1}}$, we compute $l = \lim_{n\to\infty}\left(\dfrac{n}{\sqrt{n+1}}\right)^{1/n} = \dfrac{n^{1/n}}{\sqrt{(n+1)^{1/n}}} = 1$, so that the radius of convergence is $R = 1$. Hence the series converges absolutely for $|x| < 1$ and diverges for $|x| > 1$. When $x = -1$, the series becomes $\displaystyle\sum_{n=0}^{\infty}\frac{n}{\sqrt{n+1}}$, which diverges because $\lim_{n\to\infty}\left(\dfrac{n}{\sqrt{n+1}}\right) = \infty$. When $x = 1$, the series becomes $\displaystyle\sum_{n=0}^{\infty}(-1)^n\,\frac{n}{\sqrt{n+1}}$, which diverges for the same reason. Therefore we determine that the interval of convergence is $(-1,\, 1)$.

19. Given that $\left|a_n\right| = \dfrac{2^n}{n!}$, we compute $l = \lim_{n\to\infty}\left(\dfrac{2^{n+1}/(n+1)!}{2^n/n!}\right) = \lim_{n\to\infty}\left(\dfrac{2}{n+1}\right) = 0$, so that the radius of convergence is $R = \infty$. Therefore we determine that the interval of convergence is $(-\infty,\, \infty)$.

21. Given that $\left|a_n\right| = \dfrac{1}{0.1^n}$, we compute $l = \lim_{n\to\infty}\left(\dfrac{1}{0.1^n}\right)^{1/n} = \dfrac{1}{0.1} = 10$, so that the radius of convergence is $R = 0.1$. Hence the series converges absolutely for $|x+1| < 0.1$ and diverges for $|x+1| > 0.1$. When $x = -0.9$, the series becomes $\displaystyle\sum_{n=0}^{\infty} 1$, which diverges. When $x = -1.1$, the series becomes $\displaystyle\sum_{n=0}^{\infty}(-1)^n$, which diverges as well. Therefore we determine that the interval of convergence is $(-1.1,\, -0.9)$.

23.   The original series can be written in the form $\sum_{n=0}^{\infty} \frac{\left(x-6\sqrt{2}\right)^n}{2^{n/2}}$. Then we compute $|a_n| = \frac{1}{2^{n/2}}$,

$l = \lim_{n \to \infty} \left(\frac{1}{2^{n/2}}\right)^{1/n} = \frac{1}{\sqrt{2}}$, so that the radius of convergence is $R = \sqrt{2}$. Hence the series

converges absolutely for $\left|x - 6\sqrt{2}\right| < \sqrt{2}$ and diverges for $\left|x - 6\sqrt{2}\right| > \sqrt{2}$. When $x = 7\sqrt{2}$, the

series becomes $\sum_{n=0}^{\infty} 1$, which diverges. When $x = 5\sqrt{2}$, the series becomes $\sum_{n=0}^{\infty} (-1)^n$, which

diverges as well. Therefore we determine that the interval of convergence is $\left(5\sqrt{2},\ 7\sqrt{2}\right)$.

25.   Given that $|a_n| = \frac{1}{2n+3}$, we compute $l = \lim_{n \to \infty} \frac{1}{(2n+3)^{1/n}} = 1$, so that the radius of convergence

is $R = 1$. Hence the series converges absolutely for $|x - 3| < 1$ and diverges for $|x - 3| > 1$.

When $x = 4$, the series becomes $\sum_{n=0}^{\infty} \frac{1}{2n+3}$, which diverges by comparison with the harmonic

series $\sum_{n=1}^{\infty} \frac{1}{n}$. When $x = 2$, the series becomes $\sum_{n=0}^{\infty} \frac{(-1)^n}{2n+3}$, which converges by the Alternating

Series Test. Therefore we determine that the interval of convergence is $[2,\ 4)$.

27.   Given that $|a_n| = \frac{n^2}{3n+1}$, we compute $l = \lim_{n \to \infty} \left(\frac{n^2}{3n+1}\right)^{1/n} = 1$, so that the radius of convergence

is $R = 1$. Hence the series converges absolutely for $|x + 5| < 1$ and diverges for $|x + 5| > 1$.

When $x = -6$, the series becomes $\sum_{n=0}^{\infty} \frac{n^2}{3n+1}$, which diverges because $\lim_{n \to \infty} \left(\frac{n^2}{n+1}\right) = \infty$. When

$x = -4$, the series becomes $\sum_{n=0}^{\infty} (-1)^n \frac{n^2}{3n+1}$, which diverges for the same reason. Therefore

we determine that the interval of convergence is $(-6,\ -4)$.

Further Theory and Practice

29.   Using the geometric series expansion for $(1 - \alpha)^{-1}$ we find

$$\frac{1}{1+x} = \frac{1}{1-(-x)} = \sum_{n=0}^{\infty} (-x)^n = \sum_{n=0}^{\infty} (-1)^n x^n.$$

31. Using the geometric series expansion for $(1-\alpha)^{-1}$ we find $\dfrac{1}{1-x^4} = \sum\limits_{n=0}^{\infty} \left(x^4\right)^n = \sum\limits_{n=0}^{\infty} x^{4n}$.

33. Using the geometric series expansion for $(1-\alpha)^{-1}$ we find

$$\frac{x^2}{1+x^2} = x^2 \left(\frac{1}{1-\left(-x^2\right)}\right) = x^2 \sum_{n=0}^{\infty} \left(-x^2\right)^n = x^2 \sum_{n=0}^{\infty} (-1)^n x^{2n} = \sum_{n=0}^{\infty} (-1)^n x^{2n+2} = \sum_{m=1}^{\infty} (-1)^{m-1} x^{2m} .$$

35. Using the geometric series expansion for $(1-\alpha)^{-1}$ we find

$$\frac{3}{3-x} = \frac{1}{1-x/3} = \sum_{n=0}^{\infty} \left(\frac{x}{3}\right)^n = \sum_{n=0}^{\infty} 3^{-n} x^n .$$

37. The original series can be written in the form $\sum\limits_{n=0}^{\infty} n(n+1)(n+2)2^n \left(x+\dfrac{3}{2}\right)^n$. Then we find the radius of convergence of the power series as follows: $\left|a_n\right| = n(n+1)(n+2)2^n$, and

$$l = \lim_{n\to\infty} \left(n(n+1)(n+2)2^n\right)^{1/n} = 2 , \ R = \frac{1}{l} = \frac{1}{2} .$$

39. Given that $\left|a_n\right| = \dfrac{(n!)^2}{(2n)!}$, we compute

$$l = \lim_{n\to\infty} \left(\frac{\left|a_{n+1}\right|}{\left|a_n\right|}\right) = \lim_{n\to\infty} \left(\frac{\left((n+1)!\right)^2 \big/ (2(n+1))!}{(n!)^2 \big/ (2n)!}\right) = \lim_{n\to\infty} \left(\frac{(n+1)^2}{(2n+1)(2n+2)}\right) = \frac{1}{4} ,$$

so that the radius of convergence is $R = \dfrac{1}{l} = 4$.

41. Given that $\left|a_n\right| = \dfrac{n!}{\left(n^2\right)!}$, we compute

$$l = \lim_{n\to\infty} \left(\frac{\left|a_{n+1}\right|}{\left|a_n\right|}\right) = \lim_{n\to\infty} \left(\frac{(n+1)! \big/ \left((n+1)^2\right)!}{n! \big/ \left(n^2\right)!}\right) = \lim_{n\to\infty} \left(\frac{n+1}{\left(n^2+1\right)\left(n^2+2\right)...\left(n^2+2n+1\right)}\right) = 0 ,$$

so that the radius of convergence is $R = \infty$.

43.   Let $t = (x+1)^2$. For the series $\sum\limits_{n=0}^{\infty} (-1)^n \dfrac{t^n}{9^n}$ we compute $|a_n| = \dfrac{1}{9^n}$, $l = \lim\limits_{n \to \infty} \left(\dfrac{1}{9^n} \right)^{1/n} = \dfrac{1}{9}$, so

that the radius of convergence is $R = 9$. Hence the series converges for $|t| < 9$ and diverges for

$|t| > 9$. This implies that the original series converges for $-4 < x < 2$ and diverges for $x < -4$

and $x > 2$. When $x = -4$ and $x = 2$, the series becomes $\sum\limits_{n=0}^{\infty} (-1)^n$, which diverges. The

interval of convergence of the original series is therefore $(-4, \ 2)$.

45.   Let $t = x^2$. For the series $\sum\limits_{n=0}^{\infty} \dfrac{4^n t^n}{n!}$ we compute $|a_n| = \dfrac{4^n}{n!}$,

$$l = \lim\limits_{n \to \infty} \left(\dfrac{|a_{n+1}|}{|a_n|} \right) = \lim\limits_{n \to \infty} \left(\dfrac{4^{n+1}/(n+1)!}{4^n/n!} \right) = \lim\limits_{n \to \infty} \left(\dfrac{4}{n+1} \right) = 0,$$

so that the radius of convergence is $R = \infty$. Hence the series converges for all t. This implies

that the interval of convergence for the original series is $(-\infty, \ \infty)$.

47.   Let $\sum\limits_{n=1}^{\infty} \dfrac{(2x)^{2n+1}}{n+1} = 2x \sum\limits_{n=1}^{\infty} \dfrac{(2x)^{2n}}{n+1} = 2x \sum\limits_{n=1}^{\infty} \dfrac{4^n t^n}{n+1}$, where $t = x^2$. For the last series we compute

$|a_n| = \dfrac{4^n}{n+1}$, $l = \lim\limits_{n \to \infty} \left(\dfrac{4^n}{n+1} \right)^{1/n} = 4$, so that the radius of convergence is $R = \dfrac{1}{4}$. Hence the

series converges absolutely for $|t| < \dfrac{1}{4}$ and diverges for $|t| > \dfrac{1}{4}$. This implies that the original

series converges for $|x| < \dfrac{1}{2}$ and diverges for $|x| > \dfrac{1}{2}$. When $x = \pm \dfrac{1}{2}$, the series becomes

$\pm \sum\limits_{n=1}^{\infty} \dfrac{1}{n+1}$, which diverges by comparison with the harmonic series. The interval of

convergence for the original series is therefore $\left(-\dfrac{1}{2}, \dfrac{1}{2} \right)$.

49.   Let $\sum\limits_{n=0}^{\infty} \dfrac{(x+e)^{2n+1}}{3^n} = (x+e) \sum\limits_{n=0}^{\infty} \dfrac{(x+e)^{2n}}{3^n} = (x+e) \sum\limits_{n=0}^{\infty} \dfrac{t^n}{3^n}$, where $t = (x+e)^2$. For the last

series we compute $|a_n| = \dfrac{1}{3^n}$, $l = \lim\limits_{n \to \infty} \left(|a_n|^{1/n} \right) = \lim\limits_{n \to \infty} \left(\dfrac{1}{3^n} \right)^{1/n} = \dfrac{1}{3}$, so that the radius of

convergence is $R = 3$. Hence the series converges absolutely for $|t| < 3$ and diverges for

$|t| > 3$. This implies that the original series converges for $-e - \sqrt{3} < x < -e + \sqrt{3}$ and diverges

for $x < -e - \sqrt{3}$ and $x > -e + \sqrt{3}$. When $x = -e \pm \sqrt{3}$, the series becomes $\pm \sqrt{3} \sum\limits_{n=0}^{\infty} 1$, which

diverges. The interval of convergence for the original series is therefore $\left(-e - \sqrt{3}, \ -e + \sqrt{3} \right)$.

51. Let $t = \dfrac{x-(\alpha+\beta)/2}{(\beta-\alpha)/2}$. The geometric series $\displaystyle\sum_{n=0}^{\infty} t^n$ converges for $-1 < t < 1$. Therefore the

series $\displaystyle\sum_{n=0}^{\infty} \dfrac{(x-(\alpha+\beta)/2)^n}{((\beta-\alpha)/2)^n}$ converges for $\alpha < x < \beta$.

53. Let $t = \dfrac{x-(\alpha+\beta)/2}{(\beta-\alpha)/2}$. The series $\displaystyle\sum_{n=1}^{\infty} \dfrac{t^n}{\sqrt{n}}$ converges for $-1 \le t < 1$. Therefore the series

$\displaystyle\sum_{n=1}^{\infty} \dfrac{(x-(\alpha+\beta)/2)^n}{((\beta-\alpha)/2)^n \sqrt{n}}$ converges for $\alpha \le x < \beta$.

55. It is not possible for a power series to have an interval of convergence $(-\infty, 1)$. If R is positive and finite, the interval of convergence belongs in $[c-R,\ c+R]$ (endpoints may be removed). If R is 0, the interval of convergence is $\{c\}$. If R is infinite, the interval of convergence is $(-\infty,\ \infty)$. In any case the presence of only one infinite boundary is impossible.

57. Given that $\displaystyle\lim_{n\to\infty} a_n = \dfrac{\pi^2}{6}$, we compute $l = \displaystyle\lim_{n\to\infty}\left(\dfrac{a_{n+1}}{a_n}\right) = 1$ so that the radius of convergence is $R = 1$.

59. Let $u_n = a_n x^n$, then $\displaystyle\lim_{n\to\infty}\left|\dfrac{u_{n+1}}{u_n}\right| = \lim_{n\to\infty}\left|\dfrac{a_{n+1}x}{a_n}\right| = |x|\lim_{n\to\infty}\left|\dfrac{a_{n+1}}{a_n}\right| = |x|\lim_{n\to\infty}\left(\dfrac{|a_{n+1}|}{|a_n|}\right) = |x|l$. Suppose that $0 < l < \infty$. Using the Ratio Test we find that the series converges absolutely if $|x|l < 1$, i.e. if $|x| < \dfrac{1}{l}$; it diverges if $|x|l > 1$, i.e. if $|x| > \dfrac{1}{l}$. Therefore the radius of convergence is $R = \dfrac{1}{l}$.

For $l = \infty$, by the Ratio Test, the series converges only if $x = 0$, i.e., $R = 0$. If we set $\dfrac{1}{\infty} = 0$ and $\dfrac{1}{0} = \infty$, we have $R = \dfrac{1}{l}$.

61.   We have the series

$$\sum_{k=1}^{\infty}(2k-1)\left(x^{2k-1}+2x^{2k}\right)=\sum_{k=1}^{\infty}(2k-1)x^{2k-1}(1+2x)=(1+2x)\sum_{k=1}^{\infty}(2k-1)x^{2k-1}.$$

Let $u_k=(2k-1)x^{2k-1}$. We then compute

$$\lim_{n\to\infty}\left|\frac{u_{n+1}}{u_n}\right|=\lim_{n\to\infty}\left|\frac{(2n+1)x^{2n+1}}{(2n-1)x^{2n-1}}\right|=|x|^2\lim_{n\to\infty}\left|\frac{2n+1}{2n-1}\right|=|x|^2.$$

By the Ratio Test the series converges if $|x|^2<1$ and diverges if $|x|^2>1$. Therefore the radius of convergence is $R=1$.

Calculator/Computer Exercises

63.   a.

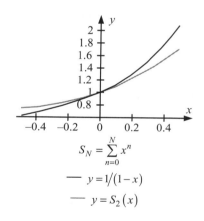

$$S_N=\sum_{n=0}^{N}x^n$$

—— $y=1/(1-x)$

—— $y=S_2(x)$

b.

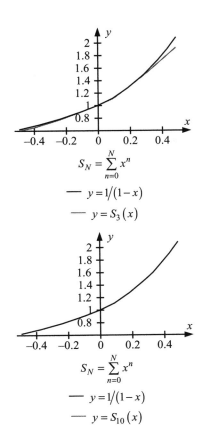

$$S_N = \sum_{n=0}^{N} x^n$$

— $y = 1/(1-x)$

····· $y = S_3(x)$

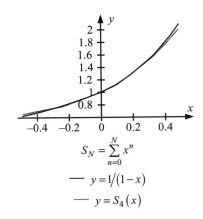

$$S_N = \sum_{n=0}^{N} x^n$$

— $y = 1/(1-x)$

····· $y = S_4(x)$

$$S_N = \sum_{n=0}^{N} x^n$$

— $y = 1/(1-x)$

····· $y = S_{10}(x)$

c.

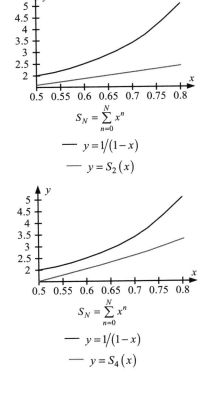

$$S_N = \sum_{n=0}^{N} x^n$$

— $y = 1/(1-x)$

····· $y = S_2(x)$

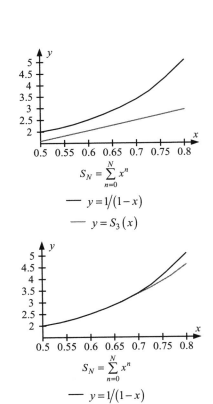

$$S_N = \sum_{n=0}^{N} x^n$$

— $y = 1/(1-x)$

····· $y = S_3(x)$

$$S_N = \sum_{n=0}^{N} x^n$$

— $y = 1/(1-x)$

····· $y = S_4(x)$

$$S_N = \sum_{n=0}^{N} x^n$$

— $y = 1/(1-x)$

····· $y = S_{10}(x)$

d.

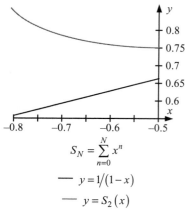

$$S_N = \sum_{n=0}^{N} x^n$$

— $y = 1/(1-x)$

— $y = S_2(x)$

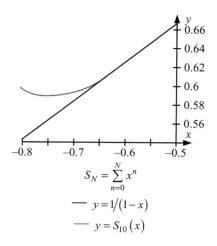

$$S_N = \sum_{n=0}^{N} x^n$$

— $y = 1/(1-x)$

— $y = S_3(x)$

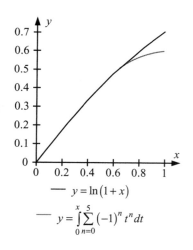

$$S_N = \sum_{n=0}^{N} x^n$$

— $y = 1/(1-x)$

— $y = S_4(x)$

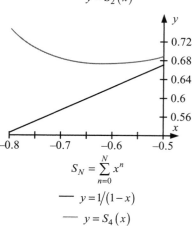

$$S_N = \sum_{n=0}^{N} x^n$$

— $y = 1/(1-x)$

— $y = S_{10}(x)$

65.   **a.**

— $y = \ln(1+x)$

— $y = \int_0^x \sum_{n=0}^{5} (-1)^n t^n \, dt$

b.

— $y = \ln(1+x)$

— $y = \int_0^x \sum_{n=0}^{10} (-1)^n t^n \, dt$

Section 10.2   Operations on Power Series

Problems for Practice

1.   Using geometric series expansion we obtain $\dfrac{1}{1-2x} = \sum\limits_{n=0}^{\infty} (2x)^n = \sum\limits_{n=0}^{\infty} 2^n x^n$. Given that

$a_n = 2^n$, the radius of convergence is $R = \dfrac{1}{2}$.

3.   Using geometric series expansion we obtain $\dfrac{1}{2-x} = \dfrac{1/2}{1-(x/2)} = \left(\dfrac{1}{2}\right)\sum\limits_{n=0}^{\infty} \left(\dfrac{1}{2}\right)^n x^n = \sum\limits_{n=0}^{\infty} \left(\dfrac{1}{2}\right)^{n+1} x^n$.

Given that $|a_n| = \left(\dfrac{1}{2}\right)^{n+1}$, the radius of convergence is $R = 2$.

5.   Using geometric series expansion we obtain

$$\frac{1}{4-x} = \frac{1/4}{1-x/4} = \left(\frac{1}{4}\right)\sum_{n=0}^{\infty}\left(\frac{x}{4}\right)^n = \sum_{n=0}^{\infty}\left(\frac{1}{4}\right)^{n+1} x^n.$$

Given that $a_n = \left(\dfrac{1}{4}\right)^{n+1}$, the radius of convergence is $R = 4$.

7.   Using geometric series expansion we obtain

$$\frac{x^2+1}{1-x^2} = \left(x^2+1\right)\sum_{n=0}^{\infty}\left(x^2\right)^n = \sum_{n=0}^{\infty}\left(x^2\right)^{n+2} + \sum_{n=0}^{\infty}\left(x^2\right)^n = 1 + x^2 + \sum_{n=2}^{\infty} 2x^{2n}.$$

Since convergence depends only on the tail, we have $\sum\limits_{n=1}^{\infty} 2x^{2n}$. This is $\sum\limits_{n=1}^{\infty} 2u^n$ for $u = x^2$.

Given that $a_n = 2$, the series converges for $|u| < 1$ and hence for $|x| < 1$, so $R = 1$.

9.   Using geometric series expansion we obtain $\dfrac{x^3}{1+x^4} = \dfrac{x^3}{1-\left(-x^4\right)} = x^3\sum\limits_{n=0}^{\infty}\left(-x^4\right)^n = \sum\limits_{n=0}^{\infty}(-1)^n x^{4n+3}$.

We know $\sum\limits_{n=1}^{\infty}(-1)^n x^{4n+3}$ converges on the same interval as $\sum\limits_{n=1}^{\infty}(-1)^n x^{4n}$ by Theorem 1c. Set

$u = x^4$, then $\sum\limits_{n=0}^{\infty}(-1)^n u^n$ has $a_n = (-1)^n$, so the series converges for $|u| < 1$, hence for $|x| < 1$,

so $R = 1$.

11. Using the series expansion for the logarithmic function we obtain

$$\ln(1+4x) = \sum_{m=1}^{\infty} (-1)^{m-1} \frac{(4x)^m}{m} = \sum_{m=1}^{\infty} (-1)^{m-1} 4^m \frac{x^m}{m}.$$

Given that $|a_m| = 4^m \frac{1}{m}$, we find $\lim_{m\to\infty} \left|\frac{a_{m+1}}{a_m}\right| = \lim_{m\to\infty} \left(\frac{4m}{m+1}\right) = 4$. The radius of convergence is thus $R = \frac{1}{4}$.

13. Using the series expansion for the logarithmic function we obtain

$$\ln(1+x^2) = \sum_{m=1}^{\infty} (-1)^{m-1} \frac{x^{2m}}{m}.$$

Given that $|a_m| = \frac{1}{m}$, we find $\lim_{m\to\infty} |a_m|^{1/m} = \lim_{m\to\infty} \left(\frac{1}{m}\right)^{1/m} = 1$. The radius of convergence is $R = 1$.

15. Using the series expansion for the logarithmic function we obtain

$$\int_0^x \ln(1+t^2)\,dt = \int_0^x \sum_{m=1}^{\infty} (-1)^{m-1} \frac{t^{2m}}{m}\,dt = \sum_{m=1}^{\infty} (-1)^{m-1} \int_0^x t^{2m} \frac{dt}{m} = \sum_{m=1}^{\infty} (-1)^{m-1} \frac{x^{2m+1}}{m(2m+1)}.$$

Given that $|a_m| = \left(\frac{1}{m(2m+1)}\right)$, we find $\lim_{m\to\infty} |a_m|^{1/m} = \lim_{m\to\infty} \left(\frac{1}{m(2m+1)}\right)^{1/m} = 1$. The radius of convergence is $R = 1$.

17. Using the series expansion for the trigonometric function we obtain

$$\arctan(x^2) = \sum_{n=0}^{\infty} (-1)^n \frac{(x^2)^{2n+1}}{2n+1} = \sum_{n=0}^{\infty} (-1)^n \frac{x^{4n+2}}{2n+1}.$$

Given that $|a_m| = \frac{1}{2m+1}$, we find $\lim_{m\to\infty} |a_m|^{1/m} = \lim_{m\to\infty} \left(\frac{1}{2m+1}\right)^{1/m} = 1$. The radius of convergence is $R = 1$.

19.   According to Exercise 10.2.17,

$$\int_0^x \arctan\left(t^2\right)dt = \int_0^x \sum_{n=0}^\infty \frac{(-1)^n}{2n+1}t^{4n+2}dt = \sum_{n=0}^\infty \frac{(-1)^n}{2n+1}\int_0^x t^{4n+2}dt = \sum_{n=0}^\infty \frac{(-1)^n}{(2n+1)(4n+3)}x^{4n+3}.$$

Given that $|a_m| = \dfrac{1}{(2m+1)(4m+3)}$, we find

$$\lim_{m\to\infty}|a_m| = \lim_{m\to\infty}\left(\frac{1}{(2m+1)(4m+3)}\right)^{1/m} = 1.$$

The radius of convergence is $R = 1$.

21.   Using a geometric series expansion around $x = -3$ we obtain

$$\frac{1}{x+1} = \left(-\frac{1}{2}\right)\cdot\frac{1}{1-(x+3)/2} = -\sum_{n=0}^\infty (-1)\cdot 2^{-(n+1)}\cdot(x+3)^n.$$

Given that $|a_m| = \dfrac{1}{2}\cdot\dfrac{1}{2^m}$, we see that $\lim_{m\to\infty}|a_m|^{1/m} = \dfrac{1}{2}$ hence the radius of convergence is

$R = 2$.

23.   Using a geometric series expansion around $x = -1$ we obtain

$$\frac{1}{2x+5} = \left(\frac{1}{3}\right)\cdot\frac{1}{1-(x+1)/(-3/2)} = \sum_{n=0}^\infty (-1)^n\cdot\left(\frac{2^n}{3^{n+1}}\right)\cdot(x+1)^n.$$

Given that $|a_m| = \dfrac{1}{3}\cdot\left(\dfrac{2}{3}\right)^m$, we see that $\lim_{m\to\infty}|a_m|^{1/m} = \dfrac{2}{3}$ hence the radius of convergence is

$R = \dfrac{3}{2}$.

25.   Using a geometric series expansion around $t = 1$ we obtain

$$\ln(x) = \int_1^x \frac{1}{t} dt = \int_1^x \left(\frac{1}{1 - (1-t)} \right) dt = \int_1^x \sum_{n=0}^{\infty} (1-t)^n dt = \sum_{n=0}^{\infty} (-1)^n \int_1^x (t-1)^n dt = \sum_{n=0}^{\infty} \frac{(-1)^n}{n+1} (x-1)^{n+1}.$$

Making the change of the index of summation $m = n+1$, we obtain

$$\ln(x) = \sum_{m=1}^{\infty} (-1)^{m-1} \frac{(x-1)^m}{m}.$$

Given that $|a_m| = \dfrac{1}{m}$, we find $\lim_{m \to \infty} |a_m|^{1/m} = \lim_{m \to \infty} \left(\dfrac{1}{m} \right)^{1/m} = 1$. The radius of convergence is $R = 1$.

Further Theory and Practice

27.   To apply geometric series expansion we first express the given function as a derivative:
$\dfrac{1}{(1-x)^2} = \dfrac{d}{dx} \left(\dfrac{1}{1-x} \right)$. Next, using the geometric series expansion we find

$$\frac{1}{(1-x)^2} = \frac{d}{dx} \sum_{n=0}^{\infty} x^n = \sum_{n=1}^{\infty} nx^{n-1} = \sum_{m=0}^{\infty} (m+1)x^m.$$

29.   Using the result of Exercise 10.2.27, we first break the function into two parts:

$$\frac{2+x}{(1-x)^2} = \frac{2}{(1-x)^2} + \frac{x}{(1-x)^2}.$$

Expanding each part as a geometric series we find $\dfrac{2+x}{(1-x)^2} = 2 \sum_{m=0}^{\infty} (m+1)x^m + x \cdot \sum_{m=0}^{\infty} (m+1)x^m$.

Finally, changing the summation indices by setting $j = m$ and $j = m+1$ in the two series, respectively, we obtain

$$\frac{2+x}{(1-x)^2} = \sum_{m=0}^{\infty} 2(m+1)x^m + \sum_{m=0}^{\infty} (m+1)x^{m+1} = 2 + \sum_{j=1}^{\infty} \left(2(j+1) + j \right)x^j = \sum_{j=0}^{\infty} (3j+2) \cdot x^j.$$

31. Using the formula for the sum of the geometric series we obtain $\sum_{n=0}^{\infty} x^{n+2} = x^2 \sum_{n=0}^{\infty} x^n = \frac{x^2}{1-x}$.

Given that $|a_n| = 1$, the radius of convergence is $R = 1$.

33. To apply geometric series expansion we first express the given function as a second

derivative: $\sum_{n=1}^{\infty} n(n+1)x^n = x\sum_{n=1}^{\infty} n(n+1)x^{n-1} = x\frac{d^2}{dx^2}\sum_{n=1}^{\infty} x^{n+1}$. Next, using the formula for the

sum of the geometric series we find

$\sum_{n=1}^{\infty} n(n+1)x^n = x\frac{d^2}{dx^2}\left(x^2\sum_{j=0}^{\infty} x^j\right) = x\frac{d^2}{dx^2}\left(\frac{x^2}{1-x}\right) = \frac{2x}{(1-x)^3}$. Given that $|a_n| = n(n+1)$, the

radius of convergence is $R = 1$.

35. First we start the summation with $n = 0$ because $a_0 = 0$. To apply the geometric series

expansion we express the given function as a derivative:

$$\sum_{n=1}^{\infty} nx^{2n+1} = \sum_{n=0}^{\infty} nx^{2n+1} = \left(\frac{x^2}{2}\right)\sum_{n=0}^{\infty} n\left(x^2\right)^{(n-1)}(2x) = \left(\frac{x^2}{2}\right)\frac{d}{dx}\sum_{n=0}^{\infty}\left(x^2\right)^n.$$

Next, using the formula for the sum of the geometric series we find that

$$\sum_{n=1}^{\infty} nx^{2n+1} = \left(\frac{x^2}{2}\right)\frac{d}{dx}\left(\frac{1}{1-x^2}\right) = \frac{x^3}{\left(x^2-1\right)^2}.$$

Given that $|a_n| = n$, the radius of convergence is $R = 1$.

37. First, we expand the function in the power series so that $f(x) = \sum_{n=0}^{\infty} a_n x^n$. Next we compute

the fifth derivative: $\frac{d^5}{dx^5}f(x) = \sum_{n=5}^{\infty} n(n-1)(n-2)(n-3)(n-4)a_n x^{n-5} = 0$. We change the

summation index $m = n-5$ as follows: $\sum_{m=0}^{\infty}(m+1)(m+2)(m+3)(m+4)(m+5)a_{m+5}x^m = 0$.

According to the Uniqueness Theorem, we have: $a_5 = 0$, $a_6 = 0$, ..., $a_n = 0$, $n \geq 5$. As this

solution does not place any restriction on the coefficients a_0, a_1, a_2, a_3, a_4, any fourth-

degree polynomial is a solution: $f(x) = a_0 + a_1 x + a_2 x^2 + a_3 x^3 + a_4 x^4$.

39. Using the expansion $y = \sum_{n=0}^{\infty} a_n x^n$, we obtain $\frac{dy}{dx} = \frac{d}{dx} \sum_{n=0}^{\infty} a_n x^n = \sum_{n=1}^{\infty} n a_n x^{n-1}$. To find solutions

for $\frac{dy}{dx} = 2y$ we equate the two series: $\sum_{n=1}^{\infty} n a_n x^{n-1} = 2 \sum_{n=0}^{\infty} a_n x^n$. We change the summation

index $m = n-1$ in the series on the left-hand side of the equation and rename this index:

$$\sum_{n=0}^{\infty} (n+1) a_{n+1} x^n = 2 \sum_{n=0}^{\infty} a_n x^n \text{ , or } a_1 + 2a_2 x + \sum_{n=2}^{\infty} (n+1) a_{n+1} x^n = 2a_0 + 2a_1 x + 2 \sum_{n=2}^{\infty} a_n x^n .$$

According to the Uniqueness Theorem, $a_1 = 2a_0$, $2a_2 = 2a_1$, $3a_3 = 2a_2$, ..., $(n+1) a_{n+1} = 2a_n$.

Using $y(0) = 3$ we obtain $a_0 = 3$, $a_1 = 6$, $a_2 = 6$, $a_3 = 4$, ..., $a_n = 3 \cdot \dfrac{2^n}{n!}$. Therefore

$$y(x) = 3 \left(1 + 2x + \sum_{n=2}^{\infty} \frac{(2x)^n}{n!} \right) = 3 \cdot e^{2x} .$$

41. Using the expansion $y = \sum_{n=0}^{\infty} a_n x^n$, we obtain $\frac{dy}{dx} = \frac{d}{dx} \sum_{n=0}^{\infty} a_n x^n = \sum_{n=1}^{\infty} n a_n x^{n-1}$. To find solutions

for $\frac{dy}{dx} = 2x - y$ we equate the two series: $\sum_{n=1}^{\infty} n a_n x^{n-1} = 2x - \sum_{n=0}^{\infty} a_n x^n$. We change the

summation index $m = n-1$ in the series on the left-hand side of the equation and rename this

index: $\sum_{n=0}^{\infty} (n+1) a_{n+1} x^n = 2x - \sum_{n=0}^{\infty} a_n x^n$, or

$$a_1 + 2a_2 x + \sum_{n=2}^{\infty} (n+1) a_{n+1} x^n = -a_0 + (2 - a_1) x - \sum_{n=2}^{\infty} a_n x^n .$$

According to the Uniqueness Theorem, $a_1 = -a_0$, $2a_2 = 2 - a_1$, $3a_3 = -a_2$, ..., $(n+1) a_{n+1} = -a_n$.

Using $y(0) = 1$ we obtain $a_0 = 1$, $a_1 = -1$, $a_2 = \dfrac{3}{2}$, $a_3 = -\dfrac{1}{2} = -\dfrac{3}{3!}$, ..., $a_n = 3 \cdot \dfrac{(-1)^n}{n!}$.

Therefore $a_n = 3 \cdot \dfrac{(-1)^n}{n!}$ for $n \geq 2$, and $y(x) = 1 - x + 3 \sum_{n=2}^{\infty} \frac{(-1)^n x^n}{n!} = 3e^{-x} + 2x - 2$.

43.   Using the expansion $y = \sum\limits_{n=0}^{\infty} a_n x^n$, we obtain $\frac{dy}{dx} = \frac{d}{dx} \sum\limits_{n=0}^{\infty} a_n x^n = \sum\limits_{n=1}^{\infty} na_n x^{n-1}$. To find solutions

for $\frac{dy}{dx} = x + xy$ we equate the two series: $\sum\limits_{n=1}^{\infty} na_n x^{n-1} = x + x \sum\limits_{n=0}^{\infty} a_n x^n$. We change the

summation index $m = n - 1$ in the series on the left-hand side of the equation and rename this

index: $\sum\limits_{n=0}^{\infty} (n+1)a_{n+1}x^n = x + \sum\limits_{n=0}^{\infty} a_n x^{n+1}$, or

$$a_1 + 2a_2 x + \sum\limits_{n=2}^{\infty} (n+1)a_{n+1}x^n = (a_0 + 1)x + a_1 x^2 + \sum\limits_{n=2}^{\infty} a_n x^{n+1}.$$

According to the Uniqueness Theorem $a_1 = 0$, $2a_2 = a_0 + 1$, $3a_3 = a_1$, ..., $(n+1)a_{n+1} = a_{n-1}$.

Using $y(0) = 0$ we obtain $a_0 = 0$, $a_1 = 0$, $a_2 = \frac{1}{2}$, $a_3 = 0$, ..., $a_{2n} = \frac{1}{(2n)!!}$, where

$$(2n)!! = 2 \cdot 4 \cdot ... \cdot 2n = 2^n \cdot n!.$$

Therefore $a_{2n} = \frac{1}{2^n n!}$ and $a_{2n+1} = 0$ for $n \geq 1$, and $y(x) = \sum\limits_{n=1}^{\infty} \frac{x^{2n}}{2^n n!} = \sum\limits_{n=1}^{\infty} \frac{\left(x^2/2\right)^n}{n!} = e^{x^2/2} - 1$.

45.   Using geometric series expansion we obtain $h(x) = \frac{1}{1-x^2} = \sum\limits_{n=0}^{\infty} x^{2n}$. On the other hand,

$$h(x) = \frac{1}{1-x^2} = \frac{1}{1-x} \cdot \frac{1}{1+x}.$$

The Cauchy product of the polynomials $g(x)$ and $f(x)$ is:

$$h(x) = \sum\limits_{n=0}^{\infty} \left(\sum\limits_{k=0}^{n} a_k b_{n-k} \right) x^n$$

where $a_k = 1$, $b_k = (-1)^k$ for all k. So:

$$\sum\limits_{k=0}^{n} a_k b_{n-k} = \sum\limits_{k=0}^{n} (-1)^{n-k} = \begin{cases} 1, & \text{if } n \text{ is even} \\ 0, & \text{if } n \text{ is odd} \end{cases}.$$

Then $h(x) = \sum\limits_{n=0}^{\infty} \left(\sum\limits_{k=0}^{n} a_k b_{n-k} \right) x^n = \sum\limits_{n=0}^{\infty} x^{2n}$, and the result is identical to the one obtained using
geometric series expansion.

47.   Let $g(x) = 1/f(x)$. Given that $f(x) = \sum_{n=0}^{\infty} a_n x^n$ and $g(x) = \sum_{n=0}^{\infty} b_n x^n$, we find that

$$f(x) \cdot g(x) = \sum_{n=0}^{\infty} \left(\sum_{k=0}^{n} a_k b_{n-k} \right) x^n = 1.$$ According to the Uniqueness Theorem, $\sum_{k=0}^{n} a_k b_{n-k} = 1$

for $n = 0$ and $\sum_{k=0}^{n} a_k b_{n-k} = 0$ for $n > 0$, or $a_0 b_0 = 1$, $b_0 = 1/a_0$; $a_0 b_1 + a_1 b_0 = 0$, $b_1 = -a_1/a_0^2$;

$a_0 b_2 + a_1 b_1 + a_2 b_0 = 0$, $b_2 = -a_2/a_0^2 + a_1^2/a_0^3$, and so on. Therefore, we find for b_n: $b_0 = 1/a_0$

and $b_n = (1/a_0) \sum_{k=0}^{n-1} (-1)^{n-k} (a_k/a_0)^{n-k}$ for $n > 0$.

49.   If $P(x) = \sum_{n=0}^{N} a_n x^n$ is a degree N polynomial and c is an arbitrary number, then one can

always find a set of coefficients $\{b_n\}$, $0 \le n \le N$ such that $P(x) = \sum_{n=0}^{N} b_n (x-c)^n$, because the

latter is a polynomial of degree N as well, and according to the Uniqueness Theorem, there are
$N+1$ linear equations to determine $N+1$ unknown coefficients b_n. These equations can be
obtained by evaluating consecutive derivatives of $P(x)$ at point c. We compute

$$P^{(k)}(x) = \sum_{n=k}^{N} \frac{n!}{(n-k)!} b_n (x-c)^{n-k}.$$

Therefore we find that $P(c) = b_0$, $P'(c) = b_1$, ..., $P^{(k)}(c) = k! \cdot b_k$, or $b_k = \dfrac{P^{(k)}(c)}{k!}$. Then

$$P(x) = \sum_{n=0}^{N} \frac{P^{(n)}(c)}{n!} (x-c)^n.$$

Calculator/Computer Exercises

51.   a.

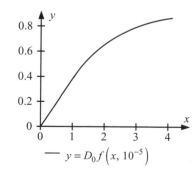

$$- \quad y = D_0 f\left(x, 10^{-5}\right)$$

b.   From the given power series, we see that

$$p(x) = \frac{2}{3} + \left(\frac{5}{27}\right)(x-2) - \left(\frac{5}{81}\right)(x-2)^2$$
$$+ \left(\frac{55}{4374}\right)(x-2)^3$$

approximates $f'(x)$.

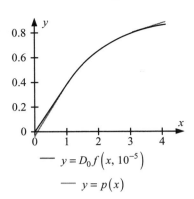

$$\quad\quad y = D_0 f\left(x, 10^{-5}\right)$$
$$\quad\quad y = p(x)$$

53.   a.   If we substitute $y(x) = \sum\limits_{n=0}^{\infty} a_n x^n$ into the differential equation, we obtain

$$\sum_{n=1}^{\infty} n a_n x^{n-1} = 2 - x - \sum_{n=0}^{\infty} a_n x^n .$$ Writing this out up to degree 5 terms we have

$$a_1 + 2a_2 x + 3a_3 x^2 + 4a_4 x^3 + 5a_5 x^4 + 6a_6 x^5 = 2 - x - a_0 - a_1 x - a_2 x^2 - a_3 x^3$$
$$- a_4 x^4 - a_5 a^5$$

or

$$\left(a_1 + a_0 - 2\right) + \left(2a_2 + a_1 + 1\right)x + \left(3a_3 + a_2\right)x^2 + \left(4a_4 + a_3\right)x^3 + \left(5a_5 + a_4\right)x^4$$
$$+ \left(6a_6 + a_5\right)x^5 = 0$$

We set each coefficient equal to 0. The initial condition gives us $a_0 = 1$. Therefore,

$$a_1 = 2 - a_0 = 1, \; a_2 = -\frac{a_1 + 1}{2} = -1, \; a_3 = \frac{-a_2}{3} = \frac{1}{3}, \; a_4 = -\frac{a_3}{4} = -\frac{1}{12}, \text{ and}$$

$$a_5 = -\frac{a_4}{5} = \frac{1}{60} .$$

From this we obtain $S_3(x) = 1 + x - x^2$ and $S_5(x) = 1 + x - x^2 + \frac{x^3}{3} - \frac{x^4}{12} + \frac{x^5}{60} .$

b.

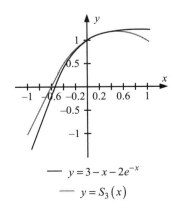

$$\quad\quad y = 3 - x - 2e^{-x}$$
$$\quad\quad y = S_3(x)$$

c.

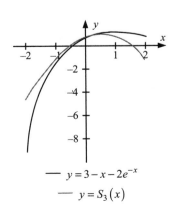

$$\quad\quad y = 3 - x - 2e^{-x}$$
$$\quad\quad y = S_3(x)$$

d.

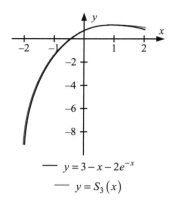

$$— \quad y = 3 - x - 2e^{-x}$$
$$— \quad y = S_3(x)$$

Section 10.3   Taylor Polynomials

Problems for Practice

1.   We compute the values of the function and its first five derivatives as follows:

$$f\left(\frac{\pi}{3}\right) = \frac{1}{2}$$

$$f'\left(\frac{\pi}{3}\right) = -\sin(x)\big|_{x=\pi/3} = -\frac{\sqrt{3}}{2}$$

$$f''\left(\frac{\pi}{3}\right) = -\cos(x)\big|_{x=\pi/3} = -\frac{1}{2}$$

$$f'''\left(\frac{\pi}{3}\right) = \sin(x)\big|_{x=\pi/3} = \frac{\sqrt{3}}{2}$$

$$f^{(4)}\left(\frac{\pi}{3}\right) = \cos(x)\big|_{x=\pi/3} = \frac{1}{2}$$

$$P_4(x) = \frac{1}{2} - \left(\frac{\sqrt{3}}{2}\right)\left(x - \frac{\pi}{3}\right) - \frac{(x - \pi/3)^2}{4} + \left(\frac{\sqrt{3}}{12}\right)\left(x - \frac{\pi}{3}\right)^3 + \frac{(x - \pi/3)^4}{48}.$$

3.   We compute the values of the function and its first five derivatives as follows:

$$f(0) = -1$$
$$f'(0) = -2\sin(2x + \pi)\big|_{x=0} = 0$$
$$f''(0) = -4\cos(2x + \pi)\big|_{x=0} = 4$$
$$f'''(0) = 8\sin(2x + \pi)\big|_{x=0} = 0$$
$$f^{(4)}(0) = 16\cos(2x + \pi)\big|_{x=0} = -16$$
$$f^{(5)}(0) = -32\sin(2x + \pi)\big|_{x=0} = 0$$

$$P_5(x) = -1 + 2x^2 - 2x^4/3.$$

5. We compute the values of the function and its first five derivatives as follows:

$$f(0)=1$$

$$f'(0)=\left(10x+e^x\right)\Big|_{x=0}=1$$

$$f''(x)=\left(10+e^x\right)\Big|_{x=0}=11$$

$$f'''(0)=f^{(4)}(0)=f^{(5)}(0)=e^x\Big|_{x=0}=1$$

$$P_5(x)=1+x+\left(\frac{11}{2}\right)x^2+\left(\frac{1}{6}\right)x^3+\left(\frac{1}{24}\right)x^4+\left(\frac{1}{120}\right)x^5.$$

7. We compute the values of the function and its first five derivatives as follows:

$$f(2)=\frac{1}{8}$$

$$f'(2)=\frac{-3}{x^4}\Big|_{x=2}=-\frac{3}{16}$$

$$f''(2)=\frac{12}{x^5}\Big|_{x=2}=\frac{3}{8}$$

$$f'''(2)=\frac{-60}{x^6}\Big|_{x=2}=-\frac{15}{16}$$

$$f^{(4)}(2)=\frac{360}{x^7}\Big|_{x=2}=\frac{45}{16}$$

$$f^{(5)}(2)=\frac{-2520}{x^8}\Big|_{x=2}=-\frac{315}{32}$$

$$P_5(x)=\frac{1}{8}-\left(\frac{3}{16}\right)(x-2)+\left(\frac{3}{16}\right)(x-2)^2-\left(\frac{5}{32}\right)(x-2)^3+\left(\frac{15}{128}\right)(x-2)^4-\left(\frac{21}{256}\right)(x-2)^5.$$

9.   We compute the values of the function and its first five derivatives as follows:

$$f(0) = \ln(3)$$

$$f'(0) = \frac{1}{x+3}\bigg|_{x=0} = \frac{1}{3}$$

$$f''(0) = \frac{-1}{(x+3)^2}\bigg|_{x=0} = -\frac{1}{9}$$

$$f'''(0) = \frac{2}{(x+3)^3}\bigg|_{x=0} = \frac{2}{27}$$

$$f^{(4)}(0) = \frac{-6}{(x+3)^4}\bigg|_{x=0} = -\frac{2}{27}$$

$$f^{(5)}(0) = \frac{24}{(x+3)^5}\bigg|_{x=0} = \frac{8}{81}$$

$$P_5(x) = \ln(3) + \frac{x}{3} - \frac{x^2}{18} + \frac{x^3}{81} - \frac{x^4}{324} + \frac{x^5}{1215}.$$

11.   We compute the values of the function and its first four derivatives as follows:

$$f(-2) = 0$$

$$f'(-2) = \frac{3}{3x+7}\bigg|_{x=-2} = 3$$

$$f''(-2) = \frac{-9}{(3x+7)^2}\bigg|_{x=-2} = -9$$

$$f'''(-2) = \frac{54}{(3x+7)^3}\bigg|_{x=-2} = 54$$

$$f^{(4)}(-2) = \frac{-486}{(3x+7)^4}\bigg|_{x=-2} = -486$$

$$P_4(x) = 3(x+2) - \left(\frac{9}{2}\right)(x+2)^2 + 9(x+2)^3 - \left(\frac{81}{4}\right)(x+2)^4.$$

13.   We compute the values of the function and its first two derivatives as follows:

$$f(-2) = -\frac{1}{2}$$

$$f'(-2) = e^{x+2} \left.\frac{x-1}{x^2}\right|_{x=-2} = -\frac{3}{4}$$

$$f''(-2) = e^{x+2} \left.\frac{x^2 - 2x + 2}{x^3}\right|_{x=-2} = -\frac{5}{4}$$

$$P_2(x) = -\frac{1}{2} - 3\frac{x+2}{4} - 5\frac{(x+2)^2}{8}.$$

15.   We compute the values of the function and its first three derivatives as follows:

$$f(0) = 0$$

$$f'(0) = \sec^2(x)\big|_{x=0} = 1$$

$$f''(0) = 2\tan(x)\sec^2(x)\big|_{x=0} = 0$$

$$f'''(0) = \left(2\sec^4(x) + 4\tan^2(x)\sec^2(x)\right)\big|_{x=0} = 2$$

$$P_3(x) = x + \frac{x^3}{3}.$$

17.   We compute the values of the function and its first two derivatives as follows:

$$f(0) = 2$$

$$f'(0) = \left(-1 + 3x^2 + 16x^3\right)\big|_{x=0} = -1$$

$$f''(0) = (6x + 48x^2)\big|_{x=0} = 0$$

$$f'''(0) = (6 + 96x)\big|_{x=0} = 6$$

$$f^{(iv)}(0) = 96\big|_{x=0} = 96$$

$$P_4(x) = 2 - x + x^3 + 4x^4, \quad R_4 = f(x) - P_4(x) \equiv 0.$$

19.   We compute the values of the function and its first four derivatives as follows:

$$f(1) = 11$$
$$f'(1) = \left(1 + 40x^3\right)\Big|_{x=1} = 41$$
$$f''(1) = 120x^2\Big|_{x=1} = 120$$
$$f'''(1) = 240x\big|_{x=1} = 240$$
$$f^{(4)}(1) = 240\big|_{x=1} = 240$$

$$P_4(x) = 11 + 41(x-1) + 60(x-1)^2 + 40(x-1)^3 + 10(x-1)^4, \quad R_4 = f(x) - P_4(x) \equiv 0.$$

21.   We compute the values of the function and its first three derivatives as follows:

$$f(0) = 49$$
$$f'(0) = \left(-2 + 6(x-2) - 12(x-2)^2\right)\Big|_{x=0} = -62$$
$$f''(0) = \left(6 - 24(x-2)\right)\big|_{x=0} = 54$$
$$f'''(0) = -24\big|_{x=0} = -24$$

$$P_3(x) = 49 - 62x + 27x^2 - 4x^3, \quad R_3 = (f)x - P_3(x) \equiv 0.$$

Further Theory and Practice

23.   We compute the values of the function and its first four derivatives at $x = 1$ as follows:

$$f(1) = \frac{1}{2}$$

$$f'(1) = -\frac{x^2 - 1}{\left(x^2 + 1\right)^2}\Bigg|_{x=1} = 0$$

$$f''(1) = 2x\frac{x^2 - 3}{\left(x^2 + 1\right)^3}\Bigg|_{x=1} = -\frac{1}{2}$$

$$f'''(1) = -6\frac{x^4 - 6x^2 + 1}{\left(x^2 + 1\right)^4}\Bigg|_{x=1} = \frac{3}{2}$$

$$f^{(4)}(1) = 24x\frac{x^4 - 10x^2 + 5}{\left(x^2 + 1\right)^5}\Bigg|_{x=1} = -3$$

Therefore $P_4(x) = \frac{1}{2} - \frac{(x-1)^2}{4} + \frac{(x-1)^3}{4} - \frac{(x-1)^4}{8}$.

25.   We compute the values of the function and its first three derivatives at $x = 0$ as follows:

$$f(0) = 1$$

$$f'(0) = e^{-x}\left(2x - x^2 - 1\right)\Big|_{x=0} = -1$$

$$f''(0) = e^{-x}\left(3 - 4x + x^2\right)\Big|_{x=0} = 3$$

$$f'''(0) = e^{-x}\left(6x - x^2 - 7\right)\Big|_{x=0} = -7$$

Therefore $P_3(x) = 1 - x + \left(\frac{3}{2}\right)x^2 - \left(\frac{7}{6}\right)x^3$.

27.   We compute the values of the function and its first three derivatives at $x = 0$ as follows:

$$f(0) = 0$$

$$f'(0) = e^x(1+x)\big|_{x=0} = 1$$

$$f''(0) = e^x(2+x)\big|_{x=0} = 2$$

$$f'''(0) = e^x(3+x)\big|_{x=0} = 3$$

Therefore $P_3(x) = x + x^2 + \dfrac{x^3}{2}$.

29.   We compute the values of the function and its first three derivatives at $x = 2$ as follows:

$$f(2) = 3$$

$$f'(2) = \frac{3x^2}{2\sqrt{1+x^3}}\Bigg|_{x=2} = 2$$

$$f''(2) = 3x\frac{x^3+4}{4\left(1+x^3\right)^{3/2}}\Bigg|_{x=2} = \frac{2}{3}$$

$$f'''(2) = -3\frac{20x^3+x^6-8}{8\left(1+x^3\right)^{5/2}}\Bigg|_{x=2} = -\frac{1}{3}$$

Therefore $P_3(x) = 3 + 2(x-2) + \dfrac{(x-2)^2}{3} - \dfrac{(x-2)^3}{18}$.

31.   Let $g(x) = f'(x)$. Then, in the expansion of $g(x)$ about base point c, the coefficient $c(g, n)$ of $(x-c)^n$ is given by

$$c(g, n) = \frac{g^{(n)}(c)}{n!} = \frac{f^{(n+1)}(c)}{n!} = (n+1)\frac{f^{(n+1)}(c)}{(n+1)!} = (n+1)c(f, n+1).$$

For example: let $f(x) = \sin(x)$, $g(x) = f'(x) = \cos(x)$, and $c = \dfrac{\pi}{2}$.

$$f\left(\frac{\pi}{2}\right) = 1 \qquad g\left(\frac{\pi}{2}\right) = 0$$

$$f'\left(\frac{\pi}{2}\right) = 0 \qquad g'\left(\frac{\pi}{2}\right) = -1$$

$$f''\left(\frac{\pi}{2}\right) = -1 \qquad g''\left(\frac{\pi}{2}\right) = 0$$

$$f'''\left(\frac{\pi}{2}\right) = 0 \qquad g'''\left(\frac{\pi}{2}\right) = 1$$

So $f(x)$ has $P_3(x) = 1 - \dfrac{(x-\pi/2)^2}{2}$, $c(f, n) = \begin{cases} \dfrac{(-1)^{n/2}}{n!} & \text{if } n \text{ is even} \\ 0 & \text{if } n \text{ is odd} \end{cases}$;

and $g(x)$ has $P_3(x) = -\left(x - \dfrac{\pi}{2}\right) + \dfrac{(x-\pi/2)^3}{6}$, $c(g, n) = \begin{cases} 0 & \text{if } n \text{ is even} \\ \dfrac{(-1)^{(n+1)/2}}{n!} & \text{if } n \text{ is odd} \end{cases}$.

We see that $c(g, n) = (n+1)c(f, n+1)$.

33.   Let $f(x) = x^4$ and $c = 1$. We compute the values of the first three derivatives at $x = c$ as follows: $f'(c) = 4c^3 = 4$, $f''(c) = 12c^2 = 12$, and $f'''(c) = 24c = 24$. Therefore,

$$P_1(x) = 1 + 4(x-1), \quad P_2(x) = 1 + 4(x-1) + 6(x-1)^2$$

and $P_3(x) = 1 + 4(x-1) + 6(x-1)^2 + 4(x-1)^3$.

Given that $f''(\xi) = 12\xi^2$ we find

$$x^4 = 1 + 4(x-1) + R_1(x) = 1 + 4(x-1) + 6\xi^2(x-1)^2,$$

or $\xi^2 = \dfrac{x^4 - 4x + 3}{6(x-1)^2} = \dfrac{x^2}{6} + \dfrac{x}{3} + \dfrac{1}{2}$. Thus, $\xi = \sqrt{\dfrac{x^2}{6} + \dfrac{x}{3} + \dfrac{1}{2}}$. Similarly, because $f'''(\xi) = 24\xi$,

we obtain $x^4 = 1 + 4(x-1) + 6(x-1)^2 + R_2(x) = 1 + 4(x-1) + 6(x-1)^2 + (24\xi)\dfrac{(x-1)^3}{3!}$, or

$\xi = \dfrac{x^4 - \left(1 + 4(x-1) + 6(x-1)^2\right)}{4(x-1)^3} = \dfrac{x}{4} + \dfrac{3}{4}$. Finally, because $f^{(4)}(\xi) = 24$ we see that

$R_3(x) = f^{(4)}(\xi)\dfrac{(x-1)^4}{4!} = (x-1)^4$. Therefore

$$x^4 = 1 + 4(x-1) + 6(x-1)^2 + 4(x-1)^3 + R_3(x)$$

becomes $x^4 = 1 + 4(x-1) + 6(x-1)^2 + 4(x-1)^3 + (x-1)^4$, an identity that is valid for all x. We may take ξ to be any value between 1 and x.

35.   Let $f(x) = \cos^2(x)$. Given that

$$f(0) = 1$$
$$f'(0) = 0$$
$$f''(0) = -2$$
$$f'''(0) = 0$$
$$f^{(4)}(0) = 8$$
$$f^{(5)}(0) = 0$$
$$f^{(6)}(0) = -32,$$

we have $P_6(x) = 1 - x^2 + \dfrac{x^4}{3} - \dfrac{2x^6}{45}$.

The degree 6 Taylor polynomial of $\cos(2x)$ is obtained by substituting $2x$ for u in the degree 6 Taylor polynomial, $1 - \dfrac{u^2}{2} + \dfrac{u^4}{24} - \dfrac{u^6}{720}$, of $\cos(u)$. The result is $1 - 2x^2 + \dfrac{2x^4}{3} - \dfrac{4x^6}{45}$. As $\cos^2(x) = \dfrac{1 + \cos(2x)}{2}$, we see that $\dfrac{1 + \left(1 - 2x^2 + 2x^4/3 - 4x^6/45\right)}{2}$, or $1 - x^2 + \dfrac{x^4}{3} - \dfrac{2x^6}{45}$, is the degree 6 Taylor polynomial of $\cos^2(x)$. This formula agrees with that of our direct calculation.

37.   Let $f'(x) = \begin{cases} -4x^3 & \text{if} \quad x \le 0 \\ 4x^3 & \text{if} \quad x > 0 \end{cases}$ and $f''(x) = \begin{cases} -12x^2 & \text{if} \quad x \le 0 \\ 12x^2 & \text{if} \quad x > 0 \end{cases}$. These formulas can be

established in the same way that we derive the formula for $f'''(x)$. It is clear that

$$f'''(x) = \begin{cases} -24x & \text{if} \quad x < 0 \\ 24x & \text{if} \quad x > 0 \end{cases}.$$

We next show that $f'''(0)$ exists. To that end, we observe that

$$\frac{f''(h) - f''(0)}{h} = \begin{cases} -12h & \text{if} \quad h < 0 \\ 12h & \text{if} \quad h > 0 \end{cases}.$$

Therefore, $\displaystyle\lim_{h \to 0} \frac{f''(h) - f''(0)}{h} = 0$ (however h approaches 0).

It follows that $f'''(0)$ exists and equals 0. Assembling all the information about f''', we see that $f'''(x) = 24|x|$. In particular, f''' is not differentiable at 0. Indeed,

$$f^{(4)}(x) = \begin{cases} -24 & \text{if} \quad x < 0 \\ \text{not defined} & \text{if} \quad x = 0 \\ 24 & \text{if} \quad x > 0 \end{cases}.$$

Given that $f(0)$, $f'(0)$, $f''(0)$, and $f'''(0)$ all exist, we can form the degree 3 Taylor polynomial $P_3(x)$ of f centered at 0. Moreover, as $f(0) = 0$, $f'(0) = 0$, $f''(0) = 0$, and $f'''(0) = 0$, we have $P_3(x) = 0$ (the zero polynomial). However, because $f^{(4)}(0)$ does not exist, we may *not* apply Taylor's Theorem with $N = 3$. Nevertheless, we can observe that Taylor's formula does hold. For $x < 0$, $f^{(4)}(\xi) = -24$ for any $\xi \in (x, 0)$. Thus,

$$-x^4 = f(x) = 0 + (-1)x^4 = P_3(x) + \frac{f^{(4)}(\xi)}{4!}(x - 0)^4.$$

Similarly, for $x > 0$, we find $f^{(4)}(\xi) = 24$ for any $\xi \in (0, x)$, and

$$x^4 = f(x) = 0 + x^4 = P_3(x) + \frac{f^{(4)}(\xi)}{4!}(x - 0)^4.$$

39.   Let $f(x)=(a+x)^N$. Then

$$f(0)=a^N$$

$$f'(0)=N(a+x)^{N-1}\Big|_{x=0}=Na^{N-1}$$

$$f''(0)=N(N-1)(a+x)^{N-2}\Big|_{x=0}=N(N-1)a^{N-2}$$

$$f^{(k)}(0)=N(N-1)\cdots(N-k+1)(a+x)^{N-k}\Big|_{x=0}=N(N-1)\cdots(N-k+1)a^{N-k},\ldots$$

$$f^{(N)}(0)=N(N-1)\cdots 1=N!$$

Given that $f^{(k)}(0)=N(N-1)\cdots(N-k+1)=N(N-1)\cdots\dfrac{1}{(N-k)(N-k-1)\ldots 1}$, and

$(N(N-1)\cdots 1)\cdot\dfrac{1}{(N-k)(N-k-1)\cdots 1}=\dfrac{N!}{(N-k)!}$, we conclude that

$$P_N(x)=\sum_{k=0}^{N}\left(\frac{N!}{(N-k)!}\right)\left(a^{N-k}\frac{x^k}{k!}\right)=\sum_{k-0}^{N}\left(\frac{N!}{k!(N-k)!}\right)a^{N-k}x^k=\sum_{k=0}^{N}\binom{N}{k}a^{N-k}x^k\,.$$

41.   We calculate $\dfrac{3\sin(t)}{2+\cos(t)}=t-\dfrac{1}{180}t^5+$ higher powers of t. For small values of t, the value of

$\dfrac{3\sin(t)}{2+\cos(t)}$ is approximately $\dfrac{t^5}{180}$ less than t.

43.   Let $f(x)=P(x)+R_N(x)$ and $f^{(n)}(c)=P^{(n)}(c)$, $0\le n\le N$. Then from Exercise 10.2.49 we

have $P(x)=\displaystyle\sum_{n=0}^{N}\dfrac{P^{(n)}(c)}{n!}(x-c)^n=P_N(x)$, i.e. $P_N(x)$ is a unique polynomial of degree N for

a given derivative up to order N at point c.

Calculator/Computer Exercises

45.   We compute the values of the function and its first five derivatives as follows:

$$f(0) = \frac{\sqrt{2}}{2}$$

$$f'(0) = \cos\left(x + \frac{\pi}{4}\right)\Big|_{x=0} = \frac{\sqrt{2}}{2}$$

$$f''(0) = -\sin\left(x + \frac{\pi}{4}\right)\Big|_{x=0} = -\frac{\sqrt{2}}{2}$$

$$f'''(0) = -\cos\left(x + \frac{\pi}{4}\right)\Big|_{x=0} = -\frac{\sqrt{2}}{2}$$

$$f^{(4)}(0) = \sin\left(x + \frac{\pi}{4}\right)\Big|_{x=0} = \frac{\sqrt{2}}{2}$$

$$f^{(5)}(0) = \cos\left(x + \frac{\pi}{4}\right)\Big|_{x=0} = \frac{\sqrt{2}}{2}$$

Therefore $P_1(x) = \frac{\sqrt{2}}{2} + \frac{\sqrt{2}x}{2}$, $P_2(x) = \frac{\sqrt{2}}{2} + \frac{\sqrt{2}x}{2} - \frac{\sqrt{2}x^2}{4}$, and

$$P_5(x) = \frac{\sqrt{2}}{2} + \frac{\sqrt{2}x}{2} - \frac{\sqrt{2}x^2}{4} - \frac{\sqrt{2}x^3}{12} + \frac{\sqrt{2}x^4}{48} + \frac{\sqrt{2}x^5}{240}.$$

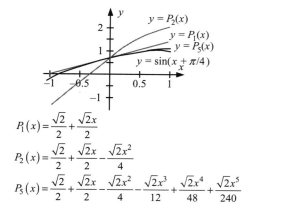

$$P_1(x) = \frac{\sqrt{2}}{2} + \frac{\sqrt{2}x}{2}$$

$$P_2(x) = \frac{\sqrt{2}}{2} + \frac{\sqrt{2}x}{2} - \frac{\sqrt{2}x^2}{4}$$

$$P_5(x) = \frac{\sqrt{2}}{2} + \frac{\sqrt{2}x}{2} - \frac{\sqrt{2}x^2}{4} - \frac{\sqrt{2}x^3}{12} + \frac{\sqrt{2}x^4}{48} + \frac{\sqrt{2}x^5}{240}$$

$$P_1(x) = \frac{\sqrt{2}}{2} + \frac{\sqrt{2}x}{2}$$

$$P_2(x) = \frac{\sqrt{2}}{2} + \frac{\sqrt{2}x}{2} - \frac{\sqrt{2}x^2}{4}$$

$$P_5(x) = \frac{\sqrt{2}}{2} + \frac{\sqrt{2}x}{2} - \frac{\sqrt{2}x^2}{4} - \frac{\sqrt{2}x^3}{12} + \frac{\sqrt{2}x^4}{48} + \frac{\sqrt{2}x^5}{240}$$

47. We compute the values of the function and its first five derivatives as follows:

$$f(0) = 1$$

$$f'(0) = \left(3x^2 - 2\sin(2x)\right)\Big|_{x=0} = 0$$

$$f''(0) = \left(6x - 4\cos(2x)\right)\Big|_{x=0} = -4$$

$$f'''(0) = \left(6 + 8\sin(2x)\right)\Big|_{x=0} = 6$$

$$f^{(4)}(0) = 16\cos(2x)\Big|_{x=0} = 16$$

$$f^{(5)}(0) = -32\sin(2x)\Big|_{x=0} = 0$$

Therefore $P_1(x) = 1$, $P_2(x) = 1 - 2x^2$, and $P_5(x) = 1 - 2x^2 + x^3 + \dfrac{2x^4}{3}$.

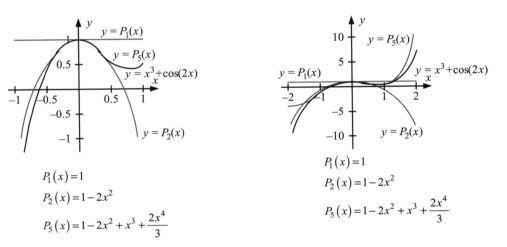

$$P_1(x) = 1$$

$$P_2(x) = 1 - 2x^2$$

$$P_5(x) = 1 - 2x^2 + x^3 + \frac{2x^4}{3}$$

$$P_1(x) = 1$$

$$P_2(x) = 1 - 2x^2$$

$$P_5(x) = 1 - 2x^2 + x^3 + \frac{2x^4}{3}$$

49.   We compute the values of the function and its first five derivatives as follows:

$$f(1) = \sin\left(1 + \frac{\pi}{4}\right)$$

$$f'(1) = \cos\left(1 + \frac{\pi}{4}\right)$$

$$f''(1) = -\sin\left(1 + \frac{\pi}{4}\right)$$

$$f'''(1) = -\cos\left(1 + \frac{\pi}{4}\right)$$

$$f^{(4)} = \sin\left(1 + \frac{\pi}{4}\right)$$

$$f^{(5)}(1) = \cos\left(1 + \frac{\pi}{4}\right)$$

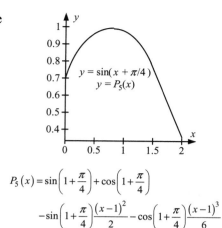

$$P_5(x) = \sin\left(1 + \frac{\pi}{4}\right) + \cos\left(1 + \frac{\pi}{4}\right)$$

$$-\sin\left(1 + \frac{\pi}{4}\right)\frac{(x-1)^2}{2} - \cos\left(1 + \frac{\pi}{4}\right)\frac{(x-1)^3}{6}$$

$$+\sin\left(1 + \frac{\pi}{4}\right)\frac{(x-1)^4}{24} + \cos\left(1 + \frac{\pi}{4}\right)\frac{(x-1)^5}{120}$$

This graph is of better quality than the ones shown in Problem 10.3.45.

Therefore

$$P_5(x) = \sin\left(1 + \frac{\pi}{4}\right) + \cos\left(1 + \frac{\pi}{4}\right)(x-1) - \sin\left(1 + \frac{\pi}{4}\right)\frac{(x-1)^2}{2}$$

$$-\cos\left(1 + \frac{\pi}{4}\right)\frac{(x-1)^3}{6} + \sin\left(1 + \frac{\pi}{4}\right)\frac{(x-1)^4}{24} + \cos\left(1 + \frac{\pi}{4}\right)\frac{(x-1)^5}{120}$$

51. We compute the values of the function and its first five derivatives:

$$f(1) = 1 + \cos(2)$$
$$f'(1) = 3 - 2\sin(2)$$
$$f''(1) = 6 - 4\cos(2)$$
$$f'''(1) = 6 + 8\sin(2)$$
$$f^{(4)}(1) = 16\cos(2)$$
$$f^{(5)}(1) = -32\sin(2)$$

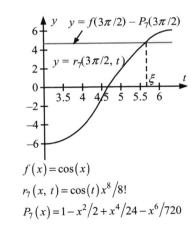

$y = x^3 + \cos(2x)$
$y = P_5(x)$

$$P_5(x) = 1 + \cos(2) + (3 - \sin(2))(x-1)$$
$$+ (3 - 2\cos(2))(x-1)^2$$
$$+ (3 + 4\sin(2))\frac{(x-1)^3}{3} + 2\cos(2)\frac{(x-1)^4}{3}$$
$$- 4\sin(2)\frac{(x-1)^5}{15}$$

This graph is of better quality than the ones shown in Problem 10.3.47.

Therefore

$$P_5(x) = 1 + \cos(2) + (3 - \sin(2))(x-1) + (3 - 2\cos(2))(x-1)^2 + (3 + 4\sin(2))\frac{(x-1)^3}{3}$$
$$+ 2\cos(2)\frac{(x-1)^4}{3} - 4\sin(2)\frac{(x-1)^5}{15}$$

53. Let

$$P_7(x) = 1 - \frac{x^2}{2} + \frac{x^4}{24} - \frac{x^6}{720}, \quad r_7(x, t) = \frac{\cos(t)x^8}{8!}.$$

The solution of the equation

$$\cos\left(\frac{3\pi}{2}\right) = P_7\left(\frac{3\pi}{2}\right) + r_7\left(\frac{3\pi}{2}, \xi\right) \text{ is}$$

$y = f(3\pi/2) - P_7(3\pi/2)$

$y = r_7(3\pi/2, t)$

$f(x) = \cos(x)$
$r_7(x, t) = \cos(t)x^8/8!$
$P_7(x) = 1 - x^2/2 + x^4/24 - x^6/720$

$$\xi = 2\pi$$
$$- \arccos\left(224\frac{-5120 + 5760\pi^2 - 1080\pi^4 + 81\pi^6}{729\pi^8}\right)$$
$$\approx 5.62323$$

and $c < \xi < x_0$ as predicted by Taylor's Theorem.

55.   Let

$$P_4(x) = 1 + 6(x-1) + 15(x-1)^2 + 20(x-1)^3$$
$$+ 15(x-1)^4,$$

$$r_4(x, t) = 720t \frac{(x-1)^5}{5!}.$$

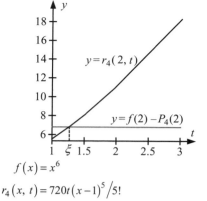

The solution of equation $2^6 = P_4(2) + r_4(2, \xi)$ is

$\xi = \dfrac{7}{6}$ and $c < \xi < x_0$ as predicted by Taylor's

Theorem.

$f(x) = x^6$

$r_4(x, t) = 720t(x-1)^5/5!$

$P_4(x) = 1 + 6(x-1) + 15(x-1)^2 + 20(x-1)^3$
$\qquad + 15(x-1)^4$

Section 10.4   **Estimating the Error Term—The Rate of Convergence of Taylor's Expansion**

Problems for Practice

Note: For problems in this section, we write $O(x^N)$ for any error term that is bounded by a constant times x^N as $x \to 0$. In particular $\lim\limits_{x \to 0} O(x^N) / x^{N-1} = 0$.

1.   **a.**   We calculate $e^{0.1} \approx 1 + (0.1) + \dfrac{(0.1)^2}{2} + \dfrac{(0.1)^3}{6} + \dfrac{(0.1)^4}{24} \approx 1.10517083$; a plot shows that $\left| f^{(5)}(x) \right|$ is maximized at $x = 0.1$.

b.   The error is bounded by $\left| f^{(5)}(0.1) \right| \dfrac{(0.1)^5}{5!} = 9.20975765 \times 10^{-8}$. (The actual error is about 8.5×10^{-8}.)

3.   **a.**   We calculate $\sin(0.3) \approx (0.3) - \dfrac{(0.3)^3}{6} + \dfrac{(0.3)^5}{120} = 0.29552025$; a plot shows that $\left| f^{(5)}(x) \right|$ is maximized at $x = 0.3$.

b.   The error is bounded by $\left| f^{(5)}(0.3) \right| \dfrac{|0.3|^5}{5!} = 2.99214209 \times 10^{-7}$. (The actual error is about 4.33×10^{-8}.)

5.   **a.**   We calculate $\cos(-0.2) \approx 1 - \dfrac{(-0.2)^2}{2} + \dfrac{(-0.2)^4}{24} = 0.9800666667$; a plot shows that $\left| f^{(5)}(x) \right|$ is maximized at $x = -0.2$.

b.   The error is bounded by $\left| f^{(5)}(-0.2) \right| \dfrac{|-0.2|^5}{5!} = 5.2978488 \times 10^{-7}$. (The actual error is about 8.89×10^{-8} .)

7.   **a.**   We calculate

$$\ln(3) = \ln\left(e + (3-e)\right) = 1 + \ln\left(1 + \frac{3-e}{e}\right)$$

$$\approx 1 + \frac{3-e}{e} - \frac{(3-e)^2}{2e^2} + \frac{(3-e)^3}{3e^3} - \frac{(3-e)^4}{4e^4} + \frac{(3-e)^5}{5e^5} = 1.098612478.$$

A plot shows that $\left| f^{(6)}(x) \right|$ is maximized at $x = e$.

b.   The error is bounded by $\left| f^{(6)}(e) \right| \dfrac{|3-e|^6}{6!} = 2.07 \times 10^{-7}$. (The actual error is about 1.90×10^{-7} .)

9.   **a.**   We calculate $\arctan(2(0.2)) \approx 2(0.2) = 0.4$; a plot shows that $\left| f^{(3)}(x) \right|$ is maximized at $x = 0$.

b.   The error is bounded by $\left| f^{(3)}(0) \right| \dfrac{(0.2-0)^3}{3!} = 0.0213$. (The actual error is about 0.01949.)

11.   **a.**   We calculate $\sqrt{1 + 8.2} \approx 3 + \dfrac{8.2-8}{6} - \dfrac{(8.2-8)^2}{216} + \dfrac{(8.2-8)^3}{3888} = 3.033150206$; a plot shows that $\left| f^{(4)}(x) \right|$ is maximized at $x = 8$.

b.   The error is bounded by $\left| f^{(4)}(8) \right| \dfrac{(8.2-8)^4}{4!} = 2.8578 \times 10^{-8}$. (The actual error is about 2.8141×10^{-8} .)

In each of Exercises 13–21, we first calculate $f^{(N+2)}(x)$ and observe that it does not vanish on the interval with endpoints c and x_0. Having done this, we have ensured that M_{N+1} is the larger of $\left|f^{(N+1)}(c)\right|$ and $\left|f^{(N+1)}(x_0)\right|$. The error estimate $R_N(x_0)$ is then $M_{N+1}\left|x_0 - c\right|^{N+1}/(N+1)!$.

13.   Given that $f^{(6)}(x) = -945\dfrac{(1+x)^{-11/2}}{64} \neq 0$ in $[0,\ 0.4]$, $f^{(5)}(x) = 105\dfrac{(1+x)^{-9/2}}{32}$, the maximum value is $M_5 = 105\dfrac{(1)^{-9/2}}{32}$, and the remainder is

$$R_4(0.4) = 105(1)^{-9/2}\frac{(0.4)^5}{32 \cdot 5!} = 0.00028.$$

15.   The maximum value is $M_3 = \left|-3\sin\left(\dfrac{\pi}{6}\right) - \left(\dfrac{\pi}{6}\right)\cos\left(\dfrac{\pi}{6}\right)\right|$, and the remainder is

$$R_2\left(\frac{\pi}{6}\right) = \left|-3\sin\left(\frac{\pi}{6}\right) - \left(\frac{\pi}{6}\right)\cos\left(\frac{\pi}{6}\right)\right|\frac{(\pi/6)^3}{3!} = 0.0467355.$$

17.   Given that $f^{(5)}(x) = e^x \neq 0$ in $[-2,\ -1.1]$, the maximum value is $M_4 = \exp(-1.1)$, and the remainder is $R_3(-1.1) = \exp(-1.1)\dfrac{\left|-2-(-1.1)\right|^4}{4!} = 9.099863251 \times 10^{-3}$.

19.   Given that $f^{(7)}(x) = \dfrac{720}{x^7} \neq 0$ in $[2.5,\ e]$, the maximum value is $M_6 = \dfrac{120}{(2.5)^6}$, and the remainder is $R_5(2.5) = 120\dfrac{\left|e-2.5\right|^6}{(2.5)^6\,6!} = 7.384382694 \times 10^{-8}$.

21.   Given that $f^{(4)}(x) = 24x\dfrac{1-x^2}{\left(1+x^2\right)^4} \neq 0$ in $(0,\ 0.6)$, $f^{(3)}(x) = 2\dfrac{3x^2-1}{\left(1+x^2\right)^3}$, the maximum value is $M_3 = \left|f^{(3)}(0)\right|$, and the remainder is $R_2(0.6) = \left|f^{(3)}(0)\right|\dfrac{(0.6)^3}{3!} = 0.072$.

Further Theory and Practice

23.   Given that $f^{(N+1)}(x) = \pm\cos(x)$ or $\pm\sin(x)$ we have $M_{N+1} = 1$ and $R_N(0.2) = \dfrac{(0.2)^{N+1}}{(N+1)!}$.

For $N = 4$ we have $R_4(0.2) = 2.67 \times 10^{-6}$ and for $N = 5$ we have $R_5(0.2) = 8.89 \times 10^{-8}$.

Thus, $N = 5$ is the smallest index N for which $R_N(0.2) < 5 \times 10^{-7}$. Our estimate is

$$P_5(0.2) = 1 - \frac{(0.2)^2}{2} + \frac{(0.2)^4}{24} = 0.9800667 .$$

25.   Given that $f^{(N+1)}(x) = \pm\cos(x)$ or $\pm\sin(x)$ we have $M_{N+1} = 1$ and $R_N(1.5) = \dfrac{(\pi/2 - 1.5)^{N+1}}{(N+1)!}$.

For $N = 4$ we have $R_N(1.5) = \dfrac{(\pi/2 - 1.5)^{4+1}}{(4+1)!} = 1.482082504 \times 10^{-8}$ and for $N = 5$ we have

$R_N(1.5) = \dfrac{(\pi/2 - 1.5)^{5+1}}{(5+1)!} = 1.748766621 \times 10^{-10}$. Thus, $N = 5$ is the smallest index N for

which $R_N(1.5) < 5 \times 10^{-9}$. Our estimate is

$$P_5(1.5) = 1 - \frac{(1.5 - \pi/2)^2}{2} + \frac{(1.5 - \pi/2)^4}{24} = 0.99749499 .$$

27.   Given that $f^{(N+1)}(x) = \pm\exp(-x)$, we have $M_{N+1} = 1$ and $R_N(0.1) = \dfrac{(0.1)^{N+1}}{(N+1)!}$. For

$N = 2$ we have $R_N(0.1) = \dfrac{(0.1)^{2+1}}{(2+1)!} = 1.666666667 \times 10^{-4}$ and for $N = 3$ we have

$R_N(0.1) = \dfrac{(0.1)^{3+1}}{(3+1)!} = 4.166666667 \times 10^{-6}$. Thus, $N = 3$ is the smallest index N for which

$R_N(0.1) < 5 \times 10^{-5}$. Our estimate is $P_3(0.1) - 1 - (0.1) + \dfrac{(0.1)^2}{2} - \dfrac{(0.1)^3}{6} = 0.9048 .$

29. We calculate

$$f'(x) = \frac{1}{x}$$

$$f''(x) = -\frac{1}{x^2}$$

$$f'''(x) = \frac{2}{x^3}$$

$$f^{(4)}(x) = -\frac{3!}{x^4}$$

$$f^{(5)} = \frac{4!}{x^5}$$

and, in general, $f^{(N)}(x) = (-1)^{N-1}(N-1)!x^{-N}$. We see that $M_N = (N-1)!$ and

$$\left|R_N(1.2)\right| \le N!\frac{(1.2-1)^{(N+1)}}{(N+1)!} = \frac{(0.2)^{N+1}}{N+1}.$$

Given that $\dfrac{(0.2)^{4+1}}{4+1} = 0.000064$ and $\dfrac{(0.2)^{5+1}}{5+1} = 1.067 \times 10^{-5}$, we conclude that $N = 5$ is the smallest index that ensures 4 decimal places of accuracy. Our approximation is

$$P_5(1.2) = (1.2-1) - \frac{(1.2-1)^2}{2} + \frac{(1.2-1)^3}{3} - \frac{(1.2-1)^4}{4} + \frac{(1.2-1)^5}{5} = 0.1823.$$

31. We will expand $\ln(x)$ about $c = e^4$, using $x_0 = 50$ and a degree N to be determined. We have

$$\ln(x_0) = 4 + \frac{\left(x_0 - e^4\right)}{e^4} - \frac{\left(x_0 - e^4\right)^2}{2e^8} + \frac{\left(x_0 - e^4\right)^3}{3e^{12}} - \frac{\left(x_0 - e^4\right)^4}{4e^{16}} + \frac{\left(x_0 - e^4\right)^5}{5e^{20}}$$

$$+ \cdots + (-1)^{N+1} \frac{\left(x_0 - e^4\right)^N}{Ne^{4N}}$$

Given that the $N+1^{\text{st}}$ derivative of $\ln(x)$ is $\dfrac{N!}{x^{N+1}}$, the absolute value of $R_N(x_0)$ is, for

some $50 < \xi < e^4$, $\dfrac{1}{(N+1)|\xi|^{N+1}}\left(e^4 - 50\right)^{N+1}$, which is less than $\dfrac{1}{(N+1)|50|^{N+1}}\left(e^4 - 50\right)^{N+1}$.

For $N = 3$ this error bound is $\dfrac{\left(e^4 - 50\right)^{3+1}}{(3+1)|50|^{3+1}} \approx 1.7881 \times 10^{-5} < 5 \times 10^{-5}$. Therefore,

$$P_3(x_0) = 4 + \frac{50 - e^4}{e^4} - \frac{\left(50 - e^4\right)^2}{2e^8} + \frac{\left(50 - e^4\right)^3}{3e^{12}} = 3.9120\ldots$$

gives the required approximation of $\ln(50)$.

33. Given that $\sin(x) = x - \dfrac{x^3}{6} + R_3(x)$ and $R_3(x)$ is bounded by a constant times x^4, we have

$$\lim_{x \to 0} \frac{\sin(x) - x}{x^3} = \lim_{x \to 0} \left(-\frac{1}{6} + \frac{R_3(x)}{x^3} \right) = -\frac{1}{6}.$$

35. $\displaystyle\lim_{x \to 0} \frac{\ln(1+x) - x}{x^2} = \lim_{x \to 0} \frac{-x^2/2 + x^3/3 + R_3(x)}{x^2} = \lim_{x \to 0} \left(-\frac{1}{2} + \frac{x}{3} \right) = -\frac{1}{2}$

37. $\displaystyle\lim_{x \to 0} \frac{\exp(x) + \exp(-x) - 2}{x^2} = \lim_{x \to 0} \frac{\left(1 + x + x^2/2 + O\left(x^3\right)\right) + \left(1 - x + x^2/2 + O\left(x^3\right)\right) - 2}{x^2}$

$$= \lim_{x \to 0} \left(1 + O\frac{x^3}{x^2} \right) = 1$$

39.
$$\lim_{x \to 0} \frac{\arctan(x) - x}{x^3} = \lim_{x \to 0} \frac{x - x^3/3 + O(x^4) - x}{x^3} = \lim_{x \to 0} \left(-\frac{1}{3} + O\frac{x^4}{x^3} \right) = -\frac{1}{3}$$

41. Given that $\sin(x) = x - \dfrac{x^3}{6} + O(x^4)$, $\cos(x) = 1 - \dfrac{x^2}{2} + O(x^4)$, and

$$\ln(1+x) = x - \frac{x^2}{2} + \frac{x^3}{3} + O(x^4),$$

we have

$$\lim_{x \to 0} \frac{\sin(x) - x}{(1 - \cos(x)) \cdot \ln(1+x)} = \lim_{x \to 0} \frac{-x^3/6 + O(x^4)}{\left(x^2/2 + O(x^4)\right)\left(x - x^2/2 + x^3/3 + O(x^4)\right)} = \lim_{x \to 0} \frac{-x^3/6 + O(x)}{x^3/2 + O(x^4)}$$

$$= -\frac{1}{3}.$$

Calculator/Computer Exercises

43. The plot shows that $M_3 \approx 10.45$, leading to the upper bound $10.45 \dfrac{(\pi/4 - \pi/12)^3}{3!} = 0.25$ for $R_2\left(\dfrac{\pi}{4}\right)$. We calculate

$$f(c) = \left(\frac{\pi^2}{144}\right) \sin\left(\frac{\pi^2}{144}\right) = 0.00469$$

$$f'(c) = \left(\frac{\pi^3}{864}\right) \cos\left(\frac{\pi^2}{144}\right) + \left(\frac{\pi}{6}\right) \sin\left(\frac{\pi^2}{144}\right) = 0.007166$$

$$\frac{f''(c)}{2!} = \left(\frac{\pi^2}{36}\right) \cos\left(\frac{\pi^2}{144}\right) + \left(\frac{\pi^2}{144}\right)\left(\cos\left(\frac{\pi^2}{144}\right) - \left(\frac{\pi^2}{72}\right) \sin\left(\frac{\pi^2}{144}\right)\right) + \sin\left(\frac{\pi^2}{144}\right) = 0.40973$$

Our approximation is

$$P_2\left(\frac{\pi}{4}\right) = 0.00469 + 0.0717\left(\frac{\pi}{4} - \frac{\pi}{12}\right)$$

$$+ 0.40973\left(\frac{\pi}{4} - \frac{\pi}{12}\right)^2 = 0.15456$$

Observe that

$$\left|f\left(\frac{\pi}{4}\right) - P_2\left(\frac{\pi}{4}\right)\right| = |0.35683 - 0.15456| = 0.20227$$

$$< 0.25$$

in agreement with our estimate.

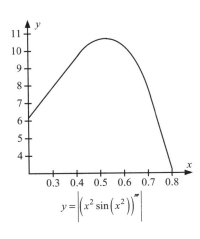

$y = \left|\left(x^2 \sin\left(x^2\right)\right)''\right|$

45.   The plot shows that $M_4 \approx 0.007215$, leading to the

upper bound $0.007215\dfrac{(4-3.5)^4}{4!} = 1.88\times10^{-5}$ for

$R_3(4)$. We calculate

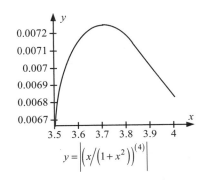

$y = \left|\left(x/\left(1+x^2\right)\right)^{(4)}\right|$

$$f(c) = \frac{3.5}{1+3.5^2} = 0.2641509$$

$$f'(c) = \frac{1-3.5^2}{\left(1+3.5^2\right)^2} = -0.0640797$$

$$\frac{f''(c)}{2!} = (3.5)\frac{3.5^2-3}{\left(1+3.5^2\right)^3} = 0.0139175$$

$$\frac{f'''(c)}{3!} = \frac{6\left(3.5^2\right)-3.5^4-1}{\left(1+3.5^2\right)^4} = -0.0025164$$

Our approximation is

$$P_3(4) = 0.2641509 - 0.0640797(4-3.5) + 0.0139175(4-3.5)^2 - 0.0025164(4-3.5)^3$$

$$= 0.2352759$$

Observe that $|f(4) - P_3(4)| = |0.2352941 - 0.2352759| = 0.0000182 < 1.88\times10^{-5}$, in agreement with our estimate.

47. The plot shows that $M_5 \approx 160.02$, leading to the upper bound

$$160.02 \cdot \frac{(5-4.4)^5}{5!} = 0.103693 \text{ for } R_4(5). \text{ We calculate}$$

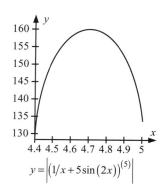

$$f(c) = 3.151859$$

$$f'(c) = -8.162583$$

$$f''\frac{(c)}{2!} = -5.837433$$

$$f'''\frac{(c)}{3!} = 5.404619$$

$$f^{(4)}\frac{(c)}{4!} = 1.950330$$

Our approximation is

$$P_4(5) = 3.151859 - 8.162583(5-4.4) - 5.837433(5-4.4)^2 + 5.404619(5-4.4)^3$$
$$+1.950330(5-4.4)^4 = -2.427006$$

Observe that $|f(5) - P_4(5)| = |-2.520106 - (-2.427006)| = 0.0931 < 0.103693$, in agreement with our estimate.

49. The plot shows that $M_3 \approx 0.071075$, leading to the

upper bound $0.071075 \cdot \dfrac{\left(\sqrt{2} - 1.4\right)^3}{3!} = 3.4 \times 10^{-8}$ for

$R_2(1.4)$. We calculate

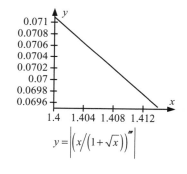

$$f(c) = 0.6459934980$$

$$f'(c) = 0.3327200914$$

$$f''\frac{(c)}{2!} = -0.04196853350$$

Our approximation is

$$P_2(1.4) = 0.6459934980 + 0.3327200914\left(1.4 - \sqrt{2}\right) - 0.04196853350\left(1.4 - \sqrt{2}\right)^2$$
$$= 0.6412558815$$

Observe that $|f(1.4) - P_2(1.4)| = |0.6412558482 - 0.6412558815| = 3.33 \times 10^{-8} < 3.4 \times 10^{-8}$, in agreement with our estimate.

Section 10.5   Taylor Series

Problems for Practice

1. We substitute $u = 2x$ in the Maclaurin series $\sum\limits_{n=1}^{\infty} (-1)^{n+1} \dfrac{u^{2n-1}}{(2n-1)!}$ of $\sin(u)$ to obtain

$$\sin(2x) = \sum_{n=1}^{\infty} (-1)^{n+1} \frac{(2x)^{2n-1}}{(2n-1)!}.$$

3. We substitute $u = -x$ in the Maclaurin series $\sum\limits_{n=0}^{\infty} \dfrac{u^n}{n!}$ of $\exp(u)$ to obtain

$$\exp(-x) = \sum_{n=0}^{\infty} (-1)^n \frac{x^n}{n!}.$$

Then $\exp(1-x) = e\exp(-x) = \sum\limits_{n=0}^{\infty} (-1)^n e \cdot \dfrac{x^n}{n!}.$

5. We substitute $u = x^2$ in the Maclaurin series $\sum\limits_{n=0}^{\infty} (-1)^n \dfrac{x^{2n}}{(2n)!}$ of $\cos(x)$ to obtain

$$\cos(x^2) = \sum_{n=0}^{\infty} (-1)^n \frac{x^{4n}}{(2n)!}.$$

Thus, $x^3 + \cos(x^2) = 1 + x^3 + \sum\limits_{n=1}^{\infty} (-1)^n \dfrac{x^{4n}}{(2n)!}.$

7. We observe that $\dfrac{1}{1+x^2} = \dfrac{1}{1-(-x^2)} = \sum\limits_{n=0}^{\infty} (-x^2)^n = \sum\limits_{n=0}^{\infty} (-1)^n x^{2n}$. We then multiply this

Maclaurin series by x to obtain $\dfrac{x}{1+x^2} = \sum\limits_{n=0}^{\infty} (-1)^n x^{2n+1}$ for $-1 < x < 1$.

9. We substitute $u = x^3$ in the binomial series $(1+u)^{1/2} = \sum\limits_{n=0}^{\infty} \binom{1/2}{n} u^n$ to obtain

$$\sqrt{1+x^3} = \sum_{n=0}^{\infty} \binom{1/2}{n} x^{3n}$$

for $-1 < x < 1$.

11.   We substitute $u = x^2$ in the Maclaurin series $\sum\limits_{n=1}^{\infty} (-1)^{n+1} \dfrac{u^n}{n}$ of $\ln(1+u)$ to obtain

$$\ln\left(1+x^2\right) = \sum_{n=1}^{\infty} (-1)^{n+1} \frac{x^{2n}}{n}$$

for $-1 < x < 1$. We then divide each term of this Maclaurin series by x to obtain

$$\frac{\ln\left(1+x^2\right)}{x} = \sum_{n=1}^{\infty} \frac{(-1)^{n+1}}{n} x^{2n-1}$$

for $-1 < x < 1$.

13.   We calculate $\sqrt{x} = \left(1+(x-1)\right)^{1/2} = \sum\limits_{n=0}^{\infty} \binom{1/2}{n}(x-1)^n$.

15.   We calculate $x^2 e^x = x^2 \sum\limits_{n=0}^{\infty} \dfrac{x^n}{n!} = \sum\limits_{n=0}^{\infty} \dfrac{x^{n+2}}{n!} \overset{m=n+2}{=} \sum\limits_{m=2}^{\infty} \dfrac{x^m}{(m-2)!}$.

17.   We calculate $\ln(x) = \ln\left(1+(x-1)\right) = \sum\limits_{n=1}^{\infty} (-1)^{n+1} \dfrac{(x-1)^n}{n}$.

19.   We calculate $\sin(x)\cos(x) = \dfrac{\sin(2x)}{2} = \left(\dfrac{1}{2}\right)\sum\limits_{n=1}^{\infty} (-1)^{n+1} \dfrac{(2x)^{2n-1}}{(2n-1)!} = \sum\limits_{n=1}^{\infty} (-1)^{n+1} 2^{2(n-1)} \dfrac{x^{2n-1}}{(2n-1)!}$.

21.   We calculate $\cos(0.2) = \sum\limits_{n=0}^{\infty} (-1)^n \dfrac{(0.2)^{2n}}{(2n)!}$. If we truncate this alternating series, the error is less than the absolute value of the first term omitted. Thus, for $n = 3$,

$$\frac{(0.2)^{2n}}{(2n)!} = 8.9 \times 10^{-8} < 5 \times 10^{-5}.$$

We calculate $\cos(0.2) = \sum\limits_{n=0}^{1} (-1)^n \dfrac{(0.2)^{2n}}{(2n)!} = 1 - \dfrac{(0.2)^2}{2} = 0.9800$, correct to four decimal places.

23. For any positive integer N, we obtain $\exp(-0.2) = \sum_{n=0}^{N-1} \dfrac{(-0.2)^n}{n!} + \varepsilon$ where $|\varepsilon| < \dfrac{(0.2)^N}{N!}$. Given

that $\dfrac{(0.2)^5}{5!} = 2.7 \times 10^{-6} < 5 \times 10^{-5}$, we find $\exp(-0.2) = \sum_{n=0}^{5-1} \dfrac{(-0.2)^n}{n!} = 0.8187$, to four decimal

places.

25. $(1+x)^{3/4} = 1 + \dfrac{3x}{4} - \dfrac{3x^2}{32} + \dfrac{5x^3}{128} + \cdots$

27. $\dfrac{1}{\sqrt{1+x}} = 1 - \dfrac{x}{2} + \dfrac{3x^2}{8} - \dfrac{5x^3}{16} + \cdots$

29. $\dfrac{1}{(1+x)^{3/2}} = 1 - \dfrac{3x}{2} + \dfrac{15x^2}{8} - \dfrac{35x^3}{16} + \cdots$

Further Theory and Practice

31. The series evaluates to $\sin(\pi/4) = 1/\sqrt{2}$, as we see by substituting $x = \pi/4$ into the Maclaurin series of $\sin(x)$.

33. The series evaluates to $\sin(\pi/2)/(\pi/2) = 2/\pi$, as we see by substituting $x = \pi/2$ into the Maclaurin series of $\sin(x)/x$ (Exercise 10).

35. The series evaluates to $1/e$, as we see by substituting $x = -1$ into the Maclaurin series of $\exp(x)$.

37. Using $\exp(x) = \sum_{n=0}^{\infty} \dfrac{x^n}{n!}$ and $\exp(-x) = \sum_{n=0}^{\infty} \dfrac{(-1)^n x^n}{n!}$, we obtain

$$\cosh(x) = \frac{\exp(x) + \exp(-x)}{2} = \sum_{n=0}^{\infty} \frac{1}{2}\left(\frac{1}{n!} + \frac{(-1)^n}{n!}\right)x^n = \sum_{k=0}^{\infty} \frac{x^{2k}}{(2k)!}.$$

39.   We calculate the integral by expanding the integrand:

$$\int_0^{1/2} e^{-x^2}\,dx = \int_0^{1/2} \sum_{n=0}^{\infty} \frac{(-1)^n}{n!} x^{2n}\,dx = \sum_{n=0}^{\infty} \frac{(-1)^n}{n!} \int_0^{1/2} x^{2n}\,dx.$$

Thus $\displaystyle\int_0^{1/2} e^{-x^2}\,dx = \sum_{n=0}^{\infty} \frac{(-1)^n}{n!\,(2n+1)} x^{2n+1} \Bigg|_{x=0}^{x=1/2} = \sum_{n=0}^{\infty} \frac{(-1)^n}{n!\,(2n+1)\cdot 2^{2n+1}}$. Given that the error in

truncating an alternating series is no greater than the first term omitted, and because

$$\frac{1}{5!\,(10+1)\cdot 2^{10+1}} = 3.7\times 10^{-7} < 5\times 10^{-7}, \text{ we calculate}$$

$$\int_0^{1/2} e^{-x^2}\,dx \approx \sum_{n=0}^{4} \frac{(-1)^n}{n!\,(2n+1)\cdot 2^{2n+1}} = 0.461281,$$

correct to six decimal places. (This accuracy is greater than that specified, but if the last summand is omitted, then the required accuracy is not guaranteed and is, in fact, off in the fifth decimal place.)

41.   We calculate $\displaystyle\int_0^{1/3} \frac{1}{1+x^5}\,dx = \int_0^{1/3} \sum_{n=0}^{\infty} (-1)^n x^{5n}\,dx = \sum_{n=0}^{\infty} (-1)^n \int_0^{1/3} x^{5n}\,dx = \sum_{n=0}^{\infty} \frac{(-1)^n}{(5n+1)\cdot 3^{5n+1}}$. Given

that $\dfrac{1}{(5n+1)\cdot 3^{5n+1}} < 5\times 10^{-6}$ for $n=2$, we see that

$$\int_0^{1/3} \frac{1}{1+x^5}\,dx = \sum_{n=0}^{1} \frac{(-1)^n}{(5n+1)\cdot 3^{5n+1}} = 0.33310.$$

43.   Suppose that f equals its Taylor Series on the interval $I = [c-h, c+h]$. Let M_3 be the maximum value of $|f'''(x)|$ on I. Then, for $x \in I$, we calculate, by Taylor's Theorem,

$$f(x) = f(c) + f'(c)(x-c) + \frac{1}{2}f''(c)(x-c)^2 + \frac{f'''(\xi)}{6}(x-c)^3 \text{ where } \xi \text{ is some number}$$

between c and x. Given that $f'(c) = 0$ the displayed equation simplifies to

$$f(x) = f(c) + (x-c)^2 \left(\frac{1}{2}f''(c) + \frac{f'''(\xi)}{6}(x-c) \right).$$

Suppose that $f''(c) > 0$. By noting that $|f'''(\xi)| < M_3$ for every $x \in I$, we see that by taking x sufficiently close to c we can force $\left| f'''(\xi) \frac{x-c}{6} \right| < f'' \frac{(c)}{2}$. We conclude that, for x sufficiently close to c, $f(x) = f(c) + \text{positive coefficient} \cdot (x-c)^2 > f(c)$. In other words, f has a local minimum at c. The case $f''(c) < 0$ is handled analogously to show that

$$f(x) = f(c) + \text{negative coefficient} \cdot (x-c)^2 < f(c).$$

In this case, f has a local maximum at c.

45.   We calculate

$$\ln\left(\frac{1+x}{1-x}\right) = \ln(1+x) - \ln(1-x) = \sum_{n=1}^{\infty} \frac{(-1)^{n+1}}{n}x^n - \sum_{n=1}^{\infty} \frac{(-1)^{2n+1}}{n}x^n$$

$$= \sum_{n=1}^{\infty} \left(\frac{(-1)^{n+1} - (-1)^{2n+1}}{n} \right)$$

Notice that $(-1)^{2n+1}$ equals -1 for every integer n. Therefore, $(-1)^{n+1} - (-1)^{2n+1}$ equals 0 for n even and 2 for n odd. The given formula follows: $\ln\left(\frac{1+x}{1-x}\right) = \sum_{k=0}^{\infty} \left(\frac{2}{2k+1} \right) x^{2k+1}$. If we set $x = \frac{1}{3}$ in this power series, then we obtain $\ln(2) = \sum_{k=0}^{\infty} \left(\frac{2}{2k+1} \right) \left(\frac{1}{3} \right)^{2k+1}$. Thus,

$$\ln(2) \approx 2 \sum_{k=0}^{N} \frac{(1/3)^{2k+1}}{2k+1}.$$

For $N = 5$, 6, and 7 we obtain 0.6931471 as our approximation.

47. Let $f(x) = (1+x)^\alpha$ and $g(x) = \sum_{n=0}^{\infty} \binom{\alpha}{n} x^n$ for $x \in (-1, 1)$. The following steps show that $f(x) = g(x)$.

   a. Beginning with the left side of the asserted identity, we calculate as follows:

$$(n+1) \cdot \binom{\alpha}{n+1} + n \cdot \binom{\alpha}{n} = (n+1) \cdot \frac{\alpha \cdot (\alpha-1) \cdots (\alpha-n+1) \cdot (\alpha-n)}{(n+1)!}$$
$$+ n \cdot \frac{\alpha \cdot (\alpha-1) \cdots (\alpha-n+1)}{n!}$$
$$= \frac{\alpha \cdot (\alpha-1) \cdots (\alpha-n+1) \cdot (\alpha-n)}{n!}$$
$$+ n \cdot \frac{\alpha \cdot (\alpha-1) \cdots (\alpha-n+1)}{n!}$$
$$= \frac{\alpha \cdot (\alpha-1) \cdots (\alpha-n+1) \cdot ((\alpha-n)+n)}{n!}$$
$$= \alpha \cdot \frac{\alpha \cdot (\alpha-1) \cdots (\alpha-n+1)}{n!} = \alpha \cdot \binom{\alpha}{n}.$$

   b. Beginning with the left side of the asserted identity, we calculate as follows:

$$(1+x) \cdot g'(x) = (1+x) \sum_{n=1}^{\infty} n \binom{\alpha}{n} x^{n-1} = \sum_{n=1}^{\infty} n \binom{\alpha}{n} x^{n-1} + \sum_{n=1}^{\infty} n \binom{\alpha}{n} x^n$$
$$= \sum_{n=0}^{\infty} (n+1) \binom{\alpha}{n+1} x^n \sum_{n=1}^{\infty} n \binom{\alpha}{n} x^n = \sum_{n=0}^{\infty} \left((n+1) \binom{\alpha}{n+1} + n \binom{\alpha}{n} \right) x^n$$
$$= \alpha \cdot g(x).$$

   c. We calculate as follows: $(1+x) \cdot f'(x) = (1+x) \cdot \alpha \cdot (1+x)^{\alpha-1} = \alpha \cdot (1+x)^\alpha = \alpha \cdot f(x)$.

   d. Clearly we calculate $f(0) = g(0) = 1$. (At this point we calculate that f and g are solutions of the same initial value problem: $(1+x) y'(x) = \alpha \cdot y(x)$, $y(0) = 1$.)

   e. The differential equation $(1+x) y' = \alpha \cdot y$ can be written as $\dfrac{y'}{y} = \dfrac{\alpha}{1+x}$, or, for $x > -1$

   and $y > 0$, $\dfrac{d}{dx} \ln(y) = \dfrac{\alpha}{1+x}$. It follows that $\ln(y) = \alpha \ln(1+x) + C$ for some constant C.

   f. If $y(x)$ satisfies the initial value problem $(1+x) y' = \alpha \cdot y$, $y(0) = 1$, then

$$\ln(1) = \alpha \ln(1+0) + C,$$

   or $C = 0$. Therefore, $\ln(y) = \alpha \ln(1+x)$, or $y(x) = (1+x)^\alpha$. In other words, the initial value problem that we studied has a unique solution: $f(x)$. Given that g is also a solution, it follows that $f = g$.

49.   Suppose that $\left(1-x^2\right)f^{(n+2)}(x)-(2n+1)xf^{(n+1)}(x)-n^2 f^{(n)}(x)=0$. If we differentiate each term with respect to x then, using the Product Rule, we have

$$f^{(n+3)}(x)-2xf^{(n+2)}(x)-x^2 f^{(n+3)}(x)-(2n+1)f^{(n+1)}(x)$$
$$-(2n+1)xf^{(n+2)}(x)-n^2 f^{(n+1)}(x)=0$$

On grouping like terms we obtain

$$\left(1-x^2\right)f^{(n+3)}(x)-(2n+3)xf^{(n+2)}(x)-(n+1)^2 f^{(n+1)}(x)=0,$$

which is the original equation with n replaced by $n+1$. When $n=0$ the equation becomes

$$\left(1-x^2\right)f''(x)-xf'(x)=0.$$

If $f(x)=\arcsin(x)$ then $\left(1-x^2\right)f''(x)=\dfrac{x}{\sqrt{1-x^2}}=xf'(x)$, which shows that

$$f(x)=\arcsin(x)$$

satisfies the original equation for $n=0$ and, hence, for all nonnegative values of n. In particular, at $x=0$, we have $f^{(n+2)}(0)=n^2 f^{(n)}(0)$. In the case of

$$f(x)=\arcsin(x),$$

we have $f^{(0)}(0)=0$ and $f^{(1)}(0)=1$. It follows that $f^{(n)}(0)=0$ for all even n, and $f^{(3)}(0)=1^2$, $f^{(5)}(0)=1^2 \cdot 3^2$, $f^{(7)}(0)=1^2 \cdot 3^3 \cdot 5^2$, and so on. We deduce that the Taylor Series of $\arcsin(x)$ is

$$x+\frac{1^2}{3!}x^3+\frac{1^2 \cdot 3^3}{5!}x^5+\frac{1^2 \cdot 3^3 \cdot 5^2}{7!}x^7+\frac{1^2 \cdot 3^3 \cdot 5^2 \cdot 7^2}{9!}x^9+\cdots.$$

Calculator/Computer Exercises

51. The plots of

$$y = \frac{f(x)}{x^5}$$

and

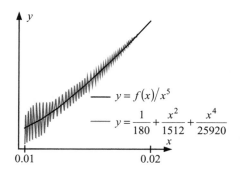

$$y = \frac{1}{180} + \frac{x^2}{1512} + \frac{x^4}{25920}$$

are given. Both were produced in the same way (apart from different choices of color). The inaccuracies in the plot of $y = \dfrac{f(x)}{x^5}$ are caused by a loss of significance. In a computer algebra system, this source of error can be eliminated by using a sufficiently large number of digits in the computations.

53. The Ratio Test shows that the Maclaurin series of $J_0(x)$ converges for all x. We may differentiate term-by-term when we use this series in the left side of the differential equation we are to verify. We obtain

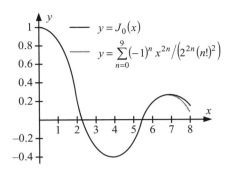

$$\sum_{n=1}^{\infty}(-1)^n \frac{2n(2n-1)x^{2n}}{2^{2n}(n!)^2} + \sum_{n=1}^{\infty}(-1)^n \frac{2nx^{2n}}{2^{2n}(n!)^2}$$

$$+ x^2 \sum_{n=0}^{\infty}(-1)^n \frac{x^{2n}}{2^{2n}(n!)^2}$$

or

$$\sum_{n=1}^{\infty}(-1)^n \frac{(2n)^2 x^{2n}}{2^{2n}(n!)^2} + \sum_{n=0}^{\infty}(-1)^n \frac{x^{2n+2}}{2^{2n}(n!)^2}.$$

If we make the change of summation $m = n+1$ in the second of these series, then we obtain

$$\sum_{n=1}^{\infty}(-1)^n \frac{(2n)^2 x^{2n}}{2^{2n}(n!)^2} - \sum_{n=1}^{\infty}(-1)^m \frac{x^{2m}}{2^{2m-2}((m-1)!)^2}.$$

On renaming the summation index of the second series to n, we see that this difference of series simplifies to

$$\sum_{n=1}^{\infty}(-1)^n \left(\frac{(2n)^2}{2^{2n}(n!)^2} - \frac{1}{2^{2n-2}((n-1)!)^2} \right) x^{2n}.$$

For every $n \in \mathbb{Z}^+$, the coefficient

$$\left(\frac{(2n)^2}{2^{2n}(n!)^2} - \frac{1}{2^{2n-2}((n-1)!)^2} \right)$$

simplifies to 0, proving that $J_0(x)$ is a solution of the given differential equation.

55.   Let n be a positive integer. By applying L'Hôpital's Rule n times we obtain

$$\lim_{u\to\infty}\frac{u^n}{e^u}=\lim_{u\to\infty}\frac{\frac{d}{du}u^n}{\frac{d}{du}e^u}$$

$$=n\lim_{u\to\infty}\frac{u^{n-1}}{e^u}$$

$$=n\lim_{u\to\infty}\frac{\frac{d}{du}u^{n-1}}{\frac{d}{du}e^u}$$

$$=n(n-1)\lim_{u\to\infty}\frac{u^{n-2}}{e^u}$$

$$\vdots$$

$$=n!\lim_{u\to\infty}\frac{u}{e^u}$$

$$=n!\lim_{u\to\infty}\frac{\frac{d}{du}u}{\frac{d}{du}e^u}$$

$$=n!\lim_{u\to\infty}\frac{1}{e^u}$$

$$=0.$$

We conclude that

$$\lim_{x\to 0}\frac{\exp\left(-1/x^2\right)}{x^{2n}}=0$$

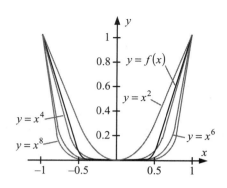

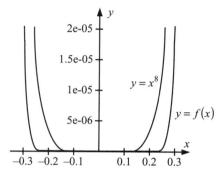

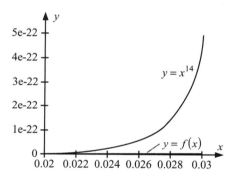

for every positive integer n. This confirms that $f(x)$ approaches 0 more rapidly than any positive power of x as x approaches 0. These calculations suggest that $f^{(n)}(0)=0$ for every nonnegative integer n, which is, in fact, true. The Taylor Series of f with base point 0 is therefore 0. Notice that the Taylor Series of $f(x)$ converges for all x, but it converges to $f(x)$ only at the base point.

ANSWERS TO QUICK QUIZ EXERCISES

Section 1.1

1. An irrational number is a number that has nonterminating, nonrepeating decimal expansion.
2. The decimal expansion of a rational number will either be terminating or repeating.
3. $(-2, 8)$
4. $\{x : |x - 0.5| < 1.5\}$

Section 1.2

1. $x^2 + (y - 2)^2 = 2^2$
2.

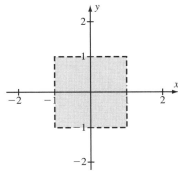

3. Parabola: $y = Ax^2 + Bx + C$ with $A \neq 0$ or $x = Ay^2 + By + C$ with $A \neq 0$

   Hyperbola: $\dfrac{(x - h)^2}{a^2} - \dfrac{(y - k)^2}{b^2} = 1$ with $a > 0, b > 0$ or

   $\dfrac{(y - k)^2}{b^2} - \dfrac{(x - h)^2}{a^2} = 1$ with $a > 0, b > 0$

   Ellipse: $\dfrac{(x - h)^2}{a^2} + \dfrac{(y - k)^2}{b^2} = 1$ with $a > 0, b > 0$
4. $(x - 6)^2 - 6^2$

Section 1.3

1. Slope signifies the steepness of the line with respect to the x-axis.
2. 0, undefined
3. $y = m(x - x_0) + y_0$
4. This is the line that deviates the least from the data points, in the sense that the sum of the squares of the vertical distances of the data point to the line is minimized.

Section 1.4

1. All the values of x in the graph
2. All the values of y in the graph

3. (Vertical line test). If every vertical line drawn through a curve intersects that curve only once, then the curve is the graph of a function.
4. $a_{n+1} = 2 \cdot a_n$ for $n \geq 1$, $a_1 = 2$

Section 1.5

1. The functions must have the same domain.
2. When we have the composition $g \circ f$, the domain of g has to contain the range of f.
3. The graph of $x \mapsto (x - 1)^2 + 4$ can be obtained by shifting the graph of $x \mapsto x^3 + 2$ to the right by 1 unit and up by 2 units.
4. The graph of $x \mapsto (x^2 + 1)/(3x^4 + 5)$ is symmetric with respect to the y-axis. None of the two graphs is symmetric with respect to the origin.
5. For $-\infty < t < 0$, the curve is the line segment $y = -x$ with positive x values. For $0 \leq t < \infty$, the curve is the line segment $y = x$ with non-negative x values.

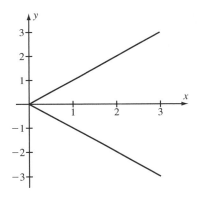

Section 1.6

1. $(-\infty, \infty)$
2. $[-1, 1]$
3. $\cos(x)$ and $\sec(x)$ are even functions. $\sin(x)$, $\csc(x)$, $\tan(x)$, and $\cot(x)$ are odd functions.
4. $2\cos(\theta)$

Section 2.1

1. Yes
2. 10
3. If $\lim\limits_{x \to c} f(x) = \ell$ exists, then given $\epsilon > 0$, we can always find a δ such that $f(x)$ stays within ϵ of ℓ when x stays within δ of c.
4. Within 0.0000249998 of 2

Section 2.2

1. We do not consider the value of x equal to c. f might not be defined at c even when $\lim_{x \to c}$ exists.

2. $3\ell + 1$

$$
\begin{aligned}
\lim_{x \to c}(3f(x) + 1) &= \lim_{x \to c} 3f(x) + \lim_{x \to c} 1 & \text{(Theorem 2 a)} \\
&= 3\lim_{x \to c} f(x) + \lim_{x \to c} 1 & \text{(Theorem 2 d)} \\
&= 3\ell + \lim_{x \to c} 1 & \text{(Given information)} \\
&= 3\ell + 1 & \text{(Theorem 3)}
\end{aligned}
$$

3. $\lim_{x \to -3^+} \sqrt{x + 3} = 0$, but $\lim_{x \to -3^-} \sqrt{x + 3}$ does not exist because the square root of negative number is not defined.

4. $\lim_{x \to c} f(x)$ does not necessarily exist. If both left limit and right limit exist, but the two limits are not equal, then $\lim_{x \to c} f(x)$ will not exist, e.g., $f(x) = \begin{cases} 1 & \text{if } x \geq 0 \\ -1 & \text{if } x < 0 \end{cases}$.

5. (See Theorem 6.)

Section 2.3

1. It means $\lim_{x \to c} f(x) = f(c)$. It means that f is continuous at every point of its domain.

2. Every point in the real line except $-2^{1/4}$ and $2^{1/4}$

3. Addition, subtraction, scalar multiplication ($\alpha \cdot f$), and multiplication ($f \cdot g$) all preserve continuity. Division (f/g) will preserve continuity if the denominator is not zero. Composition ($g \circ f$) will preserve continuity if the composed function $g \circ f$ is well-defined; that means the image of f is contained in the domain of g.

4. No, f does not necessarily have a minimum value. For example, the function $f(x) = \begin{cases} x/3, & \text{if } 0 < x \leq 3 \\ 1, & \text{if } x = 0 \end{cases}$ does not have a minimum value.

   If f is continuous, then it will have a minimum value because of the Extreme Value Theorem.

5. Yes. Since $f(2) = \dfrac{1}{4} < \dfrac{1}{\pi} < \dfrac{1}{2} = f(1)$, by the Intermediate Value Theorem, there exists a number between 1 and 2, such that $f(c) = \dfrac{1}{\pi}$. Therefore, $\dfrac{1}{f}(c) = \dfrac{1}{f(c)} = \pi$.

Section 2.4

1. f has a one-sided or two-sided infinite limit as $x \to c$.

2. $\lim_{x \to +\infty} g(x) = \alpha$ or $\lim_{x \to -\infty} g(x) = \alpha$

3. It does not have any horizontal nor vertical asymptote.

   Since $\sin(x)$ is a periodic function and its values oscillate between -1 and 1, $\lim_{x \to \pm\infty} \sin(x)$ does not exist. Therefore, it does not have any horizontal asymptote.

   Since the values of $\sin(x)$ will never be greater than 1 or less than -1, $\lim_{x \to c} \sin(x)$ will never be infinite. Therefore, it does not have any vertical asymptote.

4. Vertical asymptotes at $x = 2$ and $x = -2$. Horizontal asymptote is $y = 1/2$.

Section 2.5

1. It means that a_n is as close as we please to 3 when n is large enough.

2. It means that the limit $\lim_{n \to \infty} a_n$ does not exist.

3. $\{\sin(j\pi/2)\}_{j=1}^{\infty} = \{1, 0, -1, 0, 1, 0, -1, \ldots\}$
   The sequence diverges because $\lim_{j \to \infty} \sin(j\pi/2)$ does not exist.

4. $p > 3$

Section 2.6

1. A monotone sequence is either an increasing sequence or a decreasing sequence.

   No, if a monotone sequence is not bounded then it will not converge.

2. Let $\{a_n\}$ be a sequence of rational numbers such that $a_n \to \pi$, as $n \to \infty$. Then, we have $4^{\pi} = \lim_{n \to \infty} 4^{a_n}$.

3. e

4. Since the value of e is greater than 1, the graph of e^x is

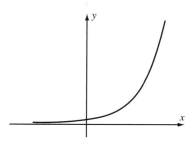

Since the value of (e/π) is less than 1, the graph of $(e/\pi)^x$ is

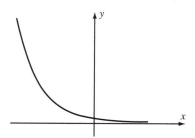

Section 3.1

1. Average velocity is calculated for a given time interval, say from $t = a$ to $t = b$, but the instantaneous velocity is calculated at one point, say at time $t = c$.

2. It is moving backward.

3. $y = -2x + 5$

4. $y = 4x - 4$

Section 3.2

1. They are almost the same except that the concept of derivative can be applied to any function with any variable, while the discussion of instantaneous velocity is restricted to distance function with time as its variable.
2. This means that the limit $\lim\limits_{\Delta x \to 0} \dfrac{f(c + \Delta x) - f(c)}{\Delta x}$ does not exist. Geometrically, this means that we cannot define a tangent line to the graph of f at $x = c$.
3. 10
4. $-\sqrt{3}/2$

Section 3.3

1. -8
2. $2x \sin(x) + x^2 \cos(x)$
3. $\dfrac{x \cos(x) - \sin(x)}{x^2}$
4. 1.000008333

Section 3.4

1. $\dfrac{d}{dx} x^p = px^{p-1}$
2. $\dfrac{d}{dx} \sin(x) = \cos(x)$, $\dfrac{d}{dx} \cos(x) = -\sin(x)$,
   $\dfrac{d}{dx} \tan(x) = \sec^2(x)$, $\dfrac{d}{dx} \cot(x) = -\csc^2(x)$,
   $\dfrac{d}{dx} \sec(x) = \sec(x) \tan(x)$, $\dfrac{d}{dx} \csc(x) = -\csc(x) \cot(x)$
3. Let $f(x) = e^x$.

   $$\lim_{x \to 0} \frac{e^x - 1}{x} = \lim_{h \to 0} \frac{e^h - e^0}{h} = f'(0) = e^0 = 1$$

4. $e^x + xe^x$

Section 3.5

1. $\dfrac{d}{dx}(g \circ f)\Big|_{x=c} = \left(\dfrac{dg}{du}\Big|_{u=f(c)} \right) \cdot \left(\dfrac{df}{dx}\Big|_{x=c} \right)$
2. $(g \circ f)'(c) = g'(f(c)) \cdot f'(c)$
3. Chain Rule is used for differentiation of composite function with $(g \circ f)'(x) = g'(f(x)) \cdot f'(x)$.
   Product Rule is for differentiation of product with $(f \cdot g)'(x) = f'(x) \cdot g(x) + f(x) \cdot g'(x)$.
4. $-\sin(x^2 + 3x) \cdot (2x + 3)$

Section 3.6

1. When it is both one-to-one and onto
2. $\dfrac{d(f^{-1})}{dt}\Big|_t = \dfrac{1}{\dfrac{df}{ds}\Big|_{s=f^{-1}(t)}}$
3. $4 + \ln(3)$

4. $2^x \ln(2)$
5. $e^{7 \ln(\pi)}$

Section 3.7

1. It tells us how the first derivative, $f'(x)$, changes with respect to the variable x.
2. $f''(x)$, $\dfrac{d^2 f}{dx^2}$, or $f^{(2)}(x)$
3. $\dfrac{d}{dt}$velocity $=$ acceleration
4. $2 \cos(x) - 4x \sin(x) - x^2 \cos(x)$

Section 3.8

1. The implicit differentiation allows us to find the derivative of y without first solving for y in terms of x.
2. We can substitute the values of x and y after differentiating both sides of the equation.
3. We have $4y^3 \dfrac{dy}{dx} + \dfrac{dy}{dx} = 2x$; therefore, $\dfrac{dy}{dx} = \dfrac{2x}{1 + 4y^3}$.
4. When $x = 3$, the equation $2x + y^2 = 10$ becomes $y^2 = 4$, and hence $y = 2$ because y is a positive variable. By differentiating the equation, we have $2\dfrac{dx}{dt} + 2y\dfrac{dy}{dt} = 0$.
   When $x = 3$, and $y = 2$, this becomes $2 \cdot 4 + 2 \cdot 2 \cdot \dfrac{dy}{dt} = 0$.
   Therefore, $\dfrac{dy}{dt}\Big|_{x=3} = -2$.

Section 3.9

1. $f(c + \Delta x) \approx f(c) + f'(c) \cdot \Delta x$
2. $\Delta f(c) \approx f'(c) \Delta x$
3. $\approx 8^{1/3} + \dfrac{1}{3} \cdot (8)^{-2/3} \cdot (-0.1) = \dfrac{239}{120}$

Section 4.1

1. (See the definition on p. 251 of the text.)
2. False. If $f'(c) = 0$, then the point c is just a candidate for a local extremum. There is no guarantee that this candidate must actually be a local extremum. For example, consider the function $f(x) = x^3$; we have $f'(0) = 0$, but 0 is not a local extremum.
3. The Mean Value Theorem says that there is a number c between a and b for which the tangent line at $(c, f(c))$ is parallel to the line ℓ passing through the points $(a, f(a))$ and $(b, f(b))$.
4. Local minimum at $x = \dfrac{1}{2}$, no local maximum

Section 4.2

1. A function f is increasing on an interval I if $f(\alpha) < f(\beta)$ whenever α and β are points in I with $\alpha < \beta$.
2. $f'(x) < 0$ for each x in I

3. The tangent line to the graph at the point $(c, f(c))$ is parallel to the x-axis.
4. Local maximum at $x = -5$. Local minimum at $x = 3$.

Section 4.3

1. At the critical points inside the interval and the end points of the closed interval
2. No
3. We need to use derivative to locate critical points which are candidates for local extrema.
4. If a function is always increasing or always decreasing, then its extrema will occur at the end points. There is no need to use derivative to locate the critical points.

Section 4.4

1. It means that as x moves from left to right on I, f' will increase.
2. It means that as x moves from left to right on I, f' will decrease.
3. Let c be the point at which $f'(c) = 0$.
   a. If $f''(c) > 0$, then c is a local minimum.
   b. If $f''(c) < 0$, then c is a local maximum.
   c. If $f''(c) = 0$, then no conclusion is possible from this test.
4. Point of inflection at $x = 3$ but not at -2

Section 4.5

1. $\pi, 2\pi, \pi$
2. $3x - 2$
3. Vertical asymptote $x = 1$; critical points $x = 0$ and 3; increasing on $(-\infty, 1)$ and $(3, \infty)$, decreasing on $(1, 3)$; concave down on $(-\infty, 0)$, concave up on $(0, 1)$ and $(1, \infty)$; local minimum at $x = 3$, point of inflection at $x = 0$; y intercept 0; since $\dfrac{2x^3}{(x-1)^2} = 2x + 4 + \dfrac{6x - 4}{x^2 - 2x + 1}$, we have skew asymptote $y = 2x + 4$.
4. In some cases in order to show the asymptotes of the graph, the scale of the actual plot has to be large, but unfortunately this may make other features of the graph, such as local extrema, inflection points, and the x-intercepts, unnoticeable in the picture. However, when we sketch the graph, we can show all features of the graph clearly visible in one viewing window without worrying about the actual scale of the picture.

Section 4.6

1. We have to make sure that the quotient has indeterminate form $\dfrac{0}{0}$ or $\dfrac{\infty}{\infty}$.

2. Rewrite it as $\displaystyle\lim_{x \to 0^+} \dfrac{\ln(x)}{\dfrac{1}{x}}$.
3. $0^0, 1^\infty$, and ∞^0
4. 0

Section 4.7

1. It is used to find an approximate solution of an equation.
2. To approximate a root of a function f, we consider $x_{j+1} = x_j - \dfrac{f(x_j)}{f'(x_j)}$, provided $f'(x_j) \neq 0$.
3. A loose rule of thumb is that the number of decimal places of accuracy doubles with each iteration.
4. We need to approximate a root of the function $f(x) = x^5 - 4$.
$$\Phi(x_n) = x_n - \frac{f(x_n)}{f'(x_n)} = x_n - \frac{(x_n)^5 - 4}{5(x_n)^4} = \frac{4(x_n)^5 + 4}{5(x_n)^4}$$

Section 4.8

1. Let f be a continuous function. If F is a differentiable function such that $F' = f$ on an open interval I, then F is an antiderivative of f for f on I.
2. The indefinite integral of f, denoted by $\int f(x)\,dx$, is the collection of all antiderivatives of f. An antiderivative $+ C$ will be the indefinite integral.
3. Antidifferentiation respects addition, subtraction and scalar multiplication.
4. Velocity $= \int$ acceleration dt
5. Distance $= \int$ velocity $dt = \int \left(\int \text{acceleration } dt \right) dt$. In other words, we have to take the antidifferentiation of acceleration twice.

Section 4.9

1. Equal
2. Minimize $1000x + 2 \cdot \dfrac{9000}{2x}$. Answer: $x = 3$.
3. \$30. When the average cost is minimized, it equals the marginal cost. Therefore, the marginal cost is $\$30000/1000 = \$30$.
4. Revenue will decrease because $E > 1$.

Section 5.1

1. $(2 \cdot 3^3 - 3^4) + (2 \cdot 4^3 - 4^4) + (2 \cdot 5^3 - 5^4) + (2 \cdot 6^3 - 6^4) + (2 \cdot 7^3 - 7^4)$
2. $3\displaystyle\sum_{j=4}^{150} j - \sum_{j=4}^{150} j^2 = 33957 - 1136261 = -1102304$
3. The subintervals are $[-1, -1/4]$, $[-1/4, 1/2]$, $[1/2, 5/4]$, and $[5/4, 2]$, with $\Delta x = 3/4$. In order for
$$\sum_{j=1}^{4} \exp(-1 + 3j/4) \cdot (3/4)$$
to be the Riemann Sum
$$\sum_{j=1}^{4} f(x_j) \cdot \Delta x,$$
we have $f(x) = e^x$.

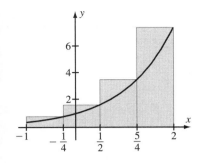

4. It is the area of the region that is bounded above by the graph of f, below by the x-axis, and laterally by the vertical lines $x = a$ and $x = b$, where a and b are the end points of the given interval.

Section 5.2

1. s_3 can be any point inside the subinterval $I_3 = [2, 2.5]$
2. (See the last paragraph on p. 359 of the text.)
3. Area of the rectangle with height 1 and base $[-1, 4]$ is 5. It should be $\int_{-1}^{4} 1 \, dx = x \big|_{x=-1}^{x=4} = 4 - (-1) = 5.$
4. $F(x) = \dfrac{x^3}{3} + \dfrac{x^2}{2}$ is an antiderivative of $x^2 + x$. Therefore,

$\int_{1}^{4} (x^2 + x) \, dx = F(4) - F(1) = \dfrac{4^3}{3} + \dfrac{4^2}{2} - \left(\dfrac{1}{3} + \dfrac{1}{2} \right) = \dfrac{57}{2}.$

Section 5.3

1. $\int_{2}^{7} 3f(x) \, dx = 3 \int_{2}^{7} f(x) \, dx$

$= 3 \left(\int_{0}^{7} f(x) \, dx - \int_{0}^{2} f(x) \, dx \right) = 3 \, (-5 - 3) = -24.$

Therefore, $\int_{7}^{2} 3f(x) \, dx = -\int_{2}^{7} 3f(x) \, dx = 24.$

2. 0
3. No
4. 3

Section 5.4

1. (See Theorem 1.)
2. $\sin(x^2)$
3. $\dfrac{d}{dx} \int_{x}^{0} (1 + t^2)^{-1} dt = -\dfrac{d}{dx} \int_{0}^{x} (1 + t^2)^{-1} dt = -(1 + x^2)^{-1}$
4. $xe^{x^2} \big|_{x=0}^{x=1} = e - 0 = e$

Section 5.5

1. (See p. 387 of the text.)
2. Use the substitution $u = \cos(x)$; the antiderivative becomes
$-\int u^3 du = -\dfrac{(\cos(x))^4}{4} + C.$

3. Use the substitution $u = x^2 - 1$; the definite integral becomes $\dfrac{1}{2} \int_{0}^{15} u^{1/2} du = 5\sqrt{15}.$
4. It is actually the same function. The substitution is basically an elegant way of applying the Chain Rule.

Section 5.6

1. If $f(x)$ is negative, then $\int_{a}^{b} f(x) \, dx$ is negative, but the area is always a positive number.
2. $\int_{0}^{\pi} \sin(x) \, dx - \int_{\pi}^{2\pi} \sin(x) \, dx = 2 - (-2) = 4$
3. $\int_{-1}^{1} (1 - y^2) \, dy = \dfrac{4}{3}$
4. $\int_{a}^{b} |f(x) - g(x)| \, dx$

Section 5.7

1. Approximate the area of the region on each subinterval by area of a rectangle with height $f(\bar{x}_j)$, where $\bar{x}_j$ is the mid-point of the subinterval. (See Figure 1c on p. 401 of the text.)
2. Approximate the area of the region on each subinterval by area of a trapezoid with the two sides $f(x_{j-1})$ and $f(x_j)$. (See Figure 4 on p. 402 of the text.)
3. Approximate the graph of f on each pair of adjacent subintervals, $I_j \cup I_{j+1}$, by the graph of a parabolic curve.
4. Order of accuracy is proportional to $\dfrac{1}{N^4}$.

Section 6.1

1. $f(x, y) = g(x) \cdot h(y)$
2. $\dfrac{dy}{dx} = F(x, y), \ y(x_0) = y_0$
3. Compute $\int \dfrac{1}{h(y)} dy = \int g(x) \, dx.$
4. $\alpha = \pi, \beta = x, C = 2$

Section 6.2

1. $\ln(x) = \int_{1}^{x} \dfrac{1}{t} \, dt, \, x > 0$
2. $\dfrac{2x}{1 + x^2}$
3. Let $u = 2x + 1$; then the integral becomes
$\dfrac{1}{2} \int \dfrac{1}{u} du = \dfrac{1}{2} \ln |2x + 1| + C.$
4. $\dfrac{a^2}{b^3}$

Section 6.3

1. Inverse of $\ln(x)$
2. $2\exp(2x)$
3. Let $u = x^2$; then the integral becomes
$$\frac{1}{2}\int \exp(u)\,du = \frac{1}{2}\exp(x^2) + C.$$
4. $e = \exp(1)$

Section 6.4

1. $\dfrac{\ln(5)}{\ln(3)}$
2. $\exp(\sqrt{2}\ln(3))$
3. $2 \cdot \ln(3) \cdot 3^{\sin(2x)} \cdot \cos(2x)$
4. $\dfrac{2}{x\ln(9)} = \dfrac{1}{x\ln(3)}$
5. $\displaystyle\int \frac{3}{\ln(6)}\ln(x)\,dx = \frac{3}{\ln(6)}(x\ln(x) - x) + C$

Section 6.5

1. Growth: $y(t) = y(0)e^{kt}$; decay: $y(t) = y(0)e^{-\lambda t}$.
   Experimental evidence suggests that the population of bacteria and the mass of a radioactive element grows/decays exponentially over time.
2. $y(t) = y_0 \cdot e^{kt}$
3. Doubling time $= \dfrac{\ln(2)}{\text{growth rate}}$
4. $\dfrac{dT}{dt} = K \cdot (\text{surrounding temperature} - T)$

Section 6.6

1. $\sin(\arcsin(t)) = t$, for all t inside the domain of the arcsine function; $\arcsin(\sin(s)) = s$, only for $s \in \left[\dfrac{-\pi}{2}, \dfrac{\pi}{2}\right]$
2. $[-1, 1]$, $[0, \pi]$
3. $\arctan(x) + C$
4. No. $\text{Tan}^{-1}(x) = \text{"inverse function of Tan}(x)\text{"} \neq \dfrac{1}{\tan(x)} = \cot(x)$.

Section 6.7

1. $x = \cosh(t)$, $y = \sinh(t)$
2. $\dfrac{\sinh(t)}{\cosh(t)}$, $\text{sech}^2(t)$
3. $\cosh(x) + C$
4. $\tanh^{-1}(x) + C$

Section 7.1

1. $\int u \cdot dv = u \cdot v - \int v \cdot du$
2. Product rule
3. $x^n v(x) - \int nx^{n-1}v(x)\,dx$
4. $x\ln(2x) - x + C$

Section 7.2

1. $A + B = 5$, $2A - 3B = -7$
2. $A = 1$, $B = 2$, $C = -1$
3. $\dfrac{A}{x} + \dfrac{B}{x^2} + \dfrac{C}{x^3} + \dfrac{D}{(x+1)}$
4. $\dfrac{A}{x} + \dfrac{B}{x-2} + \dfrac{C}{x+2}$

Section 7.3

1. $\dfrac{1}{2}x + \dfrac{1}{4}\sin(2x) + C$
2. $-\dfrac{1}{12}\sin^{11}(x)\cos(x) + \dfrac{11}{12}\int \sin^{10}(x)\,dx$
3. Let $u = \cos(x)$; then the integral becomes
$$-\int (1 - u^2)u^5\,du = -\frac{\cos^6(x)}{6} + \frac{\cos^8(x)}{8} + C.$$
4. To compute $\int \sin^n(x)\,dx$ or $\int \cos^n(x)\,dx$ when n is even, use the appropriate reduction formula.

   To compute $\int \sin^n(x)dx$ when n is odd, use the substitution $u = \cos(x)$, then $du = -\sin(x)\,dx$ and $\sin(x) = (1 - u^2)^{1/2}$. The integral becomes $\int \sin^{n-1}(x)\sin(x)\,dx = -\int (1 - u^2)^{(n-1)/2}\,du$.

   To compute $\int \cos^n(x)\,dx$ when n is odd, use the substitution $u = \sin(x)$, then $du = \cos(x)\,dx$ and $\cos(x) = (1 - u^2)^{1/2}$. The integral becomes $\int \cos^{n-1}(x)\cos(x)\,dx = \int (1 - u^2)^{(n-1)/2}\,du$.

   To compute $\int \sin^m(x)\cos^n(x)\,dx$ when m is odd, use the substitution $u = \cos(x)$; then the integral becomes $\int \sin^{m-1}(x)\sin(x)\cos^n(x)\,dx = -\int (1 - u^2)^{(m-1)/2}u^n\,du$.

   To compute $\int \sin^m(x)\cos^n(x)\,dx$ when n is odd, use the substitution $u = \sin(x)$; then the integral becomes $\int \sin^m(x)\cos^{n-1}(x)\cos(x)\,dx = \int u^m(1 - u^2)^{(n-1)/2}\,du$.

   To compute $\int \sin^m(x)\cos^n(x)\,dx$ when both m and n are even, change it to $\int (1 - \cos^2(x))^{m/2}\cos^n(x)\,dx$ or $\int \sin^m(x)(1 - \sin^2(x))^{n/2}\,dx$, then use the corresponding reduction formula for powers of cosine or sine.

Section 7.4

1. $x = \dfrac{1}{2}\sin(\theta)$
2. $x = 3\tan(\theta) - 1$
3. $x = 2\sec(\theta)$
4. $\displaystyle\int 1 + \frac{1}{1 + x^2}\,dx = x + \tan^{-1}(x) + C$

Section 7.5

1. $\dfrac{(x+1)}{(x-1)x^2} = \dfrac{A}{(x-1)} + \dfrac{B}{x} + \dfrac{C}{x^2}$,

   $\dfrac{(x+1)}{(x-1)(x^2+1)} = \dfrac{A}{x-1} + \dfrac{Bx+C}{x^2+1}$

2. $\dfrac{(x+1)}{(x^2-3x-4)^2} = \dfrac{A}{(x+1)} + \dfrac{B}{(x+1)^2} + \dfrac{C}{x-4} + \dfrac{D}{(x-4)^2}$,

$\dfrac{(x+1)}{(x^2-3x+4)^2} = \dfrac{Ax+B}{(x^2-3x+4)} + \dfrac{Cx+D}{(x^2-3x+4)^2}$

3. $\dfrac{Ax+B}{x^2+4} + \dfrac{C}{x+1}$

4. $\dfrac{Ax+B}{x^2+4} + \dfrac{Cx+D}{(x^2+4)^2} + \dfrac{E}{x+1}$

Section 8.1

1. y will be the variable because $x = g(y)$ will be the radius of each disk.

2. y will be the variable because $x = g(y)$ will be the height of each cylindrical shell.

3. $2\pi \int_0^5 y \cdot \left(11 - \dfrac{11}{5}y\right) dy$

4. 24π

Section 8.2

1. $L \approx \displaystyle\sum_{j=1}^{N} \sqrt{(x_j - x_{j-1})^2 + (f(x_j) - f(x_{j-1}))^2}$

$= \displaystyle\sum_{j=1}^{N} \sqrt{(\Delta x)^2 + (f'(\xi_j)\Delta x)^2}$

2. $f(x) = \sqrt{R^2 - x^2}$

$\displaystyle\int_{-R}^{R} \sqrt{1 + (f'(x))^2}\,dx = \int_{-R}^{R} \sqrt{\dfrac{R^2}{R^2 - x^2}}\,dx = \pi R$

3. Frustum

4. $2\pi \int_0^h mx\sqrt{1+m^2}\,dx = \pi m h^2 \sqrt{1+m^2}$

Section 8.3

1. $f_{ave} = \dfrac{1}{b-a}\int_a^b f(x)\,dx$

2. 0. Consider the region bounded between the graph $y = \sin(x)$ and the x-axis. The area of the region lying above $[0, \pi]$ equals to the area of the region lying below $[0, \pi]$.

3. $\dfrac{1}{2}$

4. $\displaystyle\int_0^2 x \cdot \dfrac{x}{2}\,dx = \dfrac{4}{3}$

Section 8.4

1. Negative

2. $c = 0$

3. (a, a) with $a < 1/2$

Section 8.5

1. Work $=$ Force $\times$ Distance

2. Work $= \displaystyle\int_a^b$ Force dx, Area $= \displaystyle\int_a^b$ Height dx

3. $\displaystyle\int_0^{0.01} kx\,dx = 0.5$, therefore $k = 10000$ N/m.

Force $= (10000)(0.01) = 100$N

4. $\displaystyle\int_0^1 1 \cdot y\,dy = \dfrac{1}{2} J$

Section 8.6

1. $\displaystyle\int_a^b f(x)\,dx$, where f is unbounded as $x \to b^-$ or $x \to a^+$

2. $\displaystyle\lim_{\epsilon \to 0^+} \int_a^{b-\epsilon} f(x)\,dx$ or $\displaystyle\lim_{\epsilon \to 0^+} \int_{a+\epsilon}^b f(x)\,dx$

3. $\dfrac{3}{4}(6^{2/3} - (-4)^{2/3})$

4. Diverges

Section 8.7

1. $\displaystyle\int_A^\infty f(x)\,dx$ or $\displaystyle\int_{-\infty}^B f(x)\,dx$

2. $\displaystyle\lim_{N \to +\infty} \int_A^N f(x)\,dx$ or $\displaystyle\lim_{M \to -\infty} \int_M^B f(x)\,dx$

3. $\displaystyle\lim_{N\to\infty} \int_1^N (1+x)^{-3}\,dx = \lim_{N\to\infty} \dfrac{(1+x)^{-2}}{-2}\Big|_{x=1}^{x=N}$

$= \displaystyle\lim_{N\to\infty} \dfrac{(1+N)^{-2}}{-2} + \dfrac{(1+1)^{-2}}{2} = \dfrac{1}{8}$

4. Diverges

Section 9.1

1. $\displaystyle\sum_{n=1}^\infty a_n = a_1 + a_2 + a_3 + \cdots + a_n + \cdots$

2. Let $S_N = \displaystyle\sum_{n=1}^N a_n$ be the partial sum. If the sequence $\{S_N\}$ converges, then the series $\displaystyle\sum_{n=1}^\infty a_n$ converges.

3. A sequence is a list of numbers, $\{a_n\} = \{a_1, a_2, \ldots, a_n, \ldots\}$. A series is a sum of numbers,
$\displaystyle\sum a_n = a_1 + a_2 + \cdots + a_n + \cdots$.

4. Diverges

Section 9.2

1. False. The series diverges.

2. True. The series converges.

3. False. Consider the harmonic series.

4. False. Consider the divergent series $\displaystyle\sum_{n=1}^\infty (-1)^n$; its partial sums are bounded.

Section 9.3

1. (See Theorem 1.)
2. Both
3. Converges if $0 \le r < 1$
4. True

Section 9.4

1. (See Theorem 1.)
2. (See Theorem 2.)
3. Converges. Use the Comparison Test for Convergence with $\sum \dfrac{1}{n^2}$ or the Limit Comparison Test with $\sum \dfrac{1}{n^2}$.
4. Diverges. Use the Comparison Test for Divergence with $\sum \dfrac{1}{n}$ or the Limit Comparison Test with $\sum \dfrac{1}{n}$.

Section 9.5

1. (See Theorem 1.)
2. If $\sum |a_n|$ converges, then we say $\sum a_n$ is an absolutely convergent series.
3. If $\sum a_n$ converges, but $\sum |a_n|$ diverges, then we say $\sum a_n$ is a conditionally convergent series.
4. True

Section 9.6

1. (See Theorem 1.)
2. True. The Ratio Test will only work for absolutely convergent series or divergent series.
3. Converges absolutely
4. Converges absolutely

Section 10.1

1. $\displaystyle\sum_{n=0}^{\infty} a_n x^n = a_0 + a_1 x + a_2 x^2 + a_3 x^3 + \cdots$
2. $\{0\}, (-R, R), [-R, R), (-R, R], [-R, R], (-\infty, \infty)$
3. This is the number R, such that the power series $\sum a_n (x - c)^n$ converges absolutely for $|x - c| < R$ and diverges for $|x - c| > R$.
4. $-3 \le x \le -1$, $R = 1$

Section 10.2

1. Addition, subtraction, and scalar multiplication

2. We can differentiate a power series term by term when the variable x is inside the interval $(c - R, c + R)$, where R is the radius of convergence of the series and c is the center.
3. We can integrate term by term when the variable x is inside the interval $(c - R, c + R)$.
4. (See Theorem 3.)

Section 10.3

1. The graph of the degree one Taylor polynomial at c is the tangent line at $(c, f(c))$.
2. $1 - \dfrac{x^2}{2}$
3. $y = -x + 1$
4. Since $\dfrac{f^{(2)}(0)}{2!} = 3$, we have $f^{(2)}(0) = 6$.

Section 10.4

1. $N + 1$
2. Error $< 5 \times 10^{-k-1}$
3. i. Find a bound, M_{N+1}, for max $\left| f^{(N+1)}(x) \right|$.
   ii. Find the smallest integer N such that
   $$M_{N+1} \cdot \frac{|x - c|}{(N + 1)!} < 5 \times 10^{-k-1}.$$
   iii. Approximate the value of $f(x)$ by $P_N(x)$, the Taylor polynomial of order N with center at c, where N is the integer obtained in part ii.
4. Approximated by $P_3(1.01) = 1.0049875\ldots$.

Section 10.5

1. (See Theorem 1.)
2. The function is not differentiable at 0; therefore, it does not have a power series expansion about 0. Yes, it will have a power series expansion about π, because f is differentiable on the interval $(\pi - 1, \pi + 1)$.
3.
$$\exp(x) = 1 + x + \frac{x^2}{2!} + \frac{x^3}{3!} + \cdots$$
$$\sin(x) = x - \frac{x^3}{3!} + \frac{x^5}{5!} - \frac{x^7}{7!} + \cdots$$
$$\cos(x) = 1 - \frac{x^2}{2!} + \frac{x^4}{4!} - \frac{x^6}{6!} + \cdots$$

4. $\displaystyle\sum_{n=0}^{\infty} \binom{\alpha}{n} x^n$. It converges to $(1 + x)^\alpha$ in $(-1, 1)$.

APPENDIX: SOLUTIONS TO STUDY GUIDE SELF-TEST EXERCISES

Section 1.1

1. Set D is described by $D = \{x \in \mathbb{R} : -7 < x < 7\}$, i.e., it is the set of real numbers that are strictly greater than -7 and strictly less than 7.

2. Set E is described by $E = \{-1, 1\}$.

3. $|-5x - 25| \leq 10$ implies that $-10 \leq -5x - 25 \leq 10$. This expression is equal to $-10 \leq -(5x + 25) \leq 10$, which implies $10 \geq (5x + 25) \geq -10$. Subtracting 25 throughout, we obtain $-15 \geq 5x \geq -35$. Now divide throughout by five to obtain the solution $-3 \geq x \geq -7$, or equivalently x in $[-7, -3]$.

4. $|5x| < 2$ implies that $-2 < 5x < 2$. Divide throughout by 5 to obtain the solution $\dfrac{-2}{5} < x < \dfrac{2}{5}$.

    Since $\dfrac{-2}{5} = -0.4$ and $\dfrac{2}{5} = +0.4$, this is equivalent to stating that the solution is $-0.4 < x < 0.4$. Thus the statement that $|5x| < 2$ implies $-0.4 < x < 0.4$ is substantiated.

5. The floating point decimal representation of 276.5589156 is $(0.27655589156) \times (10^{+3})$.

6. The floating point decimal representation of 276.55589156 is $(0.27655589156)(10^{+3})$, where $a_1 = 2$, $a_2 = 7$, $a_3 = 6$, $a_4 = 5$, $a_5 = 5$, $a_6 = 5$, $a_7 = 8$, $a_8 = 9$, $a_9 = 1$, $a_{10} = 5$, and $a_{11} = 6$. Therefore, there are eleven significant digits.

Section 1.2

1. The point $A = (.1, .5)$ is located in Quadrant I because the x-coordinate of A is equal to .1, which is greater than zero, and the y-coordinate of A is equal to .5, which is also greater than zero. (Recall that Quadrant $I = \{(x, y) | x > 0$ and $y > 0\}$.) The point $B = (-1, 2)$ is located in Quadrant II because the x-coordinate of B is equal to -1, which is less than zero, and the y-coordinate of B is 2, which is greater than zero. (Recall that Quadrant II $= \{(x, y) | x < 0$ and $y > 0\}$.)

2. Let $x_a = .1$ and $y_a = .5$.
    Let $x_b = -1$ and $y_b = 2$.
    Then
    $$|\overline{AB}| = \sqrt{(x_a - x_b)^2 + (y_a - y_b)^2}$$
    $$= \sqrt{(.1 - (-1))^2 + (.5 - 2)^2}$$
    $$= \sqrt{(1.1)^2 + (-1.5)^2}$$
    $$= \sqrt{1.21 + 2.25}$$
    $$= \sqrt{3.46}$$
    $$\approx 1.860107524 \text{ units.}$$

3. We shall use the method of "Completing the Square" to express $x^2 - 2x - 14 + y^2 + 6y = 0$ in standard form.

    $$x^2 - 2x - 14 + y^2 + 6y = 0$$
    is equal to
    $$x^2 - 2x + y^2 + 6y = 14.$$

    We shall calculate one-half of the coefficient of the x-term, $\dfrac{-2}{2}$, and square it to obtain $(-1)^2 = +1$. This value will be added to both sides of the equation to obtain

    $$(x^2 - 2x + 1) + y^2 + 6y = 14 + 1.$$

    We now calculate one-half the coefficient of the y-term, $\dfrac{6}{2}$, and square it to obtain $(3)^2 = 9$. This value is also added to both sides of the equation to obtain

    $$(x^2 - 2x + 1) + (y^2 + 6y + 9) = 14 + 1 + 9.$$

    Factoring the expressions in x and y, we obtain

    $$(x - 1)^2 + (y + 3)^2 = 24.$$

    Rewriting $+3$ as $-(-3)$, we obtain

    $$(x - 1)^2 + (y - (-3))^2 = (2\sqrt{6})^2.$$

    Now divide throughout by $(2\sqrt{6})^2$ to obtain the equation in standard form:

    $$\frac{(x - 1)^2}{(2\sqrt{6})^2} + \frac{(y - (-3))^2}{(2\sqrt{6})^2} = 1.$$

    The conic is a circle with center $(1, -3)$ and radius $r = 2\sqrt{6}$.

4. $x^2 - y^2 - \pi = 0$ is equal to the equation $x^2 - y^2 = \pi$. Divide throughout by π to obtain $\dfrac{x^2}{\pi} - \dfrac{y^2}{\pi} = 1$. Since $(\sqrt{\pi})^2 = \pi$, the equation $\dfrac{x^2}{\pi} - \dfrac{y^2}{\pi} = 1$ is equivalent to $\dfrac{(x - 0)^2}{(\sqrt{\pi})^2} - \dfrac{(y - 0)^2}{(\sqrt{\pi})^2} = 1$, which is now expressed in standard form.

    The conic is a hyperbola whose branches open to the right and left approximately 1.77 units from the origin $(0, 0)$.
    The x-intercepts of the two branches are $\approx (1.77, 0)$ and $\approx (-1.77, 0)$.

5. The graph of $y = x^2$ is an upright parabola with vertex $(0, 0)$. Since $y = x^2$ satisfies $y \leq x^2$, all the points on the upright parabola with vertex $(0, 0)$ are part of the solution. We choose, as a test point, the point $(0, -10)$. Clearly $(0, -10)$ satisfies $y \leq x^2$ since $-10 < 0^2$, so the region

"outside" the parabola $y = x^2$ contains points which are also part of the solution. Therefore a sketch of region $F = \{(x, y) : y \leq x^2\}$ is given by

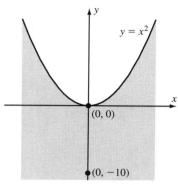

Region F includes the shaded region *and* the parabola.

6. The graph of $y = x^2$ is an upright parabola with vertex $(0, 0)$. Since $y = x^2$ satisfies $y \geq x^2$, all the points on the upright parabola with vertex $(0, 0)$ are part of the solution. If we choose, as a test point, the point $(0, -10)$ we see that $(0, -10)$ does *not* satisfy $y \geq x^2$ because $-10 \not\geq 0^2$. Therefore we choose the points "inside" the parabola as part of the solution. A sketch of region $G = \{(x, y) : y \geq x^2\}$ includes the shaded region *and* the parabola.

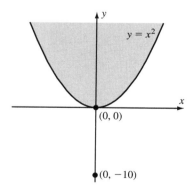

Section 1.3

1. Let $x_a = 3$ and $y_a = 4$.
   Let $x_b = -2$ and $y_b = 4$.
   The slope of line AB is defined as

$$M_{\overline{AB}} = \frac{y_b - y_a}{x_b - x_a}$$

$$= \frac{4 - 4}{-2 - 3}$$

$$= \frac{0}{-5}$$

$$= 0.$$

2. Let $x_a = 3$ and $y_a = 4$.
   Let $x_b = 3$ and $y_b = -4$.
   The slope of line AB is defined as

$$M_{\overline{AB}} = \frac{y_b - y_a}{x_b - x_a}$$

$$= \frac{-4 - 4}{3 - 3}.$$

Since the denominator of $M_{\overline{AB}}$ is equal to zero and division by zero is not permitted, we conclude that $M_{\overline{AB}}$ is undefined.

3. Let $M_{\overline{AB}}$ represent the slope of line AB and let $M_{\overline{CD}}$ represent the slope of line CD. The equation of line CD is given by $y = 3x + 10$, so recalling that the slope-intercept formula for the equation of a line, i.e., $y = mx + b$, we conclude that $M_{\overline{CD}} = 3$.

   Because lines AB and CD are perpendicular to each other, $M_{\overline{AB}} \cdot M_{\overline{CD}} = -1$, so $M_{\overline{AB}} = \dfrac{-1}{M_{\overline{CD}}} = \dfrac{-1}{3}$.

   Thus the equation of line AB can be expressed as $y = mx + b$ or $y = -\dfrac{1}{3}x + b$. But it is given that line AB passes through point $(1, -5)$, so the values $x = 1$ and $y = -5$ satisfy the equation $y = -\dfrac{1}{3}x + b$. This will allow us to solve for the value b.

   $x = 1$ and $y = -5$ and $y = -\dfrac{1}{3}x + b$ implies
   $-5 = -\dfrac{1}{3}(1) + b$, so $-5 + \dfrac{1}{3} = b$, i.e., $b = -4\dfrac{2}{3} = \dfrac{-14}{3}$

   We conclude that the equation of line AB is
   $$y = -\frac{1}{3}x - \frac{14}{3}.$$

4. The point $(0, 5)$ is the y-intercept, so in the slope intercept formula of the equation of line AB, $y = m \cdot x + b$, b is equal to 5.

   If we let $x_a = 0$, $y_a = 5$, $x_b = 2$, and $y_b = 1$, then slope $M_{\overline{AB}}$ is equal to

$$M_{\overline{AB}} = \frac{y_b - y_a}{x_b - x_a}$$

$$= \frac{1 - 5}{2 - 0}$$

$$= \frac{-4}{2}$$

$$= -2.$$

We conclude that the equation of line AB is $y = -2x + 5$.

5.

i	x_i	y_i	$x_i - x_0$	$y_i - y_0$	$(x_i - x_0)^2$	$(x_i - x_0) \cdot (y_i - y_0)$
1	1	15	$1 - 2 = -1$	$15 - (-5) = 20$	$(-1)^2 = 1$	$(-1)(20) = -20$
2	7	1	$7 - 2 = 5$	$1 - (-5) = 6$	$(5)^2 = 25$	$(5)(6) = 30$
Total					26	$+10$

$m = \dfrac{10}{26} = \dfrac{5}{13}$, so the equation of the least squares line is

$$y - (-5) = \frac{5}{13}(x - 2) \quad \text{or} \quad y + 5 = \frac{5}{13}x - \frac{10}{13}.$$

This implies that

$$y = \frac{5}{13}x - \frac{10}{13} - 5 \cdot \frac{13}{13} \quad \text{or}$$

$$y = \frac{5}{13}x - \left(\frac{10 + 65}{13}\right), \quad \text{which simplifies to}$$

$$y = \frac{5}{13}x - \frac{75}{13}.$$

Section 1.4

1. The domain of f is the set of all real numbers. So the "action of f" on each domain value is to send it to 3, i.e., if the input is a real number, then the output is 3. Since f assigns to each input value one element, 3, in its image, f is a function.
2. The domain of f is the set of real numbers such that $1 - x^2 \geq 0$, that is, all x that satisfy $-1 \leq x \leq 1$. For x in $[-1, 1]$, $f(x)$ varies from zero to one. Each x in the domain of f is assigned to $f(x) = \sqrt{1 - x^2}$ in the image of f. That is, input element x is assigned to *one* output element $f(x) = \sqrt{1 - x^2}$. Therefore, f is a function.
3. f is a function by the Vertical Line test, that is, any vertical line intersects the graph of f only once, so f is a function.
4. Because any vertical line intersects the graph of f only once, by the Vertical Line Test, f is a function.

Section 1.5

1. $(cf - g)(x) = cf(x) - g(x)$. If $f(x) = 2x^4$, $g(x) = -3x^2$, and $c = -1$, then the function $(cf - g)(x)$ is defined by $(cf - g)(x) = -2x^4 + 3x^2$.
2.
$$(cfg)(x) = c \cdot f(x) \cdot g(x)$$
$$= 10 \cdot x \cdot x^5$$
$$= 10x^6$$

   If $x = 5$, then $(cfg)(5) = 10 \cdot 5^6 = 156{,}250$.
3. $(f \circ f)(x) = f(f(x)) = f((x + 2)^2) = ((x + 2)^2 + 2)^2$ because the action of f on the input element x is to add the value of 2 to the input element, $x + 2$, and then to square the sum, $(x + 2)^2$.
4. $(f \circ g)(x) = f(g(x)) = f(3 - x) = (3 - x) - 3 = -x$ because the action of f on the input element is to subtract 3 from the input element.

5. $f(x) = x^2 + 2x + 1 = (x + 1)^2$, so the graph of $f(x)$ is an upright parabola with vertex $(-1, 0)$ and y-intercept $(0, 1)$.
   $f(x + 5) = ((x + 5) + 1)^2 = (x + 6)^2$, so the graph of $f(x + 5)$ is an upright parabola with vertex $(-6, 0)$ and y-intercept $(0, 36)$.
   In fact, the graph of $f(x + 5)$ is a horizontal translation of the graph of $f(x)$ five units to the left, i.e., five units in the negative direction.
6. $f(x) = x^2 + 2x + 1 = (x + 1)^2$
   The graph of $f(x)$ is an upright parabola with vertex $(-1, 0)$ and y-intercept $(0, 1)$.
   The function defined by $f(x) + 5 = (x + 1)^2 + 5$ describes an upright parabola with vertex $(-1, 5)$ and y-intercept $(0, 6)$. In fact, the graph of $f(x) + 5$ is a vertical translation of the graph of $f(x)$ five units in the positive direction.

Section 1.6

1.
$$\frac{57°}{180°} = \frac{x \text{ radians}}{\pi \text{ radians}} \quad \text{so}$$
$$x = \frac{57}{180}\pi \text{ radians}$$
$$\approx .317\,\pi \text{ radians}.$$

2.
$$\frac{x \text{ degrees}}{180 \text{ degrees}} = \frac{\frac{\pi}{5} \text{ radians}}{\pi \text{ radians}} \quad \text{so}$$
$$x = 180 \cdot \frac{\pi}{5} \cdot \frac{1}{\pi} \text{ degrees}$$
$$= \frac{180}{5} \text{ degrees}$$
$$= 36 \text{ degrees}.$$

3. $\sin^2(\theta) + \cos^2(\theta) = 1$ implies, for $\cos(\theta) \neq 0$,
   $\dfrac{\sin^2(\theta)}{\cos^2(\theta)} + \dfrac{\cos^2(\theta)}{\cos^2(\theta)} = \dfrac{1}{\cos^2(\theta)}$, which reduces to
   $\tan^2(\theta) + 1 = \sec^2\theta$ because $\dfrac{\sin(\theta)}{\cos(\theta)} = \tan(\theta)$ and
   $\dfrac{1}{\cos\theta} = \sec(\theta)$.
   If $\cos(\theta) = 0$, *neither* $\tan(\theta)$ *nor* $\sec(\theta)$ is defined.
4. $\sin(A + B) = \sin A \cos B + \sin B \cos A$
   If $A = B$, then
$$\sin(A + A) = \sin A \cos A + \sin A \cos A$$
   or
$$\sin 2A = 2 \sin A \cos A.$$

Section 2.1

1. The graph of $y = \dfrac{1}{x}$, for $x \neq 0$, is

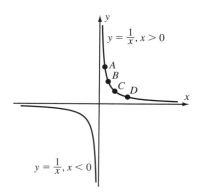

where $A = \left(\dfrac{1}{2}, 2\right)$, $B = (1, 1)$, $C = \left(2, \dfrac{1}{2}\right)$, and

$D = \left(3, \dfrac{1}{3}\right)$.

We observe that for $x > 0$, the graph of $y = \dfrac{1}{x}$ is in the first quadrant. As we approach point C from x-values larger than 2, say from point D to C, we see that the y-values approach $\dfrac{1}{2}$. As we approach point C from x-values smaller than 2, say from point B to C, we see that the y-values approach $\dfrac{1}{2}$. We conclude that the $\lim_{x \to 2} \dfrac{1}{x}$ exists and is equal to $\dfrac{1}{2}$.

2. Observe the following table that mimics the method of approaching $x = 4$ from the right.

4.0001	4.001	4.01	4.1	$4^+ \leftarrow x$
2.000025	2.000249984	2.002498439	2.024845673	$f(x) = \sqrt{x}$

4.0000001	4.000001	4.00001	$4^+ \leftarrow x$
2.000000025	2.00000025	2.0000025	$f(x) = \sqrt{x}$

Observe the following table that mimics the method of approaching $x = 4$ from the left.

$x \to 4^-$	3.9	3.99	3.999
$f(x) = \sqrt{x}$	1.974841766	1.997498436	1.999749984

$x \to 4^-$	3.9999	3.99999	3.999999	3.9999999
$f(x) = \sqrt{x}$	1.999975	1.9999975	1.99999975	1.999999975

We see that $\lim_{x \to 4^+} \sqrt{x} = 2 = \lim_{x \to 4^-} \sqrt{x}$, so $\lim_{x \to 4} \sqrt{x} = 2$.

3. In this example, we set $c = 3$, $\ell = 9$, and $\epsilon = 0.01$. We experimented with different viewing windows until we obtained the graph in the figure below. The viewing window is $[3 - .00166, 3 + .00166] \times [9 - .01, 9 + .01]$; these dimensions simplify to $[2.99834, 3.00166] \times [8.99, 9.01]$.

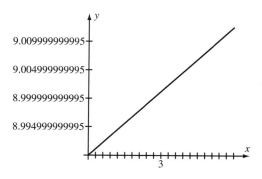

Section 2.2

1. To demonstrate that the statement $\lim_{x \to 2}(x^2 - 9) = -5$ satisfies the ϵ-δ definition of limit, we need to produce a value for $\delta > 0$ that will satisfy $|(x^2 - 9) - (-5)| < \epsilon$ whenever $0 < |x - 2| < \delta$.

Assume $\epsilon > 0$.

$|(x^2 - 9) - (-5)| < \epsilon$ whenever $0 < |x - 2| < \delta$ $\Leftrightarrow$
$|x^2 - 9 + 5| < \epsilon$ whenever $0 < |x - 2| < \delta$ $\Leftrightarrow$
$|x^2 - 4| < \epsilon$ whenever $0 < |x - 2| < \delta$ $\Leftrightarrow$
$|(x + 2)(x - 2)| < \epsilon$ whenever $0 < |x - 2| < \delta$ $\Leftrightarrow$
$|x + 2| \cdot |x - 2| < \epsilon$ whenever $0 < |x - 2| < \delta$.

If $|x - 2| < 1$, then $-1 < x - 2 < 1$, which implies $-1 + 2 < x < 1 + 2$, which implies $-1 + 2 + 2 < x + 2 < 1 + 2 + 2$. Therefore $3 < x + 2 < 5$. But $|x + 2| = x + 2$ for x satisfying $1 < x < 3$. We conclude that $|x + 2| < 5$. However, $|x + 2| < 5$ and $|x + 2| \cdot |x - 2| < \epsilon$ whenever $0 < |x - 2| < \delta \Leftrightarrow |x - 2| < \dfrac{\epsilon}{5}$ whenever $0 < |x - 2| < \delta$. Therefore δ is the smaller of the two values 1 and $\dfrac{\epsilon}{5}$. This is expressed as $\delta = \min\left(1, \dfrac{\epsilon}{5}\right)$.

2. To demonstrate that the statement $\lim_{x \to 5}(2x + 6) = 16$ satisfies the ϵ-δ definition of limit, we need to produce a value for $\delta > 0$ that will satisfy $|(2x + 6) - 16| < \epsilon, \epsilon > 0$, whenever $0 < |x - 5| < \delta$.

Assume $\epsilon > 0$.

$|(2x + 6) - 16| < \epsilon$ whenever $0 < |x - 5| < \delta$ $\Leftrightarrow$
$|2x - 10| < \epsilon$ whenever $0 < |x - 5| < \delta$ $\Leftrightarrow$
$|2(x - 5)| < \epsilon$ whenever $0 < |x - 5| < \delta$ $\Leftrightarrow$
$|2| |x - 5| < \epsilon$ whenever $0 < |x - 5| < \delta$.

But $|2| |x - 5| < \epsilon \Leftrightarrow 2 \cdot |x - 5| < \epsilon \Leftrightarrow |x - 5| < \dfrac{\epsilon}{2}$. Therefore choose $\delta = \dfrac{\epsilon}{2}$.

3.

$$\lim_{x \to 3} \left(\frac{f}{g} \right)(x) = \lim_{x \to 3} \frac{f(x)}{g(x)}$$

$$= \lim_{x \to 3} \frac{2x}{x^3 + 5}$$

$$= \frac{\lim\limits_{x \to 3} 2x}{\lim\limits_{x \to 3}(x^3 + 5)}$$

$$= \frac{6}{27 + 5}$$

$$= \frac{6}{32}$$

$$= \frac{3}{16}$$

$$= 0.1875$$

4.

$$\lim_{x \to 1} \sqrt{4x} = \lim_{x \to 1} \sqrt{4}\sqrt{x}$$

$$= 2 \lim_{x \to 1} x^{1/2}$$

$$= 2(\lim_{x \to 1} x)^{1/2}$$

$$= 2(1)^{1/2}$$

$$= 2 \cdot 1$$

$$= 2$$

Section 2.3

1. The graph of $f(x)$ is

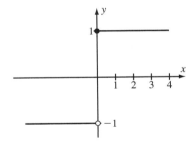

Graphical analysis shows that f is continuous on the open interval $(2, 4)$, which contains 3. Therefore

$$\lim_{x \to 3} f(x) = f(3) = 1.$$

2. The graph of $f(x)$ is

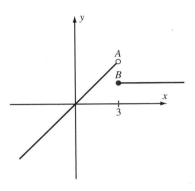

Point A is equal to $(3, 3)$ and point B is $(3, 1)$. Graphical analysis indicates that $\lim\limits_{x \to 3^+} f(x) = 1$.

3. We use a graphing calculator to produce the graph. The graph of $f(x)$ for x in $[16, 25]$ is

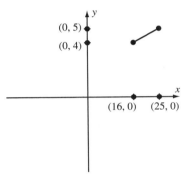

The graph shows that $f(x)$ is continuous for $16 \leq x \leq 25$. By the Extreme Value Theorem, we know that a minimum value exists and a maximum value exists because the interval $[16, 25]$ is closed. From the graph, we see that the minimum occurs at $x = 16$, with minimum value $f(16) = 4$. The maximum occurs at $x = 25$, with maximum value $f(25) = 5$.

4. With the help of a graphing calculator, we obtain the graph of $(\alpha f)(x) = 8x^{1/3}$. Clearly the graph of $(\alpha f)(x) = 8x^{1/3}$ illustrates that it is continuous for all real x.

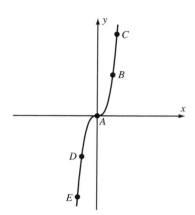

where $A = (0, 0)$, $B = (1, 8)$, $C = (8, 16)$, $D = (-1, -8)$, and $E = (-8, -16)$.

Section 2.4

1. $f(x)$ is a rational function with denominater equal to $x^2 - 5x + 6$. We observe that $x^2 - 5x + 6 = (x - 3)(x - 2)$.
$$\lim_{x \to 2^+} \frac{2}{(x - 3)(x - 2)} = -\infty \text{ or equivalently}$$
$$\lim_{x \to 2^-} \frac{2}{(x - 3)(x - 2)} = +\infty \text{ establishes that a vertical}$$
asymptote $x = 2$ exists. Similarly,
$$\lim_{x \to 3^+} \frac{2}{(x - 3)(x - 2)} = +\infty \text{ or equivalently}$$
$$\lim_{x \to 3^-} \frac{2}{(x - 3)(x - 2)} = -\infty \text{ establishes that a vertical}$$
asymptote $x = 3$ exists. This is confirmed by the graph of $f(x)$.

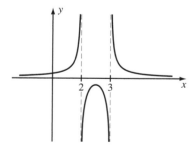

2. $\displaystyle \lim_{x \to +\infty} f(x) = \lim_{x \to +\infty} \frac{3}{x^2}$
$$= \frac{\displaystyle\lim_{x \to +\infty} 3}{\displaystyle\lim_{x \to +\infty} x^2}$$
$$= \frac{3}{\displaystyle\lim_{x \to +\infty} x^2}.$$

As $x \to +\infty$, $x^2 \to +\infty$. But because the numerator is constant and the denominator grows "without bound" as x grows "without bound," the value of $\frac{3}{x^2} \to 0$ as $x \to +\infty$.

3. $\displaystyle \lim_{x \to +\infty} \frac{1}{2x} = 0$; therefore a horizontal asymptote exists and it is the line $y = 0$. $\displaystyle \lim_{x \to 0^+} \frac{1}{2x} = +\infty$; therefore a vertical asymptote exists and it is the line $x = 0$.

Section 2.5

1. $1 + \dfrac{1}{10} + \left(\dfrac{1}{10}\right)^2 + \cdots + \left(\dfrac{1}{10}\right)^{N-1}$ is a geometric series

$1 + r + r^2 + \cdots + r^{N-1}$, where $r = \dfrac{1}{10}$. Since

$|r| = \left|\dfrac{1}{10}\right| = \dfrac{1}{10} < 1$, by Theorem 3b,

$$\lim_{N \to \infty} \left(1 + \frac{1}{10} + \left(\frac{1}{10}\right)^2 + \cdots + \left(\frac{1}{10}\right)^{N-1}\right) =$$

$$\frac{1}{1 - \dfrac{1}{10}} = \frac{1}{\dfrac{9}{10}} = \frac{10}{9}.$$

Section 2.6

1. $y = 2^{-x} = \left(\dfrac{1}{2}\right)^x$

The graph of this function is

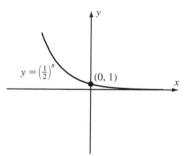

We see that $y = \left(\dfrac{1}{2}\right)^x$ is a decreasing, exponential function with a y-intercept of one.

2. In 6 years, the accumulated amount in the account is

$$A = \$1250 \left(1 + \frac{.02}{52}\right)^{(52)(6)}$$
$$= \$1409.34.$$

Therefore the total interest accumulated in 6 years is
$\$1409.34 - \$1250 = \$159.34$

3. $\dfrac{n}{n+1} = \dfrac{n}{n+1} \cdot \dfrac{\frac{1}{n}}{\frac{1}{n}} = \dfrac{1}{1 + \frac{1}{n}}.$

Therefore $\dfrac{n}{n+1} = \dfrac{1}{1 + \frac{1}{n}} \to 1$ as $n \to \infty$.

The sequence $\left\{ \dfrac{n}{n+1} \right\}_{n=1}^{\infty}$ converges to 1.

Section 3.1

1.
$$h(x) = f(x) - 2g(x)$$
$$= x^3 - 2x$$

The average rate of change of $h(x)$ from c to $(c + \Delta x)$ is

$$\dfrac{\Delta h}{\Delta x} = \dfrac{h(c + \Delta x) - h(c)}{\Delta x}$$
$$= \dfrac{[(c + \Delta x)^3 - 2(c + \Delta x)] - [c^3 - 2c]}{\Delta x}$$
$$= \dfrac{c^3 + 3c^2\Delta x + 3c(\Delta x)^2 + (\Delta x)^3 - 2c - 2\Delta x - c^3 + 2c}{\Delta x}$$
$$= \dfrac{\Delta x (3c^2 + 3c\Delta x + (\Delta x)^2 - 2)}{\Delta x}$$
$$= 3c^2 + 3c\Delta x + (\Delta x)^2 - 2.$$

If $c = 1$ and $c + \Delta x = 2$, then $\Delta x = 1$.

Thus the average rate of change of $h(x)$ between $x = 1$ and $x = 2$ is

$$\dfrac{\Delta h}{\Delta x} = 3(1)^2 + 3(1)(1) + (1)^2 - 2$$
$$= 3 + 3 + 1 - 2$$
$$= 5.$$

2.
$$h(x) = f(x) - 2g(x)$$
$$= x^3 - 2x$$

The instantaneous rate of change of $h(x)$ at $x = 1$ is

$$\lim_{\Delta x \to 0} \dfrac{h(1 + \Delta x) - h(1)}{\Delta x}$$
$$= \lim_{\Delta x \to 0} \dfrac{[(1 + \Delta x)^3 - 2(1 + \Delta x)] - [1^3 - 2(1)]}{\Delta x}$$
$$= \lim_{\Delta x \to 0} \dfrac{[1^3 + 3(1)^2\Delta x + 3(1)(\Delta x)^2 + (\Delta x)^3 - 2(1) - 2(\Delta x) - 1^3 + 2(1)]}{\Delta x}$$
$$= \lim_{\Delta x \to 0} \dfrac{[\Delta x(3(1)^2) + 3(1)\Delta x + (\Delta x)^2 - 2]}{\Delta x}$$
$$= \lim_{\Delta x \to 0} (3 + 3\Delta x + \Delta x^2 - 2)$$
$$= 3 + 0 + 0 - 2$$
$$= 1.$$

Remark: You may use the Sum Rule to do this exercise, just like example 4.

3. The instantaneous velocity at time $t = 2$ is equal to

$$\lim_{\Delta t \to 0} \dfrac{p(2 + \Delta t) - p(2)}{\Delta t}$$
$$= \lim_{\Delta t \to 0} \dfrac{[(2 + \Delta t)^2 + 3(2 + \Delta t) + 1] - 11}{\Delta t}$$
$$= \lim_{\Delta t \to 0} \dfrac{(4 + 4\Delta t + (\Delta t)^2 + 6 + 3(\Delta t) + 1) - 11}{\Delta t}$$
$$= \lim_{\Delta t \to 0} \dfrac{7\Delta t + (\Delta t)^2}{\Delta t}$$
$$= \lim_{\Delta t \to 0} 7 + \Delta t$$
$$= 7.$$

4. a.
$$m_{\text{tan}} = \lim_{\Delta x \to 0} \dfrac{f(2 + \Delta x) - f(2)}{\Delta x}$$
$$= \lim_{\Delta x \to 0} \dfrac{([-(2 + \Delta x)^2 + 4(2 + \Delta x)] - [-(2)^2 + 4(2)])}{\Delta x}$$
$$= \lim_{\Delta x \to 0} \dfrac{(-4 - 4\Delta x - (\Delta x)^2 + 8 + 4\Delta x + 4 - 8)}{\Delta x}$$
$$= \lim_{\Delta x \to 0} \dfrac{\Delta x(-\Delta x)}{\Delta x}$$
$$= \lim_{\Delta x \to 0} -\Delta x$$
$$= 0$$

The equation of the tangent line to the graph of f at point $(2, 4)$ is

$$y = 0(x - 2) + f(2)$$
$$= 0 + 4$$
$$= 4.$$

b. $m_{\text{normal}} = \dfrac{-1}{m_{\text{tan}}}$ is undefined because $m_{\text{tan}} = 0$ by part (a).

Thus the normal is a vertical line whose equation is $x = 4$.

5.
$$m_{\text{tan}} = \lim_{\Delta x \to 0} \dfrac{f\left(-\frac{1}{4} + \Delta x\right) - f\left(-\frac{1}{4}\right)}{\Delta x}$$
$$= \lim_{\Delta x \to 0} \dfrac{\left(\left[-4\left(-\frac{1}{4} + \Delta x\right) + 9\right] - \left[-4\left(-\frac{1}{4}\right) + 9\right]\right)}{\Delta x}$$
$$= \lim_{\Delta x \to 0} \dfrac{(1 - 4\Delta x + 9 - 1 - 9)}{\Delta x}$$
$$= \lim_{\Delta x \to 0} \dfrac{\Delta x(-4)}{\Delta x}$$
$$= \lim_{\Delta x \to 0} (-4)$$
$$= -4$$

The equation of the tangent line to the curve $f(x)$ at point $\left(-\dfrac{1}{4}, 10\right)$ is

$$y = (-4)\left(x - \left(-\frac{1}{4}\right)\right) + f\left(-\frac{1}{4}\right)$$

$$= -4x - 1 + 10$$

$$= -4x + 9.$$

6.
$$\lim_{\Delta x \to 0^+} \frac{f(1 + \Delta x) - f(1)}{\Delta x} = \lim_{\Delta x \to 0^+} \frac{([1 + \Delta x - 1] - [1 - 1])}{\Delta x}$$

$$= \lim_{\Delta x \to 0^+} \frac{\Delta x}{\Delta x}$$

$$= \lim_{\Delta x \to 0^+} 1$$

$$= 1$$

and

$$\lim_{\Delta x \to 0^-} \frac{f(1 + \Delta x) - f(1)}{\Delta x} = \lim_{\Delta x \to 0^-} \frac{([-(1 + \Delta x) + 1] - [1 - 1])}{\Delta x}$$

$$= \lim_{\Delta x \to 0^-} \frac{-\Delta x}{\Delta x}$$

$$= \lim_{\Delta x \to 0^-} -1$$

$$= -1$$

We conclude that $f(x) = |x - 1|$ has a corner at $(1, 0)$. This conclusion is substantiated by the graph of f.

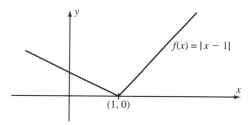

7.
$$\lim_{\Delta x \to 0} \frac{f(0 + \Delta x) - f(0)}{\Delta x} = \lim_{\Delta x \to 0} \frac{(\Delta x)^{1/5} - 0}{\Delta x}$$

$$= \lim_{\Delta x \to 0} \frac{1}{(\Delta x)^{4/5}}$$

$$= +\infty$$

Thus $f(x) = x^{1/5}$ has a vertical "tangent" at point $(0, 0)$.

8. Since $f(x) = ((x - 1)^2)^{1/3} = (x - 1)^{2/3}$, we have

$$\lim_{\Delta x \to 0^+} \frac{f(1 + \Delta x) - f(1)}{\Delta x}$$

$$= \lim_{\Delta x \to 0^+} \frac{([(1 + \Delta x - 1)^{2/3}] - [(1 - 1)^{2/3}])}{\Delta x}$$

$$= \lim_{\Delta x \to 0^+} \frac{(\Delta x)^{2/3}}{\Delta x}$$

$$= \lim_{\Delta x \to 0^+} \frac{1}{(\Delta x)^{1/3}}$$

$$= +\infty$$

and

$$\lim_{\Delta x \to 0^-} \frac{f(1 + \Delta x) - f(1)}{\Delta x}$$

$$= \lim_{\Delta x \to 0^-} \frac{([(1 + \Delta x - 1)^{2/3}] - [(1 - 1)^{2/3}])}{\Delta x}$$

$$= \lim_{\Delta x \to 0^-} \frac{(\Delta x)^{2/3}}{\Delta x}$$

$$= \lim_{\Delta x \to 0^-} \frac{1}{(\Delta x)^{1/3}}$$

$$= -\infty.$$

Since $\displaystyle\lim_{\Delta x \to 0^+} \frac{f(1 + \Delta x) - f(1)}{\Delta x} \neq \lim_{\Delta x \to 0^-} \frac{f(1 + \Delta x) - f(1)}{\Delta x}$ and at least one of the limits is infinite, we conclude that $f(x) = (x - 1)^{2/3}$ has a cusp at point $(1, 0)$. This result is substantiated by the graph of $f(x)$.

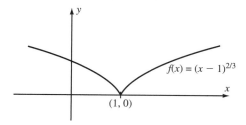

Section 3.2

1.
$$\frac{df}{dx}\bigg|_{x=2} = \lim_{\Delta x \to 0} \frac{f(2 + \Delta x) - f(2)}{\Delta x}$$

$$= \lim_{\Delta x \to 0} \frac{(2 + \Delta x - 1)^2 - (2 - 1)^2}{\Delta x}$$

$$= \lim_{\Delta x \to 0} \frac{1 + 2\Delta x + (\Delta x)^2 - 1}{\Delta x}$$

$$= \lim_{\Delta x \to 0} \frac{\Delta x(2 + \Delta x)}{\Delta x}$$

$$= \lim_{\Delta x \to 0} 2 + \Delta x$$

$$= 2$$

2.
$$\frac{d}{dt}\left(t^{1/2}\right)\bigg|_{t=4} = \lim_{\Delta t \to 0} \frac{f(4 + \Delta t) - f(4)}{\Delta t}$$

$$= \lim_{\Delta t \to 0} \frac{(4 + \Delta t)^{1/2} - (4)^{1/2}}{\Delta t}$$

$$= \lim_{\Delta t \to 0} \frac{(4 + \Delta t)^{1/2} - 2}{\Delta t} \cdot \frac{(4 + \Delta t)^{1/2} + 2}{(4 + \Delta t)^{1/2} + 2}$$

$$= \lim_{\Delta t \to 0} \frac{(4 + \Delta t) - 4}{\Delta t[(4 + \Delta t)^{1/2} + 2]}$$

$$= \lim_{\Delta t \to 0} \frac{\Delta t}{\Delta t[(4 + \Delta t)^{1/2} + 2]}$$

$$= \lim_{\Delta t \to 0} \frac{1}{(4 + \Delta t)^{1/2} + 2}$$

$$= \frac{1}{(4)^{1/2} + 2}$$

$$= \frac{1}{4}$$

3.
$$f'(x) = \lim_{\Delta x \to 0} \frac{f(x + \Delta x) - f(x)}{\Delta x}$$
$$= \lim_{\Delta x \to 0} \frac{([(x + \Delta x)^2 + 3(x + \Delta x)] - [x^2 + 3x])}{\Delta x}$$
$$= \lim_{\Delta x \to 0} \frac{(x^2 + 2x\Delta x + (\Delta x)^2 + 3x + 3\Delta x - x^2 - 3x)}{\Delta x}$$
$$= \lim_{\Delta x \to 0} \frac{\Delta x(2x + (\Delta x) + 3)}{\Delta x}$$
$$= \lim_{\Delta x \to 0} 2x + \Delta x + 3$$
$$= 2x + 3$$

4. Notice that $f(2) = \dfrac{2^2 + 6}{2^2 + 2 - 6} = \dfrac{10}{0}$ is not defined.
Therefore, f is not continuous at 2. By Theorem 1, $f'(2)$ does not exist.

5. The graph of $f(x) = \cos(x)$ in the viewing window
$$\left[-\frac{\pi}{2}, \frac{3\pi}{2} \right] \times [-1, 1] \text{ is given by}$$

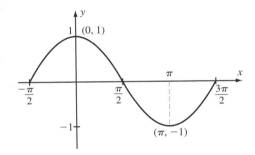

The function $f(x) = \cos(x)$ is continuous for all real x and is also a periodic function with period $p = 2\pi$. The graph of $f(x) = \cos(x)$ has no sharp corners, cusps, or vertical tangents. As a result, a tangent line exists at each and every point $(x, f(x))$ on its graph. This implies that a derivative $f'(x)$ exists at each point x in the domain of $f(x) = \cos(x)$.

6. We shall determine whether a tangent line exists from the graph of f.

Define $\phi(x) = \dfrac{f(x) - f\left(\dfrac{\pi}{4}\right)}{x - \dfrac{\pi}{4}}$ for $x \neq \dfrac{\pi}{4}$.

If $x < \dfrac{\pi}{4}$, $\phi(x) = \dfrac{\dfrac{2\sqrt{2}}{\pi}x - \dfrac{\sqrt{2}}{2}}{x - \dfrac{\pi}{4}} = \dfrac{2\sqrt{2}}{\pi}$.

If $x > \dfrac{\pi}{4}$, $\phi(x) = \dfrac{\sin(x) - \dfrac{\sqrt{2}}{2}}{x - \dfrac{\pi}{4}}$.

The graph of $\phi(x)$ in the window $\left[0, \dfrac{\pi}{2}\right] \times [0, 1]$ is given by

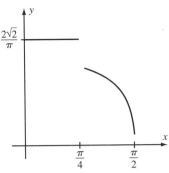

From the graph of $\phi(x)$, we see that $\phi(x)$ cannot be extended to a continuous function at $x = \dfrac{\pi}{4}$. Therefore, we believe that $f(x)$ is not differentiable at $x = \dfrac{\pi}{4}$.

Section 3.3

1.
$$f'(x) = (2x^3)' + (5\sin(x))'$$
$$= 2(x^3)' + 5(\sin(x))'$$

We have from Section 3.2 that $(x^3)' = 3x^2$ and $(\sin(x))' = \cos(x)$.
    Therefore,
$$f'(x) = 2 \cdot (3x^2) + 5 \cdot \cos(x)$$
$$= 6x^2 + 5\cos(x).$$

2.
$$D(h(x)) = h'(x)$$
$$= (3x^3)' + (5x^2)'$$
$$= 3(x^3)' + 5(x^2)'$$
$$= 3(3x^2) + 5(2x)$$
$$= 9x^2 + 10x$$
$$Dh(-3) = 9(-3)^2 + 10(-3)$$
$$= 51$$

3. Since $f(x)$ is equal to the product of two functions of x, we shall use the Product Rule.

Let $g(x) = x$ and $h(x) = \sin(x)$.
$$f(x) = g(x) \times h(x) \text{ implies that}$$
$$f'(x) = g'(x) \times h(x) + g(x) \times h'(x)$$
$$= 1 \cdot \sin(x) + x \cdot \cos x$$
$$= \sin x + x \cos x$$

4. Let $g(x) = x$ and $h(x) = \sin(x)$. Since $f(x) = \dfrac{g(x)}{h(x)}$, we shall use the Quotient Rule to calculate $f'(x)$.

$$f'(x) = \frac{h(x)g'(x) - g(x)h'(x)}{(h(x))^2}$$
$$= \frac{(\sin(x))(1) - (x)(\cos(x))}{(\sin(x))^2}$$
$$= \frac{\sin(x) - x\cos(x)}{(\sin(x))^2} \quad \text{for} \quad \sin(x) \neq 0.$$

5.

h	0.1	0.01	0.001	0.0001	0.00001	0.000001
$D_0V(0.4174, h)$	0.282924690	0.2830728	0.283074	0.283074	0.283074	0.283074

Since the leftmost four digits of $D_0V(0.4174, h)$ for $h = 0.1, 0.01, 0.001, 0.0001, 0.00001,$ and 0.000001 are equal to $0.2829, 0.2830, 0.2830, 0.2830, 0.2830,$ and $0.2830,$ we can be reasonably certain that $V'(0.4174) = 0.283$ to three significant digits.

Section 3.4

1.
$$\begin{aligned} f'(x) &= (25x^{10} - 3x + 2)' \\ &= (25x^{10})' + (-3x)' + (2)' \\ &= 25(x^{10})' - 3(x)' + (2)' \\ &= (25)(10)x^9 - 3(1) + 0 \quad \text{(by the Power Rule)} \\ &= 250x^9 - 3 \end{aligned}$$

2.
$$\begin{aligned} f(x) &= \frac{1}{\sqrt{x}} \\ &= x^{-1/2} \end{aligned}$$

Apply the Power Rule to $f(x) = x^{-1/2}$ to obtain

$$\begin{aligned} f'(x) &= \left(-\frac{1}{2}\right)(x)^{-1/2-1} \\ &= -\frac{1}{2}x^{-3/2} \\ &= -\frac{1}{2x^{3/2}} \quad \text{for} \quad x \neq 0. \end{aligned}$$

3.
$$\begin{aligned} \frac{\cos(t)\tan(t)}{\sin(t)} &= \left(\frac{\cos(t)}{\sin(t)}\right)(\tan(t)) \\ &= \left(\frac{\cos(t)}{\sin(t)}\right)\left(\frac{\sin(t)}{\cos(t)}\right) \\ &= 1 \end{aligned}$$

Thus

$$\begin{aligned} \frac{d}{dt}\left(\frac{\cos(t)\tan(t)}{\sin(t)}\right) &= \frac{d}{dt}(1) \\ &= 0. \end{aligned}$$

4. We shall use the Quotient Rule to compute the derivative of $g(x) = \dfrac{\tan(x)}{x^{1/2} - 3x}$.

$$\begin{aligned} g'(x) &= \frac{(x^{1/2} - 3x)\frac{d}{dx}(\tan(x)) - \tan(x)\frac{d}{dx}(x^{1/2} - 3x)}{(x^{1/2} - 3x)^2} \\ &= \frac{(x^{1/2} - 3x)(\sec(x))^2 - \tan(x)\left(\frac{1}{2}x^{-1/2} - 3\right)}{(x^{1/2} - 3x)^2} \end{aligned}$$

$$\begin{aligned} &= \frac{x^{1/2}(\sec(x))^2 - 3x(\sec(x))^2 - \frac{\tan(x)}{2x^{1/2}} + 3\tan(x)}{(x^{1/2} - 3x)^2} \\ &= \frac{(2x)(\sec(x))^2 - 6x^{3/2}(\sec(x))^2 - \tan(x) + 6x^{1/2}(\tan(x))}{2x^{1/2}(x^{1/2} - 3x)^2} \end{aligned}$$

5. We shall use the Quotient Rule to calculate $f'(x)$.

$$\begin{aligned} f'(x) &= \frac{(\sin(x))\frac{d}{dx}(e^x) - e^x\frac{d}{dx}(\sin(x))}{(\sin(x))^2} \\ &= \frac{(\sin(x))(e^x) - e^x(+\cos(x))}{(\sin(x))^2} \\ &= \frac{e^x(\sin(x) - \cos(x))}{(\sin(x))^2} \end{aligned}$$

Section 3.5

1. We shall use the Chain Rule to calculate $f'(x)$. The outside function is $\sin(\ )$ and the inside function is $2x$. Therefore by the Chain Rule, we obtain

$$\begin{aligned} f'(x) &= (\cos(2x))(2) \\ &= 2\cos(2x). \end{aligned}$$

2. If $n = 0$, then $\sin(2x)^0 = 1$. In this case, $f'(x) = (1)' = 0$. If $n = 1, 2, \ldots,$ then by the Chain Rule,

$$\begin{aligned} f'(x) &= n(\sin(2x))^{n-1}\frac{d}{dx}\sin(2x) \\ &= n(\sin(2x))^{n-1} \cdot \cos(2x)\cdot\frac{d}{dx}(2x) \\ &= n(\sin(2x))^{n-1} \cdot \cos(2x) \cdot 2. \end{aligned}$$

Therefore, 0 if $n = 0$.

$$2 \cdot n(\sin(2x))^{n-1} \cdot \cos(2x) \quad \text{if } n = 1, 2, \ldots$$

3. We shall use the Chain Rule to obtain $f'(x)$.
The outside function is $\cos(\ )$ and the inside function is $\sin(2x)$. Therefore by the Chain Rule,

$$\begin{aligned} f'(x) &= (-\sin(\sin(2x)))(\cos(2x))(2) \\ &= (-2\sin(\sin(2x)))(\cos(2x)). \end{aligned}$$

Section 3.6

1. Let $y = f(x)$. To obtain the inverse of $f(x)$, we shall solve for the variable x in the equation $y = 5 - \dfrac{x}{3}$.

$$\begin{aligned} y &= 5 - \frac{x}{3} \quad \text{implies that} \\ 3y &= 15 - x, \quad \text{which implies that} \\ x &= 15 - 3y. \end{aligned}$$

Thus $f^{-1}(y) = 15 - 3y$.

2. Let $x_1 = -1$ and $x_2 = +1$. $x_1 \neq x_2$.
$f(x_1) = (-1)^2 = +1$ and $f(x_2) = (+1)^2 = +1$.
Since $x_1 \neq x_2$ does not imply $f(x_1) \neq f(x_2)$, we see that f is not 1-1.

Theorem 1 states that "A function f is invertible if and only if it is both 1-1 and onto." Since $f(x)$ is not 1-1, we conclude that $f(x)$ is not invertible.

3. Let $t = f(s)$. $t = 2 + s^3$ implies that $t - 2 = s^3$ which, in turn, implies that $(t - 2)^{1/3} = s$. Thus $f^{-1}(t) = (t - 2)^{1/3}$.

$$\frac{d}{ds}(f(s)) = 0 + 3s^2$$
$$= 3s^2$$

For $s \neq 0$, $f'(s) = 3s^2 \neq 0$. Apply the Inverse Function Derivative Rule to obtain

$$\frac{d}{dt}(f^{-1}(t)) = \frac{1}{\frac{d}{ds}(f(s))}$$
$$= \frac{1}{3s^2}$$
$$= \frac{1}{3(t - 2)^{2/3}} \quad \text{for} \quad s \neq 0.$$

If $s = 0$, $f'(s) = 0$, so the Inverse Function Derivative cannot be applied.

4. We shall use the Product Rule to calculate $f'(x)$.

$$f'(x) = \left(\frac{d}{dx}(x^2 + 2)\right) \cdot \ln(x) + (x^2 + 2) \cdot \frac{d}{dx} \ln(x)$$
$$= 2x \ln(x) + (x^2 + 2)\left(\frac{1}{x}\right)$$
$$= 2x \ln(x) + \frac{x^2 + 2}{x}$$

5. We use the Chain Rule to calculate $f'(x)$. The outside function is cos( ) and the inside function is 2^x.
Therefore,

$$f'(x) = (-\sin(2^x))\frac{d}{dx}(2^x)$$
$$= (-\sin(2^x))(2^x \ln(2)).$$

Section 3.7

1. a. $f(x) = 3^x$ implies that $f'(x) = (\ln(3))(3^x)$.

$$f''(x) = \frac{d}{dx}f'(x)$$
$$= \frac{d}{dx}(\ln(3)(3^x))$$
$$= \ln(3)\frac{d}{dx}(3^x), \quad \text{since } \ln(3) \text{ is a constant}$$
$$= \ln(3)((\ln(3))(3^x))$$
$$= (\ln(3)^2(3^x))$$

b. Using Leibniz's Rule, we have

$$((1)(3^x))'' = (1)''(3^x) + (2)(1)'(3^x)' + (1)(3^x)''$$
$$= (0)(3^x) + (2)(0)((\ln(3))(3^x))$$
$$\quad + (1)((\ln 3)^2(3^x)) \quad \text{by part (a)}$$
$$= 0 + 0 + (\ln(3))^2(3^x)$$
$$= (\ln(3))^2(3^x).$$

2.
$$f'(x) = \frac{d}{dx}(3x^2 + 20x - 6)$$
$$= 6x + 20 - 0$$
$$= 6x + 20 \quad \text{by the Power Rule}$$

$$f''(x) = \frac{d}{dx}(f'(x))$$
$$= \frac{d}{dx}(6x + 20)$$
$$= 6 + 0$$
$$= 6 \quad \text{by the Power Rule}$$

$$f'''(x) = \frac{d}{dx}(f''(x))$$
$$= \frac{d}{dx}(6)$$
$$= 0$$

3. Velocity, $v(t)$, is equal to $p'(t) = 2t - 5$ by the Power Rule. Acceleration, $a(t)$, is equal to $v'(t) = 2 - 0 = 2$.

4. Acceleration, $a(t)$, is equal to $v'(t) = 4t - 0 = 4t$. Thus if $t = 3$, $a(3) = 4(3) = 12$.

Section 3.8

1. $\frac{x^2}{y^5} - xy^{2/3} + y^6 = 10$ is equal to $x^2y^{-5} - xy^{2/3} + y^6 = 10$.
Differentiate $x^2y^{-5} - xy^{2/3} + y^6 = 10$ implicitly to obtain

$$\left(x^2\frac{d}{dx}y^{-5} + y^{-5}\frac{d}{dx}x^2\right) - \left(x\frac{d}{dx}(y^{2/3}) + y^{2/3}\frac{d}{dx}(x)\right)$$
$$+ \frac{d}{dx}(y^6) = \frac{d}{dx}(10),$$

which is equal to

$$\left((x^2)\left(-5y^{-6}\frac{dy}{dx}\right) + (y^{-5})(2x)\right)$$
$$- \left((x)\left(\frac{2}{3}y^{-1/3}\frac{dy}{dx}\right) + (y^{2/3})(1)\right) + 6y^5\frac{dy}{dx} = 0.$$

Grouping all terms associated with $\frac{dy}{dx}$ on the left side of the equation and all others on the right side of the equation, we obtain

$$-\frac{5x^2}{y^6}\frac{dy}{dx} - \frac{2x}{3y^{1/3}}\frac{dy}{dx} + 6y^5\frac{dy}{dx} = -2xy^{-5} + y^{2/3}$$

Treating $\dfrac{dy}{dx}$ as if it were a common factor, we obtain

$$\left(\frac{-5x^2}{y^6} - \frac{2x}{3y^{1/3}} + 6y^5\right)\left(\frac{dy}{dx}\right) = -2xy^{-5} + y^{2/3}$$

$$\left(\frac{-15x^2}{3y^6} - \frac{2xy^{17/3}}{3y^6} + \frac{18y^{11}}{3y^6}\right)\frac{dy}{dx} = -2xy^{-5} + y^{2/3}$$

$$\left(\frac{-15x^2 - 2xy^{17/3} + 18y^{11}}{3y^6}\right)\frac{dy}{dx} = -2xy^{-5} + y^{2/3}$$

$$\frac{dy}{dx} = \frac{(-2xy^{-5} + y^{2/3})(3y^6)}{-15x^2 - 2xy^{17/3} + 18y^{11}}$$

$$= \frac{-6xy + 3y^{20/3}}{(-15x^2 - 2xy^{17/3} + 18y^{11})}.$$

2. Let D represent the diameter of the sphere and t represent time. $\dfrac{dD}{dt} = -\dfrac{1}{2}$ in/s is given.

If r represents the radius of the sphere, then $D = 2r$.

Thus $\dfrac{d(D)}{dt} = \dfrac{d(2r)}{dt} = \dfrac{2dr}{dt} = -\dfrac{1}{2}$ in/s, which implies

that $\dfrac{dr}{dt} = -\dfrac{1}{4}$ in/s.

The volume of a sphere, V, is given by $V = \dfrac{4}{3}\pi r^3$. Thus

$$\frac{dV}{dt} = \frac{4}{3}\pi \frac{d}{dt}(r^3)$$

$$= \frac{4}{3}\pi 3r^2 \frac{dr}{dt}.$$

If $r = 5''$ and $\dfrac{dr}{dt} = -\dfrac{1}{4}$ in/s, then

$$\left.\frac{dV}{dt}\right|_{r=5} = \left(\frac{4}{3}\pi\right)(3)(5)^2 \text{ in}^2 \left(-\frac{1}{4}\text{in/s}\right)$$

$$= -25\pi \text{ in}^3/\text{s}.$$

Section 3.9

1. We shall use the approximation $f(c + \Delta x) \approx f(c) + f'(c)\Delta x$ to estimate the value of $\sqrt{17}$.

If $f(x) = x^{1/2}$ then $f'(x) = \dfrac{1}{2x^{1/2}}$ by the Power Rule. Choose $c = 16$ and $\Delta x = 1$. Then

$$\sqrt{17} = f(17) = f(16 + 1) \approx f(16) + (f'(16))(\Delta x)$$

$$= \sqrt{16} + \left(\frac{1}{2\sqrt{16}}\right)(1)$$

$$= 4 + \left(\frac{1}{(2)(4)}\right)(1)$$

$$= 4 + \frac{1}{8}$$

$$= 4\frac{1}{8}$$

$$= 4.125.$$

Note that $\sqrt{17} \approx 4.123105626$.

2. The Method of Increments is represented by
$f(c + \Delta x) - f(c) = \Delta f(c) \approx f'(c)\Delta x$.

If $f(x) = x^{-1/3}$ then $f'(x) = \dfrac{-1}{3}x^{-4/3}$ by the Power Rule, so $c = 8$ implies that

$$f'(c) = f'(8) = \frac{-1}{3(8)^{4/3}} = \frac{-1}{(3)(16)} = \frac{-1}{48}.$$

If $\Delta x = .07$ then $c + \Delta x = 8 + .07 = 8.07$.

Thus $f(8.07) - f(8) \approx (f'(8))(.07)$. This implies that

$$(8.07)^{-1/3} = f(8.07) \approx \left(\frac{-1}{48}\right)(.07) + \frac{1}{8^{1/3}}$$

$$= \frac{-.07}{48} + \frac{1}{2}$$

$$\approx 0.4985416667.$$

Note that $\dfrac{1}{(8.07)^{1/3}} = \dfrac{1}{2.005816402} = 0.4985501161$.

Section 4.1

1. The graph of $f(x)$ in the interval is exhibited below.

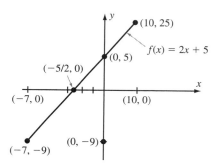

We see, by inspection of the graph of $f(x)$, that $f(x)$ is an increasing, continuous function. The values of $f(x)$ are described by $-9 \le f(x) \le 25$.

Therefore we conclude that the value -9 is an absolute minimum value and the value 25 is an absolute maximum value. For $f(x)$ in $[-7, 10]$, the local minimum coincides with the absolute minimum and the local maximum coincides with the absolute maximum.

2.

i	5	6	7	8	9	10
x	$\dfrac{7\pi}{4}$	$\dfrac{9\pi}{5}$	$\dfrac{37\pi}{20}$	$\dfrac{19\pi}{10}$	$\dfrac{39\pi}{20}$	2π
$f(x)$	$-.5$	$-.4755282582$	$-.4045084972$	$-.2938926262$	$-.1545084972$	0

Thus we see that the table values of the continuous function $f(x) = \sin(x)\cos(x)$ substantiates our observation that for x in $\left(\dfrac{3\pi}{2}, 2\pi\right)$, $\dfrac{-1}{2} \le f(x) < 0$. Because $f(x) = \sin(x)\cos(x) = \dfrac{1}{2}\sin(2x)$, we repeat that for all real x, $-\dfrac{1}{2} \le f(x) \le \dfrac{1}{2}$.

Therefore we conclude that the local minimum and global minimum values $= -\dfrac{1}{2}$ and there is no local maximum for x in $\left(+\dfrac{3\pi}{2}, 2\pi\right)$.

3. The graph of $f(x)$ in the interval $\left(\dfrac{3\pi}{2}, 2\pi\right)$ appears below.

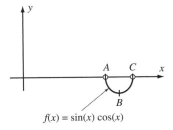

$f(x) = \sin(x)\cos(x)$

where $A = \left(\dfrac{3\pi}{2}, 0\right)$, $B = \left(\dfrac{7\pi}{4}, \dfrac{-1}{2}\right)$, and $C = (2\pi, 0)$.

By inspecting this graph, we conclude that a local minimum value $f\left(\dfrac{7\pi}{4}\right) = -\dfrac{1}{2}$ occurs at $x = \dfrac{7\pi}{4}$ and that no local maximum exists because $f\left(\dfrac{3\pi}{2}\right) = f(2\pi) = 0$ but $x = \dfrac{3\pi}{2}$ and $x = 2\pi$ are not in the open interval $\left(\dfrac{3\pi}{2}, 2\pi\right)$. Because $f(x) = \sin(x)\cos(x) = \dfrac{1}{2}\sin(2x)$, we know that $-\dfrac{1}{2} \le f(x) \le \dfrac{1}{2}$ for all real x. Therefore we can say that $f\left(\dfrac{7\pi}{4}\right) = -\dfrac{1}{2}$ is also an absolute minimum.

We shall substantiate our observation by examining values in the table that follows, where $x = \dfrac{3\pi}{2} + \dfrac{i\pi}{2}$, $i = 0, 1, 2, \ldots, 10$; we note that the first x-value is $\dfrac{3\pi}{2}$ and the last x-value is 2π. $f(x) = \sin(x)\cos(x)$.

i	0	1	2	3	4
x	$\dfrac{3\pi}{2}$	$\dfrac{31\pi}{20}$	$\dfrac{8\pi}{5}$	$\dfrac{33\pi}{20}$	$\dfrac{17\pi}{10}$
$f(x)$	0	$-.1545084973$	$-.2938926257$	$-.4045084970$	$-.4755282581$

4.
$$f(x) = x^3 + 3x^2 + 3x + 5 \Rightarrow$$
$$f'(x) = 3x^2 + 6x + 3$$
$$= 3(x^2 + 2x + 1)$$
$$= 3(x + 1)^2.$$
$$f'(x) = 0 \Rightarrow 3(x + 1)^2 = 0 \Rightarrow x = -1$$

Fermat's Theorem indicates that a local extremum may occur at $x = -1$.

If $\delta = \dfrac{1}{10}$, $x - \delta = -1.1$ and $x + \delta = -.9$.

If we compare the associated functional values in the following table,

x	-1.1	-1	$-.9$
$f(x)$	3.999	4	4.001

we see that at $x = -1$, $f(-1)$ is neither a local minimum nor a local maximum. This conclusion is substantiated by the graph of $f(x)$ in the interval $(-1.1, -.9)$.

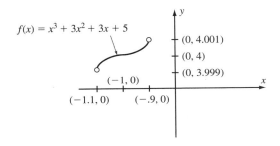

$f(x) = x^3 + 3x^2 + 3x + 5$

$(0, 4.001)$
$(0, 4)$
$(0, 3.999)$
$(-1, 0)$
$(-1.1, 0)$ $(-.9, 0)$

5. $f(x) = 2x^4 - 32$ is a polynomial; therefore it is continuous for all real x; in particular, $f(x)$ is continuous for x in $[-1, 3]$. Its derivative, $f'(x)$, is equal to $f'(x) = 8x^3$. $f'(x)$ is also a polynomial and is defined for all x in $(-\infty, \infty)$ and, therefore, for all x in $(-1, 3)$. The hypothesis that f is a function is continuous on $[a, b]$ and differentiable on (a, b), where $a = -1$ and $b = 3$, has been verified.

$$\frac{f(b) - f(a)}{b - a} = \frac{f(3) - f(-1)}{3 - (-1)}$$
$$= \frac{[2(3^4) - 32] - [2(-1)^4 - 32]}{4}$$
$$= \frac{162 - 32 - 2 + 32}{4}$$
$$= \frac{160}{4}$$
$$= 40.$$
$$f'(x) = 8x^3$$

Thus $f'(c) = 8c^3 = 40 \Rightarrow c^3 = 5 \Rightarrow c = \sqrt[3]{5} \approx 1.709975947$.

The hypothesis of continuity in $[a, b]$ and differentiability in (a, b) are satisfied because both $f(x) = 2x^4 - 32$ and $f'(x) = 8x^3$ are polynomials in x.

The conclusion of the Mean Value Theorem is that there exists a "c" in $(-1, 3)$ such that $f'(c) = \dfrac{f(3) - f(1)}{3 - (-1)}$. The value of c is $c = \sqrt[3]{5} \approx 1.709975947$. Of course in this example $a = -1$ and $b = 3$.

Section 4.2

1. $f(x) = x^4 - 1 \Rightarrow f'(x) = 4x^3$. $f'(x) > 0 \Rightarrow 4x^3 > 0 \Rightarrow x > 0$. Therefore f is increasing for all x in the interval $(0, \infty)$. $f'(x) < 0 \Rightarrow 4x^3 < 0 \Rightarrow x < 0$. Therefore f is decreasing for all x in the interval $(-\infty, 0)$.

2. $f(x) = x^4 - 1 \Rightarrow f'(x) = 4x^3$. Since $f'(x)$ is a polynomial, it is defined for all real x. Therefore, the only critical points are obtained from solving $f'(c) = 0$ and solving for c.

$$f'(c) = 4c^3 = 0 \Rightarrow c = 0.$$
$$f'(x) = 4x^3 > 0 \Rightarrow x > 0, \text{ i.e., } f'(x) > 0 \text{ for } x > c.$$
$$f'(x) = 4x^3 < 0 \Rightarrow x < 0, \text{ i.e., } f'(x) < 0 \text{ for } x < 0.$$

By the First Derivative test, we conclude that $f(x) = x^4 - 1$ has a local minimum value of $f(0) = -1$ that occurs at the critical point $x = 0$.

3. $f'(x) = 3x^2 - 24x = 3x(x - 8) = 0 \Rightarrow x = 0$ or $x = 8$. Because $f'(x)$ is a polynomial, it exists for all real x, so $x = 0$ and $x = 8$ are the only critical points.

$$f'(x) > 0 \; \Rightarrow (3x)(x - 8) > 0 \Rightarrow$$
$$x > 0 \text{ and } x > 8, \text{ i.e., } x > 8 \text{ or } x \text{ is in } (8, \infty).$$
$$\text{or} \quad x < 0 \text{ and } x < 8, \text{ i.e., } x < 0 \text{ or } x \text{ is in } (-\infty, 0).$$
$$f'(x) < 0 \; \Rightarrow (3x)(x - 8) < 0 \Rightarrow$$
$$x < 0 \text{ and } x > 8, \text{ which is impossible}$$
$$\text{or} \quad x > 0 \text{ and } x < 8, \text{ i.e., } x \text{ is in } (0, 8).$$

Thus we have the following situation.

Increasing	Decreasing	Increasing
	0	8

By the First Derivative test, we conclude that a local maximum value $f(0) = 0$ occurs at the critical point $x = 0$ and a local minimum value $f(8) = -256$ occurs at the critical point $x = 8$.

4.
$$f(x) = x(x + 2)^{1/3} \Rightarrow$$
$$f'(x) = (x)\left(\frac{1}{3}(x + 2)^{-2/3}\right) + (x + 2)^{1/3}(1)$$
$$= \frac{x + 3(x + 2)}{3(x + 2)^{2/3}}$$
$$= \frac{4x + 6}{3(x + 2)^{2/3}}$$
$$= \frac{2}{3}\left(\frac{(2x + 3)}{(x + 2)^{2/3}}\right)$$

$f'(x)$ is not defined for $x = -2$, so $x = -2$ is a critical point.

For $x \neq -2$, $f'(x) = 0 \Rightarrow 2x + 3 = 0 \Rightarrow x = -\dfrac{3}{2}$, so $x = -\dfrac{3}{2}$ is a critical point. If $\delta = \dfrac{1}{10}$ and $x = -1.5$, then $x - \delta = -1.5 - .1 = -1.6$ and $x + \delta = -1.5 + .1 = -1.4$.
$f'(-1.6) = \dfrac{\text{negative}}{\text{positive}} < 0$, $f'(-1.5) = 0$, $f'(-1.4) = \dfrac{\text{positive}}{\text{positive}} > 0$. If $\delta = \dfrac{1}{10}$ and $x = -2$, then $x - \delta = -2.1$ and $x + \delta = -1.9$. $f'(-2.1) = \dfrac{\text{negative}}{\text{positive}} < 0$, $f'(-2)$ does not exist, $f'(-1.9) = \dfrac{\text{negative}}{\text{positive}} < 0$.

We conclude that $x = -2$ and $x = -1.5$ are critical points of $f(x) = x(x + 2)^{1/3}$. By the First Derivative Test for Extrema, there is no local extremum at critical point $x = -2$, but a local minimum value $f(-1.5) \approx -1.190550789$ occurs at critical point $x = -1.5$.

Section 4.3

1. $f(x) = 2x + 5 \Rightarrow f'(x) = 2$ for all real x, in particular for x in $[30, 35]$. Thus there exists no x in $[30, 35]$ for which $f'(x) = 0$; since $f'(x)$ is a constant function equal to two, it is defined for x in $[30, 35]$. We conclude that there are no critical points for $f(x)$ in $[30, 35]$. The values of $f(x)$ at $x = 30$ and $x = 35$, the endpoints of interval $[30, 35]$, are $f(30) = 65$ and $f(35) = 75$.

Thus the absolute minimum value, $f(30) = 65$, occurs at the left endpoint $x = 30$ of interval $[30, 35]$. The absolute maximum value, $f(35) = 75$, occurs at the right endpoint $x = 35$ of interval $[30, 35]$.

2. $f(x) = \dfrac{1}{x} \Rightarrow f'(x) = \dfrac{-1}{x^2}$. $f'(x)$ does not exist for $x = 0$, but $f(x)$ is not defined at $x = 0$, either. Therefore $x = 0$ is not in the domain of $f(x)$ and there is no critical point for $f(x)$. The value of $f(x)$ at the left endpoint $x = 2$ of $[2, 5]$ is $\dfrac{1}{2}$. The value of $f(x)$ at the right endpoint $x = 5$ of $[2, 5]$ is $\dfrac{1}{5}$.

We conclude that the absolute minimum value of $f(x)$ is $\dfrac{1}{5}$ and it occurs at $x = 5$; the absolute maximum value of $f(x) = \dfrac{1}{2}$ and it occurs at $x = 2$. This is substantiated by the observation that for $x \neq 0$, $f'(x) < 0$, which implies that for $x \neq 0$, $f(x)$ is a decreasing function.

3. Let $w = $ width of the garden and let $\ell = $ length of the garden.

The perimeter, P, of the garden is $P = 2\ell + 2w$.
$$P = 200' \Rightarrow \ell + w = 100 \Rightarrow \ell = 100 - w$$

The area, A, of the garden is

$$A = \ell w = (100 - w)(w)$$
$$= 100w - w^2.$$

Since width is not negative, $w \geq 0$.

Since the total amount of fencing is 200 feet, $2w \leq 200$ or $w \leq 100$. Thus the range of values for w is $0 \leq w \leq 100$.

$A(w) = A = 100w - w^2 \Rightarrow A' = 100 - 2w$. This is a polynomial expression in w so it exists for all real values w and in particular for $0 \leq w \leq 100$.

$A' = 0 \Rightarrow w = 50$, which is the only critical point in $(0, 100)$. $w = 0$ is the left endpoint of interval $[0, 100]$ and $w = 100$ is the right endpoint of interval $[0, 100]$; we need to evaluate only $A(0)$, $A(50)$, and $A(100)$. The largest value will be the absolute maximum area.

$$A(0) = 0, \; A(50) = 2500, \text{ and } A(100) = 0$$

Thus a width of 50 feet and a length of 50 feet produces a garden with (an absolute) maximum area of 2500 square feet.

4. Let x equal the length of the stone wall in feet and let y equal the length of the wooden fence in feet.

$$\text{Total cost} = 2(5x) + 2(3y) = 5000 \Rightarrow$$
$$5x + 3y = 2500 \Rightarrow$$
$$5x = 2500 - 3y \Rightarrow$$
$$x = 500 - \frac{3}{5}y$$

If A equals the area of the garden, then

$$A = xy = \left(500 - \frac{3}{5}y\right)(y) = 500y - \frac{3}{5}y^2 \equiv A(y).$$

$A'(y) = 500 - \frac{6}{5}y$ is a polynomial, so it is defined for all y. Therefore a critical value is obtained by setting $A'(y) = 0 \Rightarrow 500 - \frac{6}{5}y = 0 \Rightarrow y = 500 \times \frac{5}{6} = \frac{2500}{6} = \frac{1250}{3}$. The length of the wooden fence cannot be negative, so $y \geq 0$.

Because $5x + 3y = 2500$, $y = \frac{2500}{3} - \frac{5}{3}x$, so $y \leq \frac{2500}{3}$.

y is a value in $\left[0, \frac{2500}{3}\right]$ because $0 \leq y \leq \frac{2500}{3}$.

Evaluating $A(y)$ at the endpoints $y = 0$ and $y = \frac{2500}{3}$ and also evaluating $A(y)$ at the critical value $y = \frac{1250}{3}$, we obtain $A(0) = 0$, $A\left(\frac{2500}{3}\right) = 0$, and

$$A\left(\frac{1250}{3}\right) = \frac{312,500}{3}.$$

The (maximum) dimension of the wooden fence y, and $\therefore$ of the garden, is $y = \frac{1250}{3} = 416\frac{2}{3}$ feet. Since

$$x = 500 - \frac{3}{5}y = 500 - \frac{3}{5}\left(\frac{1250}{3}\right) = 250 \text{ feet, we conclude}$$

that the (maximum) dimension of the stone wall x, and $\therefore$ of the garden, is $x = 250$ feet with area $= 104166^{2/3}$ ft^2.

Section 4.4

1. $f(x) = 5x^3 - 12x \Rightarrow f'(x) = 15x^2 - 12 \Rightarrow f''(x) = 30x$. $f(x)$, $f'(x)$, and $f''(x)$ are polynomials in x, so they are defined for all real x.

$f''(x) = 30x > 0 \Rightarrow x > 0$ so $f(x) = 5x^3 - 12x$ is concave up for x in $(0, \infty)$.

$f''(x) = 30x < 0 \Rightarrow x < 0$ so $f(x) = 5x^3 - 12x$ is concave down for x in $(-\infty, 0)$.

Since f is concave down to the "left" of $x = 0$, i.e., f is concave down if x is less than zero, and f is concave up to the "right" of $x = 0$, i.e., f is concave up if x is greater than zero, we conclude that the point of inflection is at $x = 0$.

To obtain critical points, set the first derivative equal to zero and solve for x.

$$f'(x) = 15x^2 - 12 = 3(5x^2 - 4) = 3(\sqrt{5}x - 2)(\sqrt{5}x + 2)$$

$$f'(x) = 0 \Rightarrow x = \frac{2}{\sqrt{5}} \text{ or } x = \frac{-2}{\sqrt{5}}$$

These are critical points of $f(x)$. We shall use the Second Derivative Test to determine whether a relative minimum or a relative maximum occurs at the critical points $x = \frac{2}{\sqrt{5}}$ or $x = \frac{-2}{\sqrt{5}}$.

$$f''(x) = 30x \Rightarrow f''\left(\frac{2}{\sqrt{5}}\right) = (30)\left(\frac{2}{\sqrt{5}}\right) > 0 \Rightarrow$$

a relative minimum occurs at critical point $x = \frac{2}{\sqrt{5}}$.

Similarly, $f''\left(\frac{-2}{\sqrt{5}}\right) = (30)\left(\frac{-2}{\sqrt{5}}\right) < 0 \Rightarrow$ a relative maximum occurs at critical point $x = \frac{-2}{\sqrt{5}}$.

Thus we conclude that the critical points are $x = \frac{2}{\sqrt{5}}$ and $x = -\frac{2}{\sqrt{5}}$. The local minimum value is $\frac{-16}{\sqrt{5}}$, which occurs at the critical point $x = \frac{2}{\sqrt{5}}$, and the local maximum value is $\frac{16}{\sqrt{5}}$, which occurs at the critical point $x = \frac{-2}{\sqrt{5}}$.

2. $f''(x) = 0 \Rightarrow (x - 1)^2(x) = 0 \Rightarrow x = 0$ or $x = 1$.

Therefore we shall identify whether $f''(x) > 0$ or $f''(x) < 0$ for x in $(1, \infty)$, for x in $(0, 1)$, and for x in $(-\infty, 0)$.

If x is in $(1, \infty)$, then x is positive and $(x - 1)^2$ is positive, so their product, $f''(x) = (x - 1)^2(x)$, is positive. We conclude that $f(x)$ is concave up for x in $(1, \infty)$.

If x is in $(0, 1)$, then x is positive and $(x - 1)^2$ is positive, so their product, $f''(x) = (x - 1)^2(x)$, is positive. We conclude that $f(x)$ is concave up for x in $(0, 1)$.

If x is in $(-\infty, 0)$, then x is negative and $(x-1)^2$ is positive, so their product, $f''(x) = (x-1)^2(x)$, is negative. We conclude that $f(x)$ is concave down. Summarizing our conclusions we have

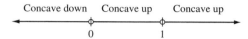

Concave down   Concave up   Concave up
0   1

We conclude that there is only one inflection point at $x = 0$ because if $x < 0$ then f is concave down and if $x > 0$, then f is concave up.

We note that the point $x = 1$ is not an inflection point because if $x < 1$, $f(x)$ is concave up and if $x > 1$, $f(x)$ is concave up.

3. Curvature, $\kappa(x)$, is defined by $\kappa(x) = \dfrac{|f''(x)|}{(1 + (f'(x))^2)^{3/2}}$ at point $(x, f(x))$ for a function f that is twice differentiable. $f(x) = x^2 - 6x$ is a polynomial so its derivative exists for all $n = 0, 1, 2, \ldots$. In particular, the nth derivative, $f^{(n)}(x)$, exists if $n = 2$.

$$f(x) = x^2 - 6x \Rightarrow f'(x) = 2x - 6 \Rightarrow f''(x) = 2.$$

Thus $f''(x)|_{x=1} = 2$, i.e., $f''(1) = 2$, and $f'(x)|_{x=1} = -4$, i.e., $f'(1) = -4$, so $(f'(1))^2 = +16$.
Thus $\kappa(x)$ at point $(1, -5)$ is equal to

$$\kappa(1) = \frac{|2|}{(1+16)^{3/2}} = \frac{2}{17\sqrt{17}} \approx 0.02853360295.$$

Section 4.5

1. $x + 5 = 0 \Rightarrow x = -5$, so $f(-5)$ is not defined. Therefore the domain of $f(x) = (-\infty, -5) \cup (-5, +\infty)$.
   The range of $f(x)$ is $(-\infty, \infty)$.
   The vertical asymptote is $x = -5$.

$$f(x) = \frac{x}{x+5} = \frac{x}{x+5} \cdot \frac{\frac{1}{x}}{\frac{1}{x}} = \frac{1}{1 + \frac{5}{x}} \cdot f(x) \to 1 \text{ as } x \to \pm\infty,$$

so the horizontal asymptote is $y = 1$.
Using the Derivative Rule for Quotients, we obtain

$$f'(x) = \frac{(x+5)(1) - x(1)}{(x+5)^2} = \frac{x+5-x}{(x+5)^2}$$

$$= \frac{5}{(x+5)^2}, \quad \text{for } x \neq -5.$$

$f'(-5)$ is not defined; however, since $x = -5$ is not in the domain of f, we conclude there is *no* critical point.

$f'(x) > 0$ for all real x, $x \neq -5$, so f is increasing for all x in the domain of f.

$$f''(x) = (5)(-2)(x+5)^{-3} = \frac{-10}{(x+5)^3}, \quad \text{for } x \neq -5$$

$f''(-5)$ is not defined, but because $x = -5$ is not in the domain of f, we conclude there is no inflection point.
If x is in $(-\infty, -5)$, $f''(x) > 0$, so $f(x)$ is concave up for these x.
If x is in $(-5, \infty)$, $f''(x) < 0$, so $f(x)$ is concave down for these x.
If $x = 0$, $y = f(0) = 0$, so the x-intercept is 0 and the y-intercept is 0.
We summarize our results:

domain f: x in $(-\infty, -5) \cup (-5, \infty)$
range f: y in $(-\infty, \infty)$
vertical asymptote: $x = -5$
horizontal asymptote: $y = 1$
critical point: none
f is increasing: for all x in domain f
inflection points: none
f is concave up: for x in $(-\infty, -5)$
f is concave down: for x in $(-5, \infty)$
x-intercept: $x = 0$
y-intercept: $y = 0$

2. The period of $g(x) = \cos(x)$ is 2π because $\cos(x + 2\pi) = \cos(x)$, as its graph and table of values demonstrate.

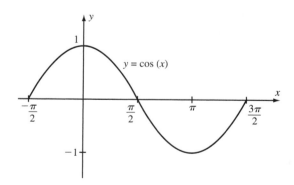

x	$-\dfrac{\pi}{2}$	0	$\dfrac{\pi}{2}$	π	$\dfrac{3\pi}{2}$
$\cos(x)$	$\cos\left(-\dfrac{\pi}{2}\right) = 0$	$\cos(0) = 1$	$\cos\left(\dfrac{\pi}{2}\right) = 0$	$\cos(\pi) = -1$	$\cos\left(\dfrac{3\pi}{2}\right) = 0$
$\cos(x + 2\pi)$	$\cos\left(\dfrac{3\pi}{2}\right) = 0$	$\cos(2\pi) = 1$	$\cos\left(\dfrac{5\pi}{2}\right) = 0$	$\cos(3\pi) = -1$	$\cos\left(\dfrac{7\pi}{2}\right) = 0$

Claim: The period of

$$f(x) = \cos(3x) = \frac{\text{period of } g(x) = \cos(x)}{3} = \frac{2\pi}{3}.$$

The graph of $f(x) = \cos(3x)$ is

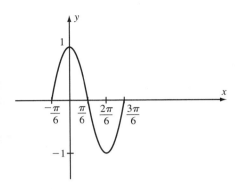

x	$\dfrac{-\pi}{6}$	0	$\dfrac{\pi}{6}$	$\dfrac{2\pi}{6}$ or $\dfrac{\pi}{3}$	$\dfrac{3\pi}{6}$ or $\dfrac{\pi}{2}$
$\cos(3x)$	$\cos\left(\dfrac{-\pi}{2}\right) = 0$	$\cos(0) = 1$	$\cos\left(\dfrac{\pi}{2}\right) = 0$	$\cos(\pi) = -1$	$\cos\left(\dfrac{3\pi}{2}\right) = 0$
$\cos\left(3x + \dfrac{2\pi}{3}\right)$	$\cos\left(\dfrac{\pi}{6}\right) = 0$	$\cos\left(\dfrac{2\pi}{3}\right) = 1$	$\cos\left(\dfrac{7\pi}{6}\right) = 0$	$\cos\left(\dfrac{5\pi}{3}\right) = -1$	$\cos\left(\dfrac{13\pi}{6}\right) = 0$

We conclude that the period of $f(x) = \cos(3x)$ is $p = \dfrac{2\pi}{3}$, since $\cos(3x) = \cos\left(3\left(x + \dfrac{2\pi}{3}\right)\right)$.

3.

$$f(x) = \frac{x^3 + 3x^2 + 5x}{x^2} = \frac{x^3}{x^2} + \frac{3x^2}{x^2} + \frac{5x}{x^2} = x + 3 + \frac{5}{x^2}$$

Notice that the remainder is $\dfrac{5}{x^2}$ and that $\displaystyle\lim_{x \to \pm\infty} \frac{5}{x^2} = 0$.

Claim that the skew asymptote, $\therefore$ is $y = x + 3$. This claim is verified because

$$\lim_{x \to \pm\infty}\left((x + 3) - \left(\frac{x^3 + 3x + 5x}{x^2}\right)\right) = \lim_{x \to \pm\infty}\frac{5}{x^2} = 0.$$

Section 4.6

1. $\displaystyle\lim_{x \to \pi}\frac{\tan(x)}{(x - \pi)} = \frac{\displaystyle\lim_{x \to \pi}\tan(x)}{\displaystyle\lim_{x \to \pi}(x - \pi)} = \frac{0}{0}$, so apply L'Hôpital's Rule.

$$\lim_{x \to \pi}\frac{\tan(x)}{(x - \pi)} = \lim_{x \to \pi}\frac{\dfrac{d}{dx}(\tan(x))}{\dfrac{d}{dx}(x - \pi)}$$

$$= \lim_{x \to \pi}\frac{\sec^2(x)}{1}$$

$$= \lim_{x \to \pi}\left(\frac{1}{\cos(x)}\right)^2$$

$$= \left(\frac{\displaystyle\lim_{x \to \pi} 1}{\displaystyle\lim_{x \to \pi}\cos(x)}\right)^2$$

$$= \left(\frac{1}{-1}\right)^2$$

$$= 1$$

2. $\displaystyle\lim_{x \to 0}(2x)^{3x} = 0^0$. If we let $f(x) = (2x)^{3x}$ and utilize logarithms, we obtain

$$\ln f(x) = \ln(2x)^{3x} = 3x\ln(2x) = \frac{\ln(2x)}{\dfrac{1}{3x}}.$$

Thus $\displaystyle\lim_{x \to 0}\ln(f(x)) = \lim_{x \to 0}\frac{\ln(2x)}{\dfrac{1}{3x}} = \frac{-\infty}{\infty}$, so we can apply L'Hôpital's Rule.

$$\lim_{x \to 0}\ln(f(x)) = \lim_{x \to 0}\frac{\ln(2x)}{\dfrac{1}{3x}} = \lim_{x \to 0}\frac{\dfrac{d}{dx}(\ln(2x))}{\dfrac{d}{dx}\left(\dfrac{1}{3x}\right)}$$

$$= \lim_{x \to 0}\frac{\dfrac{2}{2x}}{\dfrac{-1}{3x^2}}$$

$$= \lim_{x \to 0}\left(\frac{1}{x}\right)\left(\frac{-3x^2}{1}\right)$$

$$= \lim_{x \to 0}(-3x)$$

$$= 0$$

But $\displaystyle\lim_{x \to 0}(2x)^{3x} = \lim_{x \to 0}e^{\ln(2x^{3x})} = e^{\lim_{x \to 0}(\ln(2x^{3x}))} = e^0 = 1.$

Section 4.7

1. This problem is equivalent to finding the positive value where $f(x) = x^2 - 5$ vanishes.
Let $x_1 = 2.2$. $f'(x) = 2x$.

$$x_2 = x_1 - \frac{f(x_1)}{f'(x_1)} = 2.2 - \frac{[(2.2)^2 - 5]}{2(2.2)} = 2.236363636$$

$$x_3 = x_2 - \frac{f(x_2)}{f'(x_2)} = 2.236363636 - \frac{[(2.236363636)^2 - 5]}{2(2.236363636)}$$
$$= 2.236067997$$

The value of $\sqrt{5} = 2.236067977$.
Therefore the value of x_3 agrees with $\sqrt{5}$ to seven places of accuracy.

Section 4.8

1.

$$\frac{d}{dx}(\sin(x)) = \cos(x)$$

Therefore $\int \cos(x)\,dx = \sin(x) + C$.

2.

$$\frac{d}{dx}\left(\frac{x^3}{3}\right) = \frac{3x^2}{3} = x^2 \quad \text{and} \quad \frac{d}{dx}(\sin(x)) = \cos(x)$$

Therefore

$$\int (5x^2 + \cos(x))\,dx = \int 5x^2\,dx + \int \cos(x)\,dx$$

$$= 5\int x^2\,dx + \int \cos(x)\,dx$$

$$= \left(5\left(\frac{x^3}{3}\right) + C_1\right) + (\sin(x) + C_2)$$

$$= \frac{5}{3}x^3 + \sin(x) + C, \quad \text{where } C = C_1 + C_2.$$

We remark that the convention used is to not include the constants of integration C_1 and C_2 and to simply write the constant of integration C in the last statement.

3. If $p(t)$ denotes the position of an object at time t, then velocity at time t is $v(t) = p'(t)$, and acceleration at time t is $a(t) = v'(t) = p''(t)$. Therefore, given $a(t)$, we will use antidifferentiation to first find the expression for $v(t)$ and also find the expression for $p(t)$.

$$v(t) = \int v'(t)\,dt$$

$$= \int a(t)\,dt$$

$$= \int \frac{5}{2}t^2\,dt$$

$$= \frac{5}{2}\frac{t^3}{3} + C$$

$$= \frac{5}{6}t^3 + C$$

If $t = 0$, then $v(t) = 0 \Rightarrow 0 = \frac{5}{6}(0^3) + C \Rightarrow C = 0$.

Thus $v(t) = \frac{5}{6}t^3$.

$$p(t) = \int p'(t)\,dt$$

$$= \int v(t)\,dt$$

$$= \int \frac{5}{6}t^3\,dt$$

$$= \frac{5}{6}\int t^3\,dt$$

$$= \frac{5}{6}\frac{t^4}{4} + C$$

$$= \frac{5}{24}t^4 + C$$

If $t = 0$, then $p(t) = 0 \Rightarrow 0 = \frac{5}{24}(0)^4 + C = C + 0$.

Thus $p(t) = \frac{5}{24}t^4$.

Section 4.9

1. From the definition of elasticity of demand, we have

$$1.5 = E(P_0) = -\lim_{\Delta p \to 0} \frac{100\dfrac{\Delta x}{x_0}}{100\dfrac{\Delta P}{P_0}} \approx \frac{-100\dfrac{\Delta x}{x_0}}{100(.01)}.$$

This implies that $\dfrac{\Delta x}{x_0} \approx -(1.5)(.01) = -.015 = -1.5\%$.

We conclude that a 1% price increase will result in a decrease in demand of approximately 1.5%.

2. $\overline{C}(x) = \dfrac{C(x)}{x} = \dfrac{x^3 + 2x^2 - 5x + 8}{x} = x^2 + 2x - 5 + \dfrac{8}{x}$.

The graph of $\overline{C}(x)$ appears below.

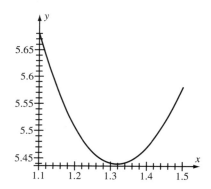

From the graph of $\overline{C}(x)$, the minimum initially appears to be in the neighborhood of $x = 1.3$. By restricting the domain to this neighborhood, we can plot a graph of $\overline{C}(x)$ with more detail. This is shown in the graph that follows.

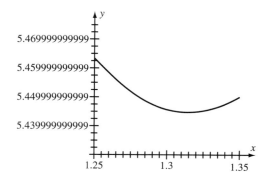

x	1.314	1.314596212	1.3147
$\overline{C}(x)$	5.442876061	5.442874454	5.442874502

We conclude that the minimum value of $\overline{C}(x)$ is approximately 5.44, which occurs when $x = 1.3$. According to the Minimum Average Cost Principle, this number equals the lowest cost. Because the company can't produce part of a certain product, they would have to produce either one or two products. $\overline{C}(1) = 6$ and $\overline{C}(2) = 7$. So if the company produces a single product, the lowest cost is achieved.

Section 5.1

1. $n = 5$, $a = 8$, and $b = 9$ implies that
$$\Delta x = \frac{b - a}{n} = \frac{9 - 8}{5} = \frac{1}{5} = 0.2. \text{ Therefore } x_0 = a = 8,$$
$x_1 = a + \Delta x = 8 + 0.2 = 8.2,$
$x_2 = a + 2 \cdot \Delta x = 8 + 0.4 = 8.4,$
$x_3 = a + 3 \cdot \Delta x = 8 + 0.6 = 8.6,$
$x_4 = a + 4 \cdot \Delta x = 8 + 0.8 = 8.8,$ and
$x_5 = a + 5 \cdot \Delta x = 8 + 1.0 = 9.$ We conclude that the uniform partition of order 5 of the closed interval $[8, 9]$ is $\{8, 8.2, 8.4, 8.6, 8.8, 9\}$.

2. $n = 2$, $a = 0$, and $b = 100$ implies that
$$\Delta x = \frac{b - a}{n} = \frac{100 - 0}{2} = 50. \text{ Therefore } x_0 = a = 0,$$
$x_1 = a + \Delta x = 0 + 50 = 50,$
$x_2 = a + 2 \cdot \Delta x = 0 + 100 = 100.$ We conclude that the uniform partition of order 2 of the closed interval $[0, 100]$ is $\{0, 50, 100\}$.

3.
$$\sum_{j=1}^{10} 2 \left(\frac{1}{4}\right)^j = 2 \sum_{j=1}^{10} \left(\frac{1}{4}\right)^j$$

$$= 2 \left[\left(\frac{1}{4}\right)^1 + \left(\frac{1}{4}\right)^2 + \left(\frac{1}{4}\right)^3 + \left(\frac{1}{4}\right)^4 + \left(\frac{1}{4}\right)^5 + \left(\frac{1}{4}\right)^6 + \left(\frac{1}{4}\right)^7 + \left(\frac{1}{4}\right)^8 + \left(\frac{1}{4}\right)^9 + \left(\frac{1}{4}\right)^{10} \right]$$

$$= 2 \left[\left(\frac{1}{4}\right) \left\{ 1 + \left(\frac{1}{4}\right)^1 + \left(\frac{1}{4}\right)^2 + \left(\frac{1}{4}\right)^3 + \left(\frac{1}{4}\right)^4 + \left(\frac{1}{4}\right)^5 + \left(\frac{1}{4}\right)^6 + \left(\frac{1}{4}\right)^7 + \left(\frac{1}{4}\right)^8 + \left(\frac{1}{4}\right)^9 \right\} \right]$$

$$= \frac{1}{2} \left\{ \frac{\left(\frac{1}{4}\right)^{10} - 1}{\frac{1}{4} - 1} \right\}$$

$$= \frac{1}{2} \left\{ \frac{\frac{1^{10}}{4^{10}} - \frac{4^{10}}{4^{10}}}{-\frac{3}{4}} \right\}$$

$$= \left(\frac{1}{2}\right) \left(\frac{-4}{3}\right) \left\{ \frac{1 - 4^{10}}{4^{10}} \right\}$$

$$= -\frac{1}{6} \left\{ \frac{1 - 4^{10}}{4^9} \right\}$$

$$= \frac{349525}{524288}$$
$$\approx 0.6666660309$$

4. $\displaystyle\sum_{k=2}^{10} (2k^2 + 3k - 7) = \sum_{k=1}^{10} (2k^2 + 3k - 7) - \sum_{k=1}^{1} (2k^2 + 3k - 7)$

First we calculate

$$\sum_{k=1}^{10} (2k^2 + 3k - 7) = 2 \sum_{k=1}^{10} k^2 + 3 \sum_{k=1}^{10} k - 7 \sum_{k=1}^{10} 1$$

$$= \frac{2(10)(11)(21)}{6} + \frac{3(10)(11)}{2} - 7(10)$$

$$= \frac{(2)(10)(11)(21)}{6} + \frac{3(10)(11)}{2} \cdot \frac{(3)}{(3)} - 7(10) \cdot \frac{6}{6}$$

$$= \frac{(2)(10)(11)(21) + (3)(3)(10)(11) - (6)(7)(10)}{6}$$

$$= \frac{(3)(10)\{(2)(7)(11) + (3)(11) - 2(7)\}}{6}$$

$$= 5\{154 + 33 - 14\}$$

$$= 5\{154 + 19\}$$

$$= 5\{173\}$$

$$= 865.$$

Then we calculate $\displaystyle\sum_{k=1}^{1} (2k^2 + 3k - 7) = 2(1)^2 + 3(1) - 7$
$= 2 + 3 - 7 = -2.$

We conclude that

$$\sum_{k=2}^{10} (2k^2 + 3k - 7) = \sum_{k=1}^{10} (2k^2 + 3k - 7) - \sum_{k=1}^{1} (2k^2 + 3k - 7)$$
$$= 865 - (-2)$$
$$= 867.$$

5. $n = 4$, $a = 0$, and $b = 4$ implies that $\Delta x = \dfrac{b - a}{n} = \dfrac{4 - 0}{4} = 1.$ So the uniform partition of order 4 is $\{0, 1, 2, 3, 4\}$. We will have four rectangles, R_1, R_2, R_3, R_4, each of width equal to one.

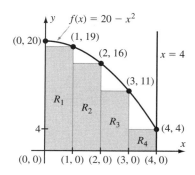

The height of R_1 is equal to $f(1) = 20 - 1^2 = 19$.
The height of R_2 is equal to $f(2) = 20 - 2^2 = 16$.
The height of R_3 is equal to $f(3) = 20 - 3^2 = 11$.
The height of R_4 is equal to $f(4) = 20 - 4^2 = 4$.
We conclude that the approximation of the area under the graph of $f(x) = 20 - x^2$ and above the integral $[0, 4]$ is

$$\text{Right Endpoint Approximation} = \sum_{i=1}^{4} f(x_i) \Delta x$$
$$= 1[f(1) + f(2)$$
$$+ f(3) + f(4)]$$
$$= 1[19 + 16 + 11 + 4]$$
$$= 1[50]$$
$$= 50 \text{ square units.}$$

6. The region whose area is to be calculated appears below.

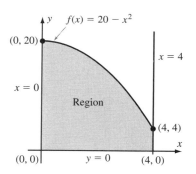

If we use the right-endpoint convention, then the area of the (shaded) Region is equal to $\text{Area} = \lim_{N \to \infty} \sum_{i=1}^{N} f(x_i) \Delta x$.

$x_i = x_0 + i \cdot \Delta x = 0 + i \Delta x$, where
$\Delta x = \dfrac{b-a}{N} = \dfrac{4-0}{N} = \dfrac{4}{N}$ and

$$f(x_i) = f(0 + i \cdot \Delta x)$$
$$= 20 - (0 + i \cdot \Delta x)^2$$
$$= 20 - i^2 \cdot (\Delta x)^2$$
$$= 20 - \left(x_0^2 + 2 \cdot i \cdot x_0 \cdot \Delta x + (i \cdot \Delta x)^2 \right)$$

$$= 20 - x_0^2 - 2 \cdot i \cdot x_0 \cdot \Delta x - i^2 \cdot (\Delta x)^2$$
$$= 20 - 0^2 - 2 \cdot i \cdot 0 \cdot \Delta x - i^2 (\Delta x)^2$$
$$= 20 - i^2 (\Delta x)^2$$

Therefore, substituting these values, we obtain

$$\text{Area} = \lim_{N \to \infty} \sum_{i=1}^{N} f(x_i) \Delta x$$

$$= \lim_{N \to \infty} \sum_{i=1}^{N} [(20 - i^2 (\Delta x)^2) \Delta x]$$

$$= \lim_{N \to \infty} \sum_{i=1}^{N} \left[\left(20 - i^2 \cdot \left(\frac{4}{N} \right)^2 \right) \cdot \frac{4}{N} \right]$$

$$= \lim_{N \to \infty} \sum_{i=1}^{N} \left[20 \cdot \frac{4}{N} - \frac{i^2 \cdot 4^3}{N^3} \right]$$

$$= \lim_{N \to \infty} \left[\sum_{i=1}^{N} \frac{80}{N} - \lim_{N \to \infty} \sum_{i=1}^{N} \frac{64}{N^3} i^2 \right]$$

$$= \lim_{N \to \infty} \frac{80}{N} \sum_{i=1}^{N} 1 - \lim_{N \to \infty} \frac{64}{N^3} \sum_{i=1}^{N} i^2$$

$$= \lim_{N \to \infty} \frac{80}{N} \cdot N - \lim_{N \to \infty} \frac{64}{N^3} \cdot \frac{(N \cdot (N+1) \cdot (2N+1))}{6}$$

$$= 80 - \frac{32}{3} \lim_{N \to \infty} \left[\left(\frac{N}{N} \right) \cdot \frac{N+1}{N} \cdot \frac{2N+1}{N} \right]$$

$$= 80 - \frac{32}{3} \cdot 1 \cdot 1 \cdot 2$$

$$= 80 - \frac{64}{3}$$

$$= \frac{240 - 64}{3}$$

$$= \frac{176}{3}$$

$$= 58 \frac{2}{3} \text{ square units}$$

$$\approx 58.6666666667 \text{ square units.}$$

Section 5.2

1. The six equal-length subintervals are $I_1 = [0, 1]$, $I_2 = [1, 2]$, $I_3 = [2, 3]$, $I_4 = [3, 4]$, $I_5 = [4, 5]$, and $I_6 = [5, 6]$. Function $f(x) = 2x$ is an increasing function, that is to say, as x increases in value, the value of $f(x)$ also increases. Therefore, the candidate for u_j, the maximum point in subinterval I_j, is the right endpoint x_j of the subinterval. We have $\Delta x = \dfrac{6-0}{6} = 1$ and $U_6\{1, 2, 3, 4, 5, 6\}$.

We conclude that

$$R(f, U_6) = \sum_{j=1}^{6} f(u_j)\Delta x$$

$$= f(1) \cdot 1 + f(2) \cdot 1 + f(3) \cdot 1 + f(4) \cdot 1$$
$$+ f(5) \cdot 1 + f(6) \cdot 1$$
$$= (2 + 4 + 6 + 8 + 10 + 12) \cdot 1$$
$$= 42.$$

Similarly, the candidate for ℓ_j, the minimum point in subinterval I_j, is the left-endpoint x_{j-1} of the subinterval. As noted earlier, $\Delta x = 1$ and $L_6 = \{0, 1, 2, 3, 4, 5\}$. We conclude that

$$R(f, L_6) = \sum_{j=1}^{6} f(\ell_j)\Delta x$$

$$= f(0) \cdot 1 + f(1) \cdot 1 + f(2) \cdot 1 + f(3) \cdot 1$$
$$+ f(4) \cdot 1 + f(5) \cdot 1$$
$$= (0 + 2 + 4 + 6 + 8 + 10) \cdot 1$$
$$= 30.$$

2. We have $\Delta x = \dfrac{2 - (-1)}{6} = \dfrac{3}{6} = \dfrac{1}{2} = 0.5$.
The six equal length subintervals are $I_1 = [-1, -0.5]$, $I_2 = [-0.5, 0]$, $I_3 = [0, 0.5]$, $I_4 = [0.5, 1]$, $I_5 = [1, 1.5]$, $I_6 = [1.5, 2]$. Function $f(x) = 4 - x^2$ is an increasing function on subintervals I_1 and I_2. It is a decreasing function on subintervals I_3, I_4, I_5, and I_6. Therefore, the maximum point in subinterval I_1 is $u_1 = -0.5$, the maximum point in subinterval I_2 is $u_2 = 0$, the maximum point in subinterval I_3 is $u_3 = 0$, and similarly we have $u_4 = 0.5$, $u_5 = 1$, and $u_6 = 1.5$. Hence $U_6 = \{-0.5, 0, 0, 0.5, 1, 1.5\}$.
We conclude that

$$R(f, U_6) = \sum_{i=1}^{6} f(u_i)\Delta x$$

$$= f(-0.5) \cdot 0.5 + f(0) \cdot 0.5 + f(0) \cdot 0.5$$
$$+ f(0.5) \cdot 0.5 + f(1) \cdot 0.5 + f(1.5) \cdot 0.5$$
$$= (3.75 + 4 + 4 + 3.75 + 3 + 1.75) \cdot 0.5 = 10.125.$$

Similarly, for the minimum points we have $\ell_1 = -1$, $\ell_2 = -0.5$, $\ell_3 = 0.5$, $\ell_4 = 1$, $\ell_5 = 1.5$, and $\ell_6 = 2$. Hence $L_6 = \{-1, -0.5, 0.5, 1, 1.5, 2\}$.
We conclude that

$$R(f, L_6) = \sum_{i=1}^{6} f(\ell_i)\Delta x$$

$$= f(-1) \cdot 0.5 + f(-0.5) \cdot 0.5 + f(0.5) \cdot 0.5$$
$$+ f(1) \cdot 0.5 + f(1.5) \cdot 0.5 + f(2) \cdot 0.5$$
$$= (3 + 3.75 + 3.75 + 3 + 1.75 + 0) \cdot 0.5$$
$$= 7.625.$$

The answers are $R(f, U_6) = 10.125$ and $R(f, L_6) = 7.625$.
3. $f(x) = 4 - x^2$ is a polynomial. Therefore it is continuous for all real x and in particular for $-1 \leq x \leq 2$. Therefore by

Theorem 2, function f is integrable on $[-1, 2]$; in other words, $\int_{-1}^{2} (4 - x^2)\, dx$ exists.

4. $f(x) = x^3 + x^2 + x + 1$ is a polynomial. Therefore it is continuous for all real x and in particular for $-100 \leq x \leq 1000$.
    Therefore by Theorem 2, function f is integrable on $[-100, 1000]$.

5. $\dfrac{d}{dx}\left(3x + \dfrac{4}{3}x^3 + C\right) = 3 + 4x^2 = f(x)$. Thus
    $f(x) = F'(x)$, where $F(x) = 3x + \dfrac{4}{3}x^3 + C$.
    $f(x) = 3 + 4x^2$ is a polynomial, so it is continuous for all real x and in particular for x in the interval $[0, 3]$. $F(x)$ is a polynomial, so it is continuous for all real x and in particular for x in interval $[0, 3]$. $F'(x) = f(x)$ for all x in $[0, 3]$. Thus the conditions to apply the Fundamental Theorem of Calculus are satisfied. Therefore,

$$\int_{0}^{3} (3 + 4x^2)\, dx = \left(3x + \dfrac{4}{3}x^3\right)\Big|_{0}^{3}$$

$$= \left(3 \cdot 3 + \dfrac{4}{3} \cdot 3^3\right) - (0 + 0)$$

$$= 9 + 36$$

$$= 45.$$

We remark that the constant of integration C is not included in the calculation for $F(b) - F(a)$ because of cancellation; i.e.,

$$F(b) = 3b + \dfrac{4}{3}b^3 + C$$

$$F(a) = 3a + \dfrac{4}{3}a^3 + C$$

$$F(b) - F(a) = \left(3b + \dfrac{4}{3}b^3 + C\right) - \left(3a + \dfrac{4}{3}a^3 + C\right)$$

$$= 3b + \dfrac{4}{3}b^3 - 3a - \dfrac{4}{3}a^3.$$

If $b = 3$ and $a = 0$, we obtain our answer

$$F(3) - F(0) = \left(9 + \dfrac{4}{3} \cdot 27\right) - 0 = 9 + 36 = 45.$$

6. $\dfrac{d}{dx}(\sin(x) + C) = \cos(x) = f(x)$. Thus $f(x) = F'(x)$,
    where $F(x) = \sin(x) + C$. $f(x) = \cos(x)$ is a continuous function for all real x and, in particular, for $x = \dfrac{\pi}{2}$.
    $F(x) = \sin(x) + C$ is a continuous function for all real x and, in particular, for $x = \dfrac{\pi}{2}$. $F'\left(\dfrac{\pi}{2}\right) = f\left(\dfrac{\pi}{2}\right)$. Thus the conditions to apply the Fundamental Theorem of Calculus are satisfied. We conclude that $\int_{\frac{\pi}{2}}^{\frac{\pi}{2}} \cos(x)\, dx = \sin(x)\Big|_{\frac{\pi}{2}}^{\frac{\pi}{2}} =$

$$\sin\left(\dfrac{\pi}{2}\right) - \sin\left(\dfrac{\pi}{2}\right) = 0.$$

Section 5.3

1. $\int_{1}^{10} f(x)\,dx = \int_{1}^{5} f(x)\,dx + \int_{5}^{10} f(x)\,dx$ by Rule (iv);

therefore $7 = \int_{1}^{5} f(x)\,dx + 3$, which implies that

$7 - 3 = 4 = \int_{1}^{5} f(x)\,dx.$

If $\int_{1}^{5} f(x)\,dx = 4$, then $+\int_{1}^{5} f(x)\,dx = -\int_{5}^{1} f(x)\,dx$

by Rule (v), so $-4 = \int_{5}^{1} f(x)\,dx.$

2.
$\int_{2}^{6} (3f(x) - 2g(x))\,dx$

$= \int_{2}^{6} (3f(x))\,dx - \int_{2}^{6} (2g(x))\,dx$ by Rule (i)

$= 3 \int_{2}^{6} f(x)\,dx - 2 \int_{2}^{6} g(x)\,dx$ by Rule (ii)

$= 3(5) - 2(-3)$

$= 15 + 6$

$= 21$

$\int_{2}^{6} (g(x) + 7)\,dx = \int_{2}^{6} g(x)\,dx + \int_{2}^{6} 7\,dx$ by Rule (i)

$= -3 + 7(6 - 2)$ by the given fact and Rule (iii)

$= -3 + 28$

$= 25.$

3. By Rule (vi), $\int_{0}^{3} 1\,dx \le \int_{0}^{3} h(x)\,dx \le \int_{0}^{3} (x^2 + 2)\,dx.$

$\int_{0}^{3} 1\,dx = 1(3 - 0) = 3$ by Rule (iii)

$\int_{0}^{3} (x^2 + 2)\,dx = \frac{x^3}{3} + 2x \Big|_{0}^{3}$

$= \left(\frac{27}{3} + 6\right) - (0 + 0)$

$= 9 + 6$

$= 15$

Therefore $3 \le \int_{0}^{3} h(x)\,dx \le 15$ gives a lower bound of 3 and an upper bound of 15 for $\int_{0}^{3} h(x)\,dx.$

Section 5.4

1. $F'(x) = \frac{d}{dx}\left(2e^{3x}\right) = 2\frac{d}{dx}\left(e^{3x}\right) = 2\left(e^{3x} \cdot \frac{d}{dx}(3x)\right) =$
$2(3e^{3x}) = 6e^{3x}.$ Thus $f(x) = 6e^{3x} = F'(x)$ establishes that
$F(x) = 2e^{3x}$ is an antiderivative of $f(x) = 6e^{3x}.$ Notice that
$\frac{1}{9}f(x) = \frac{1}{9}(6e^{3x}) = \frac{2}{3}e^{3x}.$ Therefore

$$\int_{0}^{5} \frac{2}{3}e^{3x}\,dx = \int_{0}^{5} \frac{1}{9}(6e^{3x})\,dx$$

$$= \frac{1}{9}\int_{0}^{5} 6e^{3x}\,dx$$

$$= \frac{1}{9}(2e^{3x})\Big|_{0}^{5}$$

$$= \frac{2}{9}(e^{15} - e^0)$$

$$= \frac{2}{9}e^{15} - \frac{2}{9}.$$

$$\approx 7.264480826 \cdot 10^5.$$

2. By the Fundamental Theorem of Calculus, Part Two, if

$$F(x) = \int_{2}^{x} (2\cos^2(t) + 5t + 7)\,dt, \text{ then}$$

$$F'(x) = 2\cos^2(x) + 5x + 7.$$

Section 5.5

1. Let $u = \cos(x)$. Then $du = -\sin(x)\,dx$, which implies that $-du = \sin(x)\,du$. Making these substitutions, we obtain

$$\int (\cos(x))^6 \sin(x)\,dx = \int u^6(-du)$$

$$= -\int u^6\,du$$

$$= -\frac{u^7}{7} + C$$

$$= -\frac{(\cos(x))^7}{7} + C.$$

2. Let $u = x^3 + 2$. Then $du = (3x^2 + 0)\,dx$, which implies that $\frac{1}{3}\,du = x^2\,dx$.
If $u = x^3 + 2$ and $x = 2$, then $u = 2^3 + 2 = 10$.
If $u = x^3 + 2$ and $x = 7$, then $u = 7^3 + 2 = 345$.

Thus

$$\int\limits_{x=2}^{7} \left(\sqrt{x^3 + 2}\right)(x^2)\, dx = \frac{1}{3} \int\limits_{u=10}^{345} \sqrt{u}\, du$$

$$= \frac{1}{3} \int\limits_{u=10}^{345} u^{1/2}\, du$$

$$= \frac{1}{3} \cdot \frac{2}{3} u^{3/2} \Big|_{u=10}^{345}$$

$$= \frac{2}{9}[345^{3/2} - 10^{3/2}]$$

$$\approx 1416.992847.$$

3. $\displaystyle\int\limits_{20}^{10} \left(\frac{1}{e^{3x}}\right) dx = -\int\limits_{10}^{20} e^{-3x}\, dx$

Let $u = -3x$. Then $du = -3\, dx$, which implies that $-\frac{1}{3} du = dx$.

If $x = 10$ and $u = -3x$, then $u = -30$.

If $x = 20$ and $u = -3x$, then $u = -60$.

Thus

$$-\int\limits_{10}^{20} e^{-3x}\, dx = -\int\limits_{-30}^{-60} e^{+u}\left(-\frac{1}{3} du\right)$$

$$= \frac{1}{3} \int\limits_{-30}^{-60} e^{+u}\, du$$

$$= \frac{1}{3} e^{u} \Big|_{u=-30}^{u=-60}$$

$$= \frac{1}{3}[e^{-60} - e^{-30}]$$

$$\approx -3.119207656 \cdot 10^{-14}.$$

Section 5.6

1. The graph of $f(x) = \sin(x) - \frac{1}{2}$ for $0 \le x \le 2$ is pictured below.

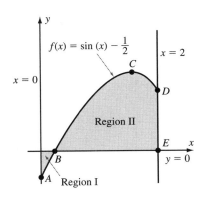

Point $A = \left(0, -\frac{1}{2}\right)$

Point $B = \left(\frac{\pi}{6}, 0\right)$

Point $C = \left(\frac{\pi}{2}, \frac{1}{2}\right)$

Point $D = (2, 0.4092974268)$

Point $E = (2, 0)$

We see there are two bounded regions, Region I, which is below the x-axis, and Region II, which is above the x-axis.

The area, A_1, of Region I is equal to

$$A_1 = \int\limits_{0}^{\frac{\pi}{6}} \left(\sin(x) - \frac{1}{2}\right) dx = -\left(-\cos(x) - \frac{1}{2}x\right)\Big|_{0}^{\frac{\pi}{6}}$$

$$= \cos\frac{\pi}{6} + \frac{\pi}{12} - 1 \approx 0.1278247918 \text{ square units.}$$

The area, A_2, of Region II is equal to

$$A_2 = \int\limits_{\frac{\pi}{6}}^{2} \left(\sin(x) - \frac{1}{2}\right) dx = -\cos x - \frac{1}{2}x \Big|_{\frac{\pi}{6}}^{2}$$

$$= -\cos 2 - 1 + \cos\frac{\pi}{6} + \frac{\pi}{12} \approx 0.543971628 \text{ square units.}$$

The total region, A, is equal to $A = A_1 + A_2 \approx 0.1278247918 + 0.543971628 = 0.6717964198$ square units.

2. The area bounded by the functions $f(x) = 3x$ and $g(x) = 3x^2$ is pictured below.

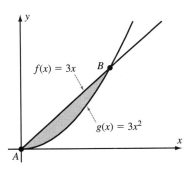

The graphs of $f(x)$ and $g(x)$ intersect at Point A and Point B. $f(x) = g(x)$ implies that $3x = 3x^2$. This, in turn, implies $3x^2 - 3x = 0$ or $3x(x - 1) = 0$. So $x = 0$ and $x = 1$ are the x-coordinates of Point A and Point B, respectively. If $x = 0$, then $y = 3(0) = 0$, so the coordinates of A are $A = (0, 0)$. If $x = 1$ and $y = 3(1)$ then the coordinates of B are $B = (1, 3)$.

The bounded region is shaded in the graph represented above.

$$\text{Area} = \int_{x=0}^{1} f(x)\,dx - \int_{x=0}^{1} g(x)\,dx$$

$$= \int_{x=0}^{1} 3x\,dx - \int_{x=0}^{1} 3x^2\,dx$$

$$= \frac{3x^2}{2}\bigg|_0^1 - \frac{3x^3}{3}\bigg|_0^1$$

$$= \frac{3}{2}(1^2 - 0^2) - (1^3 - 0^3)$$

$$= \frac{3}{2} - 1$$

$$= \frac{1}{2} \text{ square units}$$

3. The area bounded by the functions $f(x) = 3x$ and $g(x) = 3x^2$ is pictured below.

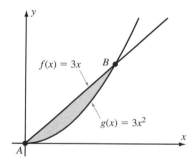

The graphs intersect at Point $A = (0, 0)$ and Point $B = (1, 3)$.

The graph of $g(x)$ is to the right of $f(x)$ and the graph of $f(x)$ is to the left of $g(x)$.

$y = 3x^2 \Rightarrow \pm\sqrt{\dfrac{y}{3}} = x$. We choose $x = +\sqrt{\dfrac{y}{3}} = G(y)$ because $x \geq 0$.

$y = 3x \Rightarrow x = \dfrac{y}{3} = F(y)$ because $x \geq 0$.

Therefore, using the "(Right-side) − (Left-side)" approach, the area of the bounded region (which is shaded) is

$$\text{Area} = \int_{y=0}^{3} \sqrt{\frac{y}{3}}\,dy - \int_{y=0}^{3} \frac{y}{3}\,dy$$

$$= \frac{1}{\sqrt{3}} \int_{y=0}^{3} y^{1/2}\,dy - \frac{1}{3} \int_{y=0}^{3} y\,dy$$

$$= \frac{1}{\sqrt{3}} \cdot \frac{2}{3} y^{3/2}\bigg|_0^3 - \frac{1}{3} \frac{y^2}{2}\bigg|_0^3$$

$$= \frac{2}{3\sqrt{3}}(3^{3/2} - 0^{3/2}) - \frac{1}{6}(3^2 - 0^2)$$

$$= \frac{2}{3\sqrt{3}}(3\sqrt{3}) - \frac{1}{6}(9)$$

$$= 2 - \frac{3}{2} = \frac{1}{2} \text{ square units.}$$

Section 5.7

1. In Example 1, $f(x)$ is defined as $f(x) = x^2$ and M_4 is calculated as $M_4 = 2.625$. In Example 2, it is calculated that the error estimate is no greater than .0417.

$$\int_0^2 f(x)\,dx = \int_0^2 x^2\,dx = \frac{x^3}{3}\bigg|_0^2 = \frac{8}{3} - 0 = \frac{8}{3} \approx 2.6667$$

$$\left| \int_0^2 f(x)\,dx - M_4 \right| \approx 2.6667 - 2.625 = 0.0417$$

Thus M_4 satisfies the error estimate obtained in Example 2.

2. $f(x)$ in Example 1 is defined as $f(x) = x^2$ and M_4 is calculated as $M_4 = 2.625$.

$$\int_0^2 x^2\,dx = \frac{x^3}{3}\bigg|_0^2 = \frac{8}{3} \approx 2.6667$$

Since $M_4 = 2.625 < 2.6667 \approx \displaystyle\int_0^2 x^2\,dx$, we conclude that

M_4 is an underestimate of $\displaystyle\int_0^2 x^2\,dx$.

3. In Example 3, $f(x)$ is defined as $f(x) = 2 - x^2$ and T_4 is calculated as $T_4 = 1.65625$.

$$\int_0^1 f(x)\,dx = \int_0^1 (2 - x^2)\,dx$$

$$= \left(2x - \frac{x^3}{3} \right)\bigg|_0^1$$

$$= \left(2 - \frac{1}{3} \right) - (0)$$

$$= 1\frac{2}{3} \approx 1.6666667$$

$$\left| \int_0^1 f(x)\,dx - T_4 \right| \approx 1.6666667 - 1.65625 = 0.0104167$$

In Example 4, the error estimate of T_4 is calculated to be less than or equal to $\dfrac{1}{96} \approx 0.0104167$, so T_4 satisfies the error estimate obtained in Example 4.

4. In Example 3, $f(x)$ is defined as $f(x) = 2 - x^2$.

$$\int_0^2 f(x)\,dx = \int_0^2 (2 - x^2)\,dx = \left(2x - \frac{x^3}{3}\right)\Big|_0^2$$

$$= \left(4 - \frac{8}{3}\right) - 0 = \frac{12 - 8}{3} = \frac{4}{3} \approx 1.333$$

$$\Delta x = \frac{b - a}{n} = \frac{2 - 0}{4} = \frac{1}{2}$$

The uniform partition of order 4 of $[0, 2]$ is $\left\{0, \frac{1}{2}, 1, \frac{3}{2}, 2\right\}$.

$$T_4 = \left(\frac{\frac{1}{2}}{2}\right)\left(f(0) + 2f\left(\frac{1}{2}\right) + 2f(1) + 2f\left(\frac{3}{2}\right) + f(2)\right)$$

$$= \left(\frac{1}{4}\right)\left(2 + 2 \cdot \frac{7}{4} + 2 \cdot 1 + 2 \cdot \left(-\frac{1}{4}\right) + (-2)\right)$$

$$= \frac{1}{4}\left(\frac{14 + 8 - 2}{4}\right)$$

$$= \frac{5}{4}$$

$$= 1.25$$

Since $\int_0^2 f(x)\,dx \approx 1.333 > 1.25 = T_4$, we conclude that T_4 is an underestimate.

5. In Example 5, $f(x)$ is defined as $f(x) = x^3$ and S_2 is calculated as $S_2 = 0.25$. In Example 6, the error estimate of S_2 is calculated to be zero.

Since $$\int_0^1 f(x)\,dx = \int_0^1 x^3 dx = \frac{x^4}{4}\Big|_0^1 = \frac{1}{4}(1^4 - 0^4)$$

$$= \frac{1}{4} = 0.25,$$ we see that the value of S_2 is equal to $\int_0^1 f(x)\,dx$, so the error estimate of zero is satisfied.

6. In Example 5, $f(x)$ is defined as $f(x) = x^3$. Therefore

$$\int_0^2 f(x)\,dx = \int_0^2 x^3 dx = \frac{x^4}{4}\Big|_0^2 = \frac{1}{4}(2^4 - 0^4) = \frac{2^4}{2^2}$$

$$= 2^2 = 4.$$

$$\Delta x = \frac{b - a}{n} = \frac{2 - 0}{4} = \frac{1}{2}$$

The uniform partition of order 4 of $[0, 2]$ is $\left\{0, \frac{1}{2}, 1, \frac{3}{2}, 2\right\}$.

Therefore,

$$S_4 = \left(\frac{\frac{1}{2}}{3}\right)\left(f(0) + 4 \cdot f\left(\frac{1}{2}\right) + 2 \cdot f(1)\right.$$

$$\left. + 4 \cdot f\left(\frac{3}{2}\right) + f(2)\right)$$

$$= \left(\frac{1}{6}\right)\left(0 + 4 \cdot \left(\frac{1}{8}\right) + 2 \cdot (1) + 4 \cdot \left(\frac{27}{8}\right) + 8\right)$$

$$= \frac{1}{6}\left(\frac{1}{2} + 2 + \frac{27}{2} + 8\right)$$

$$= \frac{1}{6}\left(\frac{1 + 4 + 27 + 16}{2}\right)$$

$$= \frac{1}{6}\left(\frac{48}{2}\right)$$

$$= \frac{1}{6}(24)$$

$$= 4.$$

Since $\int_0^2 f(x)\,dx = 4 = S_4$, we conclude that S_4 is neither an overestimate nor an underestimate.

Section 6.1

1. The fourth derivative is the highest derivative that appears in the differential equation $\frac{dy}{dx} + \frac{d^4y}{dx^4} = \frac{d^3y}{dx^3}$, so it is a fourth-order differential equation.

2. If x is greater than zero, then $|3x| = 3x$. Therefore $y(x) = \ln|3x| + C = \ln(3x) + C$ for positive x.

$y = \ln(3x) + C$ implies that

$\frac{d}{dx}(y) = \frac{d}{dx}(\ln(3x)) + \frac{d}{dx}(C)$, so $\frac{dy}{dx} = \frac{3}{3x} + 0$,

which implies that $\frac{dy}{dx} = \frac{1}{x}$ for $x > 0$.

Because $y(x) = \ln|3x| + C$ satisfies the differential equation $\frac{dy}{dx} = \frac{1}{x}$ for $x > 0$, for each value of C, it is a general solution.

3. $\frac{dy}{dx} = 1$ implies that $dy = dx$.

$\int dy = \int dx$ is equal to $y = x + C$.

If $C = 0$, then $y = x$ is a line through the origin with slope equal to one.

If $C \neq 0$, then $y = x + C$ describes parallel translates of the line $y = x$, one for every value of C.

4.

$$\frac{dy}{dx} = \frac{y}{x} \Rightarrow$$

$$\frac{dy}{y} = \frac{dx}{x} \Rightarrow$$

$$\int \frac{dy}{y} = \int \frac{dx}{x} \Rightarrow$$

$$\ln|y| = \ln|x| + C \Rightarrow$$

$$e^{\ln|y|} = e^{\ln|x| + C} \Rightarrow$$

$$|y| = C \cdot |x|, \text{ where } C = e^C$$

If $y > 0$ and $x > 0$, we have $y = C \cdot x$.

If $y > 0$ and $x < 0$, we have $y = C \cdot (-x) = C_1 x$

where $C_1 = -C$.

If $y < 0$ and $x > 0$, we have $-y = C \cdot x \Rightarrow$
$y = -C \cdot x = C_1 x$ where $C_1 = -C$.

If $y < 0$ and $x < 0$, we have $-y = C \cdot (-x) \Rightarrow y = Cx$.

In all cases, we have y is equal to a constant multiple of x.

Let the constant be represented by the letter ζ. In all cases, $y = y(x) = \zeta x$.

If $10 = y(1) = \zeta \cdot 1 \Rightarrow \zeta = 10$.

Therefore the solution to the IVP problem is
$y(x) = 10x$.

5. $P = \$1000$, $A = \$25,000$, and $r = .05$ are the given values.
$A = Pe^{rt} \Rightarrow 25,000 = 1,000e^{.05t}$, which simplifies to
$25 = e^{.05t}$. Because the natural logarithm function and the exponential function are mutually inverse functions, we take the natural logarithm of both sides to obtain
$\ln(25) = \ln(e^{.05t}) = .05t$. Solving for t, we obtain
$$t = \frac{\ln(25)}{.05} \approx 64.38 \text{ years.}$$

Section 6.2

1. Let $u(x) = 5x + 3$. Then $\dfrac{dy}{dx} = 5$.

Implementing the rule $\dfrac{d}{dx}(\ln|u(x)|) = \dfrac{1}{u(x)} \dfrac{du}{dx}$, we obtain
$$\frac{d}{dx} \ln|5x + 3| = \frac{5}{5x + 3} \text{ for } x \neq \frac{-3}{5}.$$

2.
$$\int \frac{1}{5x + 3} dx = \int \frac{1}{(5x + 3)} \cdot \frac{5}{5} dx$$
$$= \frac{1}{5} \int \frac{5}{5x + 3} dx$$
$$= \frac{1}{5} \ln|5x + 3| + C, \text{ for } x \neq \frac{-3}{5}$$

3. $\sec(x) = \dfrac{1}{\cos(x)}$ and $\cot(x) = \dfrac{\cos(x)}{\sin(x)}$, so
$$\sec(x)\cot(x) = \frac{1}{\cos(x)} \cdot \frac{\cos(x)}{\sin(x)} = \frac{1}{\sin(x)} = \csc(x).$$

$$\therefore \int_{x=\frac{\pi}{4}}^{\frac{\pi}{2}} \sec(x)\cot(x)\, dx = \int_{x=\frac{\pi}{4}}^{\frac{\pi}{2}} \csc(x)\, dx$$
$$= [-\ln|\csc(x) + \cot(x)|]_{x=\frac{\pi}{4}}^{\frac{\pi}{2}}$$
$$= \left(-\ln\left|\csc\left(\frac{\pi}{2}\right) + \cot\left(\frac{\pi}{2}\right)\right|\right)$$
$$\quad - \left(-\ln\left|\csc\left(\frac{\pi}{4}\right) + \cot\left(\frac{\pi}{4}\right)\right|\right)$$
$$= (-\ln|1 + 0|) + (\ln|\sqrt{2} + 1|)$$
$$= -\ln 1 + \ln(\sqrt{2} + 1)$$

$$= -0 + \ln(\sqrt{2} + 1)$$
$$= \ln(\sqrt{2} + 1)$$
$$\approx 0.88137$$

Section 6.3

1. Since $f(x) = x^5 \cdot \exp(5x)$ is equal to the product of two functions, we will use the Product Rule to obtain the derivative of $f(x)$.

$$\frac{d}{dx}(x^5 \cdot \exp(5x)) = \left(\frac{d}{dx}x^5\right) \cdot (\exp(5x))$$
$$+ (x^5) \cdot \left(\frac{d}{dx}\exp(5x)\right)$$
$$= 5x^4 \cdot \exp(5x) + (x^5) \cdot 5 \cdot \exp(5x)$$
$$= (5x^4 + 5x^5)\exp(5x)$$

2. We will use the method of substitution. Let $u = k \cdot x$, then
$du = k \cdot dx$.

$$\int \exp(k \cdot x)\, dx = \int (\exp(k \cdot x))\left(\frac{k}{k}\right) dx$$
$$= \frac{1}{k} \int (\exp(k \cdot x)) \cdot k\, dx$$
$$= \frac{1}{k} \int \exp(u)\, du = \frac{1}{k} \exp(u) + C$$
$$= \frac{1}{k}(\exp(k \cdot x)) + C$$

3.
$$\int_{1}^{e} \frac{1}{x} dx = \ln|x| \Big|_{x=1}^{x=e^2}$$
$$= \ln|e^2| - \ln|1|$$
$$= \ln e^2 - \ln 1$$
$$= 2 - 0 = 2$$

4. Since $f(x) = \dfrac{\exp(x)}{1 + \ln(x)}$ is the quotient of two functions of x, we shall use the Quotient Rule.

$$\frac{d}{dx}\left(\frac{\exp(x)}{1 + \ln(x)}\right)$$
$$= \frac{(1 + \ln(x))\dfrac{d}{dx}(\exp(x)) - (\exp(x))\dfrac{d}{dx}(1 + \ln(x))}{(1 + \ln(x))^2}$$
$$= \frac{(1 + \ln(x)) \cdot \exp(x) - (\exp(x)) \cdot \left(0 + \dfrac{1}{x}\right)}{(1 + \ln(x))^2}$$
$$= \frac{(\exp(x))\left(1 + \ln(x) - \dfrac{1}{x}\right)}{(1 + \ln(x))^2}$$
$$= \frac{(\exp(x))(x + x\ln(x) - 1)}{x(1 + \ln(x))^2}$$

Section 6.4

1. $\ln(5^x) = \ln(\exp(x \ln(5))) = x \ln(5)$.
Therefore $\dfrac{\ln(5^x)}{\ln(5)} = \dfrac{x \ln(5)}{\ln(5)} = x$.

2.

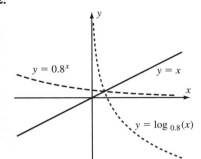

The graph of $y = 0.8^x$ is a reflection about the line $y = x$ of the graph of $y = \log_{0.8}(x)$ and conversely, the graph of $y = \log_{0.8}(x)$ is a reflection about the line $y = x$ of the graph of $y = 0.8^x$.

Thus the functions $f(x) = 0.8^x$ and $g(x) = \log_{0.8}(x)$ are inverse functions to each other.

3. Let $x_2 = x_1 + \delta$, $\delta > 0$.
$$x_1 < x_1 + \delta \Rightarrow x_1 < x_2$$
But $\left(\dfrac{1}{2}\right)^{x_1} = 2^{-x_1}$ and
$$\left(\dfrac{1}{2}\right)^{x_2} = 2^{-x_2} = 2^{-(x_1+\delta)} = 2^{-x_1} 2^{-\delta} = \dfrac{2^{-x_1}}{2^\delta}.$$
Since $\sigma > 0 \Rightarrow 2^\sigma > 2^0 = 1$ and $1 > \dfrac{1}{2^\sigma}$, hence
$$2^{-x_1} > \dfrac{2^{-x_1}}{2^\delta}, \text{ which implies } \left(\dfrac{1}{2}\right)^{x_1} > \left(\dfrac{1}{2}\right)^{x_2}.$$
Because $x_1 < x_2$ implies that $f(x_1) > f(x_2)$, we see that $f(x) = y = \left(\dfrac{1}{2}\right)^x$ is a decreasing function.

4. By Theorem 3f, $\log_{2a} 9 = \dfrac{\log_e 9}{\log_e 2a} = \dfrac{\ln 9}{\ln(2a)}$.
By Theorem 3c, $\ln(a) + \ln(2) = \ln(2a)$,
so $\left(e^{\log_{(2a)} 9}\right)^{(\ln(a)+\ln(2))} = \left(e^{\frac{\ln 9}{\ln 2a}}\right)^{\ln(2a)}$
$$= e^{\left(\frac{\ln 9}{\ln(2a)}\right)\cdot \ln(2a)}$$
$$= e^{\ln 9}$$
$$= 9.$$

5. $2^x = \left(\dfrac{1}{4}\right)^{5x}$ implies that $2^x = (2^{-2})^{5x} = 2^{-10x}$.
Take the natural logarithm of both sides to obtain $\ln(2^x) = \ln(2^{-10x})$, which implies that $x \ln(2) = (-10x) \ln 2$. Transposing $-10x \ln 2$ to the other side of the equation yields $(x + 10x) \ln(2) = 0$.

So $(11x) \ln(2) = 0 \Rightarrow x = 0$, since $\ln(2) \neq 0$ and $11 \neq 0$.

6. Since $f(x)$ is equal to the quotient of two functions of x, we shall use the quotient rule.

$$\dfrac{d}{dx}(f(x)) = \dfrac{2^x \dfrac{d}{dx}(1 - x^3) - (1 - x^3)\dfrac{d}{dx}(2^x)}{(2^x)^2}$$
$$= \dfrac{(2^x)(-3x^2) - (1 - x^3)(2^x \ln(2))}{2^{2x}}$$
$$= \dfrac{2^x[(-3x^2) - (1 - x^3)(\ln 2)]}{2^{2x}}$$
$$= \dfrac{-3x^2 - \ln 2 + x^3 \ln 2}{2^x}$$

7. Let $u = \cos(x)$. Then $du = -\sin(x)\,dx$
$\Rightarrow -du = \sin(x)\,dx$.
Therefore
$$-\int 2^u\,du = -\dfrac{1}{\ln(2)} \cdot 2^u + C$$
$$= -\dfrac{2^u}{\ln(2)} + C$$
$$= -\dfrac{2^{\cos(x)}}{\ln(2)} + C.$$

8. Let $u = 3x - 1$. Then $du = 3\,dx \Rightarrow \dfrac{1}{3}\,du = dx$.
So
$$\int 2^{3x-1}\,dx = \int 2^u \left(\dfrac{1}{3}\,du\right) = \dfrac{1}{3}\int 2^u\,du$$
$$= \dfrac{1}{3 \ln 2} \cdot 2^u + C$$
$$= \dfrac{2^{3x-1}}{3 \ln 2} + C.$$

9. $f(x) = x^x e^{\cos(x)}$ implies
$\ln(f(x)) = \ln(x^x e^{\cos(x)})$.
$\ln(f(x)) = \ln(x^x) + \ln(e^{\cos(x)})$, so
$\ln(f(x)) = x \ln(x) + \cos(x)$.

Differentiate each term with respect to x and obtain
$$\dfrac{f'(x)}{f(x)} = x \cdot \dfrac{1}{x} + (\ln(x)) \cdot 1 - \sin(x).$$

Multiply both sides of the equation by $f(x)$ to obtain
$$f'(x) = f(x) \cdot 1 + f(x) \cdot \ln(x) - f(x) \cdot \sin(x).$$

Now substitute the value $f(x) = x^x e^{\cos(x)}$ into the expression to obtain
$$f'(x) = x^x e^{\cos(x)}(1 + \ln(x) - \sin(x)).$$

Section 6.5

1. $y(t) = Ae^{kt}$, where $A = y(0)$.

It is given that $A = 5000$ and $y(3) = 5000\,e^{k \cdot 3} = 7800$.

Therefore $e^{3k} = \dfrac{7800}{5000}$. Taking the natural logarithm of each side results in $3k = \ln(7800) - \ln(5000)$, so

$k = \dfrac{1}{3}[\ln(7800) - \ln(5000)] \approx 0.1482286071$.

Thus $y(t) = 5000\,e^{0.1482286071t}$. If $t = 5$, then $y(5) = 5000\,e^{(0.1482286071) \cdot (5)} \approx 10{,}491 \cdot 66306$.

We conclude that after 5 hours there are approximately 10,492 bacteria.

2. It is given that $\lambda = .0478$ and $R(0) = 75$ g.

$R(t) = R(0)e^{-\lambda t}$, so $R(t) = 75\,e^{-.0478t}$.

If $t = 5$, $R(5) = 75e^{(-.0478) \cdot 5} \approx 59.05611618$ g. We conclude that after five years, approximately 59 g of the radioactive substance remains.

Section 6.6

1. $\vartheta = \arctan(1) \Leftrightarrow \tan(\vartheta) = 1$, where ϑ is between $-\dfrac{\pi}{2}$ and $\dfrac{\pi}{2}$. $\tan(\vartheta) = 1 \Rightarrow \vartheta = \dfrac{\pi}{4}$.

2.

$$\frac{d}{dt}(\arctan(-t)) = \frac{\frac{d}{dt}(-t)}{1 + (-t)^2} = \frac{-1}{1 + t^2}$$

3. Let $w = \dfrac{u}{a}$; then $dw = \dfrac{1}{a}\,du \Rightarrow a\,dw = du$.

$$\int \frac{du}{\left(1 + \left(\dfrac{u}{a}\right)\right)^2} = \int \frac{a}{(1 + w^2)}\,dw = a \int \frac{1}{1 + w^2}\,dw$$

$$= a \cdot \arctan(w) + C = a \cdot \arctan\left(\frac{4}{-}\right) + C$$

Section 6.7

1. We shall use the Product Rule.

$$\frac{d}{dt}(\sinh(5t) \cdot \cosh(9t))$$

$$= \sinh(5t) \cdot \frac{d}{dt}(\cosh(9t)) + \cosh(9t) \cdot \frac{d}{dt}(\sinh(5t))$$

$$= \sinh(5t) \cdot \sinh(9t) \cdot 9 + \cosh(9t) \cdot \cosh(5t) \cdot 5$$

$$= 9 \cdot \sinh(5t) \cdot \sinh(9t) + 5 \cdot \cosh(9t) \cdot \cosh(5t)$$

2.

$$\frac{d}{dt}(\operatorname{csch}(5t))^2 = 2 \cdot \operatorname{csch}(5t) \cdot \frac{d}{dt}(\operatorname{csch}(5t))$$

$$= 2 \cdot \operatorname{csch}(5t) \cdot \left(-\operatorname{csch}(5t) \cdot \coth(5t)\frac{d}{dt}5t\right)$$

$$= 2\,\operatorname{csch}(5t)(-\operatorname{csch}(5(t))\coth(5t) \cdot 5)$$

$$= -10\,(\operatorname{csch}(5t))^2 \coth(5t)$$

3. $\dfrac{d}{dx}(\tanh^{-1}(2x)) = \dfrac{1}{1 - (2x)^2} \cdot \dfrac{d}{dx}(2 \cdot x)$, for $-1 < 2x < 1$

$$= \frac{2}{1 - 4x^2}, \quad \text{for } -\frac{1}{2} < x < \frac{1}{2}$$

4. Let $u = 3x$; then $du = 3\,dx$.

$$\int \frac{dx}{\sqrt{1 + 9x^2}} = \int \frac{dx}{\sqrt{1 + (3x)^2}} = \int \frac{1}{3}\frac{1}{\sqrt{1 + u^2}}\,du$$

$$= \frac{1}{3}\sinh^{-1}(u) + C$$

$$= \frac{1}{3}\sinh^{-1}(3x) + C, \quad \text{for } x \in \mathbb{R}$$

Section 7.1

1. Let $u = x^2$ and $dv = e^{2x}\,dx$. Then $du = 2x\,dx$ and $v = \dfrac{1}{2}e^{2x}$. Using the Integration by Parts formula, we obtain

$$\int x^2 e^{2x}\,dx = \frac{1}{2}x^2 e^{2x} - \int \left(\frac{1}{2}e^{2x}\right) \cdot (2x\,dx)$$

$$= \frac{1}{2}x^2 e^{2x} - \int x e^{2x}\,dx.$$

Apply the Integration by Parts formula to $\int xe^{2x}\,dx$. Since we chose u to be the polynomial x^2 in the first application of the Integration by Parts formula, we shall make a consistent choice for u and choose $u = x$. Similarly, in a consistent fashion we choose $dv = e^{2x}\,dx$. So $du = dx$ and $v = \dfrac{1}{2}e^{2x}$. Apply the Integration by Parts formula to $\int xe^{2x}\,dx$ to obtain

$$\int x e^{2x}\,dx = \left(\frac{1}{2}e^{2x}\right)(x) - \int \frac{1}{2}e^{2x}\,dx$$

$$= \frac{1}{2}x e^{2x} - \frac{1}{4}e^{2x}.$$

Combining the results of our first and second applications of the Integration by Parts formula, we obtain

$$\int x^2 e^{2x}\,dx = \frac{1}{2}x^2 e^{2x} - \int x e^{2x}\,dx$$

$$= \frac{1}{2}x^2 e^{2x} - \left[\frac{1}{2}x e^{2x} - \frac{1}{4}e^{2x}\right]$$

$$= \frac{1}{2}x^2 e^{2x} - \frac{1}{2}x e^{2x} + \frac{1}{4}e^{2x} + C.$$

C is the constant of integration.

2. Since the integrand is equal to the product of two functions of x, we shall apply the Integration by Parts formula.

Let $u = \ln(2x)$ and $dv = x^{1/2}\,dx$. Then
$du = \dfrac{2}{2x}\,dx = \dfrac{1}{x}\,dx$ and $v = \dfrac{2}{3}x^{3/2}$.

Therefore,

$$\int_1^e \sqrt{x} \cdot \ln(2x)\,dx$$

$$= \frac{2}{3}x^{3/2} \cdot \ln(2x)\Big|_1^e - \int_1^e \frac{2}{3}x^{1/2}\,dx$$

$$= \frac{2}{3}x^{3/2} \cdot \ln(2x)\Big|_1^e - \frac{4}{9}x^{3/2}\Big|_1^e$$

$$= \frac{2}{3}[e^{3/2} \cdot \ln(2e) - 1^{3/2}\ln(2)] - \frac{4}{9}[e^{3/2} - 1^{3/2}]$$

$$= \frac{2}{3}e^{3/2}\ln(2e) - \frac{2}{9}\ln(2) - \frac{4}{9}e^{3/2} + \frac{4}{9}$$

$$= \frac{2}{3}e^{3/2}[\ln(2) + \ln(e)] - \frac{2}{3}\ln(2) - \frac{4}{9}e^{3/2} + \frac{4}{9}$$

$$= \frac{6}{9}e^{3/2}[\ln(2) + 1] - \frac{2}{3}\ln(2) - \frac{4}{9}e^{3/2} + \frac{4}{9}$$

$$= \frac{6}{9}e^{3/2}\ln(2) + \frac{6}{9}e^{3/2} - \frac{2}{3}\ln(2) - \frac{4}{9}e^{3/2} + \frac{4}{9}$$

$$= \frac{2}{3}e^{3/2}\ln(2) + \frac{2}{9}e^{3/2} - \frac{2}{3}\ln(2) + \frac{4}{9}$$

$$\approx 3.05.$$

Section 7.2

1.   Step 1: The denominator, $x^2 - x - 12$, of the rational function is of degree 2 and can be factored into the product of two linear factors, $(x - 4)$ and $(x + 3)$. We notice that the degree of the numerator is zero, which is less than the degree of the denominator.

Step 2: The basic simple linear building blocks are $\dfrac{A}{(x-4)}$ and $\dfrac{B}{(x+3)}$. Note that

$$\frac{A}{(x-4)} + \frac{B}{(x+3)} = \frac{A(x+3) + B(x-4)}{(x-4)(x+3)}$$

$$= \frac{Ax + 3A + Bx - 4B}{(x-4)(x+3)}$$

$$= \frac{(A+B)x + (3A-4B)}{(x-4)(x+3)}.$$

Step 3: Now equate the original function and this last result to obtain

$$\frac{3}{x^2 - x - 12} = \frac{0 \cdot x + 3}{x^2 - x - 12} = \frac{(A+B)x + (3A-4B)}{(x-4)(x+3)}.$$

Because the two denominators are equal, the two fractions can be equal only if the two numerators are equal. After equating the coefficients of the x-terms and the constants, we conclude that $A + B = 0$ and $3 = 3A - 4B$. Use either the

Elimination Method or the Substitution Method to solve these two equations; we obtain $A = \dfrac{3}{7}$ and $B = \dfrac{-3}{7}$.

Therefore, $\dfrac{3}{x^2 - x - 12}$ becomes $\dfrac{\frac{3}{7}}{x - 4} - \dfrac{\frac{3}{7}}{x + 3}$.
Now you can compute the original integral.

$$\int \frac{3}{x^2 - x - 12}\,dx = \int \frac{\frac{3}{7}}{x - 4}\,dx - \int \frac{\frac{3}{7}}{x + 3}\,dx$$

$$= \frac{3}{7}\ln|x - 4| - \frac{3}{7}\ln|x + 3| + C$$

$$= \ln\left|\left(\frac{(x-4)}{(x+3)}\right)^{3/7}\right| + C$$

2.   Step 1: The denominator $x^2 - 9$ of the rational function is of degree 2 and can be factored into the product of two linear factors, $(x + 3)$ and $(x - 3)$. We notice that the degree of the numerator is zero, which is less than the degree of the denominator.

Step 2: The basic simple linear building blocks are $\dfrac{A}{x+3}$ and $\dfrac{B}{x-3}$. Note that

$$\frac{A}{x+3} + \frac{B}{x-3} = \frac{A(x-3) + B(x+3)}{(x+3)(x-3)}$$

$$= \frac{Ax - 3A + Bx + 3B}{(x+3)(x-3)}$$

$$= \frac{(A+B)x + (-3A+3B)}{(x+3)(x-3)}.$$

Step 3: Now equate the original function and this last result to obtain

$$\frac{1}{x^2 - 9} = \frac{0 \cdot x + 1}{x^2 - 9} = \frac{(A+B)x + (-3A+3B)}{(x+3)(x-3)}.$$

Because the two denominators are equal, the two fractions can be equal only if the two numerators are equal. After equating the coefficients of the x-terms and the constants, we conclude that $A + B = 0$ and $(-3A + 3B) = 1$. Use either the Substitution Method or the Elimination Method to solve these two equations simultaneously; we obtain $A = \dfrac{-1}{6}$ and $B = \dfrac{1}{6}$. Therefore $\dfrac{1}{x^2 - 9}$ becomes $\dfrac{\frac{-1}{6}}{x + 3} + \dfrac{\frac{1}{6}}{x - 3}$.
Now you can compute the original integral.

$$\int \frac{1}{x^2 - 9}\,dx = \int \frac{\frac{-1}{6}}{x + 3}\,dx + \int \frac{\frac{1}{6}}{x - 3}\,dx$$

$$= \frac{-1}{6}\ln|x + 3| + \frac{1}{6}\ln|x - 3| + C$$

$$= \ln\left|\left(\frac{x-3}{x+3}\right)^{1/6}\right| + C$$

3.   Step 1:

$$p(x) = x^4 \text{ and } q(x) = (x + 2)(x^2 + 2x + 1)$$
$$= x^3 + 4x^2 + 5x + 2.$$

Because degree $p(x) = 4 > 3 = $ degree $q(x)$, divide the numerator $p(x)$ by the denominator $q(x)$ to obtain

$$\frac{p(x)}{q(x)} = x - 4 + \frac{11x^2 + 18x + 8}{x^3 + 4x^2 + 5x + 2}.$$

Step 2: Apply the Method of Partial Fractions for distinct linear fractions and repeated linear f to the remainder term $\dfrac{11x^2 + 18x + 8}{x^3 + 4x^2 + 5x + 2}$.

$$\frac{11x^2 + 18x + 8}{x^3 + 4x^2 + 5x + 2} = \frac{11x^2 + 18x + 8}{(x + 2)(x + 1)^2}$$

$$= \frac{A_1}{(x + 2)} + \frac{B_1}{(x + 1)} + \frac{B_2}{(x + 1)^2}$$

$$= \frac{A_1(x + 1)^2 + B_1(x + 1)(x + 2) + B_2(x + 2)}{(x + 2)(x + 1)^2}$$

$$= \frac{A_1(x^2 + 2x + 1) + B_1(x^2 + 3x + 2) + B_2(x + 2)}{(x + 2)(x + 1)^2}$$

$$= \frac{\{(A_1 + B_1)x^2 + (2A_1 + 3B_1 + B_2)x + (A_1 + 2B_1 + 2B_2)\}}{(x + 2)(x + 1)^2}$$

By equating coefficients of like terms, we obtain the equations

$$\begin{cases} A_1 + B_1 = 11 \\ 2A_1 + 3B_1 + B_2 = 18 \\ A_1 + 2B_1 + 2B_2 = 8. \end{cases}$$

By solving these equations simultaneously, we obtain the values $A_1 = 16$, $B_1 = -5$, and $B_2 = 1$.
Therefore,

$$\int \frac{x^4}{(x + 2)(x^2 + 2x + 1)} \, dx = \int x \, dx - 4 \int dx$$

$$+ \int \frac{16}{x + 2} \, dx - \int \frac{5}{(x + 1)} \, dx + \int \frac{1}{(x + 1)^2} \, dx.$$

$$= \frac{x^2}{2} - 4x + 16 \ln |x + 2| - 5 \ln |x + 1| - \frac{1}{(x + 1)} + C.$$

$$= \frac{x^2}{2} - 4x + \ln \left| \frac{(x + 2)^{16}}{(x + 1)^5} \right| - \frac{1}{x + 1} + C.$$

Section 7.3

1.
$$\int_0^{\frac{\pi}{6}} \sin^2(\theta) \, d\theta = \int_0^{\frac{\pi}{6}} \frac{(1 - \cos(2\theta))}{2} \, d\theta$$

$$= \frac{1}{2} \int_0^{\frac{\pi}{6}} (1 - \cos(2\theta)) \, d\theta$$

$$= \frac{1}{2} \left(\theta - \frac{1}{2} \sin(2\theta) \right) \Bigg|_{\theta=0}^{\theta=\frac{\pi}{6}}$$

$$= \frac{1}{2} \left[\left(\frac{\pi}{6} - \frac{1}{2} \sin \left(\frac{\pi}{3} \right) \right) - \left(0 - \frac{1}{2} \sin(0) \right) \right]$$

$$= \frac{\pi}{12} - \frac{1}{4} \cdot \frac{\sqrt{3}}{2}$$

$$= \frac{\pi}{12} - \frac{\sqrt{3}}{8}$$

$$\approx .045$$

2. If we apply the reduction formula

$$\int \tan^n(x) \, dx = \frac{1}{n - 1} \tan^{n-1}(x) - \int \tan^{n-2}(x) \, dx$$

with $n = 5$ to $\int \tan^5(x) \, dx$, we have

$$\int \tan^5(x) \, dx = \frac{1}{4} \tan^4(x) - \int \tan^3(x) \, dx.$$

Apply the reduction formula with $n = 3$ to $\int \tan^3(x) \, dx$.
Then we have

$$\int \tan^3(x) \, dx = \frac{1}{2} \tan^2(x) - \int \tan(x) \, dx.$$

By combining the results of the two applications of the reduction formula and the note, we obtain the solution

$$\int \tan^5(x) \, dx = \frac{1}{4} \tan^4(x) - \left[\frac{1}{2} \tan^2(x) - (- \ln |\cos(x)|) \right] + C$$

$$= \frac{1}{4} \tan^4(x) - \frac{1}{2} \tan^2(x) - \ln |\cos(x)| + C.$$

3. We shall use the identity $\cos^2(x) = 1 - \sin^2(x)$ and the Method of (Integration by) Substitution to calculate $\int \cos^3(x) \, dx$.

$$\int \cos^3(x) \, dx = \int \cos^2(x) \cdot \cos(x) \, dx$$

$$= \int (1 - \sin^2(x)) \cdot \cos(x) \, dx$$

We let $u = \sin(x)$, then $du = \cos(x) \, dx$.
Therefore,

$$\int \cos^3(x) \, dx = \int (1 - u^2) \, du$$

$$= u - \frac{u^3}{3} + C$$

$$= \sin(x) - \frac{\sin^3(x)}{3} + C_1$$

4. In this example, both powers of the sine function and the cosine function are even. We need to convert the integrand to "all sines" or "all cosines." In this case, we'll convert all even powers of sine by substituting $\sin^2(x) = 1 - \cos^2(x)$.

$$\int \sin^2(x) \cos^4(x)\, dx = \int (1 - \cos^2(x)) \cdot \cos^4(x)\, dx$$

$$= \int \cos^4(x)\, dx - \int \cos^6(x)\, dx$$

If we apply the reduction formula with $n = 4$ to $\int \cos^4(x)\, dx$, we have

$$\int \cos^4(x)\, dx = \frac{1}{4} \sin(x) \cos^3(x) + \frac{3}{4} \int \cos^2(x)\, dx.$$

If we use the identity $\cos^2(x) = \dfrac{1 + \cos(2x)}{2}$ applied to $\int \cos^2(x)\, dx$, we have

$$\int \cos^2(x)\, dx = \frac{1}{2} \int (1 + \cos(2x))\, dx$$

$$= \frac{1}{2} \left[x + \frac{1}{2} \sin(2x) \right] + C.$$

Combining these results, we obtain

$$\int \cos^4(x)\, dx = \frac{1}{4} \sin(x) \cos^3(x) + \frac{3}{4} \int \cos^2(x)\, dx$$

$$= \frac{1}{4} \sin(x) \cos^3(x) + \frac{3}{4} \cdot \frac{1}{2} \left[x + \frac{1}{2} \sin(2x) \right]$$

$$+ C_1, \quad \text{where } C_1 = \frac{3}{4} C,$$

$$= \frac{1}{4} \sin(x) \cos^3(x) + \frac{3}{8} x + \frac{3}{16} \sin(2x) + C_1.$$

We now turn our attention to $\int \cos^6(x)\, dx$ by applying the reduction formula with $n = 6$.

$$\int \cos^6(x)\, dx = \frac{1}{6} \sin(x) \cos^5(x) + \frac{5}{6} \int \cos^4(x)\, dx$$

Notice that $\dfrac{5}{6} \int \cos^4(x)\, dx$ is a multiple of the earlier integral $\int \cos^4(x)\, dx$. Combining these results, we have

$$\int \sin^2(x) \cos^4(x)\, dx = \int \cos^4(x)\, dx - \int \cos^6(x)\, dx$$

$$= \int \cos^4(x)\, dx - \left[\frac{1}{6} \sin(x) \cos^5(x) + \frac{5}{6} \int \cos^4(x)\, dx \right]$$

$$= \int \cos^4(x)\, dx - \frac{1}{6} \sin(x) \cos^5(x) - \frac{5}{6} \int \cos^4(x)\, dx$$

$$= \frac{1}{6} \int \cos^4(x)\, dx - \frac{1}{6} \sin(x) \cos^5(x)$$

$$= \frac{1}{6} \left[\frac{1}{4} \sin(x) \cos^3(x) + \frac{3}{8} x + \frac{3}{16} \sin(2x) + C_1 \right]$$

$$- \frac{1}{6} \sin(x) \cos^5(x)$$

$$= \frac{1}{24} \sin(x) \cos^3(x) + \frac{1}{16} x + \frac{1}{32} \sin(2x)$$

$$- \frac{1}{6} \sin(x) \cos^5(x) + C_2,$$

where $C_2 = \dfrac{1}{6} C_1 = \dfrac{1}{8} C$.

5. In this problem we shall use the identity $\cos^2(x) = 1 - \sin^2(x)$ and rewrite $\cos^3(x)$ as $\cos^3(x) = \cos^2(x) \cdot \cos(x) = (1 - \sin^2(x)) \cdot \cos(x)$. Therefore we can express $\int \sin^4(x) \cos^3(x)\, dx$ as

$$\int \sin^4(x) \cos^3(x)\, dx$$

$$= \int \sin^4(x) \cdot [\cos(x) - \sin^2(x) \cdot \cos(x)]\, dx$$

$$= \int \sin^4(x) \cdot \cos(x)\, dx - \int \sin^6(x) \cdot \cos(x)\, dx.$$

Using the Method of (Integration by) Substitution with $u = \sin(x)$ and $du = \cos(x)\, dx$, we obtain

$$\int \sin^4(x) \cos^3(x)\, dx = \int \sin^4(x)(1 - \sin^2(x)) \cos(x)\, dx$$

$$= \int u^4(1 - u)^2\, du$$

$$= \int u^4\, du - \int u^6\, du$$

$$= \frac{u^5}{5} - \frac{u^7}{7} + C$$

$$= \frac{\sin^5(x)}{5} - \frac{\sin^7(x)}{7} + C.$$

Section 7.4

1.

$$\int \sqrt{4 + 4x^2}\, dx = \int \sqrt{4(1 + x^2)}\, dx$$

$$= \int \sqrt{4} \cdot \sqrt{1 + x^2}\, dx$$

$$= 2 \int \sqrt{1 + x^2}\, dx$$

According to the table that summarizes trigonometric substitutions, the substitutions $x = 1 \cdot \tan\theta$ and

$dx = 1 \cdot \sec^2(\ ) \, d\theta$ should be used. Therefore we obtain

$$\int \sqrt{4 + 4x^2} \, dx = 2 \int \sqrt{1 + x^2} \, dx$$

$$= 2 \int \left(\sqrt{1 + \tan^2 \theta}\right) \cdot (\sec^2 \theta \, d\theta)$$

$$= 2 \int |\sec \theta| \cdot \sec^2 \theta \, d\theta.$$

Since $x = \tan \theta$ assumes every value in $(-\infty, \infty)$ as θ ranges from $-\dfrac{\pi}{2}$ to $\dfrac{\pi}{2}$, we can assume that θ lies inside the interval $\left(-\dfrac{\pi}{2}, \dfrac{\pi}{2}\right)$, so $|\sec \theta| = \sec \theta$. Now we have

$$\int \sqrt{4 + 4x^2} \, dx = 2 \int \sec^3(\theta) \, d\theta.$$

Use the reduction formula with $n = 3$ to obtain

$$\int \sec^3(\theta) \, d\theta = \frac{1}{2} \sec(\theta) \tan(\theta) + \frac{1}{2} \int \sec(\theta) \, d\theta$$

$$= \frac{1}{2} \sec(\theta) \tan(\theta) + \frac{1}{2}(\ln|\sec \theta + \tan \theta|) + C.$$

Therefore

$$2 \int \sec^3(\theta) \, dx = \sec(\theta) \tan(\theta) + \ln|\sec(\theta) + \tan(\theta)| + C_1,$$

where $C_1 = 2C$.

We complete the integration by resubstituting. Since $x = \tan \theta$, we have $\sec(\theta) = \sqrt{1 + x^2}$. Therefore

$$\int \sqrt{4 + 4x^2} \, dx = x\sqrt{1 + x^2} + \ln\left|\sqrt{1 + x^2} + x\right| + C_1.$$

2.

$$\int \frac{dx}{\sqrt{9 - 9x^2}} = \int \frac{dx}{\sqrt{9(1 - x^2)}}$$

$$= \int \frac{dx}{\sqrt{9}\sqrt{1 - x^2}}$$

$$= \frac{1}{3} \int \frac{dx}{\sqrt{1 - x^2}}$$

According to the table that summarizes trigonometric substitutions, $x = 1 \cdot \sin \theta$ and $dx = 1 \cdot \cos(\theta) \, d\theta$ should be used. Thus $\int \dfrac{dx}{\sqrt{1 - x^2}}$ becomes

$$\int \frac{\cos(\theta) \, d\theta}{\sqrt{1 - \sin^2 \theta}} = \int \frac{\cos(\theta) \, d\theta}{\sqrt{\cos^2 \theta}} = \int \frac{\cos \theta}{|\cos(\theta)|} \, d\theta.$$

Since $x = \sin \theta$ for $-\dfrac{\pi}{2} \le \theta \le \dfrac{\pi}{2}$, $\cos(\theta) \ge 0$, which establishes that $|\cos(\theta)| = \cos(\theta)$. Therefore

$$\int \frac{\cos(\theta)}{|\cos(\theta)|} \, d\theta = \int \frac{\cos(\theta)}{\cos(\theta)} \, d\theta$$

$$= \int 1 \, d\theta$$

$$= \theta + C.$$

But $x = \sin(\theta) \Rightarrow \sin^{-1}(x) = \theta$. So we conclude that

$$\int \frac{dx}{\sqrt{9 - 9x^2}} = \frac{1}{3} \int \frac{dx}{\sqrt{1 - x^2}}$$

$$= \frac{1}{3}\theta + C_1, \text{ where } C_1 = \frac{1}{3}C$$

$$= \frac{1}{3} \sin^{-1}(x) + C_1.$$

3. $\int \sqrt{t^2 + 4t + 3} \, dt = \int \sqrt{(t + 2)^2 - 1^2} \, dt$. If you use the substitutions $x = t + 2$ and $dx = dt$, you can write $\int \sqrt{(t + 2)^2 - 1^2} \, dt$ as $\int \sqrt{x^2 - 1^2} \, dx$. According to the table that summarizes trigonometric substitution, the substitutions $x = \sec(\theta)$ and $dx = \sec(\theta) \tan(\theta) \, d\theta$ should be used. So $\int \sqrt{x^2 - 1^2} \, dx$ becomes

$$\int \sqrt{x^2 - 1^2} \, dx = \int \sqrt{\sec^2(\theta) - 1} \cdot (\sec(\theta) \tan(\theta)) \, d\theta$$

$$= \int |\tan(\theta)| \sec(\theta) \tan(\theta) \, d\theta.$$

$x = \sec(\theta)$ for $0 < \theta < \dfrac{\pi}{2}$, so $|\tan(\theta)| = \tan(\theta)$. Therefore, $\int |\tan(\theta)| \sec(\theta) \tan(\theta) \, d\theta$ becomes

$$\int |\tan(\theta)| \sec(\theta) \tan(\theta) \, d\theta = \int \tan^2(\theta) \sec(\theta) \, d\theta.$$

Since $\tan^2(\theta) = \sec^2(\theta) - 1$, we have

$$\int \tan^2(\theta) \sec(\theta) \, d\theta = \int (\sec^2(\theta) - 1) \sec(\theta) \, d\theta$$

$$= \int \sec^3(\theta) - \sec(\theta) \, d\theta.$$

Recall that $\int \sec(\theta) \, d\theta = \ln|\sec(\theta) + \tan \theta| + C$.
To calculate $\int \sec^3(\theta) \, d\theta$, we apply the reduction formula with $n = 3$. We have

$$\int \sec^3(\theta) \, d\theta = \frac{1}{2} \sec(\theta) \tan(\theta) + \frac{1}{2} \int \sec(\theta) \, d\theta$$

$$= \frac{1}{2} \sec(\theta) \tan(\theta) + \frac{1}{2} \ln|\sec(\theta) + \tan(\theta)| + C.$$

Combining all these results, we have

$$\int \tan^2(\theta) \sec(\theta) \, d\theta = \frac{1}{2} \sec(\theta) \tan(\theta) + \frac{1}{2} \ln|\sec \theta$$

$$+ \tan \theta| - \ln|\sec(\theta) + \tan(\theta)| + C$$

$$= \frac{1}{2} \sec(\theta) \tan(\theta) - \frac{1}{2} \ln|\sec(\theta)$$

$$+ \tan(\theta)| + C_1$$

Recall our substitution $x = \sec(\theta)$. If we think of x as $\frac{x}{1}$, we can create the following right triangle.

Using the Pythagorean Theorem, we obtain that the remaining leg of the right triangle is equal to $\sqrt{x^2 - 1^2}$, so $\tan(\theta) = \sqrt{x^2 - 1^2}$. We shall use these results to obtain our final answer.

$$\frac{1}{2}\sec(\theta)\tan(\theta) - \frac{1}{2}\ln|\sec(\theta) + \tan(\theta)| + C$$
$$= \frac{1}{2}x\sqrt{x^2 - 1^2} - \frac{1}{2}\ln|x + \sqrt{x^2 - 1^2}| + C$$
$$= \frac{1}{2}(t + 2)\sqrt{t^2 + 4t + 3} - \frac{1}{2}\ln|(t + 2)$$
$$+ \sqrt{t^2 + 4t + 3}| + C$$

We have shown that

$$\int \sqrt{t^2 + 4t + 3}\, dt = \frac{1}{2}(t + 2)\sqrt{t^2 + 4t + 3}$$
$$- \frac{1}{2}\ln|(t + 2) + \sqrt{t^2 + 4t + 3}| + C.$$

4.
$$\int \frac{2}{2x^2 + 6x + 32}\, dx = \int \frac{1}{x^2 + 3x + 16}\, dx$$

Since the degree of the numerator of the integrand is zero, we shall skip steps 1 and 2 and proceed immediately to step 3.

Step 3: Using the Method of Completing the Square, the denominator of the integrand, $x^2 + 3x + 16$, is equal to $\left(x + \frac{3}{2}\right)^2 + \left(\frac{\sqrt{55}}{2}\right)^2$. If we use $u = x + \frac{3}{2}$, $du = dx$, and $a = \frac{\sqrt{55}}{2}$, the integral $\int \frac{1}{x^2 + 3x + 16}\, dx$ is equal to

$$\int \frac{1}{x^2 + 3x + 16}\, dx = \int \frac{1}{\left(x + \frac{3}{2}\right)^2 + \left(\frac{\sqrt{55}}{2}\right)^2}\, dx$$
$$= \int \frac{1}{u^2 + a^2}\, du$$
$$= \frac{1}{a}\tan^{-1}\left(\frac{u}{a}\right) + C$$
$$= \frac{1}{\left(\frac{\sqrt{55}}{2}\right)}\tan^{-1}\left(\frac{\left(x + \frac{3}{2}\right)}{\left(\frac{\sqrt{55}}{2}\right)}\right) + C$$
$$= \frac{2}{\sqrt{55}}\tan^{-1}\left(\frac{2}{\sqrt{55}}\left(x + \frac{3}{2}\right)\right) + C$$
$$= \frac{2}{\sqrt{55}}\tan^{-1}\left(\frac{2}{\sqrt{55}}x + \frac{3}{\sqrt{55}}\right) + C.$$

Section 7.5

1. The degree of the polynomial in the numerator is 2, which is less than 3, the degree of the polynomial in the denominator. Furthermore, the denominator is equal to the product of a linear factor, $(x + 1)$, and an irreducible quadratic factor, $x^2 + 1$. [$x^2 + 1$ is irreducible because for $a = 1, b = 0$, and $c = 1, b^2 - 4ac = -4 < 0$.] Therefore,

$$\frac{3x^2 + 4x}{(x + 1)(x^2 + 1)} = \frac{A}{x + 1} + \frac{Bx + C}{x^2 + 1}$$
$$= \frac{A(x^2 + 1) + (Bx + C)(x + 1)}{(x + 1)(x^2 + 1)}$$
$$= \frac{Ax^2 + A + Bx^2 + Cx + Bx + C}{(x + 1)(x^2 + 1)}$$
$$= \frac{(A + B)x^2 + (B + C)x + (A + C)}{(x + 1)(x^2 + 1)}.$$

Equating coefficients of like terms, we have the equations

$$\begin{cases} A + B = 3 \\ B + C = 4 \\ A + C = 0. \end{cases}$$

By solving these three simultaneous equations, we find $A = -\frac{1}{2}, B = \frac{7}{2}$, and $C = \frac{1}{2}$.

$$\frac{3x^2 + 4x}{(x + 1)(x^2 + 1)} = \frac{-\frac{1}{2}}{x + 1} + \frac{\frac{7}{2}x + \frac{1}{2}}{x^2 + 1}$$
$$= -\frac{1}{2}\left(\frac{1}{x + 1}\right) + \frac{7}{2}\left(\frac{x}{x^2 + 1}\right)$$
$$+ \frac{1}{2}\left(\frac{1}{x^2 + 1}\right)$$

Therefore,

$$\int \frac{3x^2 + 4x}{(x + 1)(x^2 + 1)}\, dx$$
$$= -\frac{1}{2}\int \frac{1}{x + 1}\, dx + \frac{7}{2}\int \left(\frac{x}{x^2 + 1}\right)\, dx$$
$$+ \frac{1}{2}\int \frac{1}{x^2 + 1}\, dx.$$
$$= -\frac{1}{2}\int \frac{1}{x + 1}\, dx + \frac{7}{4}\int \frac{2x}{x^2 + 1}\, dx$$
$$+ \frac{1}{2}\int \frac{1}{x^2 + 1}\, dx$$
$$= -\frac{1}{2}\ln|x + 1| + \frac{7}{4}\ln|x^2 + 1| + \frac{1}{2}\tan^{-1}(x) + C$$
$$= -\frac{1}{2}\ln|x + 1| + \frac{7}{4}\ln(x^2 + 1) + \frac{1}{2}\tan^{-1}(x) + C.$$

2. Since the degree of the numerator, $2x^2 - 4x + 2$, and the degree of the denominator, $x^2 - x + 1$, are both equal to 2, divide the denominator into the numerator.

$$\frac{2x^2 - 4x + 2}{x^2 - x + 1} = 2 - \frac{2x}{x^2 - x + 1}$$

$$= 2 - \frac{2x - 1 + 1}{x^2 - x + 1}$$

$$= 2 - \left[\frac{(2x - 1)}{x^2 - x + 1} + \frac{1}{x^2 - x + 1} \right]$$

$$= 2 - \frac{(2x - 1)}{x^2 - x + 1} - \frac{1}{\left(x^2 - x + \frac{1}{4} \right) + \frac{3}{4}}$$

$$= 2 - \frac{(2x - 1)}{x^2 - x + 1} - \frac{1}{\left(x - \frac{1}{2} \right)^2 + \left(\frac{\sqrt{3}}{2} \right)^2}.$$

Therefore,

$$\int \frac{2x^2 - 4x + 2}{x^2 - x + 1}\, dx = \int 2\, dx - \int \frac{(2x - 1)}{x^2 - x + 1}\, dx$$

$$- \int \frac{1}{\left(x - \frac{1}{2} \right)^2 + \left(\frac{\sqrt{3}}{2} \right)^2}\, dx.$$

If we let $u = x - \frac{1}{2}$, $du = dx$, and $a = \frac{\sqrt{3}}{2}$, the last integral

$$\int \frac{1}{\left(x - \frac{1}{2} \right)^2 + \left(\frac{\sqrt{3}}{2} \right)^2}\, dx$$

$$= \int \frac{1}{u^2 + a^2}\, du$$

$$= \frac{1}{a} \tan^{-1} \left(\frac{u}{a} \right) + C$$

$$= \frac{1}{\frac{\sqrt{3}}{2}} \tan^{-1} \left(\frac{x - \frac{1}{2}}{\frac{\sqrt{3}}{2}} \right) + C$$

$$= \frac{2}{\sqrt{3}} \tan^{-1} \left(\frac{2}{\sqrt{3}} \left(x - \frac{1}{2} \right) \right) + C$$

$$= \frac{2}{\sqrt{3}} \tan^{-1} \left(\frac{2}{\sqrt{3}} x - \frac{1}{\sqrt{3}} \right) + C.$$

$$\int 2\, dx - \int \frac{(2x - 1)}{x^2 - x + 1}\, dx - \int \frac{1}{\left(x - \frac{1}{2} \right)^2 + \left(\frac{\sqrt{3}}{2} \right)^2}\, dx$$

$$= 2x - \ln |x^2 - x + 1| - \frac{2}{\sqrt{3}} \tan^{-1} \left(\frac{2}{\sqrt{3}} x - \frac{1}{\sqrt{3}} \right) + C$$

Section 8.1

1. Region R is shaded in the graph that follows,

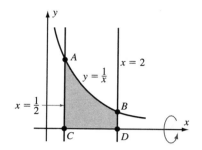

where $A = \left(\frac{1}{2}, 2 \right)$

$B = \left(2, \frac{1}{2} \right)$

$C = \left(\frac{1}{2}, 0 \right)$

$D = (2, 0)$.

The representative disk of the solid of revolution has thickness dx and area equal to πr^2 where $r = y = \frac{1}{x}$.

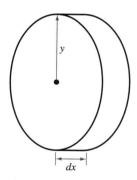

Therefore the volume of the solid of revolution is given by

$$\text{Volume} = \int_{x = \frac{1}{2}}^{2} \pi \left(\frac{1}{x} \right)^2 dx$$

$$= \pi \int_{x = \frac{1}{2}}^{2} x^{-2}\, dx$$

$$= \pi \cdot \frac{-1}{x} \Big|_{x = \frac{1}{2}}^{2}$$

$$= \pi \left[\left(-\frac{1}{2} \right) - (-2) \right] = \frac{3}{2}\pi \text{ cu. units}$$

$$\approx 4.71 \text{ cu. units}.$$

2. Region R is shaded in the graph that follows,

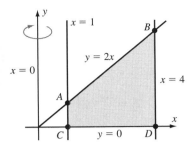

where $A = (1, 2)$
$B = (4, 8)$
$C = (1, 0)$
$D = (4, 0)$.

When region R is revolved about the y-axis [i.e., $x = 0$] we obtain a surface of revolution which looks like

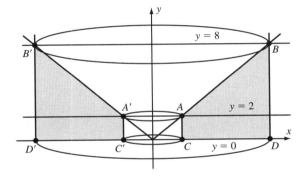

where $A = (1, 2)$ $A' = (-1, 2)$
$B = (4, 8)$ $B' = (-4, 8)$
$C = (1, 0)$ $C' = (-1, 0)$
$D = (4, 0)$ $D' = (-4, 0)$.

Since $\int\limits_{y=0}^{8} (f(y) - g(y)) \, dy = \int\limits_{y=0}^{2} (f(y) - g(y)) \, dy +$

$\int\limits_{2}^{8} (f(y) - g(y)) \, dy$, we shall deal with calculating the total
volume by considering the volume from $y = 0$ to $y = 2$ first, followed by calculating the volume from $y = 2$ to $y = 8$. Consider the region, R_1, from $y = 0$ to $y = 2$,

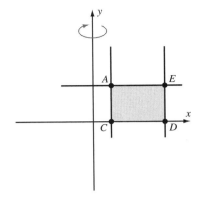

where $A = (1, 2)$
$E = (4, 2)$
$C = (1, 0)$
$D = (4, 0)$.

The two radii are $x_2 = |4 - 0| = 4$ and $x_1 = |1 - 0| = 1$. Therefore the volume of the solid of revolution from $y = 0$ to $y = 2$ is

$$\text{Volume} = \int\limits_{y=0}^{2} (\pi(4)^2 - \pi(1)^2) \, dy$$

$$= 15\pi \int\limits_{y=0}^{2} 1 \, dy$$

$$= 30 \, \pi \text{ cu. units.}$$

Now consider the region, R_2, from $y = 2$ to $y = 8$,

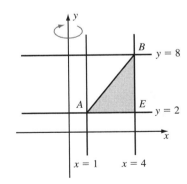

where $A = (1, 2)$
$E = (4, 2)$
$B = (4, 8)$.

The two radii are $x_2 = |4 - 0| = 4$ and $x_3 = \dfrac{y}{2}$. Therefore the volume of the solid of revolution from $y = 2$ to $y = 8$ is

$$\text{Volume} = \int_{y=2}^{8} \left[(\pi(4)^2) - \pi \left(\frac{y}{2} \right)^2 \right] dy$$

$$= \pi \int_{y=2}^{8} \left[16 - \left(\frac{y^2}{4} \right) \right] dy$$

$$= \pi \left(16y - \frac{y^3}{12} \right) \Big|_{y=2}^{y=8}$$

$$= \pi \left(16 \cdot (8) - \frac{8^3}{12} - 16 \cdot (2) + \frac{2^3}{12} \right)$$

$$= 54 \, \pi \ \text{cu. units.}$$

The total volume $= 30\pi + 54\pi = 84\pi$ cu. units.

3. Region R is shaded in the graph that follows.

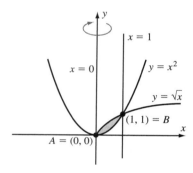

When Region R is revolved about the y-axis, we obtain a surface of revolution which looks like

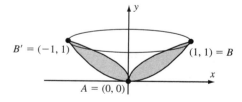

The two radii are $x_1 = y^{1/2}$ and $x_2 = y^2$. Therefore, the volume of the surface of revolution is

$$V = \pi \int_{y=0}^{1} [(y^{1/2})^2 - (y^2)^2] \, dy$$

$$= \pi \int_{y=0}^{1} [y - y^4] \, dy$$

$$= \pi \left[\frac{1}{2} y^2 - \frac{1}{5} y^5 \right] \Big|_{y=0}^{1}$$

$$= \pi \left[\frac{1}{2} - \frac{1}{5} \right]$$

$$= \frac{3}{10} \pi \ \text{cu. units} \approx .94 \ \text{cu. units.}$$

4. Region R is shaded in the graph that follows,

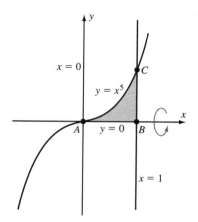

where $A = (0, 0)$
$\quad\quad\quad B = (1, 0)$
$\quad\quad\quad C = (1, 1)$.

Note that the curve $y = x^5$ is equivalent to $x = y^{1/5}$. When region R is revolved about the x-axis, we obtain a solid of revolution which looks like

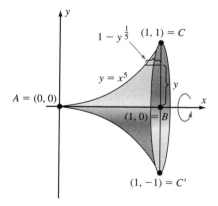

The conditions of Theorem 5 are met. Therefore the volume of the solid of revolution is

$$V = 2\pi \int_{y=0}^{1} y \cdot (1 - y^{1/5}) \, dy$$

$$= 2\pi \int_{y=0}^{1} (y - y^{6/5}) \, dy$$

$$= 2\pi \cdot \left(\frac{y^2}{2} - \frac{5}{11} y^{11/5} \right) \Bigg|_{y=0}^{y=1}$$

$$= 2\pi \cdot \left(\frac{1}{2} - \frac{5}{11} - 0 \right) = \frac{\pi}{11} \text{ cu. units.}$$

Section 8.2

1. The graph of the curve $y^2 = x^3$ is

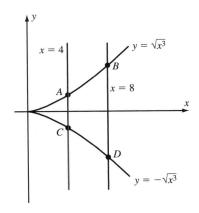

where $A = (4, 8)$
$\quad B = (8, 16\sqrt{2})$
$\quad C = (4, -8)$
$\quad D = (8, -16\sqrt{2})$.

Since the arc length from A to B is equal to the arc length from C to D, we need to calculate the arc length from A to B and to double the answer.

The arc length from A to B is given by

$L = \int\limits_{x=a}^{b} \sqrt{1 + (f'(x))^2} \, dx$, where $a = 4$ and $b = 4$. To

calculate the first derivative of $y^2 = x^3$, we observe that

$$y^2 = x^3 \Rightarrow$$
$$2y \frac{dy}{dx} = 3x^2 \Rightarrow$$
$$\frac{dy}{dx} = \frac{3x^2}{2y} \Rightarrow$$
$$\frac{dy}{dx} = \frac{3x^2}{2(\pm x^{3/2})} \Rightarrow$$
$$\frac{dy}{dx} = \pm \frac{3}{2} x^{1/2}.$$

Because we are calculating the arc length from A to B, we choose $\dfrac{dy}{dx} = +\dfrac{3}{2} x^{1/2}$.

Therefore the arc length from A to B is given by

$$L_{AB} = \int\limits_{x=4}^{8} \sqrt{1 + \left(\frac{3}{2} x^{1/2} \right)^2} \, dx$$

$$= \int\limits_{x=4}^{8} \sqrt{1 + \frac{9}{4} x} \, dx.$$

If $u = 1 + \dfrac{9}{4} x$, then $du = \dfrac{9}{4} dx \Rightarrow \dfrac{4}{9} du = dx$. If
$u = 1 + \dfrac{9}{4} x$ and $x = 4$, then $u = 1 + 9 = 10$.
If $u = 1 + \dfrac{9}{4} x$ and $x = 8$, then $u = 1 + 18 = 19$. So

$$\int\limits_{x=4}^{8} \sqrt{1 + \frac{9}{4} x} \, dx = \frac{4}{9} \int\limits_{u=10}^{19} \sqrt{u} \, du$$

$$= \frac{4}{9} \cdot \frac{2}{3} u^{3/2} \Bigg|_{u=10}^{19}$$

$$= \frac{8}{27} [19^{3/2} - 10^{3/2}]$$

$$= \frac{8}{27} [19\sqrt{19} - 10\sqrt{10}] \text{ cu. units}$$

$$\approx 15.17 \text{ cu. units.}$$

Therefore the *total* arc length of the curve $y^2 = x^3$ from $x = 4$ to $x = 8$ is equal to

$$2 \int\limits_{x=4}^{8} \sqrt{1 + \frac{9}{4} x} \, dx = \frac{16}{27} \left[19\sqrt{19} - 10\sqrt{10} \right] \text{ cu. units}$$

$$\approx 30.34 \text{ cu. units.}$$

2. Region R is shaded in the graph that follows,

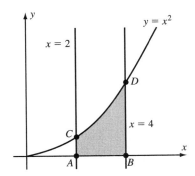

where $A = (2, 0)$
$\quad B = (4, 0)$
$\quad C = (2, 4)$
$\quad D = (4, 16)$.

The solid of revolution obtained when region R is rotated about the x-axis is

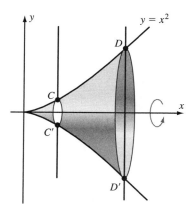

where
$$C = (2, 4)$$
$$C' = (2, -4)$$
$$D = (4, 16)$$
$$D' = (4, -16).$$

The formula for surface area is equal to

$$2\pi \int_{x=a}^{b} f(x)\sqrt{1 + (f'(x))^2}\, dx.$$

Applied to the problem, we obtain

$$\text{Surface Area} = 2\pi \int_{x=2}^{4} x^2\sqrt{1 + (2x)^2}\, dx$$
$$= 2\pi \int_{x=2}^{4} x^2\sqrt{1 + 4x^2}\, dx.$$

Section 8.3

1. Recall that $f_{ave} = \dfrac{1}{b-a} \displaystyle\int_{x=a}^{b} f(x)\, dx$. Therefore the average value of the function $f(x) = \sin(x)$ on the interval $\left[0, \dfrac{\pi}{4}\right]$ is given by

$$f_{ave} = \frac{1}{\dfrac{\pi}{4} - 0} \int_{x=0}^{\pi/4} \sin(x)\, dx$$
$$= \frac{4}{\pi} \cdot (-\cos(x)) \Big|_{x=0}^{\pi/4}$$
$$= \frac{4}{\pi} \left[\frac{\sqrt{2}}{2} - 1\right]$$
$$= \frac{4}{\pi} \left[1 - \frac{\sqrt{2}}{2}\right] \approx .37.$$

2. Let μ equal the average value (or mean) of the random variable.

$$\mu = \int_{x=a}^{b} x \cdot f(x)\, dx$$
$$= \int_{x=0}^{1} x \cdot 6x^5 dx$$
$$= 6 \int_{x=0}^{1} x^6 dx$$
$$= \frac{6}{7} x^7 \Big|_{x=0}^{1}$$
$$= \frac{6}{7} \approx .86$$

Section 8.4

1. Region R is shaded in the graph that follows,

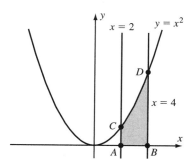

where
$$A = (2, 0)$$
$$B = (4, 0)$$
$$C = (2, 4)$$
$$D = (4, 16).$$

Let $(\bar{x}, \bar{y})$ be the center of mass of region R. Because R has uniform density δ,

$$\bar{x} = \frac{M_{x=0}}{M} = \frac{\displaystyle\int_{x=a}^{b} x \cdot f(x)\, dx}{\displaystyle\int_{x=a}^{b} f(x)\, dx}$$
$$= \frac{\displaystyle\int_{x=2}^{4} x \cdot x^2 dx}{\displaystyle\int_{x=2}^{4} x^2 dx}$$

$$= \frac{\int_{x=2}^{4} x^3 \, dx}{\int_{x=2}^{4} x^2 \, dx}$$

$$= \frac{\frac{1}{4}x^4 \Big|_{x=2}^{4}}{\frac{1}{3}x^3 \Big|_{x=2}^{4}}$$

$$= \frac{3}{4}\left(\frac{4^4 - 2^4}{4^3 - 2^3}\right)$$

$$= \frac{3}{4}\left(\frac{(2^2)^4 - 2^4}{(2^2)^3 - 2^3}\right)$$

$$= \frac{3}{4}\left(\frac{2^4(2^4 - 1)}{2^3(2^3 - 1)}\right)$$

$$= \frac{3}{4} \cdot 2 \cdot \frac{15}{7}$$

$$= \frac{45}{14} \approx 3.21.$$

Similarly,

$$\bar{y} = \frac{\frac{1}{2}\int_{x=2}^{4} (x^2)^2 \, dx}{\int_{x=2}^{4} x \, dx}$$

$$= \frac{\frac{1}{10}x^5 \Big|_{x=2}^{4}}{\frac{1}{3}x^3 \Big|_{x=2}^{4}}$$

$$= \frac{3}{10}\left[\frac{4^5 - 2^5}{4^3 - 2^3}\right]$$

$$= \frac{3}{10}\left[\frac{2^5(2^5 - 1)}{2^3(2^3 - 1)}\right]$$

$$= \frac{3}{10} \cdot 2^2 \cdot \left[\frac{31}{7}\right]$$

$$= \frac{12}{10} \cdot \frac{31}{7}$$

$$= \frac{186}{35} \approx 5.31.$$

We conclude that $(\bar{x}, \bar{y}) \approx (3.21, 5.31)$ is the center of mass of region R.

Section 8.5

1. The work done in stretching a spring from a to b is $W = \int_{x=a}^{b} k \cdot (x) \, dx$. We must first calculate k, the spring constant, by utilizing the information "4N of work is done in extending a spring 0.1 m beyond its equilibrium position."

$$4 = \int_{x=0}^{.1} k \cdot x \, dx \Rightarrow$$

$$4 = k \cdot \frac{x^2}{2} \Big|_{x=0}^{.1} \Rightarrow$$

$$8 = k\left(\frac{1}{100} - 0\right) \Rightarrow$$

$$800 = k$$

To calculate the extra work required to extend the spring an extra 0.3 m, we solve

$$W = \int_{.1}^{.4} 800 \cdot x \, dx$$

$$= \frac{800}{2} \cdot x^2 \Big|_{.1}^{.4}$$

$$= 400\left[\frac{16}{100} - \frac{1}{100}\right]$$

$$= 400 \cdot \frac{15}{100}$$

$$= 4 \cdot 15$$

$$= 60 \text{ J}.$$

Section 8.6

1. $\int_{x=0}^{2} \frac{1}{\sqrt{x}} \, dx$ is an improper integral with infinite integrand at $x = 0$. Therefore,

$$\int_{x=0}^{2} \frac{1}{\sqrt{x}} \, dx = \lim_{\epsilon \to 0^+} \int_{x=0+\epsilon}^{2} \frac{1}{\sqrt{x}} \, dx$$

$$= \lim_{\epsilon \to 0^+} \int_{x=\epsilon}^{2} x^{-1/2} \, dx$$

$$= \lim_{\epsilon \to 0^+} 2x^{1/2} \Big|_{x=\epsilon}^{2}$$

$$= \lim_{\epsilon \to 0^+} 2\left[\sqrt{2} - \sqrt{\epsilon}\right]$$

$$= 2\sqrt{2} - 2\lim_{\epsilon \to 0^+} \sqrt{\epsilon}$$

$$= 2\sqrt{2} - 0$$

$$= 2\sqrt{2}.$$

Section 8.7

1. Recall $\displaystyle\int_{-\infty}^{B} g(x)\,dx = \lim_{M\to-\infty}\int_{M}^{B} g(x)\,dx$. Therefore,

$$\int_{x=-\infty}^{-5} \frac{1}{(x)^2}\,dx = \lim_{M\to-\infty}\int_{M}^{-5}\frac{1}{x^2}\,dx$$

$$= \lim_{M\to-\infty}\int_{M}^{-5} x^{-2}\,dx$$

$$= \lim_{M\to-\infty} \frac{-1}{x}\bigg|_{M}^{-5}$$

$$= \lim_{M\to-\infty}\left[\frac{1}{5} - \frac{1}{M}\right]$$

$$= \frac{1}{5} + \left(\lim_{M\to-\infty}\frac{1}{M}\right)$$

$$= \frac{1}{5} + 0$$

$$= \frac{1}{5}.$$

2. Let μ equal the average value (or mean) of the random variable.

$$\mu = \int_{x=a}^{b} x\cdot f(x)\,dx$$

$$= \int_{x=0}^{\infty} x\cdot 3e^{-3x}\,dx$$

$$= \lim_{N\to\infty}\int_{x=0}^{N} x\cdot e^{-3x}\cdot 3\,dx$$

Using the method of integration by parts, we choose $u = x$ and $dv = e^{-3x}\cdot 3\,dx$. Therefore $du = dx$ and $v = -e^{-3x}$.

$$\int_{x=0}^{N} x\cdot e^{-3x}\cdot 3\,dx = -xe^{-3x}\bigg|_{0}^{N} - \int_{0}^{N} -e^{-3x}\,dx$$

$$= -xe^{-3x}\bigg|_{0}^{N} - \frac{1}{3}\int_{0}^{N} e^{-3x}(-3\,dx)$$

$$= -xe^{-3x}\bigg|_{0}^{N} - \frac{1}{3}e^{-3x}\bigg|_{0}^{N}$$

$$= (-N\,e^{-3N} + 0) - \frac{1}{3}\left(e^{-3N} - e^{0}\right)$$

$$= \frac{-N}{e^{3N}} - \frac{1}{3}\cdot\frac{1}{e^{3N}} + \frac{1}{3}$$

$$\lim_{N\to\infty}\int_{x=0}^{N} x\cdot e^{-3x}\cdot 3\,dx = \lim_{N\to\infty}\frac{-N}{e^{3N}} - \frac{1}{3}\lim_{N\to\infty}\frac{1}{e^{3N}} + \frac{1}{3}$$

$$= \lim_{N\to\infty}\frac{-N}{e^{3N}} - \frac{1}{3}\cdot 0 + \frac{1}{3}$$

Since $\displaystyle\lim_{N\to\infty}\frac{-N}{e^{3N}} = \frac{\infty}{\infty}$, we apply l'Hôpital's Rule to obtain $\displaystyle\lim_{N\to\infty}\frac{-N}{e^{3N}} = \lim_{N\to\infty}\frac{-1}{3e^{3N}} = 0$. We conclude that

$$\int_{x=0}^{\infty} x\cdot 3e^{-3x}\,dx = \lim_{N\to\infty}\int_{0}^{N} x\cdot 3e^{-3x}\,dx$$

$$= 0 - 0 + \frac{1}{3}$$

$$= \frac{1}{3}.$$

Section 9.1

1. $\{a_n\} = \{-1, 1, -1, 1, -1, 1, \ldots\} = \{(-1)^n\}_{n=1}^{\infty}$. Therefore, $\displaystyle\sum_{j=1}^{\infty} a_j = \sum_{j=1}^{\infty}(-1)^j$. But

$$\sum_{j=1}^{2m}(-1)^j = \underbrace{(-1+1) + (-1+1) + \cdots + (-1+1) = 0,}_{m \text{ sums of } (-1 \text{ and } +1)}$$

$m = 1, 2, \ldots$, and

$$\sum_{j=1}^{2m+1}(-1)^j = \sum_{j=1}^{2m}(-1)^j + \underset{\underset{(2m+1)^{\text{th}} \text{ term}}{\uparrow}}{(-1)} = 0 + (-1) = 1,$$

$m = 1, 2, \ldots$. Because either this sum is zero if there is an even number of terms or this sum is -1 if there is an odd number of terms, the infinite series $\displaystyle\sum_{j=1}^{\infty} a_j$ does *not* converge to a unique (i.e., one) limit, so it diverges.

2. $\displaystyle S_N = \sum_{j=1}^{N} 5\cdot j = 5\cdot\sum_{j=1}^{N} j = 5\cdot\frac{N\cdot(N+1)}{2}$.

$$\lim_{N\to\infty} S_N = \frac{5}{2}\cdot\left(\lim_{N\to\infty} N\cdot N + 1\right) = \infty;$$ therefore, the series $\displaystyle\sum_{j=1}^{\infty} 5\cdot j$ diverges.

3.

$$\sum_{k=1}^{\infty}\frac{5^k}{6^{k+1}} = \sum_{k=1}^{\infty}\frac{5^k}{6\cdot 6^k}$$

$$= \frac{1}{6}\sum_{k=1}^{\infty}\frac{5^k}{6^k}$$

$$= \frac{1}{6}\sum_{k=1}^{\infty}\left(\frac{5}{6}\right)^k$$

$$= \frac{1}{6} \cdot \left[\frac{5}{6} + \left(\frac{5}{6} \right)^2 + \cdots \right]$$

$$= \frac{1}{6} \cdot \frac{5}{6} \cdot \left[1 + \left(\frac{5}{6} \right) + \left(\frac{5}{6} \right)^2 + \cdots \right]$$

$$= \frac{5}{36} \cdot \sum_{k=0}^{\infty} \left(\frac{5}{6} \right)^k$$

$\displaystyle\sum_{k=0}^{\infty} \left(\frac{5}{6} \right)^k$ is a geometric series with ratio $|r| = \dfrac{5}{6} < 1$. By

Theorem 2, therefore, $\displaystyle\sum_{k=0}^{\infty} \left(\frac{5}{6} \right)^k$ converges to

$\dfrac{1}{1-r} = \dfrac{1}{1 - \dfrac{5}{6}} = 6$. By Theorem 3, part b,

$\dfrac{5}{36} \cdot \displaystyle\sum_{k=0}^{\infty} \left(\frac{5}{6} \right)^k$ converges to $\dfrac{5}{36} \cdot 6 = \dfrac{5}{6}$.

Section 9.2

1. In the series $\displaystyle\sum_{n=1}^{\infty} \frac{6^n}{5^{n+1}}$, $a_n = \dfrac{6^n}{5^{n+1}}$.

$$\lim_{n\to\infty} a_n = \lim_{n\to\infty} \frac{6^n}{5^{n+1}}$$

$$= \lim_{n\to\infty} \frac{1}{5} \cdot \left(\frac{6}{5} \right)^n = \infty$$

Since $\lim_{n\to\infty} a_n \neq 0$, Theorem 1 establishes that $\displaystyle\sum_{n=1}^{\infty} \frac{6^n}{5^{n+1}}$ diverges.

2. In the series $\displaystyle\sum_{n=1}^{\infty} 5$, $a_n = 5$.

$$\lim_{n\to\infty} a_n = \lim_{n\to\infty} 5 = 5 \neq 0$$

Theorem 1 establishes that $\displaystyle\sum_{n=1}^{\infty} 5$ diverges.

3.
$$\sum_{n=0}^{\infty} 3^{-n} = \sum_{n=1}^{\infty} 3^{-(n-1)}$$

$$S_N = \sum_{n=1}^{N} 3^{-(n-1)}$$

$$= \sum_{n=1}^{N} \frac{1}{3^{n-1}}$$

$$= \sum_{n=1}^{N} \left(\frac{1}{3} \right)^{n-1}$$

$$= \left(\frac{1}{3} \right)^0 + \left(\frac{1}{3} \right)^1 + \left(\frac{1}{3} \right)^2 + \cdots + \left(\frac{1}{3} \right)^{N-1}$$

S_N is a *finite* geometric series with ratio $r = \dfrac{1}{3}$. Since $|r| = \dfrac{1}{3} < 1$, S_N sums to

$$\frac{1 - r^N}{1 - r} = \frac{1 - \left(\dfrac{1}{3} \right)^N}{1 - \dfrac{1}{3}}$$

$$= +\frac{3}{2} \left(1 - \left(\frac{1}{3} \right)^N \right)$$

$$= \frac{3}{2} - \frac{3}{2} \left(\frac{1}{3} \right)^N$$

$$\leq \frac{3}{2}.$$

Since there exists a real number $U = \dfrac{3}{2}$ such that $S_N \leq U$ for all N and each $a_n = \dfrac{1}{3^{n-1}} \geq 0$ for each n, Theorem 2 establishes that $\displaystyle\sum_{n=1}^{\infty} a_n$ converges.

4. $\displaystyle\sum_{n=85}^{\infty} \frac{1}{\sqrt[3]{n^3}} = \sum_{n=85}^{\infty} \frac{1}{n^{3/3}} = \sum_{n=85}^{\infty} \frac{1}{n^1}$, so $\displaystyle\sum_{n=85}^{\infty} \frac{1}{\sqrt[3]{n^3}}$ converges if and only if $\displaystyle\sum_{n=1}^{\infty} \frac{1}{n}$ converges.

Since $\displaystyle\sum_{n=1}^{\infty} \frac{1}{n}$ is the harmonic series that diverges,

$\displaystyle\sum_{n=85}^{\infty} \frac{1}{\sqrt[3]{n^3}}$ also diverges.

Section 9.3

1. Define $f(x) = \dfrac{1}{x}$ for $x \geq 1$. For x_1, x_2 in domain f, $x_1 \neq x_2$, if $x_1 > x_2$ then $\dfrac{1}{x_1} < \dfrac{1}{x_2}$, which implies that $f(x_1) < f(x_2)$ whenever $x_1 > x_2$. Thus we see that f is a decreasing function. $f(x) > 0$ for all $x \geq 1$ and it is a continuous function. Therefore the hypotheses of the Integral Test are satisfied. We need only to establish if $\displaystyle\int_{x=1}^{\infty} f(x)\, dx$ converges or diverges.

$$\int_{x=1}^{\infty} f(x)\, dx = \lim_{b\to\infty} \int_{1}^{b} f(x)\, dx$$

$$= \lim_{b\to\infty} \int_{1}^{b} \frac{1}{x}\, dx$$

$$= \lim_{b\to\infty} \ln|x| \Big|_{1}^{b}$$

$$= \lim_{b\to\infty} \ln(x) \Big|_{1}^{b} \text{ since } x \text{ in } I = [1, \infty] \text{ is positive}$$

$$= \lim_{b \to \infty} (\ln(b) - \ln(1))$$
$$= \lim_{b \to \infty} (\ln(b) - 0)$$
$$= \infty$$

Since $\displaystyle\int_{x=1}^{\infty} f(x)\,dx$ does *not* converge, we conclude that

$$\sum_{n=1}^{\infty} f(n) \text{ does } not \text{ converge, i.e., } \sum_{n=1}^{\infty} f(n) = \sum_{n=1}^{\infty} \frac{1}{n}$$

diverges, by the Integral Test.

2.
$$\frac{(n+2)^2}{\sqrt{n^5}} = \frac{n^2 + 4n + 4}{n^{5/2}}$$
$$= \frac{n^2}{n^{5/2}} + \frac{4n}{n^{5/2}} + \frac{4}{n^{5/2}}$$
$$= \frac{1}{n^{1/2}} + 4 \cdot \frac{1}{n^{3/2}} + 4 \cdot \frac{1}{n^{5/2}}$$

Therefore $\displaystyle\sum_{n=1}^{\infty} \frac{(n+2)^2}{\sqrt{n^5}} = \sum_{n=1}^{\infty} \frac{1}{n^{1/2}} + 4 \sum_{n=1}^{\infty} \frac{1}{n^{3/2}}$

$+ 4 \displaystyle\sum_{n=1}^{\infty} \frac{1}{n^{5/2}}$. Each of the series in the preceding sum is a

p-series with $p = \frac{1}{2}$, $p = \frac{3}{2}$, and $p = \frac{5}{2}$ in the first, second, and third series, respectively. Therefore, by Theorem 2, the

first series $\displaystyle\sum_{n=1}^{\infty} \frac{1}{n^{1/2}}$ diverges and the second and third series

converge. Because the first series in the sum is *not* convergent (even though the other two are convergent), the

original series $\displaystyle\sum_{n=1}^{\infty} \frac{(n+2)^2}{\sqrt{n^5}}$ is divergent.

Section 9.4

1. For $n = 1, 2, 3, \ldots, 3^n < 3^n + 3n$. Therefore

$\frac{1}{3^n} > \frac{1}{3^n + 3} > 0$. This implies that

$$\sum_{n=1}^{\infty} \frac{1}{3^n + 3 \cdot n} < \sum_{n=1}^{\infty} \frac{1}{3^n}.$$

But

$$\sum_{n=1}^{\infty} \frac{1}{3^n} = \sum_{n=0}^{\infty} \frac{1}{3^{n+1}}$$
$$= \frac{1}{3} \sum_{n=0}^{\infty} \frac{1}{3^n}.$$

Since $\displaystyle\sum_{n=0}^{\infty} \frac{1}{3^n}$ is a geometric series with ratio $|r| =$

$r = \frac{1}{3} < 1$, it converges to $\dfrac{1}{1-r} = \dfrac{1}{1 - \dfrac{1}{3}} = \dfrac{3}{2}$. Therefore

$\displaystyle\sum_{n=1}^{\infty} \frac{1}{3^n} = \frac{1}{3} \sum_{n=0}^{\infty} \frac{1}{3^n}$ converges to $\frac{1}{3} \cdot \frac{3}{2} = \frac{1}{2}$. We conclude

that $\displaystyle\sum_{n=1}^{\infty} \frac{1}{3n + 3^n}$ converges by the Comparison Test since

$$\sum_{n=1}^{\infty} \frac{1}{3n + 3^n} < \sum_{n=1}^{\infty} \frac{1}{3^n} = \frac{1}{2}.$$

2. For $n \geq 1$, $\sqrt{n} + 1 < 9n$. This implies that $0 < \dfrac{1}{9n} <$

$\dfrac{1}{\sqrt{n}+1}$, which in turn implies $0 < \sqrt{\dfrac{1}{9n}} < \sqrt{\dfrac{1}{\sqrt{n}+1}}$.

However, $\sqrt{\dfrac{1}{9n}} = \dfrac{1}{\sqrt{9n}} = \dfrac{1}{3\sqrt{n}}$ and $\sqrt{\dfrac{1}{\sqrt{n}+1}}$

$= \dfrac{1}{\sqrt{\sqrt{n}+1}}$, so we have $0 < \dfrac{1}{3\sqrt{n}} < \dfrac{1}{\sqrt{\sqrt{n}+1}}$ for

$n \geq 1$. The series $\displaystyle\sum_{n=1}^{\infty} \frac{1}{3\sqrt{n}} = \frac{1}{3}\sum_{n=1}^{\infty} \frac{1}{\sqrt{n}} = \frac{1}{3}\sum_{n=1}^{\infty} \frac{1}{n^{1/2}}$ is a

p-series with $p = \dfrac{1}{2} < 1$, so the series $\displaystyle\sum_{n=1}^{\infty} \frac{1}{3 \cdot n^{1/2}}$ diverges.

By the Comparison Test for Divergence, the series

$\displaystyle\sum_{n=1}^{\infty} \frac{1}{\sqrt{\sqrt{n}+1}}$ also diverges.

3. For large values of n, $a_n \approx \dfrac{n^2}{n^2} = 1$. Apply the Limit

Comparison Test to the series $\displaystyle\sum_{n=1}^{\infty} b_n = \sum_{n=1}^{\infty} 1$ and

$$\sum_{n=1}^{\infty} a_n = \sum_{n=1}^{\infty} \frac{n^2}{(n^2 + 1)}.$$

$$\lim_{n \to \infty} \frac{a_n}{b_n} = \lim_{n \to \infty} \frac{\dfrac{n^2}{n^2 + 1}}{1}$$
$$= \lim_{n \to \infty} \frac{n^2}{n^2 + 1}$$
$$= \lim_{n \to \infty} \frac{n^2}{(n^2 + 1)} \cdot \frac{\dfrac{1}{n^2}}{\dfrac{1}{n^2}}$$
$$= \lim_{n \to \infty} \frac{1}{1 + \dfrac{1}{n^2}}$$
$$= 1 < \infty$$

The limit is one, a finite positive integer. But

$\displaystyle\sum_{n=1}^{\infty} b_n = \sum_{n=1}^{\infty} 1 = \infty$. Since $\displaystyle\sum_{n=1}^{\infty} b_n$ diverges, by applying

the Limit Comparison Test, we conclude that

$\displaystyle\sum_{n=1}^{\infty} a_n = \sum_{n=1}^{\infty} \frac{n^2}{n^2 + 1}$ also diverges.

Section 9.5

1. Choose $a_n = \dfrac{1}{\sqrt{2n-1}}$. We shall first show that the conditions stated in the hypotheses of the Alternating Series Test are satisfied.

For $n \geq 1$, $2n - 1 < 2n + 1$, so $\dfrac{1}{2n+1} < \dfrac{1}{2n-1}$.

But this implies that $\sqrt{\dfrac{1}{2n+1}} < \sqrt{\dfrac{1}{2n-1}}$.

Since $\sqrt{\dfrac{1}{2n+1}} = \dfrac{1}{\sqrt{2n+1}}$ and $\sqrt{\dfrac{1}{2n-1}} = \dfrac{1}{\sqrt{2n-1}}$,

we conclude that $\dfrac{1}{\sqrt{2n+1}} < \dfrac{1}{\sqrt{2n-1}}$. But $2n + 1 = 2(n+1) - 1$ establishes that

$a_{n+1} = \dfrac{1}{\sqrt{2(n+1)-1}} < \dfrac{1}{\sqrt{2n-1}} = a_n$ for $n \geq 1$, i.e.,

$a_1 > a_2 > a_3 > \ldots$. We note that each a_n, $n = 1, 2, 3, \ldots$, is nonnegative and that $a_n \to 0$ as $n \to \infty$. We have established that the hypotheses of the Alternating Series Test are satisfied. Therefore, we conclude that the series

$$\sum_{n=1}^{\infty} (-1)^{n+1} a_n = \sum_{n=1}^{\infty} \frac{(-1)^{n+1}}{\sqrt{2n-1}} \text{ converges.}$$

2.
$$\sum_{n=0}^{\infty} |a_n| = \sum_{n=0}^{\infty} \left| \frac{1}{n^3 + 5} \right|$$
$$= \sum_{n=0}^{\infty} \frac{1}{n^3 + 5}$$

For $n = 0, 1, 2, \ldots$, $(n+1)^3 < (n+1)^3 + 5$, which implies that $\dfrac{1}{(n+1)^3 + 5} < \dfrac{1}{(n+1)^3}$ and

$$\sum_{n=0}^{\infty} \frac{1}{(n+1)^3 + 5} < \sum_{n=0}^{\infty} \frac{1}{(n+1)^3}.$$

But $\displaystyle\sum_{n=0}^{\infty} \frac{1}{(n+1)^3 + 5} = \sum_{n=1}^{\infty} \frac{1}{n^3 + 5}$ and

$\displaystyle\sum_{n=0}^{\infty} \frac{1}{(n+1)^3} = \sum_{n=1}^{\infty} \frac{1}{n^3}$. Therefore $\displaystyle\sum_{n=1}^{\infty} \frac{1}{n^3 + 5} < \sum_{n=1}^{\infty} \frac{1}{n^3}$.

Because $\displaystyle\sum_{n=1}^{\infty} \frac{1}{n^3}$ is a p-series with $p = 3 > 1$, $\displaystyle\sum_{n=1}^{\infty} \frac{1}{n^3}$ is convergent.

This implies that $\displaystyle\sum_{n=1}^{\infty} \frac{1}{n^3 + 5}$ is convergent by the Comparison Test for Convergence. Since

$\dfrac{1}{5} + \displaystyle\sum_{n=1}^{\infty} \frac{1}{n^3 + 5} = \sum_{n=0}^{\infty} \frac{1}{n^3 + 5}$ and $\displaystyle\sum_{n=1}^{\infty} \frac{1}{n^3 + 5}$ is

convergent, we conclude that $\dfrac{1}{5} + \displaystyle\sum_{n=1}^{\infty} \frac{1}{n^3 + 5}$ is convergent,

i.e., $\displaystyle\sum_{n=0}^{\infty} \frac{1}{n^3 + 5}$ is convergent.

Because $\displaystyle\sum_{n=0}^{\infty} \left| \frac{1}{n^3 + 5} \right| = \sum_{n=0}^{\infty} \frac{1}{n^3 + 5}$, which is

convergent, we conclude that $\displaystyle\sum_{n=0}^{\infty} \frac{1}{n^3 + 5}$ is absolutely convergent.

We now turn our attention to

$$\sum_{n=0}^{\infty} |b_n| = \sum_{n=0}^{\infty} \left| \frac{(-1)^n}{\sqrt{n^3 + 5}} \right|$$
$$= \sum_{n=0}^{\infty} \frac{1}{\sqrt{n^3 + 5}}.$$

For $n = 1, 2, 3, \ldots$, $n^3 < n^3 + 5$. This implies that $\dfrac{1}{n^3 + 5} < \dfrac{1}{n^3}$, which, in turn, implies that

$\sqrt{\dfrac{1}{n^3 + 5}} < \sqrt{\dfrac{1}{n^3}}$. Since $\sqrt{\dfrac{1}{n^3 + 5}} = \dfrac{1}{\sqrt{n^3 + 5}}$ and

$\sqrt{\dfrac{1}{n^3}} = \dfrac{1}{\sqrt{n^3}}$, we conclude that $\dfrac{1}{\sqrt{n^3 + 5}} < \dfrac{1}{\sqrt{n^3}}$ and

$\displaystyle\sum_{n=1}^{\infty} \frac{1}{\sqrt{n^3 + 5}} < \sum_{n=1}^{\infty} \frac{1}{\sqrt{n^3}}$. Since $\displaystyle\sum_{n=1}^{\infty} \frac{1}{\sqrt{n^3}} = \sum_{n=1}^{\infty} \frac{1}{n^{3/2}}$, we

have a convergent p-series because $p = \dfrac{3}{2} > 1$.

By the Comparison Test for Convergence, because

$\displaystyle\sum_{n=1}^{\infty} \frac{1}{n^{3/2}}$ converges and $\displaystyle\sum_{n=1}^{\infty} \frac{1}{\sqrt{n^3 + 5}} < \sum_{n=1}^{\infty} \frac{1}{n^{3/2}}$, we

conclude that $\displaystyle\sum_{n=1}^{\infty} \frac{1}{\sqrt{n^3 + 5}}$ is convergent. But $\displaystyle\sum_{n=1}^{\infty} \frac{1}{\sqrt{n^3 + 5}}$

is convergent implies that $\dfrac{1}{\sqrt{5}} + \displaystyle\sum_{n=1}^{\infty} \frac{1}{\sqrt{n^3 + 5}}$ is also

convergent. Since $\dfrac{1}{\sqrt{5}} + \displaystyle\sum_{n=1}^{\infty} \frac{1}{\sqrt{n^3 + 5}} = \sum_{n=0}^{\infty} \frac{1}{\sqrt{n^3 + 5}}$, we

conclude that $\displaystyle\sum_{n=0}^{\infty} \frac{1}{\sqrt{n^3 + 5}}$ is convergent.

We have shown that $\displaystyle\sum_{n=0}^{\infty} |b_n| = \sum_{n=0}^{\infty} \frac{1}{\sqrt{n^3 + 5}}$ is

convergent, so $\displaystyle\sum_{n=0}^{\infty} \frac{(-1)^n}{\sqrt{n^3 + 5}}$ is absolutely convergent.

Since $\displaystyle\sum_{n=0}^{\infty} a_n = \sum_{n=0}^{\infty} \frac{1}{n^3 + 5}$ and $\displaystyle\sum_{n=0}^{\infty} \frac{(-1)^n}{\sqrt{n^3 + 5}}$ are both

absolutely convergent, we conclude by applying Theorem 3, part a, that $\displaystyle\sum_{n=0}^{\infty} a_n + \sum_{n=0}^{\infty} b_n = \sum_{n=0}^{\infty} (a_n + b_n)$ is also

absolutely convergent.

3. $\displaystyle\sum_{n=1}^{\infty}\left|\frac{(-1)^{n-1}}{e^{3n}}\right| = \sum_{n=1}^{\infty}\frac{1}{e^{3n}} = \sum_{n=1}^{\infty}\left(\frac{1}{e^3}\right)^n$. Since

$\displaystyle\sum_{n=1}^{\infty}\left(\frac{1}{e^3}\right)^n$ is a geometric series with $|n| = \dfrac{1}{e^3} < 1$, we

conclude that $\displaystyle\sum_{n=1}^{\infty}\left(\frac{1}{e^3}\right)^n$ converges.

Since $\displaystyle\sum_{n=1}^{\infty}\left|\frac{(-1)^{n-1}}{e^{3n}}\right| = \sum_{n=1}^{\infty}\frac{1}{e^{3n}}$ is convergent, we

conclude that $\displaystyle\sum_{n=1}^{\infty}\frac{(-1)^{n-1}}{e^{3n}}$ is absolutely convergent. By

Theorem 2, since $\displaystyle\sum_{n=1}^{\infty}\frac{(-1)^{n-1}}{e^{3n}}$ is absolutely convergent, we

can state that $\displaystyle\sum_{n=1}^{\infty}\frac{(-1)^{n-1}}{e^{3n}}$ is also covergent.

Section 9.6

1. $a_n = \dfrac{3^n}{n!}$ and $a_{n+1} = \dfrac{3^{n+1}}{(n+1)!}$. Therefore

$$\lim_{n\to\infty}\left|\frac{a_{n+1}}{a_n}\right| = \lim_{n\to\infty}\left|\frac{\dfrac{3^{n+1}}{(n+1)!}}{\dfrac{3^n}{n!}}\right|$$

$$= \lim_{n\to\infty}\left|\frac{3^{n+1}}{(n+1)!}\cdot\frac{n!}{3^n}\right|$$

$$= \lim_{n\to\infty}\left|\frac{3}{n+1}\right|$$

$$= \lim_{n\to\infty}\frac{3}{n+1}$$

$$= 0.$$

Since $L = 0 < 1$, by the Ratio Test, we conclude that the

series $\displaystyle\sum_{n=1}^{\infty}\frac{3^n}{n!}$ is absolutely convergent.

2. $a_n = \dfrac{3^n}{n^3}$. Therefore,

$$\lim_{n\to\infty}|a_n|^{1/n} = \lim_{n\to\infty}\left|\frac{3^n}{n^3}\right|^{1/n}$$

$$= \lim_{n\to\infty}\left(\frac{3^n}{n^3}\right)^{1/n}$$

$$= \lim_{n\to\infty}\frac{(3^n)^{1/n}}{(n^3)^{1/n}}$$

$$= \lim_{n\to\infty}\frac{3}{n^{3/n}}$$

$$= \lim_{n\to\infty}\frac{3}{(n^{1/n})^3}$$

$$= \frac{3}{\left(\lim_{n\to\infty}n^{1/n}\right)^3}$$

$$= \frac{3}{1^3}$$

$$= 3.$$

Since $L = 3 > 1$, by the Root Test, the series $\displaystyle\sum_{n=1}^{\infty}\frac{3^n}{n^3}$

diverges.

Section 10.1

1. $\displaystyle\sum_{n=0}^{\infty}\left(\frac{x}{5}\right)^n = 1 + \frac{x}{5} + \left(\frac{x}{5}\right)^2 + \cdots$ is a geometric series

with ratio $r = \dfrac{x}{5}$. It converges absolutely when $|r| < 1$.

But $|r| = \left|\dfrac{x}{5}\right| = \dfrac{|x|}{|5|} = \dfrac{|x|}{5}$, so $|r| < 1$ implies that $\dfrac{|x|}{5} < 1$,

i.e., $|x| < 5$. Since $|x| < 5$ implies that $-5 < x < 5$, we

conclude that the interval of convergence for the power

series $\displaystyle\sum_{n=0}^{\infty}\left(\frac{x}{5}\right)^n$ is $(-5, 5)$. The interval of convergence is

open because if $x = 5$, then $\displaystyle\sum_{n=0}^{\infty}\left(\frac{x}{5}\right)^n =$

$\displaystyle\sum_{n=0}^{\infty}\left(\frac{5}{5}\right)^n = \sum_{n=0}^{\infty}1^n = \infty$. If $x = -5$, then

$\displaystyle\sum_{n=0}^{\infty}\left(\frac{x}{5}\right)^n = \sum_{n=0}^{\infty}(-1)^n = 1 - 1 + 1 - 1 + \cdots$ does not

converge to a unique limit, so it diverges.

2. $\displaystyle\sum_{n=0}^{\infty}\frac{(x-6)^n}{3^n} = \sum_{n=0}^{\infty}\left(\frac{1}{3^n}\right)(x-6)^n$

$$a_n = \frac{1}{3^n} = \left(\frac{1}{3}\right)^n. \quad \ell = \lim_{n\to\infty}|a_n|^{1/n}$$

$$= \lim_{n\to\infty}\left|\left(\frac{1}{3}\right)^n\right|^{1/n}$$

$$= \lim_{n\to\infty}\left(\left(\frac{1}{3}\right)^n\right)^{1/n}$$

$$= \lim_{n\to\infty}\frac{1}{3}$$

$$= \frac{1}{3},$$

a positive, finite real number. Therefore, $R = \dfrac{1}{\ell} = 3$. Now

$(c - R, c + R) = (6 - 3, 6 + 3) = (3, 9)$.

We need to test the endpoints $x = 3$ and $x = 9$.

If $x = 3$, $\displaystyle\sum_{n=0}^{\infty}\frac{(x-6)^n}{3^n} = \sum_{n=0}^{\infty}\frac{-3^n}{3^n}$

$$= \sum_{n=0}^{\infty}\frac{(-1)^n(3^n)}{3^n}$$

$$= \sum_{n=0}^{\infty}(-1)^n, \text{ which diverges.}$$

If $x = 9$, $\displaystyle\sum_{n=0}^{\infty} \frac{(x-6)^n}{3^n} = \sum_{n=0}^{\infty} \frac{3^n}{3^n}$

$\displaystyle = \sum_{n=0}^{\infty} 1$, which diverges.

Therefore the interval of convergence is $(3, 9)$.

Section 10.2

1.

$$\frac{x^5}{(2+3x^2)} = \frac{x^5}{2\left(1 + \dfrac{3}{2}x^2\right)}$$

$$= \frac{1}{2} \cdot x^5 \cdot \frac{1}{\left(1 - \left(\dfrac{-3}{2}x^2\right)\right)}$$

If $u = \dfrac{-3}{2}x^2$ and $\dfrac{1}{1-u} = \displaystyle\sum_{n=0}^{\infty} u^n$,

$$\frac{1}{\left(1 + \dfrac{3}{2}x^2\right)} = \sum_{n=0}^{\infty} \left(\frac{-3}{2}x^2\right)^n$$

$= 1 - \dfrac{3}{2}x^2 + \dfrac{9}{4}x^4 - \dfrac{27}{8}x^6 + \cdots$ is a geometric series with

ratio $r = -\dfrac{3}{2}x^2$. $\displaystyle\sum_{n=0}^{\infty} \left(\frac{-3}{2}x^2\right)^n$ converges for

$|r| = \left|\dfrac{-3}{2}x^2\right| < 1.$

But

$$\left|\frac{-3}{2}x^2\right| < 1 \;\Rightarrow\; \left|\frac{-3}{2}\right| \cdot |x^2| < 1$$

$$\Rightarrow \frac{3}{2}|x^2| < 1$$

$$\Rightarrow |x^2| < \frac{2}{3}$$

$$\Rightarrow |x| < \sqrt{\frac{2}{3}}.$$

Checking endpoint $x = \sqrt{\dfrac{2}{3}}$, we have

$$\sum_{n=0}^{\infty} \left(\frac{-3}{2} \cdot \left(\sqrt{\frac{2}{3}}\right)^2\right)^n = \sum_{n=0}^{\infty} (-1)^n, \text{ which diverges.}$$

Checking endpoint $x = -\sqrt{\dfrac{2}{3}}$, we have

$$\sum_{n=0}^{\infty} \left(\frac{-3}{2} \cdot \left(-\sqrt{\frac{2}{3}}\right)^2\right)^n = \sum_{n=0}^{\infty} (-1)^n, \text{ which diverges.}$$

Thus $\displaystyle\sum_{n=0}^{\infty} \left(\frac{-3}{2}x^2\right)^n$ converges for $-\sqrt{\dfrac{2}{3}} < x < \sqrt{\dfrac{2}{3}}$.

$$\frac{x^5}{(2+3x^2)} = \frac{1}{2} \cdot x^5 \cdot \frac{1}{\left(1 - \left(\dfrac{-3}{2}x^2\right)\right)}$$

$$= \frac{1}{2} \cdot x^5 \sum_{n=0}^{\infty} \left(\frac{-3}{2}x^2\right)^n$$

$$= \sum_{n=0}^{\infty} (-1)^n \cdot \frac{3^n}{2^n} \cdot \frac{1}{2} \cdot x^5 \cdot x^{2n}$$

$$= \sum_{n=0}^{\infty} (-1)^n \cdot \frac{3^n}{2^{n+1}} \cdot x^{2n+5}$$

$$= \sum_{n=0}^{\infty} (-1)^n \cdot \frac{3^n}{2^{n+1}} \cdot (x-0)^{2n+5} \text{ for } |x| < \sqrt{\frac{2}{3}}$$

2.

$$\sum_{n=0}^{\infty} 5^{-n}(x-1)^n = \sum_{n=0}^{\infty} \frac{(x-1)^n}{5^n}$$

$$= \sum_{n=0}^{\infty} \left(\frac{x-1}{5}\right)^n$$

$$= 1 + \frac{(x-1)}{5} + \frac{(x-1)^2}{25} + \cdots$$

If $u = \dfrac{x-1}{5}$, $\displaystyle\sum_{n=0}^{\infty} 5^{-n}(x-1)^n = \sum_{n=0}^{\infty} u^n$, which converges

for $|u| < 1$. But $|u| = \left|\dfrac{x-1}{5}\right| = \dfrac{|x-1|}{|5|} = \dfrac{|x-1|}{5}$. So

$|u| < 1 \Rightarrow \dfrac{|x-1|}{5} < 1 \Rightarrow |x-1| < 5 \Rightarrow -5 < x-1 <$
$5 \Rightarrow -4 < x < 6$. If $x = -4$, then

$$\sum_{n=0}^{\infty} \left(\frac{x-1}{5}\right)^n = \sum_{n=0}^{\infty} \left(\frac{-5}{5}\right)^n$$

$$= \sum_{n=0}^{\infty} (-1)^n \left(\frac{5}{5}\right)^n$$

$$= \sum_{n=0}^{\infty} (-1)^n, \text{ which diverges.}$$

If $x = 6$, then $\displaystyle\sum_{n=0}^{\infty} \left(\frac{x-1}{5}\right)^n = \sum_{n=0}^{\infty} \left(\frac{5}{5}\right)^n$

$$= \sum_{n=0}^{\infty} 1^n, \text{ which also diverges.}$$

Thus the interval of convergence for $\displaystyle\sum_{n=0}^{\infty} 5^{-n}(x-1)^n =$
$f(x)$ is $(-4, 6)$.

By Theorem 2, part a, $f'(x)$ also converges absolutely on $(-4, 6)$.

3. Let $F(x) = \ln(1 + 2x)$.

Then $F'(x) = \dfrac{2}{1 + 2x}$, since

$$\frac{d}{dx}(\ln(1 + 2x)) = \frac{\frac{d}{dx}(1 + 2x)}{1 + 2x}.$$

Rewrite $F'(x)$ as $2 \cdot \dfrac{1}{1 - (-2x)}$.

With $u = -2x$ and $\dfrac{1}{1 - u} = \displaystyle\sum_{n=0}^{\infty} u^n$, we have

$$F'(x) = 2 \cdot \sum_{n=0}^{\infty} (-2x)^n$$
$$= 2 \cdot \sum_{n=0}^{\infty} (-1)^n (2)^n (x)^n$$
$$= \sum_{n=0}^{\infty} (-1)^n (2)^{n+1} x^4.$$

We note that

$$F'(x) = 2 \cdot \sum_{n=0}^{\infty} (-2x)^n$$
$$= 2 \cdot [1 - 2x + 4x^2 - 8x^3 + \cdots]$$

and $[1 - 2x + 4x^2 - 8x^3 + \cdots]$ is a geometric series with radius $r = -2x$. Therefore $\displaystyle\sum_{n=0}^{\infty} (-2x)^n$ converges for $|r| = |-2x| < 1$. Since

$$|-2x| < 1 \Rightarrow |-2||x| < 1 \Rightarrow 2 \cdot |x| < 1 \Rightarrow |x| < \frac{1}{2}, \text{ so}$$

$\displaystyle\sum_{n=0}^{\infty} (-2x)^n$ converges for $-\dfrac{1}{2} < x < \dfrac{1}{2}$.

We now check for convergence at the endpoints $x = -\dfrac{1}{2}$ and $x = \dfrac{1}{2}$.

If $x = -\dfrac{1}{2}$, $2 \cdot \displaystyle\sum_{n=0}^{\infty} (-2x)^n = 2 \cdot \displaystyle\sum_{n=0}^{\infty} (1)^n$, which diverges.

If $x = +\dfrac{1}{2}$, $2 \cdot \displaystyle\sum_{n=0}^{\infty} (-2x)^n = 2 \cdot \displaystyle\sum_{n=0}^{\infty} (-1)^n$, which diverges.

Therefore the interval of convergence for

$$F'(x) = 2 \sum_{n=0}^{\infty} (-2x)^n \text{ is } -\frac{1}{2} < x < \frac{1}{2}.$$

$$F(x) = \int F'(x)\, dx$$
$$= \int 2 \cdot \sum_{n=0}^{\infty} (-2x)^n\, dx$$
$$= \sum_{n=0}^{\infty} (-1) \cdot \int (-2x)^n (-2\, dx)$$
$$= \sum_{n=0}^{\infty} \left[(-1) \cdot \frac{(-2x)^{n+1}}{n + 1} + C \right]$$

$C = F(0) = \ln(1 + 2(0)) = 0$, so we have

$$F(x) = \sum_{n=0}^{\infty} \frac{(-1)}{n + 1} \cdot (-2x)^{n+1}.$$

$F(x)$ converges for x in open interval $\left(-\dfrac{1}{2}, \dfrac{1}{2} \right)$.
We need to check for convergence at the endpoints.
If $x = -\dfrac{1}{2}$, then

$$F\left(-\frac{1}{2} \right) = \sum_{n=0}^{\infty} \frac{(-1)}{n + 1} \cdot (1)^{n+1}$$
$$= \sum_{n=0}^{\infty} \frac{(-1)}{n + 1}.$$

If $x = +\dfrac{1}{2}$, then

$$F\left(\frac{1}{2} \right) = \sum_{n=0}^{\infty} \frac{(-1)}{n + 1} \cdot (-1)^{n+1}$$
$$= \sum_{n=0}^{\infty} \frac{(-1)^{n+2}}{(n + 1)}$$
$$= \sum_{n=0}^{\infty} \frac{(-1)^n}{n + 1}.$$

For $x = -\dfrac{1}{2}$, $\displaystyle\sum_{n=0}^{\infty} \frac{(-1)}{n + 1}$ is a harmonic series that diverges.

For $x = \dfrac{1}{2}$, choose $a_n = \dfrac{1}{n + 1}$ and see if the hypotheses of the Alternating Series Test are satisfied. $a_n = \dfrac{1}{n + 1} \to 0$ as $n \to \infty$.

For $n = 0, 1, 2, \ldots$, since $n + 1 < n + 2$, we have

$$a_{n+1} = \frac{1}{n + 2} < \frac{1}{n + 1} = a_n, \text{ so } a_0 > a_1 > a_2 > \cdots.$$

Because the hypotheses of the Alternating Series Test are satisfied when $x = +\dfrac{1}{3}$, we see that the interval of convergence for

$$F(x) = \ln(1 + 2x)$$
$$= \sum_{n=0}^{\infty} \frac{(-1)}{n + 1} \cdot (-2x)^{n+1}$$
$$= \sum_{n=0}^{\infty} \frac{(-1) \cdot (-1)^{n+1} \cdot (2)^{n+1} \cdot x^{n+1}}{n + 1}$$
$$= \sum_{n=0}^{\infty} \frac{(-1)^{n+2} (2)^{n+1} (x)^{n+1}}{n + 1}$$
$$= \sum_{n=0}^{\infty} \frac{(-1)^n (2)^{n+1} (x)^{n+1}}{n + 1}$$

is the interval $\left(-\dfrac{1}{2}, \dfrac{1}{2} \right]$.

4. Suppose $y \equiv y(x) = \sum_{n=0}^{\infty} a_n x^n$. Using Theorem 2, we obtain

$$\frac{dy}{dx} = \sum_{n=0}^{\infty} a_n \cdot n \cdot x^{n-1}.$$

To satisfy the given differential equation, we have

$$\sum_{n=0}^{\infty} a_n \cdot n \cdot x^{n-1} = 2x \sum_{n=0}^{\infty} a_n x^4$$

$$= \sum_{n=0}^{\infty} 2a_n x^{n+1}.$$

Expanding the series on each side and comparing coefficients of like terms, we obtain

$$0 + a_1 x^0 + 2a_2 x^1 + 3a_3 x^2 + 4a_4 x^3 + \sum_{n=5}^{\infty} a_n \cdot n \cdot x^{n-1}$$

$$= 2a_0 x^1 + 2a_1 x^2 + 2a_2 x^3 + \sum_{n=3}^{\infty} 2a_n x^{n+1}.$$

The initial condition $y(0) = 3 \Rightarrow a_0 = 3$. Comparing coefficients of like terms, we have

$$a_1 = 0$$
$$2a_2 = 2a_0$$
$$3a_3 = 2a_1$$
$$4a_4 = 2a_2$$
$$\vdots$$

i.e., for $n \geq 2$, $a_n = \frac{2}{n} a_{n-2}$, $a_1 = 0$ and from the initial condition $y(0) = 3$, $a_0 = 3$.

Listing some specific coefficient values, we have

$$a_0 = 3$$
$$a_1 = 0$$
$$a_2 = a_0 = 3$$
$$a_3 = \frac{2}{3} a_1 = 0$$
$$a_4 = \frac{1}{2} a_2 = \frac{3}{2}$$
$$a_5 = \frac{2}{5} a_3 = 0$$
$$a_6 = \frac{1}{3} a_4 = \frac{1}{3} \cdot \frac{3}{2} = \frac{1}{2}.$$

Thus we see if $n = 2k + 1$, for $k = 0, 1, 2, \ldots$, we have $a_1 = 0$,

$$a_{2k+1} = \frac{2}{2k+1} \cdot a_{2k+1-2} = \frac{2}{2k+1} a_{2k-1} \equiv 0.$$

If $n = 2k$, for $k = 1, 2, \ldots$, we have

$$a_{2k} = \frac{2}{2k} a_{2k-2} = \frac{1}{k} a_{2k-2} = \frac{1}{k(k-1)} a_{2k-4}$$

$$= \cdots = \frac{1}{k(k-1)(k-2)\ldots 1} a_0 = \frac{1}{k!} \cdot 3.$$

So in its interval of convergence, the solution is

$$y \equiv y(x) = 3 + 0 \cdot x^1 + 3 \cdot x^2 + 0 \cdot x^3$$
$$+ \frac{3}{2} x^4 + 0 + \sum_{n=6}^{\infty} \frac{2}{n} a_{n-2} x^n$$

$$= 3 + 3x^2 + \frac{3}{2} x^4 + \sum_{k=3}^{\infty} \frac{1}{k!} \cdot 3 \cdot x^{2k}.$$

Section 10.3

1. Let

$$f(x) = \ln(2x) \Rightarrow f(2) = \ln(4)$$
$$f'(x) = \frac{1}{x} \Rightarrow f'(2) = \frac{1}{2} = .5$$
$$f''(x) = \frac{-1}{x^2} \Rightarrow f''(2) = -\frac{1}{4} = -.25.$$

Therefore,

$$P_2(x) = f(2) + f'(2)(x - 2) + \frac{f''(2)}{2}(x - 2)^2$$

$$= \ln(4) + \frac{1}{2}(x - 2) - \frac{1}{8}(x - 2)^2.$$

2.
$$f(x) = x^{-4} = \frac{1}{x^4} \Rightarrow f(1) = 1$$
$$f'(x) = -4 \cdot x^{-5} = \frac{-4}{x^5} \Rightarrow f'(1) = -4$$
$$f''(x) = 20x^{-6} = \frac{20}{x^6} \Rightarrow f''(1) = 20$$
$$f'''(x) = -120x^{-7} = \frac{-120}{x^7} \Rightarrow f'''(1) = -120$$

From Theorem 1, for $N \geq 0$, $P_N(x) = \sum_{n=0}^{N} \frac{f^{(n)}(c)}{n!}(x - c)^n$ is the unique degree N polynomial such that $P_N(c) = f(c), \ldots P_N^{(N)}(c) = f^{(N)}(c)$ for N times continuously differentiable function f.

$$P_1(x) = \frac{f(1)}{0!}(x - 1)^0 + \frac{f'(1)}{1!}(x - 1)^1$$
$$= 1 - 4(x - 1) = -4x + 5$$

$$P_2(x) = \frac{f(1)}{0!}(x - 1)^0 + \frac{f'(1)}{1!}(x - 1)^1 + \frac{f''(1)}{2!}(x - 1)^2$$
$$= P_1(x) + \frac{f''(1)}{2!}(x - 2)^2$$
$$= -4x + 5 + \frac{20}{2}(x^2 - 2x + 1)$$
$$= -4x + 5 + 10x^2 - 20x + 10$$
$$= 10x^2 - 24x + 15$$

$$P_3(x) = \frac{f(1)}{0!}(x - 1)^0 + \frac{f'(1)}{1!}(x - 1)^1$$
$$+ \frac{f''(1)}{2!}(x - 2)^2 + \frac{f'''(1)}{3!}(x - 1)^3$$
$$= P_2(x) + \frac{(-120)}{6}(x^3 - 3x^2 + 3x - 1)$$
$$= 10x^2 - 24x + 15 - 20x^3 + 60x^2 - 60x + 20$$
$$= -20x^3 + 70x^2 - 84x + 35$$

3. Fix $X_0 \in I$.

$$
\begin{aligned}
f(x) &= \tan(x) \Rightarrow f(0) = \tan(0) = 0 \\
f'(x) &= \sec^2(x) \Rightarrow f'(0) = \sec^2(0) = 1 \\
f''(x) &= 2\sec^2(x)\,\tan(x) \Rightarrow f''(0) = 2\sec^2(0) \cdot \tan(0) = 0 \\
f'''(x) &= 4\sec^2(x)\,\tan^2(x) + 2\sec^2(x)(1 + \tan^2(x))
\end{aligned}
$$

Therefore, by Theorem 2,

$$
\begin{aligned}
f(x_0) &= P_2(x_0) + R_2(x_0) \\
&= \left(\frac{f(0)}{0!}(x-0)^0 + \frac{f'(\epsilon)}{1!}(x-0)^1 + \frac{f''(0)}{2!}(x-0)^2 \right) \\
&\quad + \left(\frac{f'''(\epsilon)}{3!}(x-0)^3 \right) \\
&= (0 + 1x + 0) + \frac{1}{6}(4\sec^2(\epsilon)\tan^2(\epsilon) \\
&\quad + 2\sec^2(\epsilon)(1 + \tan^2(\epsilon))x^3 \\
&= x + \frac{1}{6}(4\sec^2(\epsilon)\tan^2(\epsilon) + 2\sec^2(\epsilon)(1 + \tan^2(\epsilon)))x^3.
\end{aligned}
$$

Section 10.4

1. Let $f(x) = e^{-x}$.

$$
\begin{aligned}
f'(x) &= (-1)^1 e^{-x} \\
f''(x) &= e^{-x} = (-1)^2 e^{-x} \\
f'''(x) &= -e^{-x} = (-1)^3 e^{-x} \\
&\vdots
\end{aligned}
$$

i.e., $f^{(n)}(x) = (-1)^n e^{-x}$, for $n = 0, 1, 2, \ldots$.

Choose $x_0 = 0$. Then $f^{(n)}(x_0) = (-1)^n$.

The degree N Taylor polynomial of $f(x) = e^{-x}$ with base point zero is

$$
\begin{aligned}
P_N(x) &= f(0) + \frac{f'(0)}{1!}x + \frac{f''(0)}{2!}x^2 + \cdots + \frac{f^{(N)}(0)}{(N)!}x^N \\
&= 1 - x + \frac{x^2}{2!} - \frac{x^3}{3!} + \cdots + \frac{(-1)^N x^N}{(N)!}.
\end{aligned}
$$

To estimate the error, we note $f^{(N+1)}(x) = (-1)^{N+1} e^{-x}$. Therefore $\max \left| f^{(N+1)}(x) \right| = \max \left| e^{-x} \right| \leq 1$, for x between 0 and 0.5.

$$
|R_N(-0.5)| \leq 1 \cdot \frac{|-0.5 - 0|^{N+1}}{(N+1)!} = \frac{0.5^{N+1}}{(N+1)!}.
$$

We require four decimal places of accuracy.

Therefore we want to find the smallest N such that
$$
\frac{0.5^{N+1}}{(N+1)!} < 5 \times 10^{-5}.
$$

Substituting in some values for N in the expression $\frac{0.5^{N+1}}{(N+1)!}$, we have

$$
\text{if } N = 3, \; \frac{(0.5)^4}{4!} \approx 0.00260,
$$

$$
\text{if } N = 4, \; \frac{(0.5)^5}{5!} \approx 0.00026,
$$

$$
\text{if } N = 5, \; \frac{(0.5)^6}{6!} \approx 0.00002,
$$

and $|R_5(+0.5)| \leq 1 \cdot \dfrac{(0.5)^6}{6!} \approx 0.00002$.

Thus for $N = 5$, $\dfrac{(0.5)^6}{6!} < 0.00005$.

We shall use $P_5(+0.5)$ in the estimation and the inequality $|R_5(+0.5)| < 5 \times 10^{-5}$ assures that $P_5(+0.5)$ and $e^{-0.5}$ will agree to four decimal places.

Section 10.5

1. In section 5 of chapter 10, the Maclaurin series representing the function $y = \sin(x)$ is given. It is

$$
\begin{aligned}
\sin(x) &= x - \frac{x^3}{3!} + \frac{x^5}{5!} - \frac{x^7}{7!} + \frac{x^9}{9!} - \cdots, \; -\infty < x < \infty \\
&= \sum_{n=0}^{\infty} \frac{(-1)^n x^{2n+1}}{(2n+1)!}, \; -\infty < x < \infty.
\end{aligned}
$$

Therefore $\sin(.32) = \displaystyle\sum_{n=0}^{\infty} \frac{(-1)^n (.32)^{2n+1}}{(2n+1)!}$.

This is an alternating series, so the truncation error is less than the first term omitted (Theorem 1 of Section 5, Chapter 9). The maximum allowable error is 5×10^{-4}. The first term that is less than 5×10^{-4} is $\dfrac{(.32)^5}{5!} \approx 0.000027$.

Therefore we should use $\displaystyle\sum_{n=0}^{1} \frac{(-1)^n (.32)^{2n+1}}{(2n+1)!} =$

$.32 - \dfrac{(.32)^3}{6} \approx 0.3145386667$.

The actual value of $\sin(.32)$ is 0.3145665606.

Index

Quick Reference Guide

Geometry

Basic Formulas in Geometry

Pythagorean Theorem
$$a^2 + b^2 = c^2$$

Circle
Circumference
$= 2\pi r$
Area $= \pi r^2$

Trapezoid
Area $= \dfrac{1}{2}(a+b)c$

Cylinder
$V = \pi r^2 h$

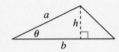

Triangle
Area $= \dfrac{1}{2}bh = \dfrac{1}{2}ab\sin(\theta)$

Sector of Circle
Area $= \dfrac{1}{2}r^2\theta$
Arc length $= r\theta$
(θ is measured in radian)

Sphere
$V = \dfrac{4}{3}\pi r^3$
$A = 4\pi r^2$

Cone
$V = \dfrac{1}{3}\pi r^2 h$
$A = \pi r\sqrt{r^2 + h^2}$

Basic Formulas in Coordinate Geometry

Distance and Midpoint Formulas (Section 1.2)

Distance between $P_1 = (x_1, y_1)$ and $P_2 = (x_2, y_2)$:

$$d = \sqrt{(x_2 - x_1)^2 + (y_2 - y_1)^2}$$

Midpoint of $\overline{P_1 P_2}$: $\left(\dfrac{x_1 + x_2}{2}, \dfrac{y_1 + y_2}{2} \right)$

Slopes and Lines (Section 1.3)

Slope of line through $P_1 = (x_1, y_1)$ and $P_2 = (x_2, y_2)$:

$$m = \frac{y_2 - y_1}{x_2 - x_1}$$

Point-slope form of the line with slope m and passes through (x_1, y_1):

$$y - y_1 = m(x - x_1)$$

Slope-intercept form of line with slope m and y-intercept b:

$$y = mx + b$$

Circles (Section 1.2)

Equation of the circle with center (h, k) and radius r:

$$(x - h)^2 + (y - k)^2 = r^2$$

Ellipses (Section 1.2)

Equation of an ellipse with vertical and horizontal axes of symmetry centered at (h, k):

$$\frac{(x - h)^2}{a^2} + \frac{(y - k)^2}{b^2} = 1$$

Hyperbolas (Section 1.2)

Equation of a hyperbola with vertical and horizontal axes of symmetry centered at (h, k):

$$\frac{(x - h)^2}{a^2} - \frac{(y - k)^2}{b^2} = 1$$

or

$$\frac{(y - k)^2}{b^2} - \frac{(x - h)^2}{a^2} = 1$$

Parabolas (Section 1.2)

Equation of a parabola with vertical axis of symmetry:

$$y = Ax^2 + Bx + C \text{ with } A \neq 0.$$

Algebra

Interval Notation

Bounded Intervals

Open interval (endpoints excluded):

$$(a, b) = \{x : a < x < b\}$$

Closed interval (both endpoints included):

$$[a, b] = \{x : a \le x \le b\}$$

Half-Open (or Half closed) interval:

$$[a, b) = \{x : a \le x < b\}$$

$$(a, b] = \{x : a < x \le b\}$$

Unbounded Intervals

$$(a, \infty) = \{x : a < x\}$$

$$[a, \infty) = \{x : a \le x\}$$

$$(-\infty, b) = \{x : x < b\}$$

$$(-\infty, b] = \{x : x \le b\}$$

$$(-\infty, \infty) = \{x : -\infty < x < \infty\}$$

Arithmetic

Factorial Notation

$$n! = n(n-1)(n-2) \cdots 3 \cdot 2 \cdot 1 \quad \text{for } n \text{ a nonnegative integer}$$
$$0! = 1$$

Sums of Powers of the First *n* Integers (Section 5.1)

$$\sum_{k=1}^{n} 1 = n$$

$$\sum_{k=1}^{n} k = 1 + 2 + 3 + \cdots + n = \frac{n(n+1)}{2}$$

$$\sum_{k=1}^{n} k^2 = 1^2 + 2^2 + 3^2 + \cdots + n^2 = \frac{n(n+1)(2n+1)}{6}$$

$$\sum_{k=1}^{n} k^3 = 1^3 + 2^3 + 3^3 + \cdots + n^3 = \frac{n^2(n+1)^2}{4}$$

$$\sum_{k=1}^{n} k^4 = 1^4 + 2^4 + 3^4 + \cdots + n^4 = \frac{n(n+1)(2n+1)(3n^2 + 3n - 1)}{30}$$

Arithmetic Operations

$$a(b + c) = ab + ac \qquad \frac{a}{b} + \frac{c}{d} = \frac{ad + bc}{bd}$$

$$\frac{a + c}{b} = \frac{a}{b} + \frac{c}{b} \qquad \frac{\dfrac{a}{b}}{\dfrac{c}{d}} = \frac{a}{b} \times \frac{d}{c} = \frac{ad}{bc}$$

Algebraic Identities

Quadratic Formula

If $ax^2 + bx + c = 0$, then $x = \dfrac{-b \pm \sqrt{b^2 - 4ac}}{2a}$.

Exponents

$$(xy)^m = x^m y^m \qquad \left(\frac{x}{y}\right)^m = \frac{x^m}{y^m} \qquad x^m x^n = x^{m+n}$$

$$\frac{x^m}{x^n} = x^{m-n} \qquad x^{-m} = \frac{1}{x^m} \qquad (x^m)^n = x^{mn}$$

Radicals

$$x^{1/n} = \sqrt[n]{x} \qquad \sqrt[n]{x^m} = (\sqrt[n]{x})^m = x^{m/n} \qquad \sqrt[n]{xy} = \sqrt[n]{x} \cdot \sqrt[n]{y} = x^{1/n} y^{1/n}$$

Binomial Formulas

$$(x + y)^2 = x^2 + 2xy + y^2 \qquad (x - y)^2 = x^2 - 2xy + y^2$$
$$(x + y)^3 = x^3 + 3x^2 y + 3xy^2 + y^3$$
$$(x - y)^3 = x^3 - 3x^2 y + 3xy^2 - y^3$$
$$(x + y)^n = x^n + nx^{n-1}y + \frac{n(n - 1)}{2}x^{n-2}y^2 + \cdots + \binom{n}{k} x^{n-k} y^k + \cdots + nxy^{n-1} + y^n$$
$$\text{where } \binom{n}{k} = \frac{n(n - 1) \cdots (n - k + 1)}{1 \cdot 2 \cdot 3 \cdots \cdot k}$$

Factoring Special Polynomials

$$x^2 - y^2 = (x + y)(x - y)$$
$$x^3 + y^3 = (x + y)(x^2 - xy + y^2)$$
$$x^3 - y^3 = (x - y)(x^2 + xy + y^2)$$

If n is a positive integer, then $x^n - y^n = (x - y)(x^{n-1} + x^{n-2}y + x^{n-3}y^2 + \cdots + xy^{n-2} + y^{n-1})$.
If n is an *odd* positive integer, then

$$x^n + y^n = (x + y)(x^{n-1} - x^{n-2}y + x^{n-3}y^2 - \cdots + x^2 y^{n-3} - xy^{n-2} + y^{n-1}).$$

Inequalities and Absolute Value (Section 1.1)

If $a < b$ and $b < c$, then $a < c$.
If $a < b$, then $a + c < b + c$.
If $a < b$ and $c > 0$, then $ca < cb$.

If $a < b$ and $c < 0$, then $ca > cb$.
If $a > 0$, then

$$|x| = a \text{ means } x = a \quad \text{or} \quad x = -a$$
$$|x| < a \text{ means } -a < x < a$$
$$|x| > a \text{ means } x > a \quad \text{or} \quad x < -a$$

Trigonometry

Angle Measurement

Radians and Degrees (Section 1.6)

$360° = 2\pi$ radians $= 1$ revolution

$$\text{angle in degree} = \left(\frac{180°}{\pi}\right) \cdot (\text{angle in radian})$$

$$\text{angle in radian} = \left(\frac{\pi}{180°}\right) \cdot (\text{angle in degree})$$

Angle measures, unless stated otherwise, are in radian measure.

Triangles

The following results hold for any plane triangle ABC with sides a, b, c and angles α, β, γ.

Pythagorean theorem
In a right triangle with right angle at C,
$c^2 = a^2 + b^2$

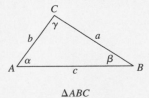

$\triangle ABC$

Heron's Formula
Let $s = \frac{1}{2}(a + b + c)$. The area of the triangle is $\sqrt{s(s-a)(s-b)(s-c)}$.

Law of cosines

$$a^2 = b^2 + c^2 - 2bc\cos\alpha \qquad b^2 = a^2 + c^2 - 2ac\cos\beta \qquad c^2 = a^2 + b^2 - 2ab\cos\gamma$$

$$\cos\alpha = \frac{b^2 + c^2 - a^2}{2bc} \qquad \cos\beta = \frac{a^2 + c^2 - b^2}{2ac} \qquad \cos\gamma = \frac{a^2 + b^2 - c^2}{2ab}$$

Law of sines

$$\frac{\sin\alpha}{a} = \frac{\sin\beta}{b} = \frac{\sin\gamma}{c}$$

Law of tangents

$$\frac{a + b}{a - b} = \frac{\tan\frac{1}{2}(\alpha + \beta)}{\tan\frac{1}{2}(\alpha - \beta)}$$

with similar relations involving the other sides and angles

Trigonometric Functions (Section 1.6)

Let θ be any angle in standard position, and let $C = (x, y)$ be any point on the terminal side of the angle a distance r from the origin ($r \neq 0$). Then

$$\cos\theta = \frac{x}{r} \quad \sin\theta = \frac{y}{r} \quad \tan\theta = \frac{y}{x}$$

$$\sec\theta = \frac{r}{x} \quad \csc\theta = \frac{r}{y} \quad \cot\theta = \frac{x}{y}$$

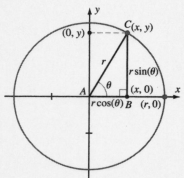

Trigonometric Graphs (Section 1.6)

Cosine $y = \cos x$

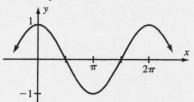

Sine $y = \sin x$

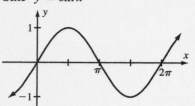

Secant $y = \sec x$

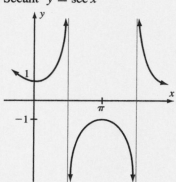

Cosecant $y = \csc x$

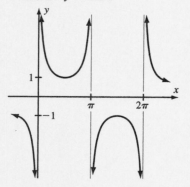

Tangent $y = \tan x$

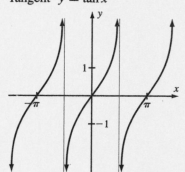

Cotangent $y = \cot x$

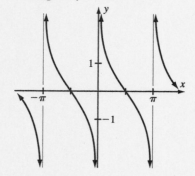

(Section 1.6)

Angle	Sine	Cosine	Tangent	Cotangent	Secant	Cosecant
0	0	1	0	undef	1	undef
$\pi/6$	$1/2$	$\sqrt{3}/2$	$1/\sqrt{3}$	$\sqrt{3}$	$2/\sqrt{3}$	2
$\pi/4$	$\sqrt{2}/2$	$\sqrt{2}/2$	1	1	$\sqrt{2}$	$\sqrt{2}$
$\pi/3$	$\sqrt{3}/2$	$1/2$	$\sqrt{3}$	$1/\sqrt{3}$	2	$2/\sqrt{3}$
$\pi/2$	1	0	undef	0	undef	1

Trigonometric Identities (Section 1.6)

Fundamental Identities

1. $\csc(\theta) = \dfrac{1}{\sin(\theta)}$

2. $\sec(\theta) = \dfrac{1}{\cos(\theta)}$

3. $\tan(\theta) = \dfrac{\sin(\theta)}{\cos(\theta)}$

4. $\cot(\theta) = \dfrac{\cos(\theta)}{\sin(\theta)}$

5. $\cot(\theta) = \dfrac{1}{\tan(\theta)}$

6. $\sin\left(\dfrac{\pi}{2} - \theta\right) = \cos(\theta)$

7. $\cos\left(\dfrac{\pi}{2} - \theta\right) = \sin(\theta)$

8. $\tan\left(\dfrac{\pi}{2} - \theta\right) = \cot(\theta)$

9. $\sin(-\theta) = -\sin(\theta)$
10. $\cos(-\theta) = \cos(\theta)$
11. $\tan(-\theta) = -\tan(\theta)$

Pythagorean Identities

12. $\cos^2(\theta) + \sin^2(\theta) = 1$
13. $1 + \tan^2(\theta) = \sec^2(\theta)$
14. $\cot^2(\theta) + 1 = \csc^2(\theta)$

Addition and Subtraction Formulas

15. $\sin(\alpha + \beta) = \sin(\alpha)\cos(\beta) + \cos(\alpha)\sin(\beta)$
16. $\sin(\alpha - \beta) = \sin(\alpha)\cos(\beta) - \cos(\alpha)\sin(\beta)$
17. $\cos(\alpha + \beta) = \cos(\alpha)\cos(\beta) - \sin(\alpha)\sin(\beta)$
18. $\cos(\alpha - \beta) = \cos(\alpha)\cos(\beta) + \sin(\alpha)\sin(\beta)$
19. $\tan(\alpha + \beta) = \dfrac{\tan(\alpha) + \tan(\beta)}{1 - \tan(\alpha)\tan(\beta)}$
20. $\tan(\alpha - \beta) = \dfrac{\tan(\alpha) - \tan(\beta)}{1 + \tan(\alpha)\tan(\beta)}$

Double-Angle Formulas

21. $\sin(2\theta) = 2\sin(\theta)\cos(\theta)$
22. $\cos(2\theta) = \cos^2(\theta) - \sin^2(\theta)$
$$= 2\cos^2(\theta) - 1$$
$$= 1 - 2\sin^2(\theta)$$
23. $\tan(2\theta) = \dfrac{2\tan(\theta)}{1 - \tan^2(\theta)}$

Half-Angle Formulas

24. $\sin\left(\dfrac{1}{2}\theta\right) = \pm\sqrt{\dfrac{1 - \cos(\theta)}{2}}$

25. $\cos\left(\dfrac{1}{2}\theta\right) = \pm\sqrt{\dfrac{1 + \cos(\theta)}{2}}$

26. $\tan\left(\dfrac{1}{2}\theta\right) = \dfrac{1 - \cos(\theta)}{\sin(\theta)}$

$\qquad\qquad = \dfrac{\sin(\theta)}{1 + \cos(\theta)}$

Sum-To-Product Formulas

27. $\sin\alpha + \sin\beta = 2\sin\left(\dfrac{\alpha + \beta}{2}\right)\cos\left(\dfrac{\alpha - \beta}{2}\right)$

28. $\sin\alpha - \sin\beta = 2\sin\left(\dfrac{\alpha - \beta}{2}\right)\cos\left(\dfrac{\alpha + \beta}{2}\right)$

29. $\cos\alpha + \cos\beta = 2\cos\left(\dfrac{\alpha + \beta}{2}\right)\cos\left(\dfrac{\alpha - \beta}{2}\right)$

30. $\cos\alpha - \cos\beta = -2\sin\left(\dfrac{\alpha + \beta}{2}\right)\sin\left(\dfrac{\alpha - \beta}{2}\right)$

Product-To-Sum Formulas

31. $\sin(\alpha)\sin(\beta) = \dfrac{1}{2}(\cos(\alpha - \beta) - \cos(\alpha + \beta))$

32. $\cos(\alpha)\cos(\beta) = \dfrac{1}{2}(\cos(\alpha - \beta) + \cos(\alpha + \beta))$

33. $\sin(\alpha)\cos(\beta) = \dfrac{1}{2}(\sin(\alpha + \beta) + \sin(\alpha - \beta))$

34. $\cos(\alpha)\sin(\beta) = \dfrac{1}{2}(\sin(\alpha + \beta) - \sin(\alpha - \beta))$

Inverse Trigonometric Functions (Section 6.6)

Inverse Function	Domain	Range
$y = \arccos(x)$, or $y = \mathrm{Cos}^{-1}(x)$	$-1 \leq x \leq 1$	$0 \leq y \leq \pi$
$y = \arcsin(x)$, or $y = \mathrm{Sin}^{-1}(x)$	$-1 \leq x \leq 1$	$-\dfrac{\pi}{2} \leq y \leq \dfrac{\pi}{2}$
$y = \arctan(x)$, or $y = \mathrm{Tan}^{-1}(x)$	All reals	$-\dfrac{\pi}{2} < y < \dfrac{\pi}{2}$
$y = \mathrm{arccot}(x)$, or $y = \mathrm{Cot}^{-1}(x)$	All reals	$0 < y < \pi$
$y = \mathrm{arcsec}(x)$, or $y = \mathrm{Sec}^{-1}(x)$	$|x| \geq 1$	$0 \leq y \leq \pi, y \neq \dfrac{\pi}{2}$
$y = \mathrm{arccsc}(x)$, or $y = \mathrm{Csc}^{-1}(x)$	$|x| \geq 1$	$-\dfrac{\pi}{2} \leq y \leq \dfrac{\pi}{2}, y \neq 0$

Inverse Trigonometric Graphs (Section 6.6)

Arccosine $y = \text{Cos}^{-1}(x)$ for $-1 \leq x \leq 1$

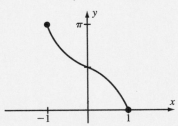

Arcsine $y = \text{Sin}^{-1}(x)$ for $-1 \leq x \leq 1$

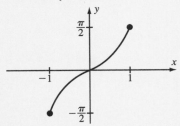

Arcsecant $y = \text{Sec}^{-1}(x)$ for $|x| \geq 1$

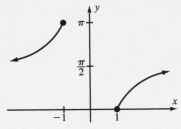

Arccosecant $y = \text{Csc}^{-1}(x)$ for $|x| \geq 1$

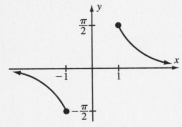

Arctangent $y = \text{Tan}^{-1}(x)$ for all x

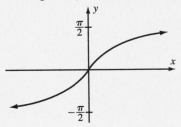

Arccotangent $y = \text{Cot}^{-1}(x)$ for all x

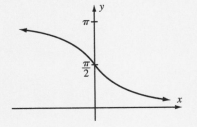

Special Functions

Exponential and Logarithmic Functions (Sections 2.6, 3.6, 6.2, 6.3 and 6.4)

$\log_a(x) = y \Leftrightarrow a^y = x$

$\ln(x) = \log_e(x), \quad \text{where} \quad \ln e = 1$

$\ln(x) = y \Leftrightarrow e^y = x$

$a^x = e^{x \cdot \ln(a)}$

$\log_a(x) = \dfrac{\ln(x)}{\ln(a)}$

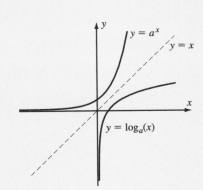

Cancellation Equations

$\ln(e^x) = x$ $e^{\ln x} = x$ (Section 3.6)

$\log_a(a^x) = x$ $a^{\log_a x} = x$ (Sections 6.3 and 6.4)

Laws of Logarithms (Sections 3.6 and 6.4)

1. $\ln(1) = 0$
2. $\log_a(1) = 0$
3. $\ln(e) = 1$
4. $\log_a(a) = 1$
5. $\ln(xy) = \ln(x) + \ln(y)$
6. $\log_a(xy) = \log_a(x) + \log_a(y)$

7. $\ln\left(\dfrac{x}{y}\right) = \ln(x) - \ln(y)$
8. $\log_a\left(\dfrac{x}{y}\right) = \log_a(x) - \log_a(y)$
9. $\ln(x^p) = p\ln(x)$
10. $\log_a(x^p) = p\log_a(x)$

Hyperbolic Trigonometric Functions

The hyperbolic sine and hyperbolic cosine are defined by

$$\sinh(t) = \frac{e^t - e^{-t}}{2}$$

and

$$\cosh(t) = \frac{e^t + e^{-t}}{2},$$

respectively. The other hyperbolic trigonometric functions are defined in terms of

$$\tanh(t) = \frac{\sinh(t)}{\cosh(t)}, \quad \coth(t) = \frac{\cosh(t)}{\sinh(t)}.$$

$$\operatorname{sech}(t) = \frac{1}{\cosh(t)}, \quad \operatorname{csch}(t) = \frac{1}{\sinh(t)}.$$

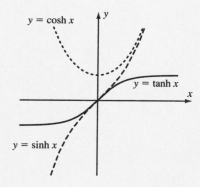

Hyperbolic Identities

1. $\cosh^2(x) - \sinh^2(x) = 1$
2. $1 - \tanh^2(x) = \operatorname{sech}^2(x)$
3. $\coth^2(x) - 1 = \operatorname{csch}^2(x)$
4. $\sinh(-x) = -\sinh(x)$
5. $\cosh(-x) = \cosh(x)$
6. $\tanh(-x) = -\tanh(x)$
7. $\sinh(x \pm y) = \sinh(x)\cosh(y) \pm \cosh(x)\sinh(y)$

8. $\cosh(x \pm y) = \cosh(x)\cosh(y) \pm \sinh(x)\sinh(y)$

9. $\tanh(x \pm y) = \dfrac{\tanh(x) \pm \tanh(y)}{1 \pm \tanh(x)\tanh(y)}$

10. $\cosh(2x) = \cosh^2(x) + \sinh^2(x)$

$\qquad = 2\cosh^2(x) - 1$

$\qquad = 1 + 2\sinh^2(x)$

11. $\sinh(2x) = 2\sinh(x)\cosh(x)$

12. $\tanh(2x) = \dfrac{2\tanh(x)}{1 + \tanh^2(x)}$

13. $\cosh\left(\dfrac{1}{2}x\right) = \pm\sqrt{\dfrac{\cosh(x) + 1}{2}}$

14. $\sinh\left(\dfrac{1}{2}x\right) = \pm\sqrt{\dfrac{\cosh(x) - 1}{2}}$

15. $\tanh\left(\dfrac{1}{2}x\right) = \dfrac{\cosh(x) - 1}{\sinh(x)} = \dfrac{\sinh(x)}{\cosh(x) + 1}$

Inverse Hyperbolic Trigonometric Functions

1. $y = \sinh^{-1}(x) \Leftrightarrow \sinh(y) = x$

2. $y = \cosh^{-1}(x) \Leftrightarrow \cosh(y) = x \quad$ and $\quad y \geq 0$

3. $y = \tanh^{-1}(x) \Leftrightarrow \tanh(y) = x$

4. $\sinh^{-1}(x) = \ln(x + \sqrt{x^2 + 1})$

5. $\operatorname{csch}^{-1}(x) = \ln\left(\dfrac{1 + \sqrt{1 + x^2}}{x}\right) \quad$ if $x > 0$

6. $\cosh^{-1}(x) = \ln(x + \sqrt{x^2 - 1})$

7. $\operatorname{sech}^{-1}(x) = \ln\left(\dfrac{1 + \sqrt{1 - x^2}}{x}\right) \quad$ if $0 < x \leq 1$

8. $\tanh^{-1}(x) = \dfrac{1}{2}\ln\left(\dfrac{1 + x}{1 - x}\right) \quad$ if $-1 < x < 1$

9. $\coth^{-1}(x) = \dfrac{1}{2}\ln\left(\dfrac{x + 1}{x - 1}\right) \quad$ if $x^2 > 1$

Differentiation

Procedural Rules of Differentiation (Sections 3.3, 3.5 and 3.6)

Let f and g be differentiable functions of x, and α and β are any real numbers.

Name of Rule	*Derivative*
Scalar Multiplication Rule	$(\alpha \cdot f)'(x) = \alpha \cdot f'(x)$
Addition and Subtraction Rule	$(f \pm g)'(x) = f'(x) \pm g'(x)$
Linear Combination Rule	$(\alpha \cdot f + \beta \cdot g)'(x) = \alpha \cdot f'(x) + \beta \cdot g'(x)$
Product Rule	$(f \cdot g)'(x) = f'(x) \cdot g(x) + f(x) \cdot g'(x)$
Quotient Rule	$\left(\dfrac{f}{g}\right)'(x) = \dfrac{g(x) \cdot f'(x) - f(x) \cdot g'(x)}{g^2(x)}$
Chain Rule	$(g \circ f)'(x) = g'(f(x)) \cdot f'(x)$
Inverse Function Derivative Rule	$\dfrac{d(f^{-1})}{dt}\bigg|_{t} = \dfrac{1}{\dfrac{df}{dx}\bigg|_{x=f^{-1}(t)}}$

Differentiation Rules

Power Rule (Section 3.4)

1. $\dfrac{d}{dx}x^p = px^{p-1}$ for any real number p

Extended-Power Rule (Section 3.5)

2. $\dfrac{d}{dx}(f(x))^p = p \cdot (f(x))^{p-1} \cdot f'(x)$ for any real number p

Trigonometric Functions (Section 3.4)

3. $\dfrac{d}{dx}\cos(x) = -\sin(x)$

4. $\dfrac{d}{dx}\sin(x) = \cos(x)$

5. $\dfrac{d}{dx}\tan(x) = \sec^2(x)$

6. $\dfrac{d}{dx}\cot(x) = -\csc^2(x)$

7. $\dfrac{d}{dx}\sec(x) = \sec(x)\tan(x)$

8. $\dfrac{d}{dx}\csc(x) = -\csc(x)\cot(x)$

Exponential and Logarithmic Functions (Sections 3.4, 6.2, 6.4)

9. $\dfrac{d}{dx}e^x = e^x$

10. $\dfrac{d}{dx}a^x = \ln(a) \cdot a^x$

11. $\dfrac{d}{dx}\ln(x) = \dfrac{1}{x}$

12. $\dfrac{d}{dx}\log_a(x) = \dfrac{1}{x\ln(a)}$

13. $\dfrac{d}{dx}\ln|x| = \dfrac{1}{x}$

14. $\dfrac{d}{dx}\log_a|x| = \dfrac{1}{x\ln(a)}$

Inverse Trigonometric Functions (Section 6.6)

15. $\dfrac{d}{dx}\text{Cos}^{-1}(x) = \dfrac{-1}{\sqrt{1-x^2}}$

16. $\dfrac{d}{dx}\text{Sin}^{-1}(x) = \dfrac{1}{\sqrt{1-x^2}}$

17. $\dfrac{d}{dx}\mathrm{Tan}^{-1}(x) = \dfrac{1}{1+x^2}$

18. $\dfrac{d}{dx}\mathrm{Cot}^{-1}(x) = \dfrac{-1}{1+x^2}$

19. $\dfrac{d}{dx}\mathrm{Sec}^{-1}(x) = \dfrac{1}{|x|\sqrt{x^2-1}}$

20. $\dfrac{d}{dx}\mathrm{Csc}^{-1}(x) = \dfrac{-1}{|x|\sqrt{x^2-1}}$

Hyperbolic Functions (Section 6.7)

21. $\dfrac{d}{dx}\sinh(x) = \cosh(x)$

22. $\dfrac{d}{dx}\cosh(x) = \sinh(x)$

23. $\dfrac{d}{dx}\tanh(x) = \mathrm{sech}^2(x)$

24. $\dfrac{d}{dx}\coth(x) = -\mathrm{csch}^2(x)$

25. $\dfrac{d}{dx}\mathrm{sech}(x) = -\mathrm{sech}(x)\cdot\tanh(x)$

26. $\dfrac{d}{dx}\mathrm{csch}(x) = -\mathrm{csch}(x)\cdot\coth(x)$

27. $\dfrac{d}{dx}\sinh^{-1}(x) = \dfrac{1}{\sqrt{1+x^2}}$

28. $\dfrac{d}{dx}\cosh^{-1}(x) = \dfrac{1}{\sqrt{x^2-1}}$

29. $\dfrac{d}{dx}\tanh^{-1}(x) = \dfrac{1}{1-x^2}$

30. $\dfrac{d}{dx}\mathrm{sech}^{-1}(x) = -\dfrac{1}{x\sqrt{1-x^2}}$

Numeric Differentiation (Section 3.3)

$$f'(c) \approx D_0 f(c,h) = \frac{f\left(c+\dfrac{h}{2}\right) - f\left(c-\dfrac{h}{2}\right)}{h}$$

Integration

Procedural Rules for Integration

In what follows, u, v, and w are functions of x, and a, b, p, q, α, β, and n are any constants, restricted if indicated. It is assumed in all cases that division by zero is excluded.

Properties of Integrals (Section 4.8)

Name of Rule   *Integral*

Scalar Multiple $\int \alpha \cdot f(x)\, dx = \alpha \cdot \int f(x)\, dx$

Addition Rule $\int [f(x) + g(x)]\, dx = \int f(x)\, dx + \int g(x)\, dx$

Subtraction Rule $\int [f(x) - g(x)]\, dx = \int f(x)\, dx - \int g(x)\, dx$

Linear Combination Rule $\int (\alpha \cdot f(x) + \beta \cdot g(x))\, dx = \alpha \cdot \int f(x)\, dx + \beta \cdot \int g(x)\, dx$

By Parts (Section 7.1) $\int u\, dv = uv - \int v\, dx$

Integration Rules

Basic Formulas

Constant Rule (Section 4.8)

1. $\int 0\, dx = C$

Power Rule (Section 4.8)

2. $\int x^p\, dx = \dfrac{x^{p+1}}{p + 1} + C$, where p is any real number except -1

   $\int x^{-1}\, dx = \ln |x| + C$

Exponential Rule (Section 4.8)

3. $\int e^x\, dx = e^x + C$

Logarithmic Rule (Section 7.1)

4. $\int \ln(x)\, dx = x \ln(x) - x + C$

Trigonometric Rules (Sections 4.8, 6.2)

5. $\int \sin(x)\, dx = -\cos(x) + C$

6. $\int \cos(x)\, dx = \sin(x) + C$

7. $\int \tan(x)\, dx = \ln(|\sec(x)|) + C$

8. $\int \cot(x)\, dx = -\ln(|\csc(x)|) + C$

9. $\int \sec(x)\, dx = \ln(|\sec(x) + \tan(x)|) + C$

10. $\int \csc(x)\, dx = -\ln(|\csc(x) + \cot(x)|) + C$

11. $\int \sec^2(x)\, dx = \tan(x) + C$

12. $\int \csc^2(x)\, dx = -\cot(x) + C$

13. $\int \sec(x) \tan(x)\, dx = \sec(x) + C$

14. $\int \csc(x) \cot(x)\, dx = -\csc(x) + C$

Exponential Rule (base *a*) (Section 6.4)

15. $\int a^x \, dx = \dfrac{a^x}{\ln(a)} + C, \quad a \neq 1$

Hyperbolic Rules (Section 6.7)

16. $\int \cosh(x) \, dx = \sinh(x) + C$

17. $\int \sinh(x) \, dx = \cosh(x) + C$

18. $\int \tanh(x) \, dx = \ln(\cosh(x)) + C$

19. $\int \coth(x) \, dx = \ln|\sinh(x)| + C$

20. $\int \operatorname{sech}(x) \, dx = 2\operatorname{Tan}^{-1}(e^x) + C$

21. $\int \operatorname{csch}(x) \, dx = \ln\left|\tanh\left(\dfrac{x}{2}\right)\right| + C$

Inverse Rules (Sections 6.6, 6.7)

22. $\displaystyle\int \frac{dx}{\sqrt{a^2 - x^2}} = \operatorname{Sin}^{-1}\left(\frac{x}{a}\right) + C$

23. $\displaystyle\int \frac{dx}{\sqrt{x^2 - a^2}} = \ln\left(x + \sqrt{x^2 - a^2}\right) + C = \cosh^{-1}\left(\frac{x}{a}\right) + C$

24. $\displaystyle\int \frac{dx}{a^2 + x^2} = \frac{1}{a}\operatorname{Tan}^{-1}\left(\frac{x}{a}\right) + C$

25. $\displaystyle\int \frac{dx}{a^2 - x^2} = \begin{cases} \dfrac{1}{a}\tanh^{-1}\left(\dfrac{x}{a}\right) + C, & \text{if } \left|\dfrac{x}{a}\right| < 1 \\ \dfrac{1}{a}\coth^{-1}\left(\dfrac{x}{a}\right) + C, & \text{if } \left|\dfrac{x}{a}\right| > 1 \end{cases}$

26. $\displaystyle\int \frac{dx}{x\sqrt{x^2 - a^2}} = \frac{1}{a}\operatorname{Sec}^{-1}\left|\frac{x}{a}\right| + C,$

27. $\displaystyle\int \frac{dx}{x\sqrt{a^2 - x^2}} = -\frac{1}{a}\ln\left|\frac{a + \sqrt{a^2 - x^2}}{x}\right| + C = -\frac{1}{a}\operatorname{sech}^{-1}\left|\frac{x}{a}\right| + C$

28. $\displaystyle\int \frac{dx}{\sqrt{a^2 + x^2}} = \ln(x + \sqrt{a^2 + x^2}) + C = \sinh^{-1}\left(\frac{x}{a}\right) + C$

29. $\displaystyle\int \frac{dx}{x\sqrt{a^2 + x^2}} = -\frac{1}{a}\ln\left|\frac{\sqrt{a^2 + x^2} + a}{x}\right| = -\frac{1}{a}\operatorname{csch}^{-1}\left|\frac{x}{a}\right| + C$

Suggested Integration Method for the Following Forms of Integral

Forms Involving Products of 1 or x^n with Exponential, Log, Sine, and Cosine (Sections 5.5 and 7.1)

1. $\int e^{ax} \, dx$: Use the substitution $u = ax$.
2. $\int \ln(ax) \, dx$: Use Integration by Parts with $u = \ln(ax)$ and $dv = dx$.
3. $\int \sin(ax) \, dx$: Use the substitution $u = ax$.
4. $\int \cos(ax) \, dx$: Use the substitution $u = ax$.
5. $\int x^n e^{ax} \, dx$: Use Integration by Parts repeatedly until the integral is reduced to $\int e^{ax} \, dx$.

6. $\int x^n \ln(ax)\,dx$: Use Integration by Parts with $u = \ln(ax)$ and $dv = x^n\,dx$.
7. $\int x^n \sin(ax)\,dx$: Use Integration by Parts repeatedly until the integral is reduced to $\int \sin(ax)\,dx$ or $\int \cos(ax)\,dx$.
8. $\int x^n \cos(ax)\,dx$: Use Integration by Parts repeatedly until the integral is reduced to $\int \sin(ax)\,dx$ or $\int \cos(ax)\,dx$.

Forms Involving Products of Exponential, Sine, and Cosine (Sections 5.5 and 7.1)

9. $\int e^{kx} \sin(ax)\,dx$, $k \neq 0$: Use Integration by Parts twice. (See Example 5 in Section 7.1 of the student's text.)
10. $\int e^{kx} \cos(ax)\,dx$, $k \neq 0$: Use Integration by Parts twice. (See Example 5 in Section 7.1 of the student's text.)
11. $\int \sin(ax) \sin(bx)\,dx$, $a \neq b$: Use the identity $\sin(ax)\sin(bx) = \frac{1}{2}(\cos((a-b)x) - \cos((a+b)x))$, and then apply the Method of Substitution.
12. $\int \cos(ax) \cos(bx)\,dx$, $a \neq b$: Use the identity $\cos(ax)\cos(bx) = \frac{1}{2}(\cos((a-b)x) + \cos((a+b)x))$, and then apply the Method of Substitution.
13. $\int \sin(ax) \cos(bx)\,dx$, $a \neq b$: Use the identity $\sin(ax)\cos(bx) = \frac{1}{2}(\sin((a+b)x) + \sin((a-b)x))$, and then apply the Method of Substitution.

Forms Involving Powers of Sine, Cosine, Tangent, and Secant (Section 7.3)

14. $\int \cos^2(x)\,dx = \dfrac{1}{2}(x + \sin(x)\cos(x)) + C$

15. $\int \sin^2(x)\,dx = \dfrac{1}{2}(x - \sin(x)\cos(x)) + C$

16. $\int \cos^n(x)\,dx = \dfrac{1}{n}\sin(x)\cos^{n-1}(x) + \dfrac{n-1}{n}\int \cos^{n-2}(x)\,dx$, n positive: Use the reduction formula repeatedly until the integral is reduced to $\int \cos(x)\,dx$ or $\int \cos^0(x)\,dx = \int 1\,dx$.

17. $\int \sin^n(x)\,dx = -\dfrac{1}{n}\sin^{n-1}(x)\cos(x) + \dfrac{n-1}{n}\int \sin^{n-2}(x)\,dx$, n positive: Use the reduction formula repeatedly until the integral is reduced to $\int \sin(x)\,dx$ or $\int \sin^0(x)\,dx = \int 1\,dx$.

18. $\int \tan^n(x)\,dx = \dfrac{1}{n-1}\tan^{n-1}(x) - \int \tan^{n-2}(x)\,dx$, $n > 1$: Use the reduction formula repeatedly until the integral is reduced to $\int \tan(x)\,dx$ or $\int \tan^0(x)\,dx = \int 1\,dx$.

19. $\int \sec^n(x)\,dx = \dfrac{1}{n-1}\sec^{n-2}(x)\tan(x) + \dfrac{n-2}{n-1}\int \sec^{n-2}(x)\,dx$, $n > 1$: Use the reduction formula repeatedly until the integral is reduced to $\int \sec(x)\,dx$ or $\int \sec^0(x)\,dx = \int 1\,dx$.

20. $\int \sin^m(x) \cos^n(x)\,dx$ when m is odd, say $m = 2k+1$: Use the substitution $u = \cos(x)$, then $\int \sin^{2k+1}(x) \cos^n(x)\,dx = \int \sin^{2k}(x)\sin(x)\cos^n(x)\,dx = -\int (1-u^2)^k u^n\,du$.

21. $\int \sin^m(x) \cos^n(x)\,dx$ when n is odd, say $n = 2k+1$: Use the substitution $u = \sin(x)$, then $\int \sin^m(x) \cos^{2k+1}(x)\,dx = \int \sin^m(x)\cos^{2k}(x)\cos(x)\,dx = \int u^m(1-u^2)^k\,du$.

22. $\int \sin^m(x) \cos^n(x)\,dx$ when both m and n are even: Change it to $\int \left(1 - \cos^2(x)\right)^{m/2} \cos^n(x)\,dx$ or $\int \sin^m(x)\left(1 - \sin^2(x)\right)^{n/2}\,dx$, and then use the reduction formula 16 or 17.

Forms Involving $a + bx$ (Sections 5.5, 7.2)

23. $\int \dfrac{1}{a + bx}\, dx$: Use the substitution $u = a + bx$.

24. $\int \dfrac{x}{a + bx}\, dx$: Use the substitution $u = a + bx$.

25. $\int \dfrac{x^2}{a + bx}\, dx$: Use long division first, and then apply the substitution $u = a + bx$.

26. $\int \dfrac{1}{x(a + bx)}\, dx$: Use partial fraction $\dfrac{A}{x} + \dfrac{B}{a + bx}$.

27. $\int \dfrac{1}{x^2(a + bx)}\, dx$: Use partial fraction $\dfrac{A}{x} + \dfrac{B}{x^2} + \dfrac{C}{a + bx}$.

Forms Involving $(a + bx)^2$ (Sections 5.5, 7.2)

28. $\int \dfrac{1}{(a + bx)^2}\, dx$: Use the substitution $u = a + bx$.

29. $\int \dfrac{x}{(a + bx)^2}\, dx$: Use the substitution $u = a + bx$.

30. $\int \dfrac{x^2}{(a + bx)^2}\, dx$: Use the substitution $u = a + bx$ first, and then apply long division.

31. $\int \dfrac{1}{x(a + bx)^2}\, dx$: Use partial fraction $\dfrac{A}{x} + \dfrac{B}{a + bx} + \dfrac{C}{(a + bx)^2}$.

32. $\int \dfrac{1}{x^2(a + bx)^2}\, dx$: Use partial fraction $\dfrac{A}{x} + \dfrac{B}{x^2} + \dfrac{C}{a + bx} + \dfrac{D}{(a + bx)^2}$.

Forms Involving Quadratic in the Denominator (Sections 7.2, 7.4, 7.5)

33. $\int \dfrac{1}{x^2 + a^2}\, dx$: Use the Method of Inverse substitution $x = a \tan(\theta)$.

34. $\int \dfrac{bx + c}{x^2 + a^2}\, dx$: Rewrite it as $\dfrac{b}{2} \int \dfrac{2x}{x^2 + a^2}\, dx + \int \dfrac{c}{x^2 + a^2}\, dx$. Compute the first integral using the substitution $u = x^2 + a^2$ and the second integral using Formula 33.

35. $\int \dfrac{1}{(x - \alpha)(x - \beta)}\, dx,\ \alpha \neq \beta$: Use partial fraction $\dfrac{A}{x - \alpha} + \dfrac{B}{x - \beta}$.

36. $\int \dfrac{cx + d}{(x - \alpha)(x - \beta)}\, dx,\ \alpha \neq \beta$: Use partial fraction $\dfrac{A}{x - \alpha} + \dfrac{B}{x - \beta}$.

37. $\int \dfrac{1}{(x - \alpha)^2}\, dx$: Use the substitution $u = x - \alpha$.

38. $\int \dfrac{1}{Ax^2 + Bx + C}\, dx$, where $Ax^2 + Bx + C$ cannot be factorized: Use completing the square in the denominator, then apply formula 33.

39. $\int \dfrac{ax + b}{Ax^2 + Bx + C}\, dx$, where $Ax^2 + Bx + C$ cannot be factorized: Rewrite the numerator $ax + b$ as a multiple of $(2Ax + B)$ plus a constant K. Integrate $\dfrac{2Ax + B}{Ax^2 + Bx + C}$ by

substitution $u = Ax^2 + Bx + C$. Integrate $\dfrac{K}{Ax^2 + Bx + C}$ by completing the square in the denominator, then applying formula 33.

Forms Involving $\sqrt{Ax^2 + Bx + C}$ (Section 7.4)

40. $\displaystyle\int \frac{1}{\sqrt{a^2 - x^2}}\,dx$ or $\int \sqrt{a^2 - x^2}\,dx$: Use Inverse (Indirect) Substitution $x = a \cdot \cos(\theta)$.

41. $\displaystyle\int \frac{1}{\sqrt{x^2 + a^2}}\,dx$ or $\int \sqrt{x^2 + a^2}\,dx$: Use Inverse (Indirect) Substitution $x = a \cdot \tan(\theta)$.

42. $\displaystyle\int \frac{1}{\sqrt{x^2 - a^2}}\,dx$ or $\int \sqrt{x^2 - a^2}\,dx$: Use Inverse (Indirect) Substitution $x = a \cdot \sec(\theta)$.

43. $\displaystyle\int \frac{1}{\sqrt{Ax^2 + Bx + C}}\,dx$ or $\int \sqrt{Ax^2 + Bx + C}\,dx$: Use completing the square on $Ax^2 + Bx + C$, then apply either formula 40, 41, or 42.

Definite Integral Formulas

Definite Integral at a point (Section 5.3)

1. $\displaystyle\int_a^a f(x)\,dx = 0$

Reversing the Direction of Integration (Section 5.3)

2. $\displaystyle\int_a^b f(x)\,dx = -\int_b^a f(x)\,dx$

Addition and Subtraction (Section 5.3)

3. $\displaystyle\int_a^b \{f(x) \pm g(x)\}\,dx = \int_a^b f(x)\,dx \pm \int_a^b g(x)\,dx$

Scalar Multiplication (Section 5.3)

4. $\displaystyle\int_a^b cf(x)\,dx = c\int_a^b f(x)\,dx$ where c is any constant

Interior Point (subdivision rule) (Section 5.3)

5. $\displaystyle\int_a^b f(x)\,dx = \int_a^c f(x)\,dx + \int_c^b f(x)\,dx$

Order Properties of Integrals (Section 5.3)

6. If $g(x) \le f(x) \le h(x)$ on $[a, b]$, then $\displaystyle\int_a^b g(x)\,dx \le \int_a^b f(x)\,dx \le \int_a^b h(x)\,dx$

Mean-Value Theorem for Integral (Section 5.3)

7. $\int\limits_{a}^{b} f(x)\,dx = (b-a)f(c)$ where c is between a and b

Fundamental Theorem of Calculus (Sections 5.2 and 5.4)

8. If f is continuous on $[a, b]$ and has an antiderivative F, then $\int\limits_{a}^{b} f(x)\,dx = F(b) - F(a)$

9. If f is continuous and $F(x) = \int\limits_{a}^{x} f(x)\,dt$, then $F'(x) = f(x)$

Numerical Integration (Section 5.7)

In the following formulas, the interval from $x = a$ to $x = b$ is subdivided into N equal parts by the points $a = x_0, x_1, x_2, \ldots, x_{N-1}, x_N = b$, and we let $\Delta x = \dfrac{(b-a)}{N}$.

Midpoint Rule

Let $\bar{x}_j = \dfrac{(x_{j-1} + x_j)}{2}$

10. $\int\limits_{a}^{b} f(x)\,dx \approx \Delta x \cdot (f(\bar{x}_1) + f(\bar{x}_2) + \cdots + f(\bar{x}_N))$

Trapezoidal Rule

11. $\int\limits_{a}^{b} f(x)\,dx \approx \dfrac{\Delta x}{2} \cdot (f(x_0) + 2f(x_1) + 2f(x_2) + \cdots + 2f(x_{N-1}) + f(x_N))$

Simpson's Rule (or parabolic formula) for *N* even

12. $\int\limits_{a}^{b} f(x)\,dx \approx \dfrac{\Delta x}{3} f(x_0) + 4f(x_1) + 2f(x_2) + 4f(x_3) + \cdots + 2f(x_{N-2}) + 4f(x_{N-1}) + f(x_N))$

Improper Integrals (Sections 8.6 and 8.7)

13. $\int\limits_{a}^{\infty} f(x)\,dx = \lim\limits_{N \to \infty} \int\limits_{a}^{N} f(x)\,dx$

14. $\int\limits_{-\infty}^{\infty} f(x)\,dx = \lim\limits_{N \to -\infty} \int\limits_{N}^{c} f(x)\,dx + \lim\limits_{M \to \infty} \int\limits_{c}^{M} f(x)\,dx$

15. $\int\limits_{a}^{b} f(x)\,dx = \lim\limits_{\epsilon \to 0^+} \int\limits_{a}^{b-\epsilon} f(x)\,dx$ if b is a singular point

16. $\int\limits_{a}^{b} f(x)\,dx = \lim\limits_{\epsilon \to 0^+} \int\limits_{a+\epsilon}^{b} f(x)\,dx$ if a is a singular point

Series and Power Series

Series $\sum a_n$

Some Special Series (Sections 9.1, 9.2 and 9.3)

1. The *geometric series* $\displaystyle\sum_{n=0}^{\infty} ar^n = a + ar + ar^2 + \cdots$ converges if and only if $|r| < 1$.
   If $|r| < 1$, the series converges to $\dfrac{a}{1-r}$.

2. The *harmonic series* $\displaystyle\sum_{n=1}^{\infty} \frac{1}{n} = 1 + \frac{1}{2} + \frac{1}{3} + \cdots$ diverges.

3. The *p-series* $\displaystyle\sum_{n=1}^{\infty} \frac{1}{n^p} = \frac{1}{1^p} + \frac{1}{2^p} + \frac{1}{3^p} + \cdots$ converges if and only if $p > 1$.

4. The alternating series $\displaystyle\sum_{n=1}^{\infty} (-1)^{n+1} \frac{1}{n} = 1 - \frac{1}{2} + \frac{1}{3} - \frac{1}{4} + \frac{1}{5} - \cdots$ converges to $\ln 2$.

5. The alternating series $\displaystyle\sum_{n=1}^{\infty} (-1)^{n+1} \frac{1}{(2n-1)} = 1 - \frac{1}{3} + \frac{1}{5} - \frac{1}{7} + \frac{1}{9} - \cdots$ converges to $\dfrac{\pi}{4}$

6. $\displaystyle\sum_{n=1}^{\infty} \frac{1}{(2n-1)(2n+1)} = \frac{1}{1 \cdot 3} + \frac{1}{3 \cdot 5} + \frac{1}{5 \cdot 7} + \cdots = \frac{1}{2}$

7. $\displaystyle\sum_{n=1}^{\infty} \frac{1}{(n)(n+2)} = \frac{1}{1 \cdot 3} + \frac{1}{2 \cdot 4} + \frac{1}{3 \cdot 5} + \cdots = \frac{3}{4}$

Tests for Convergence and Divergence of a Series

1. **The Divergence Test:** If $\displaystyle\lim_{n \to \infty} a_n \neq 0$, then $\sum a_n$ diverges.

2. **The Integral Test** (for series with positive terms only): If $a_n = f(n)$, where f is a positive, continuous, and decreasing function, then $\sum a_n$ converges if and only if $\displaystyle\int_{1}^{\infty} f(x)\, dx$ is finite.

3. **The Comparison Tests** (for series with positive terms only): Let $\sum a_n$ be a positive series, i.e., $a_n \geq 0$.
   i. If there exists a positive series $\sum b_n$ such that $0 \leq a_n \leq b_n$ and $\sum b_n$ converges, then $\sum a_n$ converges.
   ii. If there exists a positive series $\sum b_n$ such that $0 \leq b_n \leq a_n$ and $\sum b_n$ diverges, then $\sum a_n$ diverges.
   iii. If there exists a positive series $\sum b_n$ such that $\displaystyle\lim_{n \to \infty} \frac{a_n}{b_n}$ exists as a finite positive number and $\sum b_n$ converges, then $\sum a_n$ converges.
   iv. If there exists a positive series $\sum b_n$ such that $\displaystyle\lim_{n \to \infty} \frac{a_n}{b_n}$ exists as a finite positive number and $\sum b_n$ diverges, then $\sum a_n$ diverges.

4. **The Alternating Series Test** (for alternating series only): If $a_1 \geq a_2 \geq a_3 \geq \cdots$ and $\displaystyle\lim_{n \to \infty} a_n = 0$, then the alternating series $\sum (-1)^{n+1} a_n$ converges.

5. **The Ratio Test:**

   i. If $\lim\limits_{n\to\infty}\left|\dfrac{a_{n+1}}{a_n}\right| = \ell$ and $\ell < 1$, then $\sum a_n$ converges absolutely.

   ii. If $\lim\limits_{n\to\infty}\left|\dfrac{a_{n+1}}{a_n}\right| = \ell$ and $\ell > 1$, then $\sum a_n$ diverges.

6. **The Root Test:**

   i. If $\lim\limits_{n\to\infty}|a_n|^{1/n} = \ell$ and $\ell < 1$, then $\sum a_n$ converges absolutely

   ii. If $\lim\limits_{n\to\infty}|a_n|^{1/n} = \ell$ and $\ell > 1$, then $\sum a_n$ diverges.

Power Series $\sum a_n(x - c)^n$

Radius of Convergence of a Power Series (Section 10.1)

1. If $\lim\limits_{n\to\infty}\left|\dfrac{a_{n+1}}{a_n}\right| = \ell$, then the radius of convergence, R, of the power series $\sum a_n(x - c)^n$

   is $R = \dfrac{1}{\ell}$. (If $\ell = 0$, then $R = \infty$. If $\ell = \infty$, then $R = 0$.)

2. If $\lim\limits_{n\to\infty}|a_n|^{1/n} = \ell$, then the radius of convergence, R, of the power series $\sum a_n(x - c)^n$

   is $R = \dfrac{1}{\ell}$. (If $\ell = 0$, then $R = \infty$. If $\ell = \infty$, then $R = 0$.)

3. If R is the radius of convergence of a power series $\sum a_n(x - c)^n$, then the series converges absolutely for $|x - c| < R$ and diverge for $|x - c| > R$.

4. The *interval of convergence* of a power series is the set of points at which the series is convergent. If R, the radius of convergence of $\sum a_n(x - c)^n$, is a positive finite number, then the interval of convergence will be either $(c - R, c + R)$, $(c - R, c + R]$, $[c - R, c + R)$, or $[c - R, c + R]$ depending on whether the series converges at each end point of the interval.

Taylor Series of a Function (Sections 10.3, 10.4 and 10.5)

1. **(Taylor polynomial)** If f is $N + 1$ times continuously differentiable on an interval I centered at c, then for any $x_0 \in I$, we have $f(x_0) = P_N(x_0) + R_N(x_0)$, where

   $P_N(x_0) = \sum\limits_{n=0}^{N} \dfrac{f^{(n)}(c)}{n!} \cdot (x_0 - c)^n$ is called the *Taylor Polynomial of degree N* for f, and

   $R_N(x_0) = \dfrac{f^{(N+1)}(\xi)}{(N + 1)!} \cdot (x_0 - c)^{N+1}$, with ξ an unidentified number between c and x_0, is

   called the *remainder term of order N*.

2. **(Taylor series)** The Taylor series of a function f with base point c is defined to be the

   power series $\sum\limits_{n=0}^{\infty} \dfrac{f^{(n)}(c)}{n!} \cdot (x_0 - c)^n$.

3. **(Maclaurin series)** A Taylor series with base point $c = 0$ is called a Maclaurin Series.

4. **(Convergence of Taylor series)** Using the notation as above, if $\lim\limits_{N\to\infty} R_N(x) = 0$, then

   the Taylor series of f will converge to $f(x)$, i.e., $\sum\limits_{n=0}^{\infty} \dfrac{f^{(n)}(c)}{n!} \cdot (x_0 - c)^n = f(x)$.

5. If the Taylor series of f converges to f at x_0, we can use the Taylor polynomial $P_N(x_0)$ to approximate the value of $f(x_0)$ to any desired degree of accuracy.

Power Series Representations of Some Familiar Functions (Section 10.5)

1. $\sin(x) = \sum_{n=0}^{\infty} \frac{(-1)^n}{(2n+1)!} x^{2n+1} = x - \frac{x^3}{3!} + \frac{x^5}{5!} - \frac{x^7}{7!} + \cdots \quad -\infty < x < \infty$

2. $\cos(x) = \sum_{n=0}^{\infty} \frac{(-1)^n}{(2n)!} x^{2n} = 1 - \frac{x^2}{2!} + \frac{x^4}{4!} - \frac{x^6}{6!} + \cdots \quad -\infty < x < \infty$

3. $\frac{1}{1-x} = \sum_{n=0}^{\infty} x^n = 1 + x + x^2 + x^3 + \cdots \quad -1 < x < 1$

4. $e^x = \sum_{n=0}^{\infty} \frac{1}{n!} x^n = 1 + \frac{x}{1} + \frac{x^2}{2!} + \frac{x^3}{3!} + \cdots \quad -\infty < x < \infty$

5. $a^x = e^{x \ln(a)} = \sum_{n=0}^{\infty} \frac{1}{n!} (x \ln(a))^n = 1 + \frac{x \ln(a)}{1} + \frac{(x \ln(a))^2}{2!} + \frac{(x \ln(a))^3}{3!} + \cdots$
   $-\infty < x < \infty$

6. $\ln(1+x) = \sum_{n=1}^{\infty} \frac{(-1)^{n-1}}{n} x^n = x - \frac{x^2}{2} + \frac{x^3}{3} - \frac{x^4}{4} + \cdots \quad -1 < x \le 1$

7. $\ln\left(\frac{1+x}{1-x}\right) = \sum_{n=0}^{\infty} \frac{2}{(2n+1)} x^{2n+1} = 2x + \frac{2x^3}{3} + \frac{2x^5}{5} + \frac{2x^7}{7} + \cdots \quad -1 < x < 1$

8. $\sin^{-1}(x) = x + \frac{1}{2} \cdot \frac{x^3}{3} + \frac{(1)(3)}{(2)(4)} \cdot \frac{x^5}{5} + \frac{(1)(3)(5)}{(2)(4)(6)} \cdot \frac{x^7}{7} + \cdots \quad -1 < x < 1$

9. $\cos^{-1}(x) = \frac{\pi}{2} - \sin^{-1}(x) = \frac{\pi}{2} - \left(x + \frac{1}{2} \cdot \frac{x^3}{3} + \frac{(1)(3)}{(2)(4)} \cdot \frac{x^5}{5} + \cdots\right) \quad -1 < x < 1$

10. $\sinh(x) = x + \frac{x^3}{3!} + \frac{x^5}{5!} + \frac{x^7}{7!} + \cdots \quad -\infty < x < \infty$

11. $\cosh(x) = 1 + \frac{x^2}{2!} + \frac{x^4}{4!} + \frac{x^6}{6!} + \cdots \quad -\infty < x < \infty$

12. **(Binomial Series)** $(1+x)^\alpha = \sum_{n=0}^{\infty} \binom{\alpha}{n} x^n = \sum_{n=0}^{\infty} \frac{\alpha \cdot (\alpha-1) \cdots (\alpha-n+1)}{n!} x^n$
    $-1 < x < 1$